# 7th International Conference on Climate Change 2021



IOP Conference Series: Earth and Environmental Science
Volume 1016

Online
17 – 18 November 2021

ISBN: 978-1-7138-5458-6
ISSN: 1755-1307

**Printed from e-media with permission by:**

Curran Associates, Inc.
57 Morehouse Lane
Red Hook, NY 12571

**Some format issues inherent in the e-media version may also appear in this print version.**



Institute of Physics
Dirac House, Temple Back
Bristol BS1 6BE UK

Phone: 44 1 17 929 7481
Fax: 44 1 17 920 0979

techtracking@iop.org

**Additional copies of this publication are available from:**

Curran Associates, Inc.
57 Morehouse Lane
Red Hook, NY 12571 USA
Phone: 845-758-0400
Fax: 845-758-2633
Email: curran@proceedings.com
Web: www.proceedings.com

# TABLE OF CONTENTS

# THE 7th INTERNATIONAL CONFERENCE ON CLIMATE CHANGE PREFACE

## *Intro and Goal*

Covid-19 pandemic teaches us that climate change impact sometimes incredibly unpredictable. Not only the increase of heat exposure or natural hazard due to extreme weather, but the emergence of new species or the mutation of microorganism due to the changing ecology may harmful for human beings and another organism. This anthropological climate change phenomenon has negatively altered the most fundamental natural resources for living organism: air, water, and soil. Air has been polluted and warming due to human activities. The quantity of freshwater keeps on depleting, while the ground water recharge intervened with chemical and biological contamination. The sustainability of soil as the natural planting media is under threat due to land use conversion and soil quality degradation. Water acidification and temperature increase have been proved to change the salinity of seawater. Those climate change sensitive exposure may affect the immune system of the living things to survive. Those conditions grow the concern for the health status of humans, crops, farmed and wild animals.

Preparing immunity on living organism, especially human is a must to maintain our existence. Vaccine is vital to protect human health from emergence virus and disease which may affected by climate change. But does vaccine can against climate change itself? The 7th International Conference on Climate Change (ICCC) is focusing the discussion on the Covid-19 vaccine roles to against the climate change. Focusing and improving the climatic resilience strategies on immune system will be of considerably important to overcome the key climate change sensitive pathways, as well as to support the achievements some global goals in the Sustainable Development Goals (SDG).

The 7th ICCC 2021 is organized by Dept. of Master Program of Environmental Science, Post Graduate School of Sebelas Maret University (Surakarta, Indonesia), collaborating with The United Graduate School of Agriculture Science (UGSAS), and Gifu University (Japan).

This conference aims to accommodate the new related inspiration and innovation about how to minimize the climate change impact at present. Especially at the 7th ICCC 2021, the purpose is to formulate a comprehensive and efficient strategies on how to increase the effectiveness of Covid-19 Vaccine against the climate change. The e-poster of call for papers presented in Figure 1.

IOP Conf. Series: Earth and Environmental Science 1016 (2022) 011001 doi:10.1088/1755-1315/1016/1/011001

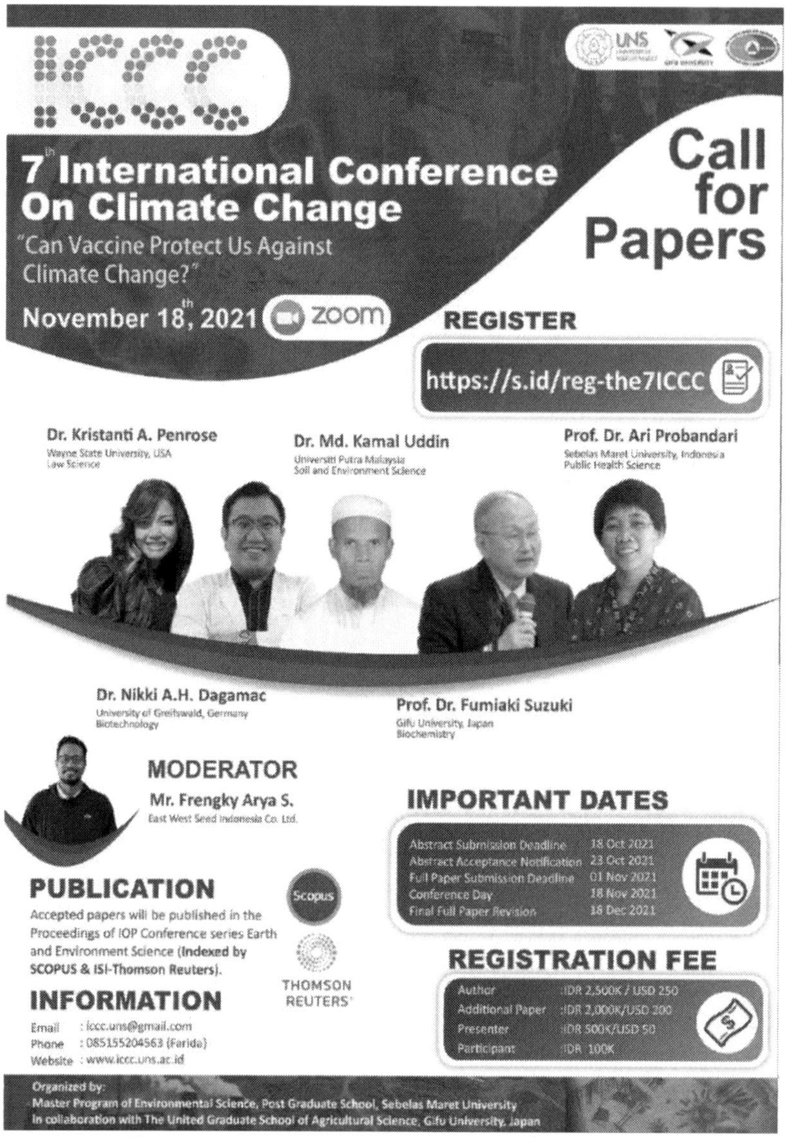


Figure 1. Call for Papers' e-poster of the 7th ICCC 2021

### *Invited Speakers and Guest Editors*

We acknowledge to all the speakers at The 7th ICCC 2020, i.e.: Prof. dr. Ari Natalia Probandari, MPH., Ph.D. from Sebelas Maret University, Indonesia; Dr. Nikki A. H. Dagamac from Dr. Nikki A. H. Dagamac, Germany; Dr. Kristanti A. Penrose from Wayne State University, USA, USA; Prof. Dr. Nurun Nabi from University of Dhaka, Bangladesh; Prof. Dr. Fumiaki Suzuki from UGSAS, Gifu University, Japan; Dr. Md. Kamal Uddin from Universiti Putra Malaysia. The sample of invited speakers presentation is presented in Figure 2 -6.

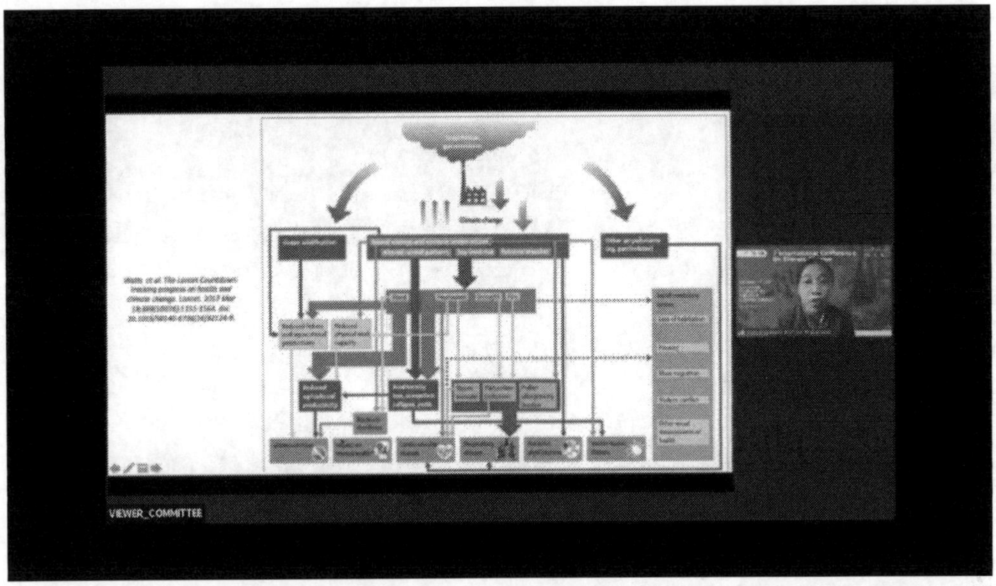


Figure 2. Presentation of invited speaker (Prof. dr. Ari Natalia Probandari, MPH., Ph.D. from Sebelas Maret University, Indonesia)

Figure 3. Presentation of invited speaker (Prof. Dr. Fumiaki Suzuki, Gifu University, Japan)

Figure 4. Presentation of invited speaker (Dr. Kristanti A. Penrose from Wayne State University, USA)

Figure 5. Presentation of invited speaker (Dr. Nikki A. H. Dagamac from University of Greifswald, Germany)

Figure 6. Presentation of invited speaker (Dr. Md. Kamal Uddin from Universiti Putra Malaysia)

### *Virtual Conference Disclaimer*

The 7th International Conference on Climate Change (ICCC) 2021 was held as a virtual conference, because the global coronavirus (Covid-19) also hit Indonesia and caused high rate of mortality. The date of the conference was November 18th, 2021. It was very interesting and challenging to organize the virtual conference, as it is becoming the demand of the technology-based future lifestyle. The virtual conference is environmentally friendly because reduce the carbon emission by minimize the global mobility. There was no significant problem regarding communication between invited speakers with participants and among participants. So, when the global pandemic has ended, the virtual conference sometimes needs to be carried out, accompanying the on-site conference.

Due to uncertain global situation due to the pandemic for unknown period, the conference should be organized virtual to support academicians and students knowledge transfer.

### *Date of conference, Location of organizer and Conference model*

The Virtual Conference Dates of was November 18th, 2021. The location of organizers: Graduate School of Sebelas Maret University, Ir. Sutami Street No. 36, Kentingan, Surakarta, Indonesia, 57126. The conference used Zoom® meeting room for both plenary and parallel sessions. Each invited speakers and participant joined the meeting room through the link, meeting ID dan password given by the organizers. The plenary session lasted for 3 hours on 18th November 2021. The Zoom® link for the speakers in the plenary session was Link Zoom: https://s.id/ICCC2021Speakers. Each participant was provided with different Zoom link. The recording of the plenary session can be seen in the Youtube link: https://youtu.be/Uy1Jym17bQQ. The parallel session was held for 6 hours (18 November 2021) and divided to 10 classrooms. Each author should make the oral presentation within 15 minutes including discussion. Total papers presented were 128 papers. Information on discussion and Question & Answer sessions. Because the parallel session was divided into 10 classrooms, other authors within the same room can ask question directly or through the chatroom to an author which was presenting his/ her paper. The example of presentation and discussion circumstance in a parallel room was recorded, one of the example is in the following Youtube link: https://www.youtube.com/watch?v=ElvznXuLGvc.

### *Location of participants and overall participant number*

Location of participants and overall participant numbers: Participants of the 7[h] ICCC was distributed worldwide: Malaysia, Japan, India, Thailand, Bangladesh, Rwanda, USA, Australia, Russia, Ethiopia, Pakistan and Korea. Overall participants including non-authors were 252 participants. The technology used was live video conference using Zoom® meeting room. The capacity of the Zoom® meeting room for the plenary session was 300 persons, where the capacity of each classroom of parallel session was 100 persons. The 7[th] ICCC 2020 was successful and was recorded with Youtube link: (https://youtu.be/Uy1Jym17bQQ). The example of authors' presentation is shown in Figure 7 and 8.

### *Committee List*

I also express my highest gratitude to the steering committees for their contributions to the 7[th] ICCC, i.e.: Prof. Dr. Ahmad Yunus and Prof. Dr. Sutarno from Sebelas Maret University, Indonesia; Prof. Dr. Ken Hiramatsu from Gifu University, Japan; Dr. James MacGregor from Eco Planet, Canada; and Dr. Anthony Kent from RMIT Univ., Australia. Also, Mr. Frenky Arya for contributed an amazing moderator during this conference. Finally, I acknowledge the organizing committee for their great valuable collaboration, namely. Prof. Dr. Maria Theresia Budiastuti, Dr. Komariah, Dr. Jauhari Syamsiyah, Dr. Andriyana Setyawati, Dr. Yuli Yanti, Dr.(c.). Sumani, Dr.(c.) Aktavia Herawati, Dr.(c.) Ganjar Herdiansyah and Mr. Sidik Pramono and all the committee member from Sebelas Maret University, Indonesia and Ms. Shigeno Kurimoto from Gifu University, Japan.

Figure 7. The parallel author presentation in room 4 of 10

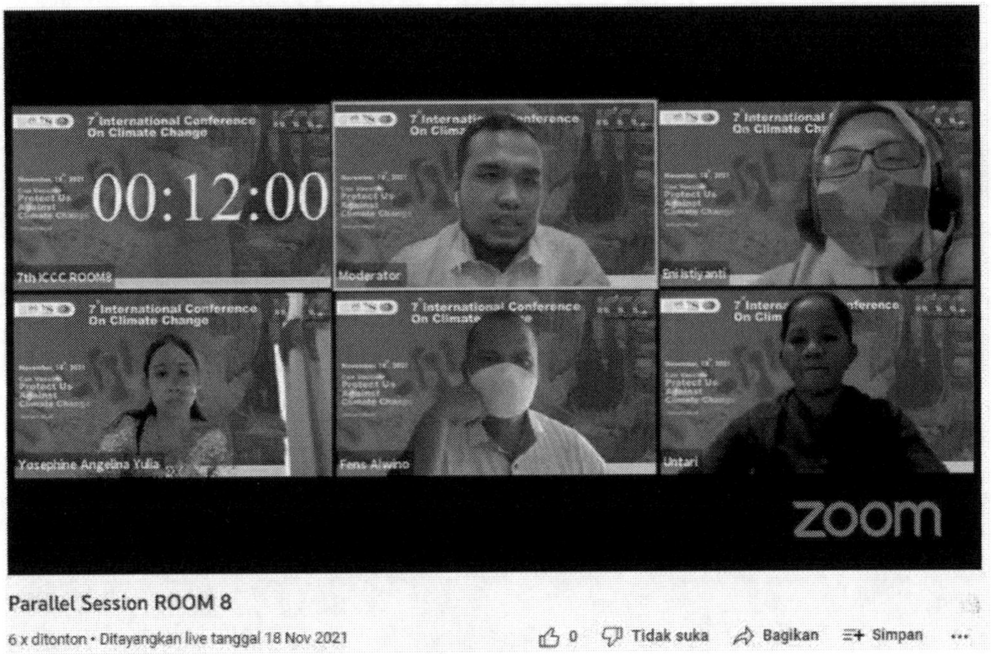


Figure 8. The parallel author presentation in room 8 of 10

***The success of delivery of the conference***

The success of the conference is well recorded at the Youtube Video, and can be seen in the following links:

- Plenary session: https://youtu.be/Uy1Jym17bQQ
- Parallel room 1: https://www.youtube.com/watch?v=ElvznXuLGvc
- Parallel room 2: https://youtu.be/RZpkICRYZhU
- Parallel room 3: https://www.youtube.com/watch?v=CJmjmsdna8g
- Parallel room 4: https://www.youtube.com/watch?v=PcaS-5DQNb8
- Parallel room 5: https://www.youtube.com/watch?v=IWwdXKDFyzY
- Parallel room 6: http://www.youtube.com/watch?v=y3rz8Jlwa5E
- Parallel room 7: https://www.youtube.com/watch?v=3GIGdutzc54
- Parallel room 8: https://www.youtube.com/watch?v=JaNZQ2asDyQ
- Parallel room 9: https://youtu.be/ed32ONyAgHQ
- Parallel room 10: https://www.youtube.com/watch?v=SohNwfIm2aQ

Thank you for your attention.

Chairman,
Prof. Dr. Maria Theresia Budiastuti

IOP Conf. Series: Earth and Environmental Science 1016 (2022) 011002 doi:10.1088/1755-1315/1016/1/011002

# Peer Review Declaration

All papers have been peer reviewed through processes administered by the Editors. Reviews were conducted by expert referees to the professional and scientific standards expected of a proceedings journal published by IOP Publishing.

- **Type of peer review: Double-blind**
- **Conference submission management system: Google Form**
- **Number of submissions received: 140 articles**
- **Number of submissions sent for review: 121 manuscripts**
- **Number of submissions accepted: 59 papers**
- **Acceptance Rate (Number of Submissions Accepted / Number of Submissions Received X 100): 42.14%**
- **Average number of reviews per paper: 2 reviews**
- **Total number of reviewers involved: 25 experts**
- **Any additional info on review process: Each paper took 2-4 weeks for revisions after review.**
- **Contact person for queries:**
  Name : Komariah, PhD.
  Affiliation: Sebelas Maret University, Indonesia
  Email : komariah@staff.uns.ac.id

# The role of land use/cover type in influencing hydrological component of a watershed in Chancho and Sorga Sub-watersheds, East Wollega Zone, Oromia, Ethiopia

**Tolera Megersa[1]* and Dessie Nedaw[2]**

[1]Department of Environmental Science, Wollega University, Ethiopia

[2]Department of Geology, Addis Ababa University, Ethiopia

*Corresponding author: toleramegersa@gmail.com

**Abstract.** Watershed hydrology can be influenced by land use/cover types of the study areas. The activities of humans exert more pressure in influencing the natural dynamism of the watersheds and hydrology through continuous and steep slope cultivation of land. The objective of this study was to assess the influences of land use/cover types and slope gradient on hydrological component of a watershed. The hydrological component of a watershed was simulated by SWAT model with arc-GIS where land use/cover map, slope map and hydrological response unit were used as an input. Farmland land has produced more surface runoff and water yield as compared to the other land use types followed by urban land, while Forest and grassland produced less surface runoff and water yield but higher in evapotranspiration, percolation, ground and lateral flow of water than cultivated and urban land. Area of farmland land is more responsive to the change of surface runoff and water yield than others land use/cover types. As farmland land size increases both surface runoff and water yield tends to decrease and had a negative correlation, whereas, other hydrological variables were less responsive to the change in land size. Slope gradient of land use/covers type contributes to the water imbalance of the watersheds. As slope increases surface runoff and water yield also decreases in farmland, grass, and forest land, whereas, increases in urban land as slope increased. The influences of land use/cover types and slope gradient on the watershed hydrology were more observed where human action is high in terms of continuous and steep slope cultivation and urban built-up expansion and appropriate land use and watershed management is paramount important for water balance in the watershed.

**Keywords:** land use/cover, soil, slope, watershed hydrology, simulation

## 1. Introduction

Flow regimes and water availability can be influenced by a change in climate, land use/land cover (LULC) [1]. Climate variability is suspected of the changes in global precipitation patterns started in the $20^{th}$ century [2], causing a shift in the global water cycle and resulting in the redistribution of water resources in time and space [3]. Humans are primarily responsible for LULC changes, which affect water partitioning among various hydrological pathways such as interception, evapotranspiration, infiltration, and runoff [4]. The climate and LULC changes are predicted to will going to increase. It will the influence hydrological process and water resources, arid and semi-arid regions will be most affected [5].

Land use has changed ecosystems more rapidly and extensively in the last few decades than in any comparable period. Agriculture and urban land expansion and forest degradation with a day to day activities of mankind resulted in temporal and spatial change in land use land cover [6]. Furthermore, powerful natural processes and phenomena such as poor climate and natural terrain have been identified as causes of land use and land cover changes. These dynamic activities have resulted in changing environments, resulting in unprecedented land degradation and natural resource depletion [7].

The effects of LULC and climate change are frequently investigated using scenario simulations and models. Despite significant progress in assessing the impacts of LULC and climate change on water resources [8] most studies have focused on individual factors (i.e., either LULC or climate); thus, the role of LULC and climate change is not well understood because their contributions are difficult to separate and vary regionally [1].

The role of land use/cover types in influencing the hydrological component of a watershed by considering the slope gradient was not yet studied. That is why this research was intended to assess and quantify the contribution of each land use/cover and slope gradient on watershed hydrology. The two Sub-watershed was taken for SWAT model calibiration and validation of hydrological parameters, whereas the Sorga Sub-watershed was used only for validation purpose. Both study areas shared a similar agro-ecology.

The study area (Chancho sub-watershed) is dominated by farmland and the Sorga Sub-watershed is a relative forest cover dominated sub-watershed The area are experienced in high rainfall resulted in more surface runoff. Farmlands are expanded at the expense of forest, grass and wetland and more runoff and reservoirs sedimentation is a widely seen problem in the study area. Therefore, this research was identified the role of land use/cover type at varying slope gradients in affecting the hydrological component of a watershed.

## 2. Materials and methods

### *2.1. Study area*

The study was carried out in the Chancho and Sorga sub-watersheds of East Wollega Zone, Oromia, Ethiopia. The study area was 328 kilometers from Addis Ababa along the main road to Asosa. (Figure 1)

*2.1.1. Chancho and Sorga Sub-watershed.* Chancho Sub-watershed is found in Diga District at 15 Km from Nekemte town to the West. It is located at 9°01' North and 36°28.2'East at about 2 km from the Nekamte. The total area of the watershed, including the water bodies, is 2,109 hectares, with water covering 250 hectares of the land. The watershed is located between 2,100 and 2260 meters above sea level (masl). Cultivated land is the most common land use/cover type in both watersheds, followed by forest and water bodies in the Sorga and Chancho Sub-watersheds, respectively (Figure 2). Sorga Sub-watershed is located in Nekemte town, which borders Diga District, East Wollega Zone, Oromia. It is situated in the western part of the region, along the main asphalt road leading to Asosa. A watershed has a total area of 3817 hectares, of which 30 and 326 are reservoir and urban land areas, respectively. The Sorga Sub-watershed is elevated around 2100-2260 masl.

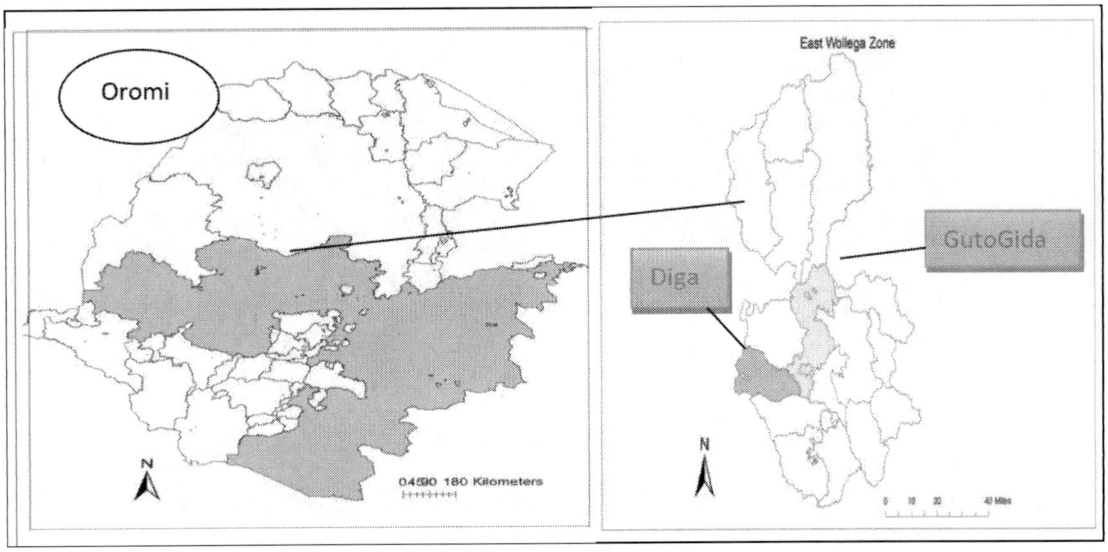


**Figure 1.** Map of the study area

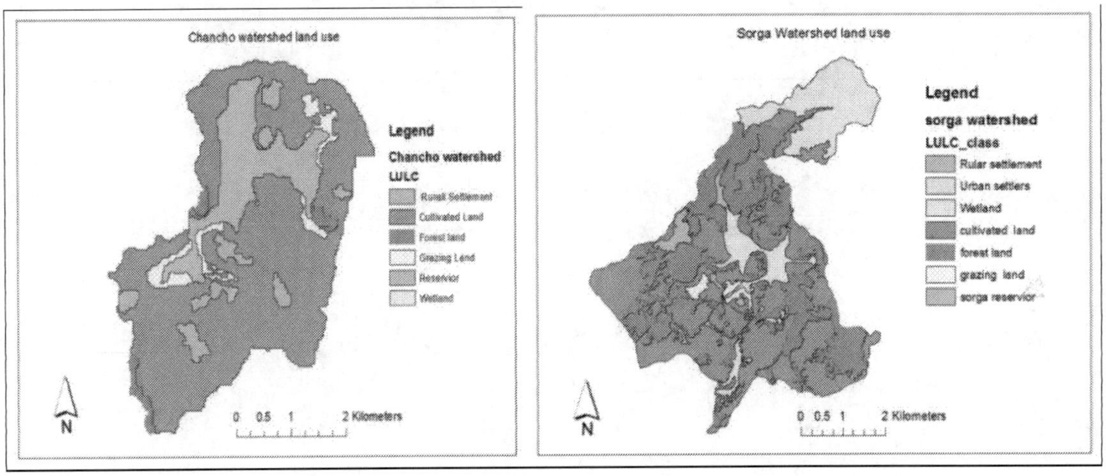


**Figure 2.** Land use map of the study watersheds

*2.1.2 Slope classification of the study area.* Chancho and Sorga Sub-watershed is categorized as steep to very steep, respectively, that is (15-30% and 8-15%) slope according to FAO classification [9] (Figure 3)

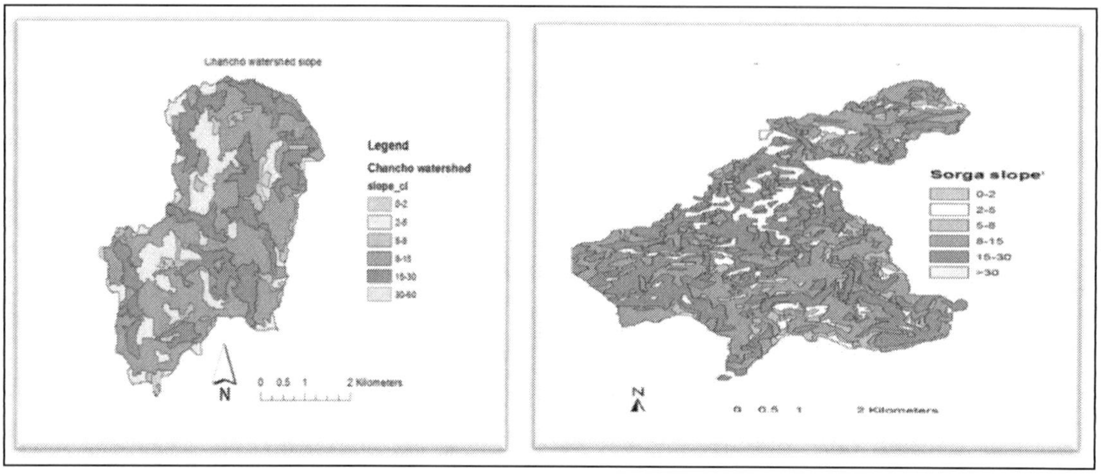


**Figure 3.** Map of slope classification of Chancho and Sorga Sub-watersheds

SWAT integrated with arc-GIS with 10%,5%, and 10% threshold percentage of land uses, soil and slope, respectively to create Hydrological Response Units (HRUs). Sorga Sub-watershed has a total of 222 and Chancho sub-watershed 301 HRU has a distinct land use, land management, soil, and slope. Simulation of the hydrological component by integrating the SWAT model with arc-GIS of the study was undertaken (Figure 4).

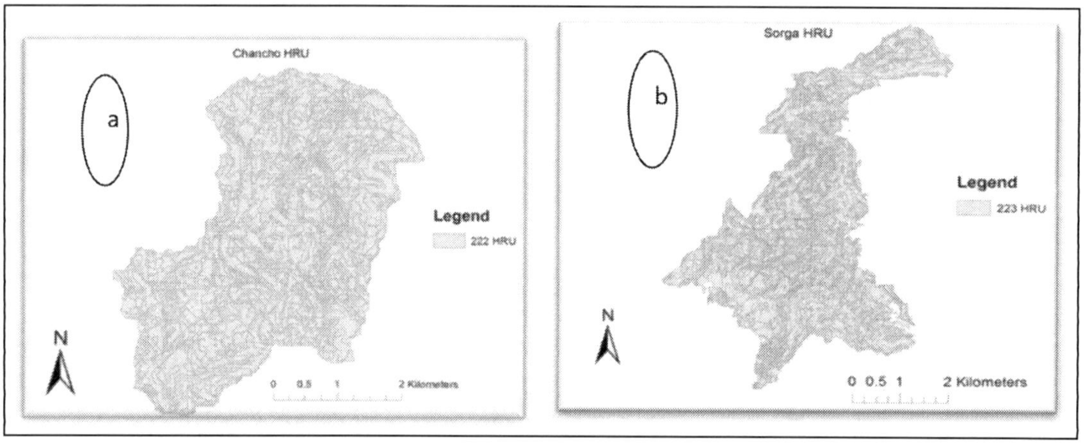


**Figure 4.** Hydrological response units of Chancho (a) and Sorga (b) Sub-watersheds

*2.1.3. Climate condition.* The study area is moderately cool, corresponding to a typical medium (moist middle altitude) agroclimatic zone. Rainfall, maximum and minimum temperatures, sun hour, wind speed, and relative humidity were calculated using data from three meteorological stations. The average annual rainfall in the study area was 2022 mm, and the average annual temperature was 19.85°C. (Figure 5). The year is divided into two seasons in the area: November to April categorized as the dry season and May to October is the rainy season.

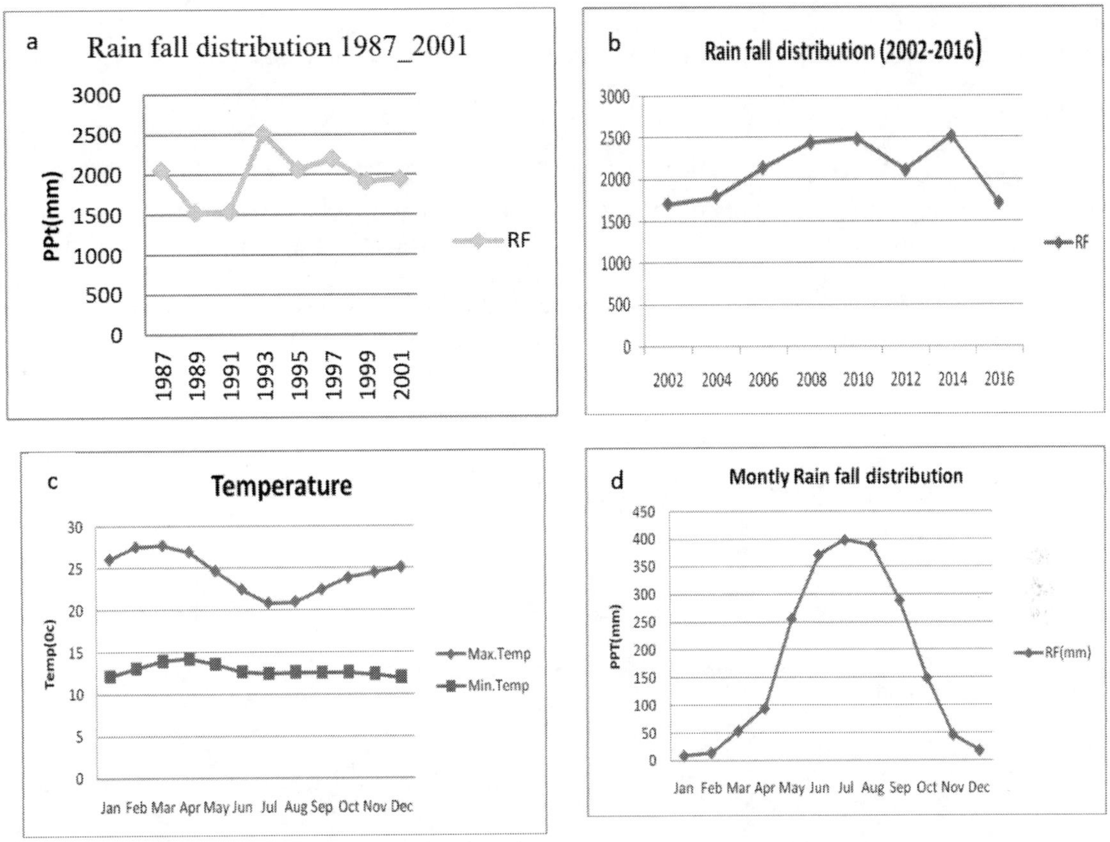


**Figure 5.** Rainfall distribution of 1987-2016 (a and b); 2016 annual air temperature (c) and rainfall (d) of the study area

*2.2. Materials*

The study methods involved Digital Elevation Model (DEM) with arc-GIS for delineation of the study Sub-watersheds and classification of soil type, slope, and land use of the study. The weather data of the year 1986– 2016 was used for this research. The Soil and Water Assessment Tool (SWAT) was used to simulate long-year hydrological parameters from HRU. This helps to predict the influences of land use/ cover type and slope gradients on the hydrological component in terms of water balance of the study areas.

*2.3. Methods*

Two Sub-watersheds, having different biophysical characteristics were selected for the study. These watersheds were selected for detailed modeling of hydrological components of a watershed. Thirteen year climatic data was used as an input for the model. Observed river flow data of (2011-2012) was used for model calibration. Sorga is a non gauged sub watershed and recorded flow data was transferred from Chancho river flow data where they share a common climatic data. They were used to estimate the general effects on watershed hydrology through simulation of hydrological components. Land classification was done by arc-GIS downloaded from Google earth Slope map and their categories were done by arc-GIS using DEM (Digital Elevation Model) of the study area

Therefore, the role of land use/cover types and slope gradient on hydrological variables was simulated by the model. The correlation between slope gradient, area of land, and hydrological variables were analyzed by using a scatter plot. These help to identify which land use/cover types are more affecting the hydrology of a watershed and need a priority intervention. This study adopted geographic information systems (GIS) and statistical (such as analysis of variance and non-parametric correlation analysis). Simulated hydrological components were compared. Simulated results of hydrological variables were subjected to correlation analysis and LSD of analysis of variance.

## 3. Result

### *3.1. SWAT parameters calibration*

The flow was simulated for 20 years including three years warm-up periods. Sensitivity analysis was carried out to identify the most sensitive parameters that determine the model output and we found all within the allowable ranges under the two sub-watersheds (Figure 6 & 7). The most influencing parameter was the soil bulk density in both sub-watersheds. Calibration results showed that there is a good agreement between simulated and gauged flows. The most sensitive to flow parameters as described in Table 1.

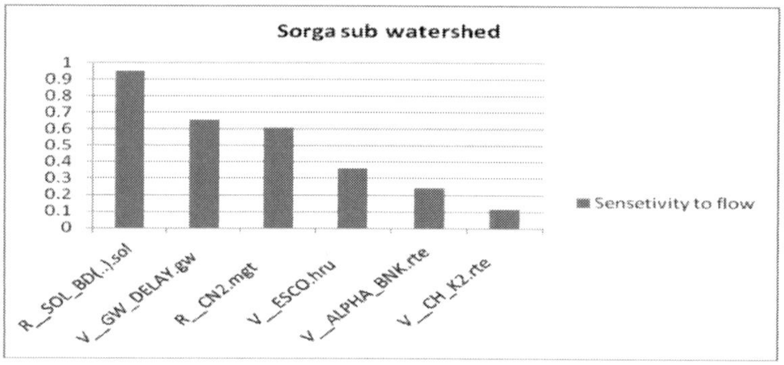


**Figure 6.** Sensitivity to flow in Sorga Sub-watershed

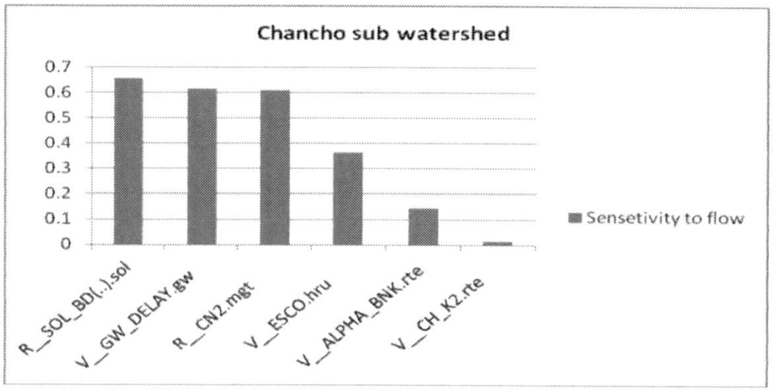


**Figure 7.** Sensitivity to flow in Chancho Sub-watershed

**Table 1.** Description of sensitive parameters to simulation flow

| Parameter Name | Description |
|---|---|
| 3:R__SOL_BD(..).sol | Soil Bulk Density |
| 2:V__GW_DELAY.gw | Groundwater delay time(days) |
| 1:R__CN2.mgt | SCS runoff Curve Number ( *Cn2*) |
| 6:V__ESCO.hru | Soil evaporation compensation factor |
| 4:V__ALPHA_BNK.rte | Baseflow alpha factor (days) |
| 5:V__CH_K2.rte | Channel effective hydraulic conductivity (mm/hr) |

The observed daily discharge recorded from existing river flow monitoring data was compared to the simulated daily flow results of the two Sub-watersheds (Sorga and Chancho). The simulated flow rate was found to be very close to the actual monitored flow rate when the default parameters were used (Table 2), with an R2=0.83, NS=0.74, and RE=0.21 for Sorga and R2=0.97, NS=0.96 and RE=0.5 for Chancho Sub-watersheds. The SWAT model parameters were calibrated until there was a satisfactory agreement between the simulated flow rate and the observed flow data. Generally, the error rates were 10% and 17% for Chancho and Sorga Sub-watershed, respectively. Less than 15%, the model is regarded to be very good for Chancho sub watershed and a good fit in Sorga Sub-watershed (Figure 8 & 9).

**Table 2.** Evaluation of the accuracy of the SWAT Model

| Watersheds | Coefficient of determination ($R^2$) | Nash-Sutcliffe coefficients (RNS) | Relative errors (RE) |
|---|---|---|---|
| Calibration (Sorga) | 0.83 | 0.74 | 0.21 |
| Calibration (Chancho) | 0.97 | 0.96 | 0.5 |

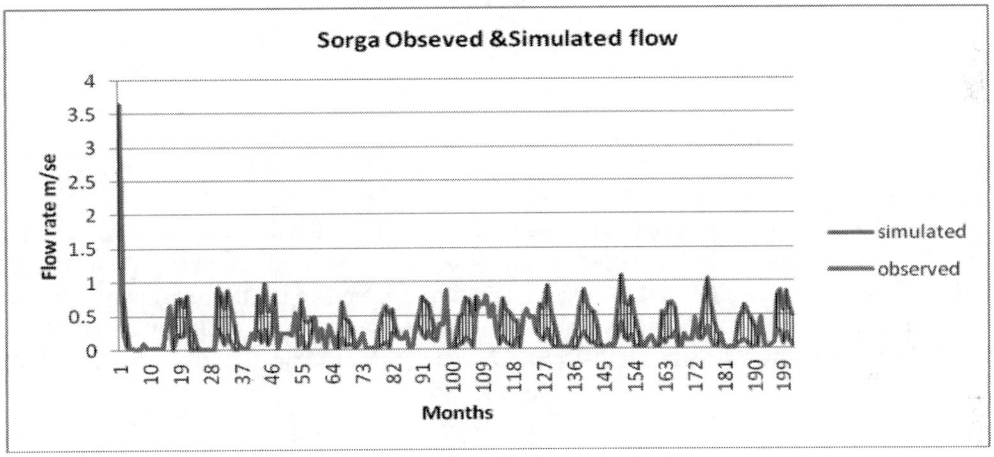


**Figure 8.** Recorded Vs simulated monthly discharge during the calibration period of the 2nd study area

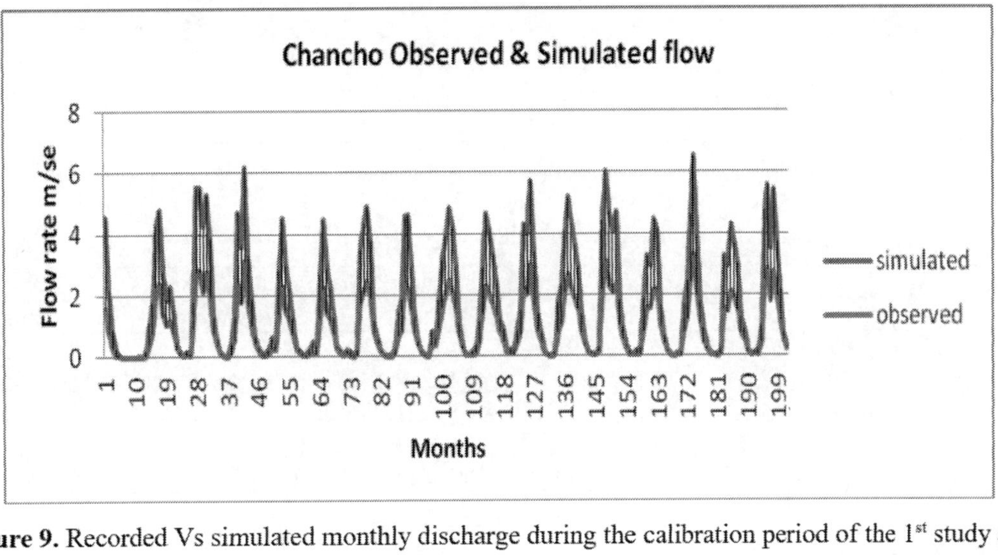


**Figure 9.** Recorded Vs simulated monthly discharge during the calibration period of the 1st study area

*Model validation.* Flow data were used for model validation. Simulated and observed were very close to each other in Sorga Sub-watershed and the model is ideal to predict the hydrological component of a sub-watershed (Figure 10).

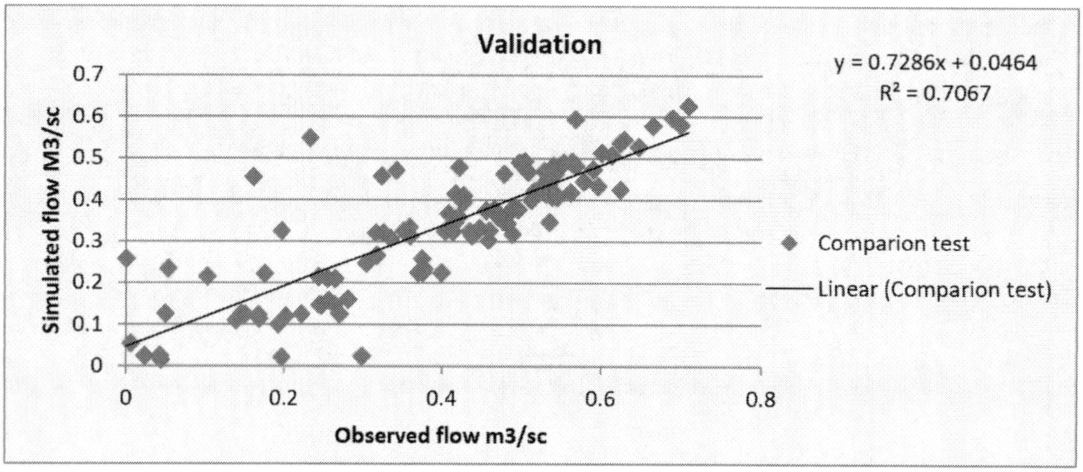


**Figure 10.** Model recorded and predicted river flow

*3.2 Role of land use/cover types on the hydrological component of a watershed*

Evapotranspiration and soil water are the two hydrological parameters that negatively affect the water balance of the watersheds. The simulated mean value of Evapotranspiration (ET) was higher in grassland(736.3mm) followed by residential land (654mm) as compared to the minimum value of urban and cultivated land (569mm and 578mm), respectively. Grassland was also higher in soil water content (131mm) than others with no significant difference between them (Table 3).

**Table 3.** Simulated values of ET and SW

| LULC | Mean and SE of hydrological parameters | |
|---|---|---|
| | ETma | SWmm |
| Cultivated | 578 ± 8.01 [a] | 84. 4 ± 3.4 [a] |
| Forest | 612 ± 8.4 [a] | 97.4 ± 3.4 [a] |
| Urban | 569 ± 7.54 [a] | 96.8 ± 4 [a] |
| Grass | 736 ± 16.7 [a] | 131 ± 1.9 [a] |
| Residential | 654 ± 15.06 [a] | 130 ± 2.77 [a] |

Notes: ET= Evapotranspiration, SW=Soil water, mm=millimeter

**Table 4.** Simulated mean values of hydrological parameters

| LULC | Mean and standard error of hydrological parameters | | | | |
|---|---|---|---|---|---|
| | PERC mm | LAT_Qmm | GWmm | SURQmm | WYLDmm |
| Cultivated | 691.6 ± 17.3 [a] | 76.5 ± 5.9 [a] | 653 ± 16.9 [a] | 641.7 ± 19.2 [a] | 1390 ± 29.4 [a] |
| Forest | 702 ± 18.6 [a] | 73 ± 5.2 [a] | 664 ± 18.1 [a] | 610 ± 14.3 [a] | 1351 ± 30.5 [a] |
| Urban | 694 ± 15.16 [a] | 79 ± 8 [a] | 661 ± 21.08 [a] | 640 ± 14.9 [a] | 1379 ± 15 [a] |
| Grass | 736 ± 20.9 [a] | 41 ± 1.6 [a] | 681 ± 19.1 [a] | 621 ± 21 [a] | 1343 ± 38 [a] |
| Residential | 736 ± 15.16 [a] | 40.7 ± 3.78 [a] | 680 ± 23.6 [a] | 629 ± 24.3 [a] | 1353 ± 45.6 [a] |

Notes: PERC=Percolation, LAT_Q=Lateral flow, GW= Ground water, SURQ= Surface runoff, WYLD= Water yield and mm=millimeter

Percolation, ground water, lateral flow and surface runoff are the important hydrological parameters contribute for the total amount of water yield produced in the watersheds. Grass land was higher in percolation, Soil water and ground water flow (736mm,131mm &681mm), than cultivated land (691mm, 84.4mm & 653mm), respectively and as well as higher than forest and urban land. PERC and GW was low in cultivated land. Surface runoff (641.7mm) of cultivated land was higher than forest and grassland (610mm & 621mm), respectively. The total water yield (1390mm) in cultivated land was greater than urban, forest and grassland (1379mm, 1351mm & 1343mm), respectively. The difference in hydrological variables were no significant difference observed between them at 95% confidence level (Table 4)

*3.2.1. Correlation between hydrological variables.* Table 5 shows the result of spearman rank correlation. Almost all simulated variables showed a positive relationship to each other except evapotranspiration and soil water. Surface runoff, Lateral flow, and ground flow are assumed to be water contributing systems whereas evapotranspiration, soil water and percolation are water losing systems. However, the findings of this study revealed that percolation is a hydrological variable that contributes water and has a strong positive correlation with total water yield and ground water. As a result, surface runoff, ground water, and percolation were strongly related to total water yield, whereas evapo-transpiration and soil water content were negatively related to water yield.

**Table 5.** Correlation among hydrological variables

| | etmm | swmm | percmm | surqmm | lat_qmm | gw_qmm | wyldmm |
|---|---|---|---|---|---|---|---|
| Etmm | 1.0000 | | | | | | |
| swmm | 0.3719* | 1.0000 | | | | | |
| percmm | -0.4039* | -0.0361 | 1.0000 | | | | |
| surqmm | -0.0557 | -0.0399 | 0.3449* | 1.000 | | | |
| lat_qmm | 0.0139 | 0.3501* | 0.1155 | 0.0052 | 1.0000 | | |
| gw_qmm | -0.4237* | -0.0499 | 0.9938* | 0.3270* | 0.1115 | 1.0000 | |
| wyldmm | -0.2965* | -0.0171 | 0.7807* | 0.7807* | 0.2348* | 0.7734* | 1.0000 |

Note: *significant correlation

*3.2.2 Role of slope on hydrological component.* Simulated hydrological variables from different slope gradient indicated that cultivated land was more responsive to difference in slope gradient than others land use/cover types. The role of slope gradient difference on hydrological variables was investigate on cultivated land. Evapotranspiration, percolation and ground water of cultivated land were very less responsive to change in slope gradient (Figure 10). Slope dependent variables identified by this study were surface runoff, Lateral flow and water yield in a cultivated land. Lateral flow in cultivated land was higher at>30% slope gradient with a significant difference ($p<0.05$) than others slope gradient. Surface runoff and water yield values were higher at 15-30% slope gradient in cultivated land and significant difference ($P<0.05$) than 0-5% and5.8% and 8-15% slope. Whereas, Evapotranspiration, Soil water, percolation and ground water has no significant difference between varying slope gradient in cultivated land (Figure 10).

*3.2.3. Spatial correlations between response variable and land use/slope areas.* This is to compare the relationships between variables and land use/cover areas. From the scatter line graph of simulated hydrological variables surface runoff and water yield values were more responsive to area difference of land use/cover types as compared to other hydrological variables. Surface runoff and water yield showed a similar trend in response to slope gradient change. Therefore, only simulated values of surface runoff and areas was subjected to correlation analysis by scatter line graph (Figure 11). The result indicates that surface runoff from cultivated, and forest land were negatively correlated to land size/area, this confirm as area of land increased surface runoff decreases. Whereas surface runoff for urban and grass land was increased as area of land use/cover land increased (Figure 11).

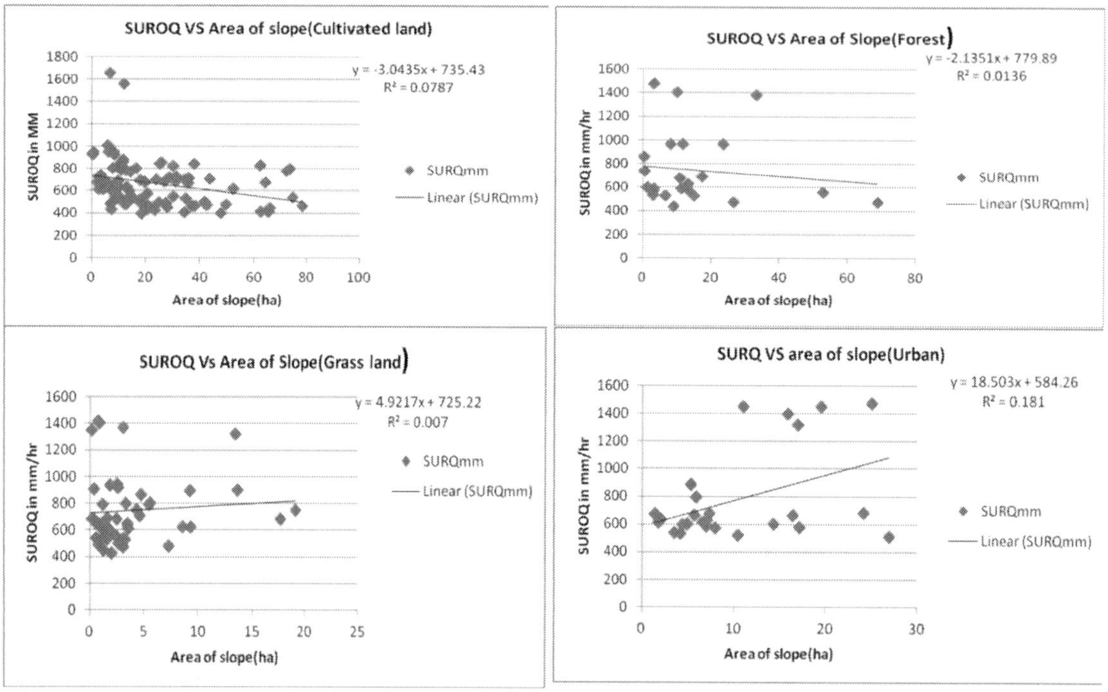


**Figure 11.** The relationship b/n surface runoff and land size

### *3.3. Discussion*

The SWAT model performance was evaluated by using observed river flow data.. The model application gave satisfactory goodness of fits levels. The variation in the values of hydrological parameters was largely due to the dominant land use/cover types and slope gradient. Land cover types contribute more in influencing the water balance in a sub-watershed. The result indicated that cultivated and urban land affects the natural water balance of the study sub watershed, whereas forest and grass land plays a great role in keeping the natural water balance.

The mean values of surface runoff and water yield of cultivated land was higher than values of forest and grassland followed by urban land, but no significant difference at among them. .The higher Surface runoff from cultivated land is also influences the hydrological component of watersheds. Continuous and steep slope cultivation of land accelerates the removal of top soil by surface runoff and resulted in onsite loss of soil and offsite water quality reduction and reservoirs sedimentation in the study sub watershed. This finding is in agreement with [10] in Rwanda, reported that the increment of surface runoff was due to the expansion of cultivated land and urbanization at the expenseof vegetation cover. Study by [11] in Lake Tana catchment and Beles watershed, Ethiopia also reported the increase of surface runoff is due to the expansion of cultivated lands and reduction of vegetative.

Forested and grass land areas are low in surface runoff and water yield, due to its capacity to control surface runoff. Both forest and grass land were higher in percolation, lateral flow, ET and ground water flow than cultivated and urban land with no significant difference at ($p>0.05$) (Table 3). The effects of the interaction between slope gradient surface runoff was plotted by using a bar graph (Figure 10). The result indicated that as slope increases from 5-15%, surface runoff also increasing almost similarly and decreases as slope gradient >15%. Slope gradient of forestland increased from 5-20% surface runoff was increased and decreased as slope going beyond 20%, this is because the area getting denser as slope gradient increased. In urban land as slope gradient increases surface runoff also increased. Grass land up to 15% slope produced almost a similar amount of surface runoff and as slope increased beyond 15% it was decreased at the increasing rate. As slope gradient increased beyond certain slope range surface runoff tend to decrease under all except urban land, this is due to the cumulative effects of surface runoff at the lower slope position and need further investigation by using runoff plot from each HRU.

The result indicated that as area of cultivated and forest land increased surface runoff decreased, this is due to the larger areas may have more potential land for percolation and evapotranspiration and vice-versa. In a grass land there was a slight increased surface runoff as the area increased, in urban land, as area of land use/cover type increased surface runoff also increased; it is due to increased impervious and paved area of urban land (Figure 11). Therefore, land use /cover types at varying slope gradient affects hydrological component of a watershed through altering the amount of water input and output from hydrological response units (HRUS) of the land.

## 4. Conclusion

Human activities in a watershed affect the watershed dynamics in the form of physical soil degradation and hydrological water imbalance in the study sub-watershed. More runoff and water yield were recorded from cultivated land and resulted in downstream reservoirs sedimentation. The natural dynamism of the watersheds was influenced through the loss of fertile topsoil and water quality reduction, whereas forest and grass cover plays a great role in regulating hydrological water balance. Dynamisms of the watershed in terms of water imbalance are mainly resulted from the use of land followed by slope gradient. The hydrologic dynamisms of the watersheds could be resulted from difference land cover and slope. Therefore, avoiding steep slope cultivation, reducing slope gradient by construction of physical structures, implementing appropriate watershed management strategies and land use according to land capability classification are important keeping the natural dynamism of the watersheds and regulate the local climate variability.

## References

[1] Wang G, Liu J, Kubota J and Chen L 2007 Effects of land-use changes on hydrological processes in the middle basin of the Heihe River, northwest China *Hydrol. Process.* **21** 1370–82

[2] IPCC 2007 Climate Change 2007: The Physical Science Basis. Contribution of Working Group I to the Fourth Assessment Report of the Intergovernmental Panel on Climate Change. Summary for Policymakers. *Climate Change 2007: The Physical Science Basis. Contribution of Working Group I to the Fourth Assessment Report of the Intergovernmental Panel on Climate Change* (United Kingdom and New York) p 996

[3] Murray S J, Foster P N and Prentice I C 2012 Future global water resources with respect to climate change and water withdrawals as estimated by a dynamic global vegetation model *J. Hydrol.* **448–449** 14–29

[4] Sterling S M, Ducharne A and Polcher J 2013 The impact of global land-cover change on the terrestrial water cycle *Nat. Clim. Chang.* **3** 385–90

[5] Vörösmarty C J, McIntyre P B, Gessner M O, Dudgeon D, Prusevich A, Green P, Glidden S, Bunn S E, Sullivan C A, Liermann C R and Davies P M 2010 Global threats to human water security and river biodiversity *Nat. 2010 4677315* **467** 555–61

[6] Rawat J S and Kumar M 2015 Monitoring land use/cover change using remote sensing and GIS techniques: A case study of Hawalbagh block, district Almora, Uttarakhand, India *Egypt. J. Remote Sens. Sp. Sci.* **18** 77–84

[7] Assen M 2011 Land Use/ Cover Dynamics and Its Implications In The Dried Lake Alemaya Watershed *J. Sustain. Dev. Africa* **13** 267–84

[8] Krysanova V and White M 2015 Advances in water resources assessment with SWAT—an overview *Hydrol. Sci. J.* **60** 771–83

[9] FAO 2002 *Major soils of the world land and water digital media series* (Rome, Italy: FAO)

[10] Karamage F, Zhang C, Fang X, Liu T, Ndayisaba F, Nahayo L, Kayiranga A and Nsengiyumva J B 2017 Modeling Rainfall-Runoff Response to Land Use and Land Cover Change in Rwanda

(1990–2016) *Water* **9** 147
[11] Woldesenbet T A, Elagib N A, Ribbe L and Heinrich J 2017 Hydrological responses to land use/cover changes in the source region of the Upper Blue Nile Basin, Ethiopia *Sci. Total Environ.* **575** 724–41

**Acknowledgements**
The research was founded by the Department of Environmental science, Wollega University. The author sending a gratitude to Addis Ababa Environmental protection Authority for the laboratory facility and Diga and Guto Gidda District for their cooperation in providing transport services during the fieldwork.

# Climate variability and the responses of crop yields to agricultural drought in the East Wollega Zone, Oromia National Regional State, Ethiopia

**Tasfaye Fayera***
Center for Studies of Environment and Society, Wollega University, P.O box 395, Nekemte, Ethiopia
Corresponding author: ftesfu2015@gmail.com

**Abstract:** Smallholder farmers in Ethiopia have been facing severe climate related hazards; droughts and floods that negatively affect their livelihoods. This study was aimed to assess climate variability and the responses of crop yields to climate induced agricultural drought. Time-series crop yields (1993-2018) and climate data (1989 to 2019) was obtained from Central Statistical Agency and NASA's portal respectively. The changes in temperature and precipitation were analyzed using Mann Kendall trend test. The agricultural drought index was analyzed using R-software. The correlation between the selected yield crops and drought indices were evaluated using Pearson correlation coefficient. Determinants on climate change adaptation strategies were assessed using a multinomial logit model. The trends of seasonal and annual minimum temperatures and spring maximum temperature results were significantly warmed up (P <0.05).However, both seasonal and annual rainfall was slightly decreased. Moderate to severe agricultural drought intensities happened 7 times and one very wet year occurred during three decades. As the result, about 0.70% of the total land area was affected by moderate drought while about 30.73% of the land area was affected by moderate wet soil moisture and 68.57% of the rest others were observed as harmless land areas. The Person's correlation coefficients showed that crop yields due to the drought interruption were 93.9, 92.4, and 92.5% for maize, wheat, and finger millet, respectively. Determinants depicted that sex, family size, extension, and education were significantly affected soil and water conservation practice and tree planting at (P <0.05). Therefore, the efforts of local government and stakeholders should be superior to build the capacity of farmers to use appropriate agricultural technologies and supply drought & wet tolerant crop varieties; improve research and development programs.

**Keywords:** agricultural drought, climate change, correlation coefficient, precipitation, temperature trends

## 1. Introduction

Agriculture is a back bone of Ethiopian total populations since its agriculture is purely rain-fed type [1]. Climate change and variability plays a great role in agricultural production having a direct impact from the start of land preparation to the final harvest [2,3]. Temperature and rainfall in particular are critical elements in determining the output of cereal crops. While non-climatic factors like availability and access to farm inputs, land quality, tenure security, infrastructure, market and policy can be managed by farmers or the government in the short or long-term, little could actually be done to respond to the influence of climate, particularly rainfall would have a large impact on agricultural production and worsen poverty level explicitly in Sub-Saharan regions [4]. But its drought patterns have become less predictable and while temperatures are rising in rural populations of developing countries [5,6]. Variation in national yields due to climate variability is common in most of the rural areas of Ethiopia [7,8]. This is largely because of poverty levels are high and capacity to adapt to climate change. In East Wollega zones, where agriculture is dependent on rainfall, the occurrence of recurrent drought in some years and the heavy rain in other times, coupled with the increasing temperature harm the Wheat, Maize and Finger millet crop productions activities of the local farmers. The problems of soil acidity, crop diseases, lack of Scientific information and land degradation (e.g landslide), which are related to the warming temperature. Erratic precipitation have been causing crop yield reduction in the study area and farmers less likely change their habitual agricultural practice along with the changing climate in order to cope with the climate bearing problems. Thus; the main objective of study was to assess climate variability and the responses of crop yields to agricultural drought, in east wollega zone Oromia national regional state, Ethiopia.

## 2. Methodology

### *2.1. Study area*

East Wollega zone is located in western part of Ethiopia (Figure 1) with an altitude ranging from 900 to 3,276 m above mean sea level. It lies between latitudes 8°30'0"to 10°20'0"N and 36°05'0" to 37°10'0"E at western Oromia. In 2010, the projected population was estimated to be 1,661,092 with an annual growth rate of 2.5%, mainly rural and engaged in rain-fed agriculture [9]. Administratively, East Wollega Zone is subdivided into 17 woredas (districts) with one special administrative city at an altitude ranging from 950 to 3276 m above mean sea level with 90% of the woredas described as midland and 10% lowland [10]. The climate is generally semi-arid with mono modal types of rainfall patterns. Annual rainfall ranges from 800-2,200 mm per year [10].

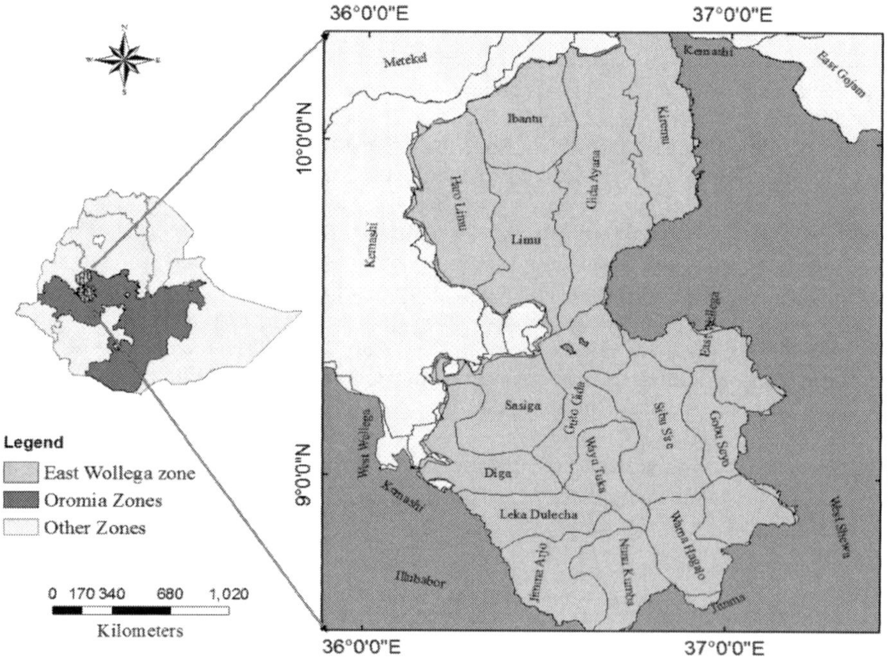


**Figure 1.** Map of study area

### *2.2. Data sources and sampling*

Survey research design and mixed research methods were used. Examination research system was used to characterize and quantify the households, climate variability pattern and local community determinants to adopt the climate change adaptation strategies while correlation methods was used to quantity responses of yield crops to climate induced agricultural drought indices. The data was mainly obtained from a survey conducted at the six villages using a structured questionnaire. Meteorological data (recorded from 1989 to 2019 was obtained from Power data access viewer, which is available at https://power.larc.nasa.gov/data-access- viewer/ at 0.50 × 0.50-degree resolution) and data for selected time series crop yield was also Obtained from Central statistical agency from 1993-2018.

Stratified sampling method with simple random was employed for survey. At first, woredas of the zone is clustered in to three eco- climatic zones. Four out of 17 woredas in the zone were purposively selected because they were Wheat, Maize & Finger millet growing areas and also vulnerable to climate change. So, Haro limu from wet, Kiramu from moderate drought, Nunukumba and Lekadulecha woreda from normal drought distribution of the zone was selected.The sample respondents were selected using simple random sampling technique from each selected woredas of village those who have lived in the area for more than ten years. Purposely six Villages based on their Agro-ecological zones and its proximity to the woredas was selected. Those Villages are namely: Haro and Bariso from Haro limu; Bado from Lekadulech and Arko; gudatu from Nunu kumba; Badessa and Boka from Kiramu Woreda. Accordingly 140 households were selected (43 farmers from the highlands, 47 farmers from midlands and 50 farmers from lowlands).

### *2.3. Data analyses*

*2.3.1. Trend analyses of temperature and precipitation.* Modified Mann Kendall Test has been used together with the Sen's Slope estimator for the determination of trend and slope magnitude to detect time series trended climatic factors was analyzed using R-software with at 0.05 ($Z\_\alpha/2 = \pm 1.96$) significance levels. It is defined as the sum of the number of positive differences minus the sum of the number of negative difference [11] and is calculated using this equation (1).

$$S = \sum_{k=1}^{n-1} \sum_{J=k+n}^{n} \operatorname{sgn}\left(Y_j - Y_k\right) \quad \text{.......... (1)}$$

Yj– Yk, where j =2, 3, 4…and k=1, 2, 3… then the function is calculated as Eq. 2.

$$\operatorname{Sgn}(\theta) = \begin{cases} 1 \text{ if } \theta > 0 \\ 0 \text{ if } \theta = 0 \\ -1 \text{ if } \theta < 0 \end{cases} \quad \text{Where; sign } (\theta) = \text{sign (Yj -YK)} \quad \text{.......... (2)}$$

*2.3.2. Analyses of agricultural drought indices.* Potential evapo-transpiration, water balance and standardized precipitation evapo transpiration index (SPEI-3) were analyzed using R- software. Drought indices such as severity, duration and intensity, were analyzed using the outputs from SPEI analyses (Table 1) [12,13]. Water balance (ω) as the difference between precipitation (PRCP) and the potential evapo-transpiration (PET) was calculated using equation (3) [14,15].

$$\omega = PRCP - PET \quad \text{.......... (3)}$$

Where: ω is water balance, PRCP is precipitation and PET is potential evapo-transpiration.

Duration is the length of period, in which the SPEI is continuously negative and it is expressed in terms of consecutive months with no rainfall, and the severity of drought was calculated using equation (4).

$$Severity(\Sigma) = \Sigma_{i=1}^{D} SPE_i \quad \text{.......... (4)}$$

Where: D is dry duration in which SPEI continuous negative. The intensity of drought is done by dividing the drought severity to its durations.

$$Intensity(I) = \frac{\Sigma\, SPEI}{D} \quad \text{.......... (5)}$$

**Table 1.** Classification of SPEI values drought [16,17]

| Anomalies | Range of SPEI value | Category |
|---|---|---|
| | 2 < SPEI ≤ Max | Extremely wet |
| **Positive** | 1.5 < SPEI ≤ 2 | Very wet |
| | 1 < SPEI ≤ 1.5 | Moderately wet |
| **None** | -1 < SPEI ≤ 1 | Normal |
| | -1.5 < SPEI ≤ -1 | Moderately dry |
| **Negative** | -2 < SPEI ≤ -1.5 | Severely dry |
| | Min < SPEI ≤ -2 | Extremely dry |

*2.3.3. Correlation coefficient analysis.* The Pearson's correlation coefficient analysis was performed to determine the crop yield and agricultural drought indices relationship was calculated using equation (6).

$$R = \frac{\sum xy}{\sqrt{\sum[\sum x^2)(\sum y^2)]}} \quad \text{.......} (6)$$

Where R is correlation coefficient, X is independent and Y is dependent variables.

The value of R is always between -1 and +1: -1 ≤ R ≤ 1. If R= -1, then, it is a perfect negative relationship between X and Y. If -0.99 < R ≤-0.5, then, it is moderately negative relationship. If 0.49 ≥ R > 0, then, it is weak negative the relationship. If R = 0, then it refers to no relationship between the two variables. If 0 < R ≤0.49, then, the relationship is weak positive. If 0.5 ≤ R ≥ 0.99, then, the relationship is moderately positive and If R= +1, then, it is a perfect positive relationship between X and Y variables. R square or coefficient of determination shows percentage variation in Y which is explained by the entire X variable together. The coefficient of determination ($R^2$) indicates proportion of the variance in Crop yields, which are predictable from the climatic variables. Higher R square value is the better and the value always lies between 0 and 1. It can never be negative since it is a squared value [9].

*2.3.4. Determinants analysis.* SPSS Version 23.0 statistical software and Multinomial logit model was used to analyse farmers determinates to adopt climate change adaptation strategies X [18].

$$prob(Yi = j) = \frac{e^{\beta' j^{x}i}}{1+\sum_{k=0}^{5} e^{\beta' k X_i}} j - 0,1,2,\ldots 5 \quad \text{.......} (7)$$

Equation (7) is normalized to remove indeterminacy in the model by assuming βo= 0 and the probabilities can be estimated using equation (8):-

$$prob(Yi = \frac{j}{X_i} = \frac{e^{\beta' j X_I}}{1+\sum_{k=0}^{j} e^{\beta' k x_i}} j = 0,1,2.\ldots J\beta_0 = 0 \quad \text{.......} (8)$$

Maximum likelihood estimation of equation (8) yields the log-odds ratio as shown in equation (9).

$$lan(\frac{pij}{pik} = xi'(\beta j - \beta k) = xi'\beta j, if K = 0 \quad \text{.......} (9)$$

According to [10], associating between βj with the $j^{th}$ outcome is interpreted by using Marginal effect of the equation (10).

$$\frac{\partial p_j}{\partial x_i} = P_j(\beta_j - \sum_{k=0}^{j} p_k \beta_j) = p_{j(\beta_j} - \beta) \quad \text{.......} (10)$$

IOP Conf. Series: Earth and Environmental Science 1016 (2022) 012002 doi:10.1088/1755-1315/1016/1/012002

The estimated coefficients of multinomial logit model provide only the direction effect of independent variables on dependent variables, but estimate neither represents the actual magnitude of change nor probabilities [19,20]. Variance inflation factor (VIF $(X_i) = (1-R_i^2)^{-1}$) where, VIF is variance inflation factor; $Ri^2$ is the adjusted square of the multiple correlation coefficients that result when one independent variable $(X_i)$ is regressed against all others which used to test the multi co linearity for continuous independent variables. The value of VIF greater than 10% is assumed to indicate model multi-co linearity problem [20]

## 3. Results and discussions

### *3.1. Climate variability*

*3.1.1. Maximum temperatures.* Maximum temperatures trends of highland and Midland of east Wollega zone were insignificantly warmed up at (P >0.05) during 1989 – 2019. The values of Mann Kendall trend tests and its rates were explained in (Table 2) below. This finding is unlikely with study by [7,21,22]. Reported that the mean and maximum temperature had revealed increasing trends from 1901 to 2014 years in North and central Ethiopia and since 1978 across almost Ethiopian country. While spring season *Maximum temperatures* trends of low lands of study were sign ificantly warmed up at (P <0.05) which was likely with study by [7,21,22]. Reported that the mean and maximum temperature had revealed increasing trends from 1901 to 2014 years in North and central Ethiopia and since 1978 across almost Ethiopian country.

*3.1.2. Minimum temperatures.* Minimum temperature of spring and summer of highland of east Wollega zone showed a significantly increasing trend (P <0.01) and annual minimum temperature also significantly warmed up at P <0.05 (Table 2). This finding is also similar with Rebetez [23], which reported that annual minimum temperature trend of Ethiopia increased. Makombe et al. [7] also reported that the mean minimum temperature trends from 1901 to 2014 in north central Ethiopia also increased. Spring and summer minimum temperature of midland of east Wollega zone were significntly increasing trend at P<0.01.While its annual temperature was notably increased at P< 0.05 (Table 2). It is line with current study, Benti and Chimdi [24] reported significant increasing trends of summer and annual minimum temperature at western parts Ethiopia too. Minimum temperature of spring and summer of east Wollega zone lowlands were showed that significantly increasing trend at P <0.01 while its annual minimum temperature was especially increased at P < 0.05 (Table 2).

*3.1.3. Precipitations.* Precipitation of seasonal and annual highland of study area revealed downward trended at insignificant levels (P >0.05) with Mann Kendall trend test (S-value) of -67, -58 and -46 for spring, summer and annual precipitation, respectively (Table 2).This is in line with the finding of [20.4, 15, and 28].Said that annual and seasonal rainfall trend in north Ethiopia was non-significant, mean annual and seasonal rains fall in Ethiopia from 1960-2006 was statistically insignificant, and most farmers noticed changes in the amount and occurrence of rainfall in their locality yet observed insignificant annual rainfall trend in Northern Ethiopia.

Precipitation of seasons and annual in the midland of east Wollega zone revealed none significant decreasing trends (Table 2). Apparently, the present study is alike to the results found by Rebetez [23]. He reported statistically non-significant trend of mean annual and seasonal rainfall from 1960-2006. Precipitation of seasons and annual in lowland of study area revealed downward trended which are statistically insignificant at P-value greater than 0.05 (Table 2).

**Table 2.** Seasonal and annual temperature and precipitation trends in three agro-ecological zones

| Climatic Variables | Highland | | | Midland | | | Low land | | |
|---|---|---|---|---|---|---|---|---|---|
| | S-score | Sen's Slope | P-value | S-score | Sen's Slope | P-value | S-score | Sen's Slope | P-value |
| Spring maximum temperature (FMAM) | 139 | 0.13 | 0.09 ns | 148 | 0.29 | 0.06 ns | 229 | 0.21 | 0.04* |
| Summer maximum temperature (JJAS) | 85 | 0.37 | 0.29 ns | 94 | 0.07 | 0.23 ns | 89 | 0.05 | 0.25 ns |
| Annual maximum temperature | 100 | 0.48 | 0.21 ns | 113 | 0.48 | 0.15 ns | 85 | 0.36 | 0.20 ns |
| Spring minimum temperature (FMAM) | 222.5 | 0.12 | 0.01** | 206.5 | 0.11 | 0.01** | 144 | 0.08 | 0.05** |
| Summer minimum temperature (JJAS) | 227 | 0.07 | 0.01** | 223 | 0.07 | 0.01** | 214 | 0.06 | 0.01** |
| Annual minimum temperature | 161 | 0.29 | 0.04** | 164 | 0.28 | 0.04** | 151 | 0.24 | 0.05* |
| Spring rainfall (FMAM) | -67 | -1.4 | 0.42 ns | -85 | -1.73 | 0.31 ns | -21 | -0.4 | 0.87 ns |
| Summer rainfall (JJAS) | -58 | -6.53 | 0.47 ns | -94 | -8.29 | 0.24 ns | -19 | -1.9 | 0.73 ns |
| Annual total rainfall | -46 | -8.36 | 0.58 ns | -82 | -11.47 | 0.31 ns | -21 | -2.77 | 0.82 ns |

Note: ** and *indicates statistically changed trend at $\alpha = 0.01$ and 0.05 respectively, **ns**= non-significant, FMAM = February, March, April and May, JJAS = June, July, August and September.

### *3.2. Characterization of agricultural drought indices*

*3.2.1. Frequency.* Three times abnormal drought events were happened for total 12 months duration in summer season, during 1989-2019 years in the highland region of east Wollega zone. Of which the drought events, the 2009 and 2015were severe dry while the 2002 was moderate. Evidence shows that the frequent of drought events is lowering Ethiopian GDP by 3 % and 6 %, in 2002 and 2015, respectively [1,13].

About 7 drought abnormal drought event was happened in the midlands of the study zone during the given time series .Of which, two of the severe drought event was occurred in 2009 while the rest of 4 drought events were moderately drought events. The year of 2010 was the very wet year while the remaining years were normal years (Table 3). This result is in lined with study conducted by Joint Government and Humanitarian Partners [1] which stated that the recurrent drought events severely declined crop productivity in Ethiopia due to climatic variability and changes. The number of drought events occurred in lowland of study area since 1989 year was 7 times abnormal drought events and of which 2 were severe ,4 were moderate drought events and one was very wet (Table 3).

*3.2.2. Severity.* The total amounts of severity indices at highland regions were -19.18. The highest severe drought indices were observed in 2015 year with the lowest index value of -7.24 followed by -6.7 index values recorded in 2009 (Table 3). This result is comparable with study result reported by Madison [25] and Akinseye et al. [26] that drought severity was achieved by heat increments and causes plant mortality in Europe and United States. The total amounts of severity indices in midland were -27.92. The highest severe drought indices were observed in the 2015 with index value of -7.21 followed by-6.30 index values recorded in the 2009 (Table 3).The total amounts of severity indices in low land were estimated to be -28.46. The highest severe drought indices were observed in 2015 with index value of -6.6 followed by-6.29 index values recorded in 2002 from the study area.

*3.2.3. Duration.* Total duration for abnormal agricultural drought events was 12 months at the highlands of east Wollega zone. The longest duration of abnormal drought events for the three years was 4 month long (Table 2).This is comparable with study results reported by Funk et al. [27] and Slater et al. [28] stated that duration of drought and temperature had direct relationships on the Pacific Ocean and in the horn of Africa. The total time covered by incidences of agricultural drought was 24 months. But the total duration for abnormal agricultural drought events was 28 months. Apparently, the longest duration of abnormal drought events with 4 month long were recorded in the, 1991, 2002, 2009, 2011, 2012, and 2015. Other 4 month (June–September) duration of very wet event was occurred in 2010 (Table 2). Droughts during summer (June–September) are largely driven by precipitation declines arising from the narrowing of the inter-tropical convergence zone (ITCZ). Increased dryness during Meher has severe consequences on the production of corn and sorghum [29].

*3.2.4. Intensity.* The total time covered by incidences of agricultural drought with its abnormality was 28 months while the longest duration of abnormal drought events with 4 month long were recorded in the, 1991, 2002, 2009, 2011, 2012 and 2015 years and lastly the average intensity of drought recorded was -1.02 drought spell. But the highest drought intensity was observed in the 2015 with lowest value of -1.65 in months June–September which indicates the extreme drought events. Evidences showed that the rising global temperature and the change in temperature in Pacific Ocean might increase the duration of agricultural drought. Meng [30] reported that when global temperature was raised from 1.5°C to 2°C, the duration drought would be increased duration from 2.9 to 3.2 months.

The average intensity of drought recorded during 1989-2019 was -1.8 drought spell. But the highest drought intensity of highland of east Wollega zone was observed in the 2015 and 2009 with lowest SPEI value of -1.81 followed -1.68 and these were labeled as severely dry soil while -1.31 moderately dry. Other most drought intensity levels of agricultural drought were also put under normal soil moisture conditions (Table 2).

The average intensity of drought recorded during 1989-2019 of midland of study area was found to be -1 drought spell. But the highest drought intensity was observed in the 2015 with lowest value of -1.8 in months June–September which indicates the severe drought events. According to Edwards et al. [16] and McKee et al. [17] , the lower values of drought indices indicate the higher intensity of drought and thus, drought intensity values -1.80 and -1.57 were labeled asseverely dry soil while -1.38, -1.35, -1.33 and -1.27 were moderately dry (Table 3). According to Edwards et al.[16], McKee et al. [17] and Wang et al. [29], the lower values of drought indices indicate the higher intensity of drought and thus, drought intensity values 1.65 and -1.57 were labeled as severely dry soil while -1.48,-1.26,-1.28,-1.24 &-1.10 were moderately dry (Table 3). Alike to this finding, Benti and Chimdi [24] reported the dry soil condition was occurred at the southern of Bako district where altitudes of the area is lower. The seasonal rainfall failure during cropping seasons devastated in causing critical food shortage for a large proportion of population [2].

**Table 3.** Frequency, severity, duration and intensity of drought in the three agro-ecological zones

| Agro-Ecological Zone | Frequency | severity | Duration | Intensity | Edward's Category Description |
|---|---|---|---|---|---|
| | 2002 | -5.24 | 4 | -1.31 | Moderately dry (June-Sept) |
| Highland | 2009 | -6.70 | 4 | -1.68 | Severely dry (June-Sept) |
| | 2025 | -7.24 | 4 | -1.81 | Severely dry (June-Sept) |
| | 1989-2019 | -19.18 | 12 | -1.8 | Severely dry |
| | 1991 | -5.39 | 4 | -1.35 | Moderately dry (June-Sept) |
| | 2002 | -5.53 | 4 | -1.38 | Moderately dry (June-Sept) |
| | 2009 | -6.30 | 4 | -1.57 | Severely dry (June-Sept) |
| Midland | 2010 | 6.92 | 4 | 1.73 | Very wet (June-Sept) |
| | 2011 | -5.08 | 4 | -1.27 | Moderately dry (June-Sept) |
| | 2012 | -5.33 | 4 | -1.33 | Moderately dry (June-Sept) |
| | 2015 | -7.21 | 4 | -1.8 | Severely dry (June-Sept) |
| | 1989-2019 | -27.92 | 28 | -1.00 | Moderately dry |
| Lowland | 1991 | -5.05 | 4 | -1.26 | Moderately dry (June-Sept) |
| | 2002 | -6.29 | 4 | -1.56 | Severely dry (June-Sept) |
| | 2009 | -5.91 | 4 | -1.48 | Moderately dry (June-Sept) |

| | | | | |
|---|---|---|---|---|
| 2010 | 4.94 | 4 | 1.2 | Moderately dry (June-Sept) |
| 2011 | -4.41 | 4 | -1.10 | Moderately dry (June-Sept) |
| 2012 | -5.14 | 4 | -1.28 | Moderately dry (June-Sept) |
| 2015 | -6.60 | 4 | -1.65 | Severely dry (June-Sept) |
| 1989-2019 | -28.46 | 28 | -1.02 | Moderately dry |

### *3.3. Spatiotemporal agricultural drought indices analyses at highland*

Three month-time scale of SPEI-3 was analyzed to identify the agricultural drought period of time (years) in the study area. The results indicated that, the values of SPEI-3 in the 2015 and 2009 were observed-1.81, -1.68, respectively which can be considered as severely dry soil condition. The value of SPEI-3 in the 2002 indicated moderately dry events. On the contrary, the drought condition in the 1989-2001, 2003-2008, 2010-2014 and 2016-2019 were normal as the values of SPEI-3 are between -1 and +1 (Figure 2).This is in sequence with study. Wagaw [31] reported that moist cool zone is more changed in climatic variability than moist warm in Ale & Adami-Tullu woreda.

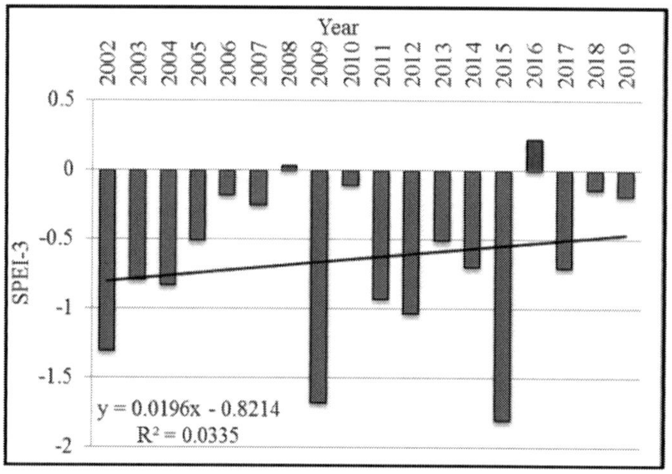


**Figure 2.** Temporal distribution of agricultural drought at highland area

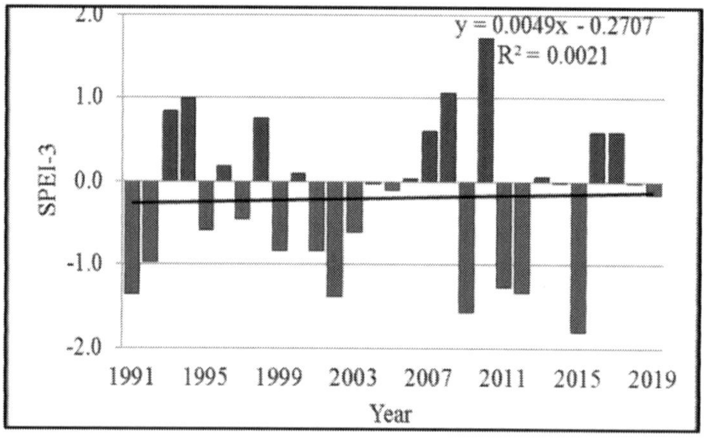


**Figure 3.** Temporal distribution of agricultural drought of midland area

### *3.4. Spatiotemporal agricultural drought indices analyses at midland*

The results indicated that, the moisture condition of the year 2009 and 2015 were estimated as severely dry. The drought condition in the 1991, 2002, 2011 and 2012 were likely moderately dry. On the contrary, the soil moisture condition in 1991-2001, 2003-2008, 2013, 2014 and 2016- 2019 were near normal and 2010 was very wet (Figure 3).

*3.5. Spatiotemporal agricultural drought indices analyses at lowland*

The results indicated that, the drought condition in the year 2002 and 2015 was estimated as severely dry. The drought condition in the 1991, 2009, 2010, 2011 and 2012 were likely moderately dry while the drought condition in the 1991-2001, 2003-2008, 2013, 2014, and 2016- 2019 were near normal (Figure 4). In 2010, the El Niño Southern Oscillation (ENSO) climate pattern swung first through its warm phase and then it's cool phase, with significant influences on seasonal climate around the globe (the Authors adapted from https://www.climate.gov/).

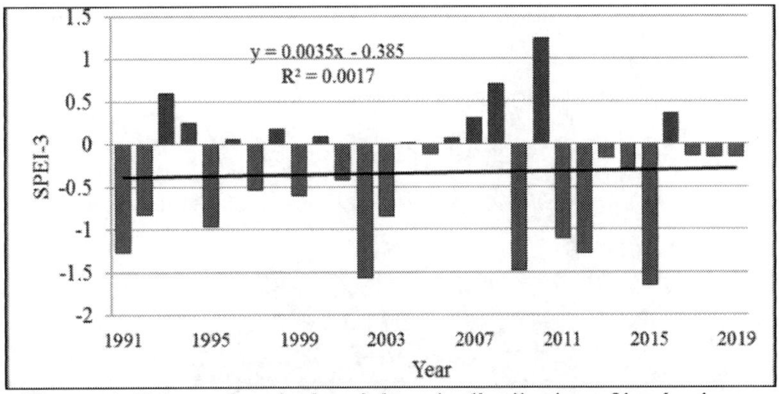


**Figure 4.** Temporal agricultural drought distribution of lowland area

*3.6. Spatial agricultural drought intensity*

The higher drought intensity ranged between -1.81 & -1.00 and covered about 94.45 square kilometer or 0.70% of the total area of the zone which is confined in the Kiramu Woreda (Table 4). The normal distribution of drought intensity values ranged between -1 and +1 was estimated to be 9209.41square kilometer or 68.57% of the total land area (Table 4). This area typically has normal soil moisture contents that an ideal suitable for crop growing and the area was likely extended from central north, to central south parts of the study area (see Figure 5).On the other hand, a very wet area with SPEI value ranged between 1 & 1 was detected in highland of HaroLimu, Ibantu, Sasiga, Limmu, GobuSayoo and BoneyaBoshe districts and accounted about 4128.13 square kilometer or 30.37% of the total area of the east Wollega zone (Table 4).

Associated to this result, many study partly recognized that drought affected regions were increasing both in geographic coverage and the number of affected populations [25,27,32] also reported that increment of drought intensity had direct relationships with heat wave, and temperatures which dramatically increase evapo-transpiration and correspondingly aggravates drought stress on cropping season of crop.

IOP Conf. Series: Earth and Environmental Science 1016 (2022) 012002 doi:10.1088/1755-1315/1016/1/012002

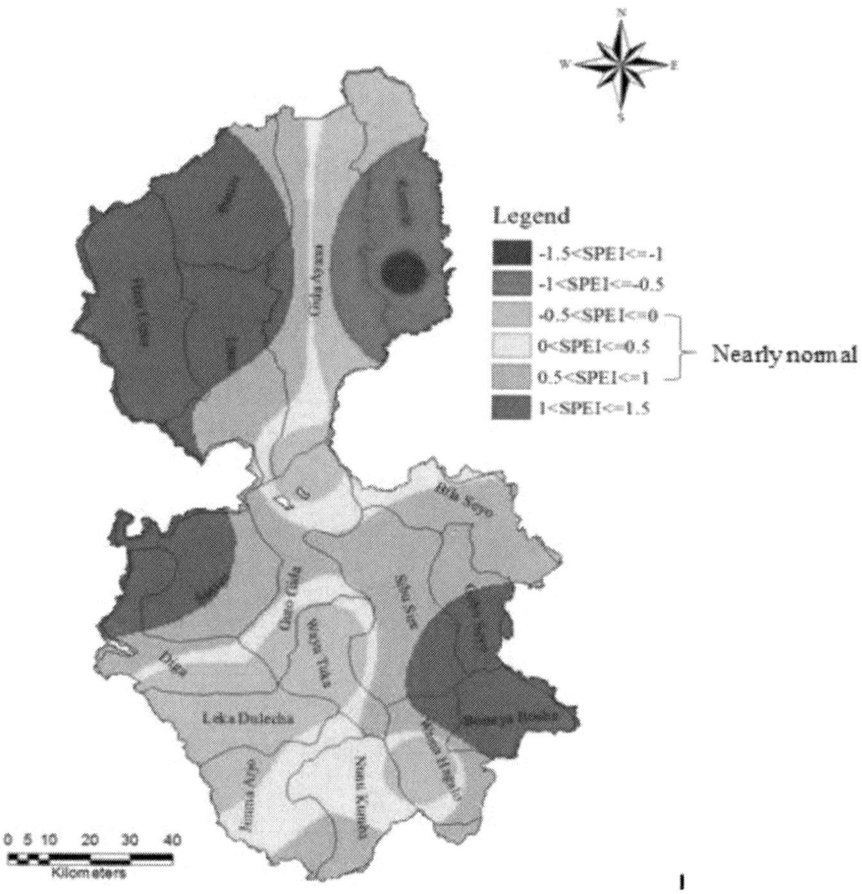


**Figure 5.** Map of spatial distribution of agricultural drought intensity

**Table 4.** Spatial distribution of agricultural drought in all provinces of East Wollega zone

| Woredas | -1.5<SPEI≤-1 (km$^2$) | -1<SPEI≤-0.5 (km$^2$) | -0.5<SPEI≤0 (km$^2$) | 0<SPEI≤0.5 (km$^2$) | 0.5<SPEI≤1 (km$^2$) | 1<SPEI≤1.5 (km$^2$) | Total (km$^2$) | Total dry (km$^2$) | Total wet (km$^2$) | %Dry | %Wet |
|---|---|---|---|---|---|---|---|---|---|---|---|
| Ibantu | 0 | 0.00 | 0.00 | 0.00 | 103.87 | 741.56 | 845.43 | 0.00 | 845.43 | 0.00 | 100.0 |
| Limu | 0 | 0.00 | 1.69 | 62.56 | 345.18 | 600.02 | 1009.5 | 64.25 | 945.19 | 6.37 | 93.63 |
| Bilaseyo | 0 | 0.00 | 1.69 | 130.44 | 490.59 | 0.00 | 622.72 | 132.13 | 490.59 | 21.22 | 78.78 |
| Sasiga | 0 | 0.00 | 0.00 | 0.00 | 382.14 | 447.35 | 829.49 | 0.00 | 829.49 | 0.00 | 100.0 |
| Gutogida | 0 | 0.00 | 252.42 | 331.41 | 257.01 | 0.00 | 840.84 | 583.83 | 257.01 | 69.43 | 30.57 |
| Wamahagalo | 0 | 0.00 | 196.86 | 98.07 | 189.14 | 76.81 | 560.88 | 294.94 | 265.95 | 52.58 | 47.42 |
| Jimmaarjo | 0 | 0.00 | 190.83 | 407.74 | 120.05 | 0.00 | 718.62 | 598.57 | 120.05 | 83.29 | 16.71 |
| Nunukumba | 0 | 0.00 | 27.05 | 417.64 | 141.55 | 0.00 | 586.25 | 444.70 | 141.55 | 75.85 | 24.15 |
| Boneyaboshe | 0 | 0.00 | 0.00 | 1.93 | 15.70 | 406.05 | 423.68 | 1.93 | 421.75 | 0.46 | 99.54 |
| Gobuseyo | 0 | 0.00 | 0.00 | 0.00 | 190.10 | 232.86 | 422.96 | 0.00 | 422.96 | 0.00 | 100.0 |
| Lekadulecha | 0 | 0.00 | 530.45 | 19.08 | 0.00 | 0.00 | 549.53 | 549.53 | 0.00 | 100.0 | 0.00 |
| Sibusire | 0 | 0.00 | 42.03 | 53.87 | 707.99 | 207.98 | 1011.8 | 95.90 | 915.97 | 9.48 | 90.52 |
| Wayptuka | 0 | 0.00 | 356.29 | 78.02 | 55.56 | 0.00 | 489.87 | 434.31 | 55.56 | 88.66 | 11.34 |
| Diga | 0 | 0.00 | 214.74 | 94.69 | 160.39 | 199.04 | 668.86 | 309.43 | 359.43 | 46.26 | 53.74 |
| Harolimu | 0 | 0.00 | 0.00 | 0.00 | 1.45 | 1095.20 | 1096.6 | 0.00 | 1096.65 | 0.00 | 100.0 |
| Gidaayana | 0 | 359.43 | 613.06 | 233.82 | 388.42 | 121.26 | 1715.9 | 1206.31 | 509.67 | 70.30 | 29.70 |
| Kiremu | 94.45 | 656.06 | 288.41 | 0.00 | 0.00 | 0.00 | 1038.9 | 1038.92 | 0.00 | 100.0 | 0.00 |
| Total | 94.45 | 1015.48 | 2715.53 | 1929.28 | 3549.12 | 4128.13 | 13431.9 | 5754.74 | 7677.25 | 42.84 | 57.16 |

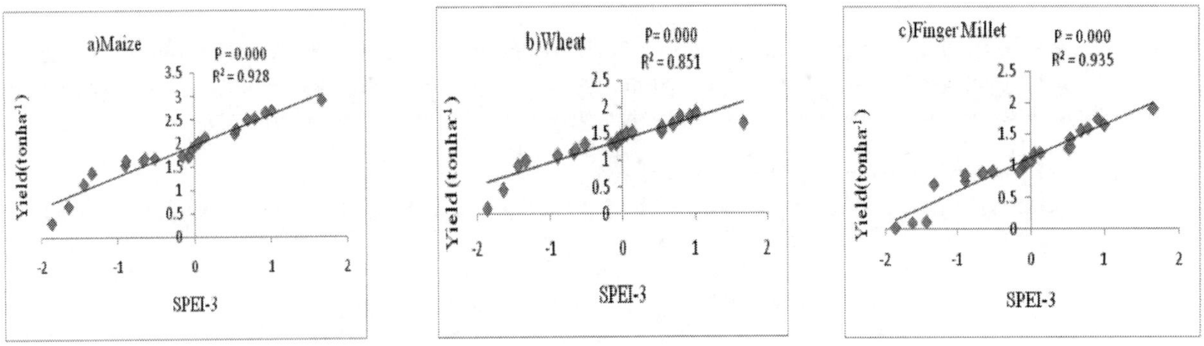


**Figure 6.** Correlation between SPEI-3 and selected crop yield wetter area

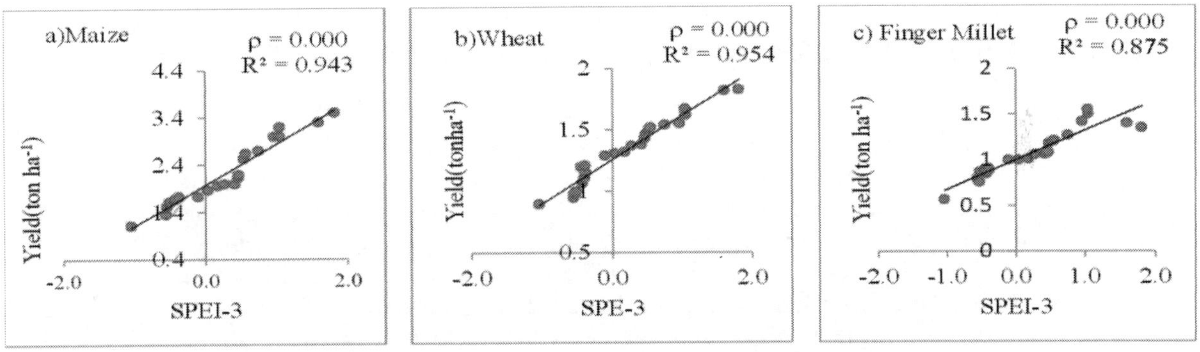


**Figure 7.** Correlation between SPEI-3 and crop yield in the normal area

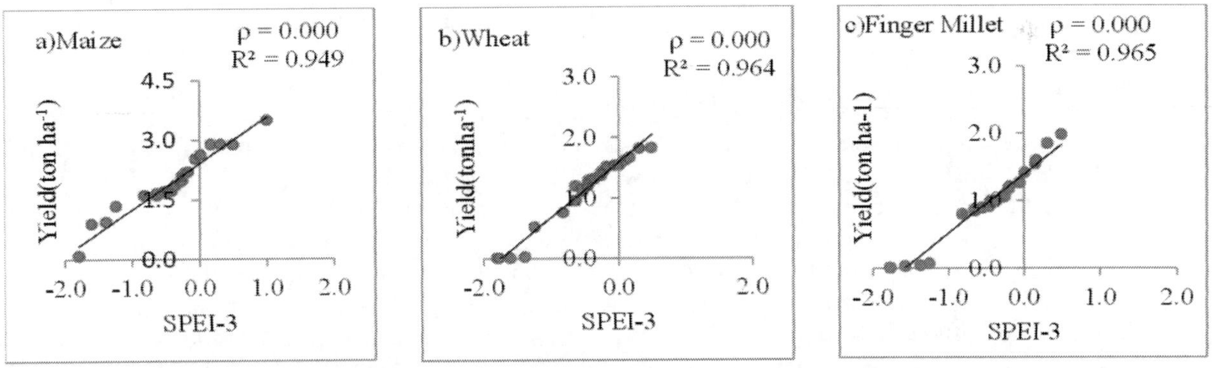


**Figure 8.** Correlation between SPEI-3 and crop yields in the drier area

### *3.7. Impacts of agricultural drought on selected crop yields at witter area*

Correlation between SPEI-3 and crop yields in the wetter area during cropping season (JJAS) was positive relationship. The values of correlation coefficient (R) between drought indices and crop yield were 0.96, 0.92 and 0.97 for maize, wheat and finger millet, respectively (Figure 6). It means that as the values of drought indices increase, the amount of crop yields increases and the vice versa. However, the decrements of crop yields tended to be as drought indices rose beyond -1 and 1.The result of maize, wheat and finger millet is significantly changed (Figure 5). The coefficients of correlation determination between agricultural drought indices and selected crop yields were 0.928, 0.851 and 0.935, for maize, wheat and finger millet, respectively. These indicated that the total yields variations in selected crop types in percent (%) were 92.8, 85.1 and 93.5 maize, wheat and finger millet, respectively. Therefore, all selected crops were significantly correlated to drought indices ($p<0.001$) (Figure 6).This finding is compatible to the information generated by [24,33,34] in which maize crop yield was highlighted as the least affected by climate change at 21st century and wheat was most vulnerable and also studied by Pant [26] stated that yield of maize was increased from1990 to 2010 in Ethiopia and Nepal.

### *3.8. Impacts of agricultural drought on selected crop yields at normal area*

Correlation between SPEI _3 and crop yields; in the normal area results were showed strong positive relationships. The values of correlation coefficient between drought indices and crop yield were 0.97, 0.97 & 0.93 for maize, wheat, and finger millet, respectively (Figure 7). It means that as the values of drought indices and crop yields had direct relationships. Yet, the decrements of crop yields tended to be as drought indices rose beyond -1&1 (Figure 7). The results of coefficients of correlation determination between agricultural drought indices and selected crop yield were 0.942, 0.954 and 0.875, respectively. These indicated that the total yields variations in selected crop types in percent (%) were 94.2, 95.4 and 87.5 maize, wheat, and finger millet, respectively. Therefore, there was a variation between chosen crop yields and drought indices in normal regions of east Wollega zone and all of them were significant ($P<0.001$) (Figure 7). This finding also parallel to [35] reported production of crop yields and seasonal temperature had strong relation in Tanzania.

### *3.9. Impacts of agricultural drought on selected crop yields at drier area*

Correlation between SPEI-3 and crop yields in the drier area results of correlation coefficient between drought indices and crop yield were 0.97, 0.98 and 0.98 for maize, wheat, and finger millet, respectively. The associations of time series drought indices and selected crop yields were strong positive (Figure 8). It means that as the values of drought indices increase, the amount of crop yields increase and the vice versa. However, the decrements of crop yields tended to be as drought indices rose beyond -1&1 and the result all is significant for maize and wheat at( $P<0.001$) (Figure 8). The coefficients of correlation determination between agricultural drought indices and selected crop yield results were 0.949, 0.964 and 0.965 for maize, wheat and finger millet, respectively. These indicated that the total yields variations in selected crop types in percent (%) were 94.9, 96.4 and 96.5 for maize, wheat, and finger millet, respectively. Therefore, the variation in yields for all was significantly influenced by the relationship ($P<0.001$) (Figure 8).

### *3.10. Characteristics of respondents*

Out of the surveyed from 140 household heads in the zone, 127 (90%) of respondents were males and 13 (10%) females. The result of the survey revealed that, age composition below 30, 31-40, 41-50 and above 51 year covered for19%, 34.3%, 18.6 and 27.9% respectively. At zonal level, the majority of farmers are between 31-40 years and above 50 years. Based on the marital status, from the total respondents, 87.9% married, 7.9% widowed, 2.9% divorced and only 1.4% are single. The survey data showed that 40.7% of the respondents achieved primary level followed by secondary level which is 12.9% of respondents. About 0.7% & 1.4% were preparatory, higher educational institutions respectively. Lastly, 42.1% and 2.1% are illiterate and none formal education respectively. The result shows there is higher percentage of illiterate in study area (Table 5). This is confirmed by [36,37], explained that farmers with more education are more likely to pursue adaptation strategies related to climate change than farmers with lower education levels.

**Table 5.** Sex, age, marital status and households' size of respondents

| Variables | | Kebeles | | | | | | Total | |
|---|---|---|---|---|---|---|---|---|---|
| | | Haro&Bariso | | Bado&Arko gudatu | | Badessa&Boka | | N=140 | |
| | | F | % | F | % | F | % | F | % |
| **Sex** | Male | 38 | 88.6 | 46 | 97.9 | 43 | 86 | 127 | 90 |
| | Female | 5 | 11.6 | 1 | 2.1 | 7 | 14. | 13 | 10 |
| **Age in year** | Below 30 | 4 | 9.3 | 12 | 25.5 | 13 | 26 | 28 | 19. |
| | 31-40 | 20 | 46.5 | 14 | 29.8 | 11 | 22 | 45 | 34.5 |
| | 41-50 | 5 | 11.6 | 11 | 23.4 | 8 | 16 | 24 | 18.6 |
| | >51 | 14 | 32.6 | 10 | 21.3 | 18 | 36 | 42 | 27.9 |
| **Marital status** | Single | - | - | - | - | 1 | 2 | 1 | 1.4 |
| | Married | 37 | 86 | 45 | 95.7 | 43 | 86 | 125 | 87.9 |
| | Divorced | - | - | 2 | 4.3 | 2 | 4 | 4 | 2.9 |
| | Widowed | 6 | 14 | - | - | 4 | 8 | 10 | 7.9 |
| **Level of education** | Illiterate | 23 | 53.5 | 16 | 32 | 17 | 34 | 56 | 42.1 |
| | Non Formal | 2 | 4.7 | - | - | - | - | 2 | 2.1 |
| | Primary | 14 | 32.6 | 23 | 46 | 19 | 38 | 56 | 40.7 |
| | Secondary | 4 | 9.3 | 11 | 22 | 11 | 22 | 26 | 12.9 |
| | Preparatory | - | - | - | - | - | - | - | - |
| | Institution | - | - | - | - | 3 | 6 | 3 | 1.4 |

Note: F =indicates Frequency, N= Number Source, Own survey, 2020

**Table 6.** Family, land holding size and average distance farm of the respondents

| Variables | | Kebeles | | | | | | Total | |
|---|---|---|---|---|---|---|---|---|---|
| | | Haro& Bariso | | Bado& Arko gudatu | | Badessa&Boka | | N=140 | |
| | | F | % | F | % | F | % | F | % |
| **Family size of respondents** | 1-3 | 5 | 11.6 | 7 | 14.9 | 12 | 24 | 24 | 15.7 |
| | 4-7 | 22 | 51.2 | 27 | 57.4 | 28 | 56 | 77 | 57.1 |
| | 8-10 | 9 | 20.9 | 11 | 23.4 | 6 | 12 | 26 | 19.3 |
| | 11-15 | 7 | 16.3 | 2 | 4.3 | 4 | 8 | 13 | 7.9 |
| **Land holding size** | <0.75 | 15 | 26.3 | 11 | 22.4 | 12 | 28.6 | 38 | 25.7 |
| | 0.75-1 | 13 | 31.0 | 12 | 31.6 | 7 | 23.3 | 32 | 21.6 |
| | 1-1.75 | 11 | 26.2 | 10 | 26.3 | 8 | 26.7 | 29 | 19.6 |
| | 1.75-2.5 | 5 | 11.9 | 6 | 15.8 | 7 | 23.3 | 18 | 12.2 |
| | 2.5-3.5 | 8 | 19.0 | 5 | 13.2 | 3 | 10.0 | 16 | 10.8 |
| | >3.5 | 1 | 2.4 | 2 | 5.3 | 4 | 3.3 | 7 | 7.4 |
| **Average distance of farm household** | 0-5minute | 1 | 2.1 | 1 | 2.1 | 1 | 10 | 3 | 1.4 |
| | 5-30minute | 43 | 91.5 | 44 | 93.6 | 47 | 48 | 134 | 94.7 |
| | 31-1hour | 2 | 4.3 | 2 | 4.3 | 2 | 2 | 6 | 2.9 |
| | >1hour | 1 | 2.1 | - | - | - | - | 1 | 1 |

Note: F =indicates Frequency, N= Number Source, Own survey, 2020

Majority of the respondents (57.1%) fell within the household size of 4 - 7 and 19.3% within the 8 -10, (15.7%), within 1-3 house hold size and followed by 7.9% of the respondents which fells in the 11 - 15 household sizes. In general the result shows that, the house hold size of the respondents fell in the range of less than 15. When respondents were asked whether they have agricultural land, majority 99.3%

of them replied yes and only 0.7% said no. The land holding size of the respondent show that majority 25.7% household has less than 0.75 hectare, and many of household have land holding size range between 0.75-1hectare which were 21.6%. 10.8% of total household possessed that 2.5-3.5 hectare in study are. Only 7.4% of household owned greater than 3.5 hectare of agricultural lands. According to this survey data the average distance of farm land from household's home 94.7% are not far from their home stead. The average distance of their farm land from their home is <30minute. 1.4% and 2.9% fell between range of 0-5minute and 31-60 minute respectively. Only 1% of households owned >1hour have average distance from home stead (Table 6).

In survey of cropping systems practiced by farmers in study area, the majority of the respondents, (about 75%) practiced crop rotation and 5% of them were practiced strip cropping farming system. About 8% of them practiced mixed cropping system and 12% practiced mono cropping. The farmers were not practiced in shifting cultivation in study area due to the shortage of cultivated lands (Figure 9).

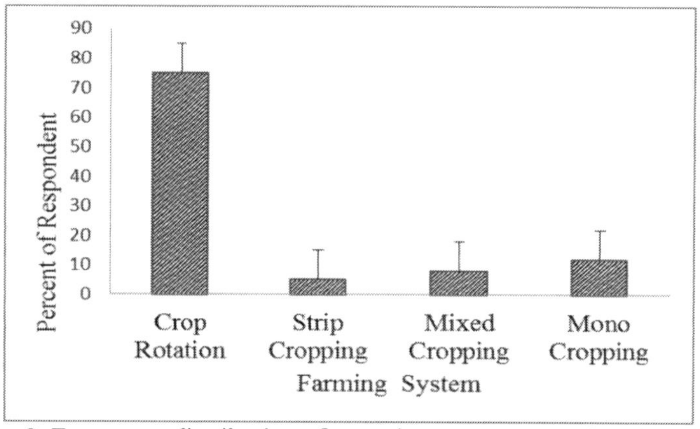


**Figure 9.** Frequency distribution of cropping systems practiced by farmers

In survey of adaptations a strategy to climate change or climate variability in study area, 21% of respondents performed physical soil water conservation strategies, 17% using of planting trees, 8%, 8%, 6% & 4 % practices irrigation, crop diversification, improved crop variety and biophysical SWC respectively (Figure 10).

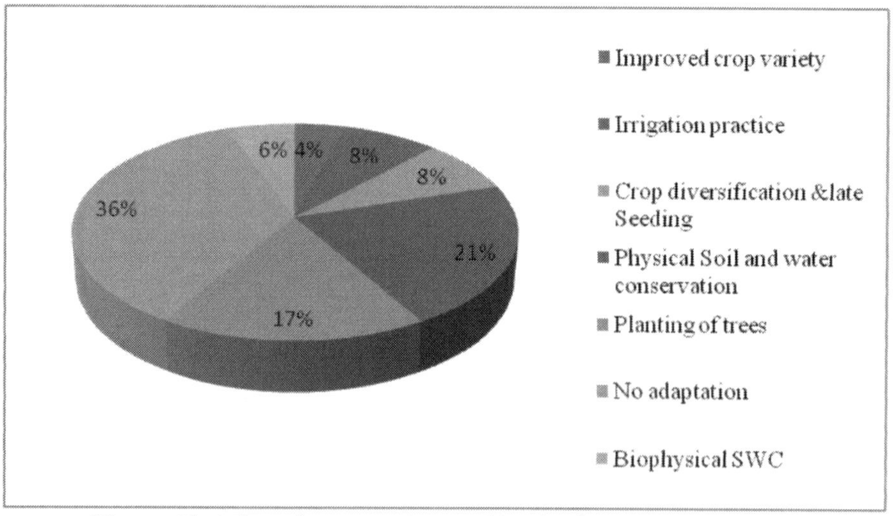


**Figure 10.** Adaptation strategies used by the farmers in crop yield

In general, a total of 64% of respondents employed at least one adaptation strategies in response to climate change. The remaining 36% of sample households did not use any mechanisms at all in response to climate change. As focus group discussion and key informant reported that households who did not undertake any of adaptations options cited shortage of sufficient financial resources, lack of climate related information and shortage of land as main reasons for not adopting. These results are similarly with, studied by [38–40] noted that crop diversification together with soil and water conservation and water harvesting practices were commonly used climate change adaptation strategies in Cameroon, South western Nigeria and eastern Ethiopia respectively. Determinants of farmers to implement climate change adaptation strategies results of MLM of the likelihood ratio statistics by chi2 statistics was (LR chi-square (54) =93.009 which are significant Prob> chi2=0.001), suggesting the model has a strong independent power (Table 7).

**Table 7**. Marginal affects variables on adaptation strategies from MLM

| Variables | ICV | | Irrigation practice | | Physical SWC | | Planting tree | | Biological SWC | | No Adaptation | |
|---|---|---|---|---|---|---|---|---|---|---|---|---|
| | CV | P | CV | P | CV | P | CV | P | CV | P | CV | P |
| Sex | 0.06 | 0.36 | 0.04 | 0.63 | -0.53 | 0.39 | 0.76 | 0.03** | 0.52 | 0.02** | 0. 00 | 0. 99 |
| Age | 0.02 | 0.44 | 0.00 | 0.60 | 0.00 | 0.57 | 0.00 | 0.91 | -0.02 | 0.76 | 0.00 | 0. 99 |
| F. Size | 0.03 | 0.30 | -0.06 | 0.23 | -0.03 | 0.46 | 0.19 | 0.05** | 0.263 | 0.020** | -5.07 | 0.99 |
| Credit | 0.59 | 0.01* | -0.02 | 0.45 | -0.02 | 0.44 | -0.46 | 0.00* | -0.07 | 0.72 | -0.02 | 0.49 |
| Extension | -0.48 | 0.07* | 0.50 | 0.07* | 0.11 | 0.21 | 0.35 | 0.05** | 0.36 | 0.06*** | -0.00 | 0.99 |
| MIF | -0.00 | 0.54 | 0.05 | 0.16** | 0.01 | 0.04** | 0.04 | 0.03** | -0.07 | 0.58 | 0.00 | 0.99 |
| Farm dis | 0.11 | 0.31 | -0.24 | 0.02** | -0.01 | 0.02** | -0.35 | 0.12** | 0. 13 | 0.35 | 0.00 | 0.99 |
| Farm size | 0.16 | 0.32 | -0.15 | 0.53 | 0.07 | 0.48 | -0.13 | 0.70 | 0.02 | 0.01* | -0.01 | 0.99 |
| Education | 0.09 | 0.02** | 0.00 | 0.11** | -0.01 | 0.03** | 0.00 | 0.03** | -0.01 | 0.02** | -0.00 | 0.99 |

Note: ***, **and * indicate the variable is significant at probability level of = 1%, 5% and 10%, respectively. CV is coefficient of variation, P- probability level, ICV – improved crop variety and MIF is meteorological information facility

*3.10.1. Sex of household's heads.* Sex of household heads has positive influence on the probability of tree planting and Soil and water conservation adaptation strategies options by 76%, 52% respectively being other variable constant. This result is similar with studied result by Asfaw and Admassie [41] and Ishaya and Abaje [42], stated that male households had get information about new technologies and undertake risky businesses than females; so that adapt to climate change than female headed households.

*3.10.2. Age of household heads.* Age has been observed positively and in significantly affects the decision to take up climate change adaptation measures in the study area. This result is in line with the findings of Madison [25], Bryan et al. [36], Ishaya and Abaje [42] and Vicente et al. [43] reported that as age of house hold increased adaptation strategies to climatic impact was increased in Ethiopia and in contrast studied by [44] reported that the age of farmers had a negative influence on adoption of soil management practices in Ethiopia and South Africa.

*3.10.3. Education level of household heads.* Education has positive and significant affects on improved crop variety, irrigation practice and soil and water conservation adaptation strategies with 0.2% and 2.6% respectively being other variable constant. This result is in line with findings by Madison [25], Meng-Ju [30] and Igodan et al [45] indicated that there was a positive relationship between the education level of the household head and adoption to climate change .In addition to this, Madison [25] and Norries and Batie [46] reported that education had direct relation with access to information on climate change, improved technologies and productivity penalty.

*3.10.4. Access to credit.* Access to credit has positive significant on using of improved crop variety in 60%, and negative significant on soil and water conservation adaptation strategies by 46% being other variable constant. This comparable with studied by WMO [47] which confirmed that access to credit had direction with climate change mitigation.

*3.10.5. Farm land size.* Farm size has positive and significant effects on using of, tree planting adaptation strategies to climate change via 1.8%, which similar with finding of Franzel [48], reported that farm size had direct relationships with adaptation measures and productivity increments.

*3.10.6. Access to extension.* Access to extension has a positive and significant effect on using of irrigation practice, soil and water conservation and tree planting near 48%, 50.9% and 35% respectively being other variable constant.

*3.10.7. Access to climate information.* Access to climate information has negative and significant affects on crop diversification adaptation strategies via 0.3% other variable being constant and has positive and significant affects on soil and water conservation adaptation strategies through 3.5% other variable being constant. This result is confirmed by finding of Legesse et al. [49] showed that, climatic information had direct relation with prefer to crop diversification and soil and water conservation as climate change adaptation options.

*3.10.8. Family size.* Family size also has significant and positive effect on probability of tree planting adaptation strategies of farmers by 13.3% keeping other variables constant. This assumption is in line with the results of similar work on climate change adaptation strategies done in Nile basin and Tanzania Bryan et al. [36] and Komba and Muchapondwa [50].

*3.10.9. Farm distance from household.* Distance to the farm is a negative and significance variable on irrigation practice, crop diversification and soil and water conservation activities near 11.4%, 24.6% and 35.6%, respectively holding other variables constant. It implies that short distance permits the use of those adaptation strategies to climate change in the study area. These results are in contrast with findings of Ishaya and Abaje [42] reported that distance for plot to farmer's residence had positive relationship with adoption of fallow, inorganic fertilizer and irrigation land management practices in Uganda.

## 4. Conclusions

This paper synthesized about of climate variability and the responses of yield crops to climate induced agricultural drought. The Mann Kendall trend of analyses clearly showed that seasonal and annual minimum temperature of all agro-ecological zones (highland, Midland and lowland) of east Wollega zone was significantly warmed up during cropping seasons (spring and summer). In addition, the climate system of the lowland strongly increased the maximum temperature. However, spring and annual maximum temperature of the highland and midland were not significantly changed. Seasonal and annual precipitation of each climatic zone was weakly declined by the climate system.. As a result, about 32% area (0.7% moderate dry and 31.3% moderate wet) of east Wollega zone was affected by abnormal drought and wet events. Subsequently, Kiremuworeda was one of the most affected provinces by moderate drought while Harolimu and Boneya Boshe were fully suffered by wet condition and Gobusayo and Sasiga, Limu and Ibantu were partly affected by abnormal or excess of water during the last three decades. The remaining 68% of the area was tended to be normal which is ideally suitable for crop growth and development. The results of correlation coefficients between drought indices and selected crop yield showed a clear variation of crop yields during the given time courses and wheat and finger millet is more vulnerable crop to agricultural drought events. Result of multinomial logistic model revealed that education and access to climate information were the prominent factors to influence the soil and water conservation practices and crop diversification at 95% confidence level. So to overcome the problems, it is suggested that strengthening agroforestry practices, crops diversification and crop intensification, increasing the awareness of farmers on the coping up of the risk of climate change, strengthening the linkage between farmers and stakeholders; agricultural research institutions, university and private sectors in order to support and build the capacity farmers with supplying new technology, providing access of credit service and agricultural drought insurance and further study should be conducted on the area of climate change with relation to soil degradation.

**Reference**

[1] Joint Government and Humanitarian Partners 2015 Ethiopia Humanitarian Requirements Document-2016 8

[2] Bekabil U T 2014 Farmers ' Perception and Adaptation To Climate Change : Heckman ' S *Ethiop. J. Environ. Stud. Manag.* **7** 832–9

[3] Beguería S, Vicente-Serrano S M, Reig F and Latorre B 2014 Standardized precipitation evapotranspiration index (SPEI) revisited: Parameter fitting, evapotranspiration models, tools, datasets and drought monitoring *Int. J. Climatol.* **34** 3001–23

[4] NMA 2007 Climate change National Adaptation Program of Action (NAPA) of Ethiopia *Minist. water Resour. Addis Ababa; North. Ghana* **Africa Por**

[5] Holmgren K and Öberg H 2006 Climate change in southern and eastern Africa during the past millennium and its implications for societal development *Environ. Dev. Sustain.* **8**

[6] Ajibefun A I, Fatuase A I and State O 2010 Reference : 0233 *Agric. Econ.* 1–29

[7] Makombe G, Namara R E, Awulachew S B, Hagos F, Ayana M and Kanjere M 2017 An analysis of the productivity and technical efficiency of smallholder irrigation in Ethiopia *Water SA* **43**

[8] Deressa T T and Hassan R M 2009 Economic impact of climate change on crop production in ethiopia: Evidence from cross-section measures *J. Afr. Econ.* **18**

[9] Ten V S, Wu E X, Tang H, Bradley-Moore M, Fedarau M V., Ratner V I, Stark R I, Gingrich J A and Pinsky D J 2004 Late measures of brain injury after neonatal hypoxia-ischemia in mice *Stroke* **35**

[10] Mugula and Mkuna 2016 Farmer ' s perceptions on climate change impacts in different rice production systems in Morogoro Tanzania *Int. J. Sci. Res. Publ.* **6**

[11] Adunya T and Benti F C 2020 The Impacts of Climate-Induced Agricultural Drought on Four Cereal Crops: A Case Study in Bako Tibe District, Oromia National Regional State, Ethiopia *Caraka Tani J. Sustain. Agric.* **35**

[12] Debalke N M 2013 Determinants of farmers' preference for adaptation strategies to climate change: evidence from north shoa zone of Amhara region Ethiopia *Munich Pers. RePEc Arch.* **2**

[13] Suryabhagavan K V. 2017 GIS-based climate variability and drought characterization in Ethiopia over three decades *Weather Clim. Extrem.* **15**

[14] Zewdu G A and Malek M 2010 *Implications of Land Policies for Rural-urban Linkages and Rural Transformation in Ethiopia* vol 15

[15] Hargreaves G H and Allen R G 2003 History and Evaluation of Hargreaves Evapotranspiration Equation *J. Irrig. Drain. Eng.* **129**

[16] Edwards, D.C. and McKee T . 1997 Characteristics of 20 the century drought in the United States at multiple times scales *Atmos. Sci. Pap.* **634** 1–30

[17] McKee T B, Nolan J and Kleist J 1993 The relationship of drought frequency and duration to time scales *Prepr. Eighth Conf. Appl. Climatol. Amer. Meteor, Soc.*

[18] Greene W W H . 2012 *Econometric analysis 7th Ed* vol 97

[19] Adhikari U, Nejadhashemi A P and Woznicki S A 2015 Climate change and eastern Africa: A review of impact on major crops *Food Energy Secur.* **4**

[20] Gujarati D N 2004 *Basic Econometrics 4th Edition*

[21] Moat J, Williams J, Baena S, Wilkinson T, Gole T W, Challa Z K, Demissew S and Davis A P 2017 Resilience potential of the Ethiopian coffee sector under climate change *Nat. Plants*

[22] Shrestha A B, Wake C P, Mayewski P A and Dibb J E 1999 Maximum temperature trends in the Himalaya and its vicinity: An analysis based on temperature records from Nepal for the period 1971-94 *J. Clim.* **12**

[23] Rebetez 2006 An analysis of the July 2006 heat wave extent in Europe compared to the record year of 2003
[24] Benti F and Chimdi A 2019 Climate Change-Induced Agricultural Drought over Moist-Cool and Moist-Warm Climatic Zones: A Case Study in Ale and Adami-Tulu Woredas, in Oromia National Regional State, Ethiopia *Caraka Tani J. Sustain. Agric.* **34**
[25] Madison D 2006 *The perception of an adaptation to climate change in Africa: CEEPA* (University of Pretoria)
[26] Akinseye F, Ajayi V and Oladitan T 2013 Assessing the impacts of climate variability on crop yield over Sudano- Sahelian zone in Nigeria *Access Int. J. Agric. Sci.* **1** 91–8
[27] Funk C, Harrison L, Shukla S, Pomposi C, Galu G, Korecha D, Husak G, Magadzire T, Davenport F, Hillbruner C, Eilerts G, Zaitchik B and Verdin J 2018 Examining the role of unusually warm Indo-Pacific sea-surface temperatures in recent African droughts *Q. J. R. Meteorol. Soc.* **144**
[28] Slater R, Peskett L, Ludi E and Brown D 2007 Climate change, agricultural policy and poverty reduction – how much do we know? *Nat. Resour. Perspect.* **109**
[29] Wang H, Stumpf A J and Kumar P 2018 Radiocarbon and Stable Carbon Isotopes of Labile and Inert Organic Carbon in the Critical Zone Observatory in Illinois *Radiocarbon* **60**
[30] Meng-Ju Lin 2011 Using Tobin's Q ratio to testing the stakeholder theory applied to the corporate social performance *AFRICAN J. Bus. Manag.* **5**
[31] Wagaw K 2019 Review on Mechanisms of Drought Tolerance in Sorghum (Sorghum bicolor (L.) Moench) Basis and Breeding Methods *Acad. Res. J. Agri. Sci. Res* **7**
[32] Asfaw A, Simane B, Hassen A and Bantider A 2018 Variability and time series trend analysis of rainfall and temperature in northcentral Ethiopia: A case study in Woleka sub-basin *Weather Clim. Extrem.* **19**
[33] IPCC I P O C C 2014 Climate change: Impacts, adaptation and vulnerability *Contrib. Work. Gr. II to 6th Assess. Rep. Intergov. panel Clim. Chang.*
[34] Adugna Lemi 2005 Rainfall Probability and Agricultural Yield in Ethiopia *East. Afr. Soc. Sci. Res. Rev.* **21**
[35] Ringler C, Shewmake S, Deressa T, Hassan R M, Admassie A, Adenew B, Tadege A, Nhemachena C, Alemu T, Yesuf M, Block P J, Strzepek K M, Rosegrant M W, Diao X, M. S K and Bluffstone R 2011 How can African agriculture adapt to climate change?: Insights from Ethiopia and South Africa *Res. briefs*
[36] Bryan E, Deressa T T, Gbetibouo G A and Ringler C 2009 Adaptation to climate change in Ethiopia and South Africa: options and constraints *Environ. Sci. Policy* **12**
[37] Mesike C S and Esekhade T U 2014 Rainfall variability and rubber production in Nigeria *African J. Environ. Sci. Technol.* **8**
[38] Schlenker W, Lobell B D, Bryan E, Ringler C, Okoba B, Roncoli C, Silvestri S, Herrero M, Nhemachena C, Rashid H, Narayanan K, Sahu S K, Mertz O, Mbow C, Reenberg A, Diouf A, Below T, Artner A, Siebert R, Seiber S and Program S 2010 Determinants of African farmers' strategies for adapting to climate change: Multinomial choice analysis *Environ. Manage.* **114**
[39] Molua E and Molua E 2008 Turning up the heat on African agriculture: The impact of climate change on Cameroonâ€™s agriculture *African J. Agric. Resour. Econ.* **02**
[40] Hicke J A and Zeppel M J B 2013 Climate-driven tree mortality: Insights from the piñon pine die-off in the United States *New Phytol.* **200**
[41] Asfaw A and Admassie A 2004 The role of education on the adoption of chemical fertiliser under different socioeconomic environments in Ethiopia *Agric. Econ.* **30**
[42] Ishaya S and Abaje I . 2008 Indigenous people's perception on climate change and adaptation strategies in Jema'a local government area of Kaduna State, Nigeria *J. Geogr. Reg. Plan.* **1**
[43] Tizale C Y 2007 The dynamics of soil degradation and incentives for optimal management in the Central Highlands of Ethiopia 263

[44] Apata T G, Samuel K D and Adeola A O 2008 Analysis of Climate Change Perception and Adaptation among Arable Food Crop Farmers in South Western Nigeria *Int. Assoc. Agric. Econ. 2009 Conf.*
[45] Igodan C O, Ohaji P E and Ekpere J A 1988 Factors associated with the adoption of recommended practices for maize production in the Kainji lake basin of Nigeria *Agric. Adm. Ext.* **29**
[46] Norris P E and Batie S S 1987 Virginia Farmers' Soil Conservation Decisions: An Application of Tobit Analysis *J. Agric. Appl. Econ.* **19**
[47] World Meteorological Organization 2012 WMO 2012 Survey on the Use of Satellite Data
[48] Franzel S 1999 Socioeconomic factors affecting the adoption potential of improved tree fallows in Africa *Agrofor. Syst.* **47**
[49] Legesse B, Ayele Y and Bewket W 2012 Smallholder Farmers' Perceptions and Adaptation to Climate Variability and Climate Change in Doba District, West Hararghe, Ethiopia *Asian J. Empir. Res.* **3**
[50] Komba C and Muchapondwa E 2018 Environment for Development Centers Adaptation to Climate Change by Smallholder Farmers in Tanzania *Adaptation to climate change by smallholder farmers in Tanzania* (Routledge) p 40

**Acknowledgements**
The author would also like to thank Dr. Fedassa Banti and Dr. Tolera Megersa for their advice and contribution towards this research work. Mr. Magarsa Abara and all field officers engaged in the research through participation in questionnaire adjustment and final enumeration are also commended. Finally, my sincere gratitude goes to Wollega University, particularly Research, Community Engagement and Technology Transfer Vice President for sponsorship.

# Prediction of drought vulnerability and its mitigation measures in the Northern Coast of Java, Indonesia

**Diah Auliyani*[1], Nining Wahyuningrum[1], Agung Budi Supangat[1] and Tyas Mutiara Basuki[1]**

[1]Watershed Management Technology Center, Ministry of Environment and Forestry, Jl. Jenderal Ahmad Yani - Pabelan, Kartasura, Central Java, Indonesia
*Corresponding author: d_auliyani@yahoo.com

**Abstract**. Climate change causes hydrometeorological disasters in Indonesia more frequently happen. One of them is drought. Drought reduces water availability and decreases agricultural production. It is necessary to select soil and water conservation techniques according to the level of vulnerability to mitigate losses. We researched Capluk Watershed which is often experiencing drought. The watershed is located on the Northern Coast of Java. The aims of this study were (1) to analyze the level of drought vulnerability and (2) to recommend drought mitigation measures. The drought vulnerability was analyzed spatially using a modification method proposed by Pramono and Savitri in 2019. The data used included rainfall, the number of dry months, actual annual evapotranspiration, geological formations, land systems, and distance to surface water sources. The analysis shows that 73% of the Capluk Watershed has a moderate vulnerability level to drought. The high level of vulnerability only exists in the upstream area, which is part of the karst landscape. We suggest that drought management focuses on the areas with moderate and high vulnerability levels. The community has been conducted water conservation by making ponds in the agricultural land and wells as water storage structures in the residential areas. In addition, we recommend several vegetative techniques such as planting annual crops with low evapotranspiration, and reforestation using Multi-purpose Tree Species (MPTS) or tree species with low evapotranspiration rates.

**Keywords** : drought, prediction, water conservation, Capluk Watershed

## 1. Introduction

Climate change has an influence on the hydrological system in watersheds. Climate change causes rainfall to fluctuate more volatile [1], so its triggers the occurrence of hydrometeorological disasters [2,3]. In Indonesia, rainfall is the utmost input in the hydrological system [4]. Rainfall which is lower than its average is the earliest sign of drought [5]. If this continues happening, it will affect soil moisture, river discharge, groundwater, ecosystems, and community welfare [6,7]. Agriculture is the first sector disturbed by drought [8,9]. Compared to other disasters, drought has the biggest impact [7,10]. It is because difficulty in predicting the beginning and the end of drought period [11]. Droughts can occur for months or even years.

Rembang is the regency on the northern coast of Central Java which is often experiencing drought [12]. In Regent Regulation Number 31 of 2020 concerning Changes in the Strategic Plan of the Regional Disaster Management Agency for 2016-2021, it is stated that the entire region of Rembang was a drought-prone area. There was a threat to the agricultural sector, therefore it is necessary to deal with drought to anticipate and reduce these losses.

Drought management must be carried out on a priority scale [13]. This priority can be determined according to the level of its vulnerability to drought. Climate change has caused a shift in rainfall patterns that have differences on drought vulnerability in the watershed. In addition, other biophysical conditions also reflect the level of drought vulnerability. As generally known, only a few watersheds have sufficient monitoring tools and available data [14,15]. Therefore, it is essential in selection of the method for determining drought vulnerability according to available data.

Pramono & Savitri [16] have proposed a modification method in calculating the level of drought vulnerability using data that are relatively easy to obtain. This method has been applied in Wonogiri Regency, and the results were able to describe the actual conditions. Based on that, this paper used their method in predicting the vulnerability level of drought in the study area. The objectives of this study are (1) to analyze the level of drought vulnerability and (2) to recommend drought mitigation measures. This information is essential in planning and managing a watershed, especially in reducing the losses due to drought.

## 2. Materials and methods

### *2.1. Materials*

This study was performed in the Capluk Watershed. Capluk is on the northern coast of Java Island, precisely in Rembang Regency. The entire area includes 218.4 sq km, expanding from 528271.89 mE to 544956.96 mE longitude and 9238128.74 mS to 9260832.87 mS latitude. Administratively, the study area located in 4 districts, Rembang, Kaliori, Sulang, Sumber, and Bulu (Figure 1). The altitude varies between 0 and 255 m asl.

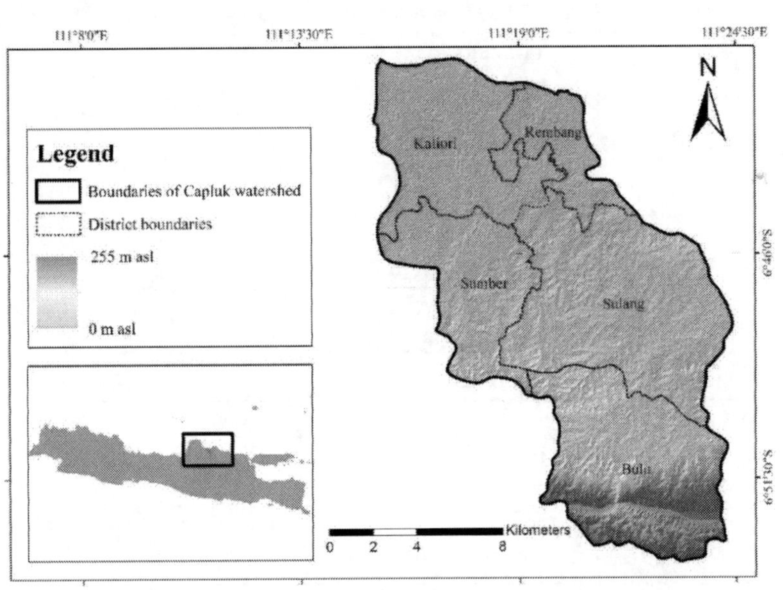


**Figure 1.** The study area map, Capluk Watershed

The data used were rainfall data and several thematic maps such as land cover, geological formation, and land systems maps. This study utilized daily rainfall data from the Climate Hazards Group InfraRed Precipitation with Station data (CHIRPS) between 1991 and 2020 [17]. The CHIRPS daily data on each grid was assessed into monthly data, then analyzed into dry month rainfall data (less than 100 mm/month).

### *2.2. Methods*

The level of drought vulnerability was evaluated spatially using the scoring technique (Table 1). The estimation refers to Pramono & Savitri method [16] (Equation 1), where DVL = drought vulnerability level; DMR = dry month rainfall (mm); AAE = actual annual evapotranspiration; GEO = geological formation; LSS = land system; DSW = distance to surface water sources. Groundchecks were undertaken to compare the result with the actual conditions.

$$DVL = 0.3\ DMR + 0.175\ AAE + 0.2\ GEO + 0.175\ LSS + 0.15\ DSW \quad (1)$$

**Table 1.** The parameters, classes, and score in Pramono & Savitri method [16]

| No | Parameters | Classes | Scores |
|---|---|---|---|
| 1 | Dry month rainfall (mm) | >60 | 1 |
| | | 45-60 | 2 |
| | | 30-45 | 3 |
| | | 15-30 | 4 |
| | | <15 | 5 |
| 2 | Actual annual evapotranspiration (mm) | <750 (Residential area) | 1 |
| | | 751-1000 (dryland agriculture) | 2 |
| | | 1001-1500 (forest/ state) | 3 |
| | | 1501-2000 (non-irrigated paddy field) | 4 |
| | | >2000 (irrigated paddy field) | 5 |
| 3 | Geological formation | Volcano | 1 |
| | | Mixed volcano & mountain folds | 2 |
| | | Mountain folds | 3 |
| | | Sediment | 4 |
| | | Limestone | 5 |
| 4 | Land system | Swamps | 1 |
| | | Alluvial valley | 2 |
| | | Plain | 3 |
| | | Alluvial fan & terrace | 4 |
| | | Mountains & hills | 5 |
| 5 | Distance to surface water sources (m) | 0-100 | 1 |
| | | 101-200 | 2 |
| | | 201-300 | 3 |
| | | 301-400 | 4 |
| | | >400 | 5 |

The drought mitigation measures were recommended on a priority scale matching the level of drought vulnerability and legal status of land in Indonesia [13,18]. The legal status of land in Indonesia were divided into forest and cultivation areas. The map belongs to the Ministry of Environment and Forestry (MoEF), the Republic of Indonesia which is freely accessible at webgis.menlhk.go.id.

## 3. Results and discussion

### *3.1. Description of Capluk Watershed*

Most of the study area is flat with steep slopes in the upstream area. The slopes range from 0 to 52%. According to MoEF, the major landcover in 2019 is agricultural area (66%) following by plantation

(17%), settlements (13%), and ponds (4%) as presented in Figure 2a and 2b. The soil types obtained from Regional Physical Planning Program for Transmigration (RePPPROT) comprise Calsiustolls, Haplustalfs, Hydraquents, Tropaquepts, and Ustropepts. The existing geological formations include the Alluvium, Bulu, Ledok, Mundu, Ngrayong, and Tuban formations (Figure 2c). Approximately 2,306.27 ha (11%) in the upstream Capluk Watershed is a karst area following the Minister of Energy and Mineral Resources No. 17 2012 concerning karst landscape.

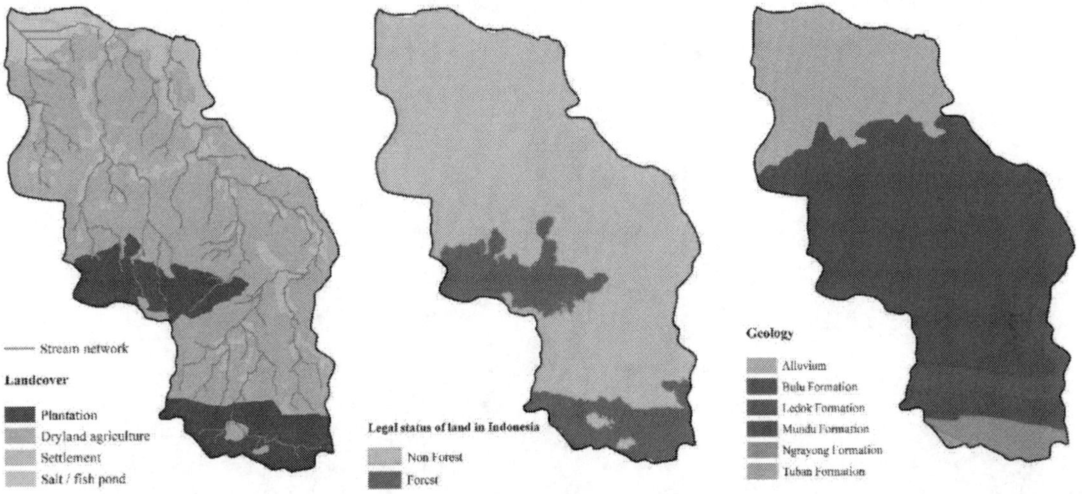


(a) (b) (c)

**Figure 2.** Landcover (a), legal status of land (b), and geological formation (c) of Capluk Watershed

*3.2. Rainfall distribution*

In the last 30 years, the average annual rainfall was 1,721 mm. The peak of the rainy season occurred between December and January, while the dry season during June to October. Figure 3 shows the rainfall temporal distribution of the study area. The dry month rainfall spatially ranged from 45.2–59.9 mm. In calculating the level of drought vulnerability, monthly rainfall has the same score for all watershed areas. The spatial distribution of rainfall is strongly influenced by the size of the watershed. The smaller the watershed, the rainfall will tend to be uniform, and vice versa.

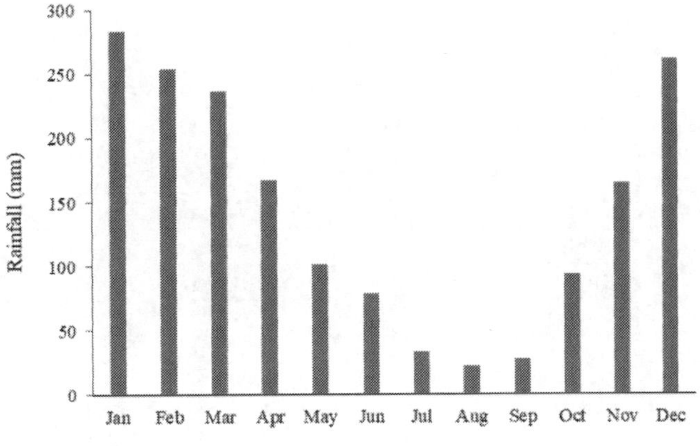


**Figure 3.** The rainfall temporal distribution of Capluk Watershed

### *3.3. The level of drought vulnerability*

Most of the study areas have a moderate level of drought vulnerability. Figure 4 shows the distribution of vulnerability to drought. The upstream areas with a high-level vulnerability are dominated Bulu Formation which consists of limestone and karst [19]. Due to the nature of karst which is composed of carbonate rock that has many crevices and is easily soluble in water, the surface drainage system of karst does not develop. Karst is dominated by the subsurface drainage system [19,20]. Thus, even though the karst is covered by forest vegetation, the karst is not able to hold water on the surface. In detail, the level of vulnerability to drought in each district is presented in Table 2.

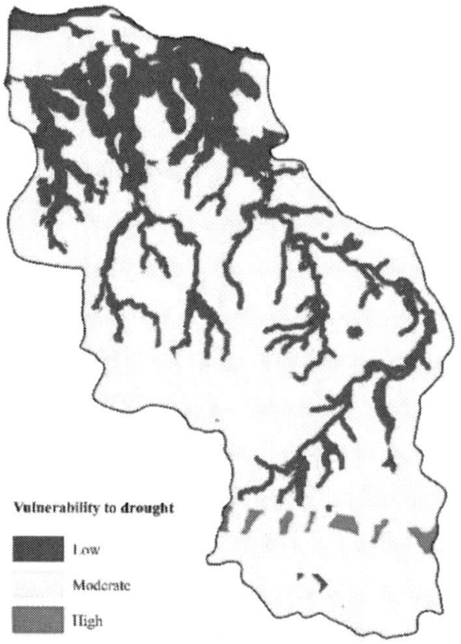


**Figure 4.** The level of vulnerability to drought in Capluk Watershed

**Table 2.** The level of vulnerability to drought in each district

| Legal status of land in Indonesia | Districts | Vulnerability level to drought (Ha) | | |
|---|---|---|---|---|
| | | Low | Moderate | High |
| Forest areas | Bulu | 5.3 | 2,146.7 | 54.0 |
| | Sulang | 15.2 | 854.5 | 0 |
| | Sumber | 11.7 | 748.3 | 0 |
| Non-Forest areas | Bulu | 339.6 | 2,834.2 | 155.0 |
| | Sulang | 1,548.6 | 4,486.7 | 0 |
| | Sumber | 599.1 | 1,924.1 | 0 |
| | Kaliori | 2,077.0 | 2,177.0 | 0 |
| | Rembang | 1,071.4 | 723.3 | 0 |
| **Total** | | **5,668** | **15,895** | **209** |
| | | **26%** | **73%** | **1%** |

Surface water sources are a determinant factor in managing drought. There were several sources of surface water in Capluk Watershed, including rivers, reservoirs, and agricultural ponds. The water level of this surface water source shrinks or even dries up during the dry season. Rivers in the watershed also experienced dry. The short stream network causes the rivers to dry up during the dry season, especially in the upper and middle parts of the watershed. The upper part did not provide optimal function as a catchment area due to the karst landscape. Karst geology is characterized by moderate to high water passing rates and scarce groundwater production, making it difficult for water to be stored and available in the soil layer. Therefore, areas with karst landscapes generally have a high level of drought vulnerability.

### *3.4. The mitigation measure to drought*

The mitigation measures to drought management in the Capluk Watershed focus on areas with moderate and high levels of vulnerability. The alternative treatment chosen needs to consider the legal status of land in Indonesia. Table 3 describes the mitigation measures to cope with drought according to the level of vulnerability and legal status of land.

**Table 3.** Drought mitigation measures in Capluk Watershed

| Legal status of land in Indonesia | Vulnerability to drought | | |
|---|---|---|---|
| | Low | Moderate | High |
| Forest | Maintenance activities | Reforestation using Multi-purpose Tree Species (MPTS) | 1. Reforestation using MPTS<br>2. Reforestation using tree species with low evapotranspiration rates, such as teak and mahagony |
| Non-Forest | | | |
| Agricultural area | Maintenance activities | 1. Alley cropping<br>2. Seasonal cropping pattern management<br>3. Planting ground cover crops such as legumes | 1. Alley cropping<br>2. Seasonal cropping pattern management<br>3. Planting ground cover crops such as legumes<br>4. Water conservation, such as ponds |
| Fishery area | Maintenance activities | Agro-silvofishery | 1. Agro-silvofishery<br>2. Water conservation, such as ponds |
| Residential area | Maintenance activities | Planting protective trees around public facilities | 1. Planting protective trees around public facilities<br>2. Water conservation, such as infiltration wells and rain harvesting instalations |

Source: Modification from Paimin et al. [13]

Drought is an annual phenomenon in the Capluk Watershed. Agricultural land in this watershed is mostly used for growing drought-tolerant seasonal crops. During field observations, part of the agricultural land was planted with tobacco, while some was abandoned (without management). The government and community build several ponds and wells to support the agricultural sector, especially during the dry season. These constructions collect rainwater to meet water needs in agricultural areas. Moreover, residents also build storage wells in their settlements to cope with drought.

## 4. Conclusions

The study area was dominated by the moderate-level vulnerability to drought, 73%. The high-level vulnerability was only found in the upper part of the watershed, the Karst area. We recommend the mitigation measures to drought focus on high and moderate-level vulnerability areas. The construction of ponds and wells are used by the community for adaptations to drought.

## References

[1] Mamuye M and Kebebewu Z 2018 Review on impacts of climate change on watershed hydrology *J. Environ. Earth Sci.* **8** 91–9
[2] Debele S E, Kumar P, Sahani J, Marti-Cardona B, Mickovski S B, Leo L S, Porcù F, Bertini F, Montesi D, Vojinovic Z and Di Sabatino S 2019 Nature-based solutions for hydro-meteorological hazards: Revised concepts, classification schemes and databases *Environ. Res.* **179** 108799
[3] Kartono K, Purwanto P and Suripin S 2020 Analysis of local rainfall characteristics as a mitigation strategy for hydrometeorology disaster in Rain-fed Reservoirs area in *Adv. Sci. Technol. Eng. Syst. J.* **5** 299–305
[4] Auliyani D and Wahyuningrum N 2021 Rainfall variability based on the Climate Hazards Group InfraRed Precipitation with Station Data (CHIRPS) in Lesti watershed, Java Island, Indonesia *IOP Conf. Ser. Earth Environ. Sci.* **874** 012003
[5] Kirono D G C, Tapper N J and McBride J L 1999 Documenting Indonesian rainfall in the 1997/1998 El Nino event *Phys. Geogr.* **20** 422–35
[6] Spinoni J, Naumann G, Carrao H, Barbosa P and Vogt J 2014 World drought frequency, duration, and severity for 1951-2010 *Int. J. Climatol.* **34** 2792–804
[7] Haile G G, Tang Q, Li W, Liu X and Zhang X 2020 Drought: Progress in broadening its understanding *Wiley Interdiscip. Rev. Water* **7** 1407
[8] Okpara J N, Afiesimama E A, Anuforom A C, Owino A and Ogunjobi K O 2017 The applicability of standardized precipitation index: Drought characterization for early warning system and weather index insurance in West Africa *Nat. Hazards* **89** 555–83
[9] Surmaini E, Susanti E, Syahputra M R and Hadi T W 2019 Exploring Standardized Precipitation Index for predicting drought on rice paddies in Indonesia *IOP Conf. Ser. Earth Environ. Sci.* **303** 012027
[10] Wilhite D A 2000 Drought as a natural hazard: Concepts and definitions *Drought A Glob. Assess.* **1** 3–18
[11] Sönmez F K, Kömüscü A Ü, Erkan A and Turgu E 2005 An analysis of spatial and temporal dimension of drought vulnerability in Turkey using the standardized precipitation index *Nat. Hazards* **35** 243–64
[12] Nurtjahjanto B D 2006 *Interactive water resources management in the implementation of the sumur resapan system in Rembang Regency* (Double Degree UGM - IHS)
[13] Paimin., Pramono I B, Purwanto and Indrawati D R 2012 *Sistem perencanaan pengelolaan daerah aliran sungai* ed H Santoso and Pratiwi (Bogor: Pusat Penelitian dan Pengembangan Konservasi dan Rehabilitasi. Badan Penelitian dan Pengembangan Kehutanan. Kementerian Kehutanan)
[14] Lufi S, Ery S and Rispiningtati R 2020 Hydrological analysis of TRMM (Tropical Rainfall Measuring Mission) data in Lesti sub watershed *Civ. Environ. Sci.* **003** 018–30
[15] Kuswanto H and Naufal A 2019 Evaluation of performance of drought prediction in Indonesia based on TRMM and MERRA-2 using machine learning methods *MethodsX* **6** 1238–51
[16] Pramono I B and Savitri E 2019 Modification method of drought vulnerability at Wonogiri District, Central Java, Indonesia *Int. J. Innov. Technol. Explor. Eng.* **8** 551–5
[17] Funk C, Peterson P, Landsfeld M, Pedreros D, Verdin J, Shukla S, Husak G, Rowland J, Harrison L, Hoell A and Michaelsen J 2015 The climate hazards infrared precipitation with stations - A new environmental record for monitoring extremes *Sci. Data* **2**

[18] Auliyani D 2020 Upaya konservasi tanah dan air pada daerah pertanian dataran tinggi di Sub-Daerah Aliran Sungai Gandul (Soil and water conservation efforts in the highland agriculture area in Gandul Sub Watershed) *J. Ilmu Pertan. Indones. (JIPI), Juli* **25** 382–7

[19] Setiawan T, Syah Alam B Y C S S, Haryono E and Hendarmawan 2020 Hydrochemical and environmental isotopes analysis for characterizing a complex karst hydrogeological system of Watuputih area, Rembang, Central Java, Indonesia *Hydrogeol. J.*

[20] Adji T N, Haryono E, Mujib A, Fatchurohman H and Bahtiar I Y 2019 Assessment of aquifer karstification degree in some karst sites on Java Island, Indonesia *Carbonates and Evaporites* **34** 53–66

**Acknowledgement**
We express our gratitude to Watershed Management Technology Center (WMTC) for accommodating the facilities to support this study. The authors would like to thank the anonymous reviewers for their valuable comments to revise this article. All authors have an equal contribution.

# Identification of Sambiloto agroecology as a basis for utilization and conservation of germplasm

**Bambang Pujiasmanto*** **and Ida Rumia Manurung**

[1]Departement of Agrotechnology, Faculty of Agriculture, Universitas Sebelas Maret, Indonesia

*Corresponding author: bambang_p56@staff.uns.ac.id

**Abstract**. Climate change can affect the diversity and growth of tropical plants, one of which is Sambiloto. Sambiloto is a wild plant with medicinal potential that is experiencing scarcity due to climate change in Indonesia. This study aims to identify the agroecological suitability of Sambiloto. The research was carried out using the survey method. The survey method includes the study of description and identification and the distribution pattern of sambiloto. The results showed that Sambiloto could grow at an altitude of 180 – 861 meters above sea level with a temperature of 20.32 $^0$C -26.93 $^0$C, relative humidity 78% - 87%, rainfall 2053.2-3555.6 mm/year. Sambiloto can grow on soils with moderate N, low P, low Mg, very low to low Ca, low to moderate organic C, while the pH is slightly acidic to acidic. The distribution pattern of sambiloto in the middle and lowlands is clustered, while in the highlands, it is uniform. The highest andrographolide content was obtained from sambiloto in the middle plains (2.27%), while in the lowlands (1.37%) and highlands (0.89%). Sambiloto is more commonly found in the middle lands (RD 76.87%, RF 10%, RK 27.75%, IVI 114.62%) than in the lowlands (RD 59.50%, RF 9.09%, RK 31.41%, IVI 99.98%), and highlands. (RD 42.20%, RF 3.88%, RK 11.62%, and IVI 78.21%).

**Keywords** : Agroecology, distribution pattern, medicinal plants, germplasm preservation

## 1. Introduction

Climate change is one of the main problems and the most significant challenge to agricultural production. Climate is a combination of geographical zone characteristics of the atmosphere that dominates and persists over time and is determined based on the variables of temperature, rainfall, humidity, wind and solar radiation [1,2]. Developing agricultural production systems with greater climate resilience is an important strategy in dealing with climate change. The potential of plants to adapt to new areas is a significant consideration in climate change in the tropics. Sambiloto (*Andrographis paniculate*) is a herbaceous plant in the family Acanthaceae and is widely distributed in subtropical areas such as India, Thailand, Vietnam, and China [3]. Sambiloto is harvested from its natural habitat by the community as a source of traditional medicinal ingredients. The plant is used as a medicine from Asian countries for the treatment of diarrhea, flu, leprosy, leptospirosis, malaria, rabies, upper respiratory infections, sinusitis, syphilis, tuberculosis and HIV/AIDS [4]. Sambiloto contains volatile chemical compounds (2-methyl-butanal, 3-methyl-butanal, 2-methyl-propanal) [5].

Sambiloto taking that is carried out continuously without proper cultivation efforts will threaten the existence of the sambiloto germplasm [6,7]. This is because Sambiloto self-pollinates, causing low genetic diversity and has been classified as an endangered species [8]. This encourages efforts to cultivate Sambiloto plants into cultivated plants. The domestication of sambiloto by considering the phytogeographical and historical evidence shows that sambiloto was domesticated in South India. Furthermore, when introduced to new environmental and ecological conditions, Sambiloto engages in diversification through natural selection [9].

Plants are characterized by morphological and chemical characters influenced by the environment and developmental stage. Sambiloto has various problems, such as highly degraded varieties, unstable active chemical content, strongly influenced by environmental conditions, single genetic traits, and limited seed supply area [10]. Various efforts can be made related to the cultivation of Sambiloto. The process from wild plants to cultivated plants through planting in new habitats is called domestication. Plant domestication is the process of genetic re-formation of wild species to meet human needs [11]. Some morphological characteristics, such as plant architecture, seed size and dispersal, have changed during domestication [12]. The first step is to evaluate the condition of the plant habitat. This study aims to identify the agroecological suitability of Sambiloto.

## 2. Materials and methods

The research was carried out using a survey method. The survey locations were selected by purposive sampling method in the lowlands (<400 m asl), middle plains (400-700 m asl), and highlands (>700 m asl). The survey location was determined based on the presence of teak trees and bitter plants in the Forest Management Units, namely the Karanganyar, Klaten and Wonogiri Forest Management Units. The sample point was determined based on purposive random sampling through a pre-survey approach where the plant could be found. The method of determining sample plots and vegetation analysis used in the survey was the quadratic method. On the transect line, an observation plot of 20 m x 20 m was made.

## 3. Results and discussion

### *3.1 Survey of the distribution of Sambiloto plants at various altitudes*

Based on the survey, sambiloto can grow at an altitude of 180 m to 861 m above sea level, temperature (20.32 $^{0}$C – 26.93 $^{0}$C), humidity (78%–87%). Rainfall ranges from 2053.2 – 3555.6 mm/yr. The light intensity above the canopy ranges from 399.74 – 456.95 lux, while it is 53.29 – 93.37 lux below the canopy. Light interception received 76.64 – 88.21%. Shade for sambiloto plants includes teak, mahogany, sengon, melinjo, coconut, lamtoro. Sambiloto can grow in the lowlands to the highlands with an altitude of 900 m above sea level, in the shade of teak, pine and gliricidia [13,14]. Sambiloto can grow on soils with the moderate value of N, low P, low Mg, very low to low Ca, low to moderate organic C, while the pH is slightly acidic to acidic.

### *3.2 Vegetation analysis*

Vegetation analysis was carried out using the quadratic method in various habitats of sambiloto. The shape of the sample plots is a square, with the sample plot sizes 20 m x 20 m (trees), 5 m x 5 m (herbs), and 1 m x 1 m for grass. The sample plots are 10, and the sample area is 4000 $m^2$ for trees, 250 m2 for herbs, and 10 m2 for grass. Based on the vegetation analysis, the highest Important Value Index (IVI) tree was teak (*Tectona grandis* L.) both in the lowlands (111.7 %), medium plains (121.75%) and highlands (122.85%). The highest IVI grass group is *Ischaemum timorense* L. in the lowlands (49.60%) and the middlelands (76.4%), while in the highlands *Digitaria longiflora* (91%). Plants that have a high IVI are the dominant species and are able to maintain their sustainability. A plant species has the potential to become dominant if it grows in a suitable location to support its growth [15]. In this study, it was found that the conditions of the research location matched the characteristics of Sambiloto which can live up to an altitude of 1300 m above sea level but is more productive at an altitude of 500-700 m above sea level, with fertile soil such as volcanic soil with an average rainfall of 1,200 mm per year [16].

### *3.3 Scatter pattern*

Based on the survey results, the distribution pattern of sambiloto in the lowlands and middle lands is clustered, while in the highlands, it is uniform. Sambiloto that grows in the highlands has lower values of dominance, relative dominance (RD), relative frequency (RF), density and relative density than those in the low and middle lands. The distribution pattern of species in nature tends to be clumped. Sambiloto is more commonly found in the middle lands (RD 76.87%, RF 10%, RK 27.75%, IVI 114.62%) than in the lowlands (RD 59.50%, RF 9.09%, RK 31.41%, IVI 99.98%), and highlands (RD 42.20%, RF 3.88%, RK 11.62%, and IVI 78.21%).

### *3.4 Andrographolide content*

Based on laboratory tests, the highest andrographolide content was in the medium land habitat (2.27%), compared to those from the lowland (1.37%), and highland (0.89%). The content of andrographolide is formed through secondary metabolism. Levels of secondary metabolites are formed in plants to defend themselves from the growing ecosystem. Therefore, the high and low levels of secondary metabolites in each plant are influenced by the environment such as altitude, rainfall and temperature [17]. Environmental factors influence andrographolide content by regulating key genes in its biosynthetic pathway including HMGR, DXS, GGPS and ApCPS. Increased accumulation of carbohydrates, especially starch, is beneficial for andrographolide biosynthesis; organic N sources decrease carbohydrate depletion by reducing N metabolism and promote plant growth and andrographolide biosynthesis synergistically [18].

## 4. Conclusion

Sambiloto grows well at an altitude of 180 m - 861 m above sea level, temperature 20.32 $^{0}$C - 26.93 $^{0}$C, air humidity 78% - 87%, rainfall ranges from 2053.2 mm/year - 3555.6 mm/year. Received light interception 76.64% - 88.21%. Sambiloto is more commonly found in the middle lands (RD 76.87%, RF 10%, KR 27.75%, IVI 114.62%) than in the lowlands (RD 59.50%, RF 9.09%, KR 31.41%, IVI 99.98%), and highlands (RD 42.20%, RF 3.88%, KR 11.62%, and IVI 78.21%). The distribution pattern of Sambiloto in the lowlands and middle lands is clumped, while in the highlands, it is uniform (uniform). The agroecological identification of Sambiloto can increase the availability of medicinal plants due to climate change.

## References

[1] Sarkar D, Kar S K, Chattopadhyay A, Shikha, Rakshit A, Tripathi V K, Dubey P K and Abhilash P C 2020 Low input sustainable agriculture: A viable climate-smart option for boosting food production in a warming world *Ecological Indicators* **115** 106412

[2] Zheng J, Wang W, Ding Y, Liu G, Xing W, Cao X and Chen D 2020 Assessment of climate change impact on the water footprint in rice production: Historical simulation and future projections at two representative rice cropping sites of China *Science of the Total Environment* **709** 136190

[3] Jiang M, Sheng F, Zhang Z, Ma X, Gao T, Fu C and Li P 2021 Andrographis paniculata (Burm.f.) Nees and its major constituent andrographolide as potential antiviral agents *Journal of Ethnopharmacology* **272** 113954

[4] Kumar S, Singh B and Bajpai V 2021 Andrographis paniculata (Burm.f.) Nees: Traditional uses, phytochemistry, pharmacological properties and quality control/quality assurance *Journal of Ethnopharmacology* **275** 114054

[5] Xia P, Li Q, Liang Z, Zhang X and Yan K 2021 Spaceflight breeding could improve the volatile constituents of Andrographis paniculata *Industrial Crops and Products* **171** 113967

[6] Ngomle S, Bhattacharya P M, Ambesh B S and Roy A 2018 Identification of promising Lentil genotypes for Terai region of West Bengal *Indian Journal of Agricultural Research* **52** 271–7

[7] Menezes Sobrinho J A de, Charchar J M and Aragão F A S 1999 Caracterização morfológica de germoplasma de alho por análises multivariada componentes principais e variáveis canônicas *Horticultura Brasileira* **17** 96–101
[8] Hiremath C, Greeshma M, Gupta N, Kuppusamy B, Shanker K and Sundaresan V 2020 Morphometric, chemotypic, and molecular diversity studies in Andrographis paniculata *Journal of Herbs, Spices and Medicinal Plants* **00** 1–14
[9] Raina A P, Gupta V, Sivaraj N and Dutta M 2013 Andrographis paniculata (Burm. f.) Wall. ex Nees (kalmegh), a traditional hepatoprotective drug from India *Genetic Resources and Crop Evolution* **60** 1181–9
[10] Wang S, Guan K, Wang Z, Ainsworth E A and Zheng T 2021 Unique contributions of chlorophyll and nitrogen to predict crop photosynthetic capacity from leaf spectroscopy *Journal of Experimental Botany* **72** 341–54
[11] Zhang Z, Liu Y, Yuan L, Weber E and van Kleunen M 2021 Effect of allelopathy on plant performance: a meta-analysis *Ecology Letters* **24** 348–62
[12] Hu M, Lv S, Wu W, Fu Y, Liu F, Wang B, Li W, Gu P, Cai H, Sun C and Zhu Z 2018 The domestication of plant architecture in African rice *Plant Journal* **94** 661–9
[13] Utami U, Nisa C, Putri A Y and Rahmawati E 2019 The potency of secondary metabolites endophytic fungi Trichoderma sp as biocontrol of Colletotrichum sp and Fusarium oxysporum causing disease in chili *AIP Conference Proceedings 2120,* **2120**
[14] Rahmawaty, Samosir J B, Batubara R and Rauf A 2019 Diversity and distribution of medicinal plants in the Universitas Sumatera Utara arboretum of deli serdang, North Sumatra, Indonesia *Biodiversitas* **20** 1457–65
[15] Prihatini R, Syarif A, Bakhtiar A and Mansyurdin 2018 Leaf epidermis and phytochemical studies of sambiloto (Andrographis paniculata (Burm. F.) Nees) *Journal of Physics: Conference Series* **1116**
[16] Kuswantoro F, Lugrayasa I N and Sujarwo W 2018 Studi ekologi kuantitatif hutan pilan sebagai dasar pengembangan Kebun Raya Gianyar *Jurnal Ilmu Kehutanan* **12** 184
[17] Nair D S and Manjula S 2020 Induction of root endosymbiosis as a highly sustainable and efficient strategy for overproduction of the medicinally important diterpenoid lactone-andrographolide in Andrographis paniculata (Burm. F.) Wall. ex Nees *Industrial Crops and Products* **156**
[18] Zhong C, Jian S F, Chen D L, Huang X J and Miao J H 2021 Organic nitrogen sources promote andrographolide biosynthesis by reducing nitrogen metabolism and increasing carbon accumulation in Andrographis paniculata *Plant Physiology and Biochemistry* **164** 82–91

# Growth of soybean (*Glycine max* L.) on various types of biochar

**A I Nurmalasari[1]*, Muji Rahayu[1] and Evita Nabila Nur Owena[2]**

[1]Departement of Agrotechnology, Faculty of Agriculture, Universitas Sebelas Maret, Indonesia
[2]Undergraduate Departement of Agrotechnology, Faculty of Agriculture, Universitas Sebelas Maret
*Corresponding author: ikeaprilia@staff.uns.ac.id

**Abstract.** Climate change especially in agriculture, have a serious impact on soil fertility. Low soil fertility can affect the quality and quantity of agricultural produce. Biochar is one of the soil-fixing materials that can be used to increase soil fertility i.e. physical, chemical and biological fertility especially in vertisol soil. The purpose of this research was to find out the influence of various types of biochar and the best dose of biochar on the growth of soybean plants. The research was conducted at Screen House, Faculty of Agriculture, Universitas Sebelas Maret, Surakarta on March to June 2021. The research was designed in randomized complete block design (RCBD) with two factors. The first factor is the type of biochar consisting of three levels, which consist of coconut shell biochar, rice husk biochar, and biochar corn cob. The second factor is the biochar dose consisting of five levels, which consist of 0 ton.ha$^{-1}$, 5 tons.ha$^{-1}$, 10 tons.ha$^{-1}$, 15 tons.ha$^{-1}$, and 20 tons.ha$^{-1}$. Each experiment was repeated three times. The results showed no interaction between the type and dose of biochar in all variables. Treatment of the type of biochar rice husk with a dose of 20 tons.ha-1 can increase the plant height by 14.6%. Treatment of the type of rice husk biochar with a dose of 15 tons.ha-1 can increase the number of leaves by 11.345%, the number of branches by 36.63%, wet weight by 17.74%, dry weight by 15.9%. Biochar treatment of coconut shell with a dose of 5 tons.ha-1 can increase the leaf area by 63.45%.

**Keywords** : climate change, growth, soybean, biochar, vertisols

## 1. Introduction

Climate change especially in agriculture, have a serious impacts. These impacts resulted can increase in extreme weather events, changes in rainfall, as well as rising temperatures. Changes in patterns and increases in extreme climate frequencies cause agricultural production to decrease significantly [1].

Soybean plants are one of the types of pods that are sensitive to climate change because they have an impact on a decrease in high production [2]. Climate change can also affect land degradation and affect the physical and chemical properties of soils, thus impacting low soil fertility rates. Soil fertility is another problem that can affect the growth of soybeans, especially in vertisol soils. Vertisol soil is a soil dominated by gray to dark blackness, formed from the accumulation of minerals 2:1, and influenced by the process of mixing the soil of the lower and upper layers caused by wet and dry conditions. This characteristic is influenced by the firm climate between the rainy season and the dry season [3].

The main obstacles to vertisol type soil are the heavy clay texture, the properties of expanding and shrinking and slow drainage [4]. Efforts to improve soil fertility in vertisol soils caused by climate change can be overcome by the addition of soil-fixing materials [5]. One of the soil fixing materials that can be obtained easily comes from agricultural waste in the form of biochar. Biochar is a soil fixing material that results from pyrolysis processes under conditions of oxygen limitation. The use of biochar can maintain soil moisture and improve the physical properties of degraded soil [6]. The utilization of biochar as agricultural waste is expected to be one of the sustainable agricultural efforts [7].

According to Endriani and Kurniawan's research [8], biochar can affect plant growth. The higher the amount of biochar applied, the better the growth of plants. The addition of biochar can increase soil moisture, soil fertility, increase water holding capacity, and can last a long day when applied in soil [9].

The Biochar can give a positive effect on semiarid vertisol soil because it is able to increase soil levels, electrical delivery power, and lower soil temperature[10]. Some studies state that the addition of rice husk biochar can increase plant productivity and nutrient availability as well as improve soil so that plant roots can absorb nutrients [11]. Based on research conducted by Zhu et al [12], the addition of biochar can produce the root system in soybeans to be fertile, increasing the absorption of areas in the soil, especially for less mobile elements such as P. Based on the information above, there needs to be research related to various types of biochar. The purpose of this study is to find out the influence of various types of biochar on the growth of soybean plants.

## 2. Materials and method

This research was conducted in March to June 2021 at the screen house of the Faculty of Agriculture, Sebelas Maret University, Surakarta, Indonesia. The soy variety used is the Grobogan variety soybean. The research was designed in randomized complete block design with two factors. The first factor is the type of biochar, which consists of coconut shell biochar, rice husk biochar, and biochar corn cob. The second factor is the dose of biochar, which consists of 0 ton.ha$^{-1}$, 5 tons.ha$^{-1}$, 10 tons.ha$^{-1}$, 15 tons.ha$^{-1}$, and 20 tons.ha$^{-1}$. Each combination treatment was repeated three times, resulting in 45 experimental units. The observed variables were plant height, number of leaves, number of branches, leaf area, wet weight of the plant, and dry weight of the plant.

## 3. Results and discussions

### *3.1 Plant height*

Based on the variety shows that the treatment of biochar type and biochar dose there is no interaction at the age of 8 week after planting. Biochar type treatment at the age of 8 week after planting hasn't real effect on plant height but the best type of biochar is rice husk with an average value of 141.60 cm plant height (Table 1.). This is because the biochar of rice husks is able to improve the physical and biological properties of the soil and can improve the soil and have a high KTK so that it can stimulate plant growth [13].

**Table 1**. The treatment various types of biochar on plant height at the age of 8 week after planting

| Age | Treatment | Dose of biochar (ton.ha$^{-1}$) | | | | | Mean |
|---|---|---|---|---|---|---|---|
| | **Type of Biochar** | **0** | **5** | **10** | **15** | **20** | |
| **8 WAP** | Coconut shell | 127.00 | 140.33 | 131.67 | 134.00 | 140.67 | 134.73a |
| | Rice husk | 138.00 | 145.67 | 138.33 | 137.33 | 148.67 | 141.60a |
| | Corn cob | 120.67 | 111.00 | 119.67 | 139.67 | 152.67 | 128.73a |
| | Mean | 128.56a | 132.33a | 129.89a | 137.00a | 147.34a | - |

Description: The average number followed by the same letter shows a result that is not real different based on Duncan's multiple range test at the level of 5%.

The treatment of biochar dose at the age of 8 week after planting doesn't have a real effect on the height of the plant at the age of 8 week after planting but the best biochar dose is 20 tons.ha$^{-1}$ with an average plant height of 147.34 cm (Table 1). This is because according to Nurida et al [14], the impact of biochar on plant productivity depends on the characteristics of biochar, the dose used and the ability to deal with soil problems. The more doses given better the growth.

*3.2 Number of leaves*

**Table 2.** The treatment various types of biochar on number of leaves at the age of 8 week after planting

| Age | Treatment / Type of biochar | Dose of biochar (ton.ha$^{-1}$) 0 | 5 | 10 | 15 | 20 | Mean |
|---|---|---|---|---|---|---|---|
| 8 WAP | Coconut shell | 35.33 | 29.00 | 39.33 | 41.00 | 29.00 | 34.73a |
| | Rice husk | 32.00 | 38.33 | 32.67 | 35.00 | 42.00 | 36.00a |
| | Corn cob | 32.67 | 39.00 | 33.67 | 35.33 | 37.00 | 35.53a |
| | Mean | 33.33a | 35.44a | 35.22a | 37.11a | 36.00a | - |

Description: The average number followed by the same letter shows a result that is not real different based on Duncan's multiple range test at the level of 5%.

Based on the variety shows that the treatment of biochar type and biochar dose there wasn't interaction at the age of 8 week after planting. The treatment of biochar type at the age of 8 week after planting hasn't real effect on the number of leaves at the age of 8 week after planting but the best type of biochar is rice husk with an average value of 36.00 leaves (Table 2.). This is because according to research Musnoi et al [15] the growth of plants that have optimal availability of nutrients will increase the process of photosynthesis, so that at the day of division, enlargement and differentiation of cells will be better. The treatment of biochar dose at the age of 8 week after planting hasn't real effect on the number of leaves at the age of 8 week after planting but the best biochar dose is 15 tons.ha$^{-1}$ with an average of 37.11 (Table 2.). This is because leaf formation is influenced by the absorption and availability of nutrients, in addition biochar is able to optimize growth and is able to reduce nutrients lost due to leaching[11].

*3.3 Number of branches*

**Table 3.** The treatment various types of biochar on number of branches at the age of 8 week after planting

| Age | Treatment / Type of biochar | Dose of biochar (ton.ha$^{-1}$) 0 | .5 | 10 | 15 | 20 | Mean |
|---|---|---|---|---|---|---|---|
| 8 WAP | Coconut shell | 12.00 | 9.67 | 13.33 | 13.67 | 9.67 | 11.66a |
| | Rice husk | 10.67 | 12.67 | 11.00 | 20.00 | 14.33 | 13.73a |
| | Corn cob | 11.00 | 13.00 | 11.67 | 12.33 | 13.00 | 12.20a |
| | Mean | 11.22a | 11.78a | 12.00a | 15.33a | 12.33a | - |

Description: The average number followed by the same letter shows a result that is not real different based on Duncan's multiple range test at the level of 5%.

Based on the variety shows that the treatment of biochar type and biochar dose there wasn't interaction at the age of 8 week after planting. The treatment of biochar type at the age of 8 week after planting hasn't real effect on the number of branches at the age of 8 week after planting but the best type of biochar is rice husk with an average value of 13.73 (Table 3.). This is because the provision of biochar used as a soil bending material is able to work well in retaining water so that the availability of macro and microdid nutrients in the soil increases, characterized by an increase in the number of plant branches better than treatment without biochar [16]. The treatment of biochar dose at the age of 8 week after planting hasn't real effect on the number of branches at the age of 8 week after planting but the best biochar dose is 15 tons.ha$^{-1}$ with an average of 15.33 (Table 3.). According to Junita et al[17] the more organic matter added in the soil, the more it will affect the increase in the ability of soil to bind water and nitrogen. Sufficient nitrogen needs make plant growth grow well.

### 3.4 *Wet weight of plant*

**Table 4**. The treatment various types of biochar on wet weight of plant at the age of 10 week after planting

| Age | Treatment | Dose of biochar (ton.ha$^{-1}$) | | | | | Mean |
|---|---|---|---|---|---|---|---|
| | **Type of biochar** | **0** | **5** | **10** | **15** | **20** | |
| **10 WAP** | Coconut shell | 6.96 | 4.54 | 5.33 | 10.14 | 5.07 | 6.40a |
| | Rice husk | 6.75 | 6.76 | 6.28 | 5.85 | 11.02 | 7.33a |
| | Corn cob | 4.39 | 6.78 | 3.33 | 5.31 | 5.12 | 4.98a |
| | Mean | 6.03a | 6.03a | 4.98a | 7.10a | 7.07a | - |

Description: The average number followed by the same letter shows a result that is not real different based on Duncan's multiple range test at the level of 5%.

Based on the variety shows that the treatment of biochar type and biochar dose there wasn't interaction at the age of 10 week after planting. Biochar type treatment at the age of 10 week after planting hasn't real effect on wet weight at the age of 10 week after planting but the best biochar type treatment is biochar rice husk with an average of 7.33 (Table 4.). This is because the biochar of rice husks has the ability to hold water and nutrients in the soil so that the provision of biochar is able to increase nutrient uptake. Biochar's ability to withstand nutrient uptake can make fertilization efficiency better and have an impact on increasing wet weight [18]. The treatment of biochar dose at the age of 10 week after planting hasn't real effect on wet weight at the age of 10 week after planting but the best biochar dose is 15 tons.ha$^{-1}$ with an average of 7.10 than the treatment of 0 ton.ha$^{-1}$ (Table 4.). According to Goenadi and Santi's research[19], biochar can last longer in the soil, so nutrients can be absorbed by the roots so that the fresh weight of the plant has a high value compared to treatment without the addition of biochar.

### 3.5 *Dry weight of plant*

**Table 5**. The treatment various types of biochar on dry weight of plant at the age of 10 week after planting

| Age | Treatment | Dose of biochar (ton.ha$^{-1}$) | | | | | Mean |
|---|---|---|---|---|---|---|---|
| | **Type of biochar** | **0** | **5** | **10** | **15** | **20** | |
| **10 WAP** | Coconut shell | 2.53 | 1.74 | 1.84 | 3.58 | 1.48 | 2.23a |
| | Rice husk | 2.31 | 2.30 | 2.05 | 2.25 | 3.95 | 2.57a |
| | Corn cob | 1.76 | 2.51 | 1.34 | 1.82 | 1.93 | 1.87a |
| | Mean | 2.20a | 2.18a | 1.74a | 2.55a | 2.45a | - |

Description: The average number followed by the same letter shows a result that is not real different based on Duncan's multiple range test at the level of 5%.

Based on the variety shows that the treatment of biochar type and biochar dose there wasn't interaction at the age of 10 week after planting. Biochar type treatment at the age of 10 week after planting hasn't real effect on dry weight at the age of 10 week after planting but the best biochar is biochar rice husk with an average of 2.57 (Table 5.). This is because organic matter given to the soil has more nutrient content than without treatment [20]. The treatment of biochar doses at the age of 10 week after planting hasn't real effect on dry weight at the age of 10 week after planting but the best biochar dose is 15 tons.ha$^{-1}$ with an average of 2.55 than the treatment of 0 ton.ha$^{-1}$ (Table 5.). According to Widowati[21] biochar have a good effect on the addition of dry weights, high dry weight values indicate that root formation is so good that roots can facilitate nutrients and water better for photosynthesis.

### 3.6 *Leaf area*

**Table 6.** The treatment various types of biochar on leaf area at the age of 4 week after planting

| Age | Treatment | Dose of biochar (ton.ha$^{-1}$) | | | | | Mean |
|---|---|---|---|---|---|---|---|
| | **Type of biochar** | **0** | **5** | **10** | **15** | **20** | |
| **4 WAP** | Coconut shell | 284.57 | 345.05 | 272.16 | 389.25 | 347.38 | 327.68a |
| | Rice husk | 266.74 | 352.03 | 146.55 | 390.80 | 354.36 | 302.09a |
| | Corn cob | 82.96 | 339.62 | 228.74 | 237.27 | 258.21 | 229.36a |
| | Mean | 211.42a | 345.57a | 215.82a | 339.11a | 319.98a | - |

Description: The average number followed by the same letter shows a result that is not real different based on Duncan's multiple range test at the level of 5%.

Based on the variety shows that the treatment of biochar type and biochar dose there wasn't interaction at the age of 4 week after planting. Biochar type treatment at the age of 4 week after planting hasn't real effect on the leaf area at the age of 4 week after planting but the best biochar type treatment is coconut shell biochar with an average of 327.68 (Table 6.). This is because the provision of biochar can increase the area of plant leaves during vegetative, with the availability of nutrients that are fulfilled able to increase the area of plant leaves, especially macro nutrients [22]. The treatment of biochar dose at 4 week after planting hasn't real effect on the leaf area at the age of 4 week after planting but the best biochar dose is 5 tons.ha$^{-1}$ (Table 6.). This is because the availability of nutrients in the soil can increase the process of photosynthesis faster to increase the number and area of leaves. If the availability of elements N, P, and K is available, the leaf area will be increas [23].

## 4. Conclusion

Climate change has an impact on the environment, especially on soybean plants. One of the impacts is on the vertisol soil if the weather is too extreme, the soil have a clay texture and will be expand in the rainy season and shrink in the dry season. The solution can use biochar as a soil fixing material. Based on the results of the research, there has been no interaction between the type and dose of biochar in all variables. Treatment of the type of biochar rice husk with a dose of 20 tons.ha$^{-1}$ can increase the height of the plant by 14.6%. Treatment of the type of rice husk biochar with a dose of 15 tons.ha$^{-1}$ can increase the number of leaves by 11.34%. Treatment of the type of rice husk biochar with a dose of 15 tons.ha$^{-1}$ can increase the number of branches by 36.63%. Treatment of the type of biochar rice husk with a dose of 15 tons.ha$^{-1}$ can increase wet weight by 17.74%. Treatment of the type of rice husk biochar with a dose of 15 tons.ha$^{-1}$ can increase dry weight by 15.9%. Treatment of coconut shell biochar type with a dose of 5 tons.ha$^{-1}$ can increase leaf area by 63.45%.

## References

[1] Surmaini E, Runtunuwu E and Las I 2011 Upaya sektor pertanian dalam menghadapi perubahan iklim *J. Litbang Pertan.* **30** 1–7
[2] Ruminta R, Irwan A W, Nurmala T and Ramadayanty G 2020 Analisis dampak perubahan iklim terhadap produksi kedelai dan pilihan adaptasi strategisnya pada lahan tadah hujan di Kabupaten Garut *Kultivasi* **19** 1089–97
[3] Prasetyo Balai B H, Penelitian B, Sumberdaya P and Pertanian L 2007 Perbedaan Sifat-Sifat Tanah Vertisol Dari Berbagai Bahan Induk *J. Ilmu-Ilmu Pertan. Indones.* **9** 20–31
[4] Nurdin, Maspeke P, Ilahude Z and Zakaria F 2009 Pertumbuhan dan Hasil Jagung yang Dipupuk N, P, dan K pada Tanah Vertisol Isimu Utara Kabupaten Gorontalo *J. Trop. SOILS* **14** 49–56
[5] Utami D N, Pertama Pada Pusat P, Reduksi T and Bencana R 2019 Kajian Dampak Perubahan Iklim Terhadap Degradasi Tanah *J. ALAMI J. Teknol. Reduksi Risiko Bencana* **3** 122–31
[6] Herviyanti H, Maulana A, Prima S, Aprisal A, Crisna S D and Lita A L 2020 Effect of biochar from young coconut waste to improve chemical properties of ultisols and growth coffee [Coffea arabica L.] plant seeds *IOP Conf. Ser. Earth Environ. Sci.* **497** 012038

[7] Nurida N L 2014 Potensi Pemanfaatan Biochar Untuk Rehabilitasi Lahan Kering Di Indonesia *J. Sumberd. Lahan* **8** 57–68

[8] Endriani E and Kurniawan A 2018 Konservasi Tanah dan Karbon Melalui Pemanfaatan Biochar Pada Pertanaman Kedelai *J. Ilm. Ilmu Terap. Univ. Jambi* **2** 93–106

[9] Andini L and Ajiningrum P S 2020 Peran Kulit Nanas Sebagai Bioaktivator Dan Penambahan Biochar Tempurung Kelapa Dalam Meningkatkan Hasil Produksi Tanaman Kedelai (Glycine max (L) Merrill) *STIGMA J. Mat. dan Ilmu Pengetah. Alam Unipa* **13** 29–34

[10] Berek A K, Berek A K, Ceunfin S, Taolin R I C ., Neonbeni E Y and Seran M J 2017 Efek Biochar Dan Teh Kompos Terhadap Pertumbuhan Dan Hasil Selada Darat (Lactuca sativa L) Di Tanah Vertisol Semiarid *J. Floratek* **12** 101–14

[11] Verdiana M A, Sebayang H T and Sumarni T 2016 Pengaruh Berbagai Dosis Biochar Sekam Padi dan Pupuk NPK terhadap Pertumbuhan dan Hasil Tanaman Jagung (Zea mays L.) *J. Produksi Tanam.* **4** 616

[12] Zhu Q, Kong L, Xie F, Zhang H, Wang H, Ao X, Zhu Q, Kong L, Xie F, Zhang H, Wang H and Ao X 2018 Effects of biochar on seedling root growth of soybeans *Chil. J. Agric. Res.* **78** 549–58

[13] Siregar D A, Lahay R R and Rahmawati N 2017 Respons Pertumbuhan Dan Produksi Kedelai (Glycine max (L. Merril) Terhadap Pemberian Biochar Sekam Padi Dan Pupuk P *J. Online Agroekoteknologi* **5** 722–8

[14] Laela Nurida N, Dariah A and Rachman A 2013 Peningkatan Kualitas Tanah dengan Pembenah Tanah Biochar Limbah Pertanian *J. Tanah dan Iklim* **37** 69–78

[15] Musnoi A, Hutapea S and Aziz R 2017 Pengaruh Pemberian Biochar Dan Pupuk Bregadium Terhadap Pertumbuhan Dan Produksi Tanaman Sawi Hijau (Brassica rapa var. parachinensis L) *Agrotekma J. Agroteknologi dan Ilmu Pertan.* **1** 160–74

[16] Sirait E E, Nelvia N and Fauzana H 2020 PertumbuhanDan Hasil Tanaman Kedelai (Glycine max L.) Terhadap Pemberian Vermikompos Dan Biochar Di Tanah Ultisol *J. Solum* **17** 29–41

[17] Junita F, Muhartini S and Kastono D 2020 Pengaruh Frekuensi Penyiraman dan Takaran Pupuk Kandang Terhadap Pertumbuhan dan Hasil Pakchoy *J. Ilmu Pertan.* **9** 37–45

[18] Ansori I, Nafi'ah H H and Nurdiana D 2021 Pengaruh Pemberian Biochar dan Pupuk Hayati Terhadap Pertumbuhan dan Hasil Pakcoy (Brassica rapa L.) *J. Agroteknologi dan Sains* **5** 394–408

[19] Goenadi D H and Laksmita S 2006 Aplikasi Bioaktivator Superdec dalam Pengomposan Limbah Padat Organik Tebu *J. Agron. Indones.* **34** 173–80

[20] Amin Sadzli M and Slamet Supriyadi 2019 Pengaruh Biochar Sekam Padi dan Kompos Paitan (Tithonia diversifolia) terhadap Pertumbuhan Tanaman Kacang Hijau (Vigna radiata L.) di Tanah Mediteran *Agrovigor J. Agroekoteknologi* **12** 102–8

[21] Widowati 2010 *Laporan Disertasi Doktor: Produksi Dan Aplikasi Biochar Atau Arang dalam Mempengaruhi Tanah Dan Tanaman*

[22] Muhammad W, Surachman S and Zulfita D 2020 Pengaruh Biochar Sekam Padi Dan Pupuk NPK Terhadap Pertumbuhan Dan Hasil Jagung Manis Di Lahan Gambut *J. Sains Mhs. Pertan.* **9** 1–10

[23] Akmal S and Simanjuntak B H 2019 Pengaruh pemberian biochar terhadap pertumbuhan dan hasil tanaman sawi pakchoy (Brassica rapa Subsp. chinensis) *J. Ilmu Pertan.* **7** 168–74

# A study of *Flickingeria angulata* (Blume) A.D.Hawkes and host trees (Phorophytes) at Wilis Mountain, Ponorogo, East Java, Indonesia

**Nina Dwi Yulia**

Purwodadi Botanic Garden, Research Center for Plant Conservation and Botanic Garden - BRIN, Surabaya Malang St. Km 65 Purwodadi, Pasuruan, East Java, Indonesia
Corresponding author: ninadwiyulia@gmail.com

**Abstract**. *Flickingeria angulata* (Blume) A.D.Hawkes is one of the species of orchids found in natural habitats in Wilis Mountain, Ponorogo, East Java. This study was aimed to explore, document, and analyze the diversity of *Flickingeria angulata* at a natural forest area in Wilis Mountain. The method used is an exploratory method by exploring the forest area in the study site. All data of *Flickingeria angulata* was recorded including its number, host trees and zone at the host tree where this orchid is attached. The result shows that orchid epiphytic *F. angulata* was found in 6 host tree species, with the two most important species being *Pinus merkusii* and *Schima wallichii*. Zone three (bottom part of the branches) was noted as the most preferred zone at host trees to be attached by *Flickingeria angulata*. Evaluation of these orchid species in natural habitats needs to be done at the same location to determine the impact of climate change.

**Keywords** : *Flickingeria angulata*, host tree, Wilis Mountain

## 1. Introduction

Orchid is an interesting plant family to study, especially epiphytic orchids. The distribution of orchids is very limited in the arctic region and abundant in the tropic [1]. The orchids epiphytic are very closely associated with their host trees, unlike terrestrial orchids [2], [3]. In tropical and subtropical ecosystems that experience anthropogenic disturbances will cause loss of biodiversity [4]. Epiphytic plants like epiphytic orchids may serve as indicators ecosystem because epiphytic orchids are one of the sensitive plants on problems that interrupt an ecosystem and microclimate and play an important role in tropical forest ecology [2]. Therefore, it is necessary to observe orchids in their natural habitat. The protected forest in Mount Wilis, Ponorogo, East Java is an interesting location to study. The natural forest area in Mount Wilis has suffered a lot of damage caused by illegal logging, forest fires, and landslides. This poses a threat to the biodiversity of the area, and in particular, the orchid species and illegal logging has also contributed to its population decline (Perhutani's forest rangers, pers. comm.).

It was explained that 10% of epiphytes of the total vascular flora grow in some humid tropical forests. Epiphytes such as epiphytes orchids that grow attached on trees, their existence is bound to their host trees and will be threatened if there is damage to tropical forests [2], [5], [6]. The communities of epiphyte as good controls to evaluate species-specific interactions, also as indicators species of habitat disturbances and orchids have a good role as indicator species that adapt to abiotic factors of the habitat [6], [7], [8]. Several research about epiphytic orchids and their host tree that have been carried out include: distribution on epiphytic orchid in *Shimma wallichii* tree [9], epiphyte host preference and host traits [3], spatial patterns of epiphytic orchid in Sempu island [10], research of host tree characteristics, distribution patterns, conditions of microclimate and diversity of epiphytic orchids [11], the role of orchids as indicator species of disturbed forest ecosystems in limestone mining areas in Georgia [8], diversity of epiphytic orchids at different elevations in the central Nepal region [12], analysis of the diversity of epiphytic orchids and their host-trees on Mount Lawu Magetan [13], diversity of epiphytic attached to the host trees [14], the species of host tree selected for the epiphytic orchid Epidendrum magnoliae Muhl (green fly orchid) in Georgia [15], types of host plants as indicators occurrence of *Grammathophyllum scriptum* in low land of Gunung Meja Natural Tourist Park [16] and there are many other research.

One of the interesting epiphytic orchids to observe in area is *Flickingeria angulata* (Blume) A.D Hawkes. *Flickingeria angulata* has the local name "anggrek tulang", a sympodial orchid of that has small fragrant flowers. This orchid is found in Java at lower foothills with elevation 250 until 500 m as [17]. *F. angulata* can be found growing well in suitable natural habitats. The purpose of this study to document and analyze the diversity of *F. angulata* and its host tree at a natural forest area in Wilis Mountain.

## 2. Materials and method

### 2.1. Materials

The activity was carried out in the Mount Willis area, precisely in the Mount Manyutan reserve forest area. The management of this area is under the West Wilis Forest Management Unit under Perum Perhutani and administratively located in Pupus Village, Ngebel District, Ponorogo Regency, East Java. This study uses an exploration method, namely tracing the forest area through several roads or paths available in the forest area. In Figure 1 is the location of the research conducted.

**Figure 1.** Location of observation area Wilis Mountain

### 2.2. Methods

Data collection on the number of *Flickingeria angulata* found at the research site and accompanied by a record of the host trees species of the host tree was this species attached. The position of the epiphytic plant attached on the host tree is divided into five zones, that is: (i) Zone 1: the bottom (1/3) of the main trunk; (ii) Zone 2: the top (2/3) of the main trunk; (iii) Zone 3: the lower part of the branch; (iv) Zone 4: the center of the branch; and (v) Zone 5: the outside of the branch [18], [19].

Observations are also equipped with environmental data including air temperature, humidity and altitude of the location.

## 3. Results and discussion

### 3.1. *Forest area*
The reserve forest area of Gunung Manyutan, Wilis Mountain (show in Figure 2) has quite good vegetation although several locations in this area have experienced fires and landslides. The environmental conditions of this area include an altitude between 1,300 - 1,413 m asl., relative humidity ranging from 60% - 80%, and a temperature of 28 - 30 C. The location point is at S 7°46,751'; E 111°39,637'. Several tree species that dominate the area include pine (*Pinus merkusii*), puspa (*Schima wallichii*), *Engelhardia spicata* and *Lithocarpus* spp

**Figure 2.** View of Wilis Mountain (credit manusialembah)

### 3.2. *Flickingeria angulata abudance*
Totally 277 individuals of *Flickingeria angulata* were found in this area and attached to five host tree species, as presented in Table 1. The distribution of this orchid species is fairly common in West Jawa and less frequently in East Java, usually growing in the lower foothills from 250 – 550 m asl [18]. In reserve forest Penanggungan Mountain East Jawa, *F. angulata* (Figure 3) was more commonly found than other orchid species, totally 119 individuals [20], and also found in Pare-pare South Sulawesi [21].

**Figure 3.** Flower of *Flickingeria angulata*

*F. angulata* grow in the range of altitude 500 m asl and the foot of the mountain with environmental humidity ranges from 80 – 90%, so that the environment of the condition in Wilis Mountain is suitable for grow and propagate of *F. angulata* found in this area, thus giving an influence positive on the population density of this species in the location. In the natural habitat, the presence a canopied of host tree is an important requirement for epiphytic orchids, that is, its role is to provide a suitable microclimate for epiphytic orchids, including shade and humidity [15].

### *3.3. Host tree and zona*

There are five species with a total 46 individuals. Species host tree and each zone of the host tree, as presented in Tabel 1. Two species of trees that are often found as host trees for *F. angulata* are *Pinus merkusii* (there are 20 individuals), and *Schima wallichii* (there are 12 individuals). Generally, the five host trees have rough bark, layered, peeling off, and the canopy of tree is not too tight. In addition, the form of branching is normal and not too dense, light sun can still penetrate the forest floor thus supporting the orchid to grow. As the inland forests of Georgia show that larger (possibly older) host trees are preferred as host trees, such as older Live Oak trees, many populations of the rare Green Fly Orchid are found [15]

Epiphytic orchids attached to the surface of the bark are strongly affected by constant or rapid successional changes, epiphytic plants must adapt to the existing conditions or they will die. These changes occur in the physical and chemical development of the bark substrate and in changes in the microclimate [14]. Vertical ecological gradients are relevant for epiphytic orchid compositions (such as moisture from forest floor to canopy surface) [5]. Orchid seeds attached to suitable trees will germinate and grow well until flowering. Although the life cycle of orchids is also very dependent on other organisms, such as pollinators and symbiotic fungi, the bond between epiphytic orchids and the host tree is very close [14].

**Table 1.** Recapitulatian of *Flickingeria angulata* at each zone of host tree

| Tree species | F of tree | Number of *F. angulata* in five zone | | | | | Total number |
|---|---|---|---|---|---|---|---|
| | | 1 | 2 | 3 | 4 | 5 | |
| *Lithocarpus sp.* | 5 | | 13 | 19 | 27 | | 59 |
| *Schima wallichii* | 12 | | 14 | 33 | 15 | | 62 |
| *Pinus merkusii* | 20 | | 57 | 40 | 5 | | 102 |
| *Lithocarpus tesymanii* | 4 | | 8 | 22 | 8 | | 38 |
| *Lithocarpus sundaicus* | 4 | | 1 | 7 | 7 | | 15 |
| *Engelhardia spicata* | 1 | | | 1 | | | 1 |
| Total number | 46 | 0 | 93 | 122 | 62 | 0 | 277 |

Naturally, epiphytic orchids will occupy and well grow in certain zones on the host tree to sunlight that suits their needs. As presented in Table 1 and Figure 4, *F. angulata* is more found in zone 3 (44.04%), the bottom part of the branches, and then in zone 2 (33.57%) and 4 (22.38). Meanwhile, this orchid not found in zone 1 and 5. Orchid epiphyte was distributed non-random in the host tree, in general, orchids found in natural habitats grow on the main stem and the bottom of the main branches [22]. In another study, it was stated that among the 97 orchid species that grow in the forest, as much as 23% in the middle of the big branch or Zone 3, and almost as many (20%) extending from the base to the middle of the big branch (Zone 4), and fewer epiphytic orchids found in another zone [18]. Although, there is a study showing that most of the epiphytic orchid species growing on *S. wallichii* trees grow scattered on almost every part of the tree and do not occupy a certain zone, although most individuals are found in the canopy [9].

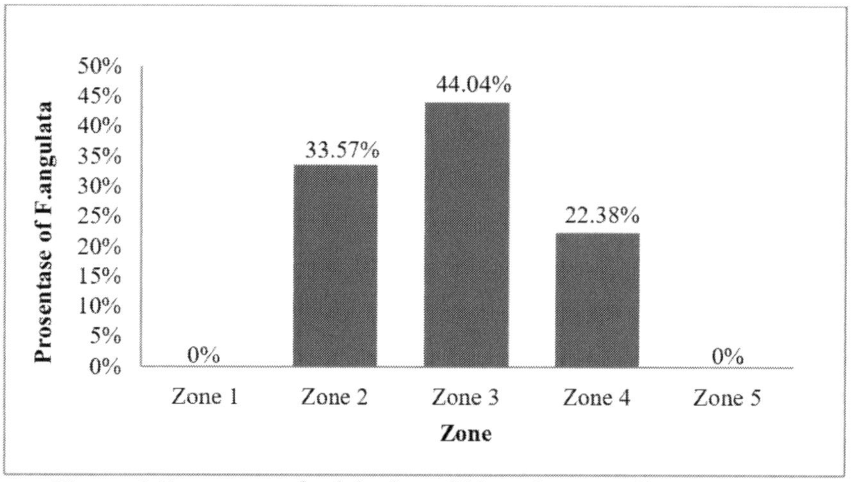


**Figure 4.** Presentage of epiphytic orchids at 5 zones of host trees

Epiphytes were distributed non-randomly in the tree canopy they were in, but the youngest part of the crown was not good for epiphytic development [14]. All conditions of environmental, things that have a direct influence on the growth of epiphytic diversity (orchids) in their natural habitat are climate (relative humidity) and tree community characteristics, while soil character has an indirect effect. The most important abiotic determinants of epiphytic diversity along tropical elevation gradients are air humidity and stand basal area as the most important biotic determinants [23]. The effect that can affect the increase in relative humidity and decrease the light intensity from the canopy surface is the microclimate gradient in the canopy [24]. Commonly, there are three things that determine the distribution of epiphytes in a rainforest, namely the microclimate within the host tree; landscape changes in the macro-climate; and characteristics of the host tree [25].

Changes in climatic conditions may not necessarily be followed by fairly rapid migration, both at the species and vegetation type levels. Epiphytes have good sensitivity to disturbances and changes in microclimate, that's why epiphytes have an important role in tropical forest ecology and serve as indicators of a well-maintained ecosystem [2]. Evaluation of *F. Angulata* in Manyutan forest reserve, Wilis Mountain needs to be done at the same location to determine the impact of climate change, because of this area is prone to environmental damage due to natural disasters and others.

## 4. Conclusions

*Flickingeria angulata* found in Manyutan forest reserve Wilis Mountian very abundance, totally 277 individuals. This orchid is attached to five host trees, a total of 46 individuals. The natural habitat is suitable for the growth of this orchid supported by suitable environmental conditions that are characteristic host tree and humidity. Considering that the Mount Wilis area is prone to disturbances such as illegal logging, forest fires, and landslides, it is necessary to evaluate the presence of F. angulata in this natural habitat.

## References

[1] Pearce N R and Cribb P J 2002 *Flora of Bhutan, including a record of plants from Sikkim and Darjeeling: The Orchids of Bhutan* (Edinburgh: Royal Botanic Garden)

[2] Hietz P 1999 Diversity and conservation of epiphytes in a changing environment *Pure Appl. Chem.* **70** 1–11

[3] Callaway R M, Reinhart K O, Moore G W, Moore D J and Pennings S C 2002 Epiphyte host preferences and host traits: mechanisms for species-specific interactions *Oecologia* **132** 221–230

[4] Adhikari Y P, Fischer S and Pauleit A 2015 Sustainable conservation perspectives for epiphytic orchids in the Central Himalayas, Nepal *Appl. Ecol. Environ. Res.* **13** 753–67
[5] Nieder J, Prosperí J and Michaloud G 2001 Epiphytes and their contribution to canopy diversity *Plant Ecol.* **153** 51–63
[6] Kress W J 1986 The systematic distribution of vascular epiphytes: An update *Selbyana* **9** 2–22
[7] Rose F 1999 Indicators of ancient woodland - the use of vascular plants in evaluating ancient woods for nature conservation *Br. Wildl.* **10** 241–51
[8] Akhalkatsi M, Arabuli G and Lorenz R 2014 Orchids as indicator species of forest disturbances on limestone quarry in Georgia (South Caucasus) *J. Eur. Orchid.* **46** 123–160
[9] Fardhani I, Torimaru T and Kisanuki H 2020 he vertical distribution of epiphytic orchids on Schima wallichiitrees in a montane forest in West Java, Indonesia *Biodiversitas* **21** 290–8
[10] Sadili A 2019 Structure, Distribution, and Spatial Patterns of Epiphytic Orchids (Orchidaceae) at Coastal Forest of the Sempu Island Nature Reserve, Malang, East Java *J. Ilmu Kehutan.* **13** 38–47
[11] Adhikari Y P 2013 *Distribution pattern, micro-site conditions, host tree characteristics and utilization of epiphytic orchids in the central Himalayas* (Technische Universität München)
[12] Timsina B, Kindlmann P, Subedi S, Khatri S and Rokaya M B 2021 Epiphytic orchid diversity along an dltitudinal gradient in Central Nepal *Plants* **10** 1381
[13] Yulia N D, Budiharta S and Yulistyarini T 2011 Analysis of epiphytic orchid diversity and its host tree at three gradient of altitudes in Mount Lawu, Java *Biodiversitas* **12** 225–8
[14] Rasmussen H N and Rasmussen F N 2018 The epiphytic habitat on a living host: reflections on the orchid–tree relationship *Bot. J. Linn. Soc.* **186** 456–472
[15] Bergstrom B J and Carter R 2008 Host-Tree Selection by an Epiphytic Orchid, Epidendrum magnoliae Muhl. (Green Fly Orchid), in an Inland Hardwood Hammock in Georgia *Southeast. Nat.* **7** 571–80
[16] Panie M B, Ungirwalu A and Wanma J F 2017 Types of host plant as indicator occurrrence of (Gramatophyllum scriptum Blume.) in the low-land forest of Gunung Meja Natural Tourist Park *J. Kehutan. Papuasia* **3** 42–151
[17] Comber J B 1990 *Orchids of Java* (Kew, London: Royal Botanic Gardens)
[18] Johansson D 1974 *Ecology of vascular epiphytes in west african rain forest: Acta Phytogeographica Suecica 59* (Sweden)
[19] Dressler R 1981 *The Orchids: Natural History and Classification* (United States: Harvard University Press)
[20] Yulia N D and Yanti R M 2010 Anggrek epifit dan pohon inangnya di kawasan gunung penanggungan, Pasuruan – Jawa Timur *Berk. Penelit. Hayati Ed. Khusus* **4A** 37–40
[21] Rahayu E M D 2018 Ex situ conservation of orchids in Jompie Botanic Gardens, Parepare, South Sulawesi *Pros. Sememinar Nas. Masy. Biodiversitas Indones.* **4** 242–7
[22] Henderson M R *The epiphytic flora of Dipterocarpus oblongifolius Bl. ("Neram")* vol 9 (Singapore: Botanical Garden)
[23] Ding Y, Liu G, Zang R, Zhang J, Lu X and Huang J 2016 Distribution of vascular epiphytes along a tropical elevational gradient: disentangling abiotic and biotic determinants *Sci. Rep.* **6** 19706
[24] Cardelús C L and Chazdon R L 2005 Inner-crown microenvironments of two emergent tree species in a lowland wet forest *Biotropica* **37** 238–44
[25] Sanger J C and Kirkpatrick J B 2015 Moss and vascular epiphyte distributions over host tree and elevation gradients in Australian subtropical rainforest *Aust. J. Bot.* **63** 696–704

# Assessing the disparities of the population exposed to flood hazards in Nigeria

**Ighile Eseosa Halima*[1] and Shirakawa Hiroaki[1]**

[1]Nagoya University, Furo-Cho, Chikusa-Ku, Nagoya 464-8601, Japan
*Corresponding author: ighileeseosa1@gmail.com

**Abstract**. The increased frequency and intensity of flooding and related disasters result from changing climatic conditions and other socio-economic factors. As flooding can be highly destructive and negatively impact human lives, this study attempts to estimate the population, capital stock and disparities in exposure to flooding hazards in Nigeria using GIS and Statistical methodologies. First, the study assessed the spatial distribution of the population and capital stock exposed to flood by utilising population and socio-economic datasets. Then, the distribution of the vulnerable groups affected is estimated by superimposing the population and socio-economic datasets onto the flood hazard maps. The results show that approximately 24.7 million (8.3%) of Nigeria's Population were exposed to floods in 2015. Most exposed groups were primarily in urban areas irrespective of the income class. Additionally, the clusters of communities within the high-risk flood hazard zones had significantly increased, evident in the number of residents exposed to flood within the 15 years (2000-2015) growing exponentially. These findings further highlight a disturbing state of localities where people are generally less responsive to climate change and natural hazards. Overall, this study provides essential information for disaster risk management and policy formation at different levels of administration and identifies areas where varied and informal strategies are needed to mitigate flood risk and climate change in regions with diverse socio-economic conditions. In addition, this study provides empirical proof of the socio-economic disparities associated with flood exposure in Nigeria and presents valuable insights into the underlying factors.

**Keywords** : Nigeria, Flood hazards, Population exposure.

## 1. Introduction

Flooding and flood-related disasters are among the significant natural hazards that have plagued Nigeria in the last decades in terms of the overall loss of lives and infrastructure [1]. In addition, the changing climatic conditions and sea-level rise have amplified the overall risk of flooding in Nigeria. Aside from climate change, flood occurrence intensity has also been heightened by the changes in the social and economic conditions such as rapid urbanisation, poor planning and land use regulations, population growth, and intensification of land conversion near coastal areas and floodplains [2,3].

Historical records of natural disasters in Nigeria have estimated over 700 flooding incidents between 1969 and 2020, affecting more than 6million people and resulting in the deaths of over 21,000 persons [4]. In addition, the estimated economic damage from these flood incidents is estimated at US$ 190 million, amounting to about US$ 6.5 million per year [5]. Thus, flooding disaster occurrence in Nigeria has produced very high human, environmental and socio-economic implications. Therefore, it is now essential to investigate and establish effective disaster risk-reduction measures to reduce the impacts. According to the Sendai Framework for disaster risk reduction (2015), prioritisation should understand disaster risks in all vulnerability dimensions, exposure, capacity, assets, hazard characteristics and the environment [6]. In addition, this framework takes the initiative to invest in the monitoring, planning, response, and recovery processes from a hazardous event to understand better the characteristics of the at-risk population, infrastructure, and physical environment [5].

In Nigeria, policies have been developed to tackle the exposure and vulnerability to natural disasters. However, these policies are designed towards structural defences, with a minimal comprehensive approach that integrates disaster preparedness, protection, response, and recovery as a tool for disaster risk reduction. To fully implement an inclusive strategy for Disaster Risk Reduction into spatial planning in Nigeria, it may require many years to reach a desirable capacity, as other factors envelop Nigeria's environmental, social, and economic characteristics. However, actions directed and designed to reduce further exposure and vulnerability to natural disasters, such as avoiding further development in high-risk zones, improving protection measures either from land use planning or land tenure practises, may be required to mitigate against further environmental and socio-economic damage in the future [7].

Several researchers have linked higher vulnerability and exposure levels as a by-product of the existing disparities within economic and societal groups, unsustainable developmental patterns, and poor governance [8]. Similar studies also tried to assess the risks to flood hazards at different spatial and geographical scales, utilising existing hazards and socio-economic data components [9–11]. Based on this existing literature, knowledge on existing hazards and vulnerable groups and exposure in an ideal developmental scenario should deter further development within at-risk areas. However, studies have shown that the population growth within and around hazard areas has been rising [12]. In sub-Saharan Africa alone, there has been a significant increase in flooding mortality and economic loss since the 1990s compared to the other regions of the world [13]. For instance, members of the lower economic groups may be disproportionately exposed to flood hazards due to the lack of basic amenities and reside in slums where property prices are usually lower and within the flood-prone areas [14]. The increased mortality rate from flooding has also been influenced by climate change, evident in the frequency of coastal and urban flooding occurrence, with research predicting that such trends would continue in the future [15,16].

Although most of the techniques used to assess flood risks, exposure and vulnerability have in most cases been geared towards hydrological modelling. However, the technicalities involved in calibrating and integration socio-economic factors into the hydrological models sometimes lead to poor and varied results, primarily due to the quality of available datasets. However, new studies have begun integrating GIS and satellite-based tools to effectively assess the changes from a hazard at different geographical scales, which hydrological models may have had difficulty processing. As for this study, flood exposure is assessed spatially by intersecting the flood hazard, population, and socio-economic distribution datasets. Similar studies have been conducted to estimate the Population and economic exposure to flood hazards in the United States and globally, applying a local regional-based assessment of urban and population exposure to flood hazards [17].

Beyond the previous studies, this study provides a nationwide local government-based assessment of the population exposed to flood hazards and their socio-economic disparities for Nigeria. We estimated the spatial distributions of the exposure population to flood hazards by intersecting the population and relevant socio-economic datasets with the flood hazards data. For this study, the spatial distribution of flood hazards was represented based on the 100-year-return flood zones and flood hazard maps produced by the UNEP Global Risk data platform and the GAR 15 global risk database to measure the population exposed to flood hazards systematically.

Finally, the flood exposure levels were quantified as the sum and ratio of population and risk levels located within the 100-year-flood zones for each state regional boundary. In addition to the total population, the disparities between the capital stock ($ U.S. million) and income groups in urban and rural areas were considered to highlight a higher percentage of exposure to flood hazards. The contributions of this study can be summarised as follows.

First, this study delivers a national-level assessment of Nigeria's total population exposed to flood hazards using open-access flood risk, socio-economic and population data. Second, we apply a quantitative assessment of the disparities in Nigeria's Population, capital stock, and geographical locations exposed to flood hazards. The result uncovers the trends of the population exposed to flood between and within the urban, rural, and different socio-economic groups. Finally, this study confirms the disparities between socio-economic groups associated with flood exposure and presents valuable insights into the underlying factors surrounding flooding in Nigeria.

## 2. Materials and methods

### *2.1. Study area*

Nigeria is in Sub-Saharan West Africa. The total area of the country is approximately 923,768 $km^2$ (**figure 1**). Furthermore, it has a population of about 200 million. Since the 1960s, there have been significant changes in the natural, social, and economic environment. Among the social and economic changes, rapid urbanisation and other anthropogenic factors have significantly influenced the physical landscape and indirectly impacted the risk of flooding [18]. Nigeria has two distinct seasons- wet and dry. The rainy season is usually long, occurring between April and October. During each of the rainy seasons in the last few years, many regions of Nigeria are prone to flooding, resulting from the continuous urban development along flood plains, inadequate drainage systems, low maintenance of the existing drainage systems, and an increase in run-offs due to uncontrolled urban expansion [19]. With cities such as Lagos ranked 15th globally in terms of the total population exposed to flooding due to climate change and the annual changes in the pattern and intensity of rainfall, Nigeria has a high risk of flooding.

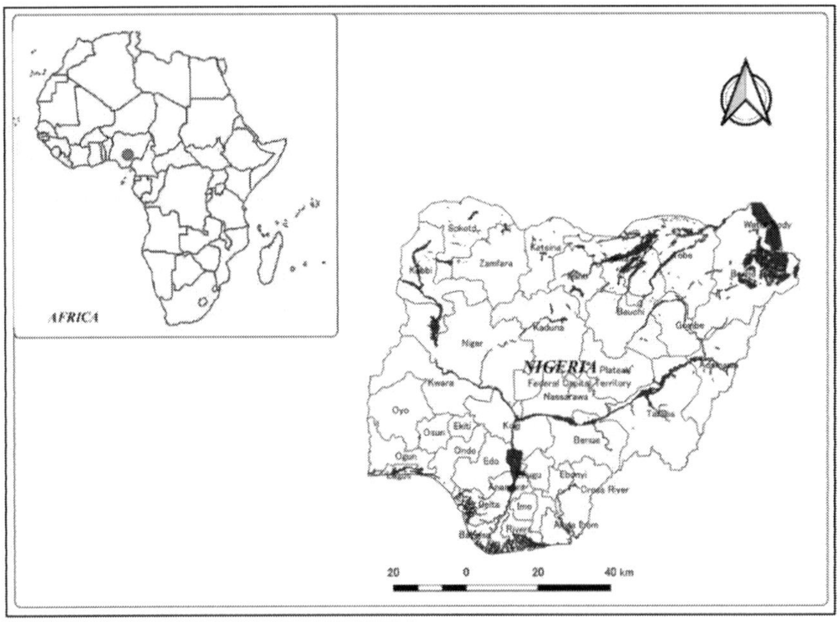


**Figure 1** Location of Nigeria

### *2.2. Data*

This study applies economic and demographic parameters to determine the disparities between exposure to flood hazards. Population and demographic data at 1km resolution from 2000-2015 were obtained from worldpop (https://www.worldpop.org). The Map of flood hazards was from the UNEP GRID Global Risk Data Platform, which shows the classification of flood hazards levels based on five categories (low, low/medium, medium, high, and extreme). Supplementary data required to estimate the socio-economic disparities from flood hazards were obtained from Nigeria's GAR15 global exposure database, which comprises information on income groups, industry and infrastructure at a national level

(2016); (https://data.humdata.org/dataset/gar15-global-exposure-dataset-for-nigeria) at 5 X 5 meters grid.

*2.2.1. Methodology.* Three analyses were conducted to estimate the total population and disparities of the exposed population to flood hazards in Nigeria. The analysis was carried out by utilising Geographic Information System (GIS) spatial analysis tools.

*2.2.2. Calculating exposed population.* The first analyses estimate the total population exposed to flood hazards between 2000 and 2015 by intersecting the total population distribution (P) with the flood hazard maps within each region's administrative boundaries [20]. The total population in flood zones ($T_p$) is the sum of the total population in each flood hazard ($F_h$) zone (equation 1).

$$T_p = P \times F_h \quad (1)$$

Where $T_p$ is the sum of the total population within the flood zones, $P$ is the total population within the administrative boundary of Nigeria and $F_h$; is the flood hazard map.

After estimating the total population exposed to flood, we further estimated the ratio of flood distribution based on two flood hazard levels. The flood hazard data from the UNEP were reclassified into two groups (**figure 2**): low and high from the original five categories (low, low/medium, medium, high, extreme) provided in the dataset. First, we highlighted the number of persons exposed to low and high-risk floods between the 15-year study intervals. Next, we estimated the ratio of the population exposed to flood as a proportion of the total population. Finally, the ratio of the population exposed to flooding is analysed to provide valuable insights on the areas within local boundaries where flood management prioritisation is needed.

The ratio of exposed population is estimated by;

$$TP_R = \frac{T_p}{P} \quad (2)$$

Where $TP_R$ is the ratio of population in the flood zone is, $T_p$ Is the total population in the flood zones, and P is the total population within the study area.

*2.2.3. Calculating the disparities between the exposed population.* The goal is to estimate and compare the disparities between the exposed population in rural and urban areas. The assumption is that observing the distribution of exposed population between the rural and urban areas would highlight what areas are most exposed and create awareness on the areas that require prioritisation for flood mitigation. The presumption is that if the percentage of population exposed to flood hazards are in urban areas exceeds that of rural areas, then the responsiveness to flood hazards is low, and there is no avoidance of flood zones for development since the urban areas have a higher population density than rural areas [4]. Similarly, the difference in per capita income between flood zones and exposed populations was calculated and compared, based on the current GDP ($ U.S. 2015 estimate); the total income distribution within flood zones is divided into low and medium, respectively, across all flood zones.

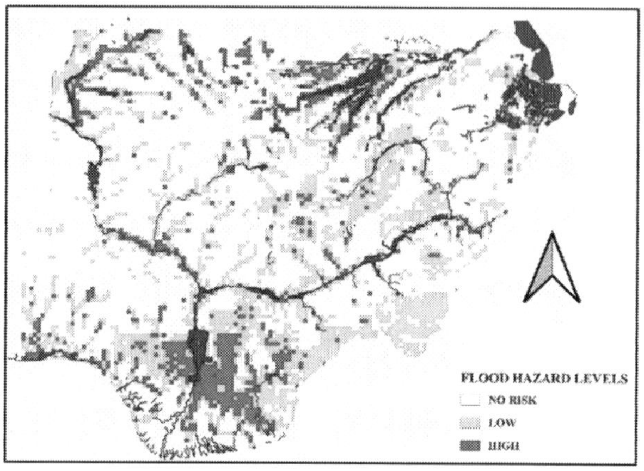


**Figure 2.** Map of flood hazard levels

## 3. Results and discussion

### *3.1. Exposed population*

The estimation of population, land use, and income distribution analyses yielded the following results.

*3.1.1. Exposure of population to flood hazard.* The findings show that the number of people exposed to flood hazards increased between 2000 and 2015. The total number of persons exposed to flooding rose from 29.7 million in 2000 to 118.8 million (53.8% of Nigeria's Population) in 2015 (**figure 3**), with an estimated 13.4 million (2000) and 24.7 million (2015) persons residing within the high-risk flood zones. An increment of 11.3 million (8.3% of Nigeria's population) over 15 years. The massive increase in the number of persons exposed (83.3 million) explains low responsiveness to flood hazards following development expansion within flood zones.

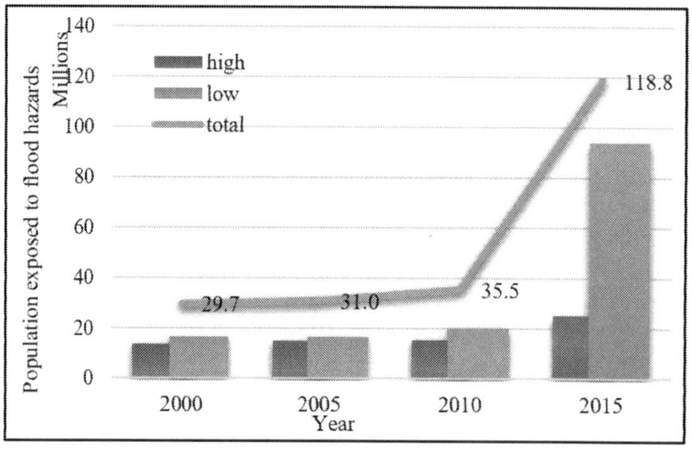


**Figure 3.** Distribution of population at risk of flooding between 2000 and 2015

Over time, the total estimated population in flood zones has remained concentrated in states with big cities and clusters of mid-sized urban areas. The top 10 states with the highest number of exposed persons as of the 2015 estimates were Bauchi, Borno, Kaduna, Katsina, Lagos, Niger, Ogun, Osun, and Oyo. Lagos had the more significant share of the exposed population (27.9%) due to its high density and proximity to the coast and other water bodies. Similarly, the same trend among the nine other states shows that regions close to a water body and the increase in precipitation levels resulting from climate change significantly impact exposure to flood hazards (**Table 1**). Therefore, we assumed that the rise in exposed populations could be attributed to various factors, including increased population growth, increase in migration due to booming economic activities, and weather conditions resulting from climate change.

The spatial distribution maps (**figure 4**) highlight the proportion of flood exposure and the changes in exposure within the last 15 years at the regional scale. The advantage of the spatial analysis of the exposure distribution will enable policymakers and governmental agencies to effectively understand the magnitude and intensity of flood exposure and proffer appropriate solutions for flood hazard mitigation.

**Table 1** Top 10 percentage distribution of exposed population to flood hazards

| State | Exposure Ratio | | | |
|---|---|---|---|---|
| | 2000 | 2005 | 2010 | 2020 |
| Bauchi | 5.90 | 6.10 | 6.16 | 7.50 |
| Borno | 9.87 | 10.07 | 9.61 | 7.07 |
| Kaduna | 12.09 | 13.30 | 11.70 | 9.68 |
| Kano | 11.81 | 11.74 | 13.73 | 14.21 |
| Katsina | 5.46 | 5.68 | 5.91 | 7.23 |
| Lagos | 38.63 | 38.67 | 37.92 | 27.90 |
| Niger | 3.30 | 3.54 | 3.78 | 5.85 |
| Ogun | 4.83 | 3.90 | 4.17 | 6.14 |
| Osun | 5.37 | 4.43 | 4.42 | 6.46 |
| Oyo | 2.75 | 2.58 | 2.59 | 7.96 |
| | 100.00 | 100.00 | 100.00 | 100.00 |

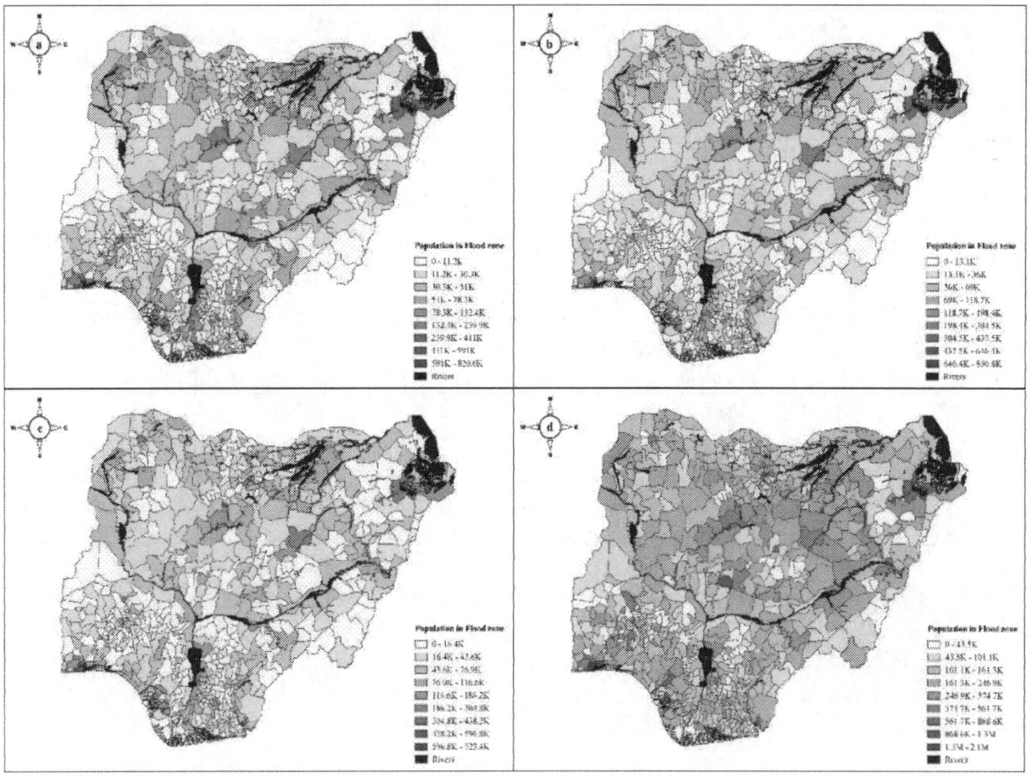

**Figure 4** Spatial distribution of population exposed to high-risk flood hazards (a); 2000 (b); 2005 (c); 2010 (d); 2015

According to the findings (**figure 4**), the total number of people exposed to high-risk flood hazards continues to rise following the current socio-economic and development trends. The increased population exposed to high-risk flooding indicates two factors: the lack of adequate knowledge and planning surrounding flood hazards or the lack of responsiveness. If the increase in the number of exposed people results from inadequate information and planning techniques, then the exposed population would continue to increase, impacting Nigeria's natural, social, and economic development [4]. However, if the exposed population increase results from the lack of responsiveness to existing hazards and climate change, the results will effectively provide tangible information that can persuade disaster risk and climate change policymakers to mitigate against significant future loss.

*3.1.2. Disparities of population exposed to flood hazards based on income distribution.* After successfully measuring the total population exposed to flood hazards, further analysis was performed to differentiate urban and rural areas exposed. Firstly, the disparities between per capita income in rural and urban areas. Then, we classified the socio-economic data into two categories (low and middle income) based on existing economic data trends for Nigeria.

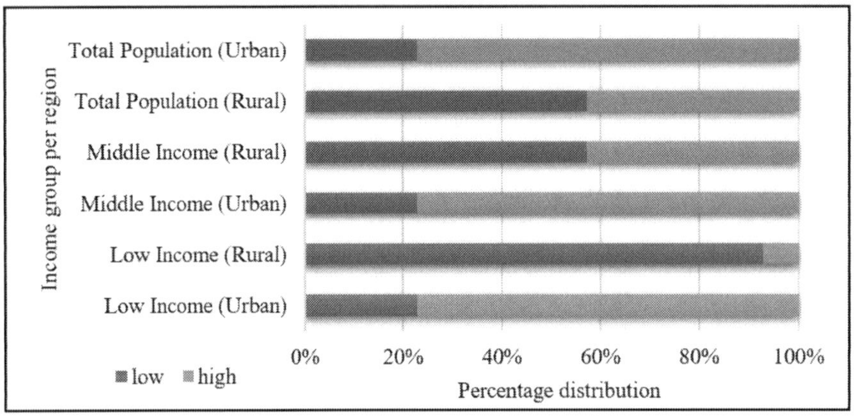


**Figure 5.** differences in the income disparity between the urban and rural population at risk of flood

The results (**figure 5**) shows that urban areas have a higher proportion (72%) of residents in the high-risk flood zones than low-risk zones. However, about 42% of residents were in the high-risk zones in the rural areas. The result explains that the residents living in urban areas are disproportionately more exposed to flooding than their rural counterparts.

Similarly, the ratio of the population within the low-risk zones was significantly higher in the rural areas than their urban counterparts. A simple explanation for the trend is that, due to the high population density in urban areas, the concentration of residents within the flood zones are higher than in rural areas. Further observation of the per capita income distribution in the urban and rural areas indicates that; in the urban areas, irrespective of the income category, they had a higher level of exposure to flooding when compared to their rural counterparts within the same socio-economic groups (**figure 5**).

*3.1.3. Disparities in capital stock exposed to flood hazards.* The ratio of urban stock in high-risk flood zones is significantly higher than in rural areas (**Figure 6**). In the low-income urban groups, the capital stock exposed to low-risk flood hazards is estimated at $US2634.6 million, while the high-risk flood zones capital stock is around $US9831.4 million. However, the middle-income group capital stock is around $US3878.12 and $US14471.66 million for low and high-risk zones, respectively.

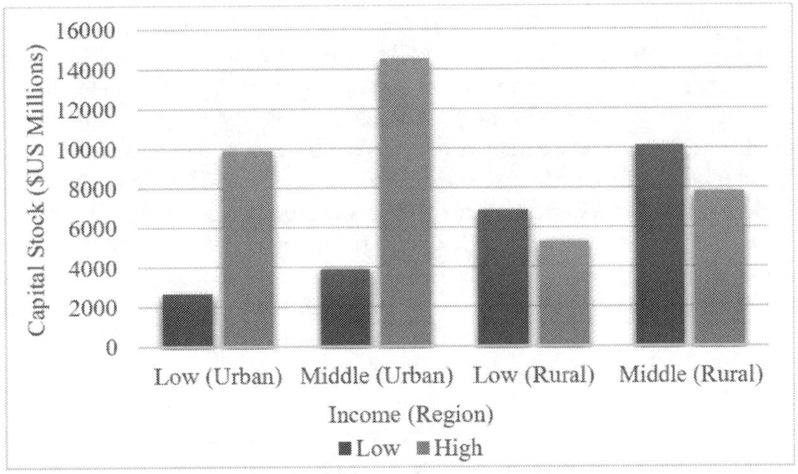


**Figure 6.** Income disparity between the per capital stock exposed to flood in rural and urban areas ($ U.S. millions).

Nonetheless, the results from the analysis are expected as a large portion of Nigeria's population, and infrastructure is in urban areas. Therefore, a significant proportion of capital stock exposed will also be in the urban areas. Similarly, the result shows that in urban areas, regardless of the per capita income, economic loss from flood hazards in a disaster is significantly higher than in rural areas. The result highlights low responsiveness to hazards and the need to improve planning and mitigation against flooding and changing climate conditions, especially in urban areas.

### *3.2. Advantages and limitations of the study*

This study utilised socio-economic and flood hazard maps to estimate the disparity in the population exposed to floods. Previous studies on flood hazards in Nigeria were segmented to a single geographical region and utilised mainly qualitative techniques to address flood hazards. The disadvantages of those studies were their inability to clearly define regions exposed to floods at both the national and local levels. The studies also failed to consider how each socio-economic group within and outside a given geographical region may differ in exposure to flood hazards.

However, in this study, we successfully showcased the regions with the highest exposure rate, the overall capital stock in monetary values, and the differences in the vulnerability to flooding hazards based on different income classifications. Secondly, by estimating the exposure population at a spatial scale, it is easier for policymakers to visualise and define the extent of flood risks, areas for prioritisation, thereby leading to better-informed decisions.

For future studies, we would recommend: (1) estimating exposure levels based on future socio-economic, climatic and land use scenarios, (2) predicting the areas with the highest risk of flooding based on collected historical flooding datasets.

## 4. Conclusions

This study aimed to assess the disparities of Nigeria's population exposed to flood hazards. The study successfully demonstrated how the population and income groups differ in exposure to flood hazards. In addition, our analysis successfully identified the most vulnerable areas and examined past trends of flood hazard exposure, which could serve as a foundation for flood risk mitigation in Nigeria. In the 15-year study period, the number of persons exposed to flooding grew from about 35.5 million in 2000 to 118.8 million (53.8%) in 2015. Similarly, the total population exposed to high-risk floods had risen to 24.7million (8.3% of Nigeria's population).

Regarding the disparity between the exposed population, we observed that urban areas, irrespective of the per capita income, had significantly higher exposure to flooding hazards than their rural counterparts within the same income class. Similarly, the exposed total carbon stock to flood hazards is about $US 60,801.9 million, with $US 30,815.8 million in urban areas and $US 29,986.1 million located in rural areas distributed between low and low middle-income groups. Based on this study approach of integrating flood hazard and socio-economic maps, it was possible to analyse (1) the population exposed to flood hazards at the national level, (2) the disparities between the exposed population in urban and rural areas, (3) the differences in hazard distribution based on per capita income in urban and rural areas, and (4) the capital stock of infrastructure exposed across all regions. The study findings highlight how increasing population and settlement development in flood-prone areas, particularly urban areas, raises economic and infrastructure loss. From the viewpoint of the effects of climate change on the existing flood hazards, this would mean a greater level of exposure to rising sea levels, storms and high levels of flooding incidents. As Nigeria continues to experience increased population growth and urbanisation, the greater risk of flooding for many communities, especially around coastal and riverine areas, is expected to rise. The rise in flooding incidents would be a repercussion of the increase in the number of rainy days, heavier torrential rainfalls, storms surges and sea-level rise propagated by climate change. The findings also highlight the importance of strictly integrating and enforcing climate change and spatial planning policies that emphasise the need to limit further settlement development in the regions identified as risk zones. Furthermore, these policies would protect human lives and existing built-up stock in high and low-risk flood areas while ensuring that new developments and assets are directed away from the flood hazard zones. As a result, assessing the population distribution and disparities to flood hazards is an innovative way to promote effective decision making for mitigating and prioritising the most vulnerable locations against flood risks. Finally, the study emphasises the importance of combining spatial policy approaches to investigate environmental problems and the need for additional and more detailed research on the impact of climatic and land use changes on flood hazards. The proposed studies should combine climatic, hydrological modelling and spatial planning to forecast future changes in flood risk and reduce the overall potential for economic and human losses.

**References**

[1] Nkwunonwo UC, Malcolm W and Brian B 2015 Flooding and flood risk reduction in Nigeria: cardinal gaps *Journal of Geography and Natural Disasters* **05** 136

[2] Oladokun V O and Proverbs D 2016 Flood risk management in Nigeria: a review of the challenges and opportunities *International Journal of Safety and Security Engineering* **6** 485–97

[3] Adelekan I O and Asiyanbi A P 2015 Flood risk perception in flood-affected communities in Lagos, Nigeria *Natural Hazards* **80** 445–69

[4] Olanrewaju C C, Chitakira M, Olanrewaju O A and Louw E 2019 Impacts of flood disasters in Nigeria: A critical evaluation of health implications and management *Jamba* **11** 557

[5] UNISDR 2015 *Disaster risk reduction and resilience in the 2030 agenda for sustainable development* (New York)

[6] Guha-Sapir D, Hoyois P and Below R 2016 *Annual disaster statistical review 2015: the numbers and trends* (Belgium)

[7] de Assis Dias M C, Saito S, Alvalá R C dos S, Stenner C, Pinho G, Nobre C A, Fonseca M R de S, Santos C, Amadeu P, Silva D, Lima C O, Ribeiro J, Nascimento F and Corrêa C de O 2018 Estimation of exposed population to landslides and floods risk areas in Brazil, on an intra-urban scale *International Journal of Disaster Risk Reduction* **31** 449–59

[8] Oruonye E D 2012 The challenges of urban flood disaster management in Nigeria: a case stufy of Jalingo LGA, Taraba State Nigeria *International Journal of Recent Scientific Research* **3** 37–42

[9] Nelson R, Kokic P, Crimp S, Meinke H and Howden S M 2010 The vulnerability of Australian rural communities to climate variability and change: Part I—Conceptualising and measuring vulnerability *Environmental Science & Policy* **13** 8–17

[10] Qiang Y, Lam N S N, Cai H and Zou L 2017 Changes in Exposure to Flood Hazards in the United States *Annals of the American Association og Geographers* **107** 1332–50

[11] Jongman B, Ward P J and Aerts J C J H 2012 Global exposure to river and coastal flooding: Long term trends and changes *Global Environmental Change* **22** 823–35
[12] Maantay J and Maroko A 2009 Mapping urban risk: flood hazards, race, & environmental justice in New York *Applied Geography* **29** 124
[13] Nkeki F N, Henah P J and Ojeh V N 2013 Geospatial techniques for the assessment and analysis of flood risk along the Niger-Benue Basin in Nigeria *Journal of Geographic Information System* **5** 123–35
[14] Tanoue M, Hirabayashi Y and Ikeuchi H 2016 Global-scale river flood vulnerability in the last 50 years *Scientific Reports* **6** 1–9
[15] Tellman B, Sullivan J A, Kuhn C, Kettner A J, Doyle C S, Brakenridge G R, Erickson T A and Slayback D A 2021 Satellite imaging reveals increased proportion of population exposed to floods *Nature* **596** 80–6
[16] Wing O E J, Bates P D, Smith A M, Sampson C C, Johnson K A, Fargione J and Morefield P 2018 Estimates of present and future flood risk in the conterminous United States *Environmental Research Letters* **13** 034023
[17] Chen Y-R and Yu B 2013 Impacts of climate and land-use changes on floods in an urban catchment in southeast Queensland, Australia *Climate and Land Surface Changes in Hydrology* pp 1–7
[18] Adelekan I O 2010 Vulnerability of poor urban coastal communities to flooding in Lagos, Nigeria: *http://dx.doi.org/10.1177/0956247810380141* **22** 433–50
[19] Sojobi A O, Balogun I I and Salami A W 2015 Climate change in Lagos state, Nigeria: what really changed? *Environmental monitoring and assessment* **188** 556
[20] Qiang Y 2019 Disparities of population exposed to flood hazards in the United States *Journal of Environmental Management* **232** 295–304

# Determination of zinc speciation in soil using synchrotron radiation and its effects on cellular organelles of *Phragmites australis*

**Tatiana Minkina, Dina Nevidomskaya, Grigoriy Fedorenko, Aleksei Fedorenko, Vishnu D. Rajput*, Victoria Shuvaeva, Victoria Tsitsuashvili, Victor Chaplygin and Priyadarshani Rajput**

Southern Federal University, 105 Bolshaya Sadovaya Street, Rostov-on-Don 344006, Russia
*Corresponding author: rajput.vishnu@gmail.com

**Abstract**. It is necessary to apply modern approaches to prevent the spread/toxic effects of pollution caused by a changing climate; especially metal pollutants to the soil. Thus, the present investigation was aimed to examine chemical speciation of soil samples collected from the River floodplain and its effects on native plant species; *Phragmites australis*. Synchrotron radiation (SR) tools for Zn speciation in soil and scanning electron microscopy for anatomical changes in plants were applied. Speciation of Zn was a dominant pollutant in Technosols. Two local surroundings were observed for Zn corresponding to its coordination by O and by S, close to ZnS. Whereas, the mixed local surrounding was noted with Zn–S and Zn–O bonds. The analysis of X-ray absorption fine structure results revealed that Technosols contained 70 and 30% of Zn–S and Zn–O bonds, respectively. SR results could be summarized that the main contribution in Technosols was from the authigenic minerals of metals with S: würtzite (hexagonal ZnS), sphalerite (cubic ZnS), bornite ($Cu_5FeS_4$), and covellite (CuS). The investigation on growth and modifications in the ultrastructure of *P. australis* roots and stems showed changes in cellular tissues such as in epidermis and mesoderm, and in subcellular organelles: mitochondria, chloroplast, cell membrane, etc. It is established that the soil contamination with exceeded limits slows down the ontogenetic developments of *P. australis*. Our findings could provide insight into the physiochemical characteristics of polluted soil as well as information on the toxic effects of pollutants on plants based on non-destructive approaches.

**Keywords** : pollution, morphology, plants, biomass, non-destructive approaches.

## 1. Introduction

The chemical contamination due to anthropogenic activities of soil with potentially toxic elements, so-called heavy metals (HMs) is s a serious threat to the environment and defines various stress for plant growth and developments. Once metal enters into plant tissues, it interacts with various compounds functional groups and causes abundant metabolic disturbances. A large number of plant species accumulate high content of HMs in roots than the aboveground organs due to direct contact with soil elements [1–3]. However, it is very important to understand the speciation of metals, which is directly linked to bio accessibility and lability of contaminant [4]. The physicochemical properties of soil, i.e., pH, organic matter, and metal storage capacity could also affect bio accessibility and mobility of elements in soil [5,6]. Heavy metals are persistent and could be more toxic after combining with another pollutants of soils, and can enter into the human body through food chain [7]. The synchrotron radiation-based approaches such as X-ray absorption spectroscopy are applying for complex sample matrices to provide information on the structural organization and speciation of soil metal compounds [2,8–10].

Several elements including Zn in soils polluted with organic compounds of urban wastes was feasible to determine the content, even at ppm level by using synchrotron radiation approaches [11]. The X-ray absorption near edge structure (XANES) and extended X-ray absorption fine structure (EXAFS) methods made it feasible to determine the geometric structure of metal compounds and revealed their real phases in soils, and establish chemical bond types. It was noted that the X-ray powder diffraction (XRD) applied to identify the crystalline phase of molecules [12].

The epigeous organs of common reed (*Phragmites australis*) due to metal accumulation capability is employing for biomonitoring of HMs, especially technogenic soils [13,14], and could reflects the cumulative effects of environmental pollution from water and soil. Thus, the present work investigates effect of persistent chemical pollutants of soils and its effects on *P. australis* grown near the sediment pond of the chemical plant by applying nondestructive approaches such as XRD, XANES and EXAFS and for plant cellular damages; scanning electron microscopy (SEM).

## 2. Materials and methods

### *2.1. Study area*

The highly polluted soils (Spolic Technosols) of Severskii Donets floodplain (48°20′42.15"N; 40°14′14.46"E) adjacent to Atamanskoe Lake, Rostov-on-Don, Russia was selected for chemical speciation, and 2 km away from this side, the Fluvisol was collected and used as a background (control) [15]. The selected site was dumping reservoir for an industrial waste generated by Khimvolokno chemical plant in the beginning of 1960s upto mid of 1990s. The Lake represents the monodominant groups of common reeds (*Phragmites australis* Cav.).

### *2.2. Methods of study*

*2.2.1. Soil.* The soil samples were collected from 0 to 20 cm as per method defined by ISO 10381-1 [16] and determined by X-ray fluorescent scanning spectrometer “SPECTROSCAN MAKC-GV” for total metal contents (Zn, Mn, Ni, Pb, Cu, Cr, Cd) in studied soil. The Zn was dominant pollutant among other determined metals; thus, it was selected for chemical speciation. The degree of soil pollution with HMs was assessed by comparing with lithosphere clarkes [17] and the MPCs: maximum permissible concentrations was considered for total contents [18].

The extended X-ray absorption fine structure (EXAFS) and X-ray absorption near edge structure (XANES) spectra were obtained at the Zn K-edge (9659 eV) on the beamline and XRD was done at X-ray structural analysis beamline “Belok/XSA” of Kurchatov Synchrotron Radiation Source (Kurchatov Institute, National Research Center, Moscow, Russia) [19]. Linear combination fittings of the XANES spectra of soil samples were performed using several reference compounds' spectra.

*2.2.2. Plant.* The 8–10 individual samples of fully developed (vegetative and generative organs reach maximum) P. australis samples were collected from each monitoring plots. The dry ash method was used to determine mineralization of plant samples [20] and atomic absorption spectroscopy (AAS) was used to estimate HMs extracted from ash of samples digested with 20% HCl. Obtained results of HMs contents in plant tissues was justified the limits defined as maximum permissible level (MPL) [21].

The samples for anatomical observations of root and stem tissues were collected; for root, it was sampled 3 and 6 cm from the growth apex due to arbitrary adventitious roots. The samples were fixed for SEM as per methods described earlier [22].

*2.2.3. Statistical analysis.* The results of analysis were presented as a mean of at least three replicates. Ten images were obtained from different parts of plant tissues and analyze for anatomical damages.

Descriptive data was carried out with STATISTICA v. 13.0 software, and differences were considered not significant at values of $P > 0.05$.

## 3. Results and discussion

The results indicate that the metal contents of soils used as a background was within the MPCs limits and largely correspond to their clarkes. However, the clarkes of Technosols samples was exceeded 100-folds for Zn and Cd, 10-folds for Pb. This exceedance was also monitored for other studied elements (Mn, Cr, Cu, Cd, Ni) (Table 1).

**Table 1.** The total metal contents (Mn, Cr, Cu, Zn, Pb, Cd, Ni) in Fluvisol (background plot) and Technosol (contamination plot), and in *P. australis* plants, mg/kg.

| Monitoring plot | | Mn | Cr | Cu | Zn | Pb | Cd | Ni |
|---|---|---|---|---|---|---|---|---|
| *Content of HMs in soil, (0- to 20-cm layer), mg/kg* | | | | | | | | |
| Background | | 230±6.6 | 97.9±7.3 | 90.0±7.0 | 90.0±7.6 | 26.1±1.7 | 0.2±0.003 | 40.9±2.6 |
| Contamination | | 1167±105 | 142.0±11.9 | **375.0[a]±40.5** | **66075±7095** | **223.4±24.6** | **8.6±0.6** | **152.4±14.8** |
| *Lithosphere clark* | | *1000* | *83* | *47* | *83* | *16* | *0.13* | *58* |
| *MPC* | | *1500* | *90* | *55* | *100* | *32* | *0.5* | *85* |
| *Content of HMs in P. australis plants, mg/kg* | | | | | | | | |
| Background | stems | 23±2.5 | 0.7±0.2 | 2.7±0.4 | 23.3±4.6 | 1.7±0.2 | 0.05±0.0[b] | 0.6±0.03 |
| | roots | 10±1.5 | 0.8±0.2 | 4.3±0.7 | 55.4±8.4 | 2.2±0.3 | 0.08±0.0 | 1.2±0.1 |
| Contamination | stems | 22±2.0 | **1.9±0.3** | **1.9±0.3** | **493.9±67.1** | 2.3±0.3 | 0.30±0.01 | 1.1±0.1 |
| | roots | 21±2.5 | **2.4±0.5** | **2.4±0.5** | **893.1±103.6** | 6.1±0.8 | **1.40±0.1** | 3.6±0.4 |
| *MPL for grasses* | | - | *0.5* | *30.0* | *50.0* | *5.0* | *0.3* | *3.0* |

[a] *Exceedances of the clark, MPC and the limit exceed MPL are marked bold.*
[b] *Permissible interval of the mean <0.001.*

The contents of elements in *P. australis* tissues from background plot was below the MPLs. Though, it was exceeded for Zn, Cd, Pb, and Cr in samples collected from the contamination plot (Table 1). Zn content in the aboveground tissues of *P. australis* crossed the MPL limits more than 10-folds. Whereas, contents of Mn, Pb and Ni in monodominant *P. australis* groups correspond to their lower threshold concentrations in plant fodder (Table 1).

Synchrotron XRD analysis of Technosol samples was performed and mineral components is shown in Figure 1. Results showed that the quartz was dominant components with diagnosed intense peaks corresponding to würtzite (hexagonal ZnS) (Figure 1). During the analysis of each reflection in X-ray diffractograms, other sulphur-containing metal compounds were determined such as sphalerite (cubic ZnS), covellite (CuS), and bornite ($Cu_5FeS_4$). Diagnosed minerals from other groups: wollastonite ($Ca_3(Si_3O_9)$), montmorillonite ($(Ca,Na...)(Mg,Al,Fe)_2 [(Si,Al)_4O_{10}] (OH)_2 \cdot nH_2O$) and anhydrite ($CaSO_4$).

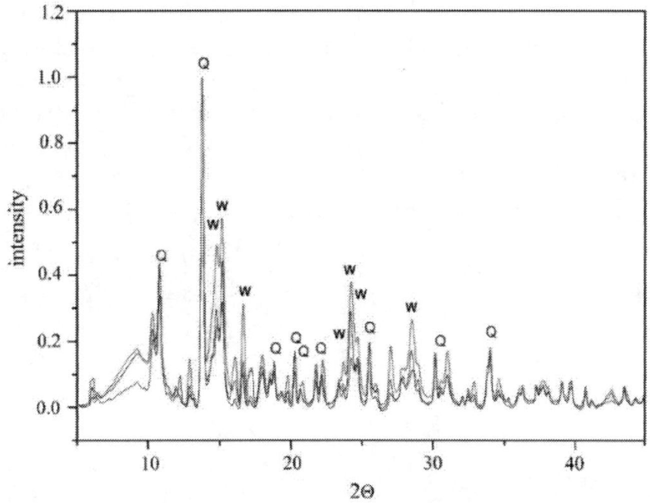


**Figure 1.** Synchrotron X-ray diffractograms of Technosols for mineral components

K-edge EXAFS spectra of Zn show speciation's of Zn in Technosol samples, and main peaks corresponded to sphalerite (ZnS). This is an indication of the dominance of Zn–S bonds (Figure 2). While, an additional parameter was an indicative of the presence of Zn–O bonding. The best-fit results were achieved, assuming mixed Zn–S and Zn–O bonding (Table 2).

**Table 2.** Best-fit results of Zn K-edge EXAFS data for Zn in Technosol (R (Å) is the interatomic distance; $\sigma^2$ (Å$^2$) is the Debye-Waller factor; R-factor (%), k range (Å$^{-1}$) and R range (Å) is the fitting quality function).

| Shell | R (Å) | $\sigma^2$ (Å$^2$) | CN | ΔE (eV) | R-factor (%) | k range (Å$^{-1}$) | R range (Å) |
|---|---|---|---|---|---|---|---|
| Zn–O1 | 2.01 | 0.008 | 2.25 | 2.16 | 3.3 | 2.66–13.25 | 1.34–2.33 |
| Zn–S | 2.34 | 0.007 | 4.6 | 2.16 | | | |

The estimation based on the Zn–S and Zn–O coordination numbers for Technosol samples show approx. 70% Zn–S bonds and 30% Zn–O bonds. In another work, Zn's dominant fixation for the upper horizon of highly contaminated soils in Palmerton (Pennsylvania, USA) was obtained due to sulfide (ZnS) [23].

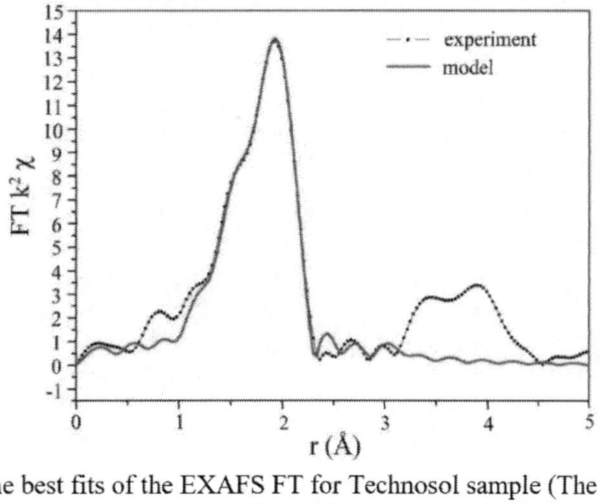


**Figure 2.** The best fits of the EXAFS FT for Technosol sample (The black line shows the experimental data and the red line indicates the fits to the data).

The light optical observations of *P. australis* showed the roots comprises single-row epidermis, well-developed cortex consists exoderm and endoderm layers, endoderm stick closely to the axis cylinder and parenchymal thin-wall cells arranged as a regular radial row (Figure 3a) in the background plot samples. The disturbance in symmetrical arrangement of exodermal cells were found in polluted soil plant samples. The significant reduction was visual in the size of intercellular air space, and swollen endodermal cells (Figure 3b).

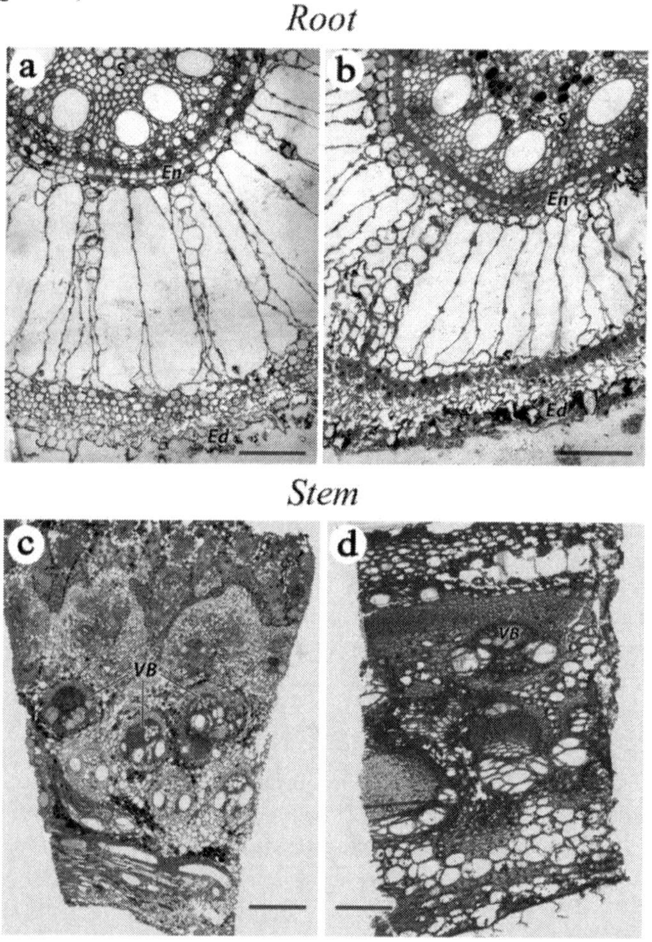


**Figure 3.** Light micrograph of the reed (*P. australis*) root and stem: *a, c* – background; *b, d* – contamination; Ep – epidermal cells; Ed – endodermal cells; S – central cylinder; Vb – vascular bundle. The scale bar *a* and *b* is 100 µm; the scale bar *c* and *d* is 200 µm.

The chlorenchyma and cortical layer was loosely arranged and formed big intercellular space for gas (Figure 3c) in background plot samples whereas greater compactness of chlorenchyma cells was found at the high contamination level (Figure 3b). The anatomical results can be concluded that soil pollution with high content resulted abnormalities in root epidermis, lessening in air space, and increase in the size of cells. Similar changes were noted in other study [24]. Apparently, these cellular-subcellular modifications in the root tissues evidently let plants to acclimatize hostile environmental circumstances.

The subcellular observations of background samples revealed that the root cells had a normal structure with unharmed cell wells, cytoplasmic membrane, and mitochondria (Figure 4a). Whereas, violations in sub-cellular organelles were noted such as disturbed integrity of tonoplast and cytoplasm contents penetrated the central vacuole (Figure 4b).

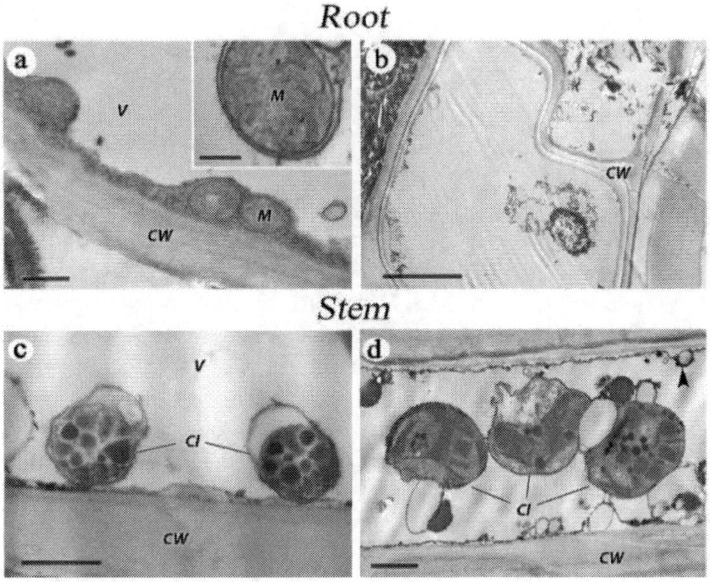


**Figure 4.** SEM micrographs of reed roots and stem (*P. australis*) (a), (c) – background; and (b), (d) – contamination. Note: M = mitochondria; CW= cell wall; (a)=0.5 μm, (b)=2 μm, (c)=2 μk, (d)=1 μm.

Chlorenchyma, parenchyma, and vascular bundles and predominantly round shape chloroplast were noted in the stem of the background reed samples (Figure 4c). Ellipsoid mitochondria contained dense matrix; elongated nuclei separated from the cytoplasm changes were identified in the stem structure of reed plants grown on contamination plot (Figure 4d). The increase bioavailability of metals in soil is one of the reasons for the disturbance in nutrients uptake, causing structural changes in cellular organelles of roots: membranes, cytoplasm, vacuoles. The disturbance of chloroplast structure could be the reason for the low content of chlorophyll pigments and which may reduce photosynthetic rate [25]. According to the previous reports under the effects of high concentrations of HMs, plastoglobules' content increases [26], indicating an enhanced organelles degradation.

## 4. Conclusions

The high concentration of chemical compounds was determined in studied technogenically polluted monitoring plots. The hundred-folds higher concentration of Zn and Cd was found. Two types of local surroundings are determined for Zn (the dominant pollutant metal) corresponding to its coordination by O and by S (close to ZnS). The mixed local surrounding with Zn–S and Zn–O bonds is identified. The EXAFS data's fitting results on the metal state in Technosols showed 70% of Zn–S bonds and 30% of Zn–O bonds. It was found that the main contribution in the composition of polluting compounds the floodplain landscapes for a long time are the authigenic minerals of metals with S: sphalerite (cubic ZnS), würtzite (hexagonal ZnS), covellite (CuS), and bornite ($Cu_5FeS_4$). The excess metal contamination decelerates ontogenetic development of *P. australis*. The changes observed in epidermis, endo- and mesoderm under the impact of metals hampered the translocation of minerals in the root from the peripheral parts. The increase density of plastoglobules probably is a part of defiance mechanisms to avoid damages in photosynthetic apparatus under chemical contamination by HMs. The results based on non-destructive approaches could helpful to rapid observation of metal speciation in soil and state of plants in stress conditions.

**References**

[1] Smolinska B and Rowe S 2015 The potential of Lepidium sativum L. for phytoextraction of Hg-contaminated soil assisted by thiosulphate *Journal of Soils and Sediments* **15** 393–400

[2] Wan B, Yan Y, Huang R, Abdala D B, Liu F, Tang Y, Tan W and Feng X 2019 Formation of Zn-Al layered double hydroxides (LDH) during the interaction of ZnO nanoparticles (NPs) with γ-Al 2 O 3 *The Science of The Total Environment* **650** 1980–7

[3] Rajput V, Minkina T, Semenkov I, Klink G, Tarigholizadeh S and Sushkova S 2021 Phylogenetic analysis of hyperaccumulator plant species for heavy metals and polycyclic aromatic hydrocarbons *Environmental Geochemistry and Health* **43** 1629–54

[4] Reeder R J, Schoonen M A A and Lanzirotti A 2006 Metal speciation and its role in bioaccessibility and bioavailability *Reviews in Mineralogy and Geochemistry* **64** 59–113

[5] Roba C, Roşu C, Piştea I, Ozunu A and Baciu C 2016 Heavy metal content in vegetables and fruits cultivated in Baia Mare mining area (Romania) and health risk assessment *Environmental Science and Pollution Research International* **23** 6062–73

[6] Chaplygin V A, Minkina T M, Mandzhieva S S, Nazarenko O G, Zimulina I v., Bauer T v., Litvinov Y A and Rajput V 2021 Heavy metals in agricultural crops of Rostov region through the example of soft wheat (Triticum aestivum) *IOP Conference Series: Earth and Environmental Science* **624** 012204

[7] Rajput V D, Minkina T M, Behal A, Sushkova S N, Mandzhieva S, Singh R, Gorovtsov A, Tsitsuashvili V S, Purvis W O, Ghazaryan K A and Movsesyan H S 2018 Effects of zinc-oxide nanoparticles on soil, plants, animals and soil organisms: A review *Environmental Nanotechnology, Monitoring & Management* **9** 76–84

[8] Porcaro F, Roudeau S, Carmona A and Ortega R 2018 Advances in element speciation analysis of biomedical samples using synchrotron-based techniques *Trends in Analytical Chemistry* **104** 22–4

[9] Tsitsuashvili V S, Minkina T M, Soldatov A v. and Nevidomskaya D G 2021 On synchrotron radiation for studying the transformation of toxic elements in the soil–plant system: a review *Journal of Surface Investigation: X-ray, Synchrotron and Neutron Techniques* **15** 814–22

[10] Minkina T, Nevidomskaya D, Bauer T, Shuvaeva V, Soldatov A, Mandzhieva S, Zubavichus Y and Trigub A 2018 Determining the speciation of Zn in soils around the sediment ponds of chemical plants by XRD and XAFS spectroscopy and sequential extraction. *The Science of the Total Environment* **634** 1165–73

[11] dos Anjos M J, Lopes R T, de Jesus E F O, Assis J T, Cesareo R and Barradas C A A 2000 Quantitative analysis of metals in soil using X-ray fluorescence *Spectrochimica Acta Part B: Atomic Spectroscopy* **55** 1189–94

[12] Vishwakarma V 2019 Environmental impact of sustainable green concrete *Smart nanoconcretes and cement-based materials : Properties, modelling and applications* ed M S Liew, P Nguyen-Tri, T A Nguyen and S Kakooei (Amsterdam: Elsevier) p 726

[13] Baldantoni D and Bellino A 2021 On the capability of the epigeous organs of Phragmites australis to act as metal accumulators in biomonitoring studies *Sustainability* **13** 7745

[14] Minkina T, Fedorenko G, Nevidomskaya D, Fedorenko A, Chaplygin V and Mandzhieva S 2018 Morphological and anatomical changes of Phragmites australis Cav. due to the uptake and accumulation of heavy metals from polluted soils *Science of The Total Environment* **636** 392–401

[15] Linnik V G, Minkina T M, Bauer T v., Saveliev A A and Mandzhieva S S 2020 Geochemical assessment and spatial analysis of heavy metals pollution around coal-fired power station *Environmental Geochemistry and Health* **42** 4087–100

[16] ISO 2002 ISO 10381-1:2002 - Soil quality — Sampling — Part 1: Guidance on the design of sampling programmes *ISO*

[17] Vinogradov A P 1957 *The geochemistry of rare and dispersed chemical elements in soils* (New York: Consultants Bureau)

[18] MPCs 1990 *Maximum permissible concentrations (MPCs) of chemical substances in soils* (Moscow: Goskompriroda SSSR no. 02-2333)
[19] Svetogorov R D, Dorovatovskii P v. and Lazarenko V A 2020 Belok/XSA diffraction beamline for studying crystalline samples at kurchatov synchrotron radiation source *Crystal Research and Technology* **55** 1900184
[20] du Laing G, Tack F and Verloo M 2003 Performance of selected destruction methods for the determination of heavy metals in reed plants (Phragmites australis) *Analytica Chimica Acta* **497** 191–8
[21] State Agroindustrial Committee of the USSR 1987 Provisional Maximum Permissible Levels (MPL) for some chemical elements and gossypol in forage for farm animals and feed additives
[22] Fedorenko G M, Fedorenko A G, Minkina T M, Mandzhieva S S, Rajput V D, Usatov A v. and Sushkova S N 2018 Method for hydrophytic plant sample preparation for light and electron microscopy (studies on Phragmites australis Cav.) *MethodsX* **5** 1213–20
[23] Scheinost A C, Kretzschmar R, Pfister S and Roberts D R 2002 Combining selective sequential extractions, X-ray absorption spectroscopy, and principal component analysis for quantitative zinc speciation in soil *Environmental science & technology* **36** 5021–8
[24] Ratushnyak A Y, Chakhirev I v., Andreeva M G and Trushin M v. 2014 Features of the formation of morphological structures and production of Typha angustifolia under load on lead *Biosciences Biotechnology Research Asia* **11** 365–9
[25] Molas J 1998 Changes in morphological and anatomical structure of cabbage (Brassica oleracea L.) outer leaves and in ultrastructure of their chloroplasts caused by an in vitro excess of nickel *Photosynthetica* **34** 513–22
[26] Rajput V D, Minkina T, Fedorenko A, Mandzhieva S, Sushkova S, Lysenko N D V, Azarov A and Chokheli V 2019 Destructive effect of copper oxide nanoparticles on ultrastructure of chloroplast, plastoglobules and starch grains in spring barley (Hordeum sativum) *International Journal of Agriculture and Biology* **21** 171–4

**Acknowledgement**
The research was financially supported by the Ministry of Science and Higher Education of the Russian Federation, no. project 0852-2020-0029) and the Russian Foundation for Basic Research, project no. 19-29-05265.

# Application of GA3 and PGPRs on growth and antioxidant content of Parijoto (*Medinilla verrucosa*) in peat soil

**A T Sakya[1,2], Sulandjari[1,2]* J Purnomo[1], and D A Bima[1]**

1 Department of Agrotechnology, Agricultural Faculty, Sebelas Maret University, Jl. Ir. Sutami 36 A, Kentingan, Surakarta, 57126, Indonesia
2 Research and Development Center for Biodiversity and Biotechnology, Sebelas Maret University, Jl. Ir. Sutami 36 A, Kentingan, Surakarta, 57126, Indonesia
*Corresponding author: sulandjari@staff.uns.ac.id

**Abstract**. Climate change is indicated to result in drought on agricultural land. In addition, increasing dryness can also cause changes in soil acidity, and these conditions can affect the growth and content of secondary metabolites of plants. Parijoto (Medinilla veruccosa) has health benefits, including lowering blood glucose levels, inhibiting the growth of cancer cells, and is traditionally believed to increase fertility during pregnancy. GA3 and PGPRs (Plant Growth Promoting Rhizobacteria) as elicitors are expected to induce and increase the production of phytoalexins and secondary metabolites. This study aimed to obtain the interaction between GA3 and PGPRs to increase the growth and antioxidant content of parijoto in acid soil. The study used a completely randomized design with two factors (GA3 25 ppm; 50 ppm; 75 ppm/plant and PGPR 2.5 g; 5.0 g; 7.5 g/l). In this study, Rowo Pening peat soil media after being given dolomite increased its pH with very high CEC, high Nitrogen, low P, and K. The results showed that the GA3 and PGPRs had no significant effect on growth. There was a positive correlation between plant height, number of leaves, and root volume. The concentration of GA3 75 ppm and PGPRs 2.5 g/l gave higher antioxidant content than other treatments.

**Keywords**: Parijoto, Medinilla verrucosa, elicitor, Antioxidant, peat soil

## 1. Introduction

Climate change will affect the soil properties, including pH as the main variable that affects all other properties of an ecosystem. Specific responses such as changes in temperature and rainfall patterns on soil and ecosystem properties vary greatly depending on the initial ecosystem. Increased rainfall due to La Nina anomalies allows for increased soil problems due to the infiltration of alkaline cations then occurs the transfer of alkalinity from the soil exchange complex to the surface and leaves the soil immersed. Likewise, drought conditions due to reduced rainfall in El Niño anomalies can increase soil pH, the problem drops to neutral or alkaline even though these changes do not go quickly. Soil acidity is too high or too low, resulting in a nutrient deficiency in plants [1].

*Medinilla verucosa* (Parijoto) is one of the forest plants that contains antioxidants from tannins, flavonoids, saponins, and glycosides [2]. Flavonoids act as inhibitors of the growth of cancer cells, saponins that have antitumor activity and tannins that act as antibacterials. Ethanol extract from Medinilla fruit lowered the blood glucose levels of white mice and lowered glucose levels of n-Hexan fraction [3]. This plant has not been cultivated outside its natural habitat so it requires research for environmental and growth factors to get high yields and antioxidant content. Plant growth and chemical content can be used as an indicator of plant tolerance to conditions. In this study, the planting media used peat soil (acidic) with a fairly high fertility character but very low nutrient availability.

Elicitor is defined as a substance that can increase the biosynthesis of certain compounds when introduced in small concentrations into living cell systems. Besides that, elicitors can be considered molecules that activate signal transduction and cause the activation and expression of genes related to the biosynthesis of secondary metabolite compounds [4] consist of biotic and abiotic elicitors. As n abiotic elicitor, PGPRs have been shown to increase Artemisia and artemisinin content [5]. In addition, PGPRs have also been shown to produce phytohormone, including gibberellins, auxins, and cytokinins that can induce growth and plant secondary metabolites [6]. Giving exogenous GA3 to plants can be interpreted as an abiotic elicitor that functions to increase flowering and fruit yields [7].

From the background, there are problems and has not been done research before. Therefore, the study wants to know whether peat soil media affects the growth and dynamics of antioxidants, whether the interaction of GA3 and PGPR can increase the growth and content of antioxidants of Parijoto in acidic soils.

## 2. Materials and method

### *2.1. Materials*

Parijoto was 3 months old, planted in 30 x 40 cm polybags, and placed under 25% paranet shade. Peat soil of Rawa Pening drained and sifted to remove dirt material and plant remains. Then mixed with 10 g of dolomite and basic fertilizer, incubated for 2 weeks. After 2 weeks, measurement of soil pH was taken, then put into a polybag. Initial analysis of soil chemistry is performed consisting of total N, available P, K available, CEC, and soil pH.

### *2.2. Methods*

The study used a completely randomized design with two factors GA3 (G) and PGPRs (P) with concentration are (GA3 25 ppm; 50 ppm; 75 ppm/plant and PGPRs 2.5 g; 5.0 g; 7.5 g/l). In this study, using a media peat soil from Rowo Pening, the pH increased (5.07 to 6,55), with very high CEC, high Nitrogen, low Phosphorous, and Potassium after being given dolomite.

### *2.3. Treatment application and observation*

The treatment was given twice, each at 1 month and 2 months after planting with the appropriate concentration. PGPRs contains Bacillus sp. 3 x 109 cfu/g and Pseudomonas sp. 6.25 x 107 cfu/g dissolved in water. GA3 treatment was sprayed on the leaves evenly after calibrating the volume/plant requirement, while PGPRs sprayed on the root area as much as 1 L/plant. Observations consisted of the height of the plant, the number of leaves, root volume, fresh weight and dry weight of plants, shoot root ratio, and antioxidant activity.

Testing of antioxidant activity is performed using the free radical DPPH 1,1- diphenyl-2-pikrilhydrazil (α, α-diphenyl-β-picrylhydrazyl). Testing of antioxidant activity with this method is suitable using ethanol and methanol solvents. Sample preparation was carried out by crushing the fresh and undamaged Parijoto leaves. Then dissolved in methanol and left overnight. After that, 5 ml of DPPH solution was added. The sample solution mixture which had been added with DPPH solution was incubated at room temperature for 40 minutes. After incubation, the absorbance was measured with a maximum of 517 nm using a UV-Vis spectrophotometer.

### *2.4. Data analysis*

The data obtained were analyzed using variance analysis ($\alpha = 5\%$) and continued with the Duncan test if there is a significant effect.

## 3. Results and discussion

The experiment was conducted in the village of Kenayan, Wedomartani, Sleman, Yogyakarta. The soil used in this research is peat. According to [8] peat soil is known as soil that has high porosity, water content, and organic matter content. The peat soil in the study used Rowo Pening peat with a low initial pH, very high organic matter, very high total N, low available phosphorus and potassium, and a clay texture.

The results of the peat soil analysis showed that the initial pH was 5.07 and after being mixed with dolomite and incubated for 2 weeks the pH increased to 6.55. Peat soil is acidic due to very high levels of organic acids resulting from weathering of organic matter. Peat with a low pH is thought to cause plant roots to be unable to absorb several nutrients, including phosphorus (P). The content of Potassium (K) available in the peat Rowo Pening soil is low because it is fixed due to the low pH [9]

### *3.1. Application of Ga3 and PGPRs to Parijoto's growth*

The correlation coefficient between growth variables in PGPR and GA3 applications is presented in Table 1. The correlation coefficient between 0.28 to 0.89, indicates that the relationship of growth parameters increases from very small to those with strong relationships. Exceptions are shown in plant heights with root shoot ratio that has negatively correlated, but the relationship is very small. The height of the plant is positively correlated moderately with the number of leaves, fresh weight, and dry weight. This explains the higher the plant, the number of leaves, the fresh weight and dry weight will be more and more. Likewise, the leaves are positively correlated and strong on root volume, fresh weight, and dry weight. The more leaves, the greater the activity of photosynthesis that runs so that more photosynthates are formed and distributed for leaf development and rooting.

**Table 1.** Correlation of growth parameters

| | **Plant height** | **Leaf number** | **Root volume** | **Root/ Shoot ratio** | **Fresh Weight** |
|---|---|---|---|---|---|
| **Leaf number** | 0.502 | | | | |
| **Root volume** | 0.282 | 0.775 | | | |
| **Root/ Shoot ratio** | -0.096 | 0.366 | 0.588 | | |
| **Fresh Weight** | 0.474 | 0.886 | 0.839 | 0.454 | |
| **Dry Weight** | 0.573 | 0.914 | 0.829 | 0.344 | 0.898 |

Analysis of variance showed no interaction on all growth parameters (Figure 1). This describes the height of the plant, the number of leaves, the fresh weight, the dry weight, the root volume, and the root shoot ratio giving the same response to GA3 applications with increasing dosage of PGPRs. The results of the analysis of variance also showed no effect on both PGPRs and GA3 individually. It is suspected that the planting media used can support the growth of Parijoto. The development of PGPR is closely related to the living conditions of microorganisms to grow. Factors that affect the growth of microorganisms include pH, humidity, temperature, and nutrients [10]. The ineffectiveness of the PGPR is also suspected to be competition between PGPR and pathogens derived from the media because the soil used is not sterilized first to allow other microorganisms to be folded and result in competition in the utilization of nutrients and space to grow. In addition, it is also suspected because the concentration of PGPR used is less.

In this study, Parijoto growth did not show any effect of GA3. Phytohormones play an important role in regulating plant growth and development. However, the effect of exogenous GA3 also depends on the endogenous GA content and the gibberellin produced by PGPRs. Some researchers claim PGPR makes it possible to produce phytohormones in both auxin, gibberellin, and cytokinin to increase root absorption or manipulate hormonal balance in plants thereby increasing growth [11,12]. Some PGPR strains produce a considerable amount of gibberellin and promote plant growth [13]. In optimal GA conditions in plants, GA application may have a positive, negative, or no effect on plant growth. [14] states that phytohormones also serve as molecular signals in achieving environmental conditions that limit plant growth and may give a fatal condition if uncontrolled.

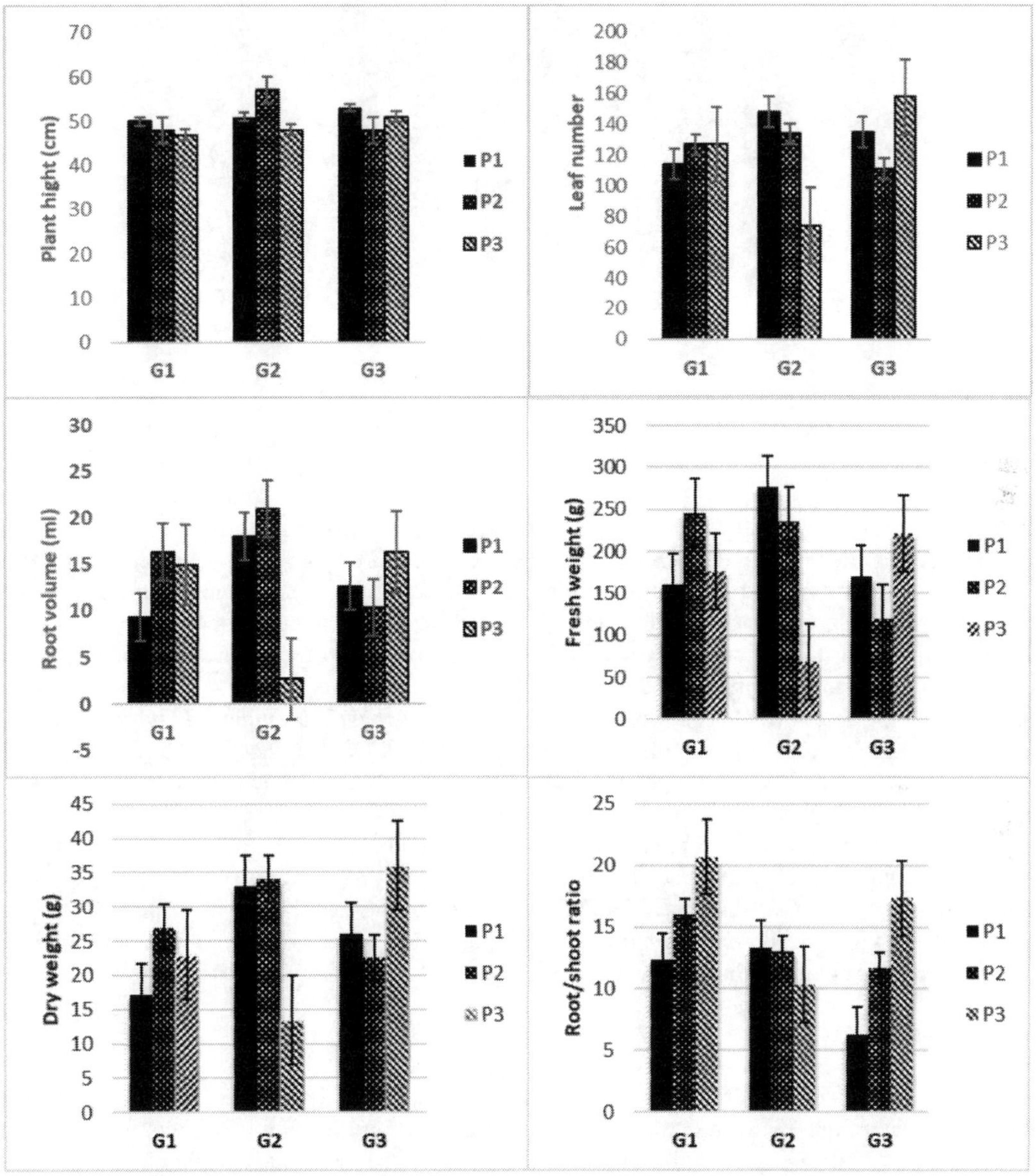


**Figure 1.** The effect of GA3 (G) and PGPRs (P) concentration on growth parameter of parijoto (G1, G2 and G3 = 25; 50; 75 ppm/plant); P1, P2 and P3 = 2.5 g; 5.0 g; 7.5g/l

### *3.2. Application of Ga3 and PGPRs to the antioxidant content of Parijoto*

Nutrient absorption, especially NPK, will increase with the administration of PGPRs. Rowo Pening peat soil media provides very low essential elements, then the ability of Rhizobacterium in PGPRs to increase nutrient absorption and production of secondary metabolites. Apart from being an elicitor by producing various phytohormones, PGPRs can activate antioxidant enzymes under abiotic stress [15]

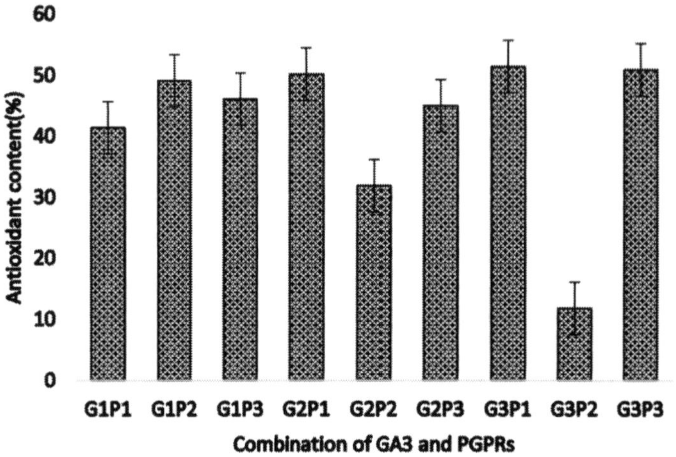


**Figure 2.** The effect of GA3 (G) and PGPRs (P) concentration on growth parameter of parijoto (G1, G2 and G3 = 25; 50; 75 ppm/plant); P1, P2 and P3 = 2.5 g; 5.0 g; 7.5g/l

Nutrient absorption, especially NPK, will increase with the administration of PGPRs. Rowo Pening peat soil media provides very low essential elements, then the ability of Rhizobacterium in PGPRs to increase nutrient absorption and production of secondary metabolites. Apart from being an elicitor by producing various phytohormones, PGPRs can activate antioxidant enzymes under abiotic stress [15].

Figure 2. shows that the combination of GA3 and PGPRs showed an increase in antioxidant content. The antioxidant activity ranged from 11.7% to 51.4%. This suggests free radical capture activity in Parijoto varies from very low to moderate. The highest antioxidant activity is found in leaves with application of GA3 75 ppm and PGPRs of 2.5 g/l. Variations that occur may be that antioxidant activity is highly dependent on secondary metabolite compounds such as flavonoids, saponin, polypenol and others. It is suspected that the addition of PGPRs and GA affects the content of these metabolites. The possibility of the addition of elicitors will change the metabolic pathway of the formation of secondary metabolites, so it will affect the content of existing compounds. [16] stated that the administration of GA3 increased Vitamin C as part of the antioxidant in grapes. Meanwhile, [17] stated that PGPRs were able to improve the quality of plant growth so that the dominant metabolism result was in the form of primary metabolites so that the increase was not too high.

## 4. Conclusions

Climate change is the main abiotic factor that changes environmental factors, such as soil acidity, drought or flooding conditions and indirectly affects the growth, yield and biosynthesis of secondary metabolites of important medicinal plants, one of which is antioxidants. Parijoto is one of the medicinal plants that allegedly contains antioxidants. The results of the study using elicitor GA3 and PGPRs showed no interaction between the two factors to Parijoto growth in peat soil. The application of GA3 and PGPRs has not been able to increase the growth of Parijoto but can increase the antioxidant content. The concentration of GA3 75 ppm and PGPRs 2.5 g/l gave higher antioxidant content than other treatments.

## References

[1] Qafoku N . 2014 *Overview of different aspects of climate change effects on soils* (Richland, Washington.: Pacific Northwest National Laoratory, Richland, Washington)

[2] Sa'adah N N, Nurhayati A P D and Purwani K I 2018 Antihyperlipidemic and anti-obesity effects of the methanolic extract of parijoto (Medinilla speciosa) *AIP Conference Proceedings* vol 2002 p 020046

[3] Vifta R L and Luhurningtyas F P 2020 Nanoparticle from Medinilla speciosa with Various of Encapsulating Agent and Their Antioxidant Activities Using Ferric Reducing Assay *Indones. J. Cancer Chemoprevention* **11** 22–9
[4] Zhao J, Davis L C and Verpoorte R 2005 Elicitor signal transduction leading to production of plant secondary metabolites *Biotechnol. Adv.* **23** 283–333
[5] Patil A and Lamnganbi M 2018 Impact of climate change on Soil Fertility: A review *Int. J. Chem. Stud.* **6** 2399–404
[6] Biswas J C, Ladha J K and Dazzo F B 2000 rhizobia inoculation improves nutrient uptake and growth of lowland rice *Soil Sci. Soc. Am. J.* **64** 1644–50
[7] Paroussi G, Voyiatzis D G, Paroussis E and Drogoudi P D 2002 Growth, flowering and yield responses to GA3 of strawberry grown under different environmental conditions *Sci. Hortic. (Amsterdam).* **96** 103–13
[8] Mochtar N E, Yulianto F E and S T R 2014 Pengaruh Usia Stabilisasi pada Tanah Gambut Berserat yang Distabilisasi dengan Campuran CaCO3 dan Pozolan *J. Tek. Sipil* **21** 57–64
[9] Gunawan, Wijayanto N and Budi S W 2019 Karakteristik Sifat Kimia Tanah dan Status Kesuburan Tanah pada Agroforestri Tanaman Sayuran Berbasis Eucalyptus Sp. *J. Silvikultur Trop.* **10** 63–9
[10] Diaz-Garza A M, Fierro-Rivera J I, Pacheco A, Schüßler A, Gradilla-Hernández M S and Senés-Guerrero C 2020 Temporal Dynamics of Rhizobacteria Found in Pequin Pepper, Soybean, and Orange Trees Growing in a Semi-arid Ecosystem *Front. Sustain. Food Syst.* **4**
[11] Gupta G, Parihar S S, Ahirwar N K, Snehi A K and Singh V 2015 Plant Growth Promoting Rhizobacteria (PGPR): Current and Future Prospects for Development of Sustainable Agriculture *J. Microb. Biochem. Technol.* **07** 96–102
[12] Kumar S, Chauhan P S, Agrawal L, Raj R, Srivastava A, Gupta S, Mishra S K, Yadav S, Singh P C, Raj S K and Nautiyal C S 2016 Paenibacillus lentimorbus inoculation enhances tobacco growth and extenuates the virulence of cucumber mosaic virus *PLoS One* **11** 1–23
[13] Jha C K and Saraf M 2015 Plant Growth Promoting Rhizobacteria (PGPR): a review *J. Agric. Res. Dev.* **5** 2227–35
[14] Fahad S, Hussain S, Bano A, Saud S, Hassan S, Shan D, Khan F A, Khan F, Chen Y, Wu C, Tabassum M A, Chun M X, Afzal M, Jan A, Jan M T and Huang J 2014 Potential role of phytohormones and plant growth-promoting rhizobacteria in abiotic stresses: consequences for changing environment *Environ. Sci. Pollut. Res.* **22** 4907–21
[15] Heidari M and Golpayegani A 2012 Effects of water stress and inoculation with plant growth promoting rhizobacteria (PGPR) on antioxidant status and photosynthetic pigments in basil (Ocimum basilicum L.) *J. Saudi Soc. Agric. Sci.* **11** 57–61
[16] Gougoulias N and Masheva L 2010 Effect of gibberillic acid (ga3) on polyphenols content and antioxidative activity of some table grape varieties of vitis vinifera l *Oxid. Commun.* **33** 652–60
[17] Chiappero J, Cappellari L del R, Palermo T B, Giordano W, Khan N and Banchio E 2021 Antioxidant status of medicinal and aromatic plants under the influence of growth-promoting rhizobacteria and osmotic stress *Ind. Crops Prod.* **167** 113541

**Acknowledgement**
The author greatly acknowledgments Mandiri grants, LPPM Sebelas Maret University for funding this research

# The role of agricultural extension workers' competency in climate scenarios through farm services centers (FSCS) in improving the yield and performance of targeted farming community (A case study of Khyber Pakhtunkhwa, Pakistan)

**Muhammad Zafarullah Khan*, Ayesha Khan[1], Urooba Pervaiz[1], Arshad Farooq[2] and Raheel Saqib[1]**

[1]Department of Agricultural Extension Education and Communication, The University of Agriculture Peshawar, Khyber Pakhtunkhwa, Pakistan
[2]Principal Scientific Officer Horticulture, Pakistan Agriculture Research Council Islamabad-Pakistan
*Corresponding author: drzafar@aup.edu.pk

**Abstract.** Pakistan is sensitive to climate change and has started billion tsunami trees project in forest and many other approaches in Agriculture Extension like privatization, decentralization, and public-private partnership which has received more attention for competencies of extension workers to educate farmers for longer summer to adjust for crop. Farm Services Centers (FSCs) motivate agriculture extension agents to enhance competency in present less precipitation for improving crops' growth initiated in Province of Khyber Pakhtunkhwa. Member farmers of 400 from five districts named Swat, Dir Lower, Swabi, Mansehra and Dera Ismail Khan of province were selected for data collection. Findings revealed that land-hold size of majority (94%) was less than 100 acres registered with FSCs from five years while limited respondents got membership for more than 10 years. Agricultural extension agents were the main source of awareness and motivation followed by fellow farmers for FSC membership. Cultivator, land leveler, boom sprayer, mold board plough, and wheat, rice, maize and tomato were the most commonly provided inputs to member farmers to improve their competency for getting higher yield with by competent extension workers in present rise of temperature. A Significant increase in yield of major crops such as wheat, rice, maize, gram and tomato were observed before and after registration in FSCs. It is recommended that non-member farmers should be motivated for membership in FSCs as majority of member farmers increased yield of major crops after registration to observe climatic factors which affected all stages of crop growth along with lands suitable agriculturally with proper precipitation and temperature.

**Keywords:** Farm services center, competency of extension workers, climate scenario. yield performance and Khyber Pakhtunkhwa-Pakistan

## 1. Introduction

Pakistan is sensitive to climate change as glaciers are affected due to an increase in temperature. Agriculture is destroyed due to more industries and population which increased the temperature and environmental pollution. Any variations in a climate with the passage of time either due to variability in nature or due to human activity [1]. According to IPCC [2], the composition of global atmosphere and human activities are directly attributed to the changes in climate over comparable period of time. Temperature, cover of clouds, precipitation and vapor pressure are the various parameters of climate where changes occur due to variation in these parameters [3]. Green House Gases (GHG) emissions are increased due to human activities along with additional warming of the earth's surfaces with more ruinous shocks [4] [5]. Khan et al [6] revealed that the length summer period increased due to increase in temperature and less rainfall over the last 20 years while winter in Pakistan turned short. They recommended that agriculture research should develop climate extreme resistant varieties by building the adaptive capacity of the farmers with the use of cultural practices for mitigation of climate change.

Agriculture is the main source of income generation for a vast majority of population in Pakistan and contributes about 18.9% to national Gross Domestic Product (GDP) [7]. Agriculture improvement, poverty alleviation and rural development in the country have been supported by extension services through the public-private sector under the ministry of agriculture. During the last few decades significant improvement in extension services has been observed in Pakistan. Currently the government is focusing to modernize traditional farming practices by transferring improved agricultural technologies to develop rural life in general and enhance crop production in particular. Agricultural extension system in developing countries is faced with numerous problems and challenges such as lack of skills in administrative capabilities and supervisory services to produce more efficiently, lack of information regarding local markets, unavailability of sufficient budget, high cost of transportation and poor linkages among the related entities which have dwarfed significantly the production and productivity of the stockholders. It is thus imperative to analyze and properly address these problems. During the last decades of the 20$^{th}$ century some of the agricultural developmental programs in region were criticized as being deficient, ineffective, utilizing top-down approach and less targeted to poorer farmers.

Most of these agriculture related problems are not only limited to this particular region as in most other parts of the world numerous such problems are faced by farming community. As world's population is growing with time at a rapid pace, cultivable land is decreasing by global industrialization. Global warming in the form of climatic changes is further aggravating the concerns as it has been reported to negatively affect the average yield of some main crops. There is a tremendous pressure on the growers to produce more and more from the same or smaller fields. As the core objectives of extension work consist of transferring and conveying empirical information [8] and properly making the farmers update of new agricultural technologies and finally motivating them for adaptation of agricultural innovation [9]. In such scenario there is an immense need for a comprehensive agriculture supporting system where every grower combat and develop mutual contacts with researchers, competent extensionists and policy makers.

In order to improve the extension system, government of Khyber Pakhtunkhwa province adopted a new agricultural extension approach, called the Farm Services Centers (FSCs) in the year 2000. Available literatures from World Bank [10] and Feder et al. [11] indicate that local extension work will be more productive if it is strongly supported by the local government with the competent extension workers. Hence the basic perceptions of these centers are the awareness of local farmers about the improved farming system and to ultimately uplift their socio-economic status in the community. This approach was purely launched for extension services and the competent extension workers and is based on the public-private partnership to enhance agricultural yield and create linkages among farming community and different other related departments. The described centers have the aim to exacerbate hidden capabilities of tenants in the community as well as nurture their relationship with Government and Non-Government organizations. This was first executed in the province of Khyber in the country with the goal of providing physical and financial resources, technical support along with provision of farm machinery and inputs to the farmers by enhancing their competencies for higher yield by the capable staff of agriculture extension.

Agricultural land or livestock farms possess by 18 years or above can have registration access with minimum charges of Rs. 100. While lifetime membership fee for these FSCs is Rs.500. After membership FSCs provide various farm inputs such as improved varietal seed, fertilizers and agricultural machinery, plus advisory services and guidance regarding different usage of inputs, enhance farmers' knowledge and skills in farm management related activities. The main objectives of this study was to determine and examine services provided by these FSCs existing farming system of the study area, analyze the available extension services under FSCs and evaluate the impact of these FSCs on the overall performance of farming community in the present scenario of climate change. Furthermore, registration of the farmers in these FSCs and overall changes and impact due to these FSCs in achieving the desired improvements of the targeted farming community were also critically analyzed in this study.

## 2. Materials and methods

### *2.1. Universe of the study*

The whole Khyber Pakhtunkhwa province (shown in figure 1) was the universe for the current study. Khyber Pakhtunkhwa occupies the northwestern parts of Pakistan, adjoining Afghanistan in the west, Gilgit-Baltistan (Northern Areas) in the northeast, the semi autonomous Azad Kashmir in the east, and the provinces Punjab and Baluchistan in the southeast and southwest, respectively. Khyber Pakhtunkhwa Province covers 74,521 $Km^2$, where the Federally Administered Tribal Areas (FATA) contains 27,220$km^2$. They cover a total area of 101,741 $km^2$, which makes up 12.8 % of Pakistan's total area [12]. Agriculture especially livestock is one of the main occupations. In addition, small enterprises like village shops, fuelwood trading and fertilizer dealers also exist in the region to facilitate the local dwellers. The land is naturally suitable for growing different kinds of vegetables, fruits and some major agronomic crops. Most of the farmers have small landholdings with tenancy status of owner and tenant in mix.

### *2.2. Sampling technique and sample size*

Five districts namely Swat, Dir lower, Swabi, Mansehra and Dera Ismail Khan were purposely selected as these have diversified type of cropping systems. From these five districts, an equal number of member farmers of FSCs i.e. 80 were randomly selected. In this way a sample set of 400 respondents were selected for interview schedule with the assumption to represent the whole population.

### *2.3. Analysis of the data*

A comprehensive interview schedule was prepared for data collection in the light of the objectives of the study for respondents. Data were collected through face to face meetings with the respondents. During filling the interview schedules each and every question and its purpose was explained to respondents so that accurate and reliable information could be obtained. The collected data were fed to computer through SPSS package. After arranging data according to nature and objective of the study descriptive statistics techniques like frequencies, averages, their comparison and percentages were analyzed using SPSS package. In addition Paired Sample T-Test was also used to determine the mean differences at 95% level of significance before and after registration of the farmers with FSCs.

## 3. Results and discussion

### *3.1. Information of the respondents regarding income source and land holding size*

Data regarding different aspects of the registered respondents of these centers were analyzed and arranged. Due consideration was given to the sources of income as these reflect a situation that the farmer has better opportunity for increasing their competencies to adopt the appropriate technology. Data in figure 2 comprise of information regarding available sources of income of the respondents and their land holding size in acres. It has been reported that on farm income and related sources improve the living standards of the farmers [13]. Agriculture was found to be the supreme source of income of majority (75.5%) of the respondents followed by business i.e. 11.5%. Only 10.2% of the respondents reported that they had government jobs as the prime source of income whereas 2.8% of the respondents had other sources of income. Our results are in conformity with that of [8] who also reported that majority of the respondents i.e. 53.3% had agriculture sector as their major source of income. Generally it is believed that the farmers having large holding are expected to take risks.

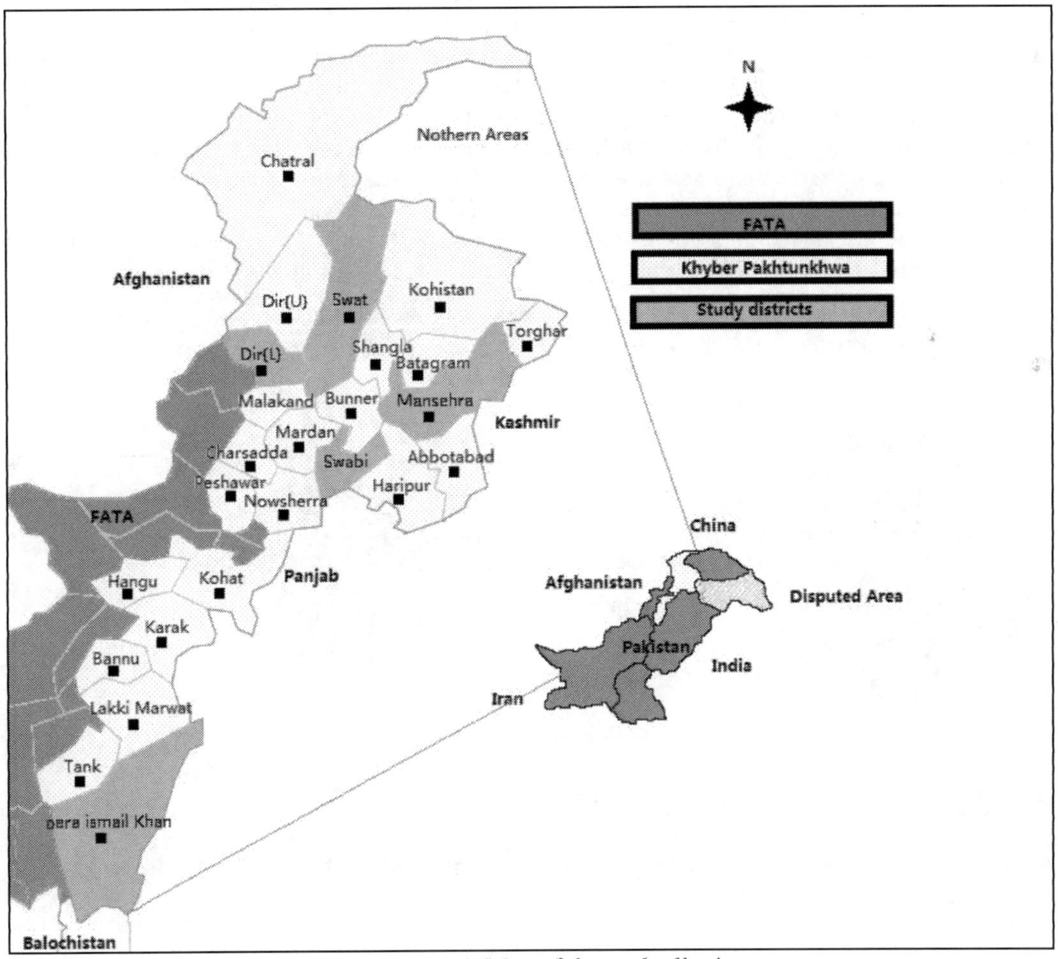


**Figure 1.** Map of the study districts

Yaseen at al. [14] stated that larger the size of landholding, the greater would be the adoption of agricultural innovations. It looks that the farmers cultivating more land have more exposure and more in need of information regarding better farming practices [15]. Different studies conducted on farm-size showed that the rate of adoption and diffusion is positively correlated with the farm size [16]. Data in the figure also depicted that majority (93.8%) of the respondents had landholdings of less than 100 acres followed by respondents who had a landholding of 101-300 acres i.e. 5.2%. The respondents with 301-500 acres and above 500 acres of landholding were meager in percentage i.e. 0.5% each as depicted in figure 2. Pervaiz [17] observed landholding size for farmers is equal or less than 12 acres for 41% farmers and more than 12 acres for 59% of farmers in his study.

*3.2 Information of the respondents regarding duration and source of membership*

Duration of registration was thought deem necessary variables because it might have impact on other variables in the study (per unit production, skill learned from FSCs, inputs obtained etc.) and thus respondents were investigated about their registration duration with FSCs. The data regarding registration duration of respondents with FSCs were collected and presented in figure 3. It was observed that greater portion (55%) of the respondents was registered with FSCs from less than 5 years. This shows that the trend of getting registered with FSCs had increased in recent years. It might be due to the fact that idea of FSCs flourished more in the recent years and thus more farmers got registered their selves at FSCs to get the training and suggestions of the competent extension workers, which ultimately boosted the public-private partnership approach. Similarly 30% of the sample respondents received FSCs memberships from last 6-10 years whereas 15% of the sample respondents obtained FSCs memberships from more than 10 years. Those respondents which had been registered with FSCs from longer duration might have the opportunity to build terms and in return got better assistance from FSCs staff. The recent increase in FSCs memberships might be on account of awareness in the farmers about FSCs from various sources.

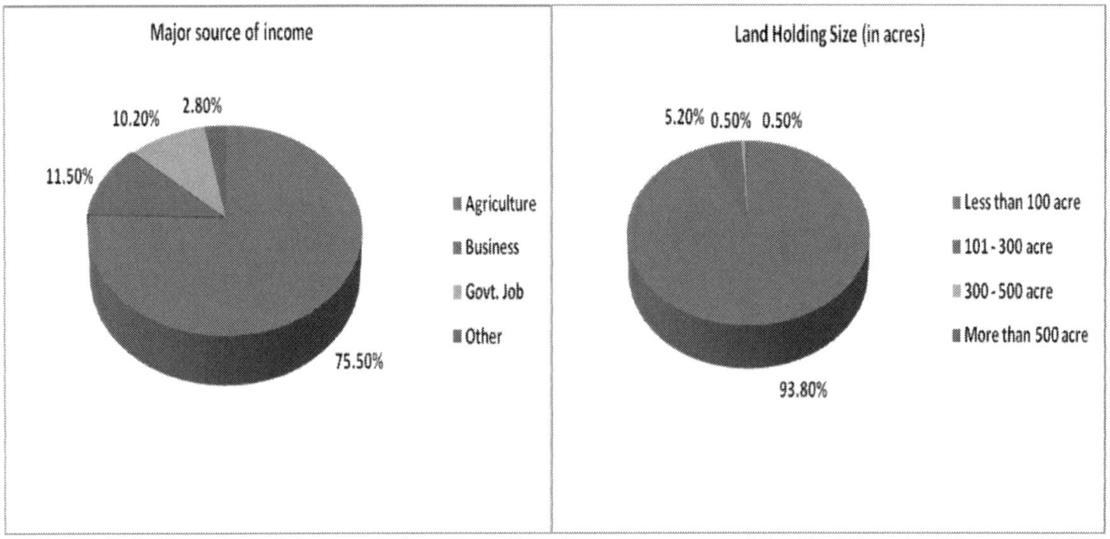


**Figure 2.** Information of the respondents regarding income source and landholding size (Source: Field Data)

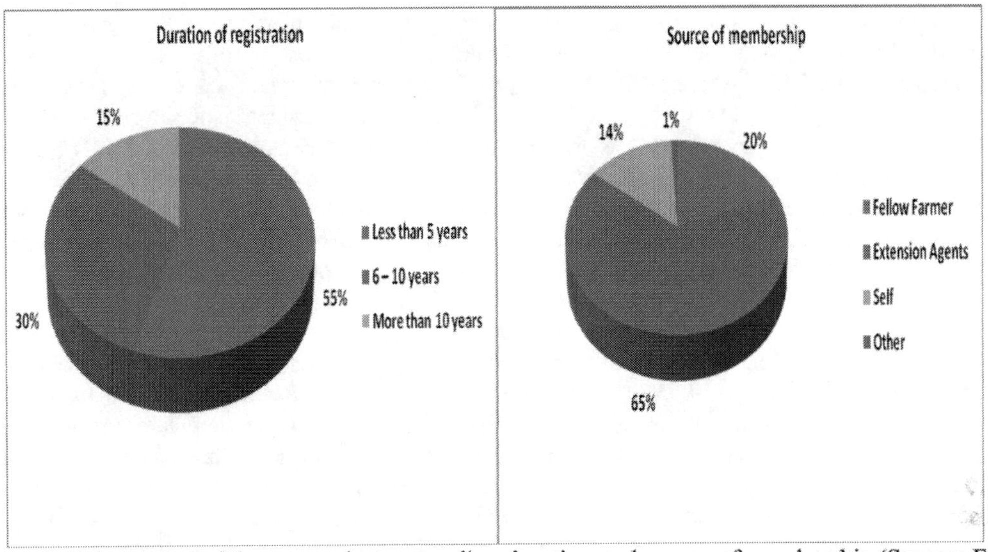


**Figure 3.** Information of the respondents regarding duration and source of membership (Source: Field Data)

Awareness of farmers about FSCs is due to popularity and sign of success among the farmers of competent extension workers. Awareness is the first step in the adoption process. The existence of FSCs can be learned through variety of sources. In order to determine the sources of awareness concerning FSCs, respondents were questioned about their awareness sources as in Figure 3 which reveals that extension agents were leading source of awareness about FSCs as 65.5 % demonstrating the success and correct achievement of extension agents. Similarly, 19.8 % of respondents learned about FSCs through fellow farmers. Only 1 % got knowledge through other sources and 14 % of them got self-awareness about FSCs. In order to support this goal, agricultural extension is one of the tools available to assist farmers in strengthening their capacity in the current climate. It is one of a kind of unique service in a way that provides access to small farmers and rural poor living outside of metropolitan regions along with the transfer of innovative technologies to cope with variation in temperature. The result showed that extension agents had the best communication fidelity of the agriculture extension agents is that they not only informed farmers about the existence of FSCs, but also encouraged them to join FSCs for better facilitation and learning of agricultural practices with a particular focus on the climate scenario, which is hot and burning issue in this province of Khyber Pakhtunkhwa.

### *3.3. Sources of Agricultural Machinery of the FSCs Respondents*

Among agricultural operations in the present climatic scenario, mechanization holds a prominent position because without proper seedbed preparation and proper mechanization the chances of better production are reduced predominantly. Farm machinery provision to registered farmers on rent is the responsibility of FSCs to fulfill their crop-related machines for agriculture farming. The mechanized implements given to the farmers who are registered with FSCs include threshers, tractors with trolleys, sheller of maize, laser leveler, cultivator etc. In this connection to know the performance of FSCs in the facilitation of farmers regarding machinery provision, they were probed. The responses of the registered farmers were noted and presented in Table No.1. The performance of FSCs regarding provision of the cultivator was much better i.e. 63.25% of the respondents reported that they got cultivators for FSCs for field preparation. While 26.75% of the sample respondents hired the cultivator form private agricultural services providers. This might be recognized to the reason that FSCs were at sufficient distance from the villages and thus some of the respondents preferred to utilize private sources for cultivators. Almost 6% of the respondents had their own cultivators whereas 4.5% of the respondents utilized the cultivator of fellow farmers on a rental basis.

Data presented in Table 1 also indicate that 277 respondents utilized land leveler among which majority (68.59%) of the respondents was facilitated by competent extension workers in FSCs. About 24.55% of the respondents utilized fellow farmer source for land levelers whereas 4.33% of the respondents utilized private sources. Only 2.53% of the respondents reported that they had their own land levelers. Data in table also reflected that the rotavator was utilized by 164 respondents out of 400. The most important tool for the provision of the fine degree of pulverization of soil is the rotary cultivator which is considered necessary for rapid soil mixing. All farmers consider the Rotavator as an important and basic need and therefore, registered farmers with FSCs were asked about Rotavator and observed that majority (41.46%) of them hired Rotavator from their respective FSCs sources. Private Sources were the second majority as revealed by (39.63%) The second major source for Rotavator was private sources i.e. 39.63%. Similarly, 15.24% of the respondents had hired Rotavator from fellow farmers whereas only limited number of registered farmers i.e. 3.66% had their own Rotavator. From the instant results it can be concluded that irrespective of sources the farmers were devoted to preparing the best land for crops cultivation to cope with present climate, thus they hired Rotavator from various sources. The reason might be that due to the shortage of Rotavator implements in FSCs where the implements cannot reach to all member farmers timely so the FSCs members hired Rotavator from other sources in the study area. In the same way information regarding hand spray machine, thresher, boom spray machine, maize sheller, laser leveler, tractor trolley and moldboard plough are provided in Table 1.

**Table 1.** Sources of agricultural machinery

| Implements | Source | % | Implements | Source | % |
|---|---|---|---|---|---|
| **Cultivator** | Own | 5.5 | **Land leveler** | Own | 2.53 |
| | FSC | 63.25 | | FSC | 68.59 |
| | Fellow farmer | 4.5 | | Fellow farmer | 24.55 |
| | Private | 26.75 | | Private | 4.33 |
| **Rotavator** | Own | 3.66 | **Hand spray machine** | Own | 37.16 |
| | FSC | 41.46 | | FSC | 31.8 |
| | Fellow farmer | 15.24 | | Fellow farmer | 19.92 |
| | Private | 39.63 | | Private | 11.11 |
| **Thresher** | Own | 6.25 | **Boom sprayer** | Own | - |
| | FSC | 9.5 | | FSC | 100 |
| | Fellow farmer | 10.5 | | Fellow farmer | - |
| | Private | 73.75 | | Private | - |
| **Maize sheller** | Own | 8.99 | **Laser leveler** | Own | - |
| | FSC | 44.94 | | FSC | 100 |
| | Fellow farmer | 10.49 | | Fellow farmer | - |
| | Private | 35.58 | | Private | - |
| **Tractor trolley** | Own | 6 | **Mold board plough** | Own | 15.52 |
| | FSC | 5.5 | | FSC | 59.48 |
| | Fellow farmer | 11.25 | | Fellow farmer | 6.9 |
| | Private | 77.25 | | Private | 18.1 |

Source: Field Data

*3.4 Sources of seeds of the FSCs respondents*

FSCs are also responsible for provision of seed of higher yield to their registered farmers by keeping in view the present climatic changes, thus the farmers were investigated about the provision of seeds. Their response about provision of seed by FSCs in presented in table 2. It was found that 385 respondents cultivated wheat among which majority (56.62%) of the respondents received wheat seed from FSCs followed by home seed 27.79%, village markets 10.13%, agricultural research 2.86% and fellow farmers 2.6%. A vast majority (49.21%) of sample respondents grow their own sugarcane seed in the form of cuttings called setts. While FSCs provided sugarcane setts to 28.57% of the respondents whereas the rest of the respondents utilized other sources for sugarcane growing i.e. Fellow Farmers (11.11%), Agriculture research station (7.94%) and village market i.e. 3.17%. More than fifty percent of the sample respondents obtained rice seed from FSCs followed by home seed 18.18%, while rest of the sample farmers received rice seed from other sources. Similarly, majority (35.58%) of the sample farmers received maize seed from followed by home seed 34.46% whereas the rest of the sample respondents utilized other sources i.e.: Village Market (27.72%), Agriculture Research (1.50%) and Fellow Farmers (0.75%). Onion was cultivated by 76 respondents among which majority (39.47%) of the respondents obtained onion seed from village market. Almost 26.32% of the respondents received onion seed from fellow farmers. Only 11.84% of the respondents got onion seed from FSCs. Home seed was cultivated by 19.74% of the respondents in the study area. More than fifty percent of the sample farmers purchased tomato seed from FSCs, while the seed of minor crops was not provided by the FSCs to their member farmers in the study area. The data regarding sources of seeds of the FSCs respondents is shown in Table 2.

**Table 2.** Distribution of respondents regarding source of seed

| Crops | Categories | % | Crops | Categories | % |
|---|---|---|---|---|---|
| | Home seed | 27.79 | | Home seed | 21.05 |
| | Fellow farmers | 2.6 | | Fellow farmers | 15.79 |
| **Wheat** | FSC | 56.62 | **Mung** | FSC | - |
| | Agriculture research | 2.86 | | Agriculture research | - |
| | Village market | 10.13 | | Village market | 63.16 |
| | Home seed | 49.21 | | Home seed | 16.67 |
| | Fellow farmers | 11.11 | | Fellow farmers | 10.26 |
| **Sugarcane setts** | FSC | 28.57 | **Tomato** | FSC | 52.56 |
| | Agriculture research | 7.94 | | Agriculture research | - |
| | Village market | 3.17 | | Village market | 20.51 |
| | Home seed | 18.18 | | Home seed | 19.74 |
| | Fellow farmers | 2.27 | | Fellow farmers | 26.32 |
| **Rice** | FSC | 52.27 | **Onion** | FSC | 11.84 |
| | Agriculture research | 2.27 | | Agriculture research | 2.63 |
| | Village market | 25 | | Village market | 39.47 |
| | Home seed | 34.46 | | Home seed | 53.66 |
| | Fellow farmers | 0.75 | | Fellow farmers | 26.83 |
| **Maize** | FSC | 35.58 | **Gram** | FSC | 12.2 |
| | Agriculture research | 1.5 | | Agriculture research | 2.44 |
| | Village market | 27.72 | | Village market | 4.88 |

### *3.5. Yield of various crops before and after fsc registration*

To investigate the differences in yield of various crops before and after registration with FSCs, paired sample T-test was applied. The results presented in Table 3 showed highly significant difference ($P \leq 0.01$) of wheat before and after FSCs registration with mean difference of 4.12 maunds/acre. Similarly highly significant ($P \leq 0.01$) difference was also observed for sugarcane yield with mean difference of 15.22 maunds/acre. In rice 2.9 maunds/acre difference was observed before and after registration with FSCs at probability of 0.000. Similarly highly significant ($P \leq 0.01$) differences were also observed in maize and onion with mean difference of 3.32 and 3.9 maunds/acre. Furthermore, significant ($P \leq 0.05$) differences were also observed in tomato and gram with mean difference of 2.58 and 1.4 mounds/acre respectively. The instant results might be attributed to the fact that FSCs performed much better in provision of technical advices, empowerment in a various activities and inputs of the appropriate type and at the appropriate time of the crop. According to research yield was increased when inputs were made available [18]. In the rest of the crop/vegetables the yield was increased but was non-significant as shown in Table 3.

**Table 3.** Paired sample T-test of yield before and after FSC registration

| Pairs | Yield before and after FSC registration | Mean (add unit of measurement) | | N | P-value |
|---|---|---|---|---|---|
| Pair 1 | Yield of wheat before FSC | 19.31 | 4.12 | 385 | 0.000 |
| | Yield of wheat after FSC | 23.43 | | 385 | |
| Pair 2 | Yield of sugarcane before FSC | 635.67 | 15.22 | 63 | 0.000 |
| | Yield of sugarcane after FSC | 650.89 | | 63 | |
| Pair 3 | Yield of rice before FSC | 22.4 | 2.9 | 44 | 0.000 |
| | Yield of rice after FSC | 25.3 | | 44 | |
| Pair 4 | Yield of gram before FSC | 14.1 | 1.4 | 41 | 0.05 |
| | Yield of gram after FSC | 15.5 | | 41 | |
| Pair 5 | Yield of mung before FSC | 4.44 | 1.2 | 38 | 0.07 |
| | Yield of mung after FSC | 5.64 | | 38 | |
| Pair 6 | Yield of tomato before FSC | 105.13 | 2.58 | 78 | 0.047 |
| | Yield of tomato after FSC | 107.71 | | 78 | |
| Pair 7 | Yield of maize before FSC | 18.46 | 3.32 | 267 | 0.000 |
| | Yield of maize after FSC | 21.78 | | 267 | |
| Pair 8 | Yield of pea before FSC | 38.13 | 27 | 32 | 0.101 |
| | Yield of pea after FSC | 45.13 | | 32 | |
| Pair 9 | Yield of onion before FSC | 108.3 | 3.9 | 76 | 0.000 |
| | Yield of onion after FSC | 112.2 | | 76 | |
| Pair 10 | Yield of *shaftal* before FSC | 23.67 | 5 | 61 | 0.423 |
| | Yield of *shaftal* after FSC | 28.67 | | 61 | |

## 4. Conclusion and recommendations

It was observed from the results that agriculture was the main occupation of the majority respondents for their livelihoods with landholding size of less than 100 acres in the areas. The duration of membership with FSCs of the greater portion (55%) of respondents was less than five years followed by 30% who obtained FSCs memberships from last 6-10 years. Major source of information regarding FSCs were extension agents (65%) about FSCs and climate change showing the competency and proper achievements of them as only 20 % registered farmers knew about FSCs through fellow farmers. This shows that the trend of getting registered with FSCs of improving their know how about climate change with the help of competent extension staff had increased with passage of time for their higher production because of variation in seasons and temperature. It is observed that the yield of various crops like wheat, sugarcane, rice, tomato, maize and onion was significantly increased after FSC registration. Therefore, it should be suggested that extension workers in FSCs also need to focus on competencies on agronomics practices along with orchard and livestock improvement in the areas by focusing on variation in climate. In addition, awareness of the community about how to modernize agriculture and provision of good quality inputs and supervisory services through these FSCs are imperative steps having potential of improvement for overall agriculture production. Adopting the practical suggestions for agricultural production given in this research will definitely increase the prosperity of the whole farming community.

**References**

[1] Ozor N 2009 Understanding Climate Change.Implications for Nigerian Agriculture, Policy and Extension *National Conference on Climate Change and the Nigerian Environment (29th June – 2nd July)* (Nssuka: Department of Geography, University of Nigeria)
[2] IPCC 2007 Climate Change 2007: The Physical Science Basis. Contribution of Working Group I to the Fourth Assessment Report of the Intergovernmental Panel on Climate Change. Summary for Policymakers. *Climate Change 2007: The Physical Science Basis. Contribution of Working Group I to the Fourth Assessment Report of the Intergovernmental Panel on Climate Change* (United Kingdom and New York) p 996
[3] DFID (Department for International Development) 2009 *Impact of Climate Change on Nigeria's Economy*
[4] Harris F 1999 Nutrient management strategies of small-holder farmers in a short-fallow farming system in north-east Nigeria *Geogr. J.* 275–85
[5] UKAID D for I D (R4D) 1998 Relevance of Nigerian farmers' responses to dryland farming systems in Southern Africa and India - Sécheresse *Dep. Geogr. Univ. Cambridge*
[6] Zafarullah Khan M, Khan P, Nawab K and Pakistan P 2018 Farmers' perception of climate change and adaptation in Khyber Pakhtunkhwa Pakistan: A case study of Nowshera district *Gontor Agrotech Sci. J.* **4** 107–24
[7] Wasti E 2011 | Ministry of Finance | Government of Pakistan | *Fianance Minist. Pakistan*
[8] Okunade E O 2007 Effectiveness of extension teaching methods in acquiring knowledge, skill and attitude by women farmers in Osun State *Eur. J. Soc. Sci.* **4** 22–30
[9] Khan G A, Muhammad S, Chaudhry K M and Khan M A 2010 Present status and future preferences of electronic media as agricultural information sources by the farmers *Pakistan J. Agric. Sci.* **47** 166–72
[10] Bank T W 2000 *Decentralizing Agricultural Extension: Lessons and... - Google Scholar*
[11] Feder G, Willett A and Zijp W 2001 Agricultural Extension: Generic Challenges and the Ingredients for Solutions *Knowledge Generation and Technical Change* (Springer US) pp 313–53
[12] Abid S, Shah N A, Hassan A and Farooq A 2014 Growth and Trend in Area , Production and Yield of Major Crops of Khyber Pakhtunkhwa , Pakistan *Asian J. Agric. Rural Dev.* **4** 149–55
[13] McGuire S 2015 FAO, IFAD, and WFP. The state of food insecurity in the world 2015: meeting the 2015 international hunger targets: taking stock of uneven progress. Rome: FAO, 2015
[14] Yaseen M, Xu S, Yu W, Luqman M, Hassan S, Ameen M, Yaseen M, Xu S, Yu W, Luqman M, Hassan S and Ameen M 2016 Factors Inhabiting ICTs usage among Farmers: Comparative Analysis from Pakistan and China *Open J. Soc. Sci.* **4** 287–94
[15] Ahmad M, Akram M, Rauf R, Khan A and Pervez U 2007 Interaction of extension worker with farmers and role of radio and television as sources of information in technology transfer: a case

study of four villages of district Peshawar and Charsadda *Sarhad J. Agric* **23**
[16] Chaudry S M and Kamal S 2010 *Introduction to Statistical Theory Part I* (Lahore: Urdu Bazar)
[17] Pervaiz U 2009 *An Investigation in to the Causes of Slow Diffusion of Tube Well Irrigation Technology in NWFP-Pakistan* (The University of Agriculture, Peshawar, Pakistan)
[18] Švarc J and Dabić M 2019 The Croatian path from socialism to European membership through the lens of technology transfer policies *J. Technol. Transf.* **44** 1476–504

**Acknowledgments**
My Special thanks and sincere appreciations to the Organizers, sponsors and facilitators of the 7th International Conference on Climate Change for their generous help, constant guidance, encouragement, valuable suggestions, advices, useful comments and timely correction throughout the whole process. Our Higher Education Commission of Pakistan is highly acknowledged by sponsoring this project 6903 for collection of this huge data.

# Nitrogen uptake in fully organic, semi-organic and conventional rice fields in Madiun Regency

**Ganjar Herdiansyah*, Suntoro, Muhammad Farid, Mahafid Aziz, Febridita sari Dewi and Adin Slamet Rahayu,**

Study Program of Soil Science, Faculty of Agriculture, Universitas Sebelas Maret
Corresponding author: ganjarh@staff.uns.ac.id

**Abstract**. Nitrogen is one of the essential nutrients for rice plants. However, the inappropriate use of nitrogen fertilizers can pollute the environment caused by the loss of N to groundwater and the atmosphere so that it can increase the nitrate content in the soil and greenhouse gas emissions. Therefore, optimal management of N nutrients is needed so that fertilization is in accordance with the needs of rice plants. This research aims to determine the estimation of nitrogen uptake in full organic, semi organic, and conventional rice fields. This research was conducted in Kab. Madiun, East Java. The research was conducted using qualitative, descriptive, and comparative methods as well as survey methods. Soil sampling used a stratified random sampling approach. Nitrogen uptake analysis was carried out by the Kjeldahl method. Data analysis was carried out by T-test. The results showed that the application of a fully organic system had a higher response than a semi organic system to plant biomass. The increase in plant biomass in full organic reached 20.36%, while semi-organic reached 2.91%. The highest tissue N content was in semi-organic. The application of semi-organic was able to increase the N uptake by 11.67% compared to the conventional system, while the full organic system was able to increase the N uptake by 1.87%.

**Keywords:** nitrogen, organic agriculture system, plant biomass

## 1. Introduction

Nutrients are important factors for plants. One of them is nitrogen. Nitrogen is a macronutrient that is needed by plants for their growth. However, the use of nitrogen fertilizers must be appropriate because improper use of nitrogen fertilizers can pollute the environment caused by the loss of N to groundwater and the atmosphere so that it can increase the nitrate content in the soil and greenhouse gas emissions [1]. Nitrogen nutrient management is needed so that fertilization is in accordance with the needs of plants, especially rice, one alternative is through an organic farming system. According to [2] one of the factors that affect rice production is the cropping system.

Currently, alternative farming systems that use low-input technology are believed to be able to maintain soil fertility and at the same time preserve the environment and increase soil productivity. Organic agriculture is a farming system that prioritizes organic matter as a source of nutrients for plants, namely the use of plant residues, manure, green manure and biological pesticides [3]. The farming system in the community, especially in Madiun Regency, consists of fully organic, semi-organic and conventional agriculture. Fertilization treatment in each agricultural system is different, as well as the absorption of nitrogen by rice plants is also different. This is done to achieve efficiency in fertilization.

To achieve optimal fertilization efficiency, it must be supported by the right nutrient sources. According to [3] high fertilization efficiency needs to be supported by the selection of the right nutrient sources, the right grain size, the right way of giving according to the needs of plant growth. Nitrogen fertilization given to the soil must be appropriate so that the efficiency of its absorption is obtained. Nitrogen is an important nutrient for rice plants. An optimal nitrogen management strategy is needed so as to reduce N loss and increase N uptake. This study aims to determine the estimation of nitrogen uptake in fully organic, semi-organic, and conventional rice plants.

## 2. Materials and method

This research was conducted in Madiun Regency, East Java. This research was conducted from May to August 2021. Chemical analysis was carried out at the Laboratory of Chemistry and Soil Fertility, Faculty of Agriculture, Universitas Sebelas Maret. Nitrogen uptake analysis was carried out using the Kjeldahl method.

The research was conducted using qualitative, descriptive, and comparative methods as well as survey methods. Plant sampling was carried out using a stratified random sampling approach. Plant samples were taken in rice fields with full organic, semi-organic, and conventional farming systems. Data analysis was carried out with the T-test, to compare the results of the organic farming system with conventional farming systems. The plant samples observed were 36 samples.

## 3. Results and discussion

### *3.1. Plant growth*

The addition of organic matter to planting growth indicated by the wet weight and dry weight of the plant (plant biomass) showed an increase from conventional to organic (Figure 1). For plant dry weight parameters, the increase in the full organic application was higher than that of semi organic. The increase in plant biomass in the application of full organic can reach 20.36%, while for semi organic it can increase 2.91%. This shows how important soil organic matter is to participate in providing plant nutrients, either directly or indirectly. Farmers in the Madiun rice production center have started to implement organic farming by using manure as a source of organic matter. The results of [4] research showed that organic treatment gave the highest yield on rice plant growth, compared to the conventional and SRI treatments.

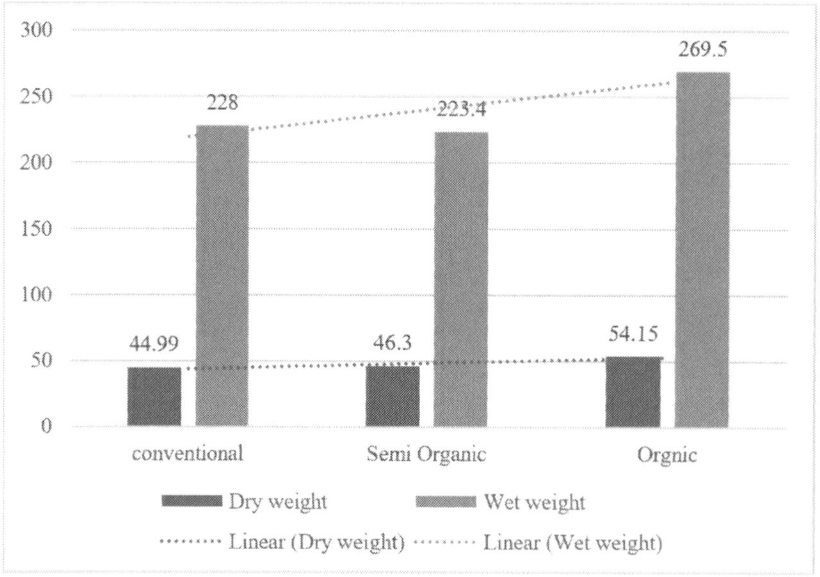


**Figure 1**. Effect of agricultural system on plant growth (plant biomass)

Basically, dry weight is a parameter to see the increase in biomass. Photosynthate accumulation will accumulate in the parts needed by plants. In the vegetative period, photosynthate is accumulated in leaves and stems, while in the generative period, it is accumulated in flowers and seeds. According to [4] organic farming systems are able to produce the heaviest plant dry weight, this is influenced by the photosynthate accumulated in the growth components. The results of the research by Supandji and Junaidi [5] and [6] stated that the use of fertilizers, especially organic fertilizers, had an effect on growth in the generative phase.

The results of the T-test analysis, to compare the growth parameters between organic farming systems and conventional farming systems, showed no significant differences. This proves that in plant growth, the absorbed fertilizer, especially N, comes from organic fertilizers and inorganic fertilizers.

### *3.2. Nitrogen content in plant*

The results showed that the highest levels of N in plant tissue were in Semi Organic, followed by Conventional, and the lowest was in Full Organic (Figure 2). The results of the T-test analysis, organic and conventional farming systems showed results that were not significantly different. This shows that at the initial growth for semi-organic, in addition to providing N from inorganic fertilizers, it is able to supply N directly to plants as well as from residues of organic matter which will also supply N nutrients, and also soil conditions will increase the ability to supply N for plants. For the conventional system, the source of N only comes from inorganic fertilizers that farmers add. As for the full organic system, the nutrient release is slow release so the response is not as fast as inorganic fertilizers.

The research of [3], showed that organic farming systems can provide higher N than semi-organic farming systems and conventional farming systems, because organic fertilizers are a source of N, so that the application of organic fertilizers increases N in plant tissues. The greater the application of organic fertilizers used can increase the N content in the roots, this is due to an improvement in physical properties and the availability of nitrogen sources so that the contact of root hairs in semi-organic farming systems is closer so that more elements are absorbed. According to [7], nitrogen content in plant tissues is influenced by the uptake of nitrate (NO3-) and ammonium (NH4+) ions by plants. This is due to the rapid movement of Nitrogen in the form of NH4+ in the soil solution.

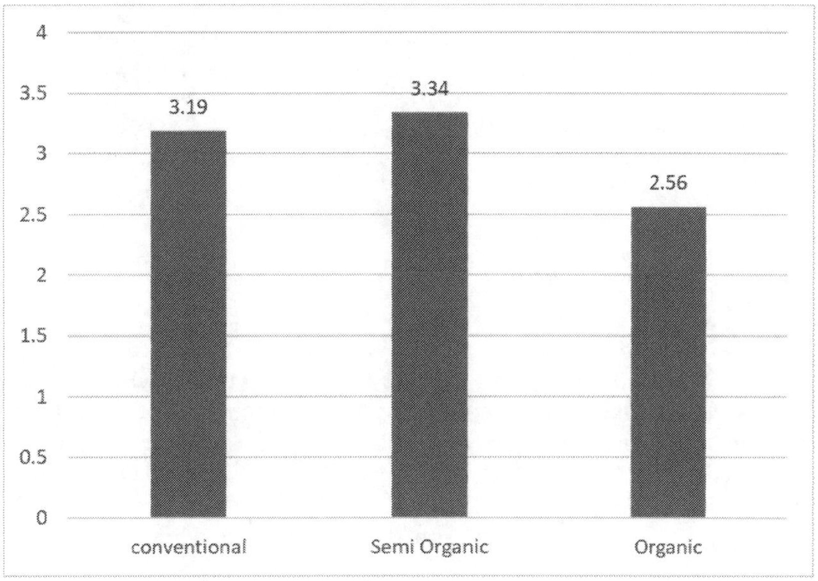


**Figure 2**. Nitrogen content in Plant tissue

### *3.3. Nitrogen uptake*

In line with the response of plants to the N content in plant tissues, the application of a semi-organic system has the highest effect on nutrient uptake, followed by the application of a full organic system, and the lowest is the conventional system. Nutrient N uptake in the semi-organic system had the highest effect so that it was able to increase N uptake by 11.67% compared to the conventional system. Meanwhile, for full organic, it is able to increase N uptake by 1.87% (Figure 3). This shows that the semi-organic system is a system that is able to increase the nutrient content in plants and provide a good response to plant growth so that it is expected to provide good results on crop yields. Nitrogen is a key nutrient for plant growth which is very important for plant metabolic processes, especially in the process of formation of amino acids, protein formation, and nitrogen is the main constituent of chlorophyll formation, so it is very decisive in the photosynthesis process.

The results of the T test showed that the organic farming system was not significantly different from the semi-organic and conventional farming systems. Nutrient uptake is influenced by soil conditions, nutrients in the soil in the form of ions that are absorbed by the roots through the mechanism of cation exchange between ions in the soil solution and roots, diffusion and mass flow. Research by [7] and [8] showed that the application of organic fertilizers was able to increase nitrogen uptake by cultivated plants.

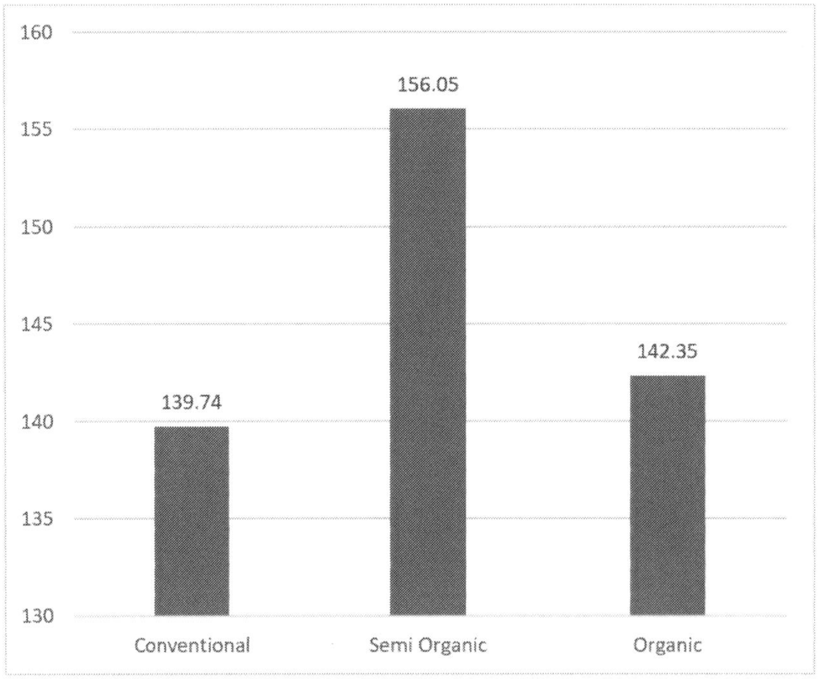


**Figure 3**. Nitrogen Uptake

Management of nitrogen fertilizers must be optimal so that fertilization according to plant needs can be met and reduce N loss and increase N uptake by plants. Improper use of N fertilizers can cause excess N fertilizers so that efficiency decreases and endangers plants and the environment. Environmental damage due to excessive N fertilization is caused by the emission of $N_2O$ gas in the ammonification, nitrification and denitrification processes. According to [9], $N_2O$ gas emissions are influenced by the dose of N fertilizer given, the higher the N dose, the greater the $N_2O$ gas emission.

## 4. Conclusions

The application of organic farming systems has a higher response to plant biomass than semi-organic. The increase in plant biomass in full organic application can reach 20.36%, while for semi-organic it can increase 2.91%. Nitrogen nutrient uptake in semi-organic has the highest effect so that it can increase N uptake by 11.67% compared to conventional systems. While for the full organic can increase 1.87%. This is in line with the highest nitrogen content of plant tissue in semi-organic. Efficient nitrogen fertilization is urgently needed to reduce the impact of climate change due to increased N2O gas emissions from the Nitrogen fertilization process.

## References

[1] Huang S, Miao Y, Zhao G, Yuan F, Ma X, Tan C, Yu W, Gnyp M L, Lenz-Wiedemann V I S, Rascher U and Bareth G 2015 Satellite remote sensing-based in-season diagnosis of rice nitrogen status in Northeast China *Remote Sens.* **7**

[2] Mergono Adi Ningrat, Carolina Diana Mual and Yohanis Yan Makabori 2021 Pertumbuhan dan Hasil Tanaman Padi (Oryza sativa L.) pada Berbagai Sistem Tanam di Kampung Desay, Distrik Prafi, Kabupaten Manokwari *Pros. Semin. Nas. Pembang. dan Pendidik. Vokasi Pertan.* **2**

[3] Nuryani H.U. S, Haji M and Widya Y. N 2010 Serapan Hara N, P, K Pada Tanaman Padi Dengan Berbagai Lama Penggunaan Pupuk Organik Pada Vertisol Sragen *Ilmu Tanah dan Lingkung.* **10**

[4] Lita T N, Soekartomo S and Guritno B 2013 Pengaruh perbedaan sistem tanam terhadap pertumbuhan dan hasil tanaman padi ( Oryza sativa L .) di lahan sawah. *Produkdi Tanam.* **1**

[5] Supandji S and Junaidi J 2020 Pengaruh pupuk urea dan pupuk organik sapi terhadap pertumbuhan dan produksi tanaman padi varietas IR. 64 (Oryza sativa L) *J. Agrinika J. Agroteknologi dan Agribisnis* **3**

[6] Veronica N T, Setiawan A and Tyasmoro S Y 2019 Respon varietas lokal dan varietas unggul nasional terhadap kombinasi pupuk organik dan anorganik pada pertumbuhan dan hasil tanaman padi (Oryza sativa L.) *J. Produksi Tanam.* **7**

[7] Yuniarti A, Damayani M and Nur D M 2019 Efek pupuk organik dan pupuk n,p,k terhadap c-organik, n-total, c/n, serapan n, serta hasil padi hitam pada inceptisols *J. Pertan. Presisi* **3** 58–66

[8] Kubát J, Klír J and Pova D 2003 The dry matter yields, nitrogen uptake, and the efficacy of nitrogen fertilisation in long-term field experiments in Prague *Plant, Soil Environ.* **49**

[9] Balasubramanian V, Morales A C, Cruz R T and Abdulrachman S 1998 On-farm adaptation of knowledge-intensive nitrogen management technologies for rice systems *Nutr. Cycl. Agroecosystems* **53**

# The effect of pottery material composition as subsurface irrigation on the growth of chilli (*Capsicum annum L.*) in Alfisols and Entisol

**Rahayu[1]*, H Aktavia[1], Rivaldi A S[2] and Komariah[1]**

[1]Department of Soil Science, Faculty of Agriculture, Sebelas Maret University, Ir. Sutami Street 36A, 57126, Jebres, Surakarta, Central Java, Indonesia
[2]Graduate School of Soil Science, Faculty of Agriculture Sebelas Maret University, Ir. Sutami Street 36A, 57126, Jebres, Surakarta, Central Java, Indonesia
*Corresponding author: rahayu_uns@yahoo.co.id

**Abstract.** Climate change requires the agricultural sector to be more precise in water management. Rainfed lands and irrigated lands with no year-round water supply require an efficient way of using water in the dry season. One of the efforts to utilize irrigation efficiently is to develop subsurface seepage irrigation techniques using pottery. This study aims to determine the composition of the pottery material, the ratio of sand, silt and clay which is optimal for the growth of chilli plants. The experiment was carried out with plots arranged in a nested design with a treatment of five pottery materials with different compositions of clay, soil and sand, and tested on entisol and alfisol soils. The results showed that the pottery irrigation with a combination of pottery material composition on two types of soil affected the weight of chilli fruit. Pottery with a composition of 70% clay and 30% sand is the most effective pottery for chilli and yields on surface growth irrigation.

**Keywords** : hydraulic conductivity, water use

## 1. Introduction

Climate change requires the agricultural sector to be more precise and efficient in water management. One of the efforts to use irrigation efficiently is to develop a subsurface irrigation system. One of the subsurface irrigation systems is using pottery, which utilizes capillary irrigation water seepage from inside the pottery out to the soil surface and root zone. According to Pariante [1], pottery irrigation is classified as water-efficient because relatively little water evaporates from the pottery. After all, it is protected by the pottery body [2]. Imanudin reported that subsurface irrigation has 100% irrigation efficiency, i.e. water is used fully to increase capillary water that wets the entire soil surface around plant roots [2].

The amount of seepage on the pottery wall is a very important measure of an earthenware irrigation system because it will determine the ability of the system to meet crop water needs. The ratio of water diffusion through the pottery is influenced by the proportion of the material mixed with clay during the pottery making process [3,4].

Chilli plant is economical plants and is generally cultivated in the dry season because they are relatively resistant to disease and the productivity of chilli plants increase when coupled with adequate irrigation. Nurfalach reported that the chilli plant did not provide optimal results in conditions of continuous heavy rain which resulted in leaf damage, flower loss and root disease [5]. Although planting in the dry season is more productive, chilli plants are sensitive to water shortages due to their shallow root system [6]. Water supply from rainfall and irrigation that is less efficient does not necessarily meet the water need of plants. Lack of water can reduce plant productivity so lack of water requires irrigation innovation by using water as efficiently as possible. This research is an experimental plot for the growth of chilli plants by using irrigation pots with different compositions of pottery materials. The subsurface irrigation system using pottery is expected to be an effective and efficient irrigation system in helping to optimize the use of dry land seasonal crops.

The purpose of this study was to determine the effect of pottery material composition in providing seepage irrigation water. To provide recommendations for appropriate pottery material composition for subsurface irrigation suitable for chilli cultivation.

## 2. Material and methods

This research was conducted during the dry season from June to November 2019, at the Experimental Field of the Faculty of Agriculture, Sebelas Maret University, located in Karanganyar Regency, Central Java, Indonesia. This research is an experimental plot using a variety of pottery materials mixed with sand and clay as the independent variable and chilli (*Capsicum annum L.*) for the dependent variable with Alfisols and Entisols soil planting media. The experiment was arranged in a nested design with two treatment factors, namely the treatment of soil types which included Alfisols (T1) and Entisols (T2), as well as the treatment of five different pottery material compositions (K1, K2, K3, K4 and K5), to obtain 10 different treatments with 3 replications.

The pottery used is pottery from the centre of the pottery silt tries in Bayat District, Klaten Regency with different material compositions. All pottery used is tubular and has the same size, which is 20 cm in diameter, 25 cm in height and has a pottery mouth with a diameter of 10 cm. Variations in the composition of pottery materials are K1: 70% clay -15% silt soil -15% sand; K2: 70% clay - 10% silt soil and 20% sand; K3: 70% clay and 30% sand; K4: 80% Clay and 20% Sand; K5: 80% clay, 10% silt soil and 10% sand. The bottom of the pottery is sealed by providing waterproof so that water does not seep down but to the side

This experiment was started by planting pottery in soil that had been given manure until it reached the neck of the pottery. Each bed on the ground is placed 5 units of pottery with a distance of 50 cm each. Each pot was planted with 4 chilli plants around the pottery with the distance of each plant from the pottery being 15 cm. Watering chilli plants are done manually before going through the seepage of pottery for 7 days after planting so that the plants can adapt first to the conditions in the land. After 7 days, irrigation was carried out through the seepage of the pottery by filling the water into the pottery every 3 days with a volume of up to 7 litres. Every 10 days after planting, additional fertilization was carried out using NPK 16:16:16 fertilizer at a dose of 100 kg/ha. Fertilization is done by immersing fertilizer around the pottery, which is between the pottery and the plants.

Plant height measurements were carried out every week. Crop yields are harvested periodically based on the maturity of the chillies and are calculated accumulatively. Analysis of the amount of water seepage is carried out every day by measuring the level of water loss in the pottery. Soil moisture was measured every day in the field using a ground moisture meter. The COLE (Coefficient of Linear Extensibility) value of post-harvest soil was observed using the oven method. The value of hydraulic conductivity is obtained by means of analysis using the constant head method. Calculation of hydraulic conductivity using the constant head method [7,8]. Soil samples were also taken and measured soil pH with a pH meter, N-total. T-test analysis was conducted to determine the difference in the results of the two different plots. To find out the relationship between the observed variables, a correlation test was conducted.

## 3. Results and discussion

From table 1. it can be seen that the highest plant was found in K5 (T2) as high as 49.83 cm, and the lowest was in treatment K3 (T2) as high as 37.66 cm. The highest plant was 221 grams and the weight was 152.6 gr in the K2(T1) the lowest fresh weight was 113.3 grams and plant dry weight was 79.33 g in K1(T2). The highest chilli weight was 221.3 g K3(T1), and the lowest chilli weight was 49 g in K1(T2). The-T test showed that there was a significant difference in plant fresh and dry weight between Alfisols and on Entisols (significance < 0.05). The composition of the pottery material had a significant effect on the weight of the chilli produced in the K1(T2) and K3(T1). The T-test also showed a significant difference in weight of chilli between Alfisols and Entisols. There was a positive correlation between plant height, chilli weight, plant fresh and dry weight.

**Table 1.** Mean growth and yield of chilli plants after the treatment is applied

| Treatment | Plant Height (cm) | Weight (gram) | Weight (gram) | Weight of Chillies (gram) |
|---|---|---|---|---|
| K1(T1) | 43.66a | 192.6a | 139.8a | 135.3ab |
| K2(T1) | 45.40a | 221.0a | 152.6a | 111.0ab |
| K3(T1) | 39.13a | 173.3a | 119.3a | 221.3b |
| K4(T1) | 38.70a | 194.3a | 136.0a | 136.6ab |
| K5(T1) | 40.20a | 198.0a | 135.1a | 167ab |
| K1(T2) | 40.66a | 113.3a | 79.33a | 49.00a |
| K2(T2) | 48.90a | 178.0a | 124.6a | 113.3ab |
| K3(T2) | 37.66a | 154.0a | 107.8a | 100.3ab |
| K4(T2) | 40.66a | 121.6a | 85.16a | 59.00a |
| K5(T2) | 49.83a | 133.0a | 93.10a | 123.0ab |

Remarks: K1; Composition of 70% clay, 15% soil and 15% sand, K2; Composition of clay 70%, soil 10% and sand 20%, K3; Composition of clay 70% and sand 30%, K4; Composition of 80% clay and 20% sand, K5; Composition of 80% clay, 10% soil and 10% sand, T1; Alfisols, T2; Entisols. The mean in column followed by the same letters is not significant base on the DMRT level of 5%.

**Table 2.** Correlation analysis

| Variable | Weight of Chillies | Plant Fresh Weight | Plant Dry Weight | Plant Height | C-Organic |
|---|---|---|---|---|---|
| Plant height | .417 | .611 | .620 | 1 | - |
| Soil Moisture (August) | - | - | - | .363 | - |
| Soil Moisture (September) | .410 | .413 | .43 | - | - |
| Soil pH | - | - | - | - | -.408 |

Remarks : The data above is only data from variables that correlate with other variables

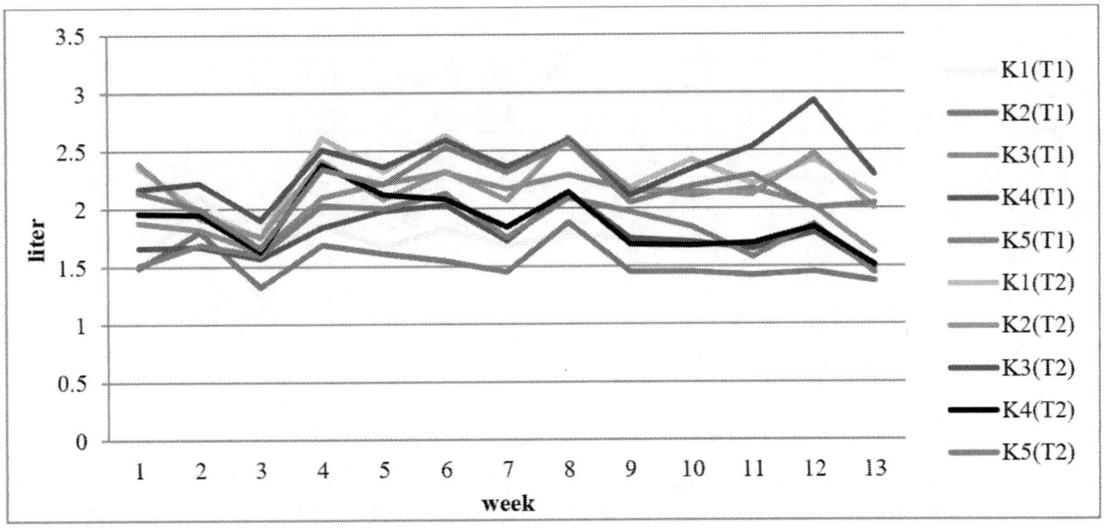


**Figure 1.** Water use dynamic

Remarks : K1; Composition of 70% clay, 15% soil and 15% sand, K2; Composition of clay 70%, soil 10% and sand 20%, K3; Composition of clay 70% and sand 30%, K4; Composition of 80% clay and 20% sand, K5; Composition of 80% clay, 10% soil and 10% sand, T1; Alfisols, T2; Entisols.

Figure 1. showed water seepage at week 3 decreased from week 2, week 4 increased from week 3 in all treatments. Water seepage at weeks 7 to 9 fluctuated in all. Water seepage at week 5, week 6 and weeks 10 to 13 showed different fluctuations in each treatment. Based on Table 3, the highest water seepage in pottery was in August with 2.20 litres per day, in September with 2.44 litres per day and in October with 2.49 litres per day obtained from the K3(T2). The lowest water seepage in pottery was in August by 1.57 litres per day, in September by1.60 litres per day and in October at 1.42 litres per day obtained from K2(T1) treatment.

**Table 3.** Mean water use after the treatment is applied

| Treatment | Water Use (litres per day) | | | |
|---|---|---|---|---|
| | August | September | October | Total |
| K1(T1) | 1.74a | 1.75a | 1.66ab | 1.72ab |
| K2(T1) | 1.57a | 1.60a | 1.42a | 1.53a |
| K3(T1) | 1.87a | 2.25a | 2.01ab | 2.04ab |
| K4(T1) | 1.72a | 1.92a | 1.66ab | 1.76ab |
| K5(T1) | 1.72a | 2.01a | 1.69ab | 1.80ab |
| K1(T2) | 2.17a | 2.42a | 2.33ab | 2.31ab |
| K2(T2) | 2.11a | 2.27a | 2.18ab | 2.19ab |
| K3(T2) | 2.20a | 2.44a | 2.49b | 2.38b |
| K4(T2) | 2.00a | 1.97a | 1.69ab | 1.88ab |
| K5(T2) | 2.03a | 2.38a | 2.26ab | 2.22ab |

Remarks : K1; Composition of 70% clay, 15% soil and 15% sand, K2; Composition of clay 70%, soil 10% and sand 20%, K3; Composition of clay 70% and sand 30%, K4; Composition of 80% clay and 20% sand, K5; Composition of 80% clay, 10% soil and 10% sand, T1; Alfisols, T2; Entisols. The mean in column followed by the same letters is not significant base on the DMRT level of 5%.

**Table 4.** Hydraulic conductivity in various pottery material compositions

| Potery | Hydraulic Conductivity of Pottery (cm/s) |
|---|---|
| K1 | $8.70263 \times 10^{-9}$ |
| K2 | $7.74746 \times 10^{-9}$ |
| K3 | $10.09314 \times 10^{-9}$ |
| K4 | $7.21681 \times 10^{-9}$ |
| K5 | $10.41683 \times 10^{-9}$ |

Remarks : K1; Composition of 70% clay, 15% soil and 15% sand, K2; Composition of clay 70%, soil 10% and sand 20%, K3; Composition of clay 70% and sand 30%, K4; Composition of 80% clay and 20% sand, K5; Composition of 80% clay, 10% soil and 10% sand, T1; Alfisols, T2; Entisols.

The highest soil water content was in August, which was 19% in the K3(T2) (Table 5). While the lowest soil water content was in September, by 11.6% in the K1(T2) and K5(T2). The highest soil porosity value was 5.71% in K4(T2), while the lowest soil porosity was 2.20% in K1(T2). The treatment of pottery material composition had a significant effect in September, at K1(T1) and K1(T2), and in October at K4(T1) and K5(T2). The results of the t-test showed significant differences in soil porosity between Alfisols and Entisols.

Based on Table 2, soil moisture in August was positively correlated with plant height. The table also showed there was a positive correlation between moisture to chilli weight, plant fresh and dry weight. They indicate that the higher soil moisture resulted in, higher plant height, fresh plant weight and dry weight, and also plant fresh and dry weight.

**Table 5.** Mean soil moisture and soil porosity after the treatment is applied

| Treatment | Soil Moisture (%) | | | Soil Porosity (%) |
|---|---|---|---|---|
| | August | September | October | |
| K1(T1) | 16.4a | 15.6c | 16.8b | 4.80a |
| K2(T1) | 15.8a | 14.3bc | 16.8b | 3.43a |
| K3(T1) | 16.3a | 15.6c | 16.2ab | 3.16a |
| K4(T1) | 15.9a | 14.7bc | 18.0b | 4.52a |
| K5(T1) | 16.1a | 14.8bc | 15.9ab | 3.71a |
| K1(T2) | 18.4a | 11.6a | 14.5ab | 2.20a |
| K2(T2) | 17.6a | 12.1a | 16.0ab | 3.89a |
| K3(T2) | 19.0a | 12.9ab | 16.1ab | 4.83a |
| K4(T2) | 18.8a | 11.8a | 13.7ab | 5.71a |
| K5(T2) | 18.8a | 11.6a | 12.2a | 5.48a |

Remarks : K1; Composition of 70% clay, 15% soil and 15% sand, K2; Composition of clay 70%, soil 10% and sand 20%, K3; Composition of clay 70% and sand 30%, K4; Composition of 80% clay and 20% sand, K5; Composition of 80% clay, 10% soil and 10% sand, T1; Alfisols, T2; Entisols.

The highest soil pH value was 6.36 which was found in the K5(T1) and the lowest soil pH value of 6.02 was found in K2(T1) (Table 6). Treatments K3(T1), K4(T1), K5(T1), K1(T2), K2(T2) and K3(T2) had the same N-Total levels of 0.01%, where this value was lower than the levels of N-Total soil in treatments K1(T1), K2(T1), K4(T2) and K5(T2). The highest soil C-organic content value of 3.53% was found in K2(T1), while the lowest soil C-Organic content value of 1.30% was found in K3(T1). The treatment of pottery material composition significantly affected the C-organic soil as happened in the K2(T1) and K3(T1).

Based on Table 2. the results of the correlation analysis showed that soil pH was negatively correlated with soil organic C with a significance value of 0.025 and a Pearson Correlation value of -0.408. This means that lower soil pH may be caused by higher C-organic soil.

**Table 6.** Mean pH, N-total and C-Organic of the soil after the treatment is applied

| Treatment | Soil pH | Soil N-Total (%) | Soil C-Organic (%) |
|---|---|---|---|
| K1(T1) | 6.16a | 0.02a | 1.71ab |
| K2(T1) | 6.02a | 0.02a | 3.53b |
| K3(T1) | 6.07a | 0.01a | 1.30a |
| K4(T1) | 6.20a | 0.01a | 2.08ab |
| K5(T1) | 6.36a | 0.01a | 1.54ab |
| K1(T2) | 6.26a | 0.01a | 1.34a |
| K2(T2) | 6.26a | 0.01a | 1.60ab |
| K3(T2) | 6.09a | 0.01a | 1.78ab |
| K4(T2) | 6.21a | 0.02a | 1.52ab |
| K5(T2) | 6.21a | 0.02a | 1.77ab |

Remarks : K1; Composition of 70% clay, 15% soil and 15% sand, K2; Composition of clay 70%, soil 10% and sand 20%, K3; Composition of clay 70% and sand 30%, K4; Composition of 80% clay and 20% sand, K5; Composition of 80% clay, 10% soil and 10% sand, T1; Alfisols, T2; Entisols.

The combination of pottery material composition and soil type had a significant effect on chilli yields. The highest chilli yields in this study were obtained from pottery with the highest use of water. The highest yield of chilli is obtained in full irrigation without a water deficit, and a water deficit will reduce yields by 5 - 20% [9]. The results of this study showed a positive correlation between chilli yields and plant height, plant fresh and dry weight, and this is in line with the results of Ganefianti [10] in Rofidah [11]which showed that tall chilli plants with many primary branches will produce fruit which is a lot per plant.

The difference in the use of water from pottery in this study was caused by the composition of different materials. This causes the hydraulic conductivity of each pottery used to be also different (Table 4). The conductivity value has an important role in the velocity of the pottery in passing air. The greater conductivity resulted in faster passing of water because it has larger pores [12,13]. This is supported by the statement of Setiawan in Agustina [7], that the amount of conductivity produced is caused by the suction of water by the soil matrix around the pottery walls. The dry condition of the soil around the pottery pulls out of the water through the pottery wall to fill the empty soil pores. Water seeps little by little through the pores of the walls of the porous media into the root zone due to hydrostatic pressure and suction of the soil matrix and the hydraulic conductivity of the pottery. In this study, the combination of pottery material composition and soil type affects the weight of chilli fruit produced by chilli plants in the subsurface irrigation system. Pottery with a composition of 70% clay and 30% sand is the most effective pottery for the growth and yield of chilli plants in the subsurface irrigation

Soil moisture content in August and September correlated with chilli plant growth, while in October there was no correlation. This can happen because the vegetative phase of chilli plants ends in early September and continues to the generative phase. Romadona [14] reported that if planting is done in the nursery first, the vegetative phase usually ends when the plant is 30-40 days after planting the generative phase is the productive period of chilli plants

The correlation between soil pH and organic C is caused by differences in water use. According to Surya [15], the application of organic fertilizer means that there is an input of organic matter. Organic matter affects the size of the soil absorption capacity of the water. The more water in the soil caused, the more soil reactions that release $H^+$ ions, so the soil is more acidic.

## 4. Conclusion

The combination of pottery material composition and soil type had a significant effect on chilli yields. The highest chilli yield in this study was obtained from pottery with the highest use of water. The higher soil moisture resulted in higher plant high, weight of chilli, plant fresh and dry weight. Pottery with a composition of 70% clay and 30% sand is the most effective pottery for the growth and yield of chilli plants in subsurface irrigation.

## References

[1] Pariante I W 2019 *Pengaruh Diameter Kendi Terhadap Daya Resapan Air Sebagai Irigasi Tanaman Kopi Di Lahan Kering* (Mataram)
[2] Imanudin M S and Prayitno 2015 Pengembangan Irigasi Bawah Tanah untuk Irigasi Mikro Melalui Metoda Kapilaritas Tanah *Prosiding Seminar Nasional Pengembangan Teknologi Pertanian* 376–81
[3] Mathai M P and Simon A 2006 Water Diffusion Through Pottery Discs of Varying Porosity *Journal of Tropical Agriculture* **42** 63–5
[4] Agung R B H 2014 *Analisis Pengaruh Pengurangan Bobot Kendi Dengan Cara Pengamplasan Terhadap Konduktivitas Hidraulik Dan Laju Rembesan Air Di Permukaan Tanah Ultisol* (Medan)
[5] Nurfalach D R 2010 *Di UPTD Perbibitan Tanaman Hortikultura Desa Pakopen Kecamatan Bandungan Kabupaten Semarang* (Surakarta: UNS (Sebelas Maret University))
[6] González-Dugo V, Orgaz F and Fereres E 2007 Responses of Pepper to Deficit Irrigation for Paprika Production *Scientia Horticulturae* **114** 77–82
[7] Agustina H, Saleh E, Desi S M K and Anggraini N 2017 Rembesan Air pada Bahan Kendi Aplikasi Irigasi Kendi dengan Berbagai Kondisi Media Tanam *Prosiding Seminar Nasional Lahan Suboptimal 2017* (Palembang: Universitas Lampung) pp 395–404
[8] Abu-Zreig M, Khdair A and Alazba A 2015 Factors Affecting Water seepage Rate of Clay Pitchers in Arid lands *University of Sharjah Journal of Pure & Applied Sciences* **6** 59–80
[9] Yang H, Liu H, Zheng J and Huang Q 2017 Effects of Regulated Deficit Irrigation on Yield and Water Productivity of Chili Pepper (Capsicum annuum L.) in The Arid Environment of Northwest China *Irrigation Science* **36** 61–74
[10] Ganefianti D W, Tunjung P, Alnopri and Hasanudin 2006 *Uji Daya Hasil Pendahuluan Galur-Galur Cabai Hasil Persilangan Talang Semut/ Tit Super* (Bengkulu)
[11] Rofidah N I, Yulianah I and Respatijarti 2018 Korelasi Antara Komponen Hasil dengan Hasil pada Populasi F6 Tanaman Cabai Merah Besar (Capsicum annuum L.). *Jurnal Produksi Tanaman* **6** 230–5
[12] Reskiana, Setiawan B I, Saptomo S K and Mustatiningsih P R D 2015 Uji Kinerja Emiter Cincin *Jurnal Irigasi* **9** 63–74
[13] Ariandi L M, Abdullah S H and Putra G M D 2018 Analisis Komposisi Serbuk Gergaji Terhadap Konduktivitas Hidrolik Pipa Mortari Irigasi Tetes Bawah Permukaan Tanah (Analysis of Sawdust Ratio on Hydraulic Conductivity in Subsurface Mortari Pipe of Drip Irrigation ) *Jurnal Ilmiah Rekayasa Pertanian dan Biosistem* **6** 39–52
[14] Romadona Z 2019 *Analisis Terhadap Nilai Kadar Air dan Pertumbuhan Vegetatif Tanaman Cabai pada Kompos Blok Berbagai Limbah Organik* (Jember: Fakultas Teknologi Pertanian)
[15] Romadona Z 2017 *Kajian Porositas Tanah Pada Pemberian Beberapa Jenis Bahan Organik di Perkebunan Kopi Robusta* (Jember: Fakultas Teknologi Pertanian)

# Growth response of *Echinacea purpurea* (L) Moench to biochar types and hormone doses

**Oentari Prilaningrum Sutanto[1], Bambang Puji Asmanto[2,3] and Ahmad Yunus[2,3*]**,

[1]Department of Agronomy, Graduated School, Universitas Sebelas Maret Surakarta, Indonesia
[2]Lecturer staff of the Agrotechnology Study Program, Faculty of Agriculture, Universitas Sebelas Maret Surakarta, Indonesia
[3]Center of Biotechnology and Biodiversity, Research and Development, Universitas Sebelas Maret Surakarta, Indonesia
*Corresponding author: yunus.uns7@yahoo.com

**Abstract.** Purple Cone Flower (*Echinacea purpurea*) is one of the introduced medicinal plants originating from America which has its native habitat growing in subtropical countries with optimal growth conditions of low temperature and full sun exposure. Problems that occur in climate change between subtropics and tropics, especially in Indonesia will have an impact on the less than optimal growth and development of plants so that it will reduce the production and quality of ekinase yields. One way to improve ekinase can suitable in Indonesia (especially in the lowlands) is by adding biochar as a soil enhancer and hormones. The aims to study the interaction between the addition of biochar and the level of hormone dose on the growth rate of ekinase. The research method used a Split-Plot Completely Randomized Block Design. The results show the interaction between the type of biochar and hormone dose. The best results were shown in the treatment of giving husk charcoal and 6 ml/L hormones of 27.67 cm for plant height and 89.67 $cm^2$ for leaf area. The formation of the highest number of leaves was 26 in the treatment of husk charcoal and hormone 2 ml/L. The best results were wood charcoal and hormone treatment 6 ml/L of 0.1467 for the rate of photosynthesis and 0.02937 for the rate of transpiration. Treatment with a dose of 6 ml/L hormone can optimize the growth rate of ekinase.

**Keywords:** *Echinacea purpurea*, Plant growth , Soil repairer, Hormone

## 1. Introduction

Purple Cone Flower Plant (*Echinacea purpurea* (L.) Moench) is one of the introduced medicinal plants originating from North America. Parts that can be used as raw materials for immune-enhancing drugs are extracts from all parts of *E. purpurea* such as leaves, stems, roots, and flowers [1]. Ekinase has benefits for humans as immunoregulation, anti-inflammatory, and antioxidant and has no side effects [2]. This species contains secondary metabolites including phenols, caffeic acid derivatives, alkamides, flavonoids, and polysaccharides and glycoproteins [3].

*Echinacea purpurea* has a native habitat growing in subtropical countries with optimal growth conditions of low temperatures and full sun exposure. The problem faced is that the climate difference between the subtropics and the tropics, especially in Indonesia, will affect the growth and yield of ekinase. This climate change will affect three elements of the microclimate that are closely related to agriculture, namely rising air temperature, humidity, and atmospheric dynamics. Climate change triggers environmental changes that cause changes in plant responses [4]. Air temperature caused by differences in subtropical and tropical climates is one of the climatic factors that greatly affects the phase of plant growth. Phenology is strongly influenced by temperature so that it can be used as a sensitive and accurate indicator in studying the effects of climate change, especially the increase in air temperature, on plant growth such as shifts in flowering time and so on [5].

The results of previous research by Rahardjo [6] showed that ekinase can adapt in the tropics in which the altitude range is 400-1200 meters above sea level (masl). *Echinacea purpurea* grows well in loose moor soil, good aeration and drainage, with a soil pH range between 5.5 – 7.5, and a high level of organic matter [7]. Seeing the various benefits and opportunities of this ekinase, it has the potential to be developed in Indonesia by adding biochar as a soil enhancer and hormone. Based on the research results of Gholami et al. [8] the addition of high doses of rice husk biochar can increase the portion of carbon in plant growth media to support plant vegetative growth. Wahyudi [9] explained the advantages and benefits of hormones, including: stimulating the acceleration of root release, extending and multiplying root fibers and shoots, stimulating the growth process, increasing the size of cells and tissues, stimulating the enlargement of rhizomes and tubers in root crops. Giving hormones to ekinase can stimulate the formation of phytohormones that are produced by plants so that they can encourage biochemical activity, both primary and secondary metabolites [10]. This study aims to study the adaptation of ekinase cultivation by adding the type of biochar and the level of hormone dose to the response to its growth rate.

## 2. Materials and method

### *2.1 Research time and place*

The research was carried out in August-September 2021 at the Experimental Garden Field of the Faculty of Agriculture, Universitas Sebelas Maret (UNS), Sukosari Village, Jumantono District, Karanganyar Regency. The type of soil in the experimental garden includes alfisols. Alfisol soil classification is: (a) molic epipedon because it has a soil solum 18 cm (23 cm), C-organic content 0.6% (1.29%), base saturation 50% (55.72%) (b) the lower characteristic horizon found is argillic because it contains < 15% clay, then the argillic horizon is > 3% more clay (c) The appropriate management for alfisol soils is conventional tillage [11].

### *2.2 Research materials and tools*

The tools used are 65% shading net, bamboo poles supporting shading net, ruler, millimeter block, and plant photosynthesis meter. The materials used are ekinase seeds (sourced from Medical Plant and Traditional Medicines Research and Development Center B2P2TOOT Tawangmangu, Indonesia), biochar (husk charcoal and wood charcoal), hormones (with the trademark "Hormax") and basic organic fertilizers (leaf compost fertilizers).

### *2.3 Research design*

This research was conducted using a Completely Randomized Block Design (RAKL) Splitplot which was arranged in a factorial manner with two factors. The first factor is the type of biochar (B) with 3 levels (B1: Control, B2: Husk charcoal, B3: Wood charcoal). The second factor is hormone dose (H) with 4 levels (H1: hormone dose 0 ml/L, H2: hormone dose 2 ml/L, H3: hormone dose 4 ml/L, H4: hormone dose 6 ml/L). Each treatment combination was repeated 3 times to obtain 32 experimental units, where each block plot consisted of 24 plants. The research data were analyzed by Analysis of Variance (ANOVA) using SPSS (Statistical Product and Service Solution) version 24.0. If there are significant differences, then proceed with the Dunca's Multiple Range Test (DMRT) level of 5%. Data presentation is presented descriptively and presented in tabular form.

## 3. Result and discussion

### *3.1 Plant height*

Height is a parameter in the process of plant growth. Plants can provide positive or negative growth responses to changes in the microclimate [12]. Table 1 explains that the control treatment (without biochar or hormone dose 0 ml/L) had higher plant height compared to other treatments. The provision of biochar has an effect on the plant height of ekinase where the highest plant height is in the treatment of husk charcoal biochar. A soil enhancer made from organic biochar which is zeolite, a containing active carbon and humic compounds, which are effective in increasing land productivity and can improve several basic components of soil chemistry, available P, pH, and total N [13].

**Table 1.** Effect of biochar type and hormone dosage on plant height of *Echinacea purpurea*

| Plant Height (cm) | | | | |
|---|---|---|---|---|
| | Hormon Dosage (ml/L) | | | |
| Biochar Type | H1 : 0 ml/L | H2 : 2 ml/L | H3 : 4 ml/L | H4 : 6 ml/L |
| No Biochar (Control) | 18.50abc | 22.50abc | 23.00abc | 26,33cd |
| Rice Husk Charcoal | 15.67a | 24.67bcd | 24,67abc | 27.67d |
| Wood Charcoal | 17.50ab | 24.33bcd | 22.00abcd | 26.00bcd |

Note: Numbers followed by different letters in the same column indicate a significant difference in DMRT at the 5% level.

The addition of hormones (with the trademark "Hormax") had an effect on the height of ekinase plants. The 6 ml/L concentration treatment gave the highest value compared to other treatments. This is due to the concentration of hormax 6-9 ml/l of air is the right concentration, so that the physiological processes of plants run well. Indole Acetate Acid (IAA) contained in homax is an auxin that plays a role in cell elongation by influencing cell wall flexibility, so that it can stimulate plant height growth. The gibberellins contained in Hormax were positively correlated with auxin in the elongation of plant cells. Pelia [14] suggests that the IAA hormone functions to play and differentiate cells, stimulate the synthesis of RNA, protein and enzymes. Auxin plays a very important role in tissue formation in the vegetative plant growth phase. Gibberellins play a role in cell elongation and division so that they have a high influence on plants [9].

### *3.2. Number of leaves*

Leaves are an important component for plant growth because they are related to physiological processes, namely photosynthesis, respiration and transpiration. Adaptability in plants is characterized by changes in plant structure such as the number of leaves_and leaf area [15].

**Table 2.** Effect of biochar type and hormone dosage on number of leaves *Echinacea purpurea*

| Number of Leaves | | | | |
|---|---|---|---|---|
| | Hormon Dosage (ml/L) | | | |
| Biochar Type | H1 : 0 ml/L | H2 : 2 ml/L | H3 : 4 ml/L | H4 : 6 ml/L |
| No Biochar (Control) | 13.00ab | 14,67abc | 19.00abc | 23.00abc |
| Rice Husk Charcoal | 10.67a | 26.00c | 23.00abc | 24.33bc |
| Wood Charcoal | 13.00ab | 20.67abc | 21.33abc | 23.67bc |

Note: Numbers followed by different letters in the same column indicate a significant difference in DMRT at the 5% level.

Table 2 explains that the control treatment (without biochar or hormone dose 0 ml/L) had the lowest amount compared to other treatments. The provision of biochar has an effect on the number of ekinase leaves where the highest number of leaves is in the rice husk charcoal biochar treatment. The addition of biochar to the soil increases the availability of major cations and phosphorus, total N and soil cation exchange capacity (CEC) which in turn increases yield [16].

The treatment of giving hormone 2 ml/L gave the highest number of leaves compared to other treatments. Hormone indications of 2 ml/L have an effect on the formation of the most optimal leaves. The use of hormones at the right dose can increase various physiological processes of plants, while at the wrong dose it can result in stunted or abnormal growth. Effectiveness of plant growth regulators is influenced by the concentration given, because different concentrations will cause different activities [17].

### *3.3 Leaf area*

Leaf area is a parameter that determines the amount of light interception, gas exchange, and transpiration rate in plants. Increasing concentrations of CO2 in the atmosphere affect the global climate, average annual temperature and rainfall patterns. An increase in atmospheric CO2 will also affect the growth rate and plant behavior, one of which is leaf formation [18].

**Table 3.** Effect of biochar type and hormone dosage on leaf area *Echinacea purpurea*.

| Leaf Area ($cm^2$) | | | | |
|---|---|---|---|---|
| | Hormon Dosage (ml/L) | | | |
| Biochar Type | H1 : 0 ml/L | H2 : 2 ml/L | H3 : 4 ml/L | H4 : 6 ml/L |
| No Biochar (Control) | 37.50a | 52.83ab | 54.33ab | 61.33ab |
| Rice Husk Charcoal | 42.33ab | 39.50ab | 67.17bc | 89.67b |
| Wood Charcoal | 49.50ab | 44.17ab | 42.67ab | 60.00ab |

Note: Numbers followed by different letters in the same column indicate a significant difference in DMRT at the 5% level.

Table 3 explains that the control treatment (without biochar or hormone dose 0 ml/L) has the lowest leaf area compared to other treatments. The application of biochar has an effect on the ekinase leaf area where the highest leaf area is in the husk charcoal biochar treatment. Research from [19] also explains that the addition of biochar can increase plant height, number of leaves, number of tillers per sample, leaf area, flowering time, fresh weight of bulbs per sample, fresh weight of bulbs per plot of sabrang onion (*Eleutherine bulbosa*).

Treatment with 6 ml/L of hormone gave the highest leaf area compared to other treatments. Hormone indications of 6 ml/L give the most optimal effect on leaf area. Research [20] used organic hormones containing auxins, cytokinins and organic gibberellins which are needed to produce vegetative and generative growth of plants. This study also showed the amount of influence of organic hormones can increase the proportion of life, shoot length, leaves, shoot diameter and volume of plant seeds.

### *3.4 Rate of photosynthesis and transpiration*

Based on the type of photosynthesis, ekinase is a group of C3 plants. The saturation level of photosynthesis in the C3 plant group was achieved at a lower light intensity than the C4 plant. Aarthi et al [21] describes C3 plants, enzymes that combine CO with substrates for the formation of carbohydrates (photosynthesis) at the same time can bind O2 for the process of photorespiration (the breakdown of carbohydrates to produce energy that occurs during the day).

Based on Table 4, it is explained that the administration of wood charcoal biochar with a hormone dose of 6 ml/L gave the highest photosynthetic rate of 0.1467 CO2/m2/s. An increase in the rate of photosynthesis indicates an increase in atmospheric CO2 so that the ambient temperature becomes higher. Metabolite storage is carried out from the accumulation of the photoassimilation process from leaf mesophyll cells (source) in which the amount of assimilate distribution is highly dependent on one of them from the rate of photosynthesis [22]. The rice husk charcoal treatment and the 2 ml/L dose had the lowest transpiration rate, which was 0.0078 mol CO2/m2/s. In hot areas growth is inhibited because the enzyme RuBisCO (Ribulose-1,5-bisphosphate carboxylase oxygenase) binds a lot of oxygen with increasing temperature so as to stimulate photorespiration which causes loss of carbon and nitrogen thus inhibiting growth [23]. In hot and dry conditions, stomata close to reduce water loss, but inhibit the exchange of CO2 thereby reducing the rate of photosynthesis.

**Table 4.** Effect of biochar type and hormone dosage on rate of photosynthesis and transpiration *Echinacea purpurea*

| Rate of Photosynthesis (μmol $CO_2/m^2/s$) | | | | |
|---|---|---|---|---|
| | Hormon Dosage (ml/L) | | | |
| Biochar Type | H1 : 0 ml/L | H2 : 2 ml/L | H3 : 4 ml/L | H4 : 6 ml/L |
| No Biochar (Control) | 0.0228a | 0.0474ab | 0.0705ab | 0.0644ab |
| Rice Husk Charcoal | 0.0199a | 0.0300ab | 0.0477ab | 0.0615ab |
| Wood Charcoal | 0.0465ab | 0.0518ab | 0.1085ab | 0.1467b |
| Rate of Transpiration (μmol $CO_2/m^2/s$) | | | | |
| | Hormon Dosage (ml/L) | | | |
| Biochar Type | H1 : 0 ml/L | H2 : 2 ml/L | H3 : 4 ml/L | H4 : 6 ml/L |
| No Biochar (Control) | 0.0198a | 0.0216a | 0.0362a | 0.0078a |
| Rice Husk Charcoal | 0.0225a | 0.01723a | 0.01756a | 0.02437a |
| Wood Charcoal | 0.01726a | 0.01907a | 0.0148a | 0.02937a |

Note: Numbers followed by different letters in the same column indicate a significant difference in DMRT at the 5% level.

## 4. Conclusion

The conclusions of this study is that microclimate affects plant responses which can be seen from the adaptability, growth response, physiological processes and plant phenotypes. Treatment of husk charcoal biochar and hormone 6 ml/L was effective in encouraging the adaptability of ekinase in the lowlands. The results on the variables of plant height, leaf area and photosynthetic rate showed that the treatment was optimal in supporting the growth response and adaptability to environmental changes in ekinase.

## References

[1] Senica M, Mlinsek G, Veberic R and Mikulic-Petkovsek M 2019 Which plant part of purple coneflower (Echinacea purpurea (L.) Moench) should be used for tea and which for tincture? *J. Med. Food* **22** 102–8

[2] Hryhoryshyn Y.V., Pospelov S.V. H O F 2020 Efficiency of pre-sowing treatment of echinacea pale seeds on morphometric indicators of second year of vegetation plants efficiency of pre-sowing treatment of echinacea pale seeds on morphometric indicators of second year of vegetation plants *Int. Symp. Biotech. Syst. Eng.* **7** 123–30

[3] Bruni R, Brighenti V, Caesar LK, Bertelli D, Cech NB P F 2018 Analytical method for the study of bioactive compounds from medicinally used Echinacea species *J. Pharm. Biomed Anal* **160** 443–77

[4] Van Alfen N K 2014 *Encyclopedia of Agriculture and Food Systems* **3**(2) 266-283

[5] Hutabarat B, Setiyanto A K R and S T 2012 Conjecturing production, imports and consumption of horticulture in indonesia in 2050: a gams simulation through changes in yields induced by climate change *J. Agro Ekon.* **30** 1–23

[6] Rahardjo M, Sudiarto S, SMD R and Sukarman S 2020 Pola pertumbuhan dan serapan hara Echinacea purpurea *J. Penelit. Tanam. Ind.* **7**

[7] Rahardjo M 2005 Peluang Pembudidayaan Tanaman Echinacea ( Echinacea purpurea ) di Indonesia *J. Perspekt.* **4** 1–10

[8] Gholami H, Saharkhiz MJ, Fard FR, Ghani A N F 2018 Humic acid and vermicompost increased

bioactive components, antioxidant activity and herb yield of Chicory (Cichorium intybus L) *Biocatal Agric Biotechnol* **14** 286–92
[9] Wahyudi A, Ernita R T 2015 Application of KCl fertilizer and hormax on Curcuma alba L *J. Din. Pertan.* **30** 125–32
[10] Pratomo G S 2016 Pengaruh Jenis Media dengan Hormon Tumbuh NAA-BAP terhadap Pertumbuhan dan Kandungan Flavonoid Kalus Daun Echinaceae purpurea (L.) Moench *J. Surya Med.* **1** 50–3
[11] Setiawan J, Arabia T and Karim A 2020 Karakteristik, Klasifikasi, dan Pengelolaan Tanah yang Terbentuk di Daerah Gunung Api Jaboi Kota Sabang *J. Ilm. Mhs. Pertan.* **5** 283–92
[12] Taufiq A 2014 Respons Tanaman Kedelai Terhadap Lingkungan Tumbuh *Bul. Palawija* **0** 13–26
[13] Sulfani S, Rahim I I N 2020 Prosiding Seminar Nasional Sinergitas Multidisiplin Ilmu Pengetahuan Dan Teknologi 128–31
[14] Pelia L, Riadi M and Rafiuddin 2020 Growth response and yield of several local yam plants with hormax applications *IOP Conference Series: Earth and Environmental Science* vol 486 012128
[15] Muriefah S S A 2015 Effects of Silicon on membrane characteristics, photosynthetic pigments, antioxidative ability, and mineral element contents of faba bean (Vicia faba L.) plants grown under Cd and Pb stress *Int. J. Adv. Res. Biol. Sci* **Coimbatore** 81–93
[16] Situmeang, Y. P., Adnyana, I. M., Subadiyasa, I. N. N., & Merit I N 2015 Effect of Dose Biochar Bamboo, Compost, and Phonska on Growth of Maize (Zea mays L.) in Dryland *Int. J. Adv. Sci. Eng. Inf. Technol.* **5** 433–9
[17] Ibrahim M, Agarwal M, Yang J O, Abdulhussein M, Du X, Hardy G and Ren Y 2019 Plant growth regulators improve the production of volatile organic compounds in two rose varieties *Plants* **8**(2) 35-38
[18] Shifa Usodri K, Prakoswo Widiyani D, Dedi Supriyatdi Jurusan Budidaya Tanaman Perkebunan dan, Negeri Lampung Jl Soekarno Hatta No P and Bandar Lampung K 2021 Identifikasi Unsur Iklim Mikro di Tanaman Karet (Hevea brasiliensis) pada Beberapa Umur Tanaman Identification of Microclimates in Rubber Plant (Hevea brasiliensis) at Several Plant Ages *J. Agric. Sci.* **2021** 115–21
[19] Sianturi D 2021 Universitas Sumatera Utara Poliklinik Universitas Sumatera Utara *J. Pembang. Wil. Kota* **1** 82–91
[20] Bocianowski J, Liersch A and Nowosad K 2020 Genotype by environment interaction for alkenyl glucosinolates content in winter oilseed rape (Brassica napus L.) using additive main effects and multiplicative interaction model *Curr. Plant Biol.* **21**
[21] Aarthi S, Suresh J, Leela N K and Prasath D 2020 Multi environment testing reveals genotype-environment interaction for curcuminoids in turmeric (Curcuma longa L.) *Ind. Crops Prod.* **14**(5) 112090
[22] Lenka N K, Lenka S, Thakur J K, Yashona D S, Shukla A K, Elanchezhian R, Singh K K, Biswas A K and Patra A K 2020 Carbon dioxide and temperature elevation effects on crop evapotranspiration and water use efficiency in soybean as affected by different nitrogen levels *Agric. Water Manag.* **2**(11) 105936
[23] Dusenge M E, Duarte A G and Way D A 2019 Plant carbon metabolism and climate change: elevated CO2 and temperature impacts on photosynthesis, photorespiration and respiration *New Phytol.* **2**(21) 32-49

**Acknowledgment**
The authors are grateful to the Research Center and Development of Medicinal Plants and Traditional Medicines (B2P2TOOT), Tawangmangu, Indonesia which has provided seeds. I also thanks you to Universitas Sebelas Maret, Surakarta, Indonesia for finance support.

# The effectiveness of organic fertilizer from natural dyes waste on the growth of *Indigofera tinctoria* L.

**M T S Budiastuti*[1], S Thaidy[1], T D Sulistyo[1], I R Manurung[1] and D Setyaningrum[2]**

[1] Departement of Agrotechnology, Faculty of Agriculture, Universitas Sebelas Maret, Indonesia
[2] Doctoral Program of Agricultural Science, Faculty of Agriculture, Universitas Sebelas Maret, Indonesia
*Corresponding author: mariatheresia@staff.uns.ac.id

**Abstract.** *Indigofera tinctoria* natural dye extraction waste has not been utilized become environmental pollution can encourage climate change. Climate change causes land degradation and decreased crop productivity. Management of the waste into organic fertilizer and used in plant cultivation need to mitigate climate change. This study aimed to examine the role of organic fertilizer from natural dye waste in obtaining the optimum dose in supporting the growth of *Indigofera tinctoria*. The study used a Randomized Completely Block Design (RCBD) with one factor, namely the dose of organic fertilizer from the extraction of *Indigofera tinctoria* L. with 11 levels (0 g/plant, 50 g/plant, 100 g/plant, 150 g/plant, 200 g/plant, 250 g/plant, 300 g/plant, 350 g/plant, 400 g/plant, 450 g/plant, and 500 g/plant). The results showed that the natural dye extraction waste could be used as organic fertilizer to support the growth of *Indigofera tinctoria*. Application of organic fertilizer at a dose of 50 g/plant showed an increase in the plant height of 14% and number of leaves by 124%. The dose of organic fertilizer of 350 g/plant is the optimum dose in producing the high number of leaves, leaf area index, and plant biomass.

**Keywords**: biodegradable, plant extracts, remediation

## 1. Introduction

The natural dye extraction process of *Indigofera tinctoria* produces 90% organic waste and only 10% natural dye [1]. The waste has not been managed properly by the farmers. Waste management is done by burning. The process of burning waste produces carbon emissions that cause air pollution and have a negative impact on climate change [2]. Climate change causes land degradation and decreased crop productivity [3].

The problem of managing waste from the extraction of natural dyes encourages the use of waste as an organic fertilizer because of the high chemical content of the extracted waste. Waste from extraction of natural dyes is organic waste consisting of the structure of the *Indigofera tinctoria* plant, namely stems, leaves and pods [4]. Composting plant waste is an environmentally friendly effort by providing product values that have a positive impact on soil and plants. Several studies have reported the management of waste into organic fertilizer as a source of nutrition in agriculture such as industrial waste with a mixture of chicken manure, sawdust and green waste [5,6]. The extracted waste contains N, $CH_4$, $NH_3$, $CO_2$, $O_2$ [7]. Waste management into organic fertilizer can have a positive impact on soil and plants.

Organic fertilizers from the composting process can overcome the degradation of agricultural land resources [8] and help the growth of horticultural crops [9] Organic fertilizers can increase soil density on dry land and increase $N_2$ fixation and the number of nodules in legumes [10] The application of a combination of organic fertilizer and biofertilizer can increase indican content, the yield of *Indigofera tinctoria* herbs and N and K content in the soil [11] This study aims to determine the effectiveness of natural dye extraction waste organic fertilizer on the growth of *Indigofera tinctoria* plants.

## 2. Materials and method

### 2.1. Materials

The research was conducted from May to September 2020, in Puron Village, Bulu District, Sukoharjo, Central Java, Indonesia. The research location is at di 110$^0$ 51'49.44'' E and 7$^0$48' 54.3'' S. The research soil had an acid pH of 5.95, 0.33% nitrogen (N), 3.94 ppm phosphorous (P), 12.16 mg/100g of soil potassium (K), C-organic was low at 2.45%, and organic matter was low at 4.23%. Materials used in this research included *Indigofera tinctoria* L. seeds and organic fertilizer from natural dyes waste. The instruments were measures to measure plant height, scale to measure leaf weight and plant biomass, and oven to dry the plant.

### 2.2. Methods

The research used a Randomized Completely Block Design (RCBD) with one factor, namely the dose of organic fertilizer from the extraction of *Indigofera tinctoria* L. with 11 levels (0 g/plant, 50 g/plant, 100 g/plant, 150 gplant, 200 g/plant, 250 g/plant, 300 g/plant, 350 g/plant, 400 g/plant, 450 g/plant, and 500 g/plant). Every experimental unit was repeated 3 times until 33 experimental units were got. Every experimental unit was consist of 4 plants. Observation variables included the plant height, number of leaves, leaf area index, and plant biomass. Research data were analysed using analysis of variance with α 5% test level 95% confidence level. If it had a significant effect, further analysis was done using Duncan's Multiple Range Test (DMRT).

## 3. Results and discussion

### 3.1. The plant height

The results showed that the dose of organic fertilizer from the extraction of natural dyes had a significant effect on plant height. Plant height increased with an increasing dose of organic fertilizer (Table 1). This is because well-decomposed organic fertilizers can provide nutrients needed by plants, especially N and K elements so that the photosynthesis process will be optimum [12].

**Table 1.** The effect of organic fertilizer from natural dyes waste on plant height

| Treatment (gram/plant) | Plant Height |
|---|---|
| 0 | 82.67 a |
| 50 | 96.83 b |
| 100 | 105.33 bc |
| 150 | 106.33 bcd |
| 200 | 112.67 cde |
| 250 | 105.67 bc |
| 300 | 113.67 cde |
| 350 | 114.67 cde |
| 400 | 115.33 cde |
| 450 | 120.33 de |
| 500 | 126.33 e |

Note: Number followed by the same columns showed no significant difference based on the DMRT level of 5%.

Fertilization is an effort to provide nutrients needed by plants in the soil. Plant height that was given organic fertilizer from the extraction of natural dyes at a dose of 500 grams/plant was able to reach a height of 120 cm. This is because organic fertilizers increase the regulation of soil microbial communities thereby increasing the content of C organic and soil nitrogen [13] which can further promote plant growth [14].

The application of organic fertilizer at a dose of 50 grams/plant has shown an increase in plant height and is significantly different from that without the application of organic fertilizer. This is supported by the chemical content of the organic fertilizer. The content of organic fertilizers can support the structure and function of microbial communities that provide available nutrients by mineralizing organic matter [15].

### *3.2. The number of leaves*

The treatment of organic fertilizer from natural dye extraction waste at various doses had a significant effect on the number of leaves (Table 2). *Indigofera tinctoria* plants treated with a dose of organic fertilizer from the extraction of natural dyes of 350 grams/plant and 500 grams/plant had an average number of leaves that were not significantly different, namely 969 leaves. This indicates that the administration of a dose of 350 grams/plant of organic fertilizer from the extraction of natural dyes has been able to affect the number of leaves of the *Indigofera tinctoria* plant. This is because the content of nutrients, C-organic and organic matter in the organic fertilizer is high so that it can support the growth of plant leaves. Leaf growth was correlated with the uptake of nitrogen and phosphate of plants which was influenced by the addition of nitrogen and phosphate to the soil [16] The addition of N can significantly affect the concentration of N and P in green and old leaves and the pattern of nutrient resorption. Based on the results of this study, shows that organic fertilizers have complete nutrition and contain growth regulators such as hormones for plant growth [17].

**Table 2.** The effect of organic fertilizer from natural dyes waste on number of leaves

| Treatment (gram/plant) | Number of Leaves |
|---|---|
| 0 | 197.33 a |
| 50 | 634.33 b |
| 100 | 668.67 b |
| 150 | 700.33 b |
| 200 | 690.33 b |
| 250 | 768.67 bc |
| 300 | 861 bc |
| 350 | 968.67 c |
| 400 | 879.33 bc |
| 450 | 879.33 bc |
| 500 | 968.67 c |

Note: Number followed by the same columns showed no significant difference based on the DMRT level of 5%.

### *3.3. The leaves area index*

The results showed that the dose of organic fertilizer from the extraction of natural dyes had a significant effect on the leaf area index (Table 3). The leaf area index of *Indigofera tinctoria* with treatment with doses of organic fertilizer from the extraction of natural dyes of 350 grams/plant and 500 grams were not significantly different. This shows that the treatment dose of organic fertilizer extraction waste of 350 grams/plant has been able to increase the leaf area index. A high leaf area index value results in a higher yield weight [18]. Organic fertilizer increases the photosynthetic efficiency of leaves so that the leaf area index increases [19]. Leaf Area Index (LAI) increased along with plant growth and reached its maximum when the canopy growth was tight. External factors also affect the optimal LAI value such as spacing (plant density) and cropping system. Leaf area index and light energy determine the rate of photosynthesis of plants [20]

**Table 3.** The effect of organic fertilizer from natural dyes waste on leaves area index

| Treatment (gram/plant) | Leaves Area Index |
|---|---|
| 0 | 0.2952 a |
| 50 | 1.0729 abc |
| 100 | 0.7864 ab |
| 150 | 1.069 abc |
| 200 | 1.9798 cd |
| 250 | 1.7343 bcd |
| 300 | 2.1958 d |
| 350 | 2.4398 d |
| 400 | 1.7504 bcd |
| 450 | 2.2291 d |
| 500 | 2.7310 d |

Note: Number followed by the same columns showed no significant difference based on the DMRT level of 5%.

### *3.4. The plant biomass*

The results showed that the dose of organic fertilizer from the extraction of natural dyes with various doses had a significant effect on plant biomass (Table 4). The average plant biomass with treatment with doses of organic fertilizer from the extraction of natural dyes of 350 grams/plant and 500 grams/plant was not significantly different, namely 186.483 grams and 182.683 grams. This shows that the treatment dose of organic fertilizer waste extraction of natural dyes 350 grams/plant has been able to increase plant biomass. The provision of organic fertilizer can increase root activity in absorbing water and nutrients needed by plants in the metabolic process [21].

Plant biomass is a measure of plant growth and development. Calculation of plant biomass was used to determine the accumulation of organic compounds that were successfully synthesized by plants to other organs [22]. The difference in plant biomass is caused by differences in the ability of leaves to produce reduced carbon for producing plant biomass [23].

**Table 4.** The effect of organic fertilizer from natural dyes waste on plant biomass

| Treatment (gram/plant) | Plant Biomass |
|---|---|
| 0 | 30.52 a |
| 50 | 108.25 b |
| 100 | 83.25 ab |
| 150 | 106.57 b |
| 200 | 142.90 bc |
| 250 | 140.47 bc |
| 300 | 152.97 bc |
| 350 | 186.48 c |
| 400 | 128.043 bc |
| 450 | 152.800 bc |
| 500 | 182.68 c |

Note: Number followed by the same columns showed no significant difference based on the DMRT level of 5%.

## 4. Conclusion

The natural dye extraction waste could be used as organic fertilizer to support the growth of *Indigofera tinctoria*. Application of organic fertilizer at a dose of 50 g/plant showed an increase in plant height of 14% and number of leaves by 124%. The dose of organic fertilizer of 350 g/plant is the optimum dose in producing the high number of leaves, leaf area index, and plant biomass. Utilization of waste as organic fertilizer as a step to reduce the impact of climate change and support plant growth.

## References

[1] Sihta F, Suyitno, Heru Wibowo A and Tanding R 2018 Enhancing biogas quality of indigofera plant waste through co-digestion with cow dung *MATEC Web of Conferences* vol 154 (EDP Sciences)

[2] Reyna-Bensusan N, Wilson D C and Smith S R 2018 Uncontrolled burning of solid waste by households in Mexico is a significant contributor to climate change in the country *Environ. Res.* **163** 280–8

[3] Macdonald C A, Anderson I C, Khachane A, Singh B P, Barton C V M, Duursma R A, Ellsworth D S and Singh B K 2021 Plant productivity is a key driver of soil respiration response to climate change in a nutrient-limited soil. *Basic Appl. Ecol.* **50** 155–68

[4] Budiastuti M T S, Supriyono, Manurung I R, Setyaningrum D, Nurmalasari A I and Arista N I D 2021 The role of organic fertilizer from natural dye waste and mycorrhizal inoculation on the growth of Indigofera tinctoria L. *IOP Conf. Ser. Earth Environ. Sci.* **905** 012011

[5] Araujo A S F, de Pereira A P de A, Antunes J E L, Oliveira L M de S, de Melo W J, Rocha S M B, do Amorim M R, Araujo F F, Melo V M M and Mendes L W 2021 Dynamics of bacterial and archaeal communities along the composting of tannery sludge *Environ. Sci. Pollut. Res.*

[6] Cheng Y, Chon K, Ren X, Kou Y, Hwang M H and Chae K J 2021 Bioaugmentation treatment of a novel microbial consortium for degradation of organic pollutants in tannery wastewater under a full-scale oxic process *Biochem. Eng. J.* **175** 108131

[7] Suyitno, Ridwan, Arifin Z and Pranolo S H 2019 The Effect of Biogas Pressure in the Performance and Emission of Spark Ignition Engine *IOP Conference Series: Materials Science and Engineering* vol 694 (Institute of Physics Publishing)

[8] Nath C P, Hazra K K, Kumar N, Praharaj C S, Singh S S, Singh U and Singh N P 2019 Including grain legume in rice–wheat cropping system improves soil organic carbon pools over time *Ecol. Eng.* **129** 144–53

[9] Boldrin A, Andersen J K, Møller J, Christensen T H and Favoino E 2009 Composting and compost utilization: Accounting of greenhouse gases and global warming contributions *Waste Manag. Res.* **27** 800–12

[10] Agegnehu G, Bass A M, Nelson P N, Muirhead B, Wright G and Bird M I 2015 Biochar and biochar-compost as soil amendments: Effects on peanut yield, soil properties and greenhouse gas emissions in tropical North Queensland, Australia *Agric. Ecosyst. Environ.* **213** 72–85

[11] Sindhu P V., Kanakamany M T and Beena C 2016 Effect of organic manures and biofertilisers on herbage yield, quality and soil nutrient balance in Indigofera tinctoria cultivation *J. Trop. Agric.* **54** 16–20

[12] Indayana T, Sadaruddin and Nurhasanah 2019 The Effect of Organic Fertilizer to The Growth and Yield of Soybean (Glycine max L. Merrill) Cultivated on Post Coal Minning Soil *J. Agroekoteknologi Trop. Lembab* **1** 38

[13] Zhaoxiang W, Huihu L, Qiaoli L, Changyan Y and Faxin Y 2020 Application of bio-organic fertilizer, not biochar, in degraded red soil improves soil nutrients and plant growth *Rhizosphere* **16** 100264

[14] Raza W, Mei X, Wei Z, Ling N, Yuan J, Wang J, Huang Q and Shen Q 2017 Profiling of soil volatile organic compounds after long-term application of inorganic, organic and organic-inorganic mixed fertilizers and their effect on plant growth *Sci. Total Environ.* **607–608** 326–38

[15] Hafez M, Popov A I and Rashad M 2021 Integrated use of bio-organic fertilizers for enhancing

soil fertility–plant nutrition, germination status and initial growth of corn (Zea Mays L.) *Environ. Technol. Innov.* **21** 101329

[16] You C, Wu F, Yang W, Xu Z, Tan B, Zhang L, Yue K, Ni X, Li H, Chang C and Fu C 2018 Does foliar nutrient resorption regulate the coupled relationship between nitrogen and phosphorus in plant leaves in response to nitrogen deposition? *Sci. Total Environ.* **645** 733–42

[17] Ravindran B, Lee S R, Chang S W, Nguyen D D, Chung W J, Balasubramanian B, Mupambwa H A, Arasu M V, Al-Dhabi N A and Sekaran G 2019 Positive effects of compost and vermicompost produced from tannery waste-animal fleshing on the growth and yield of commercial crop-tomato (Lycopersicon esculentum L.) plant *J. Environ. Manage.* **234** 154–8

[18] Zhang G, Ming B, Shen D, Xie R, Hou P, Xue J, Wang K and Li S 2021 Optimizing grain yield and water use efficiency based on the relationship between leaf area index and evapotranspiration *Agric.* **11** 313

[19] Sorrenti G, Muzzi E and Toselli M 2019 Root growth dynamic and plant performance of nectarine trees amended with biochar and compost *Sci. Hortic. (Amsterdam).* **257** 108710

[20] Setijaningsih, Lies dan Suryaningrum L H 2015 Jurnal Ilmu-ilmu Hayati *Ber. Biol.* **14** 287–93

[21] Fatahi E, Mobasser H R and Akbarian M M 2014 Effect of organic fertilizer on wet weight, dry weight and number of leaves in cowpea *J. Nov. Appl. Sci.* 440–3

[22] Putra E, Sudirman A and Indrawati W 2016 Pengaruh pupuk organik pada pertumbuhan vegetatif tanaman tebu (Saccharum officinarum L.) varietas GMP 2 dan GMP 3 *J. Agro Ind. Perkeb.* **4** 60–8

[23] Susilo D E H 2015 Identification of Constanta Value of Leaf Shape for Leaf Area Measurement Using Length Cross Width of Leaf of Horticulture Plant in Peat Soil *Anterior J.* **14** 139–46

# Morphology and physiology of *Echinacea purpurea* in the vegetative phase result of gamma ray irradiation

**Wafa' Nur Hanifah[1], Nandariyah[2], Yuli Widiyastuti[3] and Ahmad Yunus*[2]**

[1]Postgraduate Department of Agronomy, Faculty of Agriculture, Universitas Sebelas Maret, Surakarta, Indonesia
[2]Department of Agrotechnology, Faculty of Agriculture, Universitas Sebelas Maret, Surakarta, Indonesia
[3]Medicinal Plants and Traditional Medicines Research and Development Centre (B2P2TOOT) Tawangmangu, Indonesia
*Corresponding author: yunus.uns7@yahoo.com

**Abstract**. Coneflower is a medicinal plant that was introduced from North America. This plant is useful as an immunomodulator, which can protect the human body's immune system from extreme climate change. Coneflower plants in Indonesia are generally grown in highlands that have low temperatures as in their native areas. Indonesia's tropical environmental conditions, especially in the lowlands, and increasing temperatures as part of global warming are a concern for coneflower cultivation because high climate differences can affect plant growth and physiology. One of the efforts to improve plant properties to obtain plants that are suitable for living in a high-temperature environment can be done by gamma-ray irradiation. The purpose of this study was to determine the growth and physiology of the Coneflower plant irradiated by gamma rays. The study was conducted using a simple design experiment method by planting the population per treatment into successive plots. The seeds used were Coneflower plant accession 2 seeds from the Medicinal Plants and Traditional Medicines Research and Development Centre (B2P2TOOT) Tawangmangu. The irradiation doses used were 0 (control), 5Gy, 15Gy, and 25Gy. The results showed the highest plant height, and the number of leaves on 15 Gy irradiated plants. The highest total chlorophyll content was in 5Gy irradiated plants, while the highest leaf area was in 25Gy irradiated plants. Morphology and physiology of irradiated plants showed greater growth than control plants.

**Keywords** : Medical plants, gamma-ray, growth, physiology.

## 1. Introduction

Coneflower is an herbal plant of the family Asteraceae introduced from North America. This plant is known for its efficacy as an immunomodulator and anti-inflammatory that serves in maintaining the immune system of the human body [1], especially during pandemics and extreme climate change. Various studies show coneflower extract can keep the body from preventing upper respiratory infections, such as the flu [2,3]. This plant comes from sub-tropical regions so in Indonesia coneflower plants are generally grown in the highlands. The coneflower plant commonly grown in Indonesia is a species of *Echinacea purpurea* that is known to have high adaptability [4].

The use of herbal plants as medicinal raw materials in the pharmaceutical industry is increasing [5]. Efforts to increase the production of coneflower plants can be done with the development of crop cultivation, such as land use in lowlands. Rising global temperatures due to global warming are also a challenge for plant breeding to create coneflower plants that can adapt to high temperatures. Irradiation of gamma rays becomes one way of breeding plants to obtain plants that have the desired characteristics. The impact of global climate change on agriculture is difficult to predict so there needs to be the development of new varieties that can adapt to various agro-climate conditions [6].

Mutation techniques with gamma rays can form changes in plant phenotypes [7]. Genetic changes in plants from mutations can result in plants that are superior to before [8]. Gamma-ray irradiation research on *echinacea purpurea* plants is still small. The purpose of the study was to acquire mutant plants that showed good growth and physiology in low altitudes, particularly high temperatures.

## 2. Materials and method

### *2.1. Materials*
The research was conducted at the Agricultural Land Faculty of Sebelas Maret University, Jumantono, Karanganyar. The seeds used are two accession coneflower seeds obtained from The Medicinal Plants and Traditional Medicines Research and Development Centre (B2P2TOOT) Tawangmangu. The seeds are irradiated using gamma rays at the National Nuclear Energy Agency of Indonesia, then planted in a seedling tray for a month before being moved to experimental land.

### *2.2. Methods*
The study used a simple random design experiment method by planting populations per treatment into successive swaths of experiments. The doses of irradiation used are 0 Gy (control), 5 Gy, 15 Gy, and 25 Gy. Growth observations include plant height, number of leaves, and leaf area. Physiological observations measure the levels of chlorophyll a, chlorophyll b, and total chlorophyll. The data were tested using a T-test on SPSS by comparing control plants with irradiated plants.

### *2.3. Analysis of chlorophyll content*
Extraction of chlorophyll is carried out using the Arnon method (1949) by weighing fresh leaves by 1 gr. The leaves are pounded until smooth then the pigment is dissolved with 20 ml of acetone 80% and filtered using filter paper. The content of chlorophyll is measured using a spectrophotometer with a solution absorbance of $\lambda$ 663 mn and $\lambda$ 645 mn against blank solvents. Calculation of chlorophyll content based on the formula:

Chlorophyll a = 12.7 A-663 – 2.69 A-645 (mg/l)
Chlorophyll b = 22.9 A-645 – 4.68 A-663 (mg/l)
Total Chlorophyll = 20.2 A-645 + 8.02 A-663 (mg/l)

## 3. Results and discussion

### *3.1. Plant height*
One morphological character that can be an indicator of plant growth is plant height. Plants that are irradiated by gamma rays undergo morphological changes, such as in plant height. Measurement of plant height is commonly used in determining biological effects on physical mutagens [9]. In selecting mutation plants need to determine the right dose of irradiation, M1 mutation plants need to be compared to control plants based on their growth rate [10].

**Table 1.** Plant height of *Echinacea purpurea* plant at various doses of irradiation

| Dose of Irradiation | High Growth of Plants (cm) | | |
|---|---|---|---|
| | Lowest | Highest | Average |
| 0Gy (control) | 16.9 | 20.4 | 18.67±1.75 |
| 5Gy | 18.8 | 25.2 | 22.00±3.16* |
| 15Gy | 19.2 | 25.1 | 22.14±2.97* |
| 25Gy | 13.7 | 24.7 | 19.20±5.47 |

Note: Values describe the average±standard deviation (SD). The value followed by the sign (*) shows a significant difference with control based on the T-test ($\alpha = 0.005$)

Irradiated coneflower plants have a higher plant height compared to control plants. Data in Table 1. The highest average plant height is in plants that are irradiated 15Gy (22.14 cm), then 5Gy (22 cm), and 25 Gy (19.20 cm). Low doses of gamma-ray irradiation can stimulate the growth of aerial plants [9]. 5Gy and 15Gy irradiation plants showed significant height differences over control plants. High irradiation plants decreased in doses of 25Gy. Low doses of gamma-ray irradiation can stimulate tissue meristem activity, triggering high plant growth, but high doses of irradiation can damage plant cells. Hydroxyl radicals are products produced by reactions between water and radiation that cause damage to chromosomes [11]. High doses of gamma-ray irradiated seeds can disrupt plant growth because hormone balance, protein synthesis, and enzyme activity are disrupted [12], such as reduced bioactive content of Giberelat Acid (GA) or inhibition of GA signaling [13]. Irradiated plants show a higher plant height diversity than control plants. The greatest diversity of plants is in 25Gy irradiated plants, followed by 5Gy and 15Gy.

### 3.2. *Number of leaves*

The number of leaves in individual plants is related to the ability of plants to produce energy from the process of photosynthesis that affects the growth of plants and yield. The effectiveness of the dose of irradiation can be seen from its growth, such as the number of leaves. Gamma-ray irradiation did not show a significant increase in the number of coneflower plant leaves compared to control.

**Table 2.** Number of *Echinacea purpurea* leaves at various doses of irradiation

| Dose of Irradiation | Number of Leaves | | |
|---|---|---|---|
| | Lowest | Highest | Average |
| 0Gy (control) | 8 | 15 | 11.50±3.77 |
| 5Gy | 8 | 12 | 9.90±1.99* |
| 15Gy | 9 | 15 | 12.10±3.12 |
| 25Gy | 7 | 13 | 10.27±2.98 |

Note: Values describe the average±standard deviation (SD). The value followed by the sign (*) shows a significant difference with control based on the T-test ($\alpha = 0.005$)

A significant difference in the number of leaves from the controls plants was shown in 5Gy, averaging 11.50 and 9.90 respectively. The highest average number of leaves was found in irradiated plants at doses of 15Gy (12.10) then mutant plants 25Gy (10.27). On table 2. Increased doses of gamma-ray irradiation do not affect the average number of leaves of coneflower plants. This is because gamma-ray irradiation is random in changing plant morphology, so the number of doses does not affect much/little number of leaves. Abnormalities in the leaves of mutant plants can be caused by chromosomal abnormalities, changes in energy production/distribution, breakdown of phosphate metabolism, and accumulation of free amino acids [14]. M1 mutation plants are heterogeneous which means plants carry different diversity traits [15].

### 3.3. *Leaf area*

Leaf area is concerned with the capacity of photosynthetic space, energy exchange, and transpiration that determine plant growth and yield [14, 15]. The area of the leaves has an optimum value, if the leaf area is too wide it will cause the leaves to shade each other so that the light received is not optimal [18]. The larger of leaf area thus the more optimal the absorption of sunlight [19], so that the results of photosynthesis produced by plants can support plant growth and yield.

**Table 3.** The leaf area of *Echinacea purpurea* at various doses of irradiation

| Dose of Irradiation | Leaf Area ($cm^2$) | | | |
|---|---|---|---|---|
| | Above | Middle | Below | Average |
| 0Gy (control) | 48.3 | 47.6 | 51.4 | 49.1 |
| 5Gy | 37.6 | 62.7 | 59 | 53.1 |
| 15Gy | 72.8 | 74.2 | 38.3 | 61.8 |
| 25Gy | 57.7 | 78.2 | 50.3 | 62 |

The leaf area of the irradiated coneflower plant has a leaf area greater than the leaf area of the control plant. This suggests irradiated plants have better adaptability in high-temperature environments than control plants. Low-dose irradiation alters hormonal signals in plant cells or increases the number of anti-oxidative cells to cope with environmental stress [11]. Increased dose of irradiation linear with an increase in leaf area although not significant. The leaf area of the irradiated plant of 5Gy, 15Gy, and 25Gy is 53.1 $cm^2$, 61.8 $cm^2$, and 62 $cm^2$ respectively. The increment leaf area in irradiated plants reflects the large source capacity of photo-assimilation [20]. There is a difference in leaf area between the upper, middle, and lower leaves. The area of the lower leaf is smaller because the bottom leaf is the oldest leaf that grows in the early vegetative phase, so the growth is still small. The middle leaves have a larger leaf area along with plant growth indicating the adaptation of the plant to its environment.

### 3.4. *Leaf chlorophyll levels*

Leaf chlorophyll is a leaf color pigment that plays a role in capturing light in the process of photosynthesis. Chlorophyll absorbs and converts solar radiation into stored energy, as well as being a place of exchange of energy and matter flows between the biosphere and the atmosphere [21]. The content of chlorophyll consists of two types, chlorophyll a and chlorophyll b. Chlorophyll a and b absorb light at different wavelengths, chlorophyll absorbs red-orange light, while chlorophyll b absorbs blue-violet light [22]. Gamma-ray irradiation can affect the content of chlorophyll which results in changes in the rate of photosynthesis of plants [23].

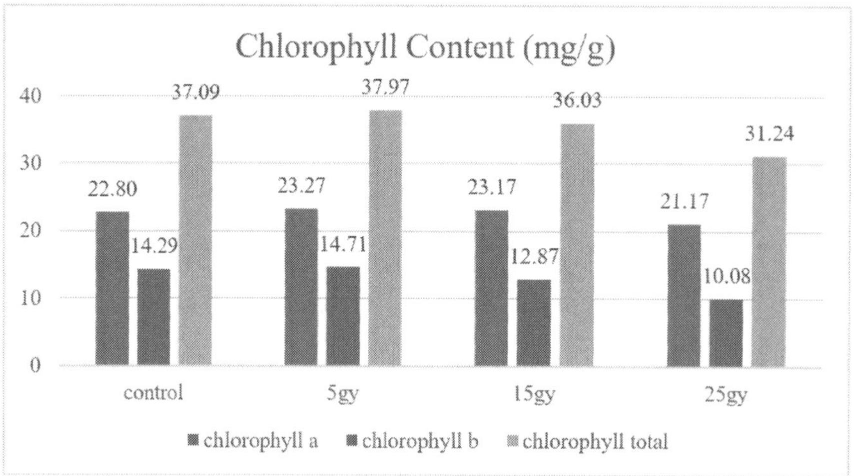


**Figure 1.** Chlorophyll content of *Echinacea purpurea* (L.) at various doses

Gamma-ray irradiation can alter plant physiology and biochemistry [7]. Based on figure 1. the total chlorophyll content of non-irradiated plants is 37.09 mg/g. Irradiated coneflower plants have chlorophyll content that is not much different from control, even tending to decrease in number as the dose of irradiation increases. The highest total chlorophyll content in 5Gy irradiated dose plants was 37.97 mg/g, followed by doses of 15Gy (36.03 mg/g), and doses of 25Gy (31.24 mg/g). The same results were also found in tomato and banana plants where the total amount of chlorophyll decreased along with increasing the dose of irradiation [19, 20]. The decrease in the total content of chlorophyll in high-dose irradiated plants is due to the suppression of the rate of chlorophyll synthesis. Low chlorophyll content will affect the growth and yield of plants. The slight absorption of light will affect the low energy produced from the process of photosynthesis.

## 4. Conclusions

Gamma-ray irradiation plants show better morphological and physiological growth than plant control, thus indicating *Echinacea purpurea* mutants are adaptable in climate change conditions of high temperature. 15Gy irradiated plants showed the highest plant height and the number of leaves. The highest chlorophyll content is present in 5Gy irradiated plants, although the difference with other doses is very small. The largest leaf area is found in 25Gy irradiated plants. This research needs to be known the yield and content of secondary metabolites so that it can be known the extent of the effect of gamma-ray irradiation on coneflower plants.

## References

[1] Ross S M 2016 Echinacea purpurea : A proprietary extract of echinacea purpurea is shown to be safe and effective in the prevention of the common cold *Holist. Nurs. Pract.* **30** 54–7

[2] Jawad M, Schoop R, Suter A, Klein P and Eccles R 2012 Safety and Efficacy Profile of Echinacea purpurea to Prevent Common Cold Episodes: A Randomized, Double-Blind, Placebo-Controlled Trial *Evid. Based. Complement. Alternat. Med.* **2012** 841315

[3] David S and Cunningham R 2019 Echinacea for the prevention and treatment of upper respiratory tract infections: A systematic review and meta-analysis *Complement. Ther. Med.* **44** 18–26

[4] Rahardjo M 2005 Peluang Pembudidayaan Tanaman Echinacea (Echinacea purpurea) di Indonesia *Perspektif* **4** 1–10

[5] Nugroho A W 2017 Review: Konservasi Keanekaragaman Hayati Melalui Tanaman Obat Dalam Hutan Di Indonesia Dengan Teknologi Farmasi: Potensi dan Tantangan *J. Sains dan Kesehat.* **1** 377–83

[6] Jan S, Parween T, Siddiqi T O and Mahmooduzzafar X 2012 Effect of gamma radiation on morphological, biochemical, and physiological aspects of plants and plant products *Environ. Rev.* **20** 17–39

[7] Hasbullah N A, Taha R M, Saleh A and Mohamed N 2012 Physiological Responses of Callus from Gerbera jamesonii Bolus ex. Hook f. to Gamma Irradiation *Brazilian Arch. Biol. Technol.* **55** 411–6

[8] Harahap F, Jusoff K, Poerwanto R, Nusyirwan, Syarifuddin and Hasruddin 2013 Mangosteen DNA analysis (Garcinia Mangostana L.) with molecular markers after gamma ray irradiation treatment *Am. J. Sustain. Agric.* **7** 37–44

[9] Silva A S da, Danielowski R, Braga E J B, Deuner S, Magalhães Junior A M de and Peters J A 2011 Development of rice seedlings grown from pre-hydrated seeds and irradiated with gamma rays *Ciência e Agrotecnologia* **35** 1093–100

[10] Ulukapi K and Ozmen S F 2018 Study of the effect of irradiation (60Co) on M1 plants of common bean (Phaseolus vulgaris L.) cultivars and determined of proper doses for mutation breeding *J. Radiat. Res. Appl. Sci.* **11** 157–61

[11] Abdullah S, Kamaruddin N Y and Harun A R 2018 The effect of gamma radiation on plant morphological characteristics of Zingiber officinale Roscoe *Int. J. Adv. Sci. Eng. Inf. Technol.* **8**

2085–91

[12] Khah M A, Ch R and Verma R 2015 Assessment of the effects of gamma radiations on various morphological and agronomic traits of common wheat ( Triticum aestivum L.) var. WH-147 *Eur. J. Exp. Biol.* **5** 6–11

[13] Ahumada-Flores S, Gómez Pando L R, Parra Cota F I, de la Cruz Torres E, Sarsu F and de los Santos Villalobos S 2021 Technical note: Gamma irradiation induces changes of phenotypic and agronomic traits in wheat (Triticum turgidum ssp. durum) *Appl. Radiat. Isot.* **167** 109490

[14] Kumari K, Dhatt K . and Kapoor M 2013 Induced mutagenesis in chrysanthemum morifolium variety 'otome pink' through gamma irradiation *Bioscan an Int. Q. J. Life Sci.* **8** 1481492

[15] Abdullah T L, Endan J and Nazir B M 2009 Changes in flower development, chlorophyll mutation and alteration in plant morphology of Curcuma alismatifolia by gamma irradiation *Am. J. Appl. Sci.* **6** 1436–9

[16] Nayak D, Patil N S, Behera L K and Jadeja D B 2015 Effects of gamma rays on germination and growth in Jatropha curcas L. *J. Appl. Nat. Sci.* **7** 964–9

[17] Yadav A, Singh B and Singh S 2019 Impact of Gamma Irradiation on Growth, Yield, and Physiological Attributes of Maize *Indian J. Exp. Biol.* **57** 116–22

[18] Awanis S, Parjanto and Purwanto E 2021 Morphophysiological characteristics and yields of black rice (Oryza sativa L. Indica) induced by gamma rays 100 gy at m3 generation *IOP Conf. Ser. Earth Environ. Sci.* **637** 012078

[19] Latifa R, Hadi S and Nurrohman E 2019 The Exploration of Chlorophyll Content of Various Plants in City Forest of Malabar Malang *Bioedukasi J. Biol. dan Pembelajarannya* **17** 50–62

[20] Singh B and Datta P S 2010 Gamma irradiation to improve plant vigour, grain development, and yield attributes of wheat *Radiat. Phys. Chem.* **79** 139–43

[21] Croft H, Chen J M, Wang R, Mo G, Luo S, Luo X, He L, Gonsamo A, Arabian J, Zhang Y, Simic-Milas A, Noland T L, He Y, Homolová L, Malenovský Z, Yi Q, Beringer J, Amiri R, Hutley L, Arellano P, Stahl C and Bonal D 2020 The global distribution of leaf chlorophyll content *Remote Sens. Environ.* **236** 111479

[22] Li Y, He N, Hou J, Xu L, Liu C, Zhang J, Wang Q, Zhang X and Wu X 2018 Factors Influencing Leaf Chlorophyll Content in Natural Forests at the Biome Scale *Front. Ecol. Evol.* **6** 64

[23] Kiong L P A, Lai A G, Hussein S and Harun A R 2008 Physiological responses of Orthosiphon stamineus plantles to gamma irradiation. *Am. J. Sustain. Agric.* **2** 135–49

[24] Chanu A M, Thokchom R, Bhaigyabati T, Devi N S and Meitei T R 2020 Effect of gamma irradiation on the chlorophyll content of tree tomato (Solanum betaceum Cav.) in M1 generation *Pharma Innov. J.* **9** 33–5

[25] Abdulhafiz F, Kayat F and Zakaria S 2018 Effect of gamma irradiation on the morphological and physiological variation from In vitro individual shoot of banana cv. Tanduk (Musa spp.) *J. Plant Biotechnol.* **45** 140–5

**Acknowledgment**

The authors thank the Centre for Research and Development of Medicinal Plants and Traditional Medicine (B2P2TOOT) Tawangmangu, Indonesia for supporting *E. purpurea* seeds and access for analysis in the laboratory and to the KEMRISTEKDIKTI for financial support.

# Study of potential plant of Biduri (*Calotropis gigantea*) with combination of bacteria, organic matter, and inorganic fertilizer for bioremediation of lead (Pb) contaminated soil

**Retno Rosariastuti*[1], Hapsari Ayu Astuti[2] and Sudadi[1]**

[1]Department of Soil Science, Faculty of Agriculture, Sebelas Maret University, Surakarta, Indonesia
[2]Department of Soil Science, Bachelor Student, Sebelas Maret University
*Corresponding author: retnobs@staff.uns.ac.id

**Abstract**. Heavy metal pollution from anthropogenic sources is a serious problem for the environment, one of these Pb. Besides industrial waste, the use of chemical fertilizers and pesticides in the agriculture sector can increase heavy metal concentration in the soil. One impact of climate change is the change of rainfall patterns. Some areas suddenly experienced a long drought, forcing farmers to irrigate their fields using industrial wastewater which contains hazardous compounds such as Pb. Bioremediation technology using the plant in combination with microorganisms and organic matter as the bioremediation agent is one way to restore polluted soil as an effort to improve the environment carried out in this study. The purpose of this study was to study the effect of inorganic fertilizers and bioremediation agents on efforts to reduce soil Pb levels with Biduri and the ability of Biduri as Pb hyperaccumulator plants. This study used a factorial experimental design with a completely randomized design with three factors and three replicates. The use of Biduri with a combination of inorganic fertilizers, bioremediation agent, and doses of Pb is able to reduce Pb levels in the soil so that it can be used as an alternative to reduce soil pollution, one of the consequences of climate change which has an impact on agriculture and Biduri in this study have potential as hyperaccumulator plants for Pb.

**Keywords** : bioremediation, lead, biduri, bacteria, organic matter, inorganic fertilizer

## 1. Introduction

Heavy metal pollution is a serious problem for the environment and human health. Heavy metals are toxic pollutants, one of which is Pb [1,2]. Heavy metal contamination in soil mainly comes from anthropogenic sources such as household activities, agrochemicals, and mining [3,4,5]. In addition, changes in rainfall patterns due to the impact of climate change can lead to long droughts in some areas. The long drought that occurred forced farmers to irrigate their fields using industrial wastewater, which contains hazardous compounds such as the Pb. Heavy metals of anthropogenic origin can be transformed, transported in the air, deposited on the soil surface, and then penetrate deep into the soil profile [6]. According to Government Regulation of Republik of Indonesia No.101 [7], the permissible level of heavy metal lead (Pb) in agricultural soil is 3ppm.

One way to restore the quality of soil contaminated with heavy metals is to remediate soil contaminated with heavy metals using bioremediation technology. One form of bioremediation is rhizoremediation, which is a process where microorganisms degrade contaminated soil in the root areas.

This process uses plant roots and a consortium of microbes to degrade environmental pollutants and restore the contaminated soil area to a condition that can be used for the intended purpose [8]. According to [9], all plants can absorb metals in varying amounts, but some plants can accumulate certain metal elements in high enough concentrations.

Hyperaccumulator plants are plants that are capable of accumulating heavy metals in high concentrations in root and shoot tissues [10] stated that the specifies a hyperaccumulator value of 1000 ppm for the Pb. Biduri is a drought-resistant, salt-tolerant plant and grows wild [11]. Based on the results of research by [12], the use of Agrobacterium sp increased Pb absorption by 84,8%-99,2% in hemp plants in Ngringo Village. *Rhizobium* sp. is a biological agent that has an important role in determining the success of plant bioremediation. Based on research by [13], the use of a combination of inorganic fertilizer treatment and Rhizobium sp. can increase the uptake of Pb in mendong plants by 52.01$\mu g.g^{-1}$ at the roots and 38.6$\mu g.g^{-1}$ at the shoots. Manure is an organic material, where the addition of manure can provide nutrients needed by plants and nutrients for microbes to allow rapid biodegradation of contaminants in the soil [14,15].

Research on Pb rhizoremediation with Biduri with a combination of inorganic fertilizers, bacteria, and organic matter has not been done much. Therefore, this research needs to be done. Pure bacterial isolates of jendi, isolates of *Rhizobium* Sp I3, *Agrobacterium* Sp I37, and manure were used as bioremediation agents to increase Biduri uptake of Pb. This study studied the use of Biduri with a combination of inorganic fertilizers, bioremediation agents, and metal doses of Pb in an effort to reduce Pb levels in the soil.

## 2. Material and method

### *2.1. Study area*
This research was conducted from June 2020-March 2021 in Greenhouse, Faculty of Agriculture, UNS. The analysis of soil and plant tissue samples was carried out in the Laboratory of Soil Biology and Biotechnology, Chemistry and Soil Fertility, and Physics and Soil Conservation.Analysis of heavy metal measurements was carried out at the Integrated Laboratory, Sebelas Maret University.

### *2.2. Materials*
The materials used in this study were *Rhizobium* sp.I3 and *Agrobacterium* sp.I37 Biduri seeds from the Bandung Textile Industry Center, inorganic fertilizer, organic matter (produced by Jatikuwung Innovation Center), The equipment needed is microscope, vortex, shaker, autoclave, Atomic Absorption Spectrophotometer (AAS), etc.

### *2.3. Experimental design*
This research using a factorial experimental design with a CRD with 3 factors: a) inorganic fertilizers(P): without inorganic fertilizers(P0) and with inorganic fertilizers(P1); b) bioremediation agent(R): without bioremediation agent (R0), with pure bacterial isolate of jendi bacteria(R1), with *Rhizobium* sp.I3(R2), with *Agrobacterium* sp.I37(R3), with organic matter(R4); c) dose of heavy metal lead(L): without metal(L0), dose 3μg.g-1(L1), and dose 233,33μg.g-1(L2). There were 30 treatment combinations with three repetitions, so there were 90 treatment units.

### *2.4. Procedure*

*2.4.1. Soil preparation.* The soil used in this study is soil originating from Ngargoyoso District, The soil to be used is sieved first, then the soil is weighed as much as 7kg.$pot^{-1}$ x 90bags. The sifted soil was then sterilized for 2 hours daily for three consecutive days using the steam method.

*2.4.2. Seed preparation.* Biduri seeds that have been sown and treated in the form of watering, when they have 4-5 leaves are transferred to polybags. Then, from the seeding results, they are ready to be planted after they are 1 month old.

*2.4.3 Bacterial inoculum*. The *Rhizobium* sp.I3 and *Agrobacterium* sp.I37 used were reinoculated from a pure isolate of *Rhizobium* sp.I3 and *Agrobacterium* sp.I37 and pure bacterial isolate of jendi bacteria were isolated from people's gold mine of Jendi Village. After growing, each bacterium was transferred to 100mL LB media. Then was shaken until the bacterial cell density reached $10^{10}$cells.mL-1. The bacterial dose was $10^{6}$cells.g$^{-1}$ soil, therefore the bacterial applied to the soil was $10^{10}$cell.mL$^{-1}$ and 7mL.

*2.4.4 Fertilizer preparation*. The dose of fertilizer used inorganic fertilizer is urea 60 kg.ha$^{-1}$, SP-36 20kg.ha$^{-1}$, KCL 30kg.ha$^{-1}$. The organic fertilizer (manure) is 10.000 kg.ha$^{-1}$. From the results after conversion, the fertilization dose for each pot was 0.47g urea, 0.195g SP-36, 0.175g KCL, and 0.035g organic matter.

*2.4.5 Heavy metal preparation*. Pb was weighed with a soil threshold dose of 3 ppm [7] as L1 and a high dose of 233.33ppm as L2. Each pot was given 0.021g.pot$^{-1}$ for L1 treatment and 0.233g.pot$^{-1}$ for L2 treatment.

### *2.5. Observation parameters and laboratory analysis*
Observed parameters were soil characteristics including a) CEC with the ammonium acetate saturation method; b) bacterial population with pH with the electrometric method; c) C-Organic levels using the Walkley & Black method; and d) Pb content with wet destruction then reading the metal content using AAS. Plant characteristics include: a) Pb levels of plants (roots and shoots) by wet destruction then reading metal content using AAS and b) dry weight of plants by direct measurement using digital scales.

### *2.6. Data analysis*
Data were analyzed by statistical using ANOVA with a 95%. The analysis is continued with an average comparison test using the DMRT.

## 3. Results and discussion

### *3.1. Dry weight of plant*
Based on the ANOVA results, all treatments had a very significant effect ($p<0.01$) on the dry weight of root and shoots. Based on Table 1, the P1R0L1 treatment had the highest average dry weight of Biduri roots at 59.77g, 61% higher, and significantly different from the control. The P1R2L1 treatment had the highest average dry biomass of Biduri shoots of 18.86 g, 41.09% higher, and significantly different from the control.

Inorganic fertilizers provide nutrients that are immediately available to plants so that they can increase plant growth and increase crop production. This study also showed that both roots and shoots had the highest dry weight of stover in the treatment with inorganic fertilizers. The combination of treatment with inorganic fertilizer, *Rhizobium* sp I3, and metal dose also had the highest dry weight of stover in the shoots. Based on the results of research by [16], Rhizobium Sp I3 can increase plant growth by synthesizing vitamins, enzymes, siderophores, and antibiotics. Rhizobium Sp I3 is a biological agent capable of releasing secondary metabolites that can change the form of heavy metals from forms that are not available to be available to plants, besides that Rhizobium Sp I3 can increase plant growth [13].

From the 3 treatment levels given the metal dose, it was also known that L2 showed the lowest growth of Biduri in terms of root and shoots dry biomass of biduri. This is possible because Biduri have started to experience Pb poisoning due to the high concentration of Pb. The higher the concentration of metals in the soil, the higher the stress level of the plant will increase, so that it can reduce plant biomass.

**Table 1.** Mean values of dry weight of plant roots and shoots

| Treatment | Root | Shoot | Total |
|---|---|---|---|
| P0R0L0 | 23.31ghijk | 11.11cdefg | 34.42 |
| P0R0L1 | 38.33bcdef | 11.02cdefg | 49.34 |
| P0R0L2 | 22.28ghijkl | 2.82i | 25.11 |
| P0R1L0 | 33.48cdefg | 9.44defgh | 42.91 |
| P0R1L1 | 22.28ghijkl | 7.25efghi | 29.54 |
| P0R1L2 | 20.36hijklm | 2.78i | 23.14 |
| P0R2L0 | 26.71fghij | 6.44efghi | 33.15 |
| P0R2L1 | 40.67bcd | 7.77efghi | 48.43 |
| P0R2L2 | 28.36defghi | 3.63hi | 31.99 |
| P0R3L0 | 15.20jklm | 13.46abc | 28.66 |
| P0R3L1 | 40.24bcde | 9.39defgh | 49.63 |
| P0R3L2 | 21.67ghijkl | 3.50hi | 25.17 |
| P0R4L0 | 43.30ab | 12.57bcde | 55.87 |
| P0R4L1 | 27.92efghi | 11.45cdefg | 39.37 |
| P0R4L2 | 13.59klm | 2.33i | 15.92 |
| P1R0L0 | 21.77ghijkl | 12.1bcdef6 | 33.93 |
| P1R0L1 | 59.77a | 14.17abcd | 73.94 |
| P1R0L2 | 17.95ijklm | 3.30hi | 21.25 |
| P1R1L0 | 40.57bcd | 15.84abc | 56.41 |
| P1R1L1 | 28.61defghi | 14.96abcd | 43.57 |
| P1R1L2 | 8.31m | 5.78ghi | 14.09 |
| P1R2L0 | 39.94bcde | 12.66abcd | 52.61 |
| P1R2L1 | 32.13cdefgh | 18.86a | 50.99 |
| P1R2L2 | 10.12lm | 3.43hi | 13.56 |
| P1R3L0 | 26.89fghij | 14.16abcd | 41.05 |
| P1R3L1 | 42.33bc | 16.99abc | 59.31 |
| P1R3L2 | 15.83ijklm | 7.44efghi | 23.27 |
| P1R4L0 | 24.74ghijk | 17.86ab | 42.60 |
| P1R4L1 | 49.67ab | 15.97abc | 65.64 |
| P1R4L2 | 16.85ijklm | 6.14fghi | 22.99 |

Note: The same small letters or no written in the same column showing no significant difference

*3.2. Soil Pb level and Pb uptake by plant*

*3.2.1. Soil Pb level.* Based on the ANOVA results, all treatments had a significant effect ($p<0.05$) on soil Pb levels. The P1R1L0 treatment had the lowest Pb content in the soil at 85.26$\mu g.g^{-1}$ or 33.71% lower and significantly different from the control treatment. P1R3L2 treatment had the highest soil Pb content of 149.91$\mu g.g^{-1}$. The P1R3L2 treatment had the highest soil Pb content of 149.91$\mu g.g^{-1}$. R1 treatment had the lowest soil Pb content compared to *Agrobacterium* sp. I37. This is presumably because the pure isolates of Jendi which include indigenous bacteria can increase the solubility and mobility of Pb so that it can be absorbed by plants. The mechanisms of bacteria with heavy metals include adsorption on extracellular polymeric substances on the cell surface, accumulation in cells, the transformation of metals into less toxic derivatives, and precipitation into carbonate compounds [17]. Secondary metabolites produced by the bacterial application can degrade heavy metal lead (Pb) via bioaccumulation [18]. Heavy metals that are accumulated in bacteria indicate that these heavy metals are still in the soil, not being absorbed by plants.

*3.2.2. Pb uptake by plant.* Based on the ANOVA results, all treatments had a very significant effect ($p<0.01$) on Pb uptake in the roots and of biduri. Based on Table 2, it is also known that the highest Pb uptake in the roots of Biduri in the P1R0L1 treatment was 2952.27μg.g$^{-1}$, 1.5x higher and significantly different from the control treatment. The highest Pb uptake in the of Biduri was the P1R2L1 treatment of 685.56μg.g$^{-1}$, 7 times higher and significantly different from the control treatment. Salt et al. [19] explained that plants release root exudates which can serve as nutrients for bacterial activity, thereby increasing the number of bacteria in the soil. Bacteria can help the process of decomposition of organic matter.

The addition of organic matter can increase heavy metal uptake because the anions in manure can bind heavy metal cations in the soil. Plants usually have different abilities for accumulating and distributing heavy metals to different parts of the plant. This can be seen from Table 2, which shows that the average Pb uptake in the roots was higher than of biduri. Some Pb hyperaccumulator plants can accumulate large amounts of Pb in plant roots contaminated with Pb, but only a small part accumulates in other plant parts. The plant's hyperaccumulator value for Pb is 1000ppm [10]. Based on Table 2, it was able to translocate Pb to its body tissues (roots to shoots) and the total Pb uptake by Biduri was more than 1000 μg.g$^{-1}$ for all treatments except the L2 treatment, so it can be said that the Biduri is a hyperaccumulator plant for Pb.

**Table 2.** Mean values of soil Pb level and Pb uptake by plant

| Treatment | Pb Soil Level (μg.g$^{-1}$) | Accumulation (μg.g$^{-1}$) | | |
|---|---|---|---|---|
| | | Root | Shoot | Total |
| P0R0L0 | 128.63abcd | 1153.27defghi | 97.46fgh | 125073 |
| P0R0L1 | 129.85abcd | 1614.60bcdefgh | 114.61fgh | 1729.21 |
| P0R0L2 | 130.98abcd | 618.82ghi | 98.25fgh | 717.07 |
| P0R1L0 | 85.59d | 1388.54bcdefghi | 81.71gh | 1470.25 |
| P0R1L1 | 100.34de | 1029.68efghi | 191.31defgh | 1220.99 |
| P0R1L2 | 103.80cde | 608.41ghi | 122.24fgh | 730.66 |
| P0R2L0 | 119.81abcde | 1159.00defghi | 97.92fgh | 1256.92 |
| P0R2L1 | 139.58abc | 1751.43bcdef | 137.77fgh | 1889.19 |
| P0R2L2 | 140.87abc | 812.05fghi | 68.91h | 880.96 |
| P0R3L0 | 111.60bcde | 807.50fghi | 326.48bcde | 1133.98 |
| P0R3L1 | 123.16abcd | 1637.81bcdefg | 199.92defgh | 1837.73 |
| P0R3L2 | 132.28abcd | 782.85fghi | 101.96fgh | 884.81 |
| P0R4L0 | 124.38abcd | 2260.67abc | 337.93bcd | 2598.61 |
| P0R4L1 | 126.14abcd | 1371.22bcdefghi | 428.75b | 1799.97 |
| P0R4L2 | 126.88abcd | 568.68ghi | 126.45fgh | 695.13 |
| P1R0L0 | 127.11abcd | 916.49efghi | 283.03bcdef | 1199.52 |
| P1R0L1 | 124.98abcd | 2952.27a | 291.89bcdef | 3244.16 |
| P1R0L2 | 131.68abcd | 612.59ghi | 81.09gh | 693.67 |
| P1R1L0 | 85.26e | 1893.24bcde | 204.40defgh | 2097.63 |
| P1R1L1 | 113.45abcde | 1229.81cdefghi | 337.15fgh | 1566.96 |
| P1R1L2 | 113.75abcde | 320.97i | 233.64cdefgh | 554.61 |
| P1R2L0 | 119.52abcde | 2330.42ab | 330.29bcde | 2660.70 |
| P1R2L1 | 122.88abcd | 1398.38bcdefghi | 685.56a | 2083.95 |
| P1R2L2 | 123.33abcd | 358.26i | 148.61defgh | 506.87 |
| P1R3L0 | 137.15abcd | 955.07efghi | 420.91bc | 1375.98 |
| P1R3L1 | 142.01ab | 1574.85bcdefgh | 442.68b | 2017.52 |
| P1R3L2 | 149.91a | 555.00ghi | 274.41bcdefg | 829.41 |
| P1R4L0 | 123.52abcd | 1085.94efghi | 437.34b | 1523.28 |
| P1R4L1 | 131.04abcd | 2173.77abcd | 333.34bcde | 2507.11 |
| P1R4L2 | 138.36abc | 550.11hi | 212.87defgh | 762.98 |

Note: The same small letters or no written in the same column showing no significant difference

### *3.3. Effectiveness of bioremediation*

The success of bioremediation in this study is a percentage of the magnitude of the decrease in Pb in the soil. However, it should be noted that the initial levels (155.35 $\mu g.g^{-1}$) of each treatment were influenced by the levels of Pb in organic matter, inorganic fertilizers, metal doses, and the initial levels of the soil itself. Plants accumulate heavy metals from the soil and distribute them to plant tissues. Ferina et al. [20] explained that the effectiveness of bioremediation or the so-called removal effectivity is the success of plants in absorbing heavy metals in soil which is calculated from the difference in initial and final levels after being given treatment. The effectiveness of bioremediation can be calculated using the following formula (equation 1):

$$\text{Effectiveness of bioremediation (\%)} = \frac{\textit{Initial Levels} - \textit{Final Grade}}{\textit{Initial Levels}} \text{ x } 100\% \quad (1)$$

The P0R1L2 treatment had the highest effectiveness of Pb by Biduri, at 73.29% (Table 3). While the effectiveness of Pb by Biduri was the highest at L0 treatment namely P1R1L0 treatment of 47.10%, at L1 treatment, namely P0R1L1 treatment of 36.63%, and at L2 treatment, that is treatment P0R1L2 at 73.29%. This shows that the pure bacterial isolate of jendi bacteria can be used as bioremediation agents for heavy metals that are not available to plants. Outside the surface of bacterial cells is a substance that functions to chelate metals, but extracellular polymeric substances derived from microbes in many types of macromolecules have not been determined to have a key role in binding metals [21, 22].

**Table 3.** Mean values of effectiveness of bioremediation

| Treatment | Effectiveness of Bioremediation (%) |
|---|---|
| P0R0L0 | 17.20 |
| P0R0L1 | 18.00 |
| P0R0L2 | 66.30 |
| P0R1L0 | 44.91 |
| P0R1L1 | 36.63 |
| P0R1L2 | 73.29 |
| P0R2L0 | 22.88 |
| P0R2L1 | 11.85 |
| P0R2L2 | 63.76 |
| P0R3L0 | 28.16 |
| P0R3L1 | 22.22 |
| P0R3L2 | 65.97 |
| P0R4L0 | 23.62 |
| P0R4L1 | 23.94 |
| P0R4L2 | 67.97 |
| P1R0L0 | 21.12 |
| P1R0L1 | 23.86 |
| P1R0L2 | 66.62 |
| P1R1L0 | 47.10 |
| P1R1L1 | 30.89 |
| P1R1L2 | 71.16 |
| P1R2L0 | 25.84 |
| P1R2L1 | 25.14 |
| P1R2L2 | 68.74 |
| P1R3L0 | 14.90 |
| P1R3L1 | 13.49 |
| P1R3L2 | 62.00 |
| P1R4L0 | 26.75 |
| P1R4L1 | 23.65 |
| P1R4L2 | 65.58 |

## 4. Conclusion

The use of Biduri with a combination of inorganic fertilizers, bioremediation agents, and doses of Pb metal is able to reduce Pb levels in the soil so that it can be used as an alternative to reduce soil pollution, one of the consequences of climate change which has an impact on agriculture and Biduri in this study have potential as hyperaccumulator plants for Pb.

## References

[1] Tchounwou P B, Yedjou C G, Patlolla A K and Sutton D J 2012 Molecular, clinical and environmental toxicicology Volume 3: Environmental Toxicology *Mol. Clin. Environ. Toxicol.* **101** 133–64
[2] Malar S, Shivendra Vikram S, JC Favas P and Perumal V 2016 Lead heavy metal toxicity induced changes on growth and antioxidative enzymes level in water hyacinths [Eichhornia crassipes (Mart.)] *Bot. Stud.* **55** 1–11
[3] Marrugo-Negrete J, Pinedo-Hernández J and Díez S 2017 Assessment of heavy metal pollution, spatial distribution and origin in agricultural soils along the Sinú River Basin, Colombia *Environ. Res.* **154** 380–8
[4] Huang Y, Deng M, Wu S, Japenga J, Li T, Yang X and He Z 2018 A modified receptor model for source apportionment of heavy metal pollution in soil *J. Hazard. Mater.* **354** 161–9
[5] Islam M S, Hossain M B, Matin A and Islam Sarker M S 2018 Assessment of heavy metal pollution, distribution and source apportionment in the sediment from Feni River estuary, Bangladesh *Chemosphere* **202** 25–32
[6] Mazurek R, Kowalska J, Gąsiorek M, Zadrożny P, Józefowska A, Zaleski T, Kępka W, Tymczuk M and Orłowska K 2017 Assessment of heavy metals contamination in surface layers of Roztocze National Park forest soils (SE Poland) by indices of pollution *Chemosphere* **168** 839–50
[7] Indonesian Goverment 2014 *PP 101 Tentang Pengelolaan Limbah Bahan Berbahaya dan Beracun (in Indonesian)* (Indonesia)
[8] Al Ameeri D K and Al Sarawi M 2017 Rhizoremediation of Contaminated Soils by Comparing Six Roots Species in Al-Wafra, State of Kuwait *J. Bioremediation Biodegrad.* **08** 1–14
[9] Rondonuwu S B 2014 Fitoremediasi Limbah Merkuri Menggunakan Tanaman Dan Sistem Reaktor *J. Ilm. Sains* **14** 52
[10] Zhang X, Li M, Yang H, Li X and Cui Z 2018 Physiological responses of Suaeda glauca and Arabidopsis thaliana in phytoremediation of heavy metals *J. Environ. Manage.* **223** 132–9
[11] Suresh Kumar P, Chezhian A, Senthil Raja P and Sathiyapriya J 2012 Computational selections of terpenes present in the plant Calotropis gigantea as mosquito larvicide's by blocking the sterol carrying protein, AeSCP-2 *Bangladesh J. Pharmacol.* **7** 1–5
[12] Rosariastuti R, Barokah U, Purwanto and Supriyadi 2018 Phytoremediation of Pb Contaminated Paddy Field Using Combination of Agrobacterium sp. I3, Compost and Ramie (Boehmeria nivea) *J. Degrad. Min. L. Manag.* **5** 1381–8
[13] Rosariastuti R, Mahardika D P, Supriyadi, Purwanto and Hartati S 2019 Soil Bioremediation of lead (Pb) polluted paddy field using Mendong (Fimbristylis globulosa), Rhizobium SpI3, compost, and inorganic fertilizer *IOP Conf. Ser. Earth Environ. Sci.* **230** 1–7
[14] Agamuthu P, Tan Y S and Fauziah S H 2013 Bioremediation of Hydrocarbon Contaminated Soil Using Selected Organic Wastes *Procedia Environ. Sci.* **18** 694–702
[15] Masciandaro G, Macci C, Peruzzi E, Ceccanti B and Doni S 2013 Organic matter-microorganism-plant in soil bioremediation: A synergic approach *Rev. Environ. Sci. Biotechnol.* **12** 399–419
[16] Pramono A, Rosariastuti M R, Ngadiman and Prijambada I D 2012 Peran Rhizobakteri dalam Fitoekstraksi Logam Berat *Ecolab* **6** 38–50
[17] Sakpirom J, Kantachote D, Nunkaew T and Khan E 2017 Characterizations of purple non-sulfur bacteria isolated from paddy fields, and identification of strains with potential for plant growth-

promotion, greenhouse gas mitigation and heavy metal bioremediation *Res. Microbiol.* **168** 266–75
[18] Narayani M and Shetty K V 2013 Critical Reviews in Environmental Science and Technology Chromium-Resistant Bacteria and Their Environmental Condition for Hexavalent Chromium Removal : A Review Chromium-Resistant Bacteria and Their Environmental Condition for Hexavalent Chromium Removal *Crit. Rev. Environ. Sci. Technol.* **43** 955–1009
[19] Salt D E, Smith R D and Raskin I 1998 Phytoremediation *Annu. Rev. Plant Physiol. Plant Mol. Biol.* **49** 643–68
[20] Ferina P, Rosariastuti R and Supriyadi S 2017 The effectiveness of Mendong plant (Fimbrystilis globulosa) as a phytoremediator of soil contaminated with chromium of industrial waste *J. Degrad. Min. Lands Manag.* **04** 899–905
[21] Paul A P A K 2008 Microbial extracellular polymeric substances : central elements in heavy metal bioremediation *Indian J. Microbiol* **48** 49
[22] Tiquia-arashiro S M 2018 Lead absorption mechanisms in bacteria as strategies for lead bioremediation *Appl. Microbiol. Biotechnol.* **102** 5437–44

**Acknowledgements**
This research is an independent research and authors would like to express their gratitude to Prof. Dr. Ir. M.M.A Retno Rosariastuti, M. Si. who has funded this independent research in 2020-2021.

# Potential lines of black rice crossing with Jeliteng variety and their reciprocals

**Nandariyah[1], E Yuniastuti [1], E Purwanto[1] and R D Astuti[2]***

[1]Lecturer of Agrotechnology Department, Faculty of Agriculture, Sebelas Maret University, Surakarta, Indonesia
[2]Undergraduate Program of Agrotechnology Department, Faculty of Agriculture, Sebelas Maret University, Surakarta, Indonesia
*Corresponding author: rizki.dyaha@student.uns.ac.id

**Abstract.** Climate change causes an increase in rice plant pests, one of them is brown planthopper. The effort through an assembly of black rice potential lines was made by mutation induction of Cempo Ireng black rice, but it's only able to shorten the age of plant without being able to improve rice resistance to brown planthopper. Jeliteng as a new high yielding variety of black rice that is resistant to brown planthopper biotype 1 makes it worthy of being a potential parent source. Reciprocal crosses needed to be carried out to get offspring that have superior traits from both parents. This research aims to get F1 generation from the potential lines of black rice crossing with the Jeliteng variety and their reciprocals to improve local varieties of black rice. The research was conducted in the experimental field of the village of Gawanan, Sukoharjo Regency from August 2020 to April 2021. The planting materials were the Jeliteng variety as male parents while 6 potential lines of black rice 8, 13, 44, 46, 51, 52, and Cempo Ireng variety as female parents. Crosses were conducted by hand crossing technique on 14 sets. The data were analyzed descriptively and using the Chi-square test. The results showed that the crossing set could be crossed and produced fertile F1 generation. The length of grain, the width of grain, and the weight of a grain of the F1 varied in all sets. The skin colour of the F1 generation in all sets was mostly straw yellow with brown spots.

**Keywords**: reciprocals crossing, potential lines, F1 generation

## 1. Introduction

Black rice (*Oryza sativa* L.) contains anthocyanin, iron, and high fiber. The fiber content of organic black rice is higher than brown rice and white rice [1]. The high economic value makes black rice has the potential to be developed. The weakness of local varieties of black rice such as the long harvesting age and susceptibility to pests and diseases. The age of black rice can reach more than 145 days [2]. In addition, local black rice is known to have high plant posture, long harvesting age, low productivity, and susceptibility to pests and diseases [3] that make farmers unwilling to cultivate it.

Climate change also has a discontinuous impact on the explosion of pests and diseases in cultivated areas [4], especially the brown planthopper (BPH). The brown planthopper will activate the biological clock so the population increases due to the extreme La Nina climate [5]. The plant breeding program through mutation induction of Cempo Ireng black rice has succeeded in assembling potential lines of black rice which have been short-lived. The selection program in the M5 generation has been obtained potential lines which have high productivity, short flowering age, and short harvesting age [6]. The yield trial has been carried out on potential lines in the M5 generation by comparison with the Cempo Ireng variety without radiation [7]. However, these potential lines are still susceptible to brown planthopper.

The government released Jeliteng as a new high-yielding variety (HYVs) in 2019. This variety has ±113 days of age, has a potential yield of ±9.87 tons/ha, a plant height of ±106 cm, and is resistant to the brown planthopper (BPH) biotype 1 makes it worthy of being a potential parent source in the next breeding programs [8]. The large diversity of germplasm is important in the creation of new varieties. A new variety can be produced from the crossing program. Plants crossing as one of the plant breeding programs are expected to have superior characteristics from both parents. However, reciprocal crosses are needed to determine the inheritance of a trait which the crosses are carried out to determine the influence of the female parent (*maternal effect*) [9]. This research aims to get F1 generation and determine the crossing success between potential lines of black rice with the Jeliteng variety and their reciprocals as a reference to improve local varieties of black rice.

## 2. Material and method

The research was conducted in the experimental field (plastic greenhouse) of the village of Gawanan Sukoharjo, Sukoharjo Regency from August 2020 to April 2021. The planting materials in the study were the Jeliteng variety as male parents and 7 female parents, including 6 potential lines of black rice numbers 8, 13, 44, 46, 51, 52, and the Cempo Ireng variety as a control. Other materials include manure, phonska NPK fertilizer, KCl, EM4, and sugar solution. The tools in this study were polybags, cultivation tools, stakes, cross tools (scissors, tweezers, toothpicks, plastic clips, magnifying glass, petridish), stationery, millimetre blocks, analytical scales, and cameras.

The arrangement of polybags is 40x20 cm between potential lines and varieties. 4 seeds were planted per polybag. Each potential line and variety was repeated 5 times, so there were 40 experimental units. This research was conducted using 2 varieties and 6 potential lines with 14 sets. There were 5 repetitions per set so there were 35 repetitions on the same crosses. The crossing technique is done by the hand crossing method. The potential lines of black rice and the Cempo Ireng variety were planted earlier because they had a long-lived other than the Jeliteng variety, while the Jeliteng variety was planted between 35-51 days later. Observations variables included the crossing success (%), the length and the width of grain, the weight of grain, and the skin colour of the grain. The data were analyzed descriptively and using the Chi-Square test. The crossing success (%) was calculated using the following formula (1).

$$\text{Percentage of crossing success} = \frac{\sum \text{Grain formed}}{\sum \text{Pollinated flower}} \text{ x } 100\% \quad (1)$$

## 3. Result and discussion

### *3.1 The crossing process*

The crossing program is carried out by taking notice of the selection of elders (selection), castration/emasculation, and hybridization. Potential lines of black rice were chosen as female parents because have short plant posture and has short-lived. The resistance to brown planthopper (BPH) biotype 1 makes the Jeliteng variety used as the male parent in the crossing program.

Castration was carried out between 02.00-05.00 pm by removing the anthers from the rice flower buds (Figure 1). Castration is carried out on the day before pollination on flowers which will use as female parents. Reciprocal crosses require that the castration process be carried out on both parents. The success of castration is influenced by the selection of rice flower buds and the castration method. The anthers which are ready for castration will be in the middle of the flower bud [10]. Castration is carried out on flowers that have not bloomed/nearly bloomed yet and are still tightly closed so the anthers can be removed intact [11]. The castration method used hand crossing which shows higher efficiency than the suction method, 70% alcohol immersion method, and absolute [12].

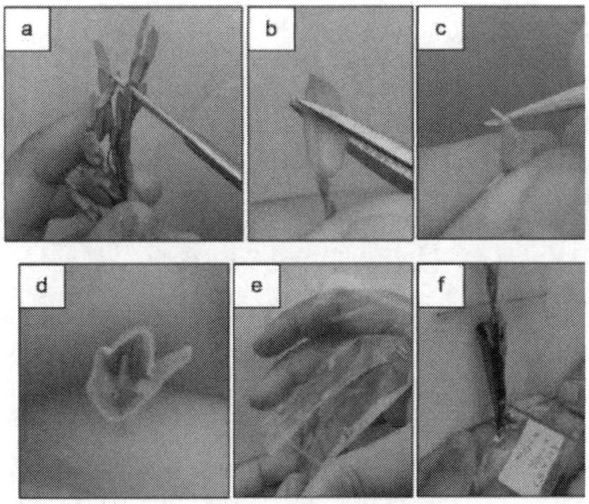

**Figure 1.** Rice flower castration process; Thinning and selection of rice flower buds (a); Cutting one-third of the rice flower buds from lemma to palea (b); Extraction of anthers from inside flower buds (c); Rice flower buds that have been cleaned from anthers (d); Closure of flower buds that have been castrated (e); Inclusion of castration date (f)

Crosses between potential lines of black rice and Jeliteng variety were carried out to produce F1 seeds which had a combination of superior traits from both parents which is short plant posture, has short-lived, and resistant to brown planthopper biotypes 1, 2, and 3. Hybridization was carried out one day after castration between 08.00-11.30 am. The success of hybridization is influenced by the accuracy of the observation time of anthesis (rupture of anthers), pollen viability, pistil receptivity, and pollination techniques. The time of anthesis is marked by the elongation of the stamens due to the expansion of the lodicule at the base of the flower. Pollen viability persisted for 5 minutes after anthesis but in contrast to the stigma which remained receptive or was able to survive waiting for pollen reception for 7 days [13]. Pollen can still germinate (viable) if it is on a receptive/compatible stigma [14]. Compatibility between pollen and the stigma will affect the success of pollen germination and the formation of ovules after pollination [15]. The pollination technique used in this research was by applying/inserting pollen using a toothpick (Figure 2). This is done to increase the success rate because the breeder can ensure in detail the pollen falls right on the stigma. Another way that can be done is by shaking the male panicles/flowers that have undergone anthesis over the female flowers [11]. However, this method is difficult to do considering the timing of anthesis between flowers is not simultaneous and cannot ensure that pollen falls on all the stigmas which are causing the low success of crossing.

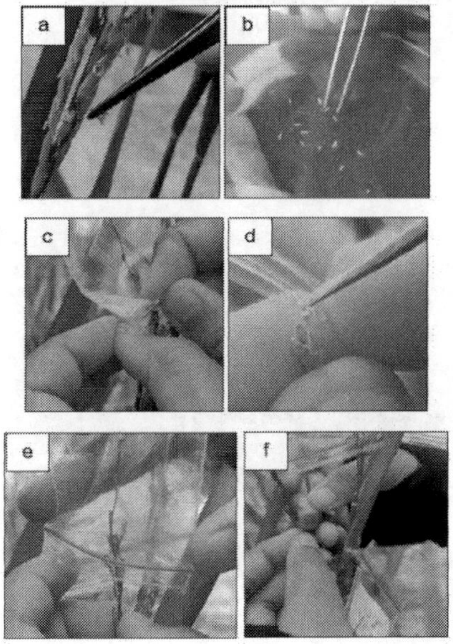

**Figure 2.** Hybridization process; Removal of anthers from male parents using tweezers (a); Collected the anthers on the petri dish (b); Opening plastic clips on the previously castrated flowers (c); Applying pollen to the top of the pistil using a toothpick (d); Closure of pollinated panicles (e); Labeling parents and date of hybridization (f)

*3.2 The crossing success (%)*

The success of the crossing showed to be high if all the cross sets show a percentage more than 30% and showed to be low if the cross set shows a percentage less than 30% [16]. The results of the Chi-Square analysis showed that the crossing between the potential lines of black rice and the Jeliteng variety was low. Table 1 shows that there were 15 repetitions of cross sets with a high percentage of success and 20 repetitions of cross sets with a low percentage of success. Although the success of the crossing is low, the seeds can still be used for further breeding programs because gene transfer in seeds still occurs. The success of the cross did not affect the gene flow between the two parents of the rice cross [17]. The close distance of flowering in the potential lines 13 with the Jeliteng variety is the cause of the high success because the flower buds are still abundant for use in crosses. The high success of the crossing was also influenced by several factors including the interaction of pollen with the pistil, pollen viability, and pollen adhesion which the state of the pollen grains was always hydrated during the introduction to the stigma [18].

**Table 1.** Percentage of the crossing success between potential lines of black rice (♀) x Jeliteng variety (♂)

| Crosses Sets | Success | | Total | P-value |
|---|---|---|---|---|
| | High | Low | | |
| 8XJ | 2 | 3 | 5 | .052 |
| | 40.0% | 60.0% | 100.0% | |
| 13XJ | 5 | 0 | 5 | |
| | 100.0% | 0.0% | 100.0% | |
| 44XJ | 3 | 2 | 5 | |
| | 60.0% | 40.0% | 100.0% | |
| 52XJ | 2 | 3 | 5 | |
| | 40.0% | 60.0% | 100.0% | |
| 51XJ | 2 | 3 | 5 | |
| | 40.0% | 60.0% | 100.0% | |
| 46XJ | 0 | 5 | 5 | |
| | 0.0% | 100.0% | 100.0% | |
| KXJ | 1 | 4 | 5 | |
| | 20.0% | 80.0% | 100.0% | |
| Total | 15 | 20 | 35 | |
| | 42.9% | 57.1% | 100.0% | |

Based on Chi-square 5% significancy; 8, 13, 44, 52, 51, 46=potential lines of black rice; K=control/Cempo Ireng variety; J=Jeliteng variety.

The results of the Chi-Square analysis showed that the reciprocal crosses between the Jeliteng variety and potential lines of black rice were low. Table 2 shows that there were 4 sets of crosses with a high percentage of success and 31 repetitions of crosses with a low percentage of success and even 2 sets of crosses that did not produce F1 seeds. The low success of the crossing is caused by the failure of anthesis found in several potential lines, especially the potential lines 8 and 46. Other factors that affected the low success of reciprocal crosses are the skill and thoroughness of the researchers and the absence of assistants in the crossing process. The determining factor for the success of the cross is one's expertise in crossing rice flowers [10]. It is necessary to experiment on various plant samples before crossing activities begin so that they have been skills during the crossing process. The assistant when the crossing process is needed to be improved so that the success of the crossing is high because many rice flowers can be crossed and the time required is short.

**Table 2.** Percentage of the crossing success between Jeliteng variety (♀) x potential lines of black rice (♂)

| Crosses Sets | Success | | Total | P value |
|---|---|---|---|---|
| | High | Low | | |
| JX8 | 0 | 5 | 5 | 1.000 |
| | 0.0% | 100.0% | 100.0% | |
| JX13 | 1 | 4 | 5 | |
| | 20.0% | 80.0% | 100.0% | |
| JX44 | 0 | 5 | 5 | |
| | 0.0% | 100.0% | 100.0% | |
| JX52 | 1 | 4 | 5 | |
| | 20.0% | 80.0% | 100.0% | |
| JX51 | 1 | 4 | 5 | |
| | 20.0% | 80.0% | 100.0% | |
| JX46 | 0 | 5 | 5 | |
| | 0.0% | 100.0% | 100.0% | |
| JXK | 1 | 4 | 5 | |
| | 20.0% | 80.0% | 100.0% | |
| Total | 4 | 31 | 35 | |
| | 11.4% | 88.6% | 100.0% | |

Based on Chi-square 5% significancy; 8, 13, 44, 52, 51, 46=potential lines of black rice; K=control/Cempo Ireng variety; J=Jeliteng variety.

*3.3 The length and the width of the grain*

One of the characters which can be inherited from the parents is the length and the width of grain. The length and the width of grain are the parameters that determine the purity of a variety [16]. Based on table 3, the cross set of the potential line 44xJeliteng variety showed the longest average grain of 7.26 mm with a range of 6.3-8 mm. The widest average grain is shown in the cross set of the potential line 8xJeliteng variety, which is 2.3 mm with a range of 2-2.5 mm. The grain size (length, width, and thickness of the grains) is influenced by the grain development process during the pre-anthesis period where this period will directly affect the availability of assimilating [19]. Rice flowers with a small size in the potential lines also affect the size of the grain.

**Table 3.** The length and the width of grain on the crossing of potential lines of black rice (♀) x Jeliteng (♂)

| Crosses Sets | The Length of Grain | | Average (mm) | The Width of Grain | | Average (mm) |
|---|---|---|---|---|---|---|
| | Min | Max | | Min | Max | |
| 8xJ | 6 | 7.3 | 6.86 | 2 | 2.5 | 2.3 |
| 13xJ | 6.3 | 7.5 | 6.81 | 2 | 2.5 | 2.19 |
| 44xJ | 6.3 | 8 | 7.26 | 2 | 2.3 | 2.12 |
| 52xJ | 6.3 | 8 | 6.73 | 2 | 2.5 | 2.11 |
| 51xJ | 6 | 8 | 7.01 | 2 | 2.5 | 2.14 |
| 46xJ | 6.3 | 7.3 | 6.81 | 2 | 2.5 | 2.22 |
| KxJ | 6 | 8 | 7.18 | 2 | 2.3 | 2.24 |

Description: 8, 13, 44, 52, 51, 46=potential lines of black rice; K=control/ Cempo Ireng variety; J=Jeliteng variety.

**Table 4.** The length and the width of grain on the crossing of Jeliteng (♀) x potential lines of black rice crossing (♂)

| Crosses Sets | The Length of Grain | | Average (mm) | The Width of Grain | | Average (mm) |
|---|---|---|---|---|---|---|
| | Min | Max | | Min | Max | |
| Jx13 | 6.3 | 8 | 7.49 | 2 | 2.5 | 2.33 |
| Jx44 | 6 | 8 | 7 | 2 | 2.3 | 2.15 |
| Jx52 | 7 | 8.5 | 7.46 | 2 | 2.7 | 2.16 |
| Jx51 | 7 | 8.5 | 7.66 | 2 | 2.7 | 2.34 |
| JxK | 5.3 | 6.5 | 6.11 | 2 | 2.7 | 2.25 |

Description: 8, 13, 44, 52, 51, 46=potential lines of black rice; K=control/ Cempo Ireng variety; J=Jeliteng variety.

Based on Table 4, the longest and widest grain sizes were shown in the reciprocal crosses of Jelitengx potential line 51 with an average of length is 7.66 mm in the range of 7-8.5 mm and average width is 2.34 mm in the range of 2-2.7 mm. The length and the width of grain are influenced by the presence of regulatory genes that play a role in the development of rice grain size [20]. Overexpression of a specific gene can also increase rice grain size to increase rice grain productivity [21]. The grain size was positively correlated with pistil mass at anthesis, grain filling rate, and seed viability [19]. Reciprocals crosses of Jelitengx potential lines 8 and 46 were failed because did not produce seeds.

*3.4 The weight of grain*

The weight of grain is the final mass of seeds after harvesting and drying. The final mass of the seed is affected by the filling period and the ripening period where is the accumulation of storage material (especially carbohydrates, proteins, and fatty acids) [19]. Based on table 5, the highest weight of 13 seeds was shown in the cross set of the Cempo Ireng (K)xJeliteng variety, which was 0.1744 g. The weight per grain will affect the weight of the seeds where the heavier the grains, the heavier the seeds will be [22]. Seeds in the cross set of potential line 44xJeliteng showed the lowest weight of 13 seeds among the other expected cross sets, which was 0.1253 g. The duration of the fast grain filling causes the accumulation of materials (carbohydrates, proteins, and fatty acids) in the seeds to be less than optimal.

**Table 5.** The weight of grain on the crossing of potential lines of black rice(♀) x Jeliteng(♂)

| Crosses Sets | Weight of 13 grains |
|---|---|
| 8xJ | 0.1313 |
| 13xJ | 0.1418 |
| 44xJ | 0.1253 |
| 52xJ | 0.1336 |
| 51xJ | 0.157 |
| 46xJ | 0.1411 |
| KxJ | 0.1744 |

**Table 6.** The weight of grain on the crossing of Jeliteng (♀) x potential lines of black rice crossing(♂)

| Crosses Sets | $\sum grain$ | The weight of all grain |
|---|---|---|
| Jx13 | 9 | 0.1176 |
| Jx44 | 2 | 0.0269 |
| Jx52 | 10 | 0.1234 |
| Jx51 | 17 | 0.1869 |
| JxK | 11 | 0.085 |

Weighing for the weight of seeds from reciprocal crosses was carried out on all F1 seeds. This is due to the small number of seeds produced through this crossing program. Based on table 6, the highest seed weight was shown in the set of reciprocal crosses Jelitengxpotential line 51, which was 0.1869 g from 17 seeds produced. The lowest weight was found in the set of reciprocal cross Jelitengxpotential line 44, which was 0.0269 g from 2 seeds. Reciprocal crosses on the Jelitengxpotential lines 8 and 46 failed because did not produce seeds.

### *3.5 The skin colour of the grain*

**Table 7.** The skin colour of grain on the crossing of potential lines of black rice(♀) xJeliteng(♂)

| Crosses Sets | The skin colour of the grain | | | | | Total of Grain |
|---|---|---|---|---|---|---|
| | a | b | c | d | e | |
| 8xJ | 2 | 0 | 11 | 12 | 16 | 41 |
| 13xJ | 8 | 0 | 36 | 39 | 10 | 93 |
| 44xJ | 6 | 0 | 38 | 22 | 11 | 77 |
| 52xJ | 4 | 0 | 42 | 26 | 2 | 74 |
| 51xJ | 6 | 0 | 37 | 20 | 11 | 74 |
| 46xJ | 0 | 0 | 6 | 6 | 1 | 13 |
| KxJ | 11 | 4 | 45 | 18 | 0 | 78 |
| Total | 37 | 4 | 215 | 143 | 51 | 450 |

Description:a=straw yellow; b= golden(golden stripe on a straw yellow background); c=straw yellow with brown spots; d=straw yellow with brown stripes; e=yellowish brown (orange brown/brown)

The true skin colour of grain is the colour of lemma and palea. The skin colour of the grain will appear after the grains are fully cooked. The characterization of grain's skin colour is done by scoring. Based on table 7, the overall results of crosses between the potential lines with Jeliteng showed that most of the skin colour of grain is straw yellow with brown spots (see Figure 3). Inheritance of colour follows the pattern of inheritance of the maternal effect [23]. Maternal effects can be detected if further selection shows the colour of the seeds from the cross according to the female parent [24].

Based on table 8, the overall results of the reciprocal crosses between Jeliteng and potential lines showed the same total number of straw yellow grain and straw yellow with brown stripes. Inheritance of colour will be controlled by a single pair of genes (monogenically inherited) which is dominant to control colour [25]. This qualitative characterization is not followed by seed colour because the seeds from the cross will be used for the next breeding program.

**Table 8.** The skin colour of grain on the crossing of Jeliteng (♀) x potential lines of black rice(♂)

| Crosses sets | The skin colour of the grain | | | | | Total of grain |
|---|---|---|---|---|---|---|
| | a | b | c | d | e | |
| Jx8 | 0 | 0 | 0 | 0 | 0 | 0 |
| Jx13 | 6 | 0 | 3 | 0 | 0 | 9 |
| Jx44 | 0 | 0 | 0 | 0 | 2 | 2 |
| Jx52 | 0 | 0 | 2 | 8 | 0 | 10 |
| Jx51 | 12 | 1 | 0 | 0 | 4 | 17 |
| Jx46 | 0 | 0 | 0 | 0 | 0 | 0 |
| JxK | 0 | 1 | 0 | 10 | 0 | 11 |
| Total | 18 | 2 | 5 | 18 | 6 | 49 |

Description:a=straw yellow; b= golden(golden stripe on a straw yellow background); c=straw yellow with brown spots; d=straw yellow with brown stripes; e=yellowish brown (orange brown/brown)

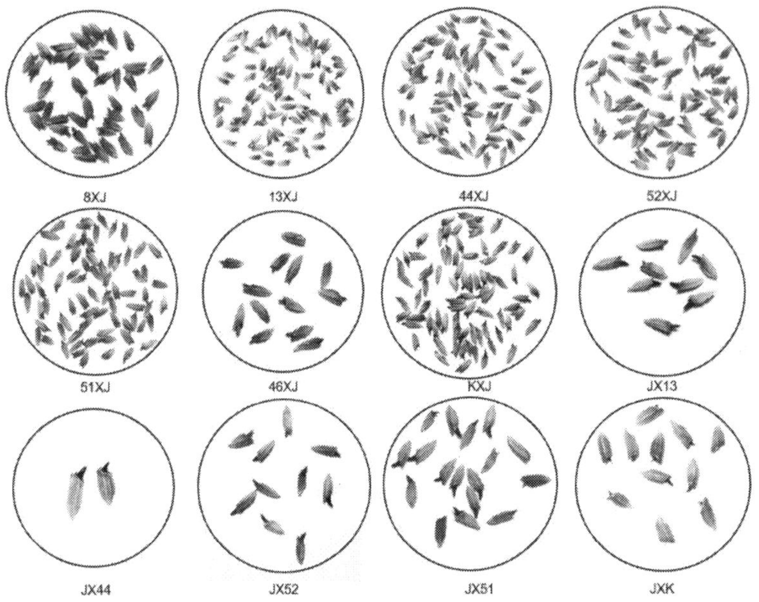


**Figure 3.** The skin colour grain on the crossing of p lines of black rice(♀) x Jeliteng(♂) and their recip

## 4. Conclusion

Based on the results of the study, it can be concluded that the potential lines of black rice and Jeliteng variety also their reciprocals can be crossed and produce a fertile F1 generation. Most of the success of the crosses in each set was low even though these cross sets produced F1 generation. The Length of grain, the width of grain, and the weight of seed on F1 progeny are varied in all sets of crosses. The skin colour of the F1 generation in all sets of crosses was mostly straw yellow with brown spots. The assembling of high-yielding variety (HYVs) is an effective effort to obtain the potential lines which resistant to brown planthoppers amid these uncertain climate change problems. However, future research must plant the F2 generation to test the resistance of planthoppers from the F1 generation.

## References

[1] Hernawan E and Meylani V 2016 Analisis Karakteristik Fisikomia Beras Putih, Beras Merah, dan Beras Hitam (Oryza sativa L., Oryza nivara dan Oryza sativa L. indica) *Jurnal Kesehatan Bakti Tunas Husada: Jurnal Ilmu-ilmu Keperawatan, Analis Kesehatan dan Farmasi* **15** 79

[2] Warman B, Suliansyah I, Swasti E and Syarif A 2016 Perbaikan Genetik Kultivar Padi Beras Hitam Lokal Sumatera Barat melalui Mutasi Induksi *Jurnal Ilmiah Aplikasi Isotop dan Radiasi* **11** 125–36

[3] Abdullah B 2017 Peningkatan Kadar Antosianin Beras Merah dan Beras Hitam melalui Biofortifikasi *Jurnal Penelitian dan Pengembangan Pertanian* **36** 91

[4] Widiarta I N 2016 Teknologi Pengelolaan Tanaman Pangan dalam Beradaptasi terhadap Perubahan Iklim pada Lahan Sawah *Jurnal Sumberdaya Lahan* **10** 91–102

[5] Susanti E, Surmaini E and Estiningtyas W 2019 Parameter Iklim sebagai Indikator Peringatan Dini Serangan Hama Penyakit Tanaman *Jurnal Sumberdaya Lahan* **13** 59

[6] Nandariyah, Devitha M, Parjanto, Suharyana, Riyatun and Sutarno 2020 Evaluation Agronomy Character of Irradiated Black Rice Cempo Ireng Mutant Strains M5 with 300 Gy of Gamma rays *International Conference on Science and Applied Science* 1–6

[7] Nandariyah, Sutarno, Riyatun and Suharyana 2020 *Pemuliaan Mutasi Padi Beras Hitam* (Surakarta: TrustMedia Publishing)

[8] Padi B B P T 2019 Inbrida Padi Sawah (Inpari) Jeliteng

[9] Napitupulu M and Damanhuri 2019 Keragaman Genetik, Fenotipe dan Heritabilitas Pada Generasi F2 Hasil Persilangan Tanaman Padi (Oryza Sativa L.) *Jurnal Produksi Tanaman* **6** 1844–50
[10] Subantoro R, Wahyuningsih S and Prabowo R 2008 Pemuliaan Tanaman Padi (Oryza sativa L.) Varietas Lokal menjadi Varietas Lokal yang Unggul *Mediagro* **4** 62–74
[11] Masniawati A, Baharuddin, Joko T and Abdullah A 2015 Pemuliaan Tanaman Padi Aromatik Lokal Kabupaten Enrekang Sulawesi Selatan *Jurnal Sainsmat* **4** 205–12
[12] Rather S A, Pandey I D, Panda G S and Kumar S 2017 Relative Effiiency of Different Emasculation Methods in Rice (Oryza sativa L.) *Environment & Ecology* **35** 2205–8
[13] Widyastuti Y, Rumanti I A and Satoto 2012 Perilaku Pembungaan Galur-galur Tetua Padi Hibrida *IPTEK Tanaman Pangan* **7** 67–78
[14] Guru T, Kunta R T and Rao P R 2016 Improving Pollen Viability and Stigma Receptivity by Boron Application as a Physiological Approach for High Yields in Rice *An International Quarterly Journal of Life Sciences* **11** 79–83
[15] Sari D K, Kendarini N and Damanhuri 2017 Studi Inkompatibilitas pada Beberapa Kombinasi Persilangan Ubi Jalar (Ipomoea batatas L.) *Jurnal Produksi Tanaman* **5** 1518–24
[16] Yanuar A D and Adiredjo A L 2018 Persilangan Beberapa Varietas Padi Gogo dan Padi Sawah (Oryza sativa L.) untuk Menghasilkan F1 *Jurnal Produksi Tanaman* **6** 2022–9
[17] Li J Q, Lu Y G, Feng J H, Zhao X J and Liu X D 2007 The Crossability and F1 Hybrid Fertility between Oryza sativa and other AA Genome Species *J. Plant Genet. Resour* **8** 1–6
[18] Patil P, Malik S K, Negi K S, John J, Yadav S, Chaudhari G and Bhat K V 2013 Pollen Germination Characteristics, Pollen-pistil Interaction and Reproductive Behaviour in Interspecific Crosses among Abelmoschus esculentus Moench and its Wild Relatives *Grana* **52** 1–14
[19] Gupta P K, Rustgi S and Kumar N 2006 Genetic and Molecular Basis of Grain Size and Grain Number and its Relevance to Grain Productivity in Higher Plants *Genome* **49** 565–71
[20] Yang X, Ren Y, Cai Y, Niu M, Feng Z, Jing R, Mou C, Liu X, Xiao L, Zhang X, Wu F, Guo X, Jiang L and Wan J 2018 Overexpression of OsbHLH107, a Member of the Basic Helix-loop-helix Transcription Factor Family, Enhances Grain Size in Rice (Oryza sativa L.) *Rice* **11** 1–12
[21] Qin M, Zhang Y, Yang Y, Miao C and Liu S 2020 Seed-Specific Overexpression of SPL12 and IPA1 Improves Seed Dormancy and Grain Size in Rice *Frontiers in Plant Science* **11** 1–7
[22] Suprayogi S, Praptiwi M A, Iqbal A and Agustono T J 2021 Keragaan Agronomik Populasi F4 Hasil Persilangan Padi IR 36 dengan Padi Merah PWR *Vegetalika* **10** 81
[23] Guo Z, Xu P, Zhang Z and Guo Y 2011 Segregation Ratios of Colored Grains in Crossed Wheat *Australian Journal of Crop Science* **5** 589–94
[24] Prastini L and Damanhuri 2017 Pengaruh Perbedaan Waktu Emaskulasi terhadap Keberhasilan Persilangan Tanaman Padi Hitam x Padi Putih (Oryza sativa L.) *Jurnal Produksi Tanaman* **5** 217–23
[25] Oktarisna F A, Soegianto A and Sugiharto A N 2013 Pola Pewarisan Sifat Warna Polong pada Hasil Persilangan Tanaman Buncis (Phaseolus vulgaris L .) Varietas Introduksi dengan Varietas Lokal *Jurnal Produksi Tanaman* **1** 81–9

# Vegetative growth of Biduri (*Calotropis Gigantea*) on mycorrhiza and azolla application

**Amalia Tetrani Sakya*, Samanhudi and M M Prakoso**

Agrotechnology Department, Faculty of Agriculture, Sebelas Maret University, Surakarta 57126, Indonesia
*Corresponding author: amaliatetrani@staff.uns.ac.id

**Abstract**. Climate change is causing many environmental problems that impact human health. Extreme environmental conditions affect the immune of the human body. It is easily exposed to the disease. The community takes various precautions to improve the immune system. One of them is to use medicinal plants as an endurance enhancer because some medicinal plants contain immunomodulators. Biduri (*Calotropis Gigantea*) is one of the medicinal plants that allegedly serves as an immune stimulator or immunosuppressive antibody production. Biduri has not been cultivated by the community, growing wildly. Research using Azolla and mycorrhizal was conducted as an effort to domesticize Biduri and prepare the source of medicinal raw material. The research aims to study the effect of Azolla and mycorrhiza on the vegetative growth of biduri. The study used 5 levels of Azolla, namely 0, 5, 7.5, 10, and 12.5 tons $ha^{-1}$, and 4 levels mycorrhizal, namely 0, 5, 10, and 15 tons $ha^{-1}$. Treatment combinations are arranged based on a Randomized Complete Block Design and repeated 6 times. The vegetative growth including the plant height, number of leaves, number of leaves, leaf width, leaf length was measured at 12 weeks after transplanting. The response of vegetative growth on Azolla and mycorrhizal application showed the same pattern, the application of mycorrhizal has not shown a difference in height or development of biduri leaves, while the application of Azolla 12.5 tons $ha^{-1}$ increases the height of plants and leaves development significantly compared to without Azolla.

**Keywords** : Azolla, mycorrhiza, Biduri

## 1. Introduction

Climate change causes extreme and unpredictable weather, such as prolonged dry seasons, heat waves that increase air temperatures severely, and heavy rains are expected. These conditions cause many environmental problems that affect human health and reduce immunity to make it easy to get sick. Therefore, maintaining the immune system becomes very important. One of the efforts that everyone can do to avoid bacterial and viral attacks is to maintain a fit and excellent endurance by utilizing herbal plants, one of which is Biduri.

Biduri or widuri (*Calotropis gigantea*) has many benefits in all parts, ranging from leaves, flower stems to sap can be used as herbal medicine to treat various wounds and diseases. The stem is used as an herbal plant that treats dizziness, asthma, and various respiratory disorders, and lately, allegedly, part of the biduri plant also contains immunomodulators [1,2,3]. In addition, biduri fiber has the potential to be a reinforcement in composites [4,5].

Biduri is a plant that grows wild in tropical and sub-tropical regions and has not been cultivated and is found in many fielded soils. Efforts to introduce biduri cultivation to the community need to be carried out. It is hoped that because of the many benefits that can be taken from biduri, the community can cultivate so that biduri can grow and develop optimally. Factors that support plants to grow and produce optimally are the availability of sufficient quantities of nutrients in the soil.

One of the efforts in domesticating biduri without damaging the environment is by using organic fertilizers. Azolla is a plant that can be used as an organic fertilizer as a source of Nitrogen. In addition, P and K content in Azolla is also high, Azolla also can help improve the physical properties, biological, soil chemistry [6,7,8,9]. Another organic fertilizer is mycorrhiza. The provision of mycorrhizal will also be expected to increase the growth of biduri. Mycorrhizal can play a role in helping in the efficiency of the absorption of water and nutrients in the soil, increase the survival of seedlings, increase growth and yield, increase resistance to abiotic or biotic stress, and increase soil aggregates [10,11,12]. In addition to growth, the formation of secondary metabolite compounds needs to be improved in medicinal plants. Plant symbiosis with mycorrhizal will not only increase nutrient uptake, especially P but will also help facilitate the signaling pathway of phytochemical compounds through upregulation of the phenylpropanoid pathway, which will improve the quality of plant production [10].

Applying Azolla and mycorrhizal fertilizers in biduri cultivation is expected to increase biduri growth, both in quality and quantity. The introduction of biduri cultivation will impact the addition of public knowledge of its benefits very much. Therefore, this study was conducted to provide an overview of the appearance of biduri growth on Azolla and mycorrhizal applications.

## 2. Materials and method

### *2.1. Materials*
The tools used in this study include polybag (10 x 15) for seedlings, polybag (30 x 35 cm) for planting, paranet 50%, planting and maintenance equipment, analytical scales, and ruler. Materials used in this study include biduri seeds, mycorrhiza and azolla.

### *2.2. Methods*
The research was conducted in the Experiment Field of the Faculty of Agriculture UNS Sukosari Jumantono, Karanganyar Central Java Indonesia (7°37'52.3"S 110°56'53.7"E). The research was carried out beginning in April until October 2021. The study used a factorial randomized block design with 2 treatment factors, namely mycorrhizal and Azolla, with 6 replicated. Azolla consisted of 5 levels, namely 0, 5, 7.5, 10 and 12.5 tons $ha^{-1}$, while the mycorrhizal factor consists of 4 levels, namely 0, 1 5, 10, and 15 tons $ha^{-1}$.

### *2.3. Stages of research*
Biduri seeds were planted in polybags and placed in the screen house of the Faculty of Agriculture. During the nursery use paranet shade. Watering of seedlings was done every day in the afternoon. Transplanting was done when the seedlings were 1.5 months old or after the plant had grown enough to have 2-4 leaves and the same height between plants. Biduri were grown in polybags that have been filled with soil as much as 5 kg. Research treatment is applied at the time of transplanting.

### *2.4. Observation and data analysis*
Observation variables include plant height, number of leaves, leaf width, leaf length at 12 weeks after transplanting. Data were analyzed using analysis of variance (ANOVA) and continued with a 5% Tukey test if there is significant effect. Data was proceed using SPSS programme.

## 3. Results and discussion

### *3.1. Result*
Analysis of variance showed no interaction between Azolla and mycorrhizal was applied to Biduri growth. Response of Biduri growth in increasing Azolla dosage is the same in different mycorrhizal applications. This means that response of Biduri growth is affected by each treatment.

**Table 1.** Mean of plant height, number, length, and width of the biduri leaves on the application of Azolla and mycorrhizal at 12 weeks of age after transplanting

| | Mycorrhiza (ton ha$^{-1}$) | | | | Mean |
|---|---|---|---|---|---|
| **Azzola (ton ha$^{-1}$)** | **0** | **5** | **10** | **15** | |
| | | **Plant height (cm)** | | | |
| 0 | 43.8 | 67.3 | 69.5 | 72.3 | 63.3 b |
| 1 | 68.7 | 81.3 | 69.0 | 73.3 | 73.1 ab |
| 7.5 | 70.0 | 76.5 | 75.5 | 64.7 | 71.7 ab |
| 10 | 65.3 | 70.7 | 81.0 | 84.3 | 75.3 ab |
| 12.5 | 81.5 | 74.7 | 75.3 | 82.8 | 78.6 a |
| Mean | 21.9 | 36.2 | 39.8 | 43.7 | (-) |
| | | **Number of leaves** | | | |
| 0 | 17 | 19 | 35 | 25 | 23.9 b |
| 1 | 33 | 37 | 44 | 48 | 40.7 b |
| 7.5 | 30 | 41 | 30 | 22 | 30.7 ab |
| 10 | 32 | 41 | 33 | 47 | 38.0 ab |
| 12.5 | 52 | 37 | 30 | 26 | 36.3. ab |
| Mean | 40.9 | 48.4 | 51.6 | 52.9 | (-) |
| | | **Leaf length** | | | |
| 0 | 14.4 | 11.8 | 17.4 | 17.5 | 15.3 a |
| 1 | 17.1 | 19.1 | 17.2 | 19.8 | 18.3 a |
| 7.5 | 18.5 | 21.9 | 20.6 | 15.7 | 19.2 a |
| 10 | 18.0 | 16.2 | 21.2 | 19.4 | 18.7 a |
| 12.5 | 19.9 | 16.2 | 18.4 | 19.7 | 18.5 a |
| Mean | 35.6 | 41.0 | 42.6 | 44.1 | (-) |
| | | **Leaf width** | | | |
| 0 | 5.5 | 5.0 | 7.4 | 7.0 | 6.2 a |
| 1 | 7.6 | 7.4 | 8.4 | 8.0 | 7.9 ab |
| 7.5 | 9.3 | 8.8 | 9.0 | 8.9 | 9.0 a |
| 10 | 7.2 | 8.1 | 8.6 | 8.8 | 8.2 a |
| 12.5 | 10.0 | 7.2 | 8.8 | 9.2 | 8.8 a |
| Mean | 29.8 | 33.9 | 35.6 | 36.8 | (-) |

Note: the number followed by the same letter is not significantly different on the 5% Tukey test. (-) there is no interaction between Azolla and mycorrhiza.

The height of the biduri in applying Azolla 12.5 tons ha$^{-1}$ is 78.6 cm and is different from the height of biduri without Azolla application but not significantly different from other Azolla applications. The application of Azolla of 12.5 tons ha$^{-1}$ increased the growth of plant height by 24.4% compared to without the administration of Azolla.

Table 1 showed that the average number of biduri leaves (40.7 leaves) is found in Azolla applications 5 tons ha-1 and differs markedly from without application which averages 23.9 leaves, while between doses Azolla applications do not show an average number of leaves significant different. Applying Azolla as much as 5 - 12.5 tons ha$^{-1}$ increases the number of leaves by 52-70% compared to without Azolla application. Although Azolla application increases the leaf length of the biduri by 19-25% compared to the non-Azolla application, it has not shown significant differences

### *3.2. Discussion*

Analysis of variance showed no interaction between Azolla and mycorrhizal was applied to biduri growth. Response of biduri growth in increasing Azolla dosage is the same in different mycorrhizal applications. This means that response of biduri growth is affected by each treatment.

In this study, the use of mycorrhizal independently did not significantly affect the parameters of observations. This is due to several things: the dose given is not appropriate, there is no compatibility of mycorrhizal with the host plant, the degree of infection is still low, or the mycorrhizal spores that have not germinated. As mentioned by other researchers, that the benefit of mycorrhizal symbiosis varies depend on the type of host plant (species and varieties) and mycorrhiza species [14, 15]. In addition, mycorrhizal symbiosis is highly dependent on environmental factors, such as soil properties, nutrient availability, or agricultural management practices [13]. Mycorrhizal in its development requires several stages before entering into the process of infection into the roots plant [15, 17,18],). Therefore, mycorrhizal symbiosis is not always beneficial, as the effect can be positive, negative, or neutral to the host plant.

The use of Azolla independently increases plant height, number of leaves and width of biduri leaves compared to without the administration of Azolla. This is due to the application of Azolla fertilizer able to improve the physical properties of the soil, so the ability of the roots to develop will also be better. It will increase the uptake of nutrients and water, which in turn will support the development of the plant's parts above ground level. In addition, the ability of Azolla to symbiosis with Anabaena in binding to free Nitrogen in the air, so the Nitrogen content in Azolla will also increase. Nitrogen is one of the macronutrients needed for the vegetative growth of plants. When the Nitrogen availability increase, the growth of the plant will be better, as other researchers have stated, Azolla application can improve the physical and chemical properties of soil and soil microbial populations [6, 8]. These improvements are significant for soil nitrogen, organic matter, and other cations (Mg, Ca and Na) that directly help improve soil fertility [8, 9,19]. Furthermore, Youssef [18] states that Azolla has the highest N, P and K content compared to other sources of organic matter, such as compost and vermicompost. Therefore, vegetative growth of biduri is better than without applying Azolla.

## 4. Conclusions

Climate change results in changes in environmental factors which are one of the main limiting factors for plant growth, including biduri. Biduri is one of the medicinal plants that have many benefits for humans but has not been widely cultivated, so attempt to cultivate biduri through mycorrhizal and Azolla was done. From the observation and analysis can be conclude that there is no interaction between the application of Azolla and mycorrhizal to the plant height, the number of leaves, the width and length of leaves the biduri; Vegetative growth of biduri shows no difference with mycorrhiza application independently and application 12.5 tons $ha^{-1}$ Azolla increases significantly on the height of the plant, the number of leaves, and the width of the leaves of biduri.

## References

[1] Chakre M B, Gaikwad P N, Medpalli K J, Birru S B, Centre W and Biotechnology F 2019 Medicinal Properties of Calotropis Gigantea *Int. J. Res. Anal. Rev.* **6** 175–81

[2] Sinha S, Raich V, Student P G and Hills S 2019 Calytropis Gigantea- a Prospective herbs in cosmetics *Int. J. Sci. Dev. Res.* **4** 348–57

[3] Sukardan M D, Natawijaya D, Prettyanti P, Cahyadi C and Novarini E 2017 Karakterisasi Serat Dari Tanaman Biduri (Calotropis Gigantea) Dan Identifikasi Kemungkinan Pemanfaatannya Sebagai Serat Tekstil *Arena Tekst.* **31** 51–62

[4] Heuvel D van den 2020 The Story of Opportunities Where Most Only Saw a Problem *J. Text. Eng. Fash. Technol.* **2** 12–4

[5] Haryati B 2015 Pengaruh pupuk bokashi azola terhadap pertumbuhan bibit tamarillo (Cypomandra betaeasenat) *AgroSainT UKI Toraja* **VI** 5–24

[6] Al-hamdawi R M N and Al-wally N S S 2020 Effect of Application Different Levels of Azolla ( Azolla Filicoides Lam .) in Some Soil Physical Properties and Barley (Hordeum Vulgare L .) *Plant Arch.* **20** 4174–82

[7] Birhane E, Sterck F J, Fetene M, Bongers F and Kuyper T W 2012 Arbuscular mycorrhizal fungi enhance photosynthesis, water use efficiency, and growth of frankincense seedlings under pulsed

water availability conditions *Oecologia* **169** 895–904
[8] Duc N H, Mayer Z, Pék Z, Helyes L and Posta K 2017 Combined inoculation of arbuscular mycorrhizal fungi, Pseudomonas Fluorescens and Trichoderma spp. For enhancing defense enzymes and yield of three pepper cultivars *Appl. Ecol. Environ. Res.* **15** 1815–29
[9] Basri A H H 2018 Kajian Peranan Mikoriza Dalam Bidang Pertanian *Agrica Ekstensia* **Vol. 12** 74–8
[10] Kaur S and Suseela V 2020 Unraveling arbuscular mycorrhiza-induced changes in plant primary and secondary metabolome *Metabolites* **10** 1–30
[11] Sikes B A, Cottenie K and Klironomos J N 2009 Plant and fungal identity determines pathogen protection of plant roots by arbuscular mycorrhizas *J. Ecol.* **97** 1274–80
[12] Fernández I, Merlos M, López-Ráez J A, Martínez-Medina A, Ferrol N, Azcón C, Bonfante P, Flors V and Pozo M J 2014 Defense Related Phytohormones Regulation in Arbuscular Mycorrhizal Symbioses Depends on the Partner Genotypes *J. Chem. Ecol.* **40** 791–803
[13] Campo S, Martín-Cardoso H, Olivé M, Pla E, Catala-Forner M, Martínez-Eixarch M and San Segundo B 2020 Effect of Root Colonization by Arbuscular Mycorrhizal Fungi on Growth, Productivity and Blast Resistance in Rice *Rice* **13** 1–14
[14] Hildebrandt U, Janetta K and Bothe H 2002 Towards growth of arbuscular mycorrhizal fungi independent of a plant host *Appl. Environ. Microbiol.* **68** 1919–24
[15] Ronsheim M L 2012 The effect of mycorrhizae on plant growth and reproduction varies with soil phosphorus and developmental stage *Am. Midl. Nat.* **167** 28–39
[16] Subedi P and Shrestha J 2015 Improving soil fertility through Azolla application in low land rice: A review *Azarian J. Agric. Rev.* **2** 35–9
[17] Putra D F and Tyasmoro S Y 2013 Pengaruh pemberian berbagai bentuk azolla dan pupuk n terhadap pertumbuhan dan hasil tanaman jagung manis ( zea mays var . Saccharata ) the effect of various forms of azolla and N fertilizer on growth and yield of sweet corn (Zea mays var . Saccharata ) *J. Produksi Tanam.* **1** 353–60
[18] Youssef M A, Al-Huqail A A, Ali E F and Majrashi A 2021 Organic amendment and mulching enhanced the growth and fruit quality of squash plants (Cucurbita pepo l.) grown on silty loam soils *Horticulturae* **7**

**Acknowledgment**
This research is funded from non APBN UNS grants 2021.

# Leaf nutrient content and corn growth with application of liquid organic fertilizer and inorganic fertilizer

**Jauhari Syamsiyah* and Ganjar Herdiansyah**

Study Program of Soil Science, Faculty of Agriculture, Universitas Sebelas Maret Jl. Ir Sutami 36 A, Kentingan, Surakarta, 57126 Indonesia
*Corresponding author: ninukts@staff.uns.ac.id

**Abstract.** Plants have an important role in mitigating climate change because plants are able to absorb carbon dioxide which is one of the greenhouse gases. Good plant growth and high yields are a reflection of its ability to absorb carbon dioxide gas, so various efforts aimed at this, such as fertilization can contribute to reducing greenhouse gas levels in the atmosphere. This research aims to determine the use of liquid organic fertilizers and inorganic fertilizers on nutrient uptake and growth of corn plants. This research was conducted in the Jumantono experimental field, Karanganyar. The research used the Randomized Block Design method, 9 treatments (A = Control, B = NPK standard, C = ¼ NPK + 1 POC, D = ½ NPK + 1 POC, E = ¾ NPK +1 POC, F = 1 NPK + 1 POC, G = ¾ NPK + ¼ POC, H = ¾ NPK + ½ POC, I = ¾ NPK + ¾ POC) with 3 replications, so we get 27 samples. The results showed that the application of liquid organic fertilizer doses of 1/4, 1/2, 3/4 and 1 combined with inorganic fertilizers gave no different results on nutrient uptake and increased plant growth, especially in the number of leaves and stem diameter, but differed in plant height. The application of 1 dose of liquid organic fertilizer and 3/4 dose of inorganic fertilizer gives higher yields compared to standard inorganic fertilizers.

**Key words:** inorganic fertilizer, liquid organic fertilizer, nutrient content

## 1. Introduction

Plants have an important role in mitigating climate change, because plants are able to absorb carbon dioxide which is one of the greenhouse gases. Good plant growth and high yields are a reflection of its ability to absorb carbon dioxide gas, so various efforts aimed at this, such as fertilization can contribute to reducing greenhouse gas levels in the atmosphere. Fertilization is an intensification effort to achieve success in agricultural production, but must be based on balanced fertilization.

The fertilizers circulating in the community are organic and inorganic fertilizers. Currently, the use of organic fertilizers is increasing in line with government programs in an effort to increase agricultural productivity without damaging the land. Excessive use of inorganic fertilizers can cause land degradation. However, the use of organic fertilizers does not directly have an impact on plant growth, therefore inorganic fertilizers with a balanced dose are still needed. The combination of organic and inorganic fertilizers is an alternative to fertilization with the aim of minimizing the use of chemical fertilizers and also keeping nutrients available for plants.

Plant response is an indicator to see how the plant absorbs nutrients. Corn plants require sufficient nutrients for growth. The provision of organic and inorganic fertilizers basically aims to meet the nutrient needs of plants [1]. Organic fertilizers can increase the availability of nutrients and increase the efficiency of nutrient uptake by plants so that fertilization efficiency increases. According to [1] Organic fertilizer added with microbes can increase nutrients and increase nutrient uptake in corn plants, so that fertilization can be carried out efficiently. The application of liquid organic fertilizer combined with inorganic fertilizer is expected to increase plant growth and increase nutrient uptake in corn plants so that it can provide the expected results. This research aims to determine the use of liquid organic fertilizers and inorganic fertilizers on nutrient uptake and growth of corn plants.

## 2. Material and methods

The research was conducted in Jumantono, Karanganyar Regency. This research was carried out using a Randomized Block Design (RAK), from 9 treatment combinations and consisted of 7 treatments with doses of organic fertilizer and 1 treatment with recommended doses of fertilizer and 1 as control and corn as a test plant (Table 1). Each treatment was repeated 3 times so that the total experimental plots were 27 plots. The doses of liquid organic fertilizer given were 1 dose (5000 ml/ha), 3/4 dose (3750 ml/ha), 1/2 dose (2500 ml/ha), and 1/4 dose (1250 ml/ha).

The data obtained were then analyzed for variance to determine the effect of each treatment. Then the DMRT test was carried out. Parameters observed were plant growth data (plant height, stem diameter, and number of leaves) and nutrient uptake (N, P, and K). Preparation of planting media, planting and fertilizing, maintenance and harvesting. carried out according to local customs

**Table 1.** The treatmens of Liquid organic fertilizar and anorganik fertilizer

| Kode | Treatments | Liquid organic fertilizer | NPK fertilizer | | |
|---|---|---|---|---|---|
| | | (ml/ha) | kg/ha | | |
| | | | Urea | SP36 | KCl |
| A | Control | 0 | 0 | 0 | 0 |
| B | NPK standar | 0 | 350 | 150 | 75 |
| C | ¼ NPK + 1 POC | 5000 | 87.5 | 37.5 | 18.75 |
| D | ½ NPK + 1 POC | 5000 | 175 | 75 | 37.5 |
| E | ¾ NPK +1 POC | 5000 | 262.5 | 112.5 | 56.25 |
| F | 1 NPK + 1 POC | 5000 | 350 | 150 | 75 |
| G | ¾ NPK + ¼ POC | 1250 | 262.5 | 112.5 | 56.25 |
| H | ¾ NPK + ½ POC | 2500 | 262.5 | 112.5 | 56.25 |
| I | ¾ NPK + ¾ POC | 3750 | 262.5 | 112.5 | 56.25 |

Note:

a. Control was treatment without organic fertilizer and without N, P, and K fertilizers.
b. Standard NP K fertilizer is inorganic fertilizer treatment at local recommended doses for maize (350 kg Urea, 150 kg SP-36, and 50 kg KCl per hectare).
c. Treatment of the recommended dose of Organic Fertilizer is given in accordance with the recommended dose of 5000 ml/ha.

## 3. Result and discussion

### *3.1. Plant growth*

*3.1.1. Plant height.* The results of measurements of plant heights can be seen in Table 2. At 14, 28, 42 and 56 DAP showed significantly different on plant height in each treatment, it mean that all fertilizer treatments gave a positive response to plant height except control.

At the age of 14, 28, 42 and 56 days, corn plants showed significantly different effects between treatments, namely the combination of NPK fertilizer with liquid organic fertilizer compared to the control treatment and standard NPK.

Table 2 shows the growth dynamics between treatments. In general, the plants with the highest stand growth rate were indicated by treatment E and the lowest was indicated by treatment A. During the maximum growth period, the increase in the dose of liquid organic fertilizer showed a significantly different effect on plant height with the control treatment, but seen from the value of standard fertilizer application, the values tended to be no different.

**Table 2.** Plant height at 14, 28, 42 and 56 days after planting

| Treatment | Plant Height | | | |
|---|---|---|---|---|
| | 14 days | 28 days | 42 days | 56 days |
| A | 4.09a | 13.77a | 33.5a | 108.98a |
| B | 5.21b | 19.56ab | 59.33bc | 169.77bc |
| C | 3.79a | 17.15ab | 56.05b | 151.83ab |
| D | 5.15b | 21.66ab | 63.73c | 165.83b |
| E | 4.59ab | 21.72ab | 65.27c | 175.93c |
| F | 4.01a | 20.21ab | 63.33c | 171.9bc |
| G | 4.52ab | 19.955ab | 49.54ab | 161.13b |
| H | 4.8ab | 18.06ab | 52.11ab | 153.93ab |
| I | 4.88ab | 20.65ab | 57.18b | 159.87b |

Note: The mean number followed by the same letter does not give a significant difference based on Duncan's Multiple Range Test at the 5% level.%.

*3.1.2. Leaf number.* The effect of liquid organic fertilizer and inorganic fertilizer showed a significant difference on the number of leaves at plant ages 14, 28 and 42 days. At the maximum vegetative phase or 56 days the combination of liquid organic fertilizer and inorganic fertilizer showed results that were not significantly different from the control and standard NPK fertilizers.

**Table 3.** Number of leaves of corn

| Treatmens | Number of leaves | | | |
|---|---|---|---|---|
| | 14 days | 28 days | 42 days | 56 days |
| A | 4.27a | 7.13a | 9.33a | 12a |
| B | 4.87ab | 8.13ab | 11.07ab | 14.2a |
| C | 4a | 7.4a | 10.6ab | 13.13a |
| D | 4.8ab | 8.33ab | 11.33ab | 13.73a |
| E | 4.6ab | 8.2ab | 11.87b | 13.87a |
| F | 4.73ab | 8.33ab | 11.33ab | 14.87a |
| G | 4.87ab | 8.13ab | 10.93ab | 14.47a |
| H | 4.6ab | 7.47a | 11.73b | 14.47a |
| I | 4.6ab | 8.27b | 11.53b | 14.27a |

Note: The mean number followed by the same letter does not give a significant difference based on Duncan's Multiple Range Test at the 5% level.

Based on Table 3. Treatment F showed the highest yield on the number of leaves that could be achieved by a combination of liquid organic fertilizer and inorganic fertilizer. While treatment A showed the lowest results.

*3.1.3. Diameter of steam.* The results showed that the effect of the combination of liquid organic fertilizer and inorganic fertilizer had no significant effect on the development of stem diameter at 14, 28 and 56 days, while at 42 days the results were significantly different.

The control treatment showed slower growth of stem diameter compared to other treatments that used a combination of liquid organic and inorganic fertilizers. The combination treatment of liquid organic and inorganic fertilizers showed results that were not different from the standard fertilizer treatment (B), because plant growth in increasing stem diameter required nutrients, especially from fertilizers.

**Table 4.** Stem Diameter

| Treatments | Stem diameter | | | |
|---|---|---|---|---|
| | 14 days | 28 days | 42 days | 56 days |
| A | 0.35a | 0.79a | 1.47a | 2.22a |
| B | 0.43a | 1.09a | 2.04ab | 2.75a |
| C | 0.31a | 0.93a | 1.8ab | 2.438a |
| D | 0.42a | 1.28a | 2.2b | 2.78a |
| E | 0.35a | 1.24a | 2.07ab | 2.784a |
| F | 0.39a | 1.18a | 2.02ab | 2.6a |
| G | 0.39a | 1.14a | 1.94ab | 2.6a |
| H | 0.3a | 0.9a | 1.78ab | 2.44a |
| I | 0.36a | 0.96a | 1.91ab | 2.48a |

Note: The mean number followed by the same letter does not give a significant difference based on Duncan's Multiple Range Test at the 5% level.

### *3.2. N, P dan K uptake.*

The results of the analysis showed that the combination of liquid organic fertilizer and inorganic fertilizer had no significant effect on the nutrient uptake of N, P and K. The uptake of these nutrients showed a proportional effect on the increase in corn yields (Table 5).

**Tabel 5.** Nutrient Uptake of N, P and K by Corn Plants

| Treatments | Nutrient uptake | | |
|---|---|---|---|
| | Nitrogen | Phosphor | Potassium |
| A | 1.87a | 2.11a | 2.73a |
| B | 2.74a | 2.66a | 2.83a |
| C | 2.98a | 2.31a | 2.92a |
| D | 3.11a | 2.16a | 2.48a |
| E | 3.14a | 2.33a | 2.09a |
| F | 3.01a | 2.24a | 2.28a |
| G | 2.95a | 2.35a | 1.99a |
| H | 2.73a | 2.31a | 2.24a |
| I | 3.32a | 2.04a | 2.04a |

Note: The mean number followed by the same letter does not give a significant difference based on Duncan's Multiple Range Test at the 5% level.

The uptake of N, P and K is in line with the dose of liquid organic fertilizer and inorganic fertilizer. Treatment I (¾ NPK + POC) was the treatment that produced the highest plant N uptake, this result confirmed that the combined application of NPK plus liquid organic fertilizer could help plants absorb nitrogen and was the best application in providing N uptake. P and K, treatment B was the best treatment in providing plant P and K nutrient uptake.

## 4. Disscusion

Giving liquid organic fertilizer with inorganic fertilizer can increase the availability of nutrients for plant growth According to [2], the increasing age of the plant, the greater the need of nutrients is higher and this situation cannot be met by the soil where it grows, so adding fertilizers can increase the availability of nutrients, especially nitrogen which is needed for plant vegetative plant growth. Nitrogen is needed by corn plants in particular to be able to grow, especially helping the growth of stems (plant height), branches and leaves. The combination of liquid organic fertilizer with standard dose of NPK fertilizer can give a positive response to the growth of corn plants (plant height, stem diameter, number of leaves, leaf length and leaf width) [3].

The number of leaves is corelated with nutrient uptaka by plant roots. In general, leaves are the main photosynthetic organ, so leaf observations are needed as supporting data to explain the process of biomass formation, as well as an indicator of plant growth. The treatments showed that at 56 days after planting no different results, this is because at this age the corn plant has entered the maximum vegetative phase which means that it has experienced aging. According to [2] corn plants that had entered the maximum vegetative phase or 56 days showed that were not significantly different on number of leaves with with organic fertilizers. it is because in this phase the leaves had experienced aging, so the fresh leaves were relatively the same in all treatments.

The application of liquid organic fertilizers combined with inorganic fertilizers generally showed no different results. This is because of the ineffectiveness of nutrient absorption by plants. Application of liquid organic fertilizer that is sprayed has a major impact on nutrient loss due to respiration. According to [4] that the application of liquid organic fertilizer does not have a significant effect due to the ineffective absorption of nutrients by plants. When viewed from the value of the stem diameter, the combination of liquid organic fertilizer and inorganic fertilizer treatment showed a higher value than the treatment of NPK fertilizer alone. The results of [5] showed that the effect of N, P, K fertilizer in supporting the growth of plant height and stem diameter was greater than the treatment without N, P, K fertilizer which showed slower growth.

The results of the analysis on nutrient uptake showed that the application of liquid organic fertilizer with inorganic fertilizer did not give significantly different results. However, there are differences in the values shown as a result of the addition of liquid organic fertilizers, especially on nitrogen uptake. According to [6] liquid organic fertilizer can increase the availability and absorption of nutrients, especially N nutrients which are needed by plants, so that plants that are given liquid organic fertilizer can stimulate their vegetative growth. Other studies have shown that the application of liquid organic fertilizer to sweet corn plants can increase nutrient uptake and can improve plant growth and produce higher production [7]. Furthermore, according to [8], the availability of sufficient nutrients during plant growth can increase metabolic activity so that it is more active and the process of elongation or cell

## 5. Conclusion

The application of liquid organic fertilizer doses of 1/4, 1/2, 3/4 and 1 combined with inorganic fertilizers gave no different results on nutrient uptake and increased plant growth, especially in the number of leaves and stem diameter, but differed in plant height. The application of 1 dose of liquid organic fertilizer and 3/4 dose of inorganic fertilizer gives higher yields compared to standard inorganic fertilizers. Fertilization with liquid organic fertilizers can mitigate climate change by reducing the use of inorganic fertilizers and energy use in the fertilizer industry. as well as increased photosynthesis and increased plant yield

## Referrence

[1] Irsyad Y M M and Kastono D 2019 Pengaruh Macam Pupuk Organik Cair dan Dosis Pupuk Anorganik terhadap Pertumbuhan dan Hasil Jagung (Zea mays L) *Vegetalika* **8** 263–75

[2] Syamsiyah J, Herdiansyah G, Aisyah N, Mentari F C and Larasati I 2021 Combined application of organic and inorganic fertilizer on Alfisol soil fertility and maize growth *IOP Conf. Ser. Earth Environ. Sci.* **824** 012025

[3] Budiono R and Sudarwati E 2019 Pengaruh Pupuk Organik Cair terhadap Pertumbuhan Tanaman Jagung Manis ( Zea mays saccarata L.) *Prosiding Semnas Fakultas Pertanian* (Purwokerto, Indonesia: Universitas Muhamadiyah Purwokerto)

[4] Rambe R D H 2014 Pengaruh Pemberian Pupuk Organik dan Pupuk Anorganik Terhadap Pertumbuhan dan Produksi Jagung Manis (Zea mays saccharata L.) *Wahana Inov.* **3** 436–43

[5] Puspadewi S, Sutari W and Kusumiyati K 2016 Pengaruh konsentrasi pupuk organik cair (POC) dan dosis pupuk N, P, K terhadap pertumbuhan dan hasil tanaman jagung manis (Zea mays L. var Rugosa Bonaf) kultivar talenta *Kultivasi* **15** 208–16

[6] Afandi F N, Siswanto B and Nuraini Y 2015 Pengaruh Pemberian Berbagai Jenis Bahan Organik Terhadap Tifat Kimia Tanah Pada Pertumbuhan dan Produksi Tanaman Ubi Jalar di Entisol Ngrangkah Pawon, Kediri *J. Tanah Dan Sumberd. Lahan* **2** 237–44

[7] Syufrin M P, Barus W A and Kurnianto H 2011 Pengaruh konsentrasi dan interval waktu pemberian pupuk organik cair (POC) nasa terhadap pertumbuhan dan produksi jagung manis (Zea mays saccharata Sturt) *J. Agrium* **17** 46–52

[8] Shaila G, Tauhid A and Tustiyani I 2019 Pengaruh dosis urea dan pupuk organik cair asam humat terhadap pertumbuhan dan hasil tanaman jagung manis *Agritrop J. Ilmu-Ilmu Pertan.* **17** 35–9

# Adaptation to climate variability of rainfed farmers in Seluma Regency, Bengkulu, Indonesia

**Gita Mulyasari[1]*, P Prawito[2] and R Yuristia[1]**

[1]Department of Agricultural Socio-Economics, University of Bengkulu, Jl. WR. Supratman, Kandang Limun, Bengkulu, 38371, Indonesia
[2]Departmen of Soil Science, University of Bengkulu, Jl. WR. Supratman, Kandang Limun, Bengkulu, 38371, Indonesia
*Corresponding author: gita.mulyasari@unib.ac.id

**Abstract**. Farmers are synonymous with poverty because of the great uncertainty about the impact and magnitude of climate variability. Adaptation efforts, in principle, minimize vulnerability to impacts caused by climate variability. This study aims to see the level of adaptation of farmers and the factors that influence it. A randomly selected 100 rainfed farmers survey was conducted in Seluma Regency, Bengkulu, from June to September 2020. The analysis used to measure the level of adaptation was to compare the number of adaptations adopted by farmers with the total adaptation used in this study and to analyze the factors that affect the level of adaptation using multiple regression models. The results showed that rainfed farmers have a low of adaptation to climate variability. Most farmers make adjustments to planting time, replace climate-resistant varieties, and take advantage of government assistance as a form of adaptation to climate variability. The level of adaptation of rainfed farmers to climate variability is influenced by experience, income, and farmer group membership.

**Keywords** : adaptation, climate variability, rainfed farmer

## 1. Introduction

Climate variability is the most complex and challenging phenomenon affecting the global community. While climate is changing and its associated anomalies are becoming more frequent, agriculture and livelihood of farm households, particularly in developing countries, are increasingly being threatened [1,2]. Climate is one of the most important factors in agricultural production, and since the climate is linked to physiological processes, it directly or indirectly influences productivity [3]. Many studies on climate variability have been carried out, especially for smallholder groups [4–6]. However, there is still minimal research on climate variability involving specific rainfed farmers [7–9]. The negative impacts of climate variability are increasingly evident and have been proven to have hit Indonesia in various environmental disasters such as floods, landslides, droughts [10,11]. These impacts pose challenges to development in terms of the social and economic environment in a sustainable manner and the achievement of Indonesia's development goals [11].

Farmers are synonymous with poverty and because of the great uncertainty about the impact and magnitude of climate variability. Research conducted by [12] showed a high correlation between hunger and climate risk, especially for the areas most affected by food insecurity. The analysis describes the impact of climate on crop yields, provides a multidimensional vulnerability analysis, and demonstrates the important role of the adaptive capacity of farming households in reducing vulnerability.

Adaptations to climate variability are strategies designed to eliminate the adverse effect of climate-related risks on agricultural activities [13,14]. Adaptation to climate variability refers to changing farming systems and on-farm operations to be in line with the climate variability [15] and is believed to be the most effective way of tackling climate impacts in agriculture [16,17]. All farmer households must carry out adaptation strategies to maintain their livelihoods. According to [18] the livelihood assets that are the key to success in adapting are land and water (natural capital); human resources, education, and health (human capital), social institutions (social capital), and financial capital. This research was conducted in Bengkulu Province, the second poorest province on the island of Sumatra [19], where the agricultural sector provides the most considerable contribution to economic growth and development. Efforts to adapt to climate variability in the agricultural sector are key factors for the success of national development, given the large number of people who work in the agricultural sector. This study aims to analyze the level of adaptation of rainfed farmers in the face of climate variability and the determinants that affect the level of adaptation of farmers.

## 2. Methodology

### *2.1. Study area*

This research was conducted in Seluma Regency, Bengkulu Province, Indonesia. Site selection was made using a purposive technique considering this area has the most extensive rain-fed rice fields in Bengkulu Province, which is 6,266 hectares [20]. Therefore, a wetland with rainfed irrigation systems is very vulnerable to the impacts of climate variability. The research locations include Talo District (Lubuk Gio Village and Kembang Seri Village) and Ilir Talo District (Penago I Village) (Figure 1).

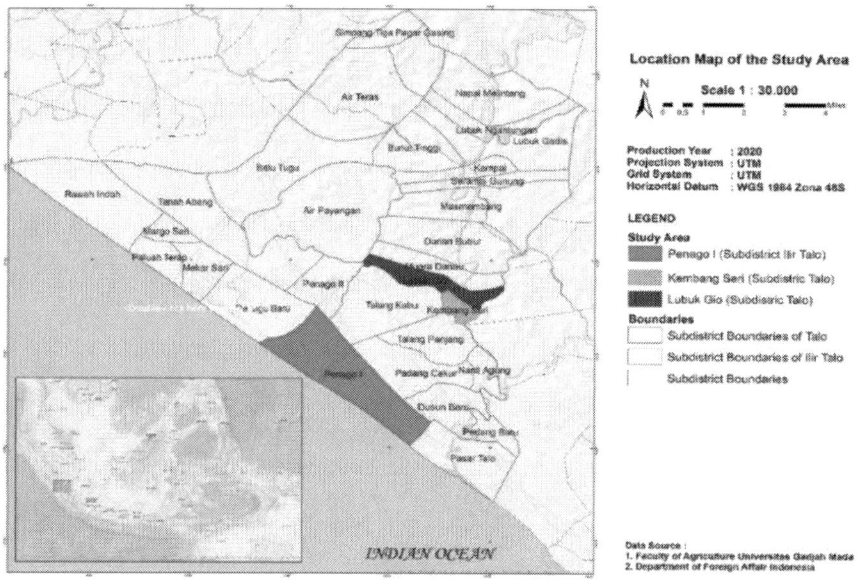


**Figure 1**. Research locations

### *2.2. Sampling and questionnaire*

In this research, the analysis unit is rainfed farmers who have their land as key informants to make decisions related to climate change adaptation. The method used in determining the respondents is purposive random sampling, with a total of 100 farmers were selected. The questionnaire was carefully designed to get information on socio-economic characteristics and climate change adaptation selected by farmers based on their knowledge, understanding, and experience.

### 2.3. *Data analysis*

The measurement of the level of adaptation to climate variability was proposed by [21] by comparing the number of adaptations adopted by farmers with the total adaptation used in this study. The value of the adaptation level ranges from 0 to 1 and is formulated as follows.

$$TA = \frac{SA_P}{SA_T} \quad (1)$$

Where TA is the level of adaptation, SAP is the number of adaptations applied by farmers, and SAT is the total number of adaptations in this study.

Data were analyzed by ordinary least square (OLS) regression analysis. OLS was used to find out important determinants affecting the farmer's level of adaptation to climate variability. Important determinants that affect farmers' level of adaptation were selected based on the previous studies, such as years of schooling [22], farming experience [22], income [23,24], perception of climate variability [22,25], membership organization [22,26], and extreme climate events [22,25].

## 3. Results and discussion

### 3.1. *Characteristic of rainfed farmer*

Age is one of the social factors that determine farmers' decisions and is related to workability. The results (Table 1) show that the average age of rainfed farmers is 49.63 years. Farmers have an average productive age, ranging from 15 to 64 years [17]. The age of farmers in the productive category can support farmers in planning, managing, and regulating their farming activities, so that farmers can work and think optimally.

The level of education is an important factor that can affect how farmers manage their farms and influence making decisions and solving problems they face [27]. Unfortunately, most of the farmers have a low level of schooling, and not a few don't have formal education. Overall, the average years of schooling of farmers have not reached the nine-year education target set by the Indonesian government (Table 1). The low education level of farmers reflects the farmers' poverty in Indonesia [28].

**Table 1**. Characteristics of farmers

| Characteristics | Min | Max | Mean | Std. deviation |
|---|---|---|---|---|
| Age (Years) | 27.00 | 76.00 | 49.63 | 11.34 |
| Years of schooling | 0.00 | 16.00 | 7.63 | 3.58 |
| Farming experience (Years) | 3.00 | 55.00 | 26.48 | 12.89 |
| Household size (Person) | 0.00 | 6.00 | 3.00 | 1.07 |
| Land area (Ha) | 0.20 | 1.00 | 0.46 | 0.24 |
| Farming income ($/month) | 105.78 | 366.71 | 194.99 | 58.66 |

Source: Primary data is processed, 2020

The study results (Table 1) show that farmers have a high average experience of 26.48 years. Overall, rice farmers have a relatively high average experience of 26.48 years. This indicates that farmers never switch jobs as farmers. The low level of education means that farmers do not have the opportunity to get other jobs outside the agricultural sector. The number of family dependents is generally related to household expenses. Family dependents are a burden that must be met every day. Therefore, the more dependents the family has, the greater the economic responsibility that must be completed in the family. The allocation of income is more significant to meet the needs of the burden. Table 1 shows that a rainfed farming family's average dependents are three people.

Most farmers have a narrow average land area of 0.46 hectares (Table 1). The land area will affect the ability of farmers to adapt to climate variability. The larger land area will require resources, significantly larger funds to take adjustment steps in rice farming to reduce the negative impact of climate variability. This narrow area of land also certainly affects the income level received by farmers as compensation for their farming services.

### *3.2. The level of adaptation to climate variability*

Rainfed rice farming is very vulnerable to climate variability because the success of this staple food production business is highly dependent on the carrying capacity of the climate, which was previously considered normal and stable. However, the study results (Figure 1) show that most farmers have a low level of adaptation to climate variability. Limited resources and financial means that farmers cannot adapt to climate variability. Regarding the adaptation capacity of farmers to climate variability, research results [29–31] reveal that farmers' ability to adapt to climate variability is primarily determined by variables such as agro-ecological conditions, socio-economic characteristics, extension, and farmer participation. Meanwhile, [32] emphasizes that if you want to be successful, adaptation strategies must consider and start from an understanding of how the experiences and thoughts of local farmers regarding the manifestations of climate change/variability are seen.

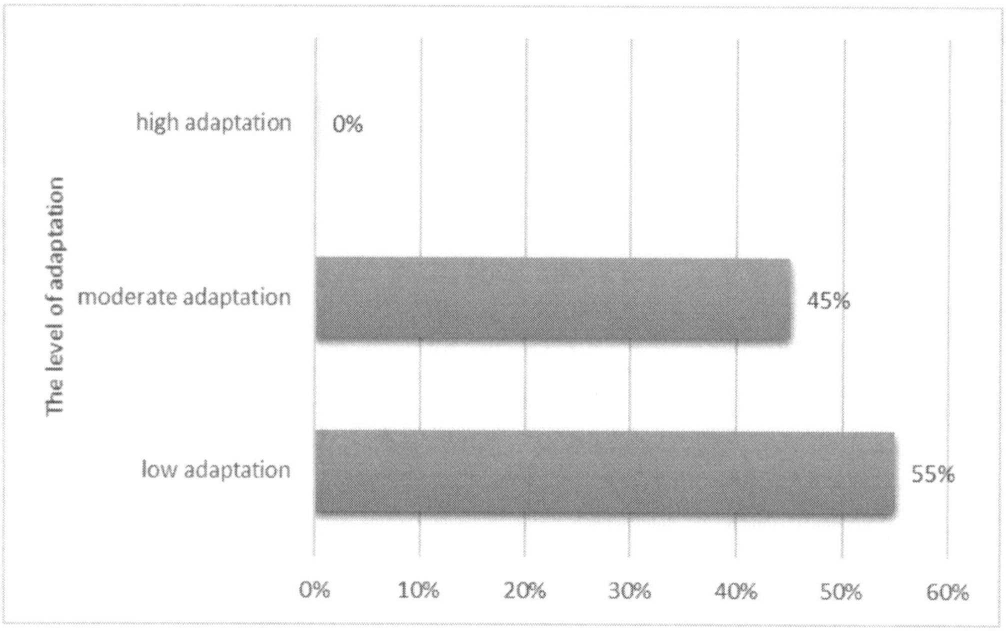


**Figure 2**. Farmer's adaptation index

The study results (Table 2) show that most rice farmers in Seluma district shift the planting season, adjust crop varieties depending on the season, and the use of government assistance as an adaptation to climate variability is the most widely used. Changing the planting season was done because farmers had to adjust to the climatic conditions to plant rice. Farmers only refer to the provisions of the coming season because the determination of the planting period is determined from the arrival of the rainy season and dry season. Syahbuddin et al. [33] revealed that farming crops often rely on habit and instinct(instinct)farmers in setting the cropping pattern in recent decades. As a result, farmers often face water shortages, especially during the long dry season. Therefore, it is necessary to adjust cropping patterns that are more adaptive to diversity and climate change.

**Table 2.** Distribution of farmers based on climate variability adaptation

| Climate variability adaptation | Distribution of farmers (%) | |
|---|---|---|
| | Already done | Have not done |
| Conduct an intercropping system | 0.00 | 100.00 |
| Shifting the planting season | 100.00 | 0.00 |
| Adjust crop varieties depending on the season | 85.00 | 15.00 |
| Pumping water from irrigation canals | 6.00 | 94.00 |
| Diversification of livelihoods off-farm | 5.00 | 95.00 |
| Mobilization of family members to participate in work | 27.00 | 73.00 |
| Use of government assistance | 56.00 | 44.00 |
| Strengthen farmer group institutions for resilience to climate variability | 26.00 | 74.00 |

Source: Primary data is processed, 2020

*3.3. Determinants of farmers' level of adaptation to climate variability*

The regression analyses were used to analyze the determinant factor adaptation to climate variability (Table 3) shows that the regression model has a value of $R^2$ by 0350, which means that the independent variable used in this study could explain adaptation to climate variability 35%. In addition, the results of the F test also show that the variables age, years of schooling, farming experience, farming income, perception of climate variability, membership organization, and extreme climate events significantly affect the adaptation level of rainfed farmers.

**Table 3**. Result estimation

| Variable | B | t | Sig. |
|---|---|---|---|
| Constant | 0.393 | 4.855 | 0.000 |
| Years of schooling | 0.003 | 0.677 | 0.500 |
| Farming experience | 0.003 | 3.500 | $0.001^{***}$ |
| Farming income | -5.119E-8 | -3.459 | $0.001^{***}$ |
| Perception on climate variability | 0.000 | -0.115 | 0.908 |
| Membership organization (dummy) | 0.093 | 4.549 | $0.000^{***}$ |
| Extreme climate events | -0.005 | -1.348 | 0.181 |
| R2 = 0.350 | | | |
| Sig-F = $0.000^{***}$ | | | |

Source: Primary data is processed, 2020
The dependent variable is the level of adaptation
$^{***}$ significance at level 0.01

The partial test (Table 3) shows that the variables farming experience, farming income, and membership organization affect the level of adaptation of farmers. Experience has a positive effect on the level of adaptation. This shows that the higher the experience of farmers, the higher the level of adaptation. Farmers have local wisdom formed from previous farmers' empirical experience and applied from generation to generation. With his experience, farmers can become more prepared to adapt to the impact of climate variability.

Farming income has a negative effect on the level of adaptation. The lower the income level, the greater farmers' adaptation in dealing with climate variability. This is because rain-fed farming is the primary income for households. If farming is vulnerable to the impacts of climate variability, farmers seek to reduce these impacts to continue to provide income for their households.

Membership organization also affects the level of adaptation and has a positive sign. This means that the more active and participating in many organizations, the more farmers will be able to improve adaptation in dealing with climate variability, which is in line with research [34]. Furthermore, by being active in farmer groups, farmers can share problems and find solutions to the farming problems they face, especially those related to climate variability.

## 4. Conclusions

The impact of climate variability creates vulnerability for farming households, especially rainfed farmers. This is because farmers are synonymous with poverty and the great uncertainty about the impact and magnitude of climate variability. The impact of climate variability threatens the sustainability of rainfed paddy farming in Seluma Regency. Most farmers have a low level of adaptation and shift the growing season, adjust crop varieties depending on the season, and use government assistance as the most widely used adaptation to climate variability. The level of adaptation of farmers is influenced by the experience of farmers, farm income, and participation in farmer groups. Adaptation efforts in the agricultural sector are critical factors for the success of national development.

## References

[1] Core Writing Team IPCC, Pachauri R K and Meyer L 2015 *Climate Change 2014: Synthesis Report. Contribution of Working Groups I, II and III to the Fifth Assessment Report of the Intergovernmental Panel on Climate Change* (Geneva, Switzerland)
[2] Dube T, Moyo P, Ncube M and Nyathi D 2016 The Impact of Climate Change on Agro-Ecological Based Livelihoods in Africa: A Review *J. Sustain. Dev.* **9** 256
[3] Ojo T O, Ogundeji A A and Belle J A 2021 Climate Change Perception and Impact of on-farm Demonstration on Intensity of Adoption of Adaptation Strategies among Smallholder Farmers in South Africa *Technol. Forecast. Soc. Change* **172** 121031
[4] Wetende E, Olago D and Ogara W 2018 Perceptions of Climate Change Variability and Adaptation Strategies on Smallholder Dairy Farming Systems: Insights from Siaya Sub-County of Western Kenya *Environ. Dev.* **27** 14–25
[5] Aniah P, Kaunza-Nu-Dem M K and Ayembilla J A 2019 Smallholder Farmers' Livelihood Adaptation to Climate Variability and Ecological Changes in the Savanna Agro Ecological Zone of Ghana *Heliyon* **5** e01492
[6] Martey E and Kuwornu J K M 2021 Perceptions of Climate Variability and Soil Fertility Management Choices among Smallholder Farmers in Northern Ghana *Ecol. Econ.* **180** 106870
[7] Sidibé Y, Foudi S, Pascual U and Termansen M 2018 Adaptation to Climate Change in Rainfed Agriculture in the Global South: Soil Biodiversity as Natural Insurance *Ecol. Econ.* **146** 588–96
[8] Wassmann R, Villanueva J, Khounthavong M, Okumu B O, Vo T B T and Sander B O 2019 Adaptation, Mitigation and Food Security: Multi-criteria Ranking System for Climate-smart Agriculture Technologies Illustrated for Rainfed Rice in Laos *Glob. Food Sec.* **23** 33–40
[9] Sarr M, Bezabih Ayele M, Kimani M E and Ruhinduka R 2021 Who Benefits from Climate-friendly Agriculture? The Marginal Returns to a Rainfed System of Rice Intensification in Tanzania *World Dev.* **138** 105160
[10] UNFCCC 2007 *Climate Change: Impacts, Vulnerabilities and Adaptation in Developing Countries* (Bonn, Germany)
[11] Opschoor J H B 2008 Fighting Climate Change - Human Solidarity in a Divided World *Dev. Change* **39** 1193–202
[12] Krishnamurthy P K, Lewis K and Choularton R J 2014 A Methodological Framework for Rapidly Assessing the Impacts of Climate Risk on National-level Food Security through a Vulnerability Index *Glob. Environ. Chang.* **25** 121–32
[13] Jiang M, Shi C L, Liu Y and Jin Z Q 2017 Response of Rice Production to Climate Change Based on Self-Adaptation in Fujian Province *J. Agric. Sci.* **155** 751–65
[14] Dang H Le, Li E, Nuberg I and Bruwer J 2019 Factors Influencing the Adaptation of Farmers in Response to Climate Change: A Review *Clim. Dev.* **11** 765–74
[15] Shaffril H A M, Krauss S E and Samsuddin S F 2018 A Systematic Review on Asian's Farmers' Adaptation Practices towards Climate Change *Sci. Total Environ.* **644** 683–95

[16] IPCC 2014 *Climate Change 2014: Impacts, Adaptation, and Vulnerability. Part A: Global and Sectoral Aspects*
[17] Masud M M, Azam M N, Mohiuddin M, Banna H, Akhtar R, Alam A S A F and Begum H 2017 Adaptation Barriers and Strategies Towards Climate Change: Challenges in the Agricultural Sector *J. Clean. Prod.* **156** 698–706
[18] Hess J J, McDowell J Z and Luber G 2012 Integrating Climate Change Adaptation into Public Health Practice: Using Adaptive Management to Increase Adaptive Capacity and Build Resilience *Environ. Health Perspect.* **120** 171–9
[19] Badan Pusat Statistik 2020 *Statistik Indonesia 2020* (Jakarta: Badan Pusat Statistik)
[20] Badan Pusat Statistik Provinsi Bengkulu 2017 *Provinsi Bengkulu dalam Angka 2017* (Bengkulu: Perum Percetakan Negara RI Cabang Bengkulu)
[21] Chukwukere A O, Udodirim U, Chidinma R O and Sulaiman J 2011 Climate Variability and Change: Perceptions and Adaptations in Subsistence Agriculture *Indian J. Anim. Res.* **45** 275–82
[22] Below T B, Mutabazi K D, Kirschke D, Franke C, Sieber S, Siebert R and Tscherning K 2012 Can Farmers' Adaptation to Climate Change be Explained by Socio-economic Household-level Variables? *Glob. Environ. Chang.* **22** 223–35
[23] Anik A R, Rahman S, Sarker J R and Al Hasan M 2021 Farmers' Adaptation Strategies to Combat Climate Change in Drought Prone Areas in Bangladesh *Int. J. Disaster Risk Reduct.* **65** 102562
[24] Getahun A B, Ayal D Y, Ture K and Zeleke T T 2021 Determinants of Climate Variability Adaptation Strategies: A Case of Itang Special District, Gambella Region, Ethiopia *Clim. Serv.* **23** 100245
[25] Dang H Le, Li E, Nuberg I and Bruwer J 2014 Understanding Farmers' Adaptation Intention to Climate Change: A Structural Equation Modelling Study in the Mekong Delta, Vietnam *Environ. Sci. Policy* **41** 11–22
[26] Abdur Rashid Sarker M, Alam K and Gow J 2013 Assessing the Determinants of Rice Farmers' Adaptation Strategies to Climate Change in Bangladesh *Int. J. Clim. Chang. Strateg. Manag.* **5** 382–403
[27] Suardika I M P, Ambarawati I G and Sukaatmadja I 2014 Analisis Perilaku Konsumen terhadap Keputusan Pembelian Sayur Organik CV Golden Leaf Farm Bali *J. Manaj. Agribisnis* **2** 1–10
[28] Mulyasari G, Irham, Waluyati L R and Suryantini A 2021 Understanding and Adaptation to Climate Change of Fishermen in the Northern Coastal of Central Java, Indonesia *IOP Conf. Ser. Earth Environ. Sci.* **724** 012094
[29] Hassan R and Nhemachena C 2008 Determinants of African Farmers' Strategies for Adapting to Climate Change: Multinomial Choice Analysis *AfJARE* **2** 83–104
[30] Ozor N and Cynthia N 2011 The Role of Extension in Agricultural Adaptation to Climate Change in Enugu State, Nigeria *J. Agric. Ext. Rural Dev.* **3** 42–50
[31] Esham M and Garforth C 2013 Agricultural adaptation to climate change: insights from a farming community in Sri Lanka *Mitig. Adapt. Strateg. Glob. Chang.* **18** 535–49
[32] Veldman R G, Szasz A and Haluza-DeLay R 2014 *How the World's Religions are Responding to Climate Change: Social Scientific Investigations* (London: Routledge)
[33] Syahbuddin H, Las I, Unadi A and Runtunuwu E 2007 *Identifikasi dan Delineasi Kalender Tanam dan Pola Tanam Pada Lahan Sawah Terhadap Anomali Iklim di Pulau Jawa* (Bogor, Indonesia: Balai Penelitian Agroklimat dan Hidrologi)
[34] Dhakal A and Rai R K 2020 Who Adopts Agroforestry in a Subsistence Economy?-Lessons from the Terai of Nepal *Forests* **11** 565

**Acknowledgment**
This article was written as a part of research "Performance of agroecosystem quality and adaptation of paddy farmer households to strengthen livelihood resilience in dealing with climate variability in Seluma district, Bengkulu Province, Indonesia". We also gratefully acknowledge the financial support of Faculty of Agriculture, University of Bengkulu, Indonesia for funding this research.

# Learning invention using satellite observations to support sustainable development goals (SDG): A use case on disaster risk reduction in Sei Serelo Indonesia

**B Setiawan[1*], E W Hastuti[1] and E Saleh[2]**

[1] Geology Engineering Study Program, University of Sriwijaya, Jl. Srijaya Negara, Bukit Besar Palembang, Indonesia
[2] Agricultural Engineering, Faculty of Agriculture, University of Sriwijaya, Jl. Raya Palembang Prabumulih Km 32 Indralaya, Indonesia
*Corresponding author: budhi.setiawan@unsri.ac.id

**Abstract**. The morphological pattern of the Sei Serelo was investigated to infer the impact of land use and climate change. Two sets of areal Landsat (1990 and 2019) identified the morphological changes to reduce the disaster risk and ideally reverse this prevailing situation. This paper presents a scalable and flexible approach to monitoring land-use change at the local level using various components of the Global Earth Observation System of Systems (GEOSS) platform. Increasing mining area has contributed to land-use change and the loss of agricultural land in many rural areas. In many cases, it worsens the poverty levels of smallholder farmers who depend on subsistence farming – an issue that Sustainable Development Goals number one seeks to address. A multi-criteria evaluation is applied using morphometric indicators, geology, and contours to identify the areas vulnerable to drainage and relief conditions. This learning invention has developed decision tools to apply GIS utilization to support disaster risk reduction. The devices are iterative and can be updated as new events occur to maximize GIS benefit, reducing disaster risk reduction and their potential consequences.

**Keywords**: morphometric, land-use change, SDGs, disaster risk reduction.

## 1. Introduction

Discussions on the relationship between Sustainable Development Goals (SDGs) and environmental assessment (EA) take various forms [1]. Most of the research on hydrological risk focuses on the risk of flooding and drought as impacts of two, so consideration is needed to design strategies and measures for reducing disaster risk (DRR) [2]. Determination of the level of integration of SDGs and EA is divided into three primary forms of integration, namely non-integration, radical integration, and partial integration, which includes six levels of integration. This research is intended to strengthen the framework of the implications for integrating the two concepts [3]. Natural conditions that have many values, which are sometimes synergistic, sometimes contradict people's lives, so that when viewed from the perspective of Sustainable Development Goals (SDGs), spatial mapping with nature's contribution to the SDGs has the potential to implement SDG strategies through the conversion of ecosystem services and land for sustainable management. However, this mapping requires a range of spatial data so that the SDGs have the potential to be in line with expectations [4].

Mapping with spatial data utilizes remote sensing data, namely satellite image data using a geographic information system (GIS) which assesses the temporal dynamics of the spatial dynamics of land cover changes in an area. It aims to understand historical events and economic policies that contributed to these changes. The analysis of land-use change aims to prevent the loss of subsistence agricultural land. This is related to the first point in the Sustainable Development Goals (SDGs) with realistic goals for people in rural areas to eradicate poverty [5]. Sustainable Development Goals (SDGs) 15.3.1. which describes progress in combating land degradation is monitored by three sub-indicators: vegetation productivity. These indicators are measured in the vegetation index using remote sensing data using a geographic information system (GIS) [6].

In the process of developing risk management, the use of geographic information systems is crucial. Geographic Information Systems (GIS) continue to build, but there are significant gaps in the risk management approach to geographic information needs, flows, and information sources. This information causes limitations on the effectiveness of management efforts in disaster risk management. Disaster risk management with geographic information systems has its concept of geographic information, namely by measuring and analyzing research capabilities in understanding, accessing, and using data on the analyzed aspects [7]. Drainage and relief parameters are very influential on the intensity of landslides. Therefore, these parameters are essential aspects that will be used in the final mapping of landslide susceptibility [8]. In this paper, the identification of areas that are vulnerable to relief and drainage conditions becomes a multi-criteria evaluation for this study.

## 2. Data dan methodology

This section provides background and analysis of morphometric indicators and methodology for morphometric change in each context. We use the watershed level as a case study; many morphometric parameters have been considered to fulfill the objective. The following characteristic of Earth Observation (EO) data can bring significant to support directly or indirectly target and indicator of the SDG Framework [9]:

a. Spatial resolution: This study uses data with a more detailed spatial resolution of 10 m × 10 m to describe the spatial characteristics of the watershed topography in more detail.
b. Temporal resolution: resolution by capturing data repeatedly at a specific time in the frequency of hours or days. This study uses a moderate temporal resolution, namely temporal resolution, within 16 days on Landsat data.
c. Scale: capacity on the size of the data spatially and in the field. The scale is used ranging from local to global scale in units of m or km.
d. Time-series: time in providing spatial data that is sustainable and continues to grow to start from 1972 (e.g., Landsat)
e. Multi-spectral: remote sensing methods in different wavelength ranges based on the need for spatial data (e.g., 0.64 – 0.67 for the Landsat infrared wavelength against the vegetation index).
f. Consistency: consistent availability of data and information at various scales and time
g. Complementary: data validation using additional sources such as sensor data.

### *2.1. Remote sensing data*

This research uses Landsat 5 TM data with band math function on band 4 and band 3 on June 26, 1996, for modeling vegetation indicators in 1996 and Landsat 8 OLI/TIRS data with band math function on band 5 and band 4 combinations of band 4, band 3, and This study utilized Landsat 5 TM on June 26, 1996, on August 31, 2019, for modeling vegetation indicators in 2019. In addition to Landsat data, it also uses spatial data in the form of SRTM DEM data with a resolution of 10 m × 10 m for modeling relief and drainage conditions. This year, we used images from Goggle Maps to assess the magnitude and impact of land change after mining operations. The location of the research is shown in Figure 1.

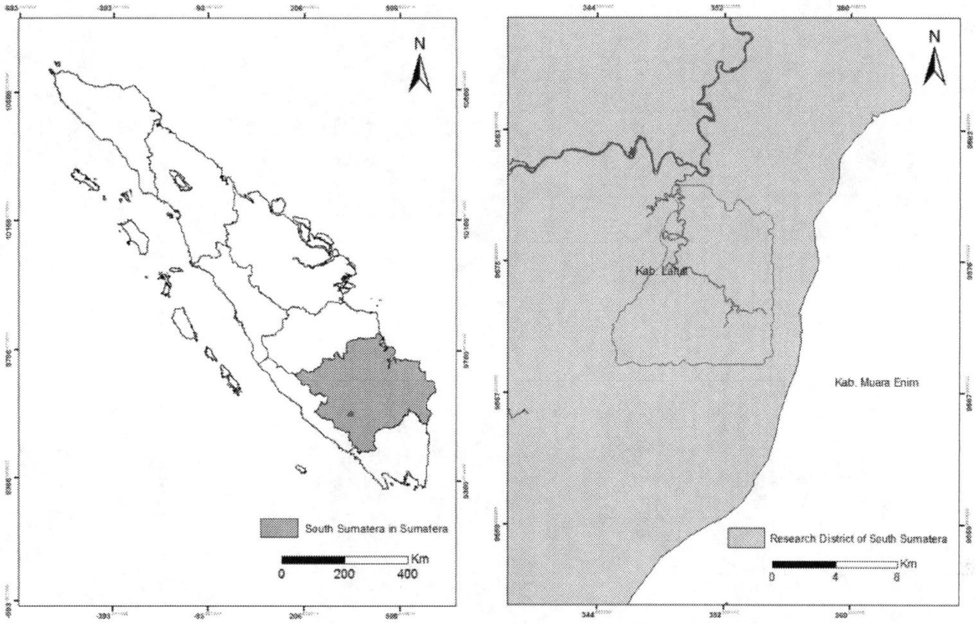


**Figure 1.** Location of the study area in Sei Serelo watershed, Merapi Selatan, Lahat, South Sumatera Province

### *2.2. Digital image enhancement and processing*

We used ArcMap software in ArcGIS version 10.3 for image data processing and SRTM DEM in this research. In addition, this study's data processing and spatial data attributes were converted geographically to the Universal Transverse Mercator (UTM) coordinate system with the World Geodetic System (WGS) 84 datum.

Extraction on drainage networks is an essential data processing in this research. The processing of the drainage system and the Sei Serelo watershed in this study uses ArcGIS 10.3 with SRTM DEM data which has a resolution of 10 m. The spatial resolution of 10 m is intended to describe the topography of the watershed in detail. In addition to processing DEM data, this study also uses Landsat data to determine the characteristics of vegetation indicators.

The purpose of this study with the result in the form of a drainage diversity map, namely, to find out which areas have drainage that plays a dominant role. Sei Serelo watershed drainage density is based on morphometric indicators, including stream frequency, drainage density, drainage intensity, length of overland flow, infiltration number, bifurcation relative relief, and ruggedness index. Meanwhile, the topographic model uses relief diversity with dissection index, relative comfort, slope in degree, lineament density.

## 3. Results and discussion

The result obtained from 12 different indices, including morphometric indicator and relief diversity for Sei Serelo, are given in Figure 2. The explanation of indicator only selected to disaster-related based on the evaluation.

IOP Conf. Series: Earth and Environmental Science 1016 (2022) 012021 doi:10.1088/1755-1315/1016/1/012021

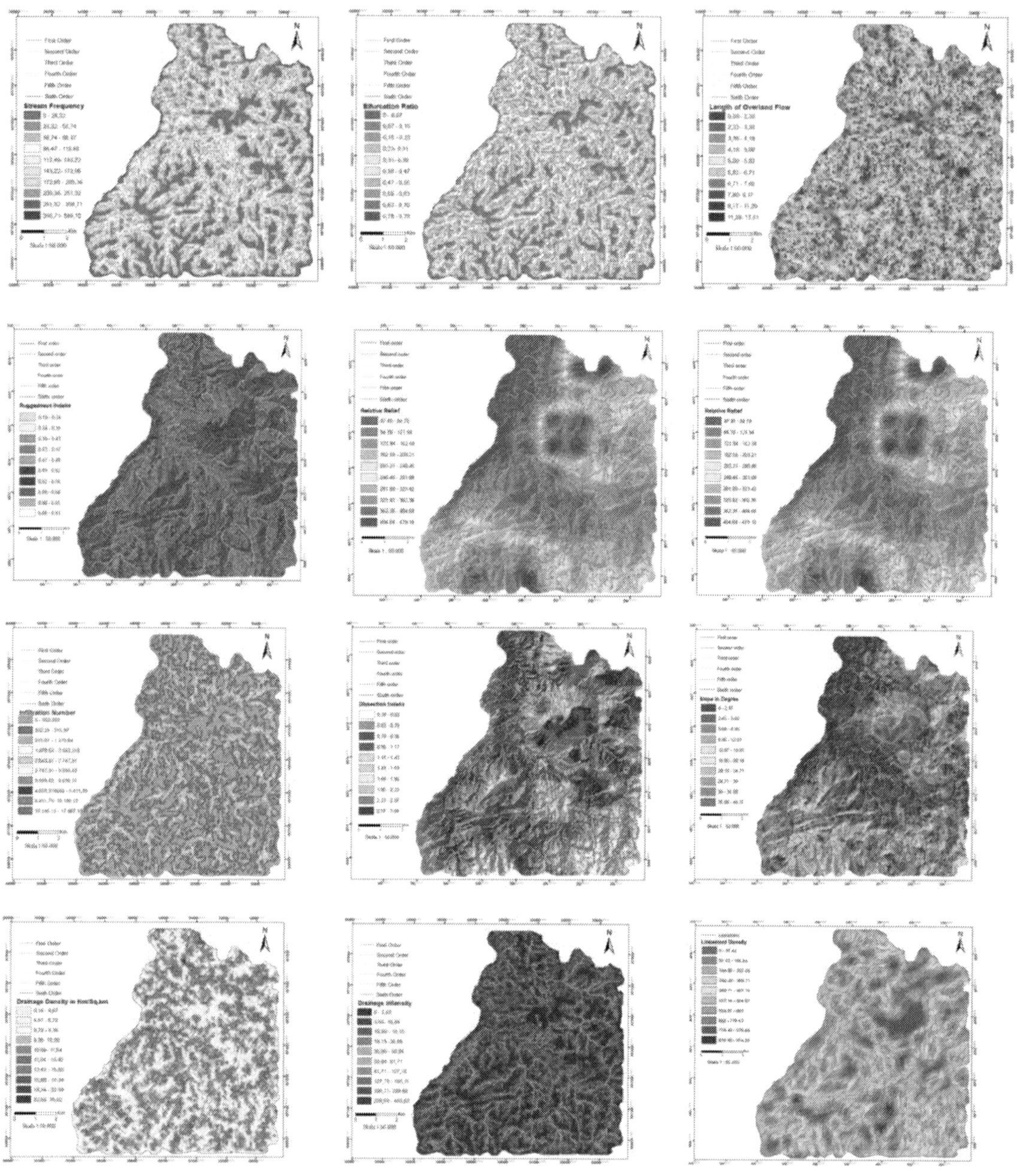

**Figure 2.** Morphometric indicator and relief density of Sei Serelo river basin

### *3.1. Morphometric indicator*

The **Stream length ratio** (Rt) indicator compares the total length of a particular river order and the total length of a subsequent river order based on the Strahler method. A high Rt value causes a high runoff rate and the possibility of flooding [10]. According to Strahler's (1952) practice, the **bifurcation ratio** (Rb) compares the number of river segments of a particular order and the number of features of a higher-order river. Therefore, while the probability of flooding is high in a watershed with a high Rb value, the rate is even lower in a low Rb value [11]. For example, the value of Rb on the Sei Serelo River has a value that varies between 0.07 to 0.90, indicating the potential for moderate flooding.

### *3.2. Relief Density*

The **Drainage density** (Dd) indicator shows the relationship between the level of permeability and the potential for flooding of an area where if the level of Drainage density (Dd) is high, it means that the area has a high runoff level so that when the permeability is low, the potential for flooding is higher [12]. Based on data analysis in the research study area, some areas have high-density values. A high Dd ratio is related to low permeability values, low vegetation index, and low relief diversity values in watershed areas. The **Ruggedness number** (Rn) indicator is one of the most significant indicators on the potential for flooding [13]. If the Rn in the drainage basin increases, the flow will increase to make an area have a low flood potential. The Rn value is low so that the probability of peak flooding decreases.

### *3.3. Land-use change*

Figure 3 below illustrates a diagrammatic representation of the classification results obtained from multitemporal satellite imagery. The figure also shows the magnitude of land change associated with each land use cover class identified in this study.

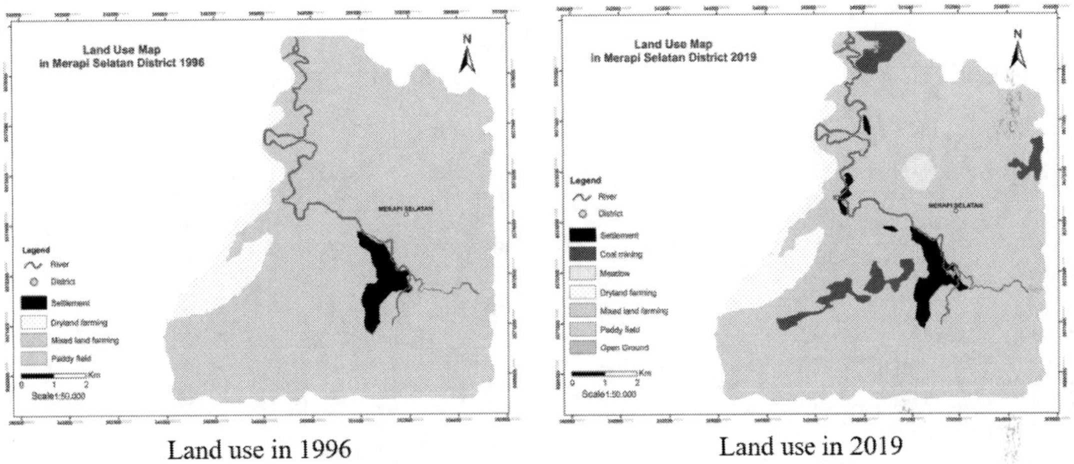


Land use in 1996 Land use in 2019

**Figure 3**. Land use and Land cover between 1996 and 2019

**Table 1**. Land use and land cover in km2 and their associated change

| Land cover | 1996 | 2007 | 2019 | Change | |
|---|---|---|---|---|---|
| | | | | 1996-2007 | 1996-2019 |
| River | 1.03 | 1.06 | 1.10 | 0.03 | 0.07 |
| Paddy field | 1.10 | 1.08 | 1.63 | -0.02 | 0.53 |
| Coal mining | 0.00 | 0.00 | 3.12 | 0.00 | 3.12 |
| Settlement | 1.96 | 2.02 | 2.34 | 0.06 | 0.38 |
| Dryland farming | 3.95 | 4.74 | 4.45 | 0.79 | 0.50 |
| Mixed land farming | 75.76 | 74.90 | 69.53 | -0.86 | -6.23 |
| Meadow | 0.00 | 0.00 | 0.88 | 0.00 | 0.88 |
| Open ground | 0.00 | 0.00 | 0.18 | 0.00 | 0.18 |

Table 1 presents the exact direction associated with the change, such as plantation, dryland farming, and mixed land farming decreased in each period. The above results show that all land cover except river bodies is characterized by only marginal change. All other land uses/covers have resulted in significant changes during the study period. Areas of coal mining, paddy fields, and settlements increased, with mixed land agriculture being the only land use/cover class that had to experience a substantial reduction in this period. These results can also be explained that after 2007, coal mining continued to increase while mixed land cover decreased.

### 3.4. Morphometric change

Various parameters of river network morphometry are seen in the previous section. Prominent river sinuses are measured for the entire watershed and straight-line segments of different lengths within the watershed. The results show that sinuosity is highly variable and that shorter segments show higher values than longer line segments, as shown in Figure 4. There is an inverse relationship between sinuosity and drainage density, although there are no data to test this.

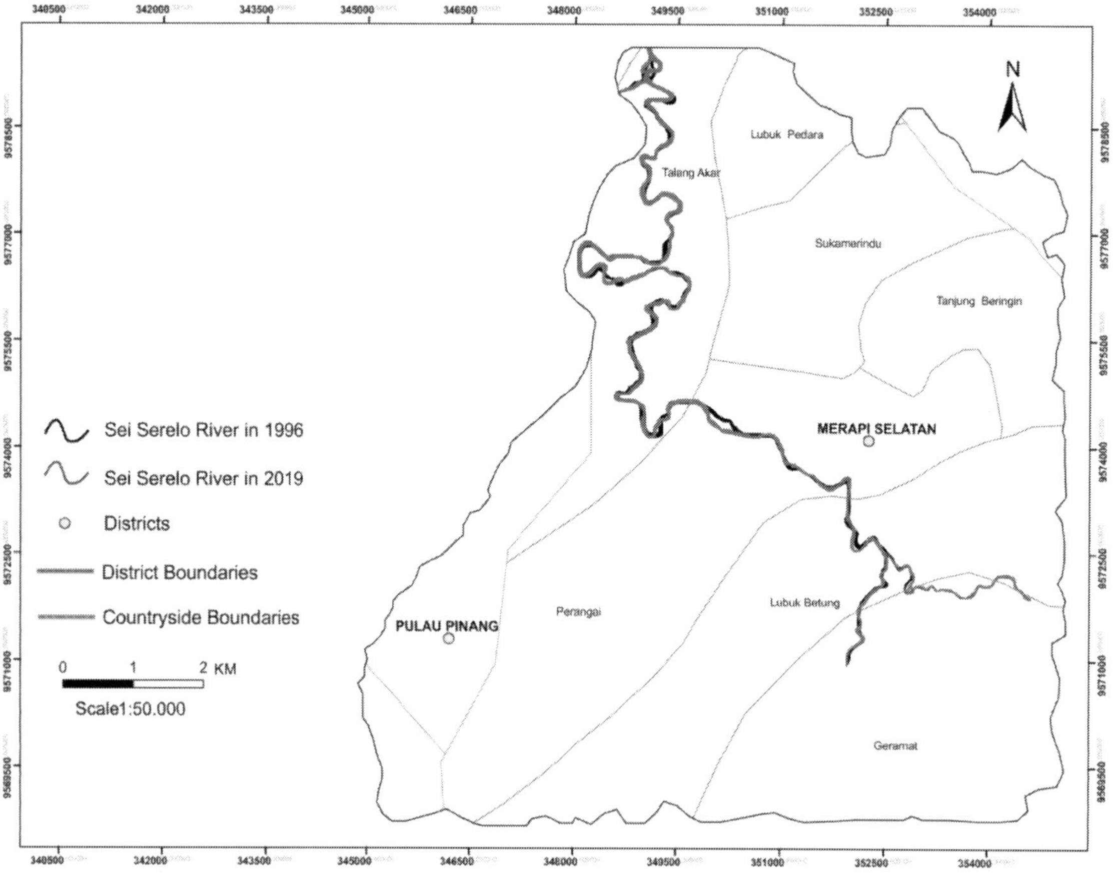


**Figure 4**. River sinuosity of Sei Serelo

The meandering river has dynamic evolutionary characteristics of lateral migration and longitudinal creep movement. The study of migration rates in meandering rivers has scientific and practical significance for understanding the evolutionary process. Unfortunately, river source areas are often underpopulated and lack long-term monitoring data, making estimating river bend migration rates challenging. However, studies have shown that topography can influence the changing trajectories of mountain streams and control the development of bedrock twists [14].

The mapped river bend patterns are also examined in detail in a smaller area of the watershed [15]. We identified several settlements living along the Sei Serelo River and assumed that each house could accommodate an average of four people. Small colonies are typical along the river, and we consider that some of these small settlements may be homes to members of the same family. The aim is to examine current river bends, including all topographic relief uniformly, or whether there is topographical evidence of river bend migration. As is known, the process of erosion and sedimentation directly impacts the communities along the banks of the Sei Serelo River. Communities experiencing a process of sedimentation, for example, tend to live far from rivers and in areas with lower river speeds (inner banks), thereby reducing their vulnerability to low inundation impacts.

## 4. Conclusions

This remarkable study uses remote sensing data from the Sei Serelo watershed in South Sumatra, which aims to show the development of settlements and mining that affect land-use changes in the area. Land-use change is an important issue that must, and land must be managed carefully to reduce disasters. In addition, this study was conducted to consider different morphometric parameters. We used morphometric parameters to describe the hydrogeomorphological characteristics of the research area. Integrated morphometric parameters to predict possible disaster vulnerability zones. This research is intended to have a deeper understanding of protecting ecosystems and an appreciation of the results of this research that can be used productively together with traditionally engineered systems for people's lives [16].

## References

[1] Boess E R, Kørnøv L, Lyhne I and Partidário M R 2021 Integrating SDGs in environmental assessment: Unfolding SDG functions in emerging practices *Environ. Impact Assess. Rev.* **90** 1–10
[2] Ward P J, de Ruiter M C, Mård J, Schröter K, Van Loon A, Veldkamp T, von Uexkull N, Wanders N, AghaKouchak A and Arnbjerg-Nielsen K 2020 The need to integrate flood and drought disaster risk reduction strategies *Water Secur.* **11** 1–14
[3] Kørnøv L, Lyhne I and Davila J G 2020 Linking the UN SDGs and environmental assessment: Towards a conceptual framework *Environ. Impact Assess. Rev.* **85** 1–10
[4] Mulligan M, van Soesbergen A, Hole D G, Brooks T M, Burke S and Hutton J 2020 Mapping nature's contribution to SDG 6 and implications for other SDGs at policy relevant scales *Remote Sens. Environ.* **239** 1–10
[5] Acheampong M, Yu Q, Enomah L D, Anchang J and Eduful M 2018 Land use/cover change in Ghana's oil city: Assessing the impact of neoliberal economic policies and implications for sustainable development goal number one–A remote sensing and GIS approach *Land use policy* **73** 373–84
[6] Prince S D 2019 Challenges for remote sensing of the Sustainable Development Goal SDG 15.3. 1 productivity indicator *Remote Sens. Environ.* **234** 1–7
[7] Tomaszewski B M, Moore E A, Parnell K, Leader A M, Armington W R, Aponte O, Brooks L, Herold B K, Meyers B S and Ruggero T 2020 Developing a geographic information capacity (GIC) profile for disaster risk management under United Nations framework commitments *Adv. Sp. Res.* **47** 1153–269
[8] Basu T and Pal S 2019 RS-GIS based morphometrical and geological multi-criteria approach to the landslide susceptibility mapping in Gish River Basin, West Bengal, India *Adv. Sp. Res.* **63** 1253–69
[9] Giuliani G, Mazzetti P, Santoro M, Nativi S, Van Bemmelen J, Colangeli G and Lehmann A 2020 Knowledge generation using satellite earth observations to support sustainable development goals (SDG): A use case on Land degradation *Int. J. Appl. Earth Obs. Geoinf.* **88** 1–12
[10] Vittala S S, Govindaiah S and Gowda H H 2004 Morphometric analysis of sub-watersheds in the Pavagada area of Tumkur district, South India using remote sensing and GIS techniques *J. Indian Soc. Remote Sens.* **32** 1–16
[11] Mahala A 2020 The significance of morphometric analysis to understand the hydrological and morphological characteristics in two different morpho-climatic settings *Appl. Water Sci.* **10** 1–16
[12] Farhan Y, Anbar A, Al-Shaikh N and Mousa R 2016 Prioritization of semi-arid agricultural watershed using morphometric and principal component analysis, remote sensing, and GIS techniques, the Zerqa River Watershed, Northern Jordan *Agric. Sci.* **8** 113–48
[13] Rai P K, Chandel R S, Mishra V N and Singh P 2018 Hydrological inferences through morphometric analysis of lower Kosi river basin of India for water resource management based

on remote sensing data *Appl. water Sci.* **8** 1–16

[14] Johnson K N and Finnegan N J 2015 A lithologic control on active meandering in bedrock channels *Geol. Soc. Am. Bull.* **127** 11–2

[15] PH N I and Setiawan B 2021 Perkembangan Morfometri Daerah Limpah Banjir Pada Sub Wilayah Aliran Sungai Sei Serelo, Kabupaten Lahat *Seminar Nasional Hari Air Sedunia* vol 3 pp 42–7

[16] Palmer M A, Menninger H L and Bernhardt E 2010 River restoration, habitat heterogeneity and biodiversity: a failure of theory or practice? *Freshw. Biol.* **55** 205–22

**Acknowledgment**

The publication of this article was funded by DIPA of Public Service Agency of University of Sriwijaya 2021. SP DIPA-023.17.2.677515/2021 on November 23, 2020. In accordance with the Rector Decree Number:010/UN9/SK.LP2M.PT/2021 on April 28, 2021. The authors also thank Nurtiara Inayah Putri Harahap and Thania Putri Firdaus for supporting fieldwork and geological mapping.

# Springshed protection planning for combating drought in Grobogan District, Central Java

**A B Supangat*[1], Purwanto[1], H Hendrayana[2] and H D Riyanto[1]**

[1]Watershed Management Technology Center (WMTC), Jl. A.Yani-Pabelan PO BOX 295, Surakarta, Indonesia
[2]Department of Geological Engineering, Faculty of Engineering, Universitas Gadjah Mada, Yogyakarta, Indonesia
*Corresponding author: maz_goenk@yahoo.com

**Abstract**. Drought occurs almost every dry season in Grobogan District, Central Java. Climate change has resulted in an atypical monthly rainfall distribution, which, along with limestone geology, has made water scarce in some months. One of the efforts to mitigate drought disasters is to revitalize water resources from existing springs. The research aims to develop a spring protection plan through a geological-spatial and participatory approach. The research location is in Kradenan Sub-District, Grogoban District, Central Java Province. The research method starts with identifying springs, delineation of springshed area, formulation of spring and springshed management, and discussion and socialization with stakeholders. The identification results show that the springs are classified as "umbul wadon" type with water discharge ranging from 0.9 to 4.8 l/minute, forming a radial pattern. The area of the springshed is 3,347.7 ha, with more than 70% of the land covered by teak forest plantations. The recommendations for spring conservation activities are organized into three zones: Zones I, II, and III, which differ in range radius, function, and proposed management. In each zone and type of land uses, the potential risk of pollution, prohibited activities, protection strategies and programs, and interested parties as program implementers have been formulated. Several official institutions can aid spring protection management. Together with water users and forest area managers, all institutions agreed to carry out activities to protect and conserve springs and springshed areas.

**Keywords :** dry season, climate change, spring water, conservation, water conservation

## 1. Introduction

Indonesia is in a disaster-prone area, one of which is drought. A drought is a form of hydrometeorological disaster induced by hydrologic and geological forces [1,2]. Climatically, parts of Indonesia have low annual rainfall, making it difficult for people to access water resources. The potential for drought is exacerbated by climate change and geology such as limestone, which is difficult to hold water, thus making it difficult for water to be available. The disproportionate shift in the distribution of monthly rainfall causes the occurrence of dry months, which leads to scarcity of water resources.

Grobogan is one district that experiences drought every year [2,3]. The type of drought in Grobogan District is a hydrological drought, where water resources are difficult to obtain in dry months [2]. The study results by [4] stated that 29.43% of the Grobogan District area had a moderate drought index, while 24.01% showed a high drought index value. [2] stated that almost 60% of the rainfall will be lost through the evapotranspiration process, while the potential annual runoff is only about 40%. However, the geological conditions of limestone/lime make it difficult for water to be stored and available in the soil layers. Tjakrawarsa and Supangat [5] has mentioned that, based on geology, more than 50% of the area is limestone with moderate to high water passing rates and aquifers with moderate groundwater production. Meanwhile, 45% of the area is marl and limestone with low water passing rates and output of scarce groundwater.

The drought in Grobogan District impacts various sectors, especially agriculture in general and domestic water needs. The study's findings concluded that, based on the parameters of the community's level of exposure, sensitivity, and adaptation capacity, the community's socioeconomic vulnerability to drought was relatively high [6].

The district government of Grobogan has carried out mitigation efforts are in the form of various water conservation buildings, the 'Pamsimas' program, and the supply of clean water for the community's domestic needs [7,8]. Some people have also adapted to drought in various ways, such as preparing alternative water sources [2], building rainwater storage infrastructure [6], or arranging rice planting schedules [4]. However, the mitigation efforts are often un-optimal in meeting the community's water needs in the dry season. Therefore, other actions are still needed to support clean water for the community by revitalizing the function of alternative water sources, namely springs.

The existence of springs in areas with karst geology usually has abundant and continuous flow water resource potential. In Grobogan District, there are quite a lot of springs scattered in almost all districts. However, some obstacles regarding the existing utilization and management of springs are still not correct. Besides inappropriate water exploitation methods, namely, directly at the edge of the spring, the management of the spring point and its recharge area has not been carried out correctly. These mistakes have caused many springs to have polluted and decreased water quantity or even be dead springs [9]. Springs are in trouble, such as being dried and polluted in many locations worldwide due to environmental deterioration, climate change, and increased demand for water, all of which put stress on these delicate and restricted natural resources [10,11].

Based on this background, it is necessary to plan well the management of the springs, which is carried out in a geospatial and participatory manner [10]. Good spring management is part of efforts to conserve groundwater resources and restore springs [12]. The research aims to develop a spring protection plan through a geological-spatial and participatory approach. The benefits of the research results are expected to provide inputs in the management by the parties, and the role of each related institution becomes focused. It is hoped that the springs will be better maintained in terms of quality, quantity, and continuity of water resources through proper management.

## 2. Materials and method

### *2.1. Research location*

The research is located in Kradenan Sub-District, Grobogan District, Central Java, Indonesia. Specifically, the spring location is located in Ngaram-Aram Hamlet, Cerewek Village. This location is in the middle area of the Lusi Watershed, which geographically is between 06°49'48,1" to 07°16'57,4" S and 110°43'41,6" to 111°35'19,8" E. (Figure 1.). In general, the research location has rainfall of 1,809 mm. year$^{-1}$, with the dry season distribution occurring from May to October. The soil type is dominated by vertisols, with geology derived from limestone sedimentary parent rock [3].

IOP Conf. Series: Earth and Environmental Science 1016 (2022) 012022 doi:10.1088/1755-1315/1016/1/012022

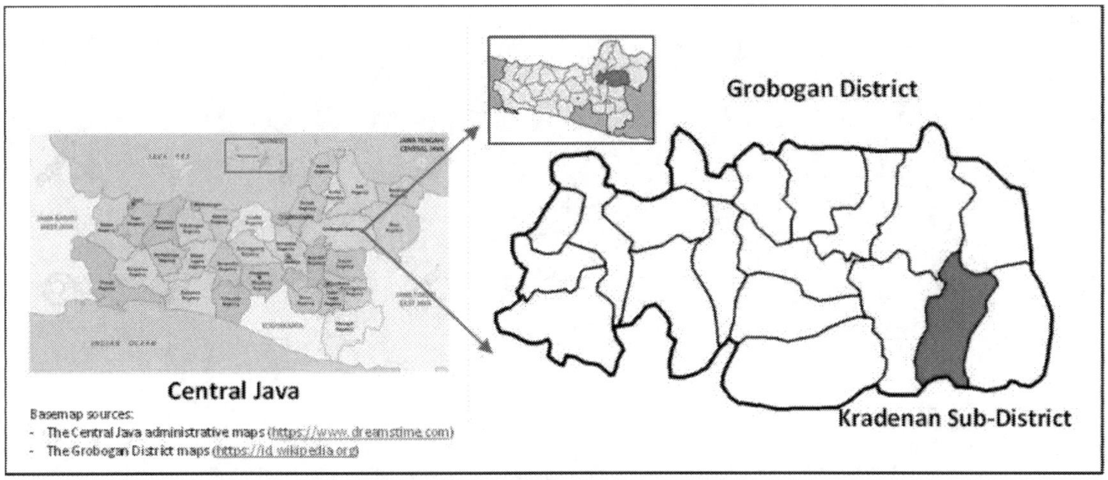


**Figure 1**. Map of research location

### *2.2. Data collection*

The stages of research begin with the activities of identifying springs, followed by delineation of the springshed area, formulation of indicative spring and springshed management, and discussion and socialization with stakeholders. Identification of springs includes activities of inventory for the presence of springs in Cerewek Village, followed by identification of the condition of the springs, evaluation of potential water resources, water use, and the conditions around the springs [9].

The delineation of the springshed area is intended to border the catchment area as a recharge area from the spring. The delineation of the springshed area is carried out using the method from Gassen et al (2013) and Smith et al. [13,14]. Springshed delineation is also used to identify land cover on the surface and map protection zones as a basis for planning best management practices in spring management [15].

The formulation of indicative spring management is carried out based on delineating the distribution of conservation zones from spring points. The preparation of directions for spring conservation activities is focused on activities that follow the objectives of each conservation zone [13,16]. Indicative springshed management is carried out based on each land cover type in the springshed area by implementing best management practices (BMP) by paying attention to soil and water conservation principles.

The institution's role was identified through stakeholder FGD activities facilitated by the regency's Bappeda (Regency Development Planning Agency). Discussion and socialization with stakeholders were carried out to formulate the parties' role in collaborative management of spring and springshed conservation activities following their respective duties. The flowchart of the research stage activities is presented in Figure 2.

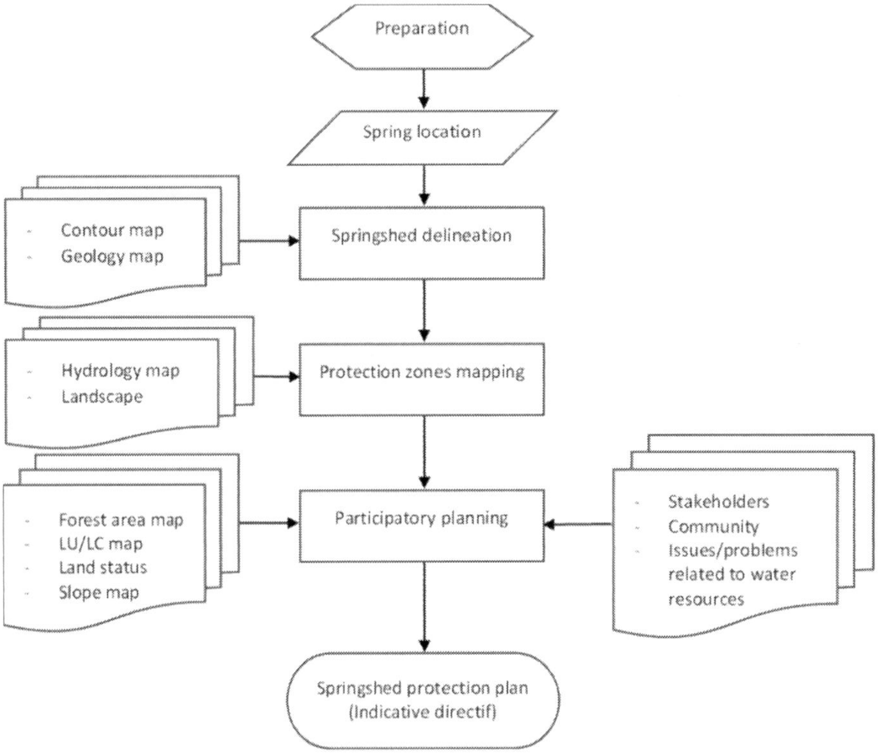


**Figure 2.** Flowchart of research activities

### 2.3. *Data analysis*

To delineate the boundaries of the springshed area, supporting maps such as contour/topographic maps and geological maps are needed. Meanwhile, mapping the zoning of the springs conservation area was carried out by analyzing overlapping maps of springs, hydrology, and landscapes/land use with the help of GIS tools. Topographic maps and geological maps are sourced from The Indonesian Geospatial Portal (https://tanahair.indonesia.go.id). While the hydrology maps and land use maps were extracted from the 2016 Earth Map (RBI) and Google Earth, respectively.

In each zone and type of land use, there has been formulated the potential risk of pollution, prohibited activities, protection strategies and programs, and the interested parties as program implementers [10,11].

## 3. Results and discussion

### 3.1. *Characteristics of spring*

In the Ngaram-Aram Hamlet, there are 4 springs, the largest being the Mindik spring, while the others are the Palem, the Waru, and the Sendang Semak Springs. The names of springs are adapted to the big trees that grow around the springs, such as the Mindik tree (*Melia azedarach*).

Based on the flow direction of the spring, Mindik spring has the gushing upwards type. That means that the spring is categorized as a spring "wadon" (female spring). The upward flowing happens because of the upward pressure from the water (aquifer) body located under the rock layers. Meanwhile, based on the location of the spring sports, they are forming a radial pattern or "nyabuk gunung" formation, with a distance between the springs of approximately 50-100m. The average spring water discharge ranges from 0.9 to 4.8 l/minute. The highest water discharge occurs in January, while the lowest is in September [9].

The springs at the research location are the main water source for the surrounding community to meet household water needs such as cooking, bathing-washing-toileting (MCK), and livestock drinking. However, the spring condition is mostly not well performed due to the wrong utilization and lack of maintenance. Taking and using water directly from the spring can cause soapy water and other impurities to enter the spring. This causes a decrease in water quality. Another way of using water is by sucking water from the spring with a water pump and following it through pipes to people's homes. The spring point is also open, so much dirt gets in, such as leaves and plastic.

(a) (b)

**Figure 3.** Spring conditions, (a) before conservation activities; (b) after broncaptering

### *3.2. Delineation of springshed and conservation zones*

The delineation of the springshed area through the overlay of several base maps showed the area of the springshed is 3,347.7 ha (Figure 4). The upstream or recharge area is in the southeast direction of the springs' point. The most upstream (in a bit of part) is in the Gabus Sub-District, Grobogan District, which also borders Sragen and Ngawi Districts. In contrast, the middle and downstream parts are in the Kradenan Sub-District, Grobogan District. There are six villages in the springshed area, namely the Cerewek, Pakai, and Rejosari Villages (Kradenan Sub-District), and the Pandan Harum, Keyongan, and Sewaktu Villages (Gabus Sub-District). The boundary of the springshed area almost coincides with the micro watershed boundary above ground level, namely the Soca River, which flows down into Cerewek Village.

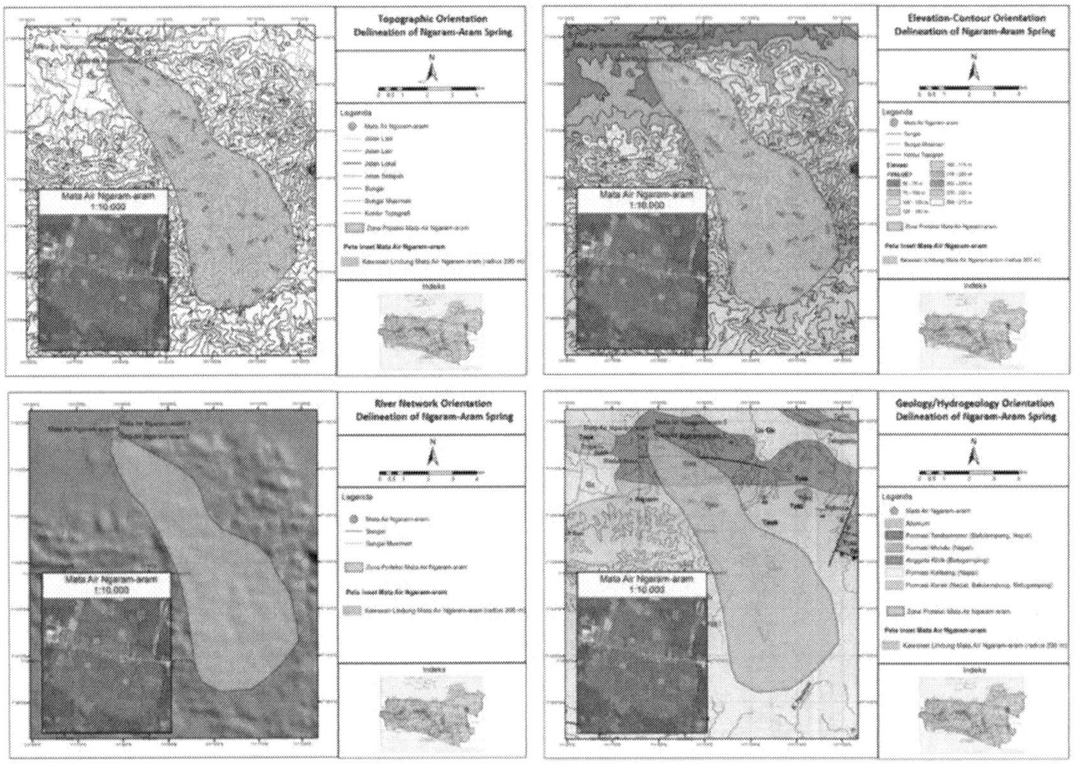


Basemap sources:
- The Topographic Map, 2015
- The Earth Map (RBI), 2016
- The Google Earth, 2016
- The Geology Map

**Figure 4.** Delineation process of springshed area

Based on the springshed area delineation, the identification of the existing land cover is then carried out as the basis for the arrangement of indicative directions for the spring protection zone management. The land cover conditions in the springshed area are presented in the following table. More than 70% of the land cover is in teak forest plantations managed by the Forestry State Enterprise (Perum Perhutani) (see Figure 5.a).

**Tabel 1.** Land use types of springshed area in Ngaram-Aram Hamlet

| Land Use Types | Area | |
|---|---|---|
| | (ha) | (%) |
| River/Water body | 0.02 | 0.00 |
| Shrubs | 31.41 | 0.94 |
| Forest/Garden | 2,580.82 | 77.09 |
| Settlement | 112.24 | 3.35 |
| Rainfed rice field | 528.89 | 15.80 |
| Dry fields | 94.32 | 2.82 |
| Total | 3,347.70 | 100.00 |

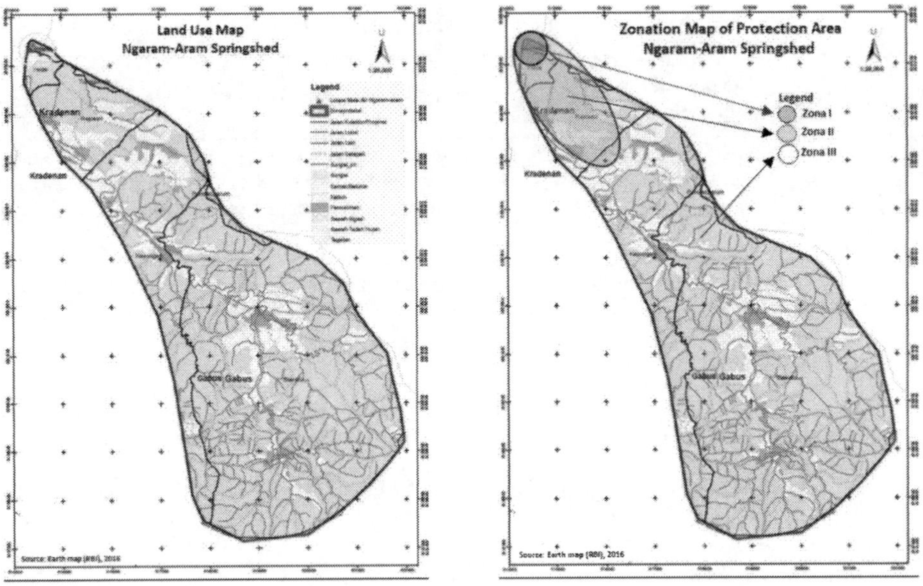


(a) (b)

**Figure 5.** Land use map of Ngaram-Aram Springshed area (a) Land Use Map; (b) Zonation Map

The protection zoning is divided into 3 zones, namely Zone I (spring protection), Zone II (protection around the springs), and Zone III (protection of springshed area). Zone I is defined as a radius of 10-15 m from a circular spring. Zone II is located 100-150 meters upstream from the outer boundary of Zone I, in an elliptical/oval shape that follows the shape of the watershed. Meanwhile, Zone III is set from the outer border of Zone II to the upstream limit of the springshed (Figure 5.b).

### *3.3. Springshed conservation plan*

The springshed management plan was divided into 2 ways, namely springs spot and springshed area protections. Based on the analysis, there were 3 zones determined, namely zones 1, II and III. Zone I (spring protection), with a radius of 10-15 m, aims to protect the springs (water source points) from all pollutant materials, either directly or indirectly, which cause water quality degradation. Zone II (protection around the springs), with a radius of 100-150 m, aims to protect the springs from contamination hazards by pathogenic bacteria, viruses, and parasites that can cause water quality degradation. While Zone III (protection of springshed area), with a radius from the boundaries of Zone II to the upstream of the spring, aims to protect the raw water sources of the springs from chemical and radioactive contamination that cannot be degraded in a short time.

The directions for spring conservation activities are carried out in each zone, and are in accordance with each existing land use. The proposed activities consist of vegetative and engineering measures in the recharge area of springs [12]. The formulation of conservation activities in each zone are presented in Tables 2 and 3.

IOP Conf. Series: Earth and Environmental Science 1016 (2022) 012022 doi:10.1088/1755-1315/1016/1/012022

**Table 2.** Directions for spring conservation activities in protection zone I

| Types of Land Cover | The potential risk of pollution | Prohibited Activities | Protection Strategy/Program |
|---|---|---|---|
| Water pond | Direct injection of harmful substances from pondside water use | Direct activities at spring point that possible to the entry of pollutants into the spring pond | - Broncaptering building<br>- Pool fence construction and repair<br>- Guard system arrangements |
| Bare land | Pollution of organic substances carried by surface runoff | Human/animal activities that can contaminate pond water | Cultivation of perennial/fruit plants that are not harvested for wood |
| Forest/shrubs | Risk of infiltration of water carrying organic substances | Human/animal activities that produce dung, such as logging | Plant and ecosystem maintenances, including enrichment planting |
| Agricultural lands | Pollution from the use of fertilizers/pesticides | Use of chemical fertilizers and pesticides | Change from agricultural land to forest land/mixed gardens |

Source: Adapted from [16], Modified

**Table 3.** Directions for spring conservation activities in protection zone II and III

| Types of Land Cover | The potential risk of pollution | Prohibited Activities | Protection Strategy/Program |
|---|---|---|---|
| Garden/Forest | Risk of infiltration water carrying organic substances into the soil | Human/animal activities that produce dung | Plant and ecosystem maintenances<br>Enrichment planting<br>Best management practices |
| Rainfed rice field | Increased nutrient content into surface water and groundwater | Excessive use of chemical fertilizers and pesticides | Best Management Practices |
| Dry-field agricultural | Pollution from the use of fertilizers/pesticides | Use of chemical fertilizers and pesticides | Best Management Practices<br>(If possible, increasing forest/mixed garden land cover) |
| Settlement | Increased contaminants from household waste (septic tanks, detergents, and other organic substances) | - Direct injection of hazardous substances into the aquifer<br>- Disposal of domestic waste into rivers/deep wells<br>- Disposal of garbage into deep pits (in Zone II) | - Repair and maintenance of well conditions (walls, floors, and drains)<br>- Repair and keeping with the integrated sanitation system<br>- The trash hole in zone II that is not too deep, or move the existing hole |
| Shrubs | Risk of infiltration water carrying organic substances from human/animal waste/dung | - Slash and burn of bushes<br>- Disposal of domestic waste to bushland<br>- Cattle grazing in zone II | - Movement of prohibited from damaging the ecosystem<br>- Movement of not littering<br>- Movement and socialization of the use of natural fertilizers<br>- Installation of signs of no livestock grazing in zone II |
| River/Water bodies | Increase in nutrients/ contaminants from various human activities | Disposal of organic and chemical waste/waste into rivers/reservoirs | - Maintenance of borders and riverbanks from potential erosion and sedimentation<br>- Movement of clean river |

| | | | |
|---|---|---|---|
| | | | - Monitoring water quality and sedimentation |

Source: Adapted from [16], Modified

### *3.4. Institutional supports*

FGDs and dialogues carried out to involve stakeholders and share roles in spring conservation initiatives, including local district institutions, central institutions in the region, community groups, and representatives of village officials are required. The results of the evaluation of the parties' roles are presented in Table 4.

**Table 4.** Identification of the institution's role in spring and springshed conservation activities

| Agencies/ Institutions | Main tasks related to water resource conservation | Potential roles that can be performed | Role realizations |
|---|---|---|---|
| Development Planning Agency at District Level (Bappeda) | District's regional development planning and coordination, including environmental and natural resource conservation activities | Allocation of funds for water resource conservation through village funds | - Dropping for conservation plant seeds<br>- Construction of water conservation buildings such as infiltration wells, biophores, and reservoirs<br>- Water conservation training for farmers |
| District Agency of Public Works and Spatial Planning (Dinas PUPR) | Handling irrigation, raw water, rivers, and reservoirs | - | - |
| District Agency of Agriculture (Dinas Pertanian) | Prevention and management of drought in agricultural areas | - | - |
| Forestry Services Branch (CDK) | Planning and operational activities in the forestry sector in the district | Allocation of funds for conservation activities | - Dropping for conservation plant seeds<br>- Water conservation training for farmers |
| District Agency of Environmental Services (Dinas LH) | Protection and preservation of natural resources (including water) | Allocation of funds for spring conservation activities | Construction of water conservation buildings such as infiltration wells, biophore, and training for farmers |
| Sub District Government (Kradenan) | Facilitation of rehabilitation and conservation programs and drought disaster management | Mobilizing villages to budget for "Pamsimas" and reservoirs through village funds | - |
| Villages Government (Banjardowo) | Facilitation of rehabilitation and conservation programs, provision of clean water, and drought management | Budgeting for conservation activities through village funds | Distribution of village funds for planting activities, constructing communal wells, and installing "Pamsimas" |
| District Disaster Management Agency (BPBD) | Disaster management in the district area, including drought | Allocation of funds for spring conservation in both mechanical and vegetative manner | Distribution of clean water to address domestic needs in drought-prone areas |

| | | | |
|---|---|---|---|
| The District Municipal Waterworks Institution (PDAM) | Organizers of drinking water management in the district areas | Maintenance of springs | - Caring for and maintaining springs as PDAM's raw water source<br>- Dropping clean water in drought-prone areas (in collaboration with BPBD) |
| Institute of Watershed Controlling and Protection Forest (BPDASHL) | Carry out rehabilitation and conservation activities in the watershed areas | Allocation of funds for spring conservation activities | - Inventory of springs as potential water resources<br>- Vegetative conservation in the recharge area of spring |
| River Basin Organization (BBWS) | Managing water resources in river basin areas | Allocation of funds for water conservation activities | - |
| Private sectors (Pertamina, Perhutani, industries) | Drought prevention and management as part of Corporate Social Responsibility (CSR) | Allocation of CSR's funds for spring conservation activities | - Constructing communal wells<br>- Distribution of clean water |

Remark: "Pamsimas" is a Community-Based Water Supply and Sanitation Program, which is one of the national mainstay programs to increase rural community access to proper drinking water and sanitation facilities with a community-based approach

According to the table above, the formal district institutions that can directly play spring protection management include The PDAM, The Dinas LH, and The BPBD Agencies. Bappeda can operate as a coordinating agency, coordinating and synchronizing activities among institutions to not overlap. The central and provincial institutions concerned with spring conservation activities include BPDASHL as a representative of the Ministry of Environment and Forestry. The CDK as a Provincial Forestry Service office branch, and Perum Perhutani as a branch the holder of forest area management rights where most of the springs are located. Some agencies do not specifically contribute to spring conservation activities but can play a role in water resource conservation activities in general, such as BBWS, Dinas PUPR, Dinas Pertanian, and village and sub-district governments. Together with water users (community and private sectors) and the forest area manager (Perum Perhutani), all institutions agreed to collaboratively carry out activities to protect and conserve springs and springsheds.

## 4. Conclusions

Springs as a provider of water resources must be adequately managed in a participatory manner and with attention to geo-spatial conditions. Springshed management can be done through spring points protection from contaminants and rehabilitation of springshed areas by applying Best Management Practices according to the protection zones and land cover type. The involvement of relevant parties in spring conservation activities is required through collaborative management. Proper spring and springshed management will be an alternative strategy for mitigating drought and adapting to climate change.

## References

[1] Adi H P 2011 Kondisi dan konsep penanggulangan bencana kekeringan di Jawa Tengah *Seminar Nasional Mitigasi dan Ketahanan Bencana* vol 26 (Universitas Islam Sultan Agung) pp 1–10
[2] Purwanto and Supangat A B 2017 Perilaku Konsumsi Air Pada Musim Kemarau Di Dusun Pamor, Kabupaten Grobogan *J. Penelit. Sos. dan Ekon. Kehutan.* **14** 157–69
[3] Supangat A B and Boediyono A 2017 Identifikasi Potensi Bencana Kekeringan di DAS Lusi *The 4th National Seminar of Coastal and Watershed Management* (Faculty of Geography, Gadjah Mada University, Yogyakarta: 24 October 2018.) pp 45–50
[4] Rahmawati A 2016 *Analisis indeks kekeringan thornthwaite-mather dan arahan jadwal tanam padi untuk adaptasi bencana kekeringan di Kabupaten Grobogan* (Sebelas Maret University)
[5] Tjakrawarsa G and Supangat A B 2015 Karakteristik Kekeringan di Kabupaten Grobogan dan

Alternatif Pemecahannya *Seminar Nasional Kemandirian Daerah dalam Mitigasi Bencana menuju Pembangunan Berkelanjutan* (Faculty of Agriculture, Sebelas Maret University, Surakarta: 19 September 2015.) pp 124–32

[6] Hastanti B W and Purwanto P 2020 Analisis Keterpaparan, Sensitivitas, dan Kapasitas Adaptasi Masyarakat Terhadap Kekeringan di Dusun Pamor, Kradenan, Grobogan (Analysis of Exposure, Sensitivity and Community Adaptation Capacity to Drought in Pamor Hamlet, Kradenan, Grobogan) *J. Penelit. Hutan dan Konserv. Alam* **17** 1689–99

[7] Anonimous 2016 Pencegahan dan penanggulangan bencana tahun 2015 (Pemerintah Kabupaten Grobogan)

[8] Hastuti D, Sarwono and Muryani C 2017 Mitigasi Kesiapsiagaan dan Adaptasi Masyarakat terhadap Bahaya Kekeringan Kabupaten Grobogan *J. GeoEco* **3** 47–57

[9] Riyanto H D and Supangat A B 2019 Teknik Konservasi Mata Air: Studi Kasus di Desa Crewek, Kecamatan Kradenan, Kabupaten Grobogan *Building Integration of Watershed Management at the Implementation Level* ed A B Supangat and I W S Dharmawan (Bogor: IPB-Press.) pp 101–11

[10] Ranjan P and Pandey P K 2021 Spring Protection: Step Towards Water Security and Sustainable Rural Water Supply *Ann. Rom. Soc. Cell Biol.* **25** 1216–22

[11] Buono J. 2019 Spring Protection and Management: Context, History and Examples of Spring Management in India *Groundwater Development and Management* ed Sikdar P. (Springer, Cham.) pp 227–41

[12] Patil S, Barola Y and Kulkarni H 2018 *Springshed Restoration in Kalsi Block of Dehradun District* (Technical report number: ACWA/Hydro/2018/H72, Under The Programme 'Strengthening State Strategies on Climate Actions Through Practicing Climate Resilient Practices'. Advanced Center for Water Resources Development and Management (ACWADAM). Maharashtra, In)

[13] Gassen N, Al-Hyari M, Hanbali B, Obaiat A, Kirsch H, Toll M and Xanke J 2013 *Delineation of groundwater protection zones for Hidan well field* (German: German-Jordanian Technical Cooperation – Water Aspects in Land-Use Planning)

[14] Smith B A, Hunt B B, Gary M O and Wierman D A 2020 Springshed delineation in a karst aquifer in Hays County, Central Texas *NCKRI Symposium* (The Sinkhole Conference) pp 255–68

[15] Goldscheider N 2010 Delineation of spring protection zones *Groundwater hydrology of springs* (Elsevier) pp 305–38

[16] Hendrayana H 2003 Zona Perlindungan Sumber Air Baku di Kota dan Kabupaten Malang - Jawa Timur *Seminar Sehari Koordinasi dalam Rangka Persiapan PERDA tentang Perlindungan Sumber Air Baku Air Minum* (UNMER Malang, 17 Juli 2003) pp 1–24

**Acknowledgment**

Special thanks to the Watershed Management Technology Center (WMTC) for financial support, the Department of Geological Engineering, Faculty of Engineering, Gadjah Mada University, Yogyakarta for collaboration research, and research assistants for technical support. Thanks are also given to the Grobogan District Government for allowing the research team to conduct their work. Appreciation is given to reviewers and editors for constructive comments.

**Contributorship**

All authors did equal contributions to conceptualization, methodology, analysis, writing, review and editing the manuscript. All authors have read and agreed to the published version of the manuscript.

# Development of a web application for climate change adaptation in Thailand

**Hiroaki Shirakawa*[1] and Weerakaset Suanpaga[2]**

[1]Nagoya University, Furo-Cho Chikusa-ku Aichi,464-8601, Japan
[2] Kasetsart University, 50 Ngamwongwan Rd, Chatuchak Bangkok 10900 Thailand
*Corresponding author: sirakawa@urban.env.nagoya-u.ac.jp

**Abstract**. There is no doubt about the warming of the climate system, and it is widely accepted by a society that adaptation measures are important to reduce the negative impact of climate change on society. However, there are socio-economic obstacles to the promotion of climate change countermeasures. In other words, to implement climate change adaptation measures, not only the economic efficiency of the measures, but also fairness, consistency with existing policies, and cultural acceptability must be considered. Moreover, future forecasts contain uncertainty. Therefore, it is very difficult to formulate adaptation measures and consider policy priorities. The purpose of the study is to create a web application that shows the risk of natural disasters and the socioeconomic status of the region, targeting Thailand, which is one of the countries affected by climate change to solve the above problems. We basically developed web applications using freeware. As the result, the user can grasp the risk situation of natural disasters in Thailand as a whole, and grasp the industrial structure, agricultural production structure, and poverty situation in each region. This application is expected to promote understanding of the impacts of climate change among stakeholders and to support discussions on policy priorities.

**Keywords** : natural disaster, adaptation, Thailand, decision making, web application

## 1. Introduction

In recent years, Thailand has been experiencing natural disasters such as large-scale floods and droughts, which have become a major social problem. In 2011, flooding mainly in the Chao Phraya River basin caused damage estimated at about 1.43 trillion baht (46.5 billion US dollars) [1]. The damage was estimated at about 1.43 trillion baht (US$46.5 billion), which is equivalent to about 13.2% of Thailand's GDP in 2010. In particular, the manufacturing industry, which is concentrated in the Chao Phraya River basin, was severely affected, accounting for about 70% of the total damage. The country also experienced a severe drought in 2015 and 2016 [2]. Landslides and flash floods in mountainous areas have been occurring every year. It is feared that these natural disasters will worsen in the future. It is believed that there are both natu
ral and social factors that contribute to the deterioration.

In Thailand, climate change is expected to increase the risk of flooding and drought due to a rise in sea level, which will reduce the number of sandy beaches and increase the variability of precipitation [3]. Social factors include urbanization and population growth in areas prone to flooding as a result of economic growth. Although urgent measures are needed, it is not easy to promote climate change adaptation measures.

There are following six obstacles to promote climate change adaptation polies; (1) transaction costs, (2) market failure and absence of markets, (3) custom, (4) social equity, (5) government failure, and (6)

forecast uncertainty [4]. Transaction costs refer to the costs of gathering information necessary for decision-making and coordination among stakeholders. Since adaptation measures often require the development of infrastructure, the project costs are very large and there are many stakeholders. In addition, it is necessary to obtain information on the return on investment. Therefore, the time and cost of information gathering and negotiation can be very high. Market failure and absence of markets are cases where the rational actions of individual economic actors result in irrational outcomes when viewed as a whole, for example, when a large amount of irrigation water is withdrawn upstream and drought damage occurs downstream. The problem of convention is when decisions are made based on past experience without carefully examining the information, even though there is a lot of published information on climate change predictions. The issue of fairness refers to the content and extent of measures taken. For example, people in poverty are less likely to live in a hazard area. In this case, the question is whether society needs to consider disaster prevention measures for that area. Government failures include cases where it is difficult to implement adaptation measures due to regulations or consistency with existing policies. Finally, uncertainty is a problem where forecasts are made over a very long period of time and the degree of uncertainty is so great that decision-making is difficult as a result.

All of the above factors can be attributed to the lack of information to compare and consider the priorities of adaptation measures, and it is very important to strengthen this knowledge base in order to promote adaptation measures. In order to overcome the problem of insufficient information on climate change, research has been conducted on the dissemination of information on climate change adaptation measures using various Web sites [5] [6] [7] [8].

In Thailand, for example, the Office of Natural Resource and Environmental Policy and Planning (ONEP), a sub-organization of the Thai Ministry of Natural Resource and Environmental Policy and Planning, which is responsible for formulating the national strategy for climate change adaptation, has been conducting research on the dissemination of information on climate change adaptation measures using various websites. Office of Natural Resource and Environmental Policy and Planning (ONEP), a subordinate body of the Thai Ministry of Natural Resource and Environmental Policy and Planning, which is responsible for formulating national strategies for climate change adaptation, has established a website called T-PLAT to disseminate information on policies and risks related to climate change adaptation measures (http://t-plat.deqp.go.th/en/home-page/). GISTDA (Geo-Informatics and Space Technology Development Agency), a subordinate organization of the Ministry of Science and Technology of Thailand, has established a website called Thailand Flood Monitoring System. It provides information on the distribution of areas inundated by past floods and the current flood situation (https://flood.gistda.or.th/).

However, most of the information dissemination on climate change adaptation measures using the web has been on disaster risks, and few of them relate disaster risks to socio-economic characteristics of the region. Local socioeconomic information is necessary when considering the contents and priorities of climate change adaptation measures. Therefore, in this study, we developed a web application that can efficiently communicate natural disaster risks and socioeconomic characteristics to support the planning of climate change adaptation measures. In recent years, the technology for viewing 3D geographic information on the web has been rapidly advancing. In this study, we also worked on the 3D display of disaster risks.

## 2. Methodology

### *2.1. System Overview of the Web Application*

Figure 1 shows the web application system. The geographic information database contains map information and statistical information. there are two types of Map servers, one providing WMS/WMTS and the other providing vector tiles, which provide two-dimensional map information and generate map information that can be used on the web based on the map information in the geographic information database. There are also servers that provide 3D topographic information. www servers access databases and map servers based on user requests, integrate the information, and display it on the homepage.

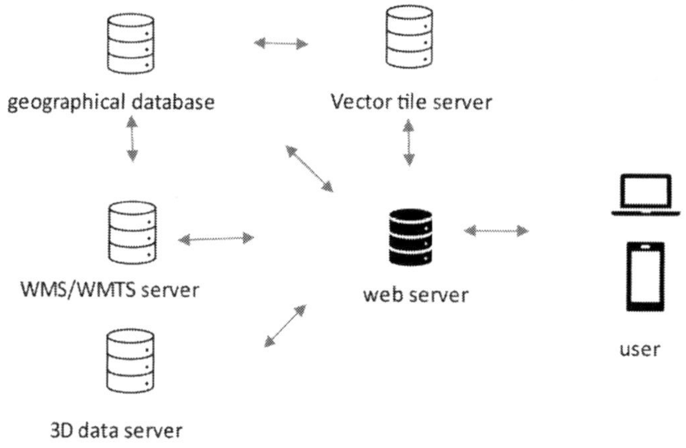


**Figure 1.** System of the web application

*2.2. Data*

For natural disaster risk, the available data include flood risk, landslide risk, drought risk, salinity risk, and loss of beaches due to sea level rise. For socio-economic data, the following data were collected in order to examine climate change adaptation measures from both efficiency and equity perspectives. From the perspective of efficiency, we collected county-level information on gross regional product by industry and production value of major agricultural crops. Secondly, from an equity perspective, information on poverty rates and household debt was collected.

*2.3. Application, libraries, and external web service*

Table 1 shows the list of libraries used in the development of the web application. First, PostgreSQL with PostGIS extension is used for a geographic information database. GeoServer and Tegola are used as the map servers, and gin written in the go language is used as the web framework. Openlayers is used as the front-end map display library. HighChart is used for graph display.

Regarding with 3D map, Cesium Js is used for 3D map display, Cesium ion is used to provide the 3D data, such as terrain data and building data.

**Table 1.** List of major application libraries and external web service which are used in the web application

| | Name | URL | Note |
|---|---|---|---|
| geographic information database | PostgreSQL | https://www.postgresql.org/ | opensource |
| | PostGIS | https://postgis.net/ | Spatial database extender for PostgreSQL, opensource |
| map server | GeoServer | http://geoserver.org/ | opensource WMS/WMTS server |
| | Tegola | https://tegola.io/ | opensource vector tile server |
| web framework | gin | https://github.com/gin-gonic/gin | opensource |
| display map | OpenLayers | https://openlayers.org/ | opensource |

| | | | |
|---|---|---|---|
| graph | Highcharts | https://www.highcharts.com/ | Non-profit organization can apply non-commercial licensg |
| display 3D map | Cesium Js | https://cesium.com/platform/cesiumjs/ | opensource |
| 3D geospatial data service site | Cesium ion | https://cesium.com/platform/cesium-ion/ | Providing 3D terrain data and 3D building data based on the OpenStreetMap |

## 3. Results and discussion

### *3.1. 2D view*

A snapshot of the application we have built is shown in Figure 2. The left side of the screen shows map information on disaster risks, and the right side shows socio-economic statistical information. Figure 1 displays the frequency of flooding and the gross regional product in Bangkok (upper panel) and the production value by industry in Bangkok (lower panel). The figure shows that the frequency of flooding is low in Bangkok compared to other regions, the gross regional product is on the rise, and the share of tertiary industries in the industrial structure is increasing, while that of the manufacturing sector, which is more economically vulnerable to flooding, is decreasing year by year.By switching the tabs at the top of the graph, one can view information on agricultural output, poverty rate, and household debt.

The type of risk can be selected from the menu that appears by pressing the button in the upper right corner of the map. When the user clicks on a point on any map, the graph will work in conjunction with it to display the information of the clicked point.

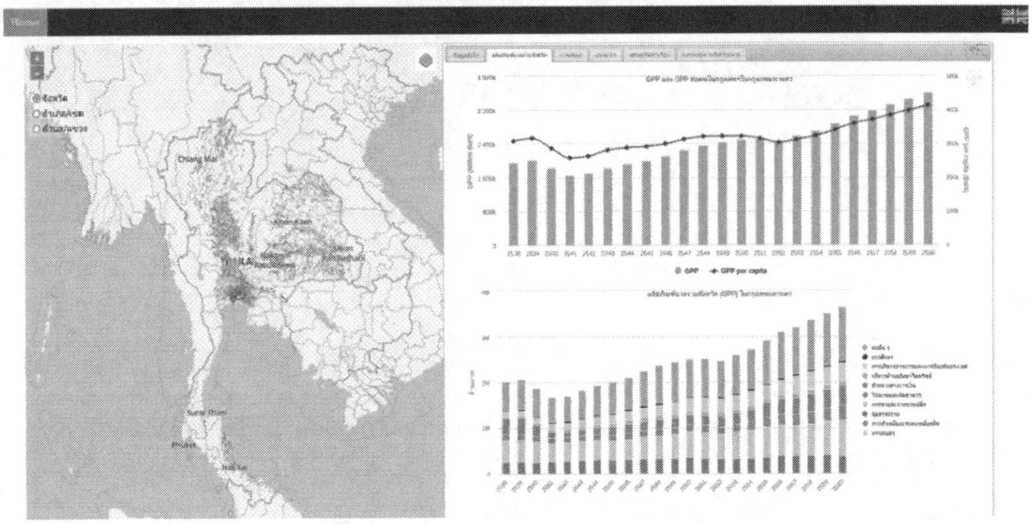

**Figure 2.** Snapshot of the web application

### *3.2. 3D view*

The results of the 3D display of disaster risks are shown in Figure 3 and Figure 4, and it is expected that the 3D display will enhance the understanding of disaster risks. It is possible to display part of a building in 3D by using information from the voluntarily maintained OpenStreetMap, but this is still limited now.

If 3D information on buildings is developed, it will be possible to convey the disaster risks of cities more realistically to users.

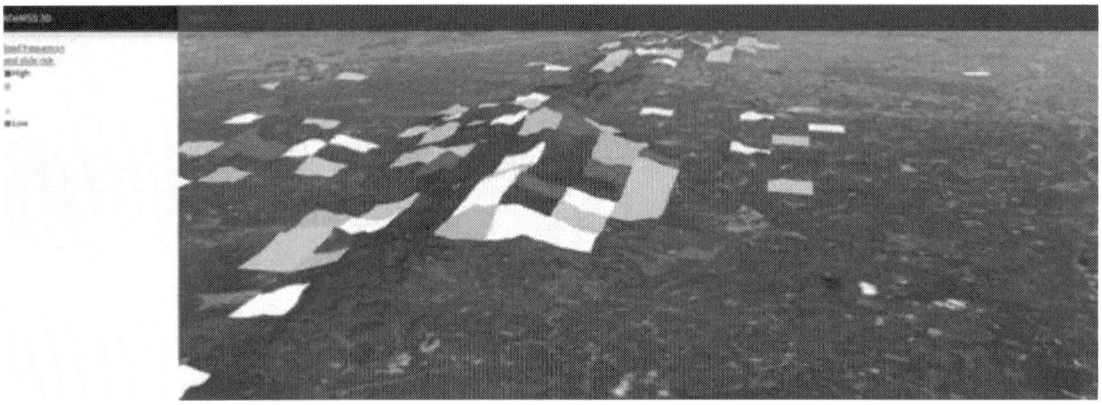

**Figure 3.** 3D representation of landslide risk in southern Thailand

**Figure 4.** 3D representation of flood risk in Bangkok

## 4. Conclusions

In this study, we have developed a web application that supports the formulation of local climate change adaptation measures by linking natural disaster risks and socioeconomic data. In the future, we plan to work with various stakeholders to discuss improvements to this application, such as data and functions that should be added.

Thailand is considered to be a country with relatively good statistical data among Southeast Asian countries. On the other hand, 3D information on roads and buildings is limited, and the development of 3D information will have a significant impact on communicating disaster risks more clearly to citizens. The future challenge is to promote the development of 3-D geospatial information in society.

## References

[1] World Bank 2012 Thai Flood 2011: Rapid Assessment for Resilient Recovery and Reconstruction Planning
[2] Kaewthongrach R, Vitasse Y, Lamjiak T and Chidthaisong A 2019 Impact of Severe Drought during the Strong 2015/2016 El Nino on the Phenology and Survival of Secondary Dry Dipterocarp Species in Western Thailand *For. 2019, Vol. 10, Page 967* **10** 967
[3] Kiguchi M, Takata K, Hanasaki N, Archevarahuprok B, Champathong A, Ikoma E, Jaikaeo C, Kaewrueng S, Kanae S, Kazama S, Kuraji K, Matsumoto K, Nakamura S, Nguyen-Le D, Noda

K, Piamsa-Nga N, Raksapatcharawong M, Rangsiwanichpong P, Ritphring S, Shirakawa H, Somphong C, Srisutham M, Suanburi D, Suanpaga W, Tebakari T, Trisurat Y, Udo K, Wongsa S, Yamada T, Yoshida K, Kiatiwat T and Oki T 2021 A review of climate-change impact and adaptation studies for the water sector in Thailand *Environ. Res. Lett.* **16** 023004

[4] Chambwera M, Heal G, Dubeux C, Hallegatte S, Leclerc L, Markandya A, McCar B A, Mechler R and Neumann J E 2014 Economics of Adaptation *Climate change 2014: impacts, adaptation and vulnerability* pp 945–78

[5] Whateley S, Walker J D and Brown C 2015 A web-based screening model for climate risk to water supply systems in the northeastern United States *Environ. Model. Softw.* **73** 64–75

[6] Glaas E, Ballantyne A G, Neset T S and Linnér B O 2017 Visualization for supporting individual climate change adaptation planning: Assessment of a web-based tool *Landsc. Urban Plan.* **158** 1–11

[7] Brown M and Bachelet D 2017 BLM Sagebrush Managers Give Feedback on Eight Climate Web Applications *Weather. Clim. Soc.* **9** 39–52

[8] Bierbaum R, Smith J B, Lee A, Blair M, Carter L, Chapin F S, Fleming P, Ruffo S, Stults M, McNeeley S, Wasley E and Verduzco L 2013 A comprehensive review of climate adaptation in the United States: More than before, but less than needed *Mitig. Adapt. Strateg. Glob. Chang.* **18** 361–406

**Acknowledgement**

This research was supported partially by the Japan Science and Technology Agency/Japan International Cooperation Agency, and the Science and Technology Research Partnership for Sustainable Development (JST/JICA, SATREPS).

# The role of soybean agroforestry in mitigating climate change in Indonesia

**Djoko Purnomo*[1], Maria Theresia Sri Budiastuti[1] and Desy Setyaningrum[2]**

[1]Departement of Agrotechnology, Faculty of Agriculture, Universitas Sebelas Maret, Indonesia
[2]Doctoral Program of Agricultural Science, Faculty of Agriculture, Universitas Sebelas Maret, Indonesia
*Corresponding author: djokopurnomo@staff.uns.ac.id

**Abstract**. Climate change is a global problem that has an impact on food production. Protein needs for people in developing countries such as Indonesia have not been fulfilled optimally. Indonesian people's protein consumption is still at 65.7%, much lower than it should be. The development of soybean cultivation as a source of vegetable protein is very potential. This article aims to examine the role of soybean agroforestry to meet human protein needs and as an effort to mitigate climate change.In Indonesia, soybean development is constrained by people's views; soybean is not the main crop and the rate of land conversion. An effort to overcome the limitation of optimum land according to the growing season is the development of soybean cultivation under forest stands as an agroforestry system. In Indonesia, there are production forests, including teak, pine, sengon, and resin, especially in teak, pine, and sengon forests, soybean cultivation with agroforestry systems has developed. The main obstacle to cultivation in the agroforestry system is low light due to the shade of the tree canopy. It requires efforts: tree spacing, especially between rows, trimming the tree canopy when the canopy is getting tighter, and low light tolerant plant varieties. Soybeans are low-light tolerant plants. The national average soybean production of about 1.5 tons ha-1 can be achieved in teak-based agroforestry systems up to 3-4 years old tree, pine-based up to 3 years old, and sengon-based up to 5 years old. The agroforestry system has the potential as land for the cultivation of organic plants in mitigating the impacts of climate change.

**Keywords**: agroforestry, climate change, soybean, vegetable protein

## 1. Introduction

The climate change phenomenon continues to threaten life, based on facts showing that an increase in atmospheric temperature has occurred from 27 in October 1981-2010 to 27.6 °C in October 2021 [1]. A threat to humans is the impact of climate change on plant growth [2]. Human nutrition depends on plants both in quantity and quality. An increase in temperature from one side is beneficial because the rate of the body's metabolic processes becomes higher, so that activity increases. However, for plants, high metabolic processes impact increasing respiration [3]. Increased respiration means photosynthate decomposes again, thereby reducing crop yields. Weather anomalies as an indicator of climate change can disrupt growth rhythm, especially tree plants [4]. The pattern of vegetative and generative growth of trees (fruit season) is very dependent on the wet and dry seasons.

Based on the discussion above, there is a change in the relationship between agriculture and seasons. Weather anomalies can cause a location to turn wetter or drier In general, wetter environmental conditions are more favorable than vice versa. To anticipate changes in environmental conditions, the selection of plants (varieties) and cultivation technology must be adjusted, so that food security is not compromised [5]. In monoculture cultivation, plants can achieve optimum results (the focus is commodity productivity) but require high energy input (high intensification) for high stability (minimum disturbanceCultivating plants with relatively low energy input, focusing on community productivity is intercropping or multiple cropping[6]. One type of intercropping cultivation is the agroforestry system by utilizes forest areas that are relatively widely available in Indonesia [7].

One of the problems with agroforestry systems is that the light under the trees is relatively low for crops. Food crops suitable for the environment below the optimum are C3 plants (e.g., rice and soybeans)[8]. In Indonesia, the main food crops are rice, corn, and soybeans. Soybean is thus potential as an intercrop in agroforestry systems[9]. In addition, the quality of human food is protein content, which is generally fulfilled from animal and vegetable protein. Protein from animal sources is still relatively unaffordable for most Indonesian people, so an alternative is vegetable protein, especially food made from soybeans [10]. This article aims to examine the role of soybean agroforestry to meet human protein needs and as an effort to mitigate climate change.

## 2. Potential and constraints of agroforestry system

Indonesia is a tropical country that has various advantages in agricultural activities. However, the current climate change is causing serious impacts. Some of the changes that occur are increasing the CO2 content in the atmosphere to increase the average temperature[11]. Indonesia's normal air temperature in the 1981-2010 period is 26.6 °C and the average air temperature in 2020 is 27.3 °C. As a whole, in Indonesia, 2016 was the hottest year with an anomaly value of 0.8 °C throughout the observation period from 1981 to 2020 [1]. An increase in temperature greatly affects plant metabolism, causing a decrease in production [12]. High temperatures can affect photosynthesis, but an increase in temperature is accompanied by the rise in air vapor pressure deficit. There is an increase in the transpiration rate simultaneously. In addition, the increased $CO_2$ concentration causes the stomata to close quickly so that the photosynthesis process can be disrupted [4,5].

Conventional agriculture i.e., monoculture systems are considered unsustainable and inadequate to address major societal challenges such as climate change. Many of these problems are related to agricultural specialization and the consequent simplification of the agroecosystems [13]. Based on conventional farming systems in the era of climate change, causing a decrease in biodiversity. Agroforestry is an agricultural approach based on the diversification of the production components of an agroecosystem (timber crops, such as trees or shrubs, plus annual crops and livestock) and the intensification of agroecological relationships between these components [14,15]. Agroforestry has been advocated to reduce the adverse environmental impacts of monoculture cultivation. This is supported by the role of trees in agroforestry systems. Trees play a role in absorbing $CO_2$ in the air and soil to help reduce the increase in $CO_2$ levels in the atmosphere [16,17]. Based on [18] research, $CO_2$ absorption by trees will reduce $CO_2$ concentrations by about 2.5% during the day. Agroforestry systems are considered the most effective carbon sequestration because they can produce zero or negative emissions [19]. In addition, trees in agroforestry can be associated with bacteria, thereby increasing microbial diversity[20,21]. Agroforestry systems play an important role as a climate change mitigation strategy [22]. Climate change mitigation through increasing soil carbon sequestration can reverse land degradation and ensure food security [23].

Agroforestry is a system of intercropping cultivation (multiple cropping) between stands and crops [24,25]. In Indonesia, this cultivation has existed since the culture of the people living permanently, in the form of a yard (a piece of land around a house). Plant cultivation in the yard is not very intensive because it is not a source of food needs [26]. Plants in the yard to support daily needs include vegetables, side dishes (especially chicken), urgent funds (coconuts), and herbal medicines. The agroforestry system (intensively) in forest areas develops from the unofficial cultivation of plants by communities around the forest [27,28]. The implementation of the agroforestry system in Indonesia has several obstacles, namely the uncertainty of land ownership, less business efficiency, expensive logistics costs, narrow cultivation area, low productivity, and centralized institutions (centralization). The main obstacle in plant cultivation is the low light received by plants. Light is one of the most important environmental factors in agricultural production [29]. Light acts as an activator of photosynthesis, which is the basis of plant growth and yield formation, and affects the structure and function of the photosynthetic apparatus [30]. Photosynthesis in agroforestry cropping systems is often limited by light due to mutual shading between intra or inter-species plants [31,32]. In general, leaves grown under shade had a higher number of chloroplast thylakoid stacks but lower chlorophyll content, Chl a/b ratio, and net photosynthetic rate than leaves grown under normal light conditions [33]. In addition, leaves under shade had lower content of electron transfer carrier and ribulose-1,5-bisphosphate carboxylase/oxygenase than those under normal light conditions [34]. Therefore, shading decreases the photosynthetic capacity of leaves during plant growth and development.

Farmers prioritize intercropping so they tend to ignore stands (even damage them) to the detriment of forest managers. Due to constraints (especially light), crops as plants between stands were shaded by the tree canopy. Competition for nutrients between stands and trees is likely to be negligible. The stand rhizosphere and intercropping rhizosphere differ in space, the stand rhizosphere is deeper than the interplant rhizosphere so that the nutrients are in different locations. Especially for N nutrients, immobilization occurs because soil microorganisms use it for the decomposition process of organic matter[32]. Based on Beule and P. Karlovsky [20] that intercropping in agroforestry systems can affect fungal and bacterial biomass. Agroforestry systems can improve soil organic matter status, microbial activity and nitrogen availability. Farmland management practices and tree species affect $CO_2$-C production, biomass-specific respiratory activity, and nitrogen mineralization . In tree-based systems, soil inorganic N levels were 8-74% higher and nitrogen mineralization by 12-37% than single planting [35].

Regarding water, in general, the agroforestry system is a rainfed area, enough water is available so that competition is not needed. C3 plants are relatively tolerant of low light, but are susceptible to photorespiration. Photorespiration occurs (interfering with the rate of photosynthesis) if in the mesophyll the concentration of $O_2$ is higher than $CO_2$, when sunlight is high intensity (high photolysis rate). This occurs because the affinity of the enzyme ribulose bisphosphate with $O_2$ is higher than with $CO_2$ [36]. In the agroforestry system it is most likely around the stand canopy that the $O_2$ concentration is high from stand respiration. High air $O_2$ concentrations then enter the mesophyll through the stomata may also cause photorespiration. The above obstacles can be overcome, among others, to increase light for intercropping by adjusting the distance of the tree, trimming the lower 1/3 of the canopy of the lower stand, and choosing low light tolerant plants N immobilization was overcome by fertilizing cultivated plants which were higher than the recommendations [37,38]. Based on the discussion above, the agroforestry system's opportunity to maintain food security is very large. The forest area in Indonesia is 66.18 million ha, if 10% is for soybean cultivation (6.6 million ha) with a yield of 1.0 tons per ha, then 6.6 million tons of soybean are obtained in one season. National soybean needs of more than 2 million tons per year (tends to increase every year) will be fulfilled and even export opportunities.

## 3. Soybean in agroforestry system

Agroforestry is indicated as a suitable agricultural practice to enhance ecosystem services. However, farmers are often reluctant to apply agroforestry because of the potential for loss of crop yields. Based on the results of [39], soybean yields decreased by 35% in the agroforestry system due to lack of light.

However, the fat and protein content of soybeans in the agroforestry system did not decrease. This is because the reduction in light availability causes a significant decrease in the net assimilation of plants. On the other hand, soil characteristics influence soybean quality because crude fat is significantly correlated with soil nutrients, pH, soil organic matter, and soil texture. Other authors found that trees might induce soybean yield stability and promote N2 fixation under water stress. Soybean in Agroforestry system as a promising solution for climate change adaptation and fulfillment of human protein [8].

Several strategies can be used to reduce the effect of the tree canopy on solar radiation, such as forest species by adjusting the canopy size, architecture and leaf geometry [40]. Several studies showed that the dynamics of solar radiation in understorey forest species differed, 41.5% in the amount of solar radiation available in understorey *Melaleuca cajuputi*, *Peltophorum dubium* and *Eucalyptus grandis* [41]. Thus, identifying the forest species that are most appropriate to compose an agroforestry system to produce a balance between the amount of solar radiation intercepted by the tree canopy and the amount transmitted downward. The application of the soybean agroforestry system with 1.5 years *Albasia chinensis* with different distances of light intensity ranging from 15,620-28,500 lux resulted in the highest soybean production (3.3 tons ha-1) can be achieved at a density of 2850 trees ha-1 with a pruned crown ( RTF: 49% equivalent to 28440 lux) compared to biomass production in open land (3.9 tonnes ha-1) [42]. In Baity et al., [43] research, soybean cultivation with an agroforestry system was carried out in areas with critical hydrological conditions. It appeared that soybean growth and yield were quite high in the agroforestry system. However, this study stated that light is an important factor and highly dependent on the character of the tree canopy, it appears that the light received by the soybean canopy is 9.42% of 49,000 lux [44]. Shade affects growth and yield characteristics and the efficiency of soybean radiation use, so it is possible to recommend the cultivation of legumes in agroforestry systems. However, in supporting soybean yields, fertilization efforts are needed to support soybean yields in low light conditions. Based on research by [45], the application of compost 10 tons ha-1 showed the highest yields of Anjasmoro, Dering I, and Grobogan, respectively 1.42, 1.56, and 1.51 tons ha-1, or increased respectively 118.46%, 102.60%, and 112.68%. Compared with no compost application. This is because soil quality and physiological parameters significantly affect the production of soybean varieties in the *M. cajuputi* agroforestry system. In addition, the application of soybean agroforestry resulted in high N2 fixation (91%), high root and nodule growth [46]. Selection of soybean varieties that are resistant to low light is also necessary. Research [47] shows that soybean varieties and soil types interact with soybean yields in agroforestry systems. The ring I planted in inceptisols and vertisols = 1.38 and 1.27 tons.ha$^{-1}$, Grobogan planted in vertisols 1.24 tons.ha$^{-1}$. Ring 1 is suitable for all soil types with agroforestry systems.

## 4. The role of soybeans in meeting protein needs

Soybean belongs to the family Leguminosae, often called legume plants, with seeds rich protein sources, essential amino acids, oils, and metabolizable energy. 32-48% of soybeans are protein, so soybeans are the main source of vegetable protein [48,49]. Protein is a key factor that determines soybeans' nutritional value and economic value. In Indonesia, soybeans are planted after rice and corn (beginning of the dry season) with relatively low intensification. Therefore, soybean yields are only about 1.5 tons per ha. Soybean seeds contain 10 g of protein per 100 g, the highest among rice and corn (7.13 and 3.4 g, respectively) [48]

Human protein needs are 0.8 g per body weight; if the average weight of an Indonesian is 62 kg, then it takes about 50 g per day, so that 18 kg of protein is needed in a year. The population of Indonesia is 270 million; it takes about 5 million tons of protein per year to be met from soybeans from the agroforestry system, even the remaining 1.6 million tons. The calculation aims to illustrate that the agroforestry system for soybean cultivation can meet the national soybean and food protein needs. The European Commission has encouraged soybean cultivation as a substitute for protein needs in recent years.

## 5. Conclusion

Soybean is a leguminous parka and a source of vegetable protein. The agroforestry system in soybean cultivation affects soybean yield but not affect soybean protein content. Soybean cultivation in agroforestry systems can support the fulfilment of human protein needs. Soybeans are low-light tolerant plants. The national average soybean production of about 1.5 tons ha-1 can be achieved in teak-based agroforestry systems up to 3-4 years old tree, pine-based up to 3 years old, and sengon-based up to 5 years old. The agroforestry system has the potential as land for the cultivation of organic plants in mitigating the impacts of climate change.

## References

[1] Ekaputra E G, Stiyanto E and Hasanah N A I 2021 Surface temperature across land-use change phenomena in Padang, Indonesia *IOP Conf. Ser. Earth Environ. Sci.* **757**

[2] Noponen M R A, Healey J R, Soto G and Haggar J P 2013 Sink or source-The potential of coffee agroforestry systems to sequester atmospheric CO2 into soil organic carbon *Agric. Ecosyst. Environ.* **175** 60–8

[3] Sharma A and Anandhi A 2021 Temperature based indicators to develop adaptive responses for crop production in Florida , USA *Ecol. Indic.* **121** 107064

[4] Hussain S, Ulhassan Z, Brestic M, Zivcak M, Weijun Zhou, Allakhverdiev S I, Yang X, Safdar M E, Yang W and Liu W 2021 Photosynthesis research under climate change *Photosynth. Res.* **150** 5–19

[5] Lee Y H, Sang W G, Baek J K, Kim J H, Shin P, Seo M C and Cho J Il 2020 The effect of concurrent elevation in CO2 and temperature on the growth, photosynthesis, and yield of potato crops *PLoS One* **15**

[6] Ma Z, Chen H Y H, Bork E W, Carlyle C N and Chang S X 2020 Carbon accumulation in agroforestry systems is affected by tree species diversity, age and regional climate: A global meta-analysis *Glob. Ecol. Biogeogr.* **29** 1817–28

[7] Smethurst P J, Huth N I, Masikati P, Sileshi G W, Akinnifesi F K, Wilson J and Sinclair F 2017 Accurate crop yield predictions from modelling tree-crop interactions in gliricidia-maize agroforestry *Agric. Syst.* **155** 70–7

[8] Lista F N, Deminicis B B, De Carvalho Almeida J C, Do Carmo Araujo S A and Zanella P G 2019 Forage production and quality of tropical forage legumes submitted to shading *Cienc. Rural* **49**

[9] Zhang S, Peng X, Zhang Y, Cai J and Jiang Z 2009 Photosynthesis, growth and yield of soybean and maize in a tree-based agroforestry intercropping system on the Loess Plateau *Agrofor. Syst.* **76** 569–77

[10] Di Giorgio L, Salgado P R and Mauri A N 2019 Encapsulation of fish oil in soybean protein particles by emulsification and spray drying *Food Hydrocoll.* **87** 891–901

[11] Javed S A, Haider A and Nawaz M 2020 How agricultural practices managing market risk get attributed to climate change? Quasi-experiment evidence *J. Rural Stud.* **73** 46–55

[12] Jumrani K and Bhatia V S 2020 Influence of different light intensities on specific leaf weight, stomatal density photosynthesis and seed yield in soybean *Plant Physiol. Reports* **25** 277–83

[13] Rosati A, Borek R and Canali S 2021 Agroforestry and organic agriculture *Agrofor. Syst.* **95** 805–21

[14] Boeckx P, Bauters M and Dewettinck K 2020 Poverty and climate change challenges for sustainable intensification of cocoa systems *Curr. Opin. Environ. Sustain.* **47** 106–11

[15] Van Noordwijk M 2021 Agroforestry-based ecosystem services: Reconciling values of humans

and nature in sustainable development *Land* **10**

[16] Nair P K R, Nair V D, Kumar B M and Haile S G 2009 Soil carbon sequestration in tropical agroforestry systems: a feasibility appraisal *Environ. Sci. Policy* **12** 1099–111

[17] Tschora H and Cherubini F 2020 Co-benefits and trade-offs of agroforestry for climate change mitigation and other sustainability goals in West Africa *Glob. Ecol. Conserv.* **22**

[18] Li J F, Zhan J M, Li Y S and Wai O W H 2013 CO2 absorption/emission and aerodynamic effects of trees on the concentrations in a street canyon in Guangzhou, China *Environ. Pollut.* **177** 4–12

[19] Feliciano D, Ledo A, Hillier J and Nayak D R 2018 Which agroforestry options give the greatest soil and above ground carbon benefits in different world regions? *Agric. Ecosyst. Environ.* **254** 117–29

[20] Beule L and Karlovsky P 2021 Tree rows in temperate agroforestry croplands alter the composition of soil bacterial communities *PLoS One* **16** 1–20

[21] Buyer J S, Baligar V C, He Z and Arévalo-Gardini E 2017 Soil microbial communities under cacao agroforestry and cover crop systems in Peru *Appl. Soil Ecol.* **120** 273–80

[22] Budiastuti M T S 2020 Agroforestri Sebagai Bentuk Mitigasi Perubahan Iklim Agroforestry as Climate Change Mitigation terkait dengan pencemaran gas , khususnya CO 2 , yang sulit terurai di atmosfer sehingga menghalangi pemantulan kembali sinar matahari yang mengakibatkan suhu b *Semin. Nas. Magister Agroteknologi Fak. Pertan. UPN "Veteran" Jawa Timur. NST Proc.* **2020** 23–9

[23] Janzen H, Bremer E and Vandenbygaart B 2007 Measuring Change in Soil Organic Carbon Storage *Soil Sampl. Methods Anal. Second Ed.*

[24] Qiao X, Sai L, Chen X, Xue L and Lei J 2019 Impact of fruit-tree shade intensity on the growth, yield, and quality of intercropped wheat *PLoS One* **14** 1–17

[25] Kaur B, Gupta S R and Singh G 2000 Soil carbon, microbial activity and nitrogen availability in agroforestry systems on moderately alkaline soils in northern India *Appl. Soil Ecol.* **15** 283–94

[26] Sofiyuddin M, Suyanto S, Kadir S and Dewi S 2021 Sustainable land preparation for farmer-managed lowland agriculture in Indonesia *For. Policy Econ.* **130** 102534

[27] Haggar J, Pons D, Saenz L and Vides M 2019 Contribution of agroforestry systems to sustaining biodiversity in fragmented forest landscapes *Agric. Ecosyst. Environ.* **283** 106567

[28] Bettles J, Battisti D S, Cook-Patton S C, Kroeger T, Spector J T, Wolff N H and Masuda Y J 2021 Agroforestry and non-state actors: A review *For. Policy Econ.* **130** 102538

[29] Yang F, Feng L, Liu Q, Wu X and Fan Y 2018 Effect of interactions between light intensity and red-to- far-red ratio on the photosynthesis of soybean leaves under shade condition *Environ. Exp. Bot.* **150** 79–87

[30] Sitompul S M and Zulfati A P 2019 Betacyanin and Growth of Beetroot ( Beta vulgaris L .) in Response to Nitrogen Fertilization in a Tropical Condition *AGRIVITA J. Agric. Sci.* **41** 40–7

[31] Reynolds P E, Simpson J A, Thevathasan N V. and Gordon A M 2007 Effects of tree competition on corn and soybean photosynthesis, growth, and yield in a temperate tree-based agroforestry intercropping system in southern Ontario, Canada *Ecol. Eng.* **29** 362–71

[32] Pontes L da S, Stafin G, Moletta J L and Porfirio-da-Silva V 2020 Performance of Purunã beef heifers and pasture productivity in a long-term integrated crop-livestock system: the effect of trees and nitrogen fertilization *Agrofor. Syst.* **94** 1713–23

[33] Li X, Takahashi T, Suzuki N and Kaiser H M 2014 Impact of Climate Change on Maize Production in Northeast and Southwest China and Risk Mitigation Strategies *APCBEE Procedia* **8** 11–20

[34] Zhu H, Li X, Zhai W, Liu Y, Gao Q, Liu J, Ren L, Chen H and Zhu Y 2017 Effects of low light on photosynthetic properties, antioxidant enzyme activity, and anthocyanin accumulation in purple pak-choi (Brassica campestris ssp. Chinensis Makino) *PLoS One* **12** 1–17

[35] Sales A, Oliveira N De, Paiva H N De and Leite H G 2021 According to Soil Management in Agroforestry Systems : A Case *diversity* **511**

[36] Kirschbaum M U F 2004 Direct and indirect climate change effects on photosynthesis and

transpiration *Plant Biol.* **6** 242–53
[37] Caron B O, Sgarbossa J, Schwerz F, Elli E F, Eloy E and Behling A 2018 Dynamics of solar radiation and soybean yield in agroforestry systems *An. Acad. Bras. Cienc.* **90** 3799–812
[38] Werner F, Balbinot Junior A A, Franchini J C, Ferreira A S and Silva M A de A e 2017 Agronomic performance of soybean cultivars in an agroforestry system1 *Pesqui. Agropecuária Trop.* **47** 279–85
[39] Mantino A, Volpi I, Micci M, Pecchioni G, Bosco S, Dragoni F, Mele M and Ragaglini G 2020 Effect of tree presence and soil characteristics on soybean yield and quality in an innovative alley-cropping system *Agronomy* **10** 1–15
[40] Sgarbossa J, Elli E F, Schwerz F, Nardini C, Miguel Knapp F, Schmidt D, Dal'Col Lúcio A and Caron B O 2021 Bean–soybean succession under full sun and in agroforestry systems: Impacts on radiation use efficiency, growth and yield *J. Agron. Crop Sci.* **207** 362–77
[41] Sgarbossa J, Elli E F, Schwerz F, Nardini C, de Cristo E, de Oliveira D and Caron B O 2020 Morphology, growth and yield of black oats cultivated in agroforestry systems in southern Brazil *Agric. Syst.* **184** 102911
[42] Affandi Z, Purnomo D and Supriyono S 2019 A Study of Light Intensity and Fertilizer on Soybean in Albizia chinensis Agroforestry System *SAINS TANAH - J. Soil Sci. Agroclimatol.* **16** 46
[43] Baity S, Purnomo D and Sulistyo T D 2015 Budidaya Organik Kedelai Pada Sistem Agroforestri Menggunakan Pupuk Hayati *Caraka Tani J. Sustain. Agric.* **30** 7
[44] Roshetko J M, Rohadi D, Perdana A, Sabastian G, Nuryartono N, Pramono A A, Widyani N, Manalu P, Fauzi M A, Sumardamto P and Kusumowardhani N 2013 Teak agroforestry systems for livelihood enhancement, industrial timber production, and environmental rehabilitation *For. Trees Livelihoods* **22** 241–56
[45] Suryanto P, Faridah E, Nurjanto H H, Supriyanta, Kastono D, Putra E T S, Handayani S, Dewi A K and Alam T 2020 Influence of siam weed compost on soybean varieties in an agroforestry system with kayu putih (Melaleuca cajuputi) *Biodiversitas* **21** 3062–9
[46] Nasielski J, Furze J R, Tan J, Bargaz A, Thevathasan N V. and Isaac M E 2015 Agroforestry promotes soybean yield stability and N2-fixation under water stress *Agron. Sustain. Dev.* **35** 1541–9
[47] Alam T, Muhartini S, Suryanto P, Ambarwati E, Kastono D, Nurmalasari A I and Kurniasih B 2020 Soybean varieties suitability in agroforestry system with kayu putih under influence of soil quality parameters *Rev. Ceres* **67** 410–8
[48] Mo W Y, Lau R S S, Kwok A C K and Wong M H 2016 Use of soybean meal and papain to partially replace animal protein for culturing three marine fish species: Fish growth and water quality *Environ. Pollut.* **219** 815–20
[49] Greenhalgh S, Chrystal P V., Selle P H and Liu S Y 2020 Reduced-crude protein diets in chicken-meat production: justification for an imperative *Worlds. Poult. Sci. J.* **76** 537–48

# Land conservation: Erosion control for sustainable agriculture (Case study in Ngargoyoso District, Karanganyar Regency, Indonesia)

**Purwo Edi Minarno*[1], Suntoro[2] and Khomariah[2]**

[1]Master Program of Environmental Science, Post Graduate School, Sebelas Maret University, Indonesia
[2]Soil Science Study Program, Faculty of Agriculture, Sebelas Maret University Surakarta, Indonesia
*Corresponding author: pewedhi@gmail.com

**Abstract**. Climate change can pose major risks to human health, global food security and economic development. The negative impact is an increase in the rate of erosion in agricultural land areas. Ngargoyoso District, Karanganyar Regency is one of the areas that have a high erosion rate. Land conservation to control the rate of erosion with a community participatory approach is carried out to realize sustainable agricultural land. The purpose of the study is to determine the level of erosion hazard in Ngargoyoso District. This research is in the form of a case study, using direct data collection methods (surveys) and laboratory tests. Analysis of the data on the level of erosion hazard (TBE) using the USLE calculation approach. The research variables were rainfall erodibility, soil erodibility, slope length and slope, crop management, and land management. The results showed that in the District of Ngargoyo: 1) areas with a very heavy level of erosion hazard covering an area of 500.66 hectares spread over Kemuning Village, Ngargoyoso Village and Nglegok Village; 2) Agricultural activities greatly affect the level of erosion hazard in the study area

**Keywords:** soil conservation, erosion hazard level, USLE

## 1. Introduction

Climate change affects the environment, economy and society [1]. Climate change can pose major risks to human health, global food security and economic development. Climate change causes erratic rainy and dry seasons. The rainy season will affect the increase in rainfall, especially in tropical countries [2]. Erosion causes the loss of a fertile layer of soil that is good for plant growth and reduces the ability of the soil to absorb and hold water [3]. Ngargoyoso Subdistrict, Karanganyar Regency has an average rainfall of 248 mm/month [4], so it has the potential to increase the rate of erosion in agricultural land areas. In addition, human activities in meeting their needs, especially food by cultivating dry land on sloping slope conditions.

Population interaction with land is one of the factors determined by the structure of the work. Based on their occupation, the population in Ngargoyoso District who works as farmers is 6,506 people (38.21%) and 5,227 farm labourers (30.69%). This condition illustrates that population pressure on land is very high because > 50% work in the agricultural sector. The sustainability of agriculture in this area affects the hydrological conditions, especially conditions downstream of the research area [5]. So that the erosion that occurs needs to be known so that actions can be taken to have a positive impact on the condition of the surrounding community and the area downstream.

The ecological conditions at the research site have been degraded and have even reached a critical point and have been degraded due to excessive land use errors. Conservation directives carried out to optimize production and reduce the risk of erosion and land landslides are carried out by considering area functions, erosion susceptibility, landslides and land suitability which are carried out mechanically and vegetatively based on the slope [6]. The purpose of the study is to determine the level of erosion hazard in Ngargoyoso District and activities that affect the level of erosion hazard in the study area. Erosion study in Ngargoyoso District, as an update on erosion data that occurred on the western slopes of Lawu.

## 2. Materials and method

### *2.1. Materials*

The research was conducted in Ngargoyoso District, Karanganyar Regency. In the period November 2020 to March 2021. Sampling with a land unit approach of 73 land units [7]. Primary data taken were slope, slope length, solum depth, land management/land conservation practices, land use observations, organic matter, cation exchange capacity (KTK), pH, Organic C, Total N, $P_2O_5$, $K_2O$, soil texture, soil structure, soil permeability (Table 1), plant species, agricultural productivity.

### *2.2. Methods*

The research method used is a survey method, through this method direct observations and measurements are made in the field of the parameters needed in the research [8]. This research is in the form of spatial descriptive, with land unit as the unit of spatial analysis. The research variables were rainfall erodibility, soil erodibility, slope length and slope, crop management, and land management. Analysis of the data on the level of erosion hazard (TBE) using the USLE calculation approach [9].

### *2.3. Subtitles*

The calculation of the amount of erosion that occurs is estimated using the General Soil Loss Equation (USLE), in equation (1):

$$\mathbf{A = R \times K \times L \times S \times C \times P} \quad (1)$$

Where :

A : the amount of land lost (ton/ha /yr)

R : the erosion factor of rain

K : soil erodibility factor

L : slope length factor

S : slope factor

C : plant management factors

P : land management factors

*2.3.1. The erosion factor of rain (R).* Erosivity in this study is calculated in equation (2):

$$EI30 = 6{,}119(RAIN)^{1{,}21}.\ (DAYS)^{-0{,}47}.\ (MAXP)^{0{,}53} \quad (2)$$

Dimana:

| | |
|---|---|
| EI30 | = Monthly rain erosion index [6] |
| RAIN | = Monthly average rainfall (cm) |
| DAYS | = Average number of rainy days per month |
| MAXP | = Daily maximum rain (24 hours) in the corresponding time in centimetre |

*2.3.2. Soil erodibility factor (K).* Soil erodibility factor using the equation matrix [10].

**Table 1.** Soil permeability to calculate K-value

| Classification | Speed (cm/hour) |
|---|---|
| Very slow | < 0.5 |
| Slow | 0.5 – 2.0 |
| Slow to Medium | 2.0 – 6.3 |
| Medium | 6.3 – 12.7 |
| Medium to Fast | 12.7 – 25.4 |
| Fast | > 25.4 |

*2.3.3. Slope factor (slope length and tilt of slope)*

$$LS= (l/22)\ mC\ (\cos\alpha)^{1,5}\ [0,5(\sin\alpha)^{1.25} + (\sin\alpha)^{2,25}] \quad (3)$$

Note:
LS = Slope Factor [6].
m = (0.5= slope ≥5%; 0.4= slope 3.5 – 4%; 0.3= slope ≤ 3.5%)
C = 34.71 (constant)
α = slope angle
l = slope length

*2.3.4. Plant management factors (C).* Plant management factor is a combination of plant species, fertility level and soil management time. The presence of plants can reduce the rate of surface erosion and erosion. Plants can influence the rate of erosion due to: 1) the interception of rainwater by leaf crowns; 2) there is an effect on surface runoff; 3) there is an influence on the physical properties of the soil; 4) an increase in the rate of water loss due to transpiration.

*2.3.5. Land management factors (P).* Land management factor is the calculation between the average soil eroded from land that received certain conservation treatment to the average eroded soil from land that is cultivated without conservation action. Soil management and conservation factors (P) are used as guidelines for forecasting P values for various conservation actions [11]. The magnitude of the Erosion Hazard Level can be found by comparing it with the depth of the soil as follows (Table 2) [12]:

**Table 2.** Classification of erosion hazard levels

| Erosion | Erosion Hazard Class (ton/ha/year) | | | | |
|---|---|---|---|---|---|
| Soil Depth (cm) | I | II | III | IV | V |
| | (<15) | (15-60) | (60-180) | (180-480) | (>480) |
| Deep (>90) | SR | R | S | B | SB |
| Medium (60-90) | R | S | B | SB | SB |
| Shallow (30-60) | SR | B | SB | SB | SB |
| Very Shallow (<30) | B | SB | SB | SB | SB |

Note: SR= very light; R = light; S= medium; B= heavy; SB= very heavy

## 3. Results and discussion

### *3.1. The erosion factor of rain (R)*

**Table 3.** Erosivity calculation of research area

| No. | Unit | Year | | | | | | | | | | R |
|---|---|---|---|---|---|---|---|---|---|---|---|---|
| | | 2009 | 2010 | 2011 | 2012 | 2013 | 2014 | 2015 | 2016 | 2017 | 2018 | |
| 1 | R= Average monthly precipitation (cm) | 25.88 | 39.02 | 27.78 | 18.48 | 23.08 | 21.42 | 26.42 | 39.67 | 32.,64 | 22.91 | 277.3 |
| 2 | D= Average number of rainy days per month. | 12.67 | 18.58 | 14.58 | 10.92 | 15.17 | 14.5 | 12.25 | 20.83 | 15.92 | 11.58 | 147 |
| 3 | MAX= Daily maximum rain (24 hours) in the relevant time in (cm) | 12.1 | 12.8 | 11.6 | 9.5 | 11.8 | 7.9 | 11.5 | 16.35 | 15.05 | 11.12 | 11.97 |
| | **EI30** | 356.44 | 504.05 | 355.42 | 223.71 | 281.37 | 212.26 | 361.38 | 554.84 | 475.93 | 306.72 | 363.21 |

Source: Rainfall Station of Ngargoyo Station and Kerjo Station

Based on Table 3 above, the Monthly Rain Erosion Index (EI30) is 363.21 cm. The greater the rainfall, the greater the ability of the rain to cause erosion. The erosivity factor is the only factor that does not use the unit of analysis of land units. The data produced by the rainfall control station is area data.

### *3.2. Soil erodibility (K)*

The erodibility value is calculated based on the structure and permeability of the soil in each land unit. The erodibility value (K) of the research area is 0.032 – 0.302. The greater the erodibility value, the worse the soil's resistance to erosion. Great erodibility values are produced from soils with low permeability. The worse the ability of the soil to pass water, the higher the possibility of erosion. The results of measurements at the research location showed permeability results with very slow, slow, slow to moderate and moderate to fast classes (Table 1).

### *3.3. Slope factor (LS)*

Based on the measurement results obtained slope length between 0.7 meters to 120 meters. The measured slope length is the length of the micro slope for each land unit. The results of the slope measurements did not show any grade I or flat slopes at the study site. This means that there are no land units with a flat slope (0-8%) at the study site. There are 22 land units with a class II slope (sloping, 8-15 %), 21 land units with a class III slope (Slightly steep, 15-25 %), 27 land units with a class IV slope (steep, 25-45%) and there is 1 land unit that has a very steep slope or class V (>45%).

### *3.4. Plant management (C)*

Based on the results of observations that are matched with the table of C values from Asdak [11], the C values are between 0.05 to 0.8. The land use in the research location is; forests, fields, gardens, and rice fields. The C value: 0.05 was obtained from the use of forest land and shrubs, while the C value: 0.8 which was the largest was obtained from the use of plantation land with cassava plants. Land management uses a terrace system, which can be found in all land units because of its easy application and has been passed down for generations using the terrace method in cultivating land.

### *3.5. Land conservation (P)*
Conservation action at the research location is making land terraces. The types of terraces carried out are; bench terraces, *gulud* terraces and traditional terraces. Making land terraces by the community according to contour lines. Land units with plantation land use carry out conservation actions using ground covers and trenches as an effort to reduce erosion.

### *3.6. Solum depth* (R)
The depth of the solum of the research area is between 20 cm – 140 cm.

### *3.7. Erosion quantity* (BE)

**Table 4.** Land area erosion quantity (BE)

| No | Area | Land area (Ha) | | | | Land area (Ha) |
|---|---|---|---|---|---|---|
| | | Very Light (SR) (0 - 15 Ton/ha/yr) | Light (R) (15 – 60 Ton/ha/yr) | Medium (S) (60 – 180 Ton/ha/yr) | Weight (B) - (180 - 480 Ton/ha/yr) | |
| 1 | Berjo | 1,433.29 | 190.58 | - | - | 1,623.87 |
| 2 | Dukuh | 318.26 | 16.70 | - | - | 334.96 |
| 3 | Girimulyo | 469.93 | 109.16 | 68.86 | - | 647.95 |
| 4 | Jatirejo | 206.58 | 10.65 | - | - | 217.23 |
| 5 | Kemuning | 326.28 | 140.63 | - | 202.39 | 669.30 |
| 6 | Ngargoyoso | 252.05 | 137.31 | 153.42 | 53.35 | 596.14 |
| 7 | Nglegok | - | 245.11 | 175.28 | 18.29 | 438.68 |
| 8 | Puntukrejo | 228.21 | - | 40.38 | - | 268.59 |
| 9 | Segorogunung | - | 1,737.23 | - | - | 1,737.23 |
| | **Total** | **3,883.53** | **1,595.29** | **704.50** | **350.63** | **6,534.00** |

Source: Analysis 2020

Based on the table 4 above, 3 villages have the largest erosion, namely Kemuning Village with an area of 202.39 Ha, Ngargoyoso Village with an area of 53.35 Ha and Nglegok Village with an area of 18.29 Ha. Overall, the area of land that is categorized as heavy erosion is 350.63 Ha, the large category of moderate erosion is 704.50 Ha, the large category of mild erosion is 1,595.29 Ha, and land area that is classified as a large category is very light erosion, which is 3,883.53 Ha.

### *3.8. Erosion Hazard Level (TBE)*
Based on Table 5 and Figure 1, the land area that has a very heavy level of erosion hazard is 3 villages, namely Kemuning Village with an area of 213.61 Ha, Ngargoyoso Village with an area of 86.98 Ha and Nglegok Village with an area of 19.42 ha, in total an area of 500.65 Ha. The level of erosion hazard in Ngargoyoso Subdistrict which is included in the category of severe erosion hazard level is 1,509.16 Ha, the category of moderate erosion hazard level is 1,351.48 Ha, the land area is included in the category of light erosion hazard level is 1,690.49 Ha, and the area of land included the large category of very light erosion is 1,482.17 Ha. The level of erosion hazard is influenced by several factors, the main one being the vegetation of the land cover. Land units with very light Erosion Hazard Levels are land units with forest land cover, gardens, and settlements.

**Table 5.** Land area erosion hazard level (TBE)

| No | Area | Land area (Ha) | | | | | Land area (Ha) |
|---|---|---|---|---|---|---|---|
| | | Very Light (SR) | Light (R) | Medium (S) | Weight (B) | Very Heavy (SB) | |
| 1 | Berjo | 608.55 | 984.27 | | 31.05 | | 1,623.87 |
| 2 | Dukuh | | 193.39 | 114.95 | 26.61 | | 334.96 |
| 3 | Girimulyo | 495.49 | 21.06 | 131.4 | | | 647.95 |
| 4 | Jatirejo | | 76.12 | 15.2 | 125.9 | | 217.23 |
| 5 | Kemuning | 8.05 | 125.8 | 111.07 | 210.77 | 213.61 | 669.3 |
| 6 | Ngargoyoso | 97.26 | 22.3 | 194.84 | 194.76 | 86.98 | 596.14 |
| 7 | Nglegok | 6.88 | | | 412.37 | 19.42 | 438.68 |
| 8 | Puntukrejo | 4.44 | 48.24 | | 215.91 | | 268.59 |
| 9 | Segorogunung | | | 1,737.23 | | | 1,737.23 |
| | **Total** | **1,482.17** | **1,690.49** | **1,351.48** | **1,509.16** | **500.65** | **6,534.00** |

Source: Analysis 2020

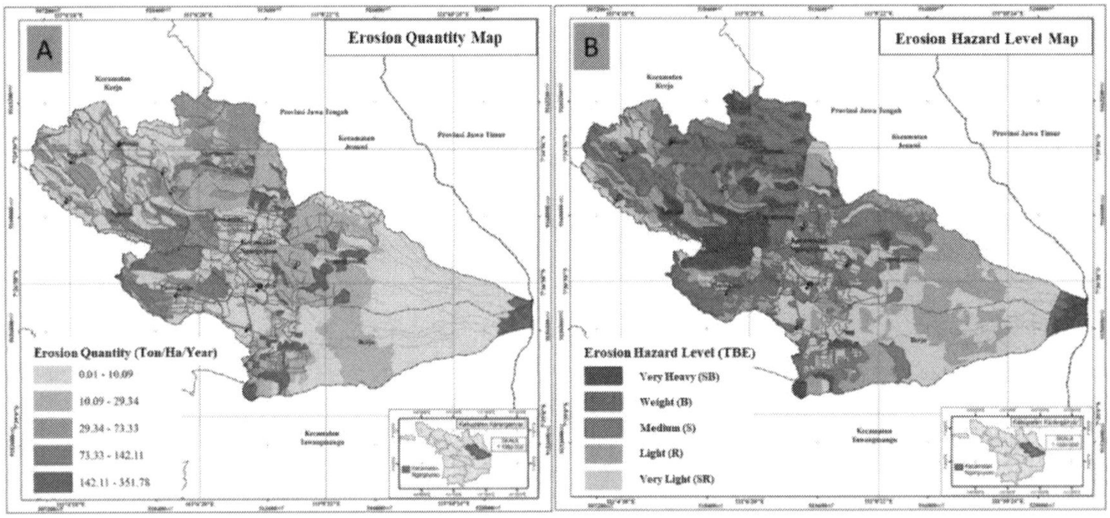


**Figure 1.** (A) Big map of erosion (Ton/ha/Year); (B) Erosion hazard level map

Efforts to overcome erosion by enriching soil fertilizing organisms, making terraces, contour farming and optimizing drainage networks can be carried out in all villages in Tawangmangu District. Enriching soil-fertilizing organisms by conserving soil and one of them is by preserving soil-fertilizing organisms such as earthworms and other types of micro-organisms to accelerate the return of soil function to its original form. Conservation by making terraces is done by making terraces by terraces like stairs on sloping land so that when it rains water does not erode the soil [13].

The next conservation step can be crop rotation which aims to preserve the nutrients contained in the soil by rotating the planting schedule for plant species, in this way substances that are useful for soil fertility are not completely absorbed by one type of plant [14]. Crop rotation land conservation priority is carried out in 3 villages, namely Kemuning Village, Ngargoyoso Village and Nglegok Village.

## 4. Conclusions

The results showed that the areas included in TBE in Ngargoyoso District were: areas that had very light TBE (SR) of 1,482.18 Ha (22.68%) and areas that included very heavy TBE (SB) of 500.66 Ha. (7.66%). Agricultural activities greatly affect the level of erosion hazard in the study area as much as 67% of the total activities. So, climate change has a negative impact on land management patterns in agricultural activities. Efforts to overcome erosion by enriching soil fertilizing organisms, making terraces, contour farming, making pair embankments, optimizing drainage, making crop rotation and reforestation can be carried out in Tawangmangu District.

**References**

[1] Arifin B 2009 Pemanasan global dan ketahanan pangan nasional *Jurnal Pangan* **18** 3–11

[2] Legionosuko T, Madjid M A, Asmoro N and Samudro E G 2019 Posisi dan Strategi Indonesia dalam Menghadapi Perubahan Iklim guna Mendukung Ketahanan Nasional *Jurnal Ketahanan Nasional* **25** 295–312

[3] Munzir T, Akbar H and Rafli M 2019 Kajian erosi tanah dan teknik konservasi tanah di Sub DAS Krueng Pirak Kabupaten Aceh Utara *Jurnal Agrium* **16** 126–34

[4] BPS 2021 Banyaknya Hari Hujan (HR) dan Curah Hujan (MM) menurut Bulan dan Tempat Pengukuran di kabupaten Karanganyar *Badan Pusat Statistik Kabupaten Karanganyar*

[5] Winarno J, Ariyanto D P and Rosariasturi R 2013 Kajian bahaya erosi pada lahan kering di Sub DAS Samin Kabupaten Karanganyar *Sains Tanah - Journal of Soil Science and Agroclimatology* **5** 101–6

[6] Christanto R E, Suryono, Mujiyo and Winarno J 2013 Pemetaan tingkat bahaya erosi lahan kering di Kecamatan Jatipuro Kabupaten Karanganyar dengan Sistem Informasi Geografi (SIG) *Sains Tanah - Journal of Soil Science and Agroclimatology* **7** 114–9

[7] Batara Y D 2016 Penentuan kesesuaian lahan berbasis Sistem Informasi Geografis studi kasus Daerah Rawa Kanamit, Kabupaten Pulang Pisau *Jurnal INTEKNA: Informasi Teknik dan Niaga* **16** 31–8

[8] al Kautsar M, Sasmito B and Hani'ah 2013 Aplikasi Echosounder Hi-Target HD 370 untuk pemeruman si perairan dangkal (Studi kasus: perairan Semarang) ed G. Balint, B Antala, C Carty, J-M A Mabieme, I B Amar and A Kaplanova *Jurnal Geodesi Undip* **2** 343–54

[9] Widodo A, Komariah and Suyana J 2015 Metode USLE untuk memprediksi erosi tanah dan nilai toleransi erosi sebuah sistem agricultural di Desa Genengan Kecamatan Jumantono Karanganyar *Agrosains : Jurnal Penelitian Agronomi* **17** 39–43

[10] Arsyad S 1989 *Konservasi Tanah dan Air* (Bogor: IPB (Bogor Agricultural University))

[11] Asdak C *Hidrologi dan pengelolaan daerah aliran sungai* (Yogyakarta: UGM Press)

[12] Hardjowigeno S 2015 *Ilmu tanah* (Jakarta: Akademika Pressindo)

[13] Arnoldus, Andawayanti U and Suhartanto E 2016 Analisis tingkat bahaya erosi dan arahan konservasi lahan dengan aplikasi GIS di DAS Manikin *Jurnal Teknik Pengairan* **7** 205–15

[14] Dhoke M E, Kusumandari A and Senawi 2020 Tingkat erosi dan rancangan teknik konservasi tanah dan air di Sub DAS Waewoki, DAS Aesesa Kabupaten Ngada Provinsi Nusa Tenggara Timur (Erosion level and soil and water conservation engineering plan in Waewoki Sub Watershed, Aesesa Watershed, Ngada Regency, East Nusa Tenggara Province) *Jurnal Manusia dan Lingkungan* **25** 7–17

**Acknowledgement**

Special thanks to Mr Setya Nugraha who has assisted in funding through joint research (fundamental grants from the Ministry of Research and Technology), as well as Prof. Suntoro, Mrs Komariah and Prof. Tuti who guided this research.

# Local food development strategy in hilly areas of Gunungkidul Indonesia

**Susanawati[*], Oki Wijaya and Muhammad Barik Rizqi**

Department of Agribusiness of Universitas Muhammadiyah Yogyakarta Indonesia, Jalan Brawijaya Tamantirto Kasihan Bantul DI Yogyakarta 55183 Indonesia
*Corresponsing author: susanawati@umy.ac.id

**Abstract.** Agriculture is one of the sectors affected by climate change. Gunungkidul is one of the areas in Indonesia vulnerable to climate change because most of its population depends on the agricultural sector. Some of the agricultural land in this region is dryland with low food crop productivity. Therefore, research aiming to analyze local food development strategies in hilly areas is necessary. The sample in this study consisted of several households and key figures in three districts of Gunungkidul Regency: Semin, Wonosari, dan Saptosari. Descriptive analysis was applied to identify the potential of local food, consumption, and processed products. The results discovered 11 food commodities in Gunungkidul, encompassing rice, brown rice, cassava, sweet potato, corn, soybean, peanuts, green beans, vegetables, fruits, and animal foods. The strategies are (1) increasing the productivity of local food ingredients, (2) farming assistance in improving the quality of harvested raw materials for local food products, (3) increasing farming technology for local raw material, (4) increasing technology in IKM local food products, (5) massive promotion of local food products, (6) creating a brand identity for superior local food products, (7) developing local food industry groups, (8) massive socialization of food diversification, (9) developing microfinance institutions, and (10) formulating regulations for the use of local food as consumption in agencies or institutions in Gunungkidul. Meanwhile, the priority strategy that should be carried out is to create a brand identity for superior local food products.

**Keywords**: local food, development strategy, hilly areas

## 1. Introduction

Climate change has developed into a global issue in recent decades [1]. The agricultural sector, especially food crops, is one sector facing major challenges from the impacts of climate change [2] because it affects agricultural infrastructure, agricultural production systems, aspects of food security and self-sufficiency, and the welfare of the community and farmers in general [3]. One of the areas in Indonesia predicted to be vulnerable to the impacts of climate change is Gunungkidul Regency because around 52.40% of its population depends on the agricultural sector for their livelihoods.

The agricultural potential in Gunungkidul Regency is dryland agriculture covering an area of 83,180 ha [4]. Dryland has limited productivity due to low soil fertility and water sources for agricultural activities depending on rainwater, thus being prone to drought hazards (Dudung 1991 in [5]). Agriculture in Gunungkidul Regency is still subsistence, that is, agriculture in which part of the output is consumed by farmers and their families [6]. The mainstay product of such agriculture is a staple food of grains with low productivity because it still uses traditional equipment and limited capital investment.

Based on data from the 2017 Energy Sufficiency Rate, the rice consumption of the people of Gunungkidul Regency is 88.3 kg/capita/year, smaller than the national rice consumption of 114 kg/capita/year. In addition, people in this regency still lack tubers, animal consumption, vegetables, and fruit. Following the food content, they consume fewer food sources of protein, vitamins, and minerals and lack variety in consuming carbohydrates. The community must be directed to diversify food by utilizing local food to overcome their lack of food content. Therefore, research aiming to formulate local food development strategies in hilly areas in Gunungkidul Regency is required to realize it.

## 2. Material and methods

The basic method used in this research is descriptive analysis. Descriptive research aims to make descriptions in a systematic, factual, and accurate way about the facts, nature, and relationships between the phenomena studied [7]. In the descriptive method, the researcher can compare certain phenomena resulting in a comparative study. The data collected are first compiled, explained, and then analyzed. The descriptive method has several advantages: (1) careful measurement of social research [8], (2) able to describe the relationship between phenomena, hypothesis testing, and policy implications [7], (3) data analysis is carried out with a descriptive-analytical approach to produce relationships or comparisons between variables [9], and (4) research results can be in the form of deductive conclusions [10].

The research location was deliberately determined according to the research objectives, namely three districts in Gunungkidul Regency, comprising (1) Semin District because it had the largest amount of peanut and soybean production in Gunungkidul in 2018, (2) Wonosari District since it had the most soybean production in Gunungkidul Regency in 2018, and (3) Saptosari District for having the largest amount of cassava production in Gunungkidul in 2018.

This study utilized both secondary and primary data. Secondary data contained the characteristics or general description of the research location collected through documentation from relevant agencies, reading literature (library research), books on the theory of the problems studied, and using the internet as a supporting medium in searching for additional information about the theory and data required. On the other hand, the primary data encompassed local food production, household consumption, and internal and external factors of local food development in the Gunungkidul Regency. The primary data were gathered using interviews, observation, and documentation.

Three essential stages were performed to determine the strategies: data collection, data matching, and decision making. Data collection revealed a picture of the local food situation in Gunungkidul. Internal and external factors employed the IFE and EFE matrices, the matching stage utilized the SWOT matrix, and the decision-making stage applied the QSPM to obtain alternative strategies in local food development to formulate STA development strategy.

In the data collection stage, the data were obtained by interviewing determined respondents. Moreover, observations were made directly on local food. The data taken comprised internal and external local food data using the IFE and EFE matrix models.

The data matching stage is where external and internal factors from the previous stage were combined to create some proper alternative strategies for the development of local food. At this stage, the analysis model used the Strengths-Weaknesses-Opportunities-Threats (SWOT) Matrix. The decision-making stage is the final stage after the data collection and matching stage. The Quantitative Strategic Planning Matrix was employed to formulate alternative strategies [11].

The first step in the SWOT analysis was to identify external and internal factors that might influence the determination of local food development strategies in Gunungkidul Regency. Identification was carried out by the External Factor Evaluation (EFE) and Internal Factor Evaluation (IFE) matrices. The factor data obtained were then grouped into internal factors (strengths and weaknesses) and external factors (opportunities and threats).

The next step was to conduct a SWOT analysis. In the SWOT matrix, the alternative strategy formulas were applied by doing pairwise comparisons. Pairwise comparison is a technique of comparing a component with other components in the same category. The SWOT matrix assisted in doing pairwise comparisons between strengths, opportunities, weaknesses, and threats.

Based on the SWOT matrix, several alternative strategies for improving the condition of food security at the village level can be developed as follows:

1. Strengths-Threats (ST) strategy employs existing strengths to avoid and overcome threats in developing local food.
2. Strengths-Opportunities (SO) strategy utilizes existing strengths to exploit existing opportunities in developing local food.
3. Weaknesses-Opportunities (WO) strategy applies the opportunities to overcome weaknesses in developing local food.
4. Weaknesses-Threats (WT) strategy is an effort to minimize weaknesses and avoid threats in developing local food.

The policy strategy was prepared based on the identification results of factors and the SWOT analysis carried out in the previous stage. The formulation of the policy strategy utilized the Quantitative Strategic Planning Matrix (QSPM) method. The QSPM is an analytical tool used to evaluate alternative strategies objectively based on key external and internal factors. Existing data were entered in a table prepared and then analyzed. Furthermore, to determine the most suitable strategies, it was continued with analysis using a strategy analysis table with the following steps:

Step 1: Listing key external opportunities or threats and internal strengths or weaknesses in the left column of the QSPM
Step 2: Assigning a value or weight to each factor (identical to the values assigned to the IFE and EFE metrics)
Step 3: Examining (matching) the matrix and identifying alternative strategies that should be considered for a setting
Step 4: Setting the attractiveness score (AS): 1=no attractiveness; 2=low traction; 3= moderate attractiveness; 4=high traction
Step 5: Calculating the total attractiveness score, which resulted from multiplying the weight by the attractiveness score in each row. The higher the total attractiveness score, the more attractive the strategy will be.

## 3. Results and discussions

The local food development strategy in Gunungkidul Regency was prepared based on the identification of its strengths, weaknesses, opportunities, and threats that will be faced. This regency has the potential strengths to develop local food products. Local food helps in increasing domestic production capacity [12]. These strengths include, among others, having a variety of local food types such as processed rice, rice, corn, cassava, sweet potatoes, soybeans, peanuts, and so on. The variety of local food types is known from generation to generation, from the ancestors to today. Some people in this regency still consume various types of local food. The reasons they consume local food vary, including dietary habits, availability of raw materials in food processing, and maintaining culture and local wisdom in consuming food.

Another strength that Gunungkidul Regency has in developing local food is the recognition of this regency as a cassava center in the Special Region of Yogyakarta (DIY). As a dryland area, cassava is suitable for farming and developing in this regency, especially as a superior local food raw material. Apart from agriculture, this regency also has strengths in natural tourism that support local food development. Gunungkidul Regency is designated as a world geopark area, increasing the number of tourist visits. Thus, Gunungkidul Regency, once known as a dry and poor area, is now known as one of the alternative beautiful DIY tourist destinations, apart from the palace and Prambanan.

Of the many potential strengths, Gunungkidul Regency also owns many weaknesses that must be improved. If not resolved with the proper strategy, these weaknesses can hinder the development of local food. One of the weaknesses due to natural resource factors is dry and hilly land. Dryland and limited water resources limit the diversity of plants in this regency. Plants that can be farmed are only those that do not require much water. In addition, most of the land in this regency is hilly. These factors cause the limitations of raw materials in processed food products. However, the limited natural resources are not the reason for the development of local food. Thus, all parties must jointly think about alternatives to dryland agriculture to support local food development in Gunungkidul.

In addition to the limited natural resources, another weakness commonly encountered is the low technology used in the production process for local food development. Most of the production still uses traditional tools, causing the low quality of processed food products. In addition, low infrastructure causes local food products developed by producers not to develop [13]. Accordingly, it will impact people's interest in buying these local food products.

Another weakness of the development of local food products is the consumption pattern of the people of Gunungkidul, which is not yet diverse. The average people of Gunungkidul currently make rice as their daily staple food. Various kinds of tubers, vegetables, grains, and fish in Gunungkidul are extremely abundant. However, the lack of public knowledge about the benefits or contents of these various commodities makes them underestimate the various processed local food products. Internal factors, strengths, and weaknesses are displayed in Table 1 below:

**Table 1**. Internal factors of local food development in Gunungkidul Regency.

| No. | Internal Factors | Weight | Rating | Score |
|---|---|---|---|---|
| **Strengths** | | | | |
| 1 | Various kinds of local food | 0.096 | 3 | 0.289 |
| 2 | People are still willing to consume local food | 0.094 | 3 | 0.282 |
| 3 | Strong community culture | 0.099 | 2 | 0.198 |
| 4 | Cassava production center | 0.089 | 2 | 0.178 |
| 5 | World geopark area | 0.099 | 2 | 0.198 |
| | | **0.4772** | | **1.145** |
| **Weaknesses** | | | | |
| 1 | Dry and hilly land | 0.102 | 3 | 0.305 |
| 2 | The low technology for local food processing | 0.086 | 3 | 0.259 |
| 3 | The low quality of processed food products | 0.071 | 2 | 0.142 |
| 4 | Limited availability of raw materials for processed food products | 0.074 | 2 | 0.147 |
| 5 | Limited institutions for local food | 0.066 | 2 | 0.132 |
| 6 | People's food consumption patterns are not yet diverse | 0.058 | 1 | 0.058 |
| 7 | Limited capital for the local food processing industry | 0.066 | 2 | 0.132 |
| | | **0.523** | | **1.175** |
| | **Total** | **1.000** | | **1.277** |

In addition to its strengths and weaknesses, Gunungkidul Regency also has opportunities that can be exploited to develop local processed food products. One of them is the existence of a central government policy in developing local food. With the government's policy on local food consumption, farmers and entrepreneurs will find it easier to market their products because they will have an easily accessible market share, such as government officials, the community, and foodservice providers such as restaurants. Moreover, the government policy regarding food diversification is also a good opportunity for farmers and entrepreneurs to develop local processed food products because to meet daily consumption, people are no longer fixated on one commodity, such as rice and other commodities (Table 2)tabl.

Another opportunity that can help increase the development of local processed food products is the presence of tourists. With an increase in the number of tourists visiting Gunungkidul, it can enhance the sales turnover of local processed food products, and the long-term effect felt by the community is an increase in the economic level. Most tourists will hunt for souvenirs being the characteristics of Gunungkidul. For example, tourists will hunt for special foods such as *gatot* and *tiwul* made from cassava, where cassava is one of the tubers widely grown in this regency.

**Table 2**. External factors of local food development in Gunungkidul Regency.

| No. | External Factors | Weight | Rating | Score |
|---|---|---|---|---|
| **Opportunities** | | | | |
| 1 | Central government policy to develop local food and food diversification | 0.081 | 1 | 0.081 |
| 2 | An increasing number of tourists | 0.074 | 2 | 0.147 |
| 3 | Increasing community income | 0.085 | 2 | 0.171 |
| 4 | Rice food prices increasingly | 0.105 | 1 | 0.105 |
| 5 | Public awareness about expected food patterns and nutritional adequacy rates | 0.155 | 3 | 0.465 |
| | | **0.500** | | **0.969** |
| **Threats** | | | | |
| 1 | Increased types of food and beverages from outside Gunungkidul | 0.143 | 2 | 0.287 |
| 2 | Import of foodstuffs | 0.136 | 2 | 0.271 |
| 3 | Climate change | 0.120 | 2 | 0.240 |
| 4 | Land function conversion | 0.101 | 2 | 0.202 |
| | | **0.500** | | **1.000** |
| | **Total** | **1.000** | | **1.969** |

Another opportunity for developing local food products is an increase in the income of the people of Gunungkidul. With public awareness about local food consumption, the community's level of welfare and income also increases because people tend to use local products from their area. In addition, as a tourist destination, local processed food products are also one of the culinary delights that tourists must try.

Currently, the market price of rice is increasing in line with the increasing public demand for the rice commodity. It can be an opportunity for the people of Gunungkidul to introduce local processed food products instead of rice. The development of local processed food products varies and has a nutritional value equivalent to rice, and the prices offered are more affordable. It is expected that people will switch from consuming rice to local processed food products. Interest in consuming local food should be driven by the belief that they are healthier [14].

Another opportunity to develop local food processing in Gunungkidul is public awareness about expected food patterns and nutritional adequacy rates. With public awareness of nutrition or diverse energy sources, the need to develop local processed food products becomes more diverse and considers the nutritional value needed. Hence, in the long term, local processed food products can meet the daily nutritional consumption needs of the Gunungkidul community, and the community certainly has a high interest in buying these products.

After understanding the various opportunities that Gunungkidul Regency has, the next factor that needs to be understood is the threats faced in developing local processed food products. If the threats are not handled properly, they will interfere with the development of local processed food products. Threats that arise include increasing types of food and beverages from outside Gunungkidul. With the increase in ready-to-eat foods and beverages increasingly varied, people will definitely be more tempted to buy these products. Besides, the affordable price also makes them more consumptive with ready-made food and beverages. It certainly affects the development of local processed food products. Hence, local government policies related to this threat are urgently required to make local food processed products remain in demand by the community.

The next threat is the import of food ingredients into Gunungkidul. As is well known, the increasing need for food and the lack of availability of food from within the country have resulted in imported food being one of the people's choices to meet their daily basic needs. In addition, affordable prices and good quality also influence the decision to buy imported food products. With this threat, local food product entrepreneurs in Gunungkidul also need to understand and improve the quality of their processed products to compete with imported products on the market today.

Climate change in Gunungkidul is also a threat to farmers. It happens because, with climate change, there will be a crop failure, causing farmers to lose money and disrupting the availability of food supplies. The climate of Gunungkidul, consisting of seven months of the rainy season and five months of a dry season, results in farmers having to understand the climate change to minimize crop failure.

**Table 3**. The SWOT matrix

| | Internal factor | |
|---|---|---|
| | Strengths (S) | Weaknesses (W) |
| | • Various kinds of local food<br>• People are still willing to consume local food<br>• Strong community culture<br>• Cassava production center<br>• World geopark area | • Dry and hilly land<br>• The low processing technology for local food<br>• The low quality of processed food products<br>• Limited availability of raw materials for processed food products<br>• Limited institutions for focal food<br>• People's food consumption patterns are not yet diverse<br>• Limited capital for the local food processing industry |
| External Factors | | |
| Opportunities (O) | | |
| • Central government policy to develop local food and food diversification<br>• Increasing number of tourists<br>• Increasing community income<br>• Rice food prices increasingly<br>• Public awareness about expected food patterns and nutritional adequacy rates | • Massive promotion of local food products<br>• Creating a superior brand identity for local food products | • Technological improvement in local food product SMEs<br>• Developing local food industry groups<br>• Massive socialization of food diversification<br>• Promoting credit schemes to the local food industry |
| Threats (T) | | |
| • Increased types of food and beverages from outside Gunungkidul<br>• Import of foodstuffs<br>• Climate change<br>• Land function conversion | • Improving local raw material farming technology<br>• Preparing regulations for the use of local food for consumption at agencies or institutions in Gunungkidul | • Increased productivity of local foodstuffs<br>• Farming assistance in improving the quality of harvested raw materials for local food products |

The next threat is the conversion of land functions. The rapid development of infrastructure in Gunungkidul is one of the factors for the conversion of agricultural land. This land conversion is feared to interfere with the management of local processed products in Gunungkidul due to limited food ingredients. Based on the analysis of internal and external factors influencing the development of local food, a policy direction and local development strategies in Gunungkidul Regency have been drawn up. The policy direction was prepared based on the main needs: (1) improving the quality and quality of local food raw materials sustainably, (2) improving the quality, quality, and scale of the local food industry, and (3) local food development based on demand-side point of view. It was then broken down into strategies and program alternatives compiled in a SWOT matrix. The above strategies (Table 3) were then compiled with the QSPM analysis to determine the most influential priorities for developing local food in Gunungkidul Regency. The results of the QSPM analysis are depicted in Table 4 below.

**Table 4**. Formulation of local food development strategies in Gunungkidul

| Priority | Score | Alternative Strategies |
|---|---|---|
| 1 | 5.252 | Creating a brand identity for superior local food products |
| 2 | 5.006 | Massive socialization of food diversification |
| 3 | 4.403 | Massive promotion of local food products |
| 4 | 4.110 | Technological improvement in local food product SMEs |
| 5 | 3.771 | Growing local food industry groups |
| 6 | 3.480 | Improving local raw material farming technology |
| 7 | 3.393 | Preparing regulations for the use of local food for consumption at agencies or institutions in Gunungkidul |
| 8 | 2.982 | Disseminating credit schemes to the local food industry |
| 9 | 2.968 | Farming assistance in improving the quality of harvested raw materials for local food products |
| 10 | 2.709 | Increased productivity of local food |

The results of the QSPM analysis revealed that the priority that should be conducted is to create a brand identity for superior local food products in Gunungkidul Regency. Other local governments have also carried out regional brands or local food. For example, Kulonprogo Regency made the slogan "Bela Beli Kulonprogo" one of the steps to create a regional brand identity aiming to increase public interest in buying local products. Moreover, Indonesia tries to increase the number of tourists with the brand identity of "Wonderful Indonesia".

## 4. Conclusions

Gunungkidul is one of the areas in Indonesia vulnerable to climate change because most of its population depends on the agricultural sector. Several strategies to develop local food in hilly areas in Gunungkidul Regency: (1) increasing the productivity of local food ingredients, (2) farming assistance in improving the quality of harvested raw materials for local food products, (3) improving farming technology for local raw materials, (4) technological improvement in local food product SMEs, (5) massive promotion of local food products, (6) creating a brand identity for superior local food products, (7) developing local food industry groups, (8) massive socialization of food diversification, (9) developing microfinance institutions, (10) preparing regulations for the use of local food for consumption at agencies or institutions in Gunungkidul. Out of these ten strategies, the priority strategy that should be immediately performed is to create a brand identity for superior local food products.

## References

[1] Kurniawan R E and Arisurya R E 2021 Kerentanan dan adaptasi rumah tangga petani terhadap perubahan iklim di Kabupaten Gunungkidul *J. Agro Ekon.* **38** 127
[2] Surmaini E, Runtunuwu E and Las I 2010 Upaya sektor pertanian dalam menghadapi perubahan iklim *J. Litbang Pertan.* **30** 1–7
[3] Rejekiningrum P, Las I, Amien I, Pujilestari N, Estiningtyas W, Surmaini E, Suciantini, Sarvina Y, Pramudia A, Kartiwi B, Muharsini S, Sudarmaji, Hardiyanto, Hermanto C, Putranto G A and Marbun O 2011 *Pedoman Umum: Adaptasi Perubahan Iklim Sektor Pertanian* (Jakarta: Badan Penelitian dan Pengembangan Pertanian)
[4] Badan Pusat Statistik 2017 *Adaptasi Perubahan Iklim Sektor Pertanian* (Gunungkidul, Indonesia: BPS Kabupaten Gunungkidul)
[5] Haryanti N, B T M and I D R 2003 Karakteristik petani lahan kering (studi pada petani lahan

kering Desa Wanadri Kabupaten Banjarnegara) *Jurnal Prosiding Hasil Litbang "Rehabilitasi Lahan Kritis"* (Banjarnegara, Indonesia) pp 74–82
[6] Todaro M P, Munandar H and Sumiharti Y 2000 *Pembangunan Ekonomi di Dunia Ketiga* (Jakarta: Erlangga)
[7] Nazir M 2009 *Metode Penelitian* ed R Sikmumbang (Bogor: Ghalia Indonesia)
[8] Singarimbun M 1982 *Metode Penelitian Survei* ed S Effendi (Jakarta: LP3ES)
[9] Mukhtar and Widodo E 2000 *Konstruksi ke Arah Penelitian Deskriptif* (Yogyakarta: Avyrouz)
[10] Subyantoro A, Suwarto F and Harjono D 2007 *Metode dan Teknik Penelitian Sosial* (Yogyakarta: Andi)
[11] David F R 2006 *Manajemen Strategis: Konsep, Buku 1* ed P Sulistio and H Mahardika (Jakarta: Salemba Empat)
[12] Kirwan J and Maye D 2013 Food Security Framings within the UK and the integration of local food systems *J. Rural Stud.* **29** 91–100
[13] Mount P 2012 Growing local food: scale and local food systems governance *Agric. Human Values* **29** 107–21
[14] Hardesty S D 2008 The growing role of local food Markets *Am. J. Agric. Econ.* **90** 1289–95

# Brand activism on the digital public sphere: campaign content analysis of #BringBackOurBottle on Instagram

**Martania Hartika[1*], Pawito[2], dan Prahastiwi Utari[3]**

[1] Master of Communication Study Program of Faculty of Social and Political Sciences of Universitas Sebelas Maret, Jl. Ir. Sutami 36 A, Kentingan, Surakarta, 57126, Indonesia
[2]Department of Communication Science of Faculty of Social and Political Sciences of Universitas Sebelas Maret, Jl. Ir. Sutami 36 A, Kentingan, Surakarta, 57126, Indonesia
[3]Department of Communication Science of Faculty of Social and Political Sciences of Universitas Sebelas Maret, Jl. Ir. Sutami 36 A, Kentingan, Surakarta, 57126, Indonesia
[*]Corresponding author: hartika@student.uns.ac.id

**Abstract**. Brand activism is one of the organizational initiative practices voluntarily conducted to promote social, economic, environmental, and cultural issues which are considered as important matter and have wide impact for society aligned with the company's vision and mission. Recently, this kind of activity mostly carried on social media as public sphere which becomes media communication, such as campaign to reduce plastic wastes conducted by The Body Shop Indonesia as effort to reduce environmental disaster and climate change impact. In this research, the researcher focused on the #BringBackOurBottle campaign carried by The Body Shop Indonesia by identifying messages through content visualization on Instagram account @thebodyshopindo.impact. Furthermore, the research used qualitative content for both photo and video post with hashtag #BringBackOurBottle. The research finding emphasized on two main messages within the content #BringBackOurBottle; the first, plastic threat for survival and the second is the drive for sustainable living. Besides, this research also showed that brands have power to create the consumer's behaviour; it can encourage the consumers to participate in taking care of the environment through the program they continue to campaign.

**Keywords** : brand activism, digital public sphere, climate change, environmental disaster

## 1. Introduction

Nowadays, we can see that climate change is an issue which continues to be discussed around the world and it gives huge impact in any part of the world in both land and sea, without exception in Indonesia. The problem of climate change in Indonesia have not been seriously responded yet and it is viewed as general problem due to other more pressing issues such as state economy and poverty [1]. Climate change can potentially cause disaster. It can be caused by natural events or humans (man-made disasters). The disaster caused by humans is an impact of humans' ignorance toward environment and ecosystem sustainability; hence they only increase quality and accessibility of life without having responsible to preserve the environment.

Human life now cannot be separated from plastic. Plastics are increasingly used because of their elastic properties, easy to shape, and easy and inexpensive production processes. Plastic is main material to make packaging, building, transportation, medical equipment, etc, because plastic has high flexibility. Besides, plastic also has benefits and easy to be processed [2]. Therefore, the use of plastic is very high and productive, because plastic is very useful and durable.

The other facts, plastic has nature which is not easily discomposed. A plastic bag has a decomposition time of 10 -12 years. Plastic bottle packaging takes 20 years to decompose, which is longer than plastic bag, due to complex polymer and thicker layer [3]. Hence, it can be concluded that humans are victim of their success. Currently, humans are destroying the earth with plastics which have been used for the last 100 years.

According to Greenpeace, since the 1950s plastic has been produced for 8.5 billion metric tons. Based on that enormous amount, only 9% is recycled, the remaining 79% only ends up in landfills or the natural environment, includes in the oceans. The marine ecosystem is disturbed by the increase of plastic waste; it means that there is an increase in the prevalence of plastic in society [3]. The problems will get worse if this method continues without any more serious handling, it is estimated that as much as 12 billion tonnes of plastic waste will be found in landfills, natural environments and oceans by 2050 [4]. Plastics cause a risk for organisms in the natural environment, for example; through consumption or entanglement in plastics, the chemicals in plastics can accumulate in the food chain as well [5].

The problems which have been elaborated above finally have made some parties start to move to take part in effort to control the development of plastic waste. In the effort to control the use of plastic, there are several difficulties faced, for example; the difficulties and solutions to bring plastic packaging into the mainstream. The next big problem is the lack of access to refill products, in order to the packaging can be used sustainably to reduce the use of single-use plastic products [6]. The Body Shop has succeeded becoming main actor in action of packaging return through program #BringBackOurBottle, in which The Body Shop Indonesia calls for consumers to return The Body Shop product packaging to near shops. The packaging which have been returned to the shops will be recycled and the processing results are used to community development.

The Body Shop Indonesia voices their actions through social media Instagram with account @thebodyshopindo.impact. Social media is very effective to promote everything because it is easy to be operated, free, and it can also reach wide market share [7]. Media social is able to become dialog place among users and it can encourage being supporter for social change. In persuading public, in positive ways, media social is considered to have high ability, i.e. The Body Shop Indonesia which finally becomes the first pioneer of waste return program in Indonesia.

The Body Shop has established since 1976 in Brighton by Anita Roddick, who believes in a revolutionary idea that business is able to be power to do kindness. Therefore, through Through the Body Shop Indo Impact the struggle begins for fairer and more beautiful world and tries to make positive impact on our planet & community.

This research focused on the #BringBackOurBottle campaign which has been carried out by The Body Shop Indonesia with message identification through content visualization on Instagram account @thebodyshopindo.impact. Credibility can be realized in visual dramatic through belief and expert themes [8]. Symbolic Convergence Theory (SCT) was used to guide the analysis and explain the research finding in this research. The previous study had used SCT to explore messages and images shared through Twitter in context of reputation management, politic, and culture [9]. This research used SCT lens in the campaign of environment issues and brand activity held on digital public sphere.

## 2. Materials and method

This research used qualitative content method. We examine all images posted on Instagram account @thebodyshopindo.impact with caption #BringBackOurBottle. Based on the method of previous research [10], this research will analyze messages through content visualization on the images, videos, and narration with indicator of the number of "likes" on the uploaded content.

Symbolic Convergence Theory (SCT) introduced by Ernest Broomman explained about messages exchange process to be able to increase group awareness in similarity of meanings, motives and patterns. This theory assumed that through communication process "humans are able to unify their individual fantasy, dream, and meaning into shared symbol system" [11]. SCT by Ernest Bormann also explained the concept that shared awareness can be formed through message exchange [12].

## 3. Results and discussion

### 3.1. *The plastic threat for survival*

The first visualization theme is about description of the terrible threat of plastic for living things survival on the earth due to the impact of climate change crisis which can be potentially caused natural disaster. Presented visualization within the images describes how terrible is the environmental impact if humans cannot manage plastic waste properly. For example, the garbage that keeps piling up in the landfills to wastes which pollute the beach. The following is The Body Shop Indonesia's post on account @thebodyshopindo.impact with #BringBackOurBottle.

Figure 1 was posted on October 13$^{th}$, 21021, liked by 521 accounts with 11 comments. The post showed a shoreline is filled with plastic bottles of cosmetics and drinks, on this picture the rubbishes is carried by the waves along with seaweeds and washed up on the beach. As mentioned by the following narration: "Every year, around the world, there are about 120 billion beauty product packaging which piles up into rubbish, you know! Even, about 70% of them cannot be recycled and end up as rubbish in landfills. That's why I am committed to being responsible for used packaging through #BringBackOurBottles! You can look good, feel good, and do good! Sumber: Zero Waste Week". The narration emphasizes that every human has responsible to plastic packaging waste from their consumed product. By mentioning the fact based on the data obtained from Zero Waste Week, The Body Shop Indonesia calls audiences for being responsible to the environment through #BringBackOurBottle program. Visually, the presented picture is very terrible; the beach should be clean and natural yet looks the opposite, and it becomes serious threat for sustainability of marine ecosystems.

**Figure 1.** Plastic threat (Oct 13$^{th}$, 2021 | 521 Likes | 11 Comments)
Source: Instagram
@thebodyshopindo.impact

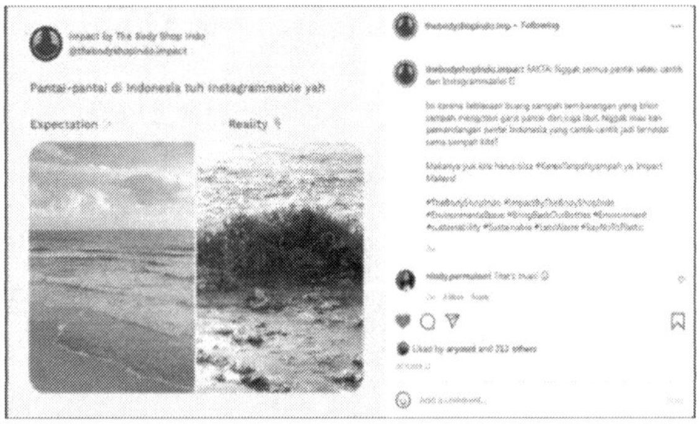


**Figure 2.** Plastic threat (Oct 12th, 2021 | 213 Likes | 1 Comment)
Source: Instagram
@thebodyshopindo.impact

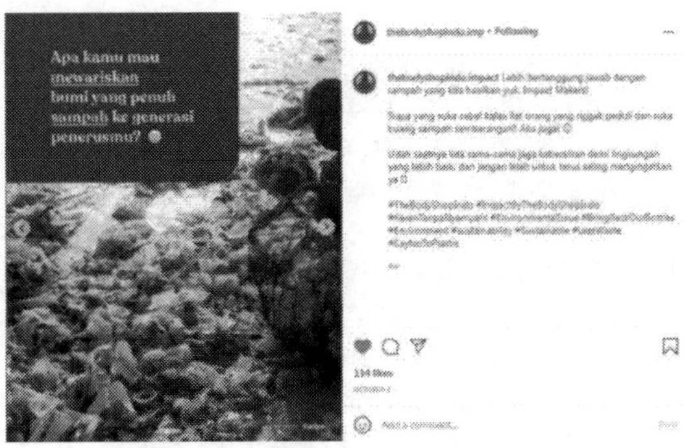


**Figure 3.** Plastic threat (Oct 2nd, 2021 | 114 Likes)
Source: Instagram @thebodyshopindo.impact

**Figure 4.** Plastic threat (June 5th, 2021 | 94 Likes)
Source: Instagram @thebodyshopindo.impact

Figure 2 was posted on October $12^{th}$, 2021, liked by 213 accounts with 1 comment. It is not different from previous image, its presented visualization showed the beach condition which we currently encounter a lot. An expectation of beach which is clean from rubbish and looks so beautiful is now inversely proportional; the waves also wash away plastic waste, which is increasingly difficult to control. The narration written on the post asks for us to give good impact for our environment. "FACT: Not all beaches are always beautiful and 'Instagrammable'! This is because of the habit of littering, which makes rubbish pollute the coastline and also the sea. You don't want the beautiful scenery of Indonesia's beaches to be stained with our rubbish, right? So let's be able to #KerenTanyampah, Impact Makers!"

Figure 3 described a future if we remain to have the same lifestyle. Rubbishes become heritage for future generations, "Let's be more responsible with the waste we produce, Impact Makers! Who likes to be annoyed when seeing people who don't care and like to litter? Me too! It's time for us to keep clean for a better environment, and don't be tired to keep reminding each other." The picture was uploaded on October $2^{nd}$, 2021 and liked by 114 accounts. The Body Shop Indonesia keeps reminding that the environmental damage threat caused by plastic wastes is not simple matter, it needs big commitment to maintaining cleanliness for the sake of better environment.

Figure 4 showed an increasingly terrifying visualization, the condition of landfill which continues to pile up. The content was uploaded on World Environment Day, June $5^{th}$, 2021 and received 94 likes. The picture has narration: "Happy World Environment Day, Impact Makers! Coinciding with today, I want to invite all of you to reduce the pile of rubbishes in the landfill, let's go! One of the ways is to return the empty packaging of The Body Shop to the store via #BringBackOurBottles. By doing this you can show how responsible you are to our earth. Who has routinely returned the packaging?" Through image visualization we are rebuked that plastic wastes are getting out of control, and small things have huge impacts. Thus, what we can do is have participated in #BringBackOurBottle program to recycle the packages becoming useful goods, and then it can reduce pile of rubbishes in landfill.

The environmental impact caused by plastic waste is really big; not only on land but the consequences also threaten marine life. Some old plastics decomposed/degraded in the sea without cytotoxic are able to release estrogenic compounds in the marine environment and dangerous for the sustainability of the oceans [13]. Therefore, due to marine plastic wastes becoming major concern, this research is mainly focused on the effect of plastic chemicals released to marine flora and fauna [14].

### 3.2. *The drive for sustainable living*

The second visualization theme is the drive for sustainable living; The Body Shop Indonesia calls it Sustainable Living. This action encourages us to recycle secondhand in order to, once more, make them be useful goods, and it emphasizes that the use of single-use items only end up being rubbish. In addition, this action is part of the #BringBackOurBottle program persuading people to change their sustainable lifestyle and give good impact for the environment. The following is the post of The Body Shop Indonesia on account @thebodyshopindo.impact with #BringBackOurBottle.

Figure 5 was uploaded on October 7th, 2021, got 117 likes with 7 comments. The narration of the post is "Crate old wood ➡ becomes stage ➡ becomes table, chair, and white board. This is one of implementations of circular economy which we are trying to do: make use of wood and empty package to be stage and it is transformed again into table, chair, and white board for @sekolahbisa students. So, nothing is wasted and is thrown into trash! In your opinion, the circular economy concept can be applied at home, right?" The Body Shop Indonesia invites us to do circular economy in which stuff which is not used can be recycled to be useful again. Plastic conventional wastes is main contributor of waste crisis, hence industries switch to circular economy model to reduce plastic wastes [15]. It is not only plastic bottles but other stuff that existed at home can also be modified in to other stuffs so that it has economic value again. This program, surely, become the program line of #BringBackOurBottle for the sake of better environmental and ecosystem sustainability.

Figure 6 got 119 likes with 7 comments and was uploaded on September 23th, 2021. Visually, the picture shows plastic waste products of The Body Shop cosmetic products. It includes narration "' After returning the empty packaging to store, where are they going?' Who are wondering? Well, I want to tell you the answer! I am with @ecollabo8 @ecobalirecycle and @waste4change processing the empty packaging that you return into Recycled Soap Dish and Recycled Pocket Mirror! You can find them again in the store! Yes, together we process and provide second chance for the empty packaging that you return to be a new item again. Psst... you can get it for free by shopping at least IDR 750,000, you know!" The Body Shop collaborates with other parties to recycle its product plastic package. It aims to let the consumers know where the returned cosmetics package bottles go. Besides, the Recycled Mirror and Recycled Soap Dish as the processed products are given as a reward with a minimum spend. The implementation of sustainable living is not only reducing the accumulation of waste in the landfill but can also encourage circular economy and increase the economic value of good/stuffs.

Figure 7 is a form of sustainable living implementation conducted by The Body Shop in one of its store located in Bandung. This plastic bottle wastes usage also have high artistic value. The content was posted on April 30th, 2021, liked by499 accounts with 11 comments. "We have a new outlet store in Bandung! Anyone know about Eco Stop? Here you can shop for your favorite products at stores which apply the concept of sustainability! There, you can see bench made of empty bottles returned through #BringBackOurBottles.

💚 There is electric power bike that you can try.

💚 Simple garbage sorting place that you can make at home.

💚 This store also uses solar power as its energy source!

Are you curious to directly see while shopping? Let's visit The Body Shop Eco Stop in Eco Camp Bandung, Jl. Pakar Barat no. 3, Bandung. If you've been there, don't forget to tag me and @thebodyshopindo! See you." The long narrative is part of the shop's photo upload with the concept of sustainability, not only chairs but this shop also has various environmentally friendly facilities and usages of second hand goods. This concept certainly gives inspiration to keep applying sustainable living started with simple things such as selecting garbage and joining #BringBackOurBottle program.

**Figure 5**. Sustainable living (Oct 7$^{th}$, 2021 | 117 Likes | 7 Comments)
Source: Instagram
@thebodyshopindo.impact

**Figure 6.** Sustainable living (Sept 23rd, 2021 | 199 Likes | 7 Comments)
Source: Instagram
@thebodyshopindo.impact

**Figure 7.** Sustainable living (April 30th, 2021 | 499 Likes | 11 Comments)
Source: Instagram
@thebodyshopindo.impact

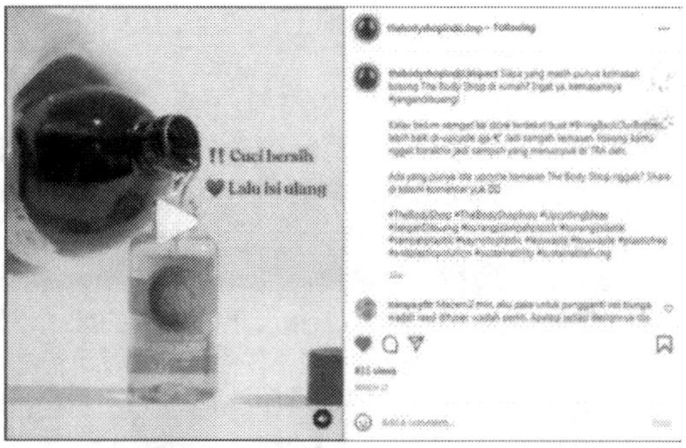


**Figure 8.** Sustainable living (March 17th, 2021 | 811 Likes | 2 Comments)
Source: Instagram
@thebodyshopindo.impact

Figure 8 is a beautiful packaged video containing one of simple ways which we can apply at home. It was uploaded on March 17$^{th}$, 2021, this content has succeeded to be liked by 811 accounts with 2 comments. The video explains that apart from joining the #BringBackOurBottle program, we can also directly apply the Sustainable Living concept by refilling used packages. "Who do still have The Body Shop empty packaging at home? Remember, don't throw the package away #jangandibuang! If you haven't had time to go to the nearest store for #BringBackOurBottles, it's better to up cycled it. So your empty packaging waste doesn't end up as garbage that piles up in the landfill. Does anyone have an idea for The Body Shop packaging up cycle? Or not? Let's share it on comment column." The narration of the content brings solution, for they who have not participated yet in the #BringbackOurBottle program, without adding the rubbishes to the landfill. Using secondhand stuffs does not mean unable to purchase new stuffs, yet it is a form to create better survival in the future by suppressing the development of plastic wastes which is getting bigger.

The Body Shop Indonesia, consistently, keeps holding #BringBackOurBottle program with unlimited time and keeps voicing it through social media. The campaign, step by step, is able to create environmentally friendly behavior of its social media followers; everything is started with increasing knowledge, the next is the growing sense of responsibility, and it leads to real action [16]. Management of plastic wastes can be conducted through three wide strategies; they are reducing plastic wastes, wastes management, and environment restoration [17]. Moreover, The Body Shop is making effort to keep implementing it.

## 4. Conclusions

The findings of this research showed that The Body Shop Indonesia makes plan through #BringBackOurBottle program by suppressing dominant message and educating the consumers to build climate crisis awareness which can cause disaster due to human action (man-made disaster). The program holds through social media Instagram account @thebodyshopindo.impact. The first message theme is Plastic threat for survival, and the second theme is the drive for sustainable living. This research also provided outlook about how campaign through social media can bring directly feedback and has a wide range, not only for the consumers but also for publics. A message, which sticks and becomes The Body Shop Indonesia value derived by Anita Roddick as the founder is "The Core of a Business should not be about Money and Profit. It is about Responsibility. A Business should Become Kindness for All people; It is not for Personal Satisfaction."

## References

[1] Aryanto V D W and Paramitadevi Y V 2018 Climate Change Social Marketing on Internet: Review of Interactivity to Promote Environmental Sustainability *E3S Web Conf.* **74** 1–4
[2] Wahyuni T 2015 Jenis Sampah dan Lama Proses Penghancurannya
[3] Vegter A C, Barletta M, Beck C, Borrero J, Burton H, Campbell M L, Costa M F, Eriksen M, Eriksson C, Estrades A, Gilardi K V K, Hardesty B D, Ivar do Sul J A, Lavers J L, Lazar B, Lebreton L, Nichols W J, Ribic C A, Ryan P G, Schuyler Q A, Smith S D A, Takada H, Townsend K A, Wabnitz C C C, Wilcox C, Young L C and Hamann M 2014 Global Research Priorities to Mitigate Plastic Pollution Impacts on Marine Wildlife *Endanger. Species Res.* **25** 225–47
[4] Strauss R 2018 14 Ways Plastic is Harming Us and the Environment
[5] Li W C, Tse H F and Fok L 2016 Plastic Waste in the Marine Environment: A Review of Sources, Occurrence and Effects *Sci. Total Environ.* **566–567** 333–49
[6] Pitt A 2021 Global Research and Innovation in Plastics Sustainability (GRIPS) 2021
[7] Purbohastuti A W 2017 Efektivitas Media Sosial sebagai Media Promosi *Tirtayasa Ekon.* **12** 212–31
[8] Page J T and Duffy M E 2018 What Does Credibility Look like? Tweets and Walls in U.S. Presidential Candidates' Visual Storytelling *J. Polit. Mark.* **17** 3–31
[9] Paul N and Perreault G 2018 The First Lady of Social Media: The Visual Rhetoric of Michelle Obama's Twitter Images *Atl. J. Commun.* **26** 164–79
[10] Perreault M F and Perreault G 2019 Symbolic Convergence in the 2015 Duggar Scandal Crisis Communication *J. Media Relig.* **18** 85–97
[11] Bormann E G 1980 *Communication Theory* (New York)
[12] McCabe J 2009 Resisting Alienation: The Social Construction of Internet Communities Supporting Eating Disorders *Commun. Stud.* **60** 1–16
[13] Kedzierski M, D'Almeida M, Magueresse A, Le Grand A, Duval H, César G, Sire O, Bruzaud S and Le Tilly V 2018 Threat of Plastic Ageing in Marine Environment. Adsorption/Desorption of Micropollutants *Mar. Pollut. Bull.* **127** 684–94
[14] Gunaalan K, Fabbri E and Capolupo M 2020 The Hidden Threat of Plastic Leachates: A Critical Review on Their Impacts on Aquatic Organisms *Water Res.* **184** 1–15
[15] Payne J, McKeown P and Jones M D 2019 A Circular Economy Approach to Plastic Waste *Polym. Degrad. Stab.* **165** 170–81
[16] Krisyanti K, Situmeang I V and Priliantini A 2020 Pengaruh Kampanye #PantangPlastik terhadap Sikap Ramah Lingkungan (Survei pada Pengikut Instagram @GreenpeaceID) *J. Komunika J. Komunikasi, Media dan Inform.* **9** 40
[17] Borrelle S B, Ringma J, Law K L, Monnahan C C, Lebreton L, McGivern A, Murphy E, Jambeck J, Leonard G H, Hilleary M A, Eriksen M, Possingham H P and Rochman C M 2020 Mitigate Plastic Pollution *Science (80-. ).* **369** 1515–8

# Optimization of the utilization of blood meal with palm oil fronds absorbance fermented as feed stuff to changes of nutrition content

**Afnur Imsya [1]*, Riswandi [1] and Yaqub[2]**

[1]Animal science departement Agriculture Faculty Sriwijaya University, Palembang Prabumulih Street Inderalaya Ogan Ilir District, Indonesia
[2]Agronomy Departement Agriculture Faculty Sriwijaya University Jln Raya Palembang Prabumulih Inderalaya Ogan Ilir Distric
*Corresponding author: aimsya@yahoo.com

**Abstract**. The utilization of agricultural and livestock waste by-products, apart from being an alternative feed ingredient, can also reduce environmental pollution. Palm fronds as a by-product of oil palm plantations are rarely utilized and most of them will be burned and couraging climate change phenomenon. Cattle blood is a by-product of abattoirs that directly impact the environment. Most of the cattle blood not used efficiently, and pollutes the environment. The research aims to increase the use of blood meal with the absorbance of palm fronds fermented as feedstuff on changes in nutritional content. The study used a completely randomized design with a factorial pattern. Factor A is the inoculant treatment, namely A1: *Bacillus amylolytic flavacient* bacteria A2: *Lactobacillus plantarum* bacteria, factor B is the incubation time, namely: B1 = 0 hours; B2 = 60 hours and B3 = 120 hours., each treatment was repeated 3 times respectively. Parameters observed were nutrient content based on proximate analysis The results showed that there was interaction between the type of inoculant and the incubation time used in fermenting blood meal with the absorbance of palm fronds. Incubation time had a significant effect ($P<0.05$) on the content of dry matter, organic matter, crude fiber and crude protein, while the incubation time had no significant effect on ($P>0.05$) crude fat content. Based on the results of the study, it was found that in blood meal with absorbent palm oil fronds, *Lactobacillus plantarum* inoculants with an incubation times of 120 days gave the best results in blood meal with absorbance palm oil fronds fermented.

**Key words:** Blood meal, fermented, palm oil frond, Lactobacillus plantarum, Amillolitic flavacient, waste

## 1. Introduction

In Indonesia, like most other developing countries in the tropics, the use of by-products from agricultural waste and livestock waste as alternative feed ingredients has been widely used. The utilisation of agricultural and live stock waste by-products, apart from being an alternative feed ingredient, can also reduce environmental pollution. Palm fronds as a by-product of oil palm plantations are rarely utilized and most of them will be burned and couraging climate change phenomenon. Cattle blood is a by-product of abattoirs that directly impact the environment. Most of the cattle blood not used efficiently, and pollutes the environment. Likewise, cattle blood is a by-product of abattoirs that has a direct impact on the environment. Most of the blood from large numbers of beef cattle is wasted, not used efficiently, and pollutes the environment [1].

Palm fronds have the potential as a source of fiber feed for ruminants, with nutritional content consisting of dry matter (BK) 88.14%, crude protein (PK) 5.28%, Neutral Detergent Fiber (NDF) 65.59%, Acid Detergent Fiber (ADF) 52.72 %, Hemicellulose 12.87%, Cellulose 27.79%, and Lignin 25.42%, the use in rations can reach 40% substitution of native grass after fermenting with *Phanerochaetae chrysosporium* [2]. Blood meal, obtained from cattle blood after drying, contains 80% - 90% crude protein, with a high content of essential amino acids, especially lysine [3]. Constraints on the use of blood meal are the difficulty in the drying process, it will be a good medium for microbial growth when humidity is high, while drying using sunlight takes a long time for 3 days [4]. Several previous studies have applied absorbent technology derived from agricultural waste to speed up the process of drying cattle blood. Research on the processing of blood meal by using absorbent corn waste and obtained the results of the drying process of cattle blood faster and increasing the nutritional content of corn waste as animal feed [5].

The use of oil palm fronds as absorbent for processing cattle blood has never been studied, therefore this study will examine the role of oil palm fronds as absorbent of blood meal and the fermentation process with different inoculants as feed ingredients for ruminants. The purpose of this study was to examine the fermentation process with different inoculants and incubation times on blood meal with palm frond absorbance on changes in nutrients content.

## 2. Materials and method

### *2.1 Materials*

Material of the research: Palm oil frond, blood meal, *Bacillus Amilolityc flavacient, Lactobacillus plantarum,* rice brands, material for proximate analysed

### *2.2 Methods*

The study used a completely randomized design with a factorial pattern. Factor A is the inoculant treatment, namely A1: *Bacillus amylolytic flavacient* bacteria A2: *Lactobacillus plantarum* bacteria, factor B is the f incubation time, namely: B1 = 0 hours; B2 = 60 hours and B3 = 120 hours., each treatment was repeated 3 times respectively. Parameters observed were nutrient content based on proximate analysis (Dry matter, organic matter, crude protein, crude fiber and crude fiber)

### *2.3 Research implementation*

*2.3.1. Inoculant preparation.* The inoculant used was 100g rice bran, sterilized using an autoclave for 15 minutes at 121oC, 1 atm pressure, then cooled to room temperature of 24°C. Ten millilitres of distilled water was put into a petri dish that had been grown with pure cultures of *Bacillus Amilolityc flavacient* and *Lactobacillus plantarum* bacteria and shaken to mix. The mixture of distilled water and bacteria was entered to an Erlenmeyer tube which already contained 190 ml of aquadest and then it mixed into the sterilized rice bran media and incubated for 24 hours. The inoculants were dried at 60°C.

*2.3.2 Processing of blood meal with palm oil fronds absorbent.* Processing of blood meal with absorbent refers to Makinde and Sonaiya [5] (modified method). The palm fronds were chopped and air-dried under the sun for 24 hours. The absorbent was mixed with fresh blood in a ratio of 1:1 (w/w). To prevent coagulation, the blood is mixed with 18g/litre of blood salt. The blood and absorbent mixture was dried for 3-4 hours. The resulting mixture is then mixed with blood in a ratio of 5: 4 (w/w) and fermented.

*2.3.3. Fermentation process.* Three percent inoculant was mixed with blood meal and absorbent, incubated according to the treatment, namely 0 hours, 60 hours and 120 hours at 40°C. After the incubation period was complete, the fermented material was dried at 60°C for 24 hours. Fermentation results in the analysis of nutritional content with proximate analysis. The result of Fermentation analysed with proximate analysis.

*2.4 Statistical analysis*
Data were analyzed as completely randomized design using Factorial Pattern the General Linear Means procedure of SAS (2000) for analysis of variance (ANOVA). Differences between treatment were resolved by Duncan's multiple range test of the SAS statistical package.

## 3. Results and discussion
Based on statistical analysis, it was found that there was an interaction (P<0.05) between the kind of inoculant and the incubation time on changes in the nutritional contents of blood meal and absorbance of palm fronds fermented, namely the content of dry matter, organic matter, crude protein and crude fiber. Based on further tests on dry matter content, it showed that incubation time had a significant effect (P<0.05) with the using *of Bacillus Amilolityc flavacient*, while incubation time had no effect (P>0.05) for *Lactobacillus plantarum* on dry matter content (Figure 1).

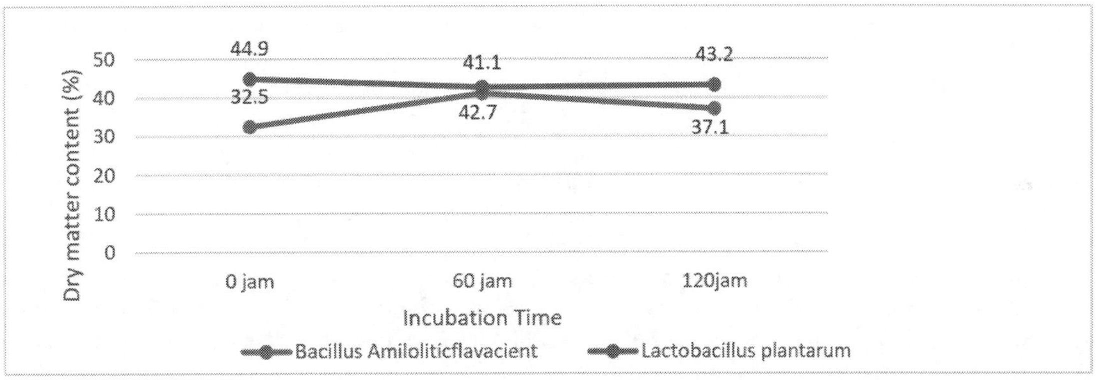


**Figure 1.** Dry matter content of blood meal with palm frond absorbance fermented

The dry matter content increased by 31.39% with the use of Bacillus Amilolityc flavacient inoculants and an incubation time of 60 hours compared to before fermentation. The dry matter content at an incubation time of 120 hours decreased by 13.11% compared to the dry matter content at an incubation time of 60 hours.

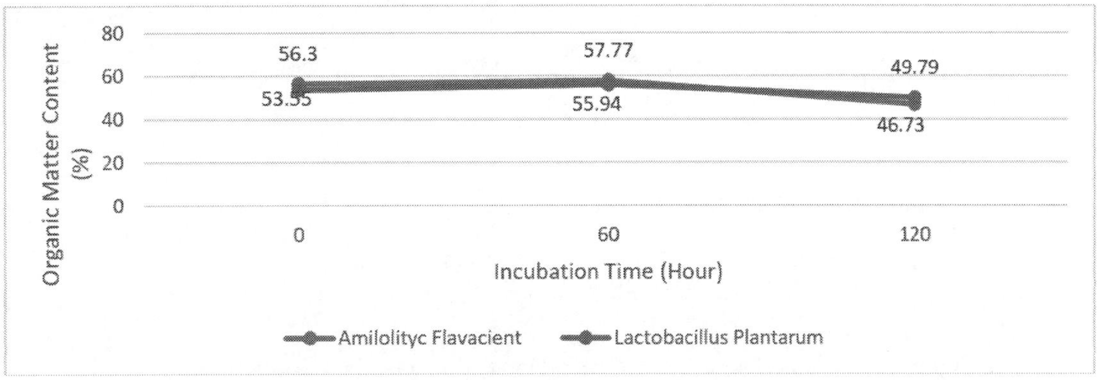


**Figure 2.** Organic matter content of blood meal with absorbent palm fronds fermented

Incubation time had a significant effect (P<0.05) on the organic matter content of blood meal with absorbant palm oil fronds in both kind of inoculants (Figure 2). During the incubation period of 60 hours, there was an increase in organic matter by 2.61% for *Amilolityc flavacient* inoculants and 4.46% for *Lactobacillus plantarum* inoculants compared to incubation time of 0 hours. At the incubation time 120 hours there was a significant decrease in organic matter (P<0.05) 7.02% with the using of

*Lactobacillus plantarum* and 16.99% for the using of *Amilolityc flavacient* inoculant compared to the incubation time of 0 hours. The kind of inoculant not significant effect (P>0.05) on changes of the organic matter content of blood meal with the absorbant palm oil fronds fermented. Changes in the content of dry matter and organic matter in all treatments were caused by the fermentation process of blood meal with the absorbance palm fronds resulting in changes of nutrients such as crude protein and crude fiber. Crude protein and crude fiber are part of dry matter and organic matter, as in the nutrient scheme based on proximate analysis [6]. Organic matter is part of the dry matter, while organic matter is composed of proteins, carbohydrates, fats and vitamins [7].

Crude protein in blood meal with absorbant palm oil fermented increased significantly (P<0.05) until an incubation time of 120 hours in both *Amilolityc flavacient* and *Lactobacillus plantarum* inoculants (Figure 3). The increase of crude protein was due to the higher inoculant activity with increasing incubation time. The Increasing of inoculant activity resulted in increased enzymes, thereby increasing crude protein. The research of [8] showed that at an incubation time of 120 hours, fermented blood meal with *Amilolityc flavacient* with coconut pulp absorbent resulted in an increase in crude protein. The use of blood meal with corn waste absorbance also resulted in an increase in crude protein [9].[10]stated that in the fermentation process, inoculants form like is bacteria, yeast or fungi are a source of protein for the substrate, in addition to carbohydrate degradation in the fermentation process will result in an increasing of crude protein [11]. Saxena and Singh [12] stated that there was an increase in protein content as the fermentation time increased. [13] also reported that the results of pods fermentation using *Bacillus subtilis* with 48 hours can increase protein up to 1.4 times which can be related to the synthesis time of microbial enzymes for substrate decomposition.

The use of *Lactobacillus plantarum* inoculants showed an increase in crude protein content which was higher than *Amilolityc flavacient* inoculants, especially at an incubation time of 60 hours. it was suspected that *Lactobacillus plantarum* activity was higher than *Amilolityc flavacient* on fermentation of blood meal with absorbent palm oil fronds

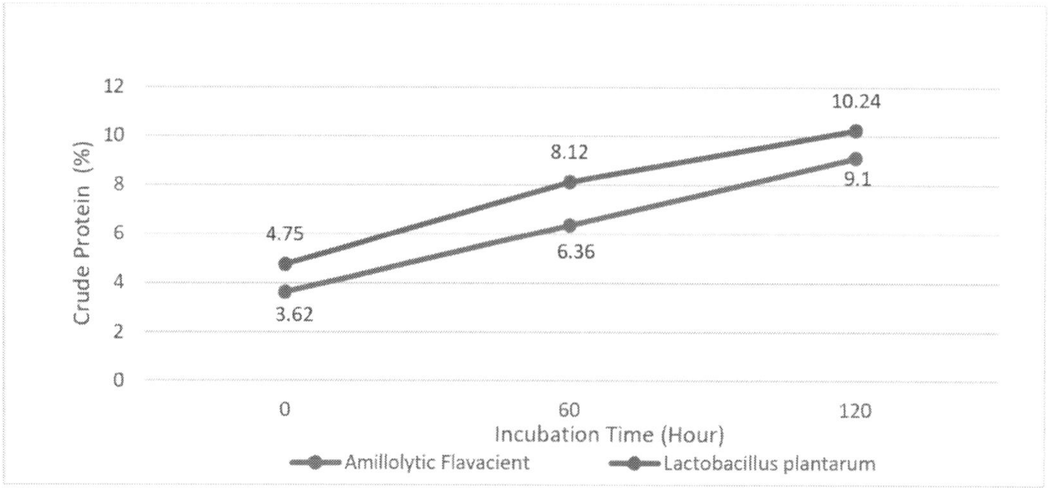


**Figure 3.** Crude protein content of blood meal with absorbent palm oil fronds fermented

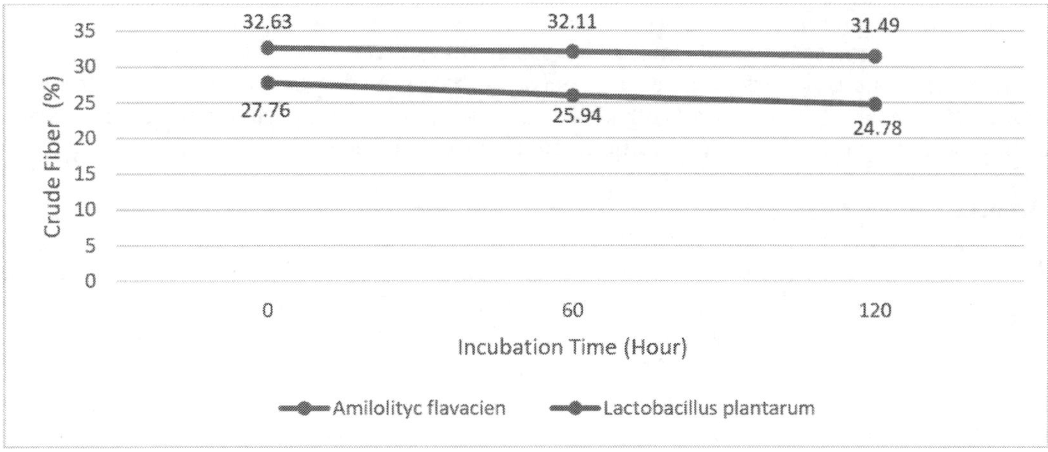


**Figure 4**. Crude fiber content of blood meal with absorbent palm oil fronds Fermented

Parameters of crude fiber content showed a significant decrease (P<0.05) until the incubation time of 120 hours (Figure 4). This decrease was due to the activity of cellulase and hemicellulose enzymes produced by *Amilolityc flavacient* bacteria and *Lactobacillus plantarum*. Cellulase and hemicellulase enzymes produced by inoculants will degrade crude fiber into simple carbohydrate compounds such as glucose. Simple carbohydrate compounds will be utilized by inoculants as an energy source. Amylolytic flavacient bacteria are cellulolytic bacteria that can degrade crude fiber because they produce cellulase and hemicellulase enzymes [14]. The longer of incubation time, the more enzymes will be produced so that the level of degradation of crude fiber will increase. Ramadhan et al. [15] stated that the length of fermentation can reduce crude fiber because the opportunity for enzymes to degrade crude fiber, especially cellulose and hemicellulose, becomes more perfect with increasing fermentation time. According to Chiang et al. [16] also stated that fermentation treatment with cellulosic bacteria can reduce crude fiber content. The decrease in crude fiber at incubation time of 120 hours for *Amilolityc flavacient* inoculants was 3.49% compared to incubation time of 0 hours. *Lactobacillus plantarum* inoculants resulted in a decrease in crude fiber of 10.73% at 120hour incubation time. Based on the level of decrease in crude fiber during the same incubation period, the activity of Lactobacillus plantarum inoculants indicated that they were more active than the *Amilolityc flavacient* inoculants.

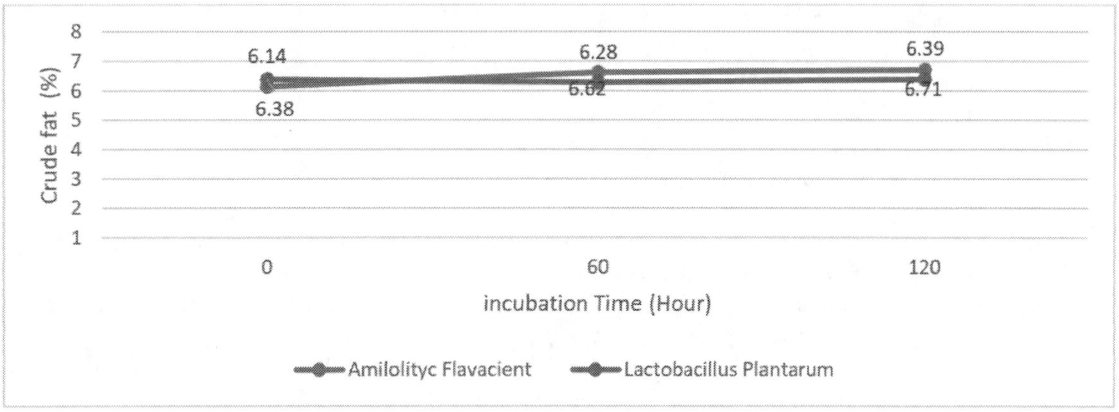


**Figure 5.** Crude fat content of blood meal with absorbent palm oil fornds fermented

There was not significant effect (P>0.05) of incubation time and type of inoculant on crude fat content (Figure 5), it was because Lactobacillus plantarum and Amilolityc flavacient inoculants did not have lipase enzyme activity to degrade fat content in blood meal with absorbent palm fronds. The average fat content of blood meal with absorbance of palm fronds fermented with Lactobacillus plantarum inoculants was 6.35% and Amilolityc flavacient inoculants were 6.71%.

## 4. Conclusion

Based on the results of the study, it was found that there was a reduction in pollution from livestock waste such as blood and agricultural waste due to increased efficiency in the use of blood meal and palm frond waste as animal feed. The blood meal with absorbent palm oil fronds, *Lactobacillus plantarum* inoculants with an incubation time of 120 days gave the best results in blood meal with absorbance palm oil fronds fermented. The nutritional contents of blood meal with absorbance of palm oil frond fermented with Lactobacillus plantarum, incubation time of 120 hours is dry matter 43.20%, organic matter 49.79%, crude protein 10.24%, crude fiber 24.78% and crude fat 6.71%.

## References

[1] Makinde O A and Sonaiya E B 2007 Determination of Water, Blood and Rumen Fluid Absorbencies of Some Fibrous Feedstuffs *Livest. Res. Rural Dev.* **19**

[2] Imsya A, Laconi E B, Wiryawan K G and Widyastuti Y 2013 In Vitro Digestibility of Ration Containing Different Level of Palm Oil Frond Fermented with Phanerochaetae chrysosporium *Media Peternak.* **36** 131–6

[3] NRC (National Research Council) 2015 *Nutrient Requirements of Beef Cattle, 8th Revised Edition* (Washington, DC: National Academy Press)

[4] Odunsi A A, Akingbade A A and Farinu G O 2004 Effect of Bovine Blood-Rumen Digesta Mixture on Growth Performance, Nutrient Retention and Carcass Characteristics of Broiler Chickens *J. Anim. Vet. Adv.* **3** 662–6

[5] Makinde O A and Sonaiya E B 2010 A Simple Technology for Production of Vegetable-Carried Blood or Rumen Fluid Meals from Abattoir Wastes *Anim. Feed Sci. Technol.* **162** 12–9

[6] AOAC 2019 *Official Methods of Analisys* ed G W and L Jr

[7] Tilman, H.D., H. Haetadi, S. Reksohardiprojo S P and S L 2006 *Livestock Food Science Basis* (Yogyakarta: Mada University Press)

[8] Ramadhan R F, Wizna, Marlida Y and Mirzah 2016 Fermentation of Blood Meal with Bacillus Amyloliquefaciens as Broiler Feeding *Asian J. Anim. Vet. Adv.* **11** 840–6

[9] Makinde O A and Sonaiya E B 2011 Utilization of Sun-Dried Maize Offal with Blood Meal in Diets for Broiler Chickens *Open J. Anim. Sci.* **01** 106–11

[10] Esonu B O, Azubuike A B ., Udedibie O ., Emenalom T C and Odoemenam I 2011 Evaluation of the Nutritive Value of Mixture of Fermented Bovine Blood and Rumen Digesta for Broiler Finisher *J. Nat. Sci. Res* **1** 12–26

[11] Xu F, Li L, Xu J, Qian K, Zhang Z and Liang Z 2011 Effects of Fermented Rapeseed Meal on Growth Performance and Serum Parameters in Ducks *Asian-Australasian J. Anim. Sci.* **24** 678–84

[12] Saxena R and Singh R 2011 Characterization of a Metallo-Protease Produced in Solid State Fermentation by a Newly Isolated Bacillus Strain *Acta Biol. Szeged.* **55** 13–8

[13] Sarasvati S, Sujata B, Amita S and Doshi B R 2014 *Effects of Fermentation on Nutritional Quality of Prosopis Juliflora Pods as Alternative Fish Feed*

[14] Wizna, Abbas H, Rizal Y, Dharma A and Putu Kompiang I 2007 Selection and Identification of Cellulase-Producing Bacteria Isolated from the Litter of Mountain and Swampy Forest *Microbiol. Indones.* **1** 135–9

[15] Ramadhan R F, Marlida Y and Wizna D 2015 Method of Cattle Blood Processing in Producing Powder for Poultry Feed Component: A Review *J. Peternak. Indones. Februari* **17** 63–76

[16] Chiang G, Lu W Q, Piao X S, Hu J K, Gong L M and Thacker P A 2010 Effects of Feeding

Solid-State Fermented Rapeseed Meal on Performance, Nutrient Digestibility, Intestinal Ecology and Intestinal Morphology of Broiler Chickens *Asian-Australasian J. Anim. Sci.* **23** 263–71

**Acknowledgement**

This research of this article was funded by DIPA of public service agency of Universitas Sriwijaya 2021 SP-DIP-023.17.2.677515/2021 on November 23, 2020 in Accordance with the Rector's Decree number :0010/UN9/SK.LP2M.PT/2021, on April 28, 2021

# Pollution load capacity of the Larangan/Premulung River Sukoharjo Regency, Central Java Province, Indonesia in 2020

**Purwo Edi Minarno*[1], Agus Suprapto [2] and Harsono [3]**

[1]Master Program of Environmental Science, Post Graduate School, Sebelas Maret University, Indonesia
[2]Environmental Services, Sukoharjo, Indonesia
[3]PT. Karsa Buana Lestari, Jakarta, Indonesia
*Corresponding author: pewedhi@gmail.com

**Abstract**. Climate change has a negative impact on surface water quality. This is influenced by the increase in human activities in the use of water. The resulting water output is discharged directly into water bodies and can increase the pollutant load of river water. One of the recipients of the burden of surface water pollution is the Larangan/Premulung River, which is indicated to have decreased quality due to liquid waste originating from domestic, industrial, and agricultural activities. The purpose of this study was to determine the load carrying capacity of water pollution in the Larangan River in the upstream, middle, and downstream parts. The water samples taken amounted to 10 sample points. The data was collected in the form of physical parameters (flow rate, water temperature, and TSS) and chemical parameters (pH, COD, BOD, and DO). The method of calculating the carrying capacity of the pollution load uses the mass balance method. Data analysis used quantitative descriptive analysis. The results showed that the carrying capacity of the Larangan/Premulung River pollution load: (1) the upstream part could no longer accept the pollutant load in the moderate to the heavy category in the quality standard class II; (2) the middle and downstream parts for COD and BOD parameters are no longer able to accept the pollutant load in the medium to large category.

**Keywords:** river, pollution, capacity, mass balance

## 1. Introduction

Climate change has a negative impact on surface water quality [1]. This is influenced by the increase in human activities in the use of water. Water as a component of the environment will affect and be influenced by other components [2]. If water quality declines, it will reduce usability, usability, productivity, carrying capacity and capacity and will reduce the wealth of natural resources. The resulting water output is discharged directly into water bodies and can increase the pollutant load of river water [3].

One of the recipients of the burden of surface water pollution is the Larangan/Premulung River, which is administratively located in the Kartasura District, Sukoharjo Regency [4]. Human activities that reduce the water quality of the Larangan/Premulung River, are caused by disposing of liquid waste into the river body. These activities are indicated from six sources, namely: (1) home industry activities, namely the tofu making industry; (2) educational activities; (3) agricultural activities using pesticide fertilizers; (4) hospital activities; (5) livestock activities; and (6) household activities [5].

Population growth in the research location in the period 2010 to 2019 has increased along with the increase in population. In 2020 Kartasura District is the most populous sub-district in Sukoharjo district, with a density of 6,884 people per km$^2$. The increase in population will have an impact on providing more basic needs, so that it will put greater pressure on the environment, one of which is on the function of rivers as surface water containers [6].

The Larangan/Premulung River is designated as a river that has class II water quality, namely water whose designation can be used for water recreation infrastructure/facilities, freshwater fish cultivation, animal husbandry, water for irrigating crops, and/or other designations that require the same water quality as the intended use. The purpose of this study was to determine the load carrying capacity of water pollution in the Larangan River in the upstream, middle, and downstream parts. Which is useful for supporting activities according to the designation of class II river quality standards

## 2. Materials and method

### 2.1. Materials

The tools and materials used in the study of the capacity of the Larangan/Premulung River are are takers samples, including: buckets, dippers, 1 l water sampling equipment, vertical and horizontal water samples), glass bottles/glass sizes 1 – 5 liters, plastic bottles, do bottles, and cool boxes/boxes. Then measuring instruments, namely pH meter, digital burette, spectrophotometer, coliform test, conductivity/TDS meter, GPS, current meter, echo meter. The sample results are then carried out with an accredited laboratory test.

### 2.2. methods

Determination of the location and point of the river water sample as well as the sampling procedure is based on the Sampling Method of the Indonesian National Standard Water Quality Test (SNI) Number SNI 06-2421-1991 [7]. Determination of the carrying capacity of the pollution load can be determined in a simple way, namely by using the mass balance method [8].

### 2.3. Subtitles

Mathematical models that use mass balance calculations can be used to determine the average concentration of down stream flows originating from point sources and non-point sources of pollutant sources. This calculation can also be used to determine the percentage change in flow rate or pollutant load [9]. Mass balance analysis with the equation (1):

$$C_R = \frac{\sum Ci\, Qi}{\sum Qi} = \frac{\sum Mi}{\sum Qi} \tag{1}$$

Where:

| | |
|---|---|
| $C_R$ | = average concentration of constituents for combined flow |
| Ci | = constituent concentration in the i-th stream |
| Qi | = flow rate to-i |
| Mi | = mass of constituents in the i-th stream |

The data used in the preparation of the study were collected through survey methods and laboratory tests. Data analysis with quantitative descriptive analysis is to describe the description of chemical and biological testing. Samples were taken at 10 points, then 10 water samples were tested in the laboratory, the results were presented in the form of numerical values and graphs, and the characteristics of each data parameter were compared. Unit of location with other parameters/locations, or between upstream, middle, and downstream watersheds. Map of the study area can be seen in Figure 1.

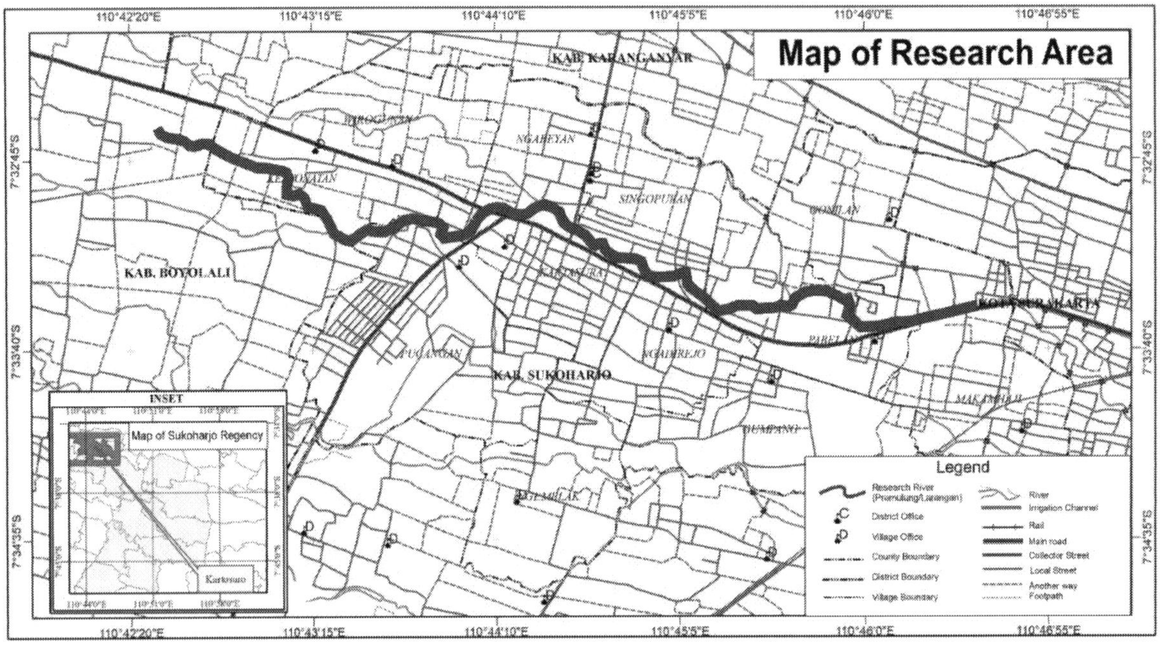


**Figure 1.** Map of the study area

## 3. Results and discussion

### *3.1. Channel debit*

The mean main river discharge was greatest in the measurement of the downstream river sample (S4) and the lowest average river discharge in the measurement of the upstream river sample (S1). Then for the average discharge channel/flow the lowest pollutant source is at sample point L6 (Domestic Waste) with an average discharge of 0.001 $m^3$/second, while the highest discharge is at sample point S4 (River Downstream) with an average discharge of 2.401 $m^3$/sec.

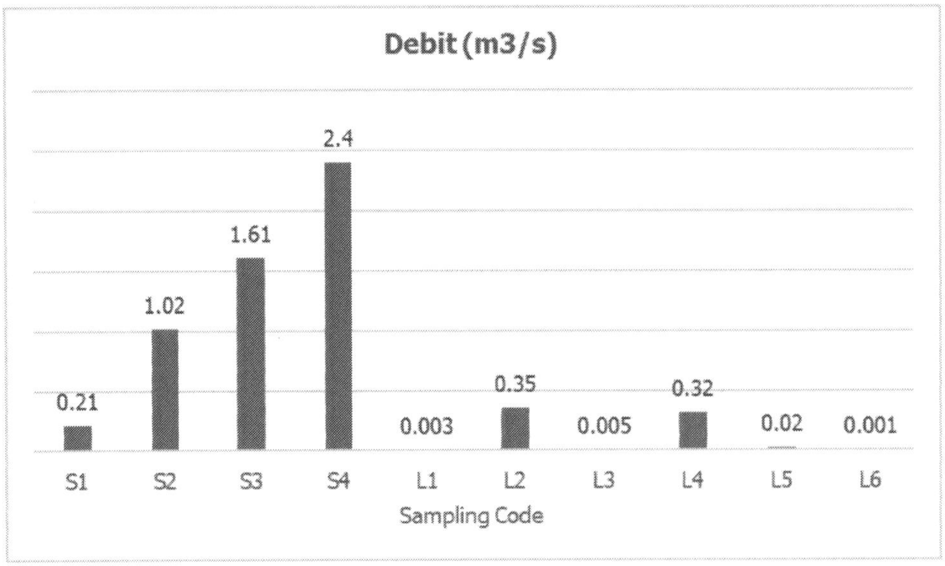


**Figure 2.** Study area water discharge

### 3.2. *Main river water quality*

Water quality measurements carried out in the main river body obtained the following results:

**Table 1.** Main river water quality.

| Parameter | Unit | Nama Lokasi | | | | Quality standards * |
|---|---|---|---|---|---|---|
| | | Upstream | Middle 1 | Middle 2 | Downstream | |
| Debit | $m^3$/sec | 0.21 | 1.02 | 1.61 | 2.40 | |
| pH | - | 8 | 7 | 7 | 8 | 6.0-9.0 |
| COD | mg/L | 27 | 51 | 37 | 61 | 25 |
| BOD | mg/L | 3.4 | 17 | 16 | 11 | 3 |
| TSS | mg/L | 51 | 12 | 10 | 104 | 50 |
| Total Coliform | Total Coliform/100 ml | 5.840x$10^2$ | 8.880x$10^2$ | 9.480x$10^2$ | 9.840x$10^2$ | 5,000 |

Note:
: Exceeding the quality standard
* : Ministerial Regulation No. 82 of 2001 Class II concerning Water Quality Management and Water Pollution Control

From table 1, the Chemical Oxygen Demand (COD) parameter for all samples has exceeded the quality standard. The condition of the samples of the highest downstream segment has exceeded the quality standard of 61 of the quality standard value of 25. The Biological Oxygen Demand (BOD) parameter for all samples has exceeded the quality standard. Parameters of Total Suspended Solid/TSS obtained the results that there are 2 sample locations that have exceeded the quality standard, namely upstream and downstream. Parameters Total coliform for all samples has exceeded the quality standard [10]. Total coliform or coliform density is an early indicator of bacteria that is used to determine whether it is safe for consumption. Total coliform includes bacteria that can be found in soil and water environments that have been affected by human and animal waste disposal.

### 3.3. *Pollution source water quality*

The quality of the pollutant load is for the pH parameter, it is known that the tofu waste industry is not in accordance with the specified pH level, with a value of 5.2. Chemical Oxygen Demand (COD) parameters obtained from measurements of pig waste and agricultural waste have exceeded the quality standard. Pig waste has exceeded the quality standard of 42.9, then for agricultural waste the result is 33.4. The Biological Oxygen Demand (BOD) parameter at the measurement of 3 locations exceeds the quality standard, including pig, agricultural and domestic 2. The Total Suspended Solids (TSS) parameter shows that in segment 1 which includes tofu waste and pig livestock waste that has exceeded the quality standard specified, for other samples have not exceeded the quality standard. Parameters of Total Coliform obtained the results of 2 activities that have exceeded the quality standard, namely domestic activities 1 and 2. The quality of wastewater at pollutant sources is presented in table 2.

**Table 2.** Pollutant source water quality.

| Measurement Location | Debit (m3/ detik) | Test Results/Parameters | | | | | | | | | |
|---|---|---|---|---|---|---|---|---|---|---|---|
| | | pH | | COD (mg/L) | | BOD (mg/L) | | TSS (mg/L) | | Total Coliform (/100 ml) | |
| | | TR | QS | TR | QS | TR | QS | TR | QS | TR | QS |
| Tofu Industry ** | 0.9 | 5.2 | 6-9 | 219 | 275 | 131 | 150 | 492 | 100 | 2.380 x$10^3$ | - |
| Pig Cattle * | 0.354 | 6.5 | 6-9 | 42.9 | 25 | 10 | 3 | 268 | 50 | 2.380 x$10^3$ | - |
| Domestic 1 *** | 0.005 | 8 | 6-9 | 75.5 | 100 | 22 | 30 | 18 | 30 | 3.390 x$10^3$ | 3000 |
| Agriculture * | 0.316 | 7.2 | 6-9 | 33.4 | 25 | 8,7 | 3 | 14 | 50 | 1.044 x$10^3$ | - |
| Hospital **** | 21 | 7.1 | 6-9 | 24 | 80 | 13 | 30 | 7 | 30 | 1.480 x$10^2$ | - |
| Domestic 2 *** | 0.001 | 7 | 6-9 | 73.9 | 100 | 32 | 30 | 26 | 30 | 8.000 x$10^2$ | 3000 |

Note:
(shaded) : Exceeding the quality standard
TR : Test results
QS : Quality standards
* : Central Java Regional Regulation No. 5 of 2012 Appendix IX, Wastewater whose quality standards have not been determined (Gol. I)
** : Central Java Regional Regulation No. 5 of 2012 Appendix I, No. 30 Tofu Industrial Wastewater
*** : LHK Ministerial Decree No. 68/Menlhk/Setjen/Kum.1/8/2016 Attachment I of the Ministerial Regulation on Domestic Wastewater Quality Standards
**** : Central Java Regional Regulation No. 5 of 2012 Appendix IV Hospital Wastewater

### *3.4. Pollutant load capacity*

Determination of the carrying capacity of direct/ban River water pollution loads using three parameters: Dissolved Oxygen (DO), Chemical Oxygen Demand (COD) and Biological Oxygen Demand (BOD). Determination of River Class is determined by Government Regulation Number 82 of 2001 concerning Water Quality Management and Water Pollution Control.

*3.4.1. Upstream river capacit.* The condition of the pollutant load carrying capacity in the upstream is that the concentrations of COD, BOD and TSS when compared with the class II river quality standard have exceeded the quality standard, while with the class III quality standard only the BOD concentration has exceeded the quality standard (Table 3).

**Table 3.** Average concentration of each parameter upstream.

| Measurement Location | Code | Debit (m/dtk) | COD (mg/L) | BOD (mg/L) | TSS (mg/L) | $C_{R\,COD}$ | $C_{R\,BOD}$ | $C_{R\,TSS}$ |
|---|---|---|---|---|---|---|---|---|
| Upstream | S1 | 0,2139 | 27 | 3,4 | 51 | 37.89 | 8.18 | 187.8668 |
| Tofu Industry | L1 | 0,0031 | 219 | 131 | 492 | | | |
| Pig Cattle | L2 | 0,3535 | 42,9 | 10 | 268 | | | |
| Middle 1 | S2 | 1,0158 | 51 | 17 | 12 | | | |
| Quality standards | | Class II | | | | 25 | 3 | 50 |
| | | Class III | | | | 50 | 6 | 400 |

Source: Calculations, 2020

*3.4.2. Middle river capacity.* The condition of the carrying capacity of the pollutant load in the middle, that the concentrations of COD and BOD when juxtaposed with the class II river quality standard have exceeded the quality standard, while with the class III quality standard only the BOD concentration has exceeded the quality standard. However, the TSS concentration has not exceeded the class II and class III quality standards (Table 4).

**Table 4.** Average concentration of each parameter middle.

| Measurement Location | Code | Debit (m/dtk) | COD (mg/L) | BOD (mg/L) | TSS (mg/L) | $C_{R\ COD}$ | $C_{R\ BOD}$ | $C_{R\ TSS}$ |
|---|---|---|---|---|---|---|---|---|
| Middle 1 | S2 | 1.0158 | 51 | 17 | 12 | 46.93 | 15.06 | 12.49 |
| Domestic 1 | L3 | 0.0049 | 75.5 | 22 | 18 | | | |
| Agriculture | L4 | 0.3158 | 33.4 | 8.7 | 14 | | | |
| Middle 2 | S3 | 1.6105 | 37 | 16 | 10 | | | |
| Quality standards | | Class II | | | | 25 | 3 | 50 |

Source: Calculations, 2020

*3.4.3. Downstream river capacity.* The condition of the carrying capacity of the pollutant load in the downstream, that the concentration of COD and BOD when juxtaposed with the river quality standard class II has exceeded the quality standard, while with the class III quality standard only the concentration of BOD has exceeded the quality standard. However, the TSS concentration has not exceeded the class II and class III quality standards (Table 5).

**Table 5.** Average concentration of each parameter downstream.

| Measurement Location | Code | Debit (m/dtk) | COD (mg/L) | BOD (mg/L) | TSS (mg/L) | $C_{R\ COD}$ | $C_{R\ BOD}$ | $C_{R\ TSS}$ |
|---|---|---|---|---|---|---|---|---|
| Middle 2 | S3 | 1.6105 | 37 | 16 | 10 | 36.86 | 15.98 | 9.98 |
| Hospital | L5 | 0.0206 | 24 | 13 | 7 | | | |
| Domestic 2 | L6 | 0.0014 | 73,9 | 32 | 26 | | | |
| Downstream | S4 | 2.4013 | 61 | 11 | 104 | | | |
| Quality standards | | Class II | | | | 25 | 3 | 50 |
| | | Class III | | | | 50 | 6 | 400 |

Source: Calculations, 2020

*3.5. Determinant parameter of pollutant load carrying capacity*

Table 6 shows the determinant parameter of pollutant load carrying capacity.

**Table 6.** Determinant parameter of pollutant load carrying capacity

| No | Segment | Determinant Parameter of Pollutant Load Carrying Capacity (Class) | | | | | |
|---|---|---|---|---|---|---|---|
| | | COD | | BOD | | TSS | |
| | | II | III | II | III | II | III |
| 1 | Upstream | -/S | +/S | -/B | -/S | -/B | +/B |
| 2 | Middle | -/S | +/S | -/B | -/B | +/B | +/B |
| 3 | Downstream | -/S | +/S | -/B | -/B | +/B | +/B |

Note:

- \- : Premulung River water no longer accepts the burden of pollution
- \+ : Premulung River water is still able to accept the pollutant load
- K : Small means the comparison of the calculated value compared to the quality standard < 50 %.
- S : Medium means that the comparison of the calculated value is compared to the quality standard 50 % - 100 %
- B : Weight/Large means the comparison of the calculated value compared to the quality standard > 100 %

## 4. Conclusions

The upstream capacity of the Larangan/Premulung River is no longer able to accept the pollutant load in the moderate to heavy category at the class II quality standard. Then for the carrying capacity of the Larangan/Premulung River in the middle and downstream for the COD and BOD parameters, it can no longer accept the pollutant load in the medium to large category. However, the TSS parameter is still able to accept large categories of pollutant loads.

This condition shows that the pollutant load carrying capacity of the upstream Premulung river is in the medium category, while the middle and downstream sections are in the poor category. If the river is not managed properly, it will worsen environmental conditions and can cause accelerated climate change. To improve water quality and increase river capacity, several alternatives can be done, namely optimizing the existing Communal IPAL and increasing community participation in waste management, domestic waste and capacity building for sustainable agricultural systems.

## References

[1] Hukom E, Limantara L M and Andawayanti U 2012 Ussy Pengaruh Perubahan Iklim Terhadap Optimasi Ketersediaan Air di Irigasi Way Mitalpropinsi Maluku *J. Tek. Pengair.* **3** 24–32

[2] Alihar F 2018 Penduduk dan Akses Air Bersih di Kora Semarang *J. Kependud. Indones.* **13** 67–76

[3] Rahayu Y, Juwana I and Marganingrum D 2018 Kajian Perhitungan Beban Pencemaran Air Sungai Di Daerah Aliran Sungai (DAS) Cikapundung dari Sektor Domestik *Rekayasa Hijau J. Teknol. Ramah Lingkung.* **2** 61–71

[4] BPS Kabupaten Sukoharjo 2020 *Sukoharjo in Figures for 2020* (Sukoharjo, Indonesia: BPS Kabupaten Sukoharjo)

[5] Heru P, Totok G and Bowo S 2020 *Study of Environmental Damage to Premulung River Waters Due to Domestic Waste Disposal in Surakarta City* (Gadjah Mada University)

[6] Arbain T 2019 Tekanan Penduduk Terhadap Masa Depan Lingkungan: Perspektif Kebijakan Publik *J. Kebijak. Publik* **9** 61–70

[7] Lutfi M 2015 *Penyusunan Daya Dukung dan Daya Tampung Lingkungan Hidup Berbasis Jasa Ekosistem Sebagai Dasar Pengendalian dan Pengelolaan Lingkungan Hidup* (Yogyakarta, Indonesia: FGMP)

[8] Effendi H 2003 *Telaah Kualitas Air Bagi Pengelolaan Sumber Daya dan Lingkungan* (Yogyakarta, Indonesia: Kanisius)

[9] Kodoatie R J and Sjarief R 2008 *Pengelolaan Sumber Daya Air Terpadu, Edisi Revisi* (Yogyakarta, Indonesia: Andi Offset)

[10] Anisafitri J, Dan K and Rasmi D A C 2020 Analisis Total Bakteri Coliform Sebagai Indikator Pencemaran Air Pada Sungai Unus Lombok *J. Pijar Mipa* **15** 266–72

## Acknowledgement

Thank you to the Sukoharjo Regency Environmental Service for giving the responsibility in carrying out a study of the carrying capacity and carrying capacity of the environment, especially the Larangan/Premulung river.

# Local climate change indications: Historical trends, multi-model projections and implications in Bilate Catchment, Southern Ethiopia

**Hailu Gisha Kuma***, **Fekadu Fufa Feyessa and Tamene Adugna Demissie**
Faculty of Civil and Environmental Engineering, Jimma Institute of Technology, Jimma University, Jimma 378, Ethiopia
*Corresponding author: hailugisha@yahoo.com

**Abstract**. Climate change is a major factor affecting hydrologic processes, water resources and agricultural productions in a catchment. This study aimed to evaluate the historical trends, future climate changes, characteristics of heat waves, dry and wet months in Bilate catchment, southern Ethiopia. The Mann-Kendal, Theil-Sen estimator were employed to analyse the trends and magnitude of trends of historical temperature and rainfall. Ensemble mean of regional climate models (RCMs) was employed for the future climate projections of this (2021-2050), mid (2051-2070), and end (2081-2100) centuries relative to baseline (1981-2010). Results revealed that the change in temperature under RCP4.5 and RCP8.5 showed an increase in the three future periods. Under the scenarios, the monthly maximum temperature is projected to increase in the range of 0.1 to 2.4 °C and 0.1 to 4.2 °C, respectively. Likewise, the minimum temperature is in the range of 0.1 to 1.9°C and 0.3 to 3.5 °C. Under the scenarios, seasonal rainfall is projected to increase in the range of 124 to 3,009.9 mm and 52.5 to 1111.0 mm; while decreases are from -19.4 to -1679.3 and -243.3 to -2,444.7 mm, respectively. Besides, dry months and heat waves are expected to increase. The combined effect of a reduction in rainfall, increase in heat waves and dry months will likely impact on the people whose livelihood depend on rain-fed subsistence agriculture in the catchment. Water and crop yield reduction are the major implications in this century. These findings raise awareness about the impacts of climate change in the community and regional authorities to endorse development policies.

**Keywords** : Bilate catchment, dry month, heat wave, regional climate models, Trend analysis

## 1. Introduction

Climate change is the departure in the normal state of individual weather events owing to likely processes within the climate structure and human persuaded activities [1] . The human effect on the climate structure, have been increasing greenhouse gas concentrations in the air, radiating forcing, and observed warming [2]. Fluctuating precipitation reformed the hydrologic systems, affecting water resources and crop yields has been attributed to climate variation [3]. The influences of climate variation are forecasted to appease development and reworking is the only option to reduce the effects [4]. Climate alteration has a substantial impact on natural resources, agricultural production [5;6]. Future rise in precipitation excesses linked to the monsoon is likely in Africa, South America, East Asia, South Asia, Southeast Asia and Australia [7]. In many East African countries drought occurrences are frequent and the population depend on the rainfall for agricultural practices [8].

Climate changes and their negative impacts had directed to the advance of different climate models to indicate the changes and respond how to give adoption strategies. Climate models make climate predictions on different time scales. The regional climate models at their better resolution let the replication of small catchment climate circumstances by incorporating impacts of mountains, water bodies, vegetation, and other local physical boundary processes. CORDEX-Africa arranged to replicate Africa climate established on the range of climate systems [9]. The fitness of individual and ensemble mean of four RCMs to replicate the rainfall pattern of the 1979-2005 Mbarali watershed, Tanzania was evaluated [10]. The evaluation presumed that the ensemble mean performed well than a single model. The ensemble average of ten RCMs in replicating the features of rainfall patterns in Eastern Africa [11;12]. In this analysis, the ensemble means executed better than a single model in replicating the rainfall patterns of Eastern Africa. The fitness of RCM-COSMO-CLM and four driving GCMs to mimic the projected climate over Africa [13]. In this projection, the seasonal temperature on both global circulation models and a regional climate model was relatively similar at a continental level, but locally there was a change. Regional models ensemble with driving GCMs from 1971 to 2100, the annual mean temperature was predicted to rise by 1.0$^0$C and 1.5$^0$C over the horn of Africa [14]. In their study, the highest temperature rise was predicted in Sudan and Northern Ethiopia. However, the lower temperature rise was projected in Tanzania. They concluded that projected rainfall has a considerable decline over Ethiopia, while the variations are uncertain across other parts of Africa. An ensemble average of regional models to evaluate climate change over Somalia was executed [15]. In this investigation, the anticipated rainfall demonstrated a diminishing pattern up to 2030 followed by an increase with 2050 and 2070, while the temperatures increased in all seasons.

Several confirmations had made on tendencies of temperature and rainfall in Ethiopia. For example, the short and long seasons rainfall decreased by 15-20 per cent in the southern Ethiopia between the 1970s and 2000s [16]. Similarly, Ethiopia has experienced decreased tendencies of the annual rainfall totals, sequential wet days, and heavy rainfall days, while sequential dry days shows an increasing tendency between 1980-2010 [17]. Besides, they indicated that temperatures feature substantially increasing tendencies, and extreme overflow and famine led to more economic losses in Ethiopia.

This study emphasis on the rain-fed and subsistence agriculture, which could be affected by climate alterations in the Bilate catchment. The purpose of this study is to raise awareness about the impacts of climate change, to urge climate-resilient managing strategies and counteract the climate changes. Besides, local climate alteration investigation becomes very important instead of national level coarse-scale analysis. Seasonal temperature projection using RCMs and GCMs featured the same at continental level, however locally there was a change by more than 1°C [13].

## 2. Materials and methods

### *2.1. Materials*

*2.1.1. Study area.* Bilate catchment covers an area of 562,560 ha and situated in the Rift Valley Lakes Basin of Ethiopia [18] in Figure 1**Error! Reference source not found.**. The annual rainfall distribution is commonly distinguished as the rainy season (Kiremt) June to September, dry season (Bega) October to January and the small rainy season (Belg) February to May. It receives about 854 to 1,039 mm rainfall annually. The mean annual maximum and minimum temperatures are 22.8 to 30.4°C and 10.7 to 16.8°C. Depending upon the landscape and topography there are different types of indigenous and non-indigenous vegetation existing in the catchment. The land-use system of the catchment is dominated by small-scale subsistence agriculture. The high lands and middle areas are known for growing different crops such as wheat, clover, barley and teff. Farmers have intensive home gardens close to their homes, where they cultivate Enset, avocado and coffee. Agroforestry and farmyard Eucalyptus plantations are mostly practiced to overcome the shortage of fuel and construction wood and to earn extra income.

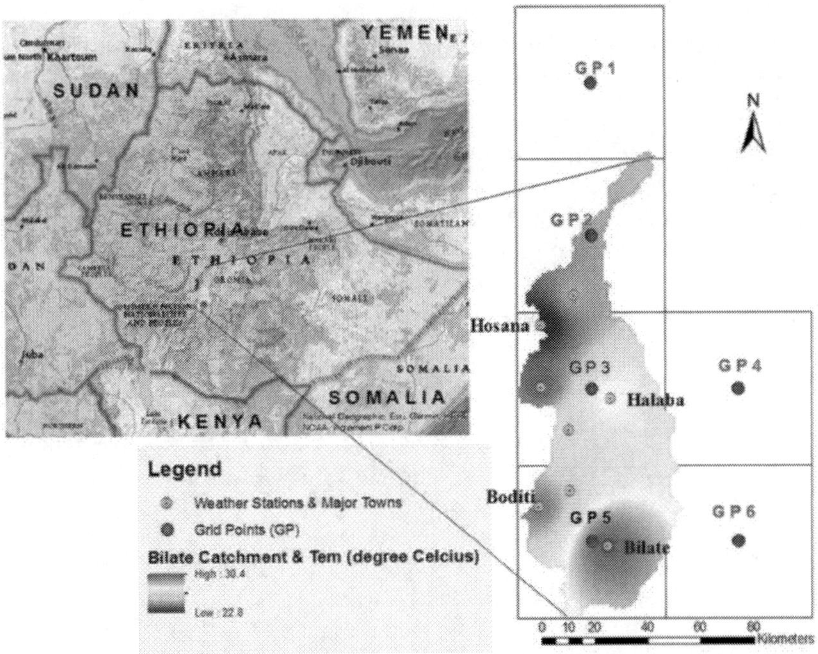


**Figure 1.** Spatial distribution of mean annual maximum temperature climatology of Bilate catchment

*2.1.2. Data.* The monthly climate data (1931-2018) were acquired from the climate research unit (CRU) database. Climate research unit gridded Time Series (CRUTS) is a broadly employed climate dataset on a 0.5° latitude by 0.5° longitude grid overall land domains except for Antarctica [19] produced by the University of East Anglia. The long-recorded CRU_ts_4.03 rainfall and temperature data from center for environmental data analysis (CEDA) are available http://data.ceda.ac.uk/badc/cru_ts_4.03/data. Temperature and rainfall data were extracted from the dataset for six data points, which are selected by including one point from each grid (0.5° by 0.5°) covering Bilate catchment. The monthly data from all data points were area-weight averaged to get areal-rainfall of Bilate catchment in Figure 1**Error! Reference source not found.**. Gridded climate dataset are employed in some studies [20;21;22;23] .

There are eight weather observation stations in Bilate catchment. Four weather stations (Bilate, Boditi, Halaba and Hosana) were selected. The selection was centred on the fullness of time series weather records and representing high, middle, and low lands of Bilate catchment; the data from 1981-2017 [24]. Figure 1, depicts the chosen weather stations, which are almost different climate distributions to the middle (Weynadega), low (Qola) and high (Dega) lands. The missing values of observed daily data were filled by using the Markov-Chain Simulation Model [25]. The simulation was executed via Instat+V 3.37 software.

The regional climate models (RCMs) data was downloaded from CORDEX-Africa domain at longitude 0.44° and latitude 0.44° and their selection based on changes in climatic means, changes climatic limits and the model ability to reproduce the past climate [26]. The regional climate models with their driving GCMs employed in the study are: CCLM4-ICHEC, RCA4-CanESM2, REMO2009-MPI, HIRHAM5-ICHEC and RACMO22T-ICHEC [27;28].

*2.2. Methods*

*2.2.1. Analysis of historical climate data and performance evaluation.* The monthly rainfall and temperature data (1931-2018) for four selected gauge stations were extracted from CRUTS. The two groups of data, CRUTS and historical observed were evaluated to indicate the altering points and tendencies. The computation of gathered irregularities to capture the potential altering points in both rainfall and temperature records [29]. The gathered irregularities at time t, $E_t$, of the generic series $X = (X_1, X_2, X_3, \ldots X_n)$ was estimated by equation (1)

$$\mathrm{Et} = t\sum_{i=1}^{t}(\mathrm{Xi} - \bar{X}),\ \ (\mathrm{t} = 1,2,3,\ldots \mathrm{n}) \quad (1)$$

Where $\bar{X}$ is the mean of X. Irregularities were calculated on monthly scale and x is month dependent. The Mann-Kendal test was executed to indicate tendencies of rainfall and temperature time-series data [30;31]. The degree of the tendency was computed by the Theil-Sen estimator [32]. Both the Mann-Kendal and Theil-Sen estimator were non-parametric methods, which are less subtle to outliers than the modest linear regression model. Evaluation was executed, on a monthly and annual basis at all gauge locations. To identify an appropriate time series model, lag-k autocorrelation was employed to get non-random data [33].

The correlation of CRU and observed data through the root mean square error (RMSE) and Pearson correlation coefficient (r) [34]. The gridded CRU data compared with observation data at four weather stations. It was found that the CRU dataset is strongly correlated with the corresponding temperature and rainfall data from the weather stations. The values of r greater than zero and less than one; and the values of RMSE greater than one was found according to the recommended ranges.

*2.2.2 Data extraction and bias correction of regional climate models.* The downloaded CORDEX-Africa regional climate model data was extracted by using Arc-GIS 10.3. Rainfall was corrected by employing power transformation algorithm. The method matches the coefficient of variations of the observed and RCM rainfall data [35;36;37], the daily rainfall sum p is transformed to a modified rainfall amount P* is given by equation (2).

$$P^* = aP^b \quad (2)$$

Where a is the transformation coefficient and b is a scaling exponent.

Temperature was corrected by variance scaling due to its almost normal distribution [38;39]. In this method, the mean and standard deviation adjustment was employed, and given by equation (3).

$$T_c = T_{obs} + \frac{Std_{obs}}{Std_{model}}(T_{model} - \bar{T}_{model}) \quad (3)$$

where Tc = bias-corrected future temperature, $\bar{T}_{obs}$ is mean observed temperature in the baseline period, $\bar{T}_{model}$ is mean RCM temperature in the baseline period and $T_{model}$ is RCM temperature in the future period, $std_{obs}$ is the standard deviation of observed temperature in the base period, $std_{model}$ is the standard deviation of regional climate model temperature in the base period.

The correlation of climate model and observed data was employed using the root mean square error (RMSE) and Pearson correlation coefficient (r) [34;40]. The values of r and RMSE were in the recommended ranges

*2.2.3 Climate change projection.* The model data downloaded under two emission scenarios of RCP4.5 and RCP8.5 was used to project future rainfall and temperature over the catchment. To account for uncertainties of climate models ensemble mean was employed in the projection.

The standardized precipitation index (SPI) used to indicate the future normal, wet and dry months in the ranges of 2.0 and above, 1.55 to 1.99, 1.00 to 1.49, 0.99 to -0.99, -1.00 to -1.49, -1.50 to -1.99 and -2.00 above are categorized extremely wet, severely wet, moderate wet, near normal, moderate drought and extremely drought, respectively [41]. Heat waves are consecutive numbers of hot days where daily mean or maximum temperatures exceed a predetermined threshold value [42]. The heat wave is a period with three or more consecutive hot days [43;44]. Projections indicate substantial increases in the frequency of hot days, which will occur on 19-40% of the days by the 2060s [45]. The SPI and heat waves are evaluated by statistical R-package [46].

## 3. Results and discussion

*3.1 Historical gathered irregularities*

The gathered irregularities were assessed for Bilate catchment using the historical monthly rainfall, minimum and maximum temperature data to detect the changing points and variations in slopes. Figure 2 shows the monthly accumulated anomalies of CRU and observed rainfall. The rainfall below and above the mean with CRU for the period 1931 to 2018 and observed rainfall for the period 1981 to 2017 analyzed were alternate, probably related to natural hydrologic cycles. The downward slopes of the gathered irregularities indicate CRU rainfall lower than the mean from the second half of the 1940s to first half of the 1950s, from second half of the 1960s to first half of the1970s. The alternating downward slopes of the accumulated anomalies indicate rainfall values lower than the mean from the second half of 1930s, 1970s, 1980s, and first half of the 1960s, 1970s, 1980s, 1990s, 2000s; then fluctuations restart until the 2010s. Both the CRU and observed rainfall showed increasing trends from 1980-1983, 1993-1995, 1998-2000, 2005-2006, and 2011-2012. While the downward trends from 1985-1987, 2002-2004 and 2013-2016 were observed.

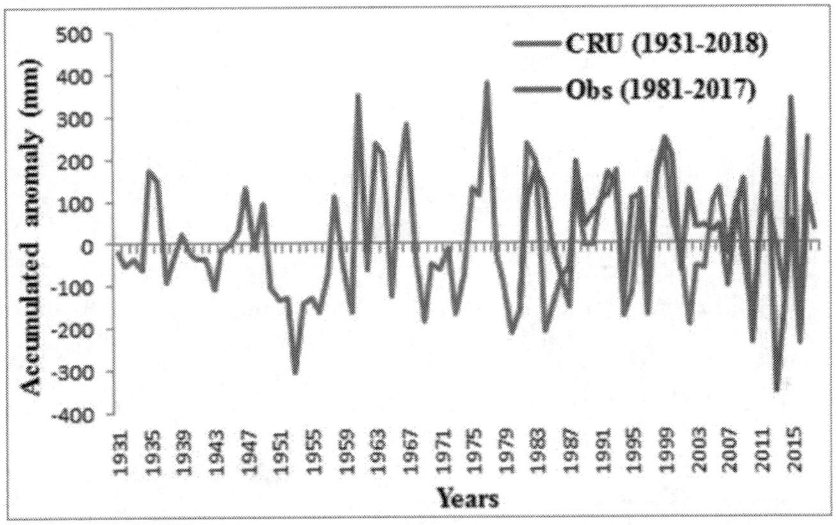


**Figure 2.** Accumulate anomalies of rainfall of Bilate catchment from 1931-2018

The CRU and observed monthly gathered irregularities of maximum and minimum temperature are shown in Figure 3. The gathered irregularities of maximum temperature in Figure 3.a show altering points around first and second half of 1980s, 1990s, 2000s and 2010s. For example, upward slopes from 1981-1984, 1989-1991, 1992-1995, 1999-2002, 2006-2009, while downward slopes from 1987-1989, 2004-2006, 2009-2010, 2015-2017 were observed by both datasets. After the 1999, the maximum temperature is above the mean value with an increase in the slope of the accumulated anomalies, i,e higher positive deviation from the mean. The accumulated anomalies of minimum temperature Figure 3.b indicate down and upward slopes from 1981 to 2017. After 2002 the minimum temperature indicates upward slopes and higher positive deviations from the mean.

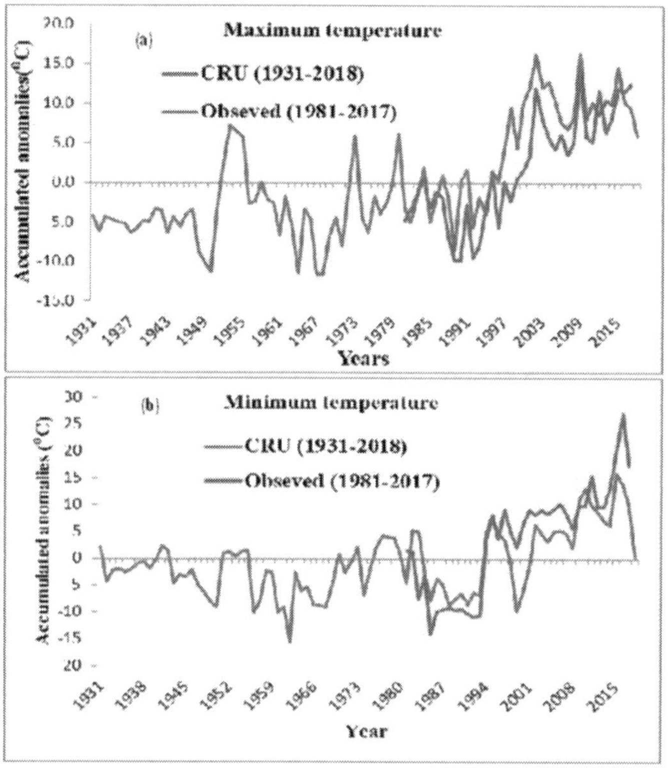


**Figure 3.** Gathered irregularities of maximum (a) and minimum (b) temperature of Bilate catchment from 1931-2018

The CRU and observed show positive and negative slope trends of rainfall from 1981-2017. However, the CRU and observed show positive slope trends of maximum temperature from 1999-2017, minimum temperature from 2002-2017. Bilate catchment becomes warmer from 1999-2017. This shows the featuring of warming in the catchment.

*3.2 Historical trend analysis*

The trend analysis on the historical data was shown inTable 1. The CRU and observed datasets for the periods 1931-2018 and 1981-2017 respectively, were chosen. The analysis was conducted at four meteorological stations and for Bilate catchment at monthly, seasonal and annual scales. The trend analysis of the historical temperature shows the gradual warming in Bilate catchment. In the period from 1981-2017, the maximum temperature presents a positive trend for all months; the trends are statistically significant except for March, April and November. The monthly positive gradients from 0.020-0.080°C/decade and the highest annual gradient 0.45°C/decade were observed in the catchment. The minimum temperature exhibits positive trends in all months, the highest annual trend of 1.072°C/decade and the tendencies are statistically significant.

Rainfall trends are represented by the box-whisker plots [47] (Figure 4). The plots represent rainfall variability between four stations, and the solid line represents the trend slope of Bilate catchment. The trends show differences in rainfall values, with increasing and decreasing magnitudes (Figure 4.a). For example, increasing rainfall trends were observed in June, July, August, October and November having 0.111, 0.121, 0.047, 0.015, 1.038mm/decade, respectively. Though, in the other months, the rainfall trends were decreasing. The seasonal rainfalls from June-September and October-January have increasing trends with the magnitude of 0.112 and 2.293 mm/decade (Figure 4.b). February- May and annual rainfall tendencies have decreasing trends with the magnitude of -3.549, -2.151 mm/decade. The short (February-May) and long (June-September) rainy seasons are required for crop production in Bilate catchment.

**Table 1.** Monthly, seasonal and annual trend slopes evaluated by Theil-Sen estimator

| Date | CRU (1931-2018) | | | Observed (1981-2017) | | |
|---|---|---|---|---|---|---|
| | Rainfall [mm] | Tmax [°C] | Tmin [°C] | Rainfall [mm] | Tmax [°C] | Tmin [°C] |
| Jan | 0.019 | 0.014 | 0.014 | -0.303 | 0.047* | 0.070* |
| Feb | -0.15 | 0.020* | 0.016* | -0.809 | 0.060* | 0.100* |
| Mar | -0.04 | 0.015* | 0.017* | -0.736 | 0.023 | 0.095* |
| Apr | -0.113 | 0.019* | 0.014* | -0.664 | 0.039 | 0.087* |
| May | 0.224* | 0.015* | 0.013* | -0.796 | 0.028* | 0.100* |
| Jun | 0.024 | 0.012* | 0.016* | 0.111 | 0.021* | 0.087* |
| Jul | 0.065 | 0.018* | 0.014* | 0.121 | 0.020* | 0.063* |
| Aug | 0.054 | 0.018* | 0.015* | 0.047 | 0.023* | 0.088* |
| Sept | -0.091 | 0.017* | 0.005* | -0.182 | 0.037* | 0.100* |
| Oct | 0.545 | 0.015* | -0.009* | 0.015 | 0.025* | 0.117* |
| Nov | -0.056 | 0.014* | -0.004* | 1.038 | 0.006 | 0.100* |
| Dec | 0.059 | 0.014 | 0.015 | -0.154 | 0.080* | 0.100* |
| Feb-May | -0.188 | | | -3.549 | | |
| Jun-Sept | 0.176 | | | 0.112 | | |
| Oct-Jan | 0.614 | | | 2.293 | | |
| Annual | 0.537 | 0.197* | 0.129* | -2.151 | 0.450* | 1.072* |

* = change trend is significant, Rainfall = mm/decade, Temperature = °C/decade

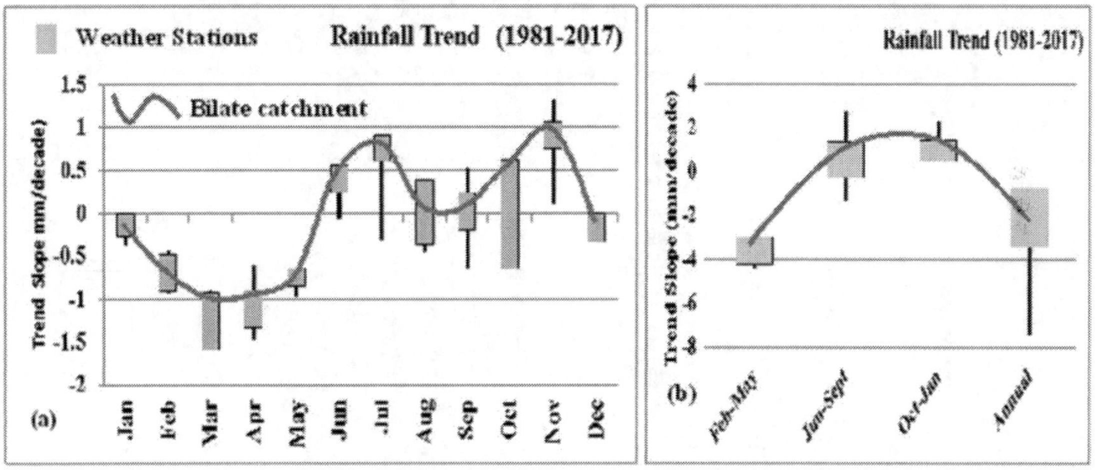


**Figure 4**. Monthly (a), seasonal and annual (b) rainfall trend slopes evaluated by Theil-Sen estimator

Historical accumulated anomalies and trends analysis of temperature and rainfall were performed by using CRU and observed data. Based on the accumulate anomalies and historical trend analysis of maximum and minimum temperatures, the tendency of positive trends was higher than negative trends since 1999 and 2002, respectively. The positive slope trends of maximum temperature started from 1999 to 2017, and that of minimum temperature from 2002 to 2017. Maximum and minimum temperatures featured the highest positive deviation from the mean after 1999 and 2002, respectively. Bilate catchment becomes warmer from 1999-2017. The warmest years have all come in the 2000s and 2010s. This shows the featuring of climate change indications in the catchment.

*3.3 Regional climate model projections*

*3.3.1 Projection of maximum temperature.* The projection of maximum temperature employed under RCP4.5 and RCP8.5 relative to the baseline (1981-2010) period over four representative weather stations in Bilate catchment. The projected change in monthly maximum temperatures was put in this (2021-2050), mid (2051-2080), and end (2081-2100) centuries under two scenarios in Figure 5. Figure 5 shows increased maximum temperature trend under two emission scenarios over Bilate catchment.

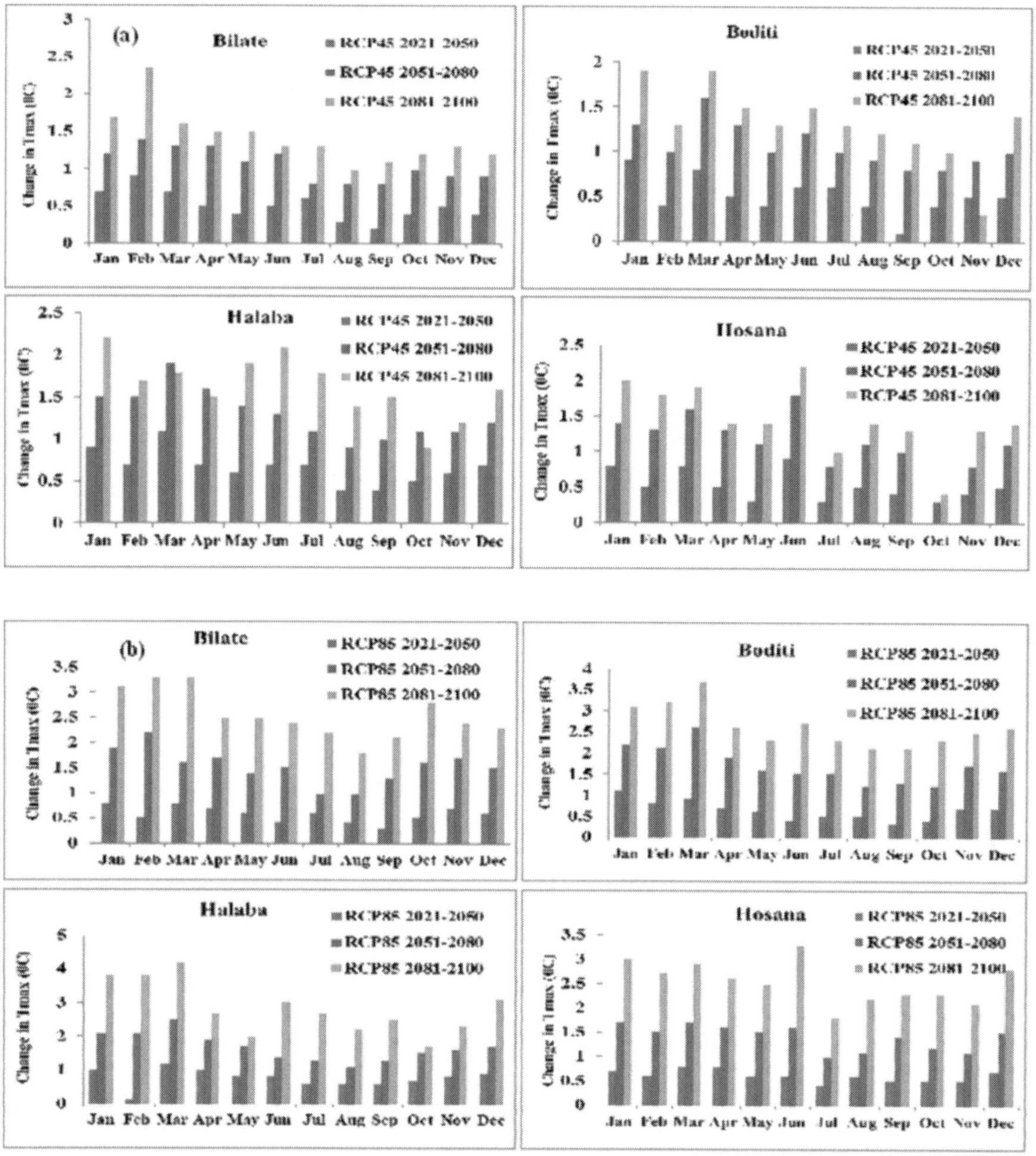


**Figure 5.** Expected change in maximum temperature across weather stations under (a) RCP4.5 and (b) RCP8.5

Outcomes indicate that projected change in temperature will vary from month to month and location to location. The maximum temperature time series in **Figure** shows an increasing trend in the current, mid and end centuries under both emission scenarios. This period monthly mean maximum temperature had increased in the range from 0.2 to 0.9 °C and 0.3 to 0.8 °C (Bilate), 0.1 to 0.9 °C and 0.3 to 1.1 °C (Boditi), 0.9 to 1.9 °C and 0.1 to 1.2 °C (Halaba), 0 to 0.9 °C and 0.4 to 0.8 °C (Hosana), respectively under RCP4.5 and RCP8.5. In the mid-century it increased from 0.3 to 1.9 °C and 1.0 to 2.6 °C, respectively under RCP4.5 and RCP8.5. Similarly, in the end century it increased from 0.3 to 2.4 °C and 1.7 to 4.2 °C, respectively under RCP4.5 and RCP8.5. It can be seen the maximum temperatures in all representative locations of Bilate catchment in the mid-century are likely to become warmer than those in this century. Besides, the end century maximum temperature features are warmer than the mid-century. In all representative locations, the maximum temperatures portray increasing trends in three future periods under both RCP4.5 and RCP8.5. The projected maximum temperature of Bilate catchment is in the line with global and regional reports that an increase in the mean temperature by 2030s and a further increase in 2050s [48].

*3.3.2 Projection of minimum temperature.* The spatial distribution of minimum temperature in this (2021-2050), mid (2051-2080) and end (2081-2100) centuries are presented in Figure 6. The time sequence of minimum temperature in Figure 6 displays an increasing tendency in the current, mid and end centuries under both emission scenarios. This is true especially the Southern lowland (Bilate), the Western Part (Boditi), Eastern land (Halaba) and Northern highland (Hosana) are expected to feature increased minimum temperature under two emission scenarios.The monthly mean minimum temperature had increased from 0.3 to 1.8 °C and 0.4 to 3.5 °C (Bilate), 0.4 to 1.9 °C and 0.5 to 3.5 °C (Boditi), 0.4 to 1.8 °C and 0.5 to 3.3 °C (Halaba), 0.1 to 2.3 °C and 0.3 to 3.1 °C (Hosana) under two emissions in the future centuries. It can be seen that the minimum temperatures in all representative locations in the end century are likely to become warmer than those in this and mid-centuries.

The expected temperatures in the catchment show a steady increase throughout the projection periods under two emission scenarios. The mean global temperature would rise by 1.4-5.8 °C in the future periods reported [49]. This study is further supported by the findings of some studies conducted in Ethiopia. For example, the annual mean temperature is predicted to rise from 1.4 to 4.1 °C by 2080s in the middle Rift Valley of Ethiopia reported [50]. Similarly, the expected changes in temperature further upheld [45] on climate alteration country profiles, shows that mean annual temperature is expected to increase by 1.1 to 3.1 °C by the 2060s and 1.5 to 5.1 °C by the 2090s. This study found pronounced warming in the weather locations of Bilate catchment which may have potential impacts on agricultural practices and water resources.

*3.3.3 Projection of seasonal rainfall.* The change in seasonal rainfall during this (2021-2050), mid (2051-2080) and end (2081-2100) centuries relative to baseline (1981-2010) under two scenarios of the catchment are shown in Figure 7. The selected representative weather stations of the catchment are likely to feature increased and decreased seasonal rainfall during the periods under two emissions. The expected rainfall in seasonal time series are of February to May (short rainy season: FMAM), June to September (long rainy season: JJAS), and October to January (dry season: ONDF).

The seasonal rainfall time series in Figure 7 show increasing and decreasing trends in this, mid and end centuries under both emission scenarios. The seasonal rainfall over the catchment features an increasing trend from 124 to 3009.9 mm and 52.1 to 1111 mm, respectively under RCP4.5 and RCP8.5. However, the decreasing seasonal rainfall features in the range of -19.4 to -1679.5 mm and -247.3 to -2444.7 mm, respectively under RCP4.5 and RCP8.5. The highest rainfall decline by -2444.7 mm is expected to be featured at Halaba in the end century under RCP8.5. The highest rainfall rise by 3009.9 mm is expected to be featured at Hosana in the end century under RCP4.5. This result shows decreasing seasonal rainfall is more pronounced over the catchment in the future.

IOP Conf. Series: Earth and Environmental Science 1016 (2022) 012030 doi:10.1088/1755-1315/1016/1/012030

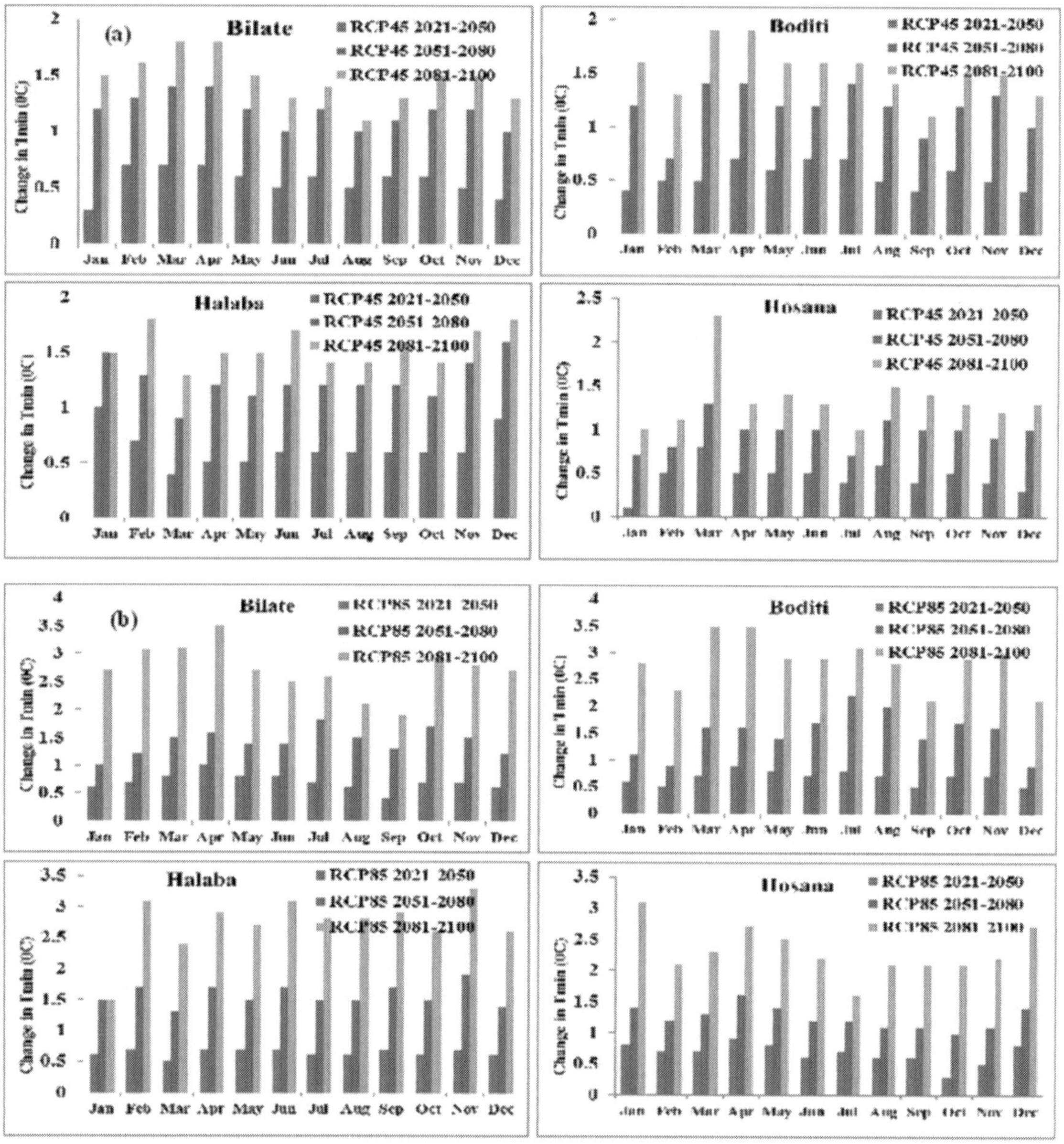


**Figure 6.** Expected change in minimum temperature across weather stations under RCP4.5 (a) and RCP8.5 (b)

### *3.4 Implications of climate change*

#### *3.4.1 Future dry, wet and months*

Monthly rainfall data of the regional climate models (RCMs) ensemble mean was used to indicate future characteristics of dry, wet and normal months. The projected dry, wet and normal months in this (2021-2050) century showed in Figure 8. For example, on January 13, 12, and 11 dry months were observed at Halaba, Bilate/ Boditi and Hosana stations respectively under RCP4.5 emission scenario. On February, 6, 5, and 4 dry months were observed at Bilate, Halaba and Boditi/ Hosana stations, respectively under the RCP4.5 emission scenario. The dry months increase over time; outstrips are higher for RCP4.5 than RCP8.5. The frequencies of dry months are increasing from October to April.

Dry months (spells) occur at increasing rates in Bilate catchment and can affect crop production because; farming is their major livelihood and food source in the catchment.

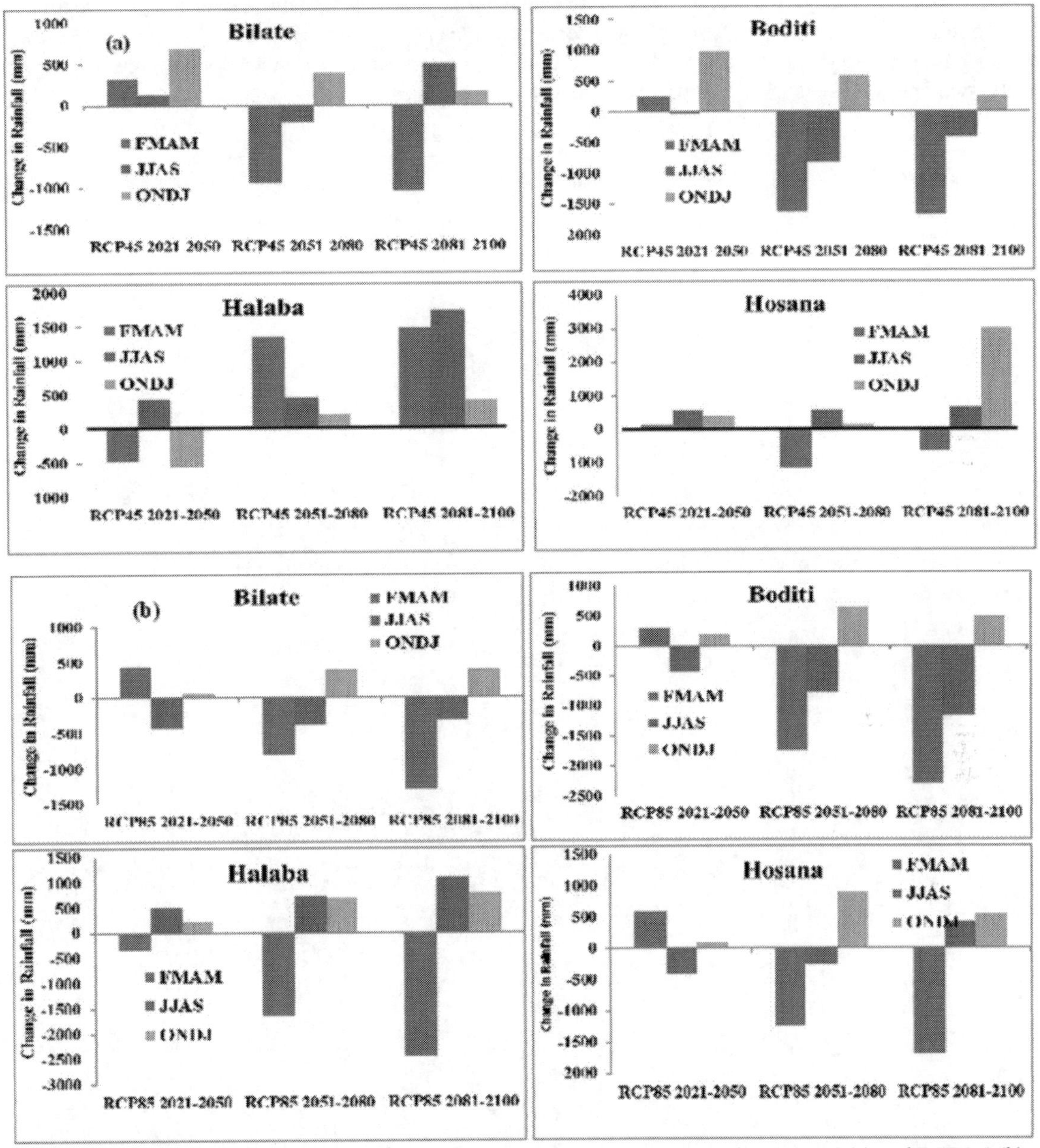


**Figure 7.** Expected change in seasonal rainfall across weather stations under RCP4.5 (a) and RCP8.5 (b)

*3.4.2 Future heat waves*

Daily maximum temperature of regional climate models ensemble was employed to detect this century changes in the heat waves. For the representative stations in Bilate catchment, a hot day was considered as a percentile of year-round daily maximum temperatures recorded in between 1981-2010. The hot days were detected for the moths of January, February, March, November and December. In the evaluation, a daily maximum temperature reaching or exceeding the threshold is considered a hot day. Thresholds corresponding to the 90th, 95th and 98th percentiles of daily maximum temperature were evaluated for each weather station. The moderate heat waves correspond to 90th and 95th percentiles, and more heat waves correspond to the 98th percentile. The thresholds vary across the stations and are above 24.0°C and below 36.5°C [44]. Table 2 shows the frequency of heat waves per year in terms of climate models mean and base period. The heat waves increase over time; the threshold exceedances are higher for the RCP8.5 than RCP4.5. The frequencies of heat waves per number of consecutive hot days of four weather stations are given in Table 2. The number of consecutive 7, 8, 9, and 10 hot days are increasing under two emission scenarios, and these can increase heat waves.

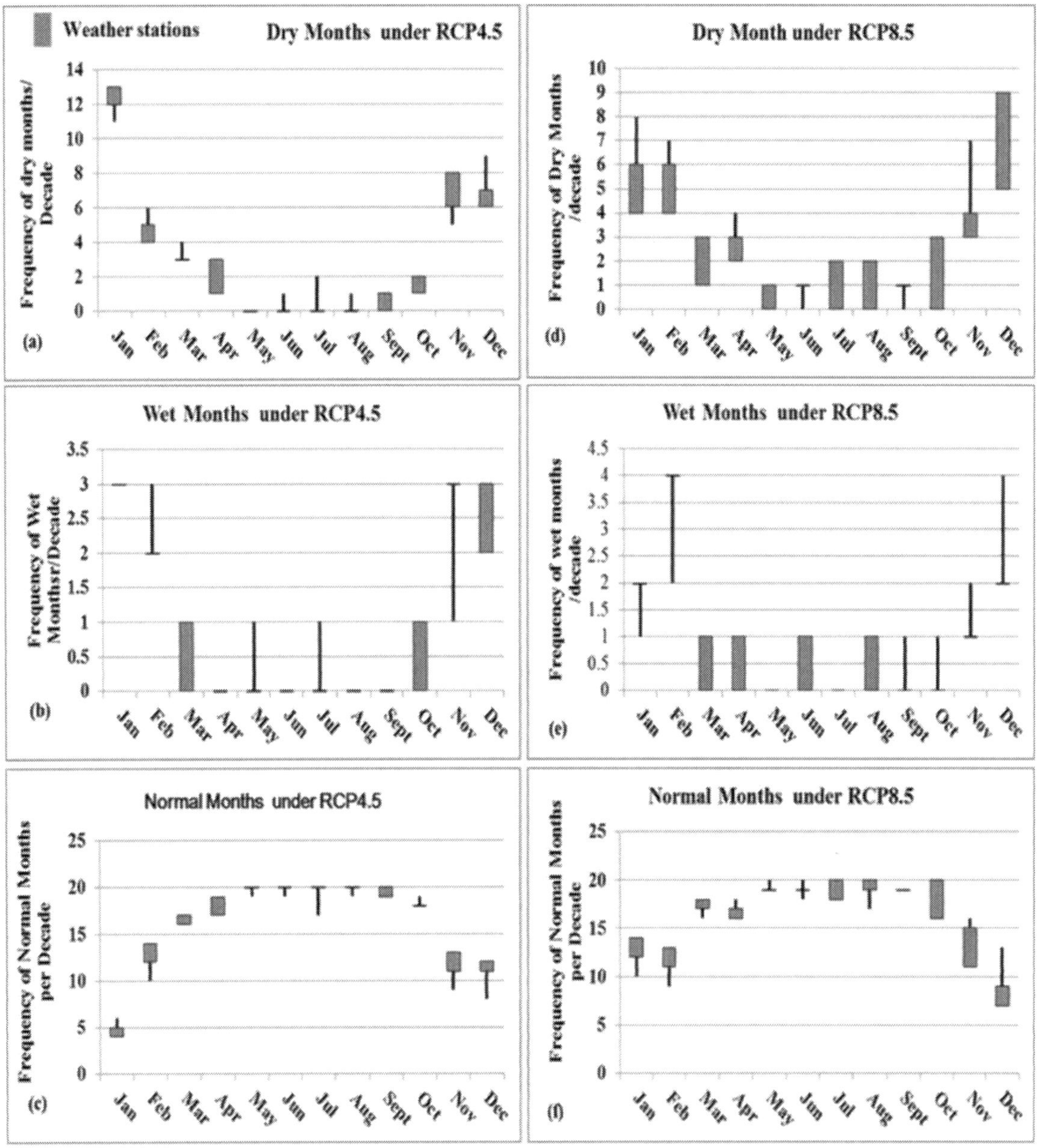


**Figure 8.** Frequency of dry, wet and normal months from 2021-2050, under RCP4.5 (a-c) and RCP8.5 (d-f)

**Table 2.** Frequencies of heat waves per year for base (1981-2010) and current (2021-2050) century

| Base Period and Scenario | Number of consecutive hot days | | | | | | | | Stations |
|---|---|---|---|---|---|---|---|---|---|
| | 3 | 4 | 5 | 6 | 7 | 8 | 9 | 10 | |
| BP | 1.48 | 1.12 | 0.68 | 0.8 | 0.84 | 0.48 | 0.24 | 2.48 | Bilate |
| RCP4.5 | 2.1 | 1.15 | 1.35 | 0.9 | 1.35 | 1.8 | 1.65 | 3.0 | |
| RCP8.5 | 2.25 | 1.75 | 1.40 | 1.15 | 1.4 | 1.85 | 2.05 | 4.0 | |
| BP | 1.6 | 1.28 | 1.12 | 1.0 | 0.68 | 0.68 | 0.4 | 1.76 | Boditi |
| RCP4.5 | 2.1 | 1.15 | 1.35 | 0.9 | 1.35 | 1.8 | 1.65 | 3.0 | |
| RCP8.5 | 2.25 | 2.05 | 1.85 | 1.55 | 1.45 | 1.9 | 1.75 | 3.4 | |
| BP | 1.88 | 1.4 | 0.84 | 1.36 | 0.68 | 0.96 | 0.88 | 1.2 | Halaba |
| RCP4.5 | 2.35 | 1.45 | 1.65 | 1.65 | 1.4 | 1.0 | 1.85 | 3.7 | |
| RCP8.5 | 2.5 | 2.0 | 1.85 | 1.75 | 1.5 | 1.95 | 2.35 | 4.5 | |
| BP | 1.76 | 1.04 | 1.04 | 0.8 | 0.4 | 0.44 | 0.64 | 1.4 | Hosana |
| RCP4.5 | 2.1 | 1.35 | 1.45 | 1.2 | 1.0 | 1.3 | 1.15 | 2.95 | |
| RCP8.5 | 2.35 | 2.2 | 1.6 | 1.75 | 1.8 | 2.15 | 1.95 | 3.75 | |

Note: BP is base period (1981-2010)

Heat waves are expected frequent and severe as climate change exists in a catchment. This result is in the line with some studies conducted in East Africa. Intense heat waves over the East Africa may have values of heat wave degree index-daily larger, but relatively weak, heat waves over West Africa [13]. The likely long-lived heat waves lasting 1.5-2 weeks was assessed in the period from 1961-1990 in Dar es Salaam and Addis Ababa [51]. The illness and mortality risk of older adults and children is due to heat pressure, which is likely to increase [52]. Heat waves can lead to water shortage, increased stress for plants, mortality and morbidity [53;54]. Heat waves occur at increasing rates and are likely to affect human health, reduce crop production, enhance water scarcity, and further affecting people depending on farming as their major livelihood and food source in Bilate catchment.

Temperature and rainfall are important climatic parameters as their changes could affect human environments. Projection of the short and long rainy seasons feature decreased in the catchment. These seasons are important, because of the practices of subsistence rain-fed agriculture in the catchment. The results are similar to the outcomes of [55;56] the long rainy season decreases over the southern part of Ethiopia. The decreasing rainfall of short and long rainy seasons in the catchment could cause problems to the subsistence and rain-fed agriculture.

Temperature Changes are reportedly linked to increasing diseases, a decline in agricultural products and shortage of water. The impact of an increasing temperature would be far-reaching for the socio-economic condition of the community. Sub-Saharan Africa becoming unsuitable for multiple major crops due to increasing temperature, this needs a shift to more heat resistant crops [20]. Sarofim et al. [57] reported that raised temperature compromises the human body's normal functioning concerning internal temperature regulation, leading to health complications.

The increasing temperature and decreasing rainfall could decline the water resources in Bilate catchment. This is further upheld by studies conducted on climate change effects on water accessibility in Eastern Africa and Ethiopia. For example, a decline of water resource by 7.43% and 11.3% for 2080s in the Anger catchment, Nile Basin of Ethiopia [58]. A similar result was reported in Dejen District, Blue Nile Basin, Northern Ethiopia [59]; and the country with the least capacity to address them [60]. Climate alteration poses challenges in Ethiopia: prolonged drought affect livestock, more extreme and erratic rainfall generating erosion and low crop production [61].

## 4. Conclusions

In this study, climate change indications in historical period, the future climate change projections, and heat waves, wet and dry months were evaluated over Bilate catchment. The evaluation was performed using Mann-Kendal, Theil-Sen and Regional Climate Models. The historical maximum and minimum temperature showed positive trends in all months except in October and November of CRU. The CRU and historical observed short rainy season showed a negative trend, while other seasons showed positive trends. The future rainfall and temperature are projected using an ensemble mean of regional climate model data under RCP4.5 and RCP8.5 over Bilate catchment. All the representative locations are projected to have increased temperatures and the rainfall was projected to be significantly increasing and decreasing. The variations in the characteristics of dry months per year are expected to gradually increase over this century. The study suggests that the evaluation of local climate change indications in Bilate catchment are well-represented by the Mann-Kendal, Sen and Regional Climate Model simulations. However, lack of data or available literature and information on past extreme rainfall and temperature effects (flood, crop failure and drought) in the catchment need urgent attention to improve our understanding of the variations in the future climate. The results indicate that the need for adaptation plans to increase resilience of the catchment under the changing climate. Sustainable catchment protection programs including planting drought resistant crops such as Enset (staple to the catchment, livestock feed and human food source), performing indigenous tree plantation based on decisions, and soil and water conservation works properly implemented to increase the groundwater recharge. Besides, this study can be a starting point but further studies must investigate water resources management strategies. Finally, local climate change studies are important to raise awareness about its impacts on the community and regional authorities to endorse development policies and sustainable mitigation measures.

**References**

[1] Field C B, Barros V, Stocker T F, Dahe Q, Jon Dokken D, Ebi K L, Mastrandrea M D, Mach K J, Plattner G K, Allen S K, Tignor M and Midgley P M 2012 Managing The Risks Of Extreme Events And Disasters To Advance Climate Change Adaptation: Special Report Of The Intergovernmental Panel On Climate Change *IPCC* vol 9781107025 (New York: University Press, Cambridge, UK and New York, NY, USA) pp 1–582

[2] Stocker, Qin D, Plattner G ., Tignor M, Allen, S K, Boschung, Nauels A, Xia Y, Bex V and Midgley P . 2013 Climate Change 2013: The Physical Science Basis *IPCC* (New York: IPCC) p 1535 pp

[3] Field C B, Barros V R, Dokke D J, Mach K J, Mastrandrea M D, Biler E T, Chatterjee M, Ebi K L, Estrada Y O, Genova R C, Girma B, Kissel E S, Levy A N, MacCracken S and Mastradrea P R and W L L 2014 Impacts, Adaptaion and Vulnerability *IPCC* (Cambridge University Press, Cambridge, UK and New York, NY, USA) pp 1-32p

[4] Luhunga P M, Kijazi A L, Chang'a L, Kondowe A, Ng'ongolo H and Mtongori H 2018 Climate Change Projections For Tanzania Based On High-Resolution Regional Climate Models From The Coordinated Regional Climate Downscaling Experiment (CORDEX)-Africa *Front. Environ. Sci.* **6** 122

[5] Abera K, Crespo O, Seid J and Mequanent F 2018 Simulating the Impact of Climate Change on Maize Production in Ethiopia, East Africa *Environ. Syst. Res.* **7**

[6] Carraro C, Bogardi J, Eulisse E, Vlachos E, Renaud F, Mostert E, Damm M, Fekete A, B M, Ridolf E, Edmunds M and Ortuno F 2010 Challenges In Water Resources Management Vulnerability, Risk And Water Resources Preservation *European Sustainable Water Goals* (Rome: 978-88-88997-57-5) p 102

[7] Christensen J H, Kumar K K, Aldrian E, An S I, Cavacanti I F ., de Castro M, Dong W, Goswami P, Hall A, Kanyang J ., Kitoh A, Kossin J, Lau N ., Renwick J, Stephenson D ., Xie S . and Zhou T 2013 Climate Phenomena and their Relevance for Future Regional Climate Change *IPCC* (Cambridge University Press, Cambridge Univer,United Kingdom and New York, NY, USA) p pp 1217-1308

[8] Nicholson S E 2017 Climate and Climatic Variability of Rainfall Over Eastern Africa *Rev. Geophys.* **55** 590–635

[9] Giorgi F, Jones C and Asrar G 2009 Addressing Climate Information Needs at the Regional Level: the CORDEX Framework *World Meteorol. Organ. Bull.* **58**
[10] Mutayoba E and Kashaigili J J 2017 Evaluation for the Performance of the CORDEX Regional Climate Models in Simulating Rainfall Characteristics Over Mbarali River Catchment in the Rufiji Basin, Tanzania *J. Geosci. Environ. Prot.* **05** 139–51
[11] Buontempo C, Mathison C, Jones R, Williams K, Wang C and McSweeney C 2015 An Ensemble Climate Projection for Africa *Clim. Dyn.* **44** 2097–118
[12] Nikulin G, Jones C, Giorgi F, Asrar G, Büchner M, Cerezo-Mota R, Christensen O B, Déqué M, Fernandez J, Hänsler A, van Meijgaard E, Samuelsson P, Sylla M B and Sushama L 2012 Precipitation Climatology in an Ensemble of CORDEX-Africa Regional Climate Simulations *J. Clim.* **25** 6057–78
[13] Dosio A and Panitz H J 2016 Climate Change Projections for CORDEX-Africa with COSMO-CLM Regional Climate Model and Differences with the Driving Global Climate Models *Clim. Dyn.* **46** 1599–625
[14] Osima S, Indasi V S, Zaroug M, Endris H S, Gudoshava M, Misiani H O, Nimusiima A, Anyah R O, Otieno G, Ogwang B A, Jain S, Kondowe A L, Mwangi E, Lennard C, Nikulin G and Dosio A 2018 Projected Climate Over the Greater Horn of Africa Under 1.5 °c and 2 °c Global Warming *Environ. Res. Lett.* **13** 1–11
[15] Ajuang Ogallo L, Omondi P, Ouma G and Wayumba G 2018 Climate Change Projections and the Associated Potential Impacts for Somalia *Am. J. Clim. Chang.* **07** 153–70
[16] Funk C, Rowland J, Eilerts G, Kebede E and Biru N 2012 *A Climate Trend Analysis of Ethiopia*
[17] Teshome A and Zhang J 2019 Increase of Extreme Drought Over Ethiopia Under Climate Warming *Adv. Meteorol.* **2019** 18
[18] Kuma H G, Feyessa F F and Demissie T A 2021 Hydrologic Responses to Climate and Land-Use/Land-Cover Changes in the Bilate Catchment, Southern Ethiopia *J. Water Clim. Chang.* **12** 3750–69
[19] Harris I, Jones P D, Osborn T J and Lister D H 2014 Updated High-Resolution Grids of Monthly Climatic Observations - the CRU TS3.10 Dataset *Int. J. Climatol.* **34** 623–42
[20] Asfaw A, Simane B, Hassen A and Bantider A 2018 Variability and time series trend analysis of rainfall and temperature in northcentral Ethiopia: A case study in Woleka sub-basin *Weather Clim. Extrem.* **19**
[21] Alhamshry A, Fenta A A, Yasuda H, Kimura R and Shimizu K 2020 Seasonal Rainfall Variability in Ethiopia and Its Long-Term Link to Global Sea Surface Temperatures *Water (Switzerland)* **12** 55
[22] Mulugeta S, Fedler C and Ayana M 2019 Analysis of Long-Term Trends of Annual and Seasonal Rainfall in the Awash River Basin, Ethiopia *Water (Switzerland)* **11**
[23] Mutti P R, Dubreuil V, Bezerra B G, Arvor D, de Oliveira C P and Santos E Silva C M 2020 Assessment of Gridded CRU TS Data for Long-Term Climatic Water Balance Monitoring over the São Francisco Watershed, Brazil *Atmosphere (Basel).* **11** 1207
[24] NMSA 2019 The National Meteorological Services Agency of Ethiopia *Natl. Meteorol. Serv. Agency*
[25] Stern R and Rijks D 2006 *Instat Climatic Guide*
[26] Castro C L, Pielke R A and Leoncini G 2005 Dynamical Downscaling: Assessment of Value Retained and Added Using the Regional Atmospheric Modeling System (RAMS) *J. Geophys. Res. D Atmos.* **110** 1–21
[27] Lutz A F, Immerzeel W W, Biemans H, ter Maat H, Veldore V and Shrestha A 2016 Selection of Climate Models for Developing Representative Climate Projections for the Hindu Kush Himalayan Region *Hi-Aware* 46
[28] Mascaro G, White D D, Westerhoff P and Bliss N 2015 Performance of the CORDEX-Africa Regional Climate Simulations in Representing the Hydrological Cycle of the Niger River Basin *J. Geophys. Res.* **120** 12,425-12,444

[29] Lozowski E P, Charlton R B, Nguyen C D and Wilson J D 1989 The Use of Cumulative Monthly Mean Temperature Anomalies in the Analysis of Local Interannual Climate Variability *J. Clim.* **2** 1059–68
[30] Mann H B 1945 Non-Parametric Test Against Trend *Econometrica* **13** 245–59
[31] Kendall M G 1975 *Rank Correlation ,ethods* (London)
[32] Sen P K 1968 Estimates of the Regression Coefficient Based on Kendall's Tau *J. Am. Stat. Assoc.* **63** 1379–89
[33] Box G E P and Jenkins G 1976 An Unexpected Route to Time Series *This Week's Cit. Class.* 575p
[34] Chai T and Draxler R R 2014 Root Mean Square Error (RMSE) or Mean Absolute Error (MAE)? -Arguments Against Avoiding RMSE in the Literature *Geosci. Model Dev.* **7** 1247–50
[35] Leander R and Buishand T A 2007 Resampling of Regional Climate Model Output for the Simulation of Extreme River Flows *J. Hydrol.* **332** 487–96
[36] Shabalova M V., van Deursen W P A and Buishand T A 2003 Assessing Future Discharge of the River Rhine Using Regional Climate Model Integrations and a Hydrological Model *Clim. Res.* **23** 233–46
[37] Min T 2006 *Assessment of the Effects of Climate Variability and Land Use Change on the Hydrology of the Meuse River Basin*
[38] Terink W, Hurkmans R T W L, Torfs P J J F and Uijlenhoet R 2010 Evaluation of a Bias Correction Method Applied to Downscaled Precipitation and Temperature Reanalysis Data for the Rhine Basin *Hydrol. Earth Syst. Sci.* **14** 687–703
[39] Teutschbein C and Seibert J Bias Correction Of Regional Climate Model Simulations For Hydrological Climate-Change Impact Studies: Review And Evaluation Of Different Methods *J. Hydrol.* **456–457** 12–29
[40] Flato G, J. Marotzke B, Abiodun P, Braconnot S C, Chou W, Collins P, Cox F, Driouech S, Emori V, Eyring C, Forest P, Gleckler E, Guilyardi C, Jakob V and Kattsov C 2013 *Evaluation of Climate Models. In: Climate Change 2013: The Physical Science Basis* vol 9781107057
[41] McKee T.B., Doesken N.J., Kleist J. 1993 1993 The Relationship of Drought Frequency and Duration to Time scales, in: Proc. 8th Conf. on Applied Climatology, Anaheim, California, 179–184. *Eighth Conf. Appl. Climatol.* 17–22
[42] Sanderson M G, Economou T, Salmon K H and Jones S E O 2017 Historical Trends and Variability in Heat Waves in the United Kingdom *Atmosphere (Basel).* **8** 191
[43] Nairn J R and Fawcett R J B 2014 The Excess Heat Factor: A Metric for Heatwave Intensity and its Use in Classifying Heatwave Severity *Int. J. Environ. Res. Public Health* **12** 227–53
[44] Perkins S E and Alexander L V. 2013 On the Measurement of Heat Waves *J. Clim.* **26** 4500–17
[45] McSweeney C, New M and Lizcano G 2008 UNDP Climate Change Country Profiles: Ethiopia *United Nations Dev. Program. 304 East 45th Street, Fl. 9 New York* p27
[46] Venables W N and Smith D M 2016 An Introduction to R Notes on R: A Programming Environment for Data Analysis and Graphics *R Core Team*
[47] Banacos P C 2011 Box and Whisker Plots for Local Climate Datasets: Interpretation and Creation using Excel 2007/2010. *East. Reg. Tech. Attach.* **Vol.1** 2–20
[48] Anyah R O and Qiu W 2012 Characteristic 20th and 21st Century Precipitation and Temperature Patterns and Changes Over the Greater Horn of Africa *Int. J. Climatol.* **32** 347–63
[49] Cooper R N, Houghton J T, McCarthy J J and Metz B 2002 Climate Change 2001: The Scientific Basis *Foreign Aff.* **81** 208
[50] Kassie B T, Rötter R P, Hengsdijk H, Asseng S, Van Ittersum M K, Kahiluoto H and Van Keulen H 2014 ClimateVariability and Change in the Central Rift Valley of Ethiopia: Challenges for Rainfed Crop Production *J. Agric. Sci.* **152** 58–74
[51] Capuano P, Sellerino M, Ruocco A, Kombe W J and Yeshitila K 2013 Climate Change Induced Heat Wave Hazard in Eastern Africa: Dar Es Salaam (Tanzania) and Addis Ababa (Ethiopia) Case Study *EGU General Assembly Conference Abstracts*

[52] Ahmadalipour A and Moradkhani H 2018 Escalating Heat-Stress Mortality Risk Due to Global Warming in the Middle East and North Africa (MENA) *Environ. Int.* **117** 215–25
[53] Shumake-Guillemot J and Ebi K 2019 A String of Deadly heat Waves has Impacted Millions Around the World in 2019 *ISSUE Br.*
[54] Hatfield J L and Prueger J H 2015 Temperature Extremes: Effect on Plant Growth and Development *Weather Clim. Extrem.* **10** 4–10
[55] Simane B, Beyene H, Deressa W, Kumie A, Berhane K and Samet J 2016 Review of Climate Change and Health in Ethiopia: Status and Gap Analysis *Ethiop. J. Heal. Dev.* **30** 28–41
[56] Li L, Li W, Ballard T, Sun G and Jeuland M 2016 CMIP5 Model Simulations of Ethiopian Kiremt-Season Precipitation: Current Climate and Future Changes *Clim. Dyn.* **46** 2883–95
[57] Fekadu K 2015 Ethiopian Seasonal Rainfall Variability and Prediction Using Canonical Correlation Analysis (CCA) *Earth Sci.* **4** 112
[58] Sarofim M C, Saha S, Hawkins M D, Mills D M, Hess J, Horton R, Kinney P, Schwartz J and Juliana A St. 2016 Temperature-Related Death and Illness *The Impacts of Climate Change on Human Health in the United States: A Scientific Assessment* p p26
[59] Amare Z Y 2018 Perceived Impacts of Climate Change and Disaster Risk Management by Rural Communities in Ethiopia *J. Degrad. Min. Lands Manag.* **5** 1181–90
[60] Cochrane L and Singh R 2018 Climate services for resilience: the changing roles of NGOs in Ethiopia Climate services for resilience p32
[61] Warner K L, Van P, Brouwer M, Van, Bodegom A. J Pluie B ., Galema F M and Buit G . . 2015 Climate Change Profile Ethiopia p24

**Acknowledgment**
The authors are pleased to thank the Center for Environmental Data Analysis (CEDA) and Earth Systems Grid Federation (ESGF) for the provision of CRU gridded and regional climate models data. Thanks to Jimma University and Wolaita Sodo University for their support.

# Paving the way for more accurate earth system modelling in Malaysia

**T A S T M Suhari[1], Eranga M Wimalasiri*[1,2] and Ebrahim Jahanshiri[1]**

[1]Crops for the Future UK, NIAB, 93 Lawrence Weaver Road, Cambridge, England, UK
[2]Department of Export Agriculture, Faculty of Agricultural Sciences, Sabaragamuwa University of Sri Lanka, Belihuloya, 70140, Sri Lanka
*Corresponding author: eranga.wimalasiri@cropsforthefutureuk.org

**Abstract**. Variations in soil structural and physiochemical properties can impact earth system models related to climate change. However, in most of the countries including Malaysia, soil data are available as low resolution and semi-detailed soil series maps, which hinder their applicability. Therefore, the objective of this study was to develop a framework to compare the conventional soil mapping with digital soil databases and to determine the accuracy of digital soil mapping. Kedah province located in the northern region of Peninsular Malaysia was selected as the study area. Observed soil (Cation exchange capacity – CEC, pH, organic matter as chemical and clay, sand and silt content as physical properties) data published by the Department of Agriculture, Malaysia was compared with SoilGrid (soillgrid.org) dataset. The depths of observed data were harmonised to international standard soil depths as; 0-5, 5-15, 15-30, s30-60, 60-100, and 100-200 cm using the equal quadratic spline approach. Global data were compared with harmonised 16 selected soil series data in Malaysia. Observed soil chemical properties (CEC, pH, and organic carbon) were within the range of SoilGrids data which allows the selected chemical properties to be used where observed data are limited or not available. In contrast, most of the soil physical properties from SoilGrids were not accurate when compared with the observed data. The developed framework in this study can be used to strengthen the accuracy of the available soil data in Earth system models, particularly those that are related to organic carbon and soil structural properties.

**Keywords** : Conventional soil mapping, SoilGrids, soil harmonization, spline tools

## 1. Introduction

Soil information plays a pivotal role in climate change studies, particularly those that are related to soil carbon storage [1] microbiome [2], geomorphological process and ecosystem health [3] and agricultural land use and land degradation [4] [5]. In Malaysia close to 500 soil series were identified and mapped as an extensive variety of soils of mountainous, hilly, rolling, undulating, level and swampy terrain landscapes are [6]. The association of mapping units is very helpful in determining general soil types across the country and for regional and national planning activities. However, such data are rarely useful for detailed analysis such as a change in carbon stocks and other climate change related studies.

The digital soil mapping technique for mapping of soil properties also has not yet been implemented in Malaysia. A major reason for this is that soil profile data are rarely made available publicly and the availability of such data depends solely on conventional survey methods for specific purposes which are often very limited in Malaysia. Per-property soil maps are available at a global scale, alas with limited detail and they are already being used for studies in Malaysia [7].

A harmonised data of soil property profiles in Malaysia can pave the way for more accurate mapping of important soil properties such as soil organic carbon and soil carbon stock and carbon pool that is directly related to the climate change studies [8] [9]. Therefore, it is crucial to compare data that are traditionally provided based on conventional soil series maps and those that are available globally through digital soil mapping techniques. Therefore, the objective of this study was to develop a framework to compare the conventional soil mapping with digital soil databases and to determine the accuracy of digital soil mapping which will lead to accurate earth system modelling in Malaysia.

## 2. Materials and methods

### 2.1. Study area

Kedah province which is located in the northern region of Peninsular Malaysia was selected as the study area. The province covers approximately 10,430 km$^2$ area. The area has a tropical climate with generally flat terrain (elevation 60 m above sea level). The land use map of Kedah is shown in Figure 1. Paddy is the major crop while different types of land use systems that include rubber, oil palm and some fruit crops such as pineapple are observed in the area. A major part of the area is covered by Lowland Forest.

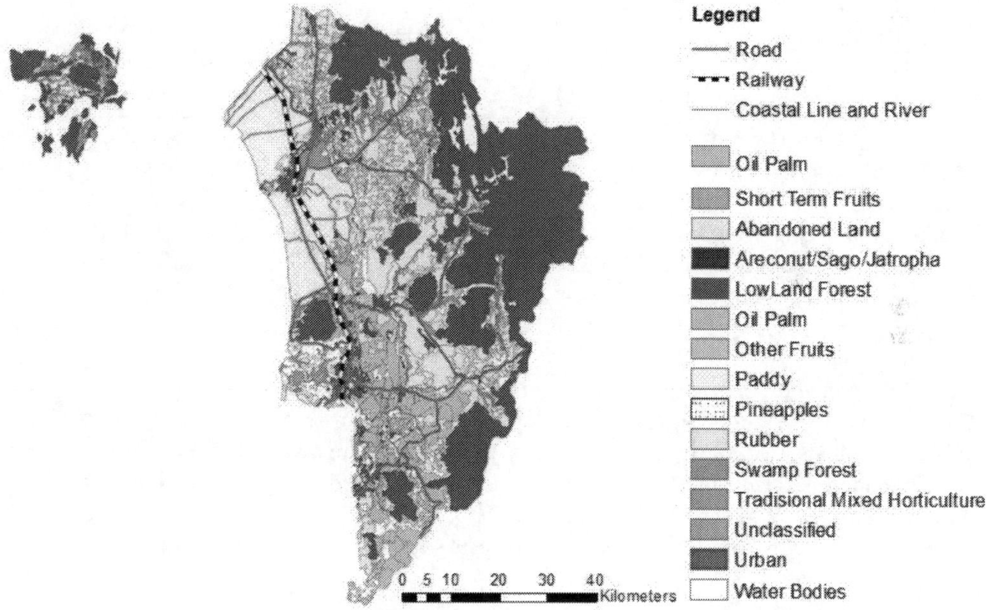


**Figure 1.** Land Use Map of Kedah

### 2.2. Data

To cover both physical and chemical soil properties, pH, clay (%), sand (%) and silt content (%), organic carbon (%), and cation exchange capacity (cmol/kg) were selected. The semi detailed soil map developed by the Department of Agriculture, Malaysia was used as observed data. The Kedah province consists of 16 soil series as shown in Table 1.

**Table 1.** Available soil series and the respective major soil groups in the study region

| No | Soil ID | Mapping Unit Reference | Soil Series | Major Soil Group |
|---|---|---|---|---|
| 1 | A | 37 | Beriah-Tanah Liat Organan & Muck | Marine Alluvial Soil |
| 2 | B | 26 | Durian-Melaka-Tavy | Sedentary Soil |
| 3 | C | 18 | Holyrood-Lunas | Riverine Alluvial Soil |
| 4 | D | 15 | Hutan-Semberin | Riverine Alluvial Soil |
| 5 | E | 2 | Keranji | Marine Alluvial Soil |
| 6 | F | 37 | Langkawi | Limestone Material |
| 7 | G | 28 | Muncung-Serdang | Sedentary Soil |
| 8 | H | 31 | Rengan-Jerangau | Sedentary Soil |
| 9 | I | 1 | Rudua-Rusila | Marine Alluvial Soil |
| 10 | J | 3 | Sedu-Parit Botak-Linau | Marine Alluvial Soil |
| 11 | K | 30 | Serdang-Bungor-Muncung' | Sedentary Soil |
| 12 | L | 16 | Sogomana-Sitiawan-Manik | Riverine Alluvial Soil |
| 13 | M | 11 | Telemung-Akob-Lanar Tempatan | Riverine Alluvial Soil |
| 14 | N | 27 | Muncung-Seremban | Sedentary Soil |
| 15 | O | 47 | Rengam-Bukit Temiang | Sedentary Soil |
| 16 | P | 25 | Gajah Mati-Muncung-Melaka | Reworked Soil |

Note: Soil ID was given by authors for easy identification while Soil Mapping Unit refers to the code of identification given by the Department of Agriculture, Malaysia.

Global soil data were retrieved from SoilGrids (soillgrid.org) dataset which is available at 250 m resolution [10]. Around 150,000 profile data points from the World Soil Information Service (WoSIS) and other sources were used to develop Soil Grids database. The SoilGrids database has been used in various earth systems and agricultural related applications in different geographical scales [7] [11] [12].

### *2.3. Harmonization of attributes and mapping*

Soil attribute varies continuously with depth in a soil profile which hinders proper comparison and application [13]. Also, the data in semi-detailed soil maps are available for different depths that are vary among locations. Therefore, soil properties were harmonised using the equal quadratic spline approach into international standard soil depths as; 0-5, 5-15, 15-30, 30-60, 60-100, and 100-200 cm. The SplineTool v2 (available at: http://www.asris.csiro.au) was used for harmonisation.

### *2.4. Data analysis*

Global soil data from SoilGrids were compared with 16 selected soil series (mapping units) in the Kedah province (Table 1). The data of 0-30 cm depth were compared and other data were omitted due to lack of information for all the locations. The averages of the standard depth intervals (0-30 cm) were obtained using the integration method.

## 3. Results and discussion

### *3.1. Mapping the smoothing spline soil depth function*

Three standardised datasets were created the equal-area spline depth function; the actual data, the spline curve and the standardised depth data. The actual data, for example, soil organic carbon content are real observations based on soil horizon that were determined by the surveyors based on the profile morphological features. The best fitted equal area spline was derived and used in the harmonisation of depths for each soil property. The lambda (λ) value of 0.1 reported the best fitted splines on test data.

Estimates of the quality of the predictions were obtained by the calculation of Root Mean Square Error (RMSE) values which measure the difference between the true and fitted function. Figure 2 shows a fitted spline of sand content (%) as an example. Please note that this spline interpolation has been done for all tested soil properties in each location.

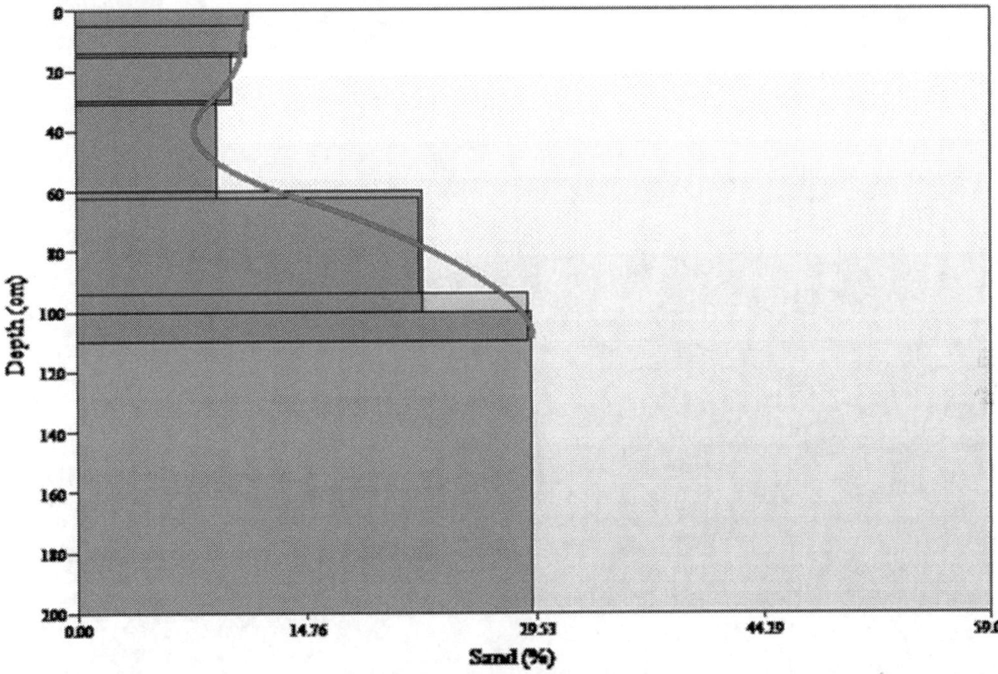


**Figure 2.** Fitted spline of sand content (%) for different soil depths. The original sampled soil profiles are in blue rectangles, the predicted six standard depths in purple rectangles and the fitted spline with 1 cm interval is in red line.

### *3.2. Comparison between conventional and digital soil mapping approaches*

Comparison of observed and global soil data of selected properties is shown in Figures 3 and 4. Observed soil chemical properties (CEC, pH and organic carbon) were within the range of SoilGrids data which allows selected chemical properties to be used where observed data are limited or not available. Soil pH data obtained from SoilGrids were used in crop suitability assessment in Malaysia and resulted in a higher (89%) median accuracy of estimation [11].

In contrast to chemical properties, most of the data points of observed soil physical properties (sand, silt and clay content) were beyond the global soil data which hinders their applicability in gap filling of observed data. However, these uncertainties may also be associated with measurement errors during the original soil survey and prediction errors during the data integration.

Having both dataset with different periods of data acquisition could also give dissimilarities in both values in soil chemical and physical characteristics as in Figure 3. The soil condition could be changed due to several factors such as urbanization, land degradation, climate change, and natural disaster, and many more.

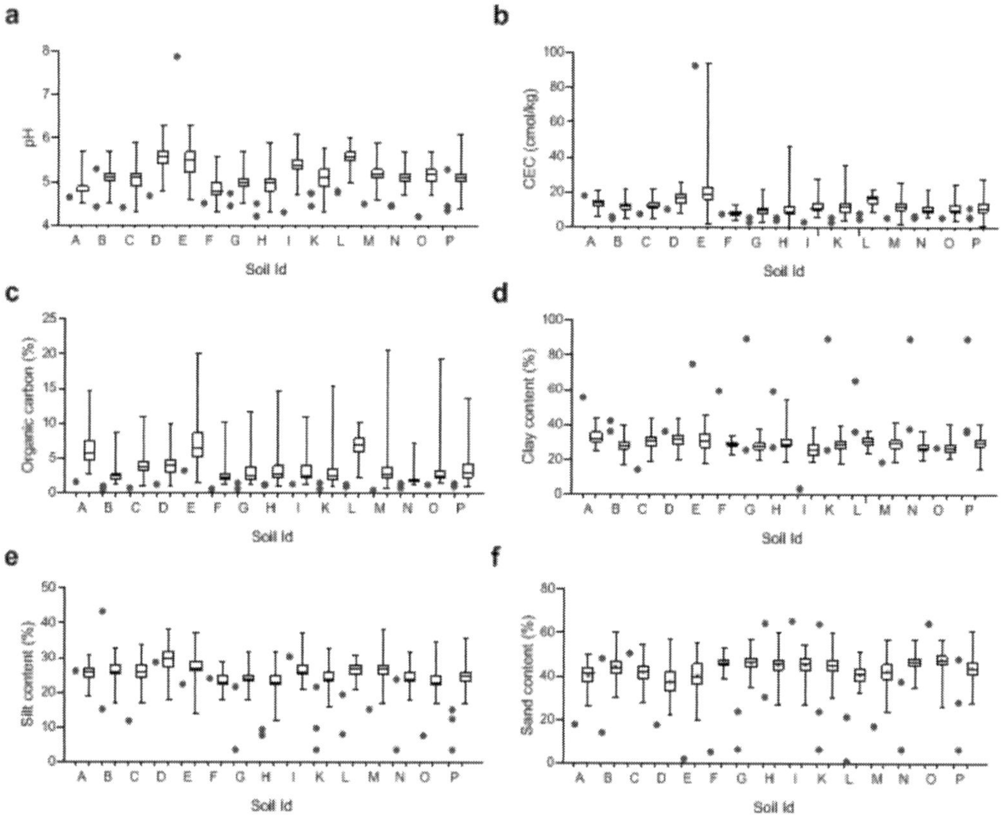


**Figure 3.** Comparison between observed (conventional) and predicted (global) data for 6 soil pH (a); cation exchange capacity (b); organic carbon (c); clay (d); silt (e); sand (f)

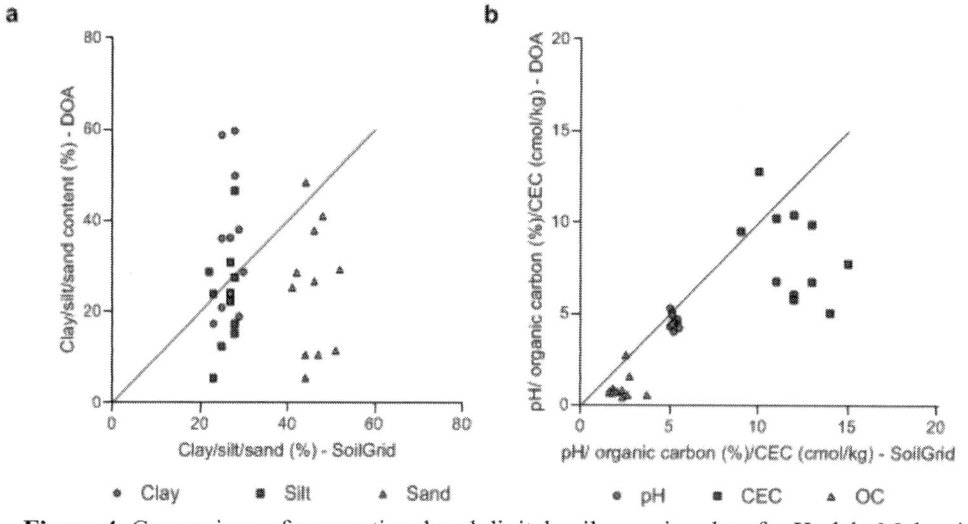


**Figure 4**. Comparison of conventional and digital soil mapping data for Kedah, Malaysia

When consider soil texture, the SoilGrids data showed overestimation compared to silt and clay on which a fairly good estimation was reported for them (Figure 4). In chemical properties, soil pH values had the best relationship between both datasets (Figure 4). Previous comparisons of soil organic carbon between regional databases and SoilGrids showed that SoilGrids was the closest to the field measurements in several geographic scales [14]. The uncertainties and errors could be due to measurement errors, prediction during the original soil survey or data integration.

Although SoilGrids data provide worldwide soil information, the samples of observations for some areas are lacking particularly for Malaysia. Wider selection of soil observations and more profile observations will increase the quality assessment and improve the consistency and standardisation across the different points datasets would give better accuracies for global available dataset in the future with no doubt [15].

### 3.3. *Applicability of the work and future work*

Although the statistical analysis determined that both data have significant differences, it shows that both conventional soil mapping and digital soil mapping have the capability in providing a source of soil information for environmental research such as monitoring the soil changes and earth systems and environmental modelling. Conventional soil mapping has the potential in generating detailed soil information on a regional scale. Figure 5 shows the schematic process flow chart developed in this study.

However, this study should be extended and more work is needed in larger geographical scales including the additional datasets and covariates that would be necessary to test these possibilities systematically.

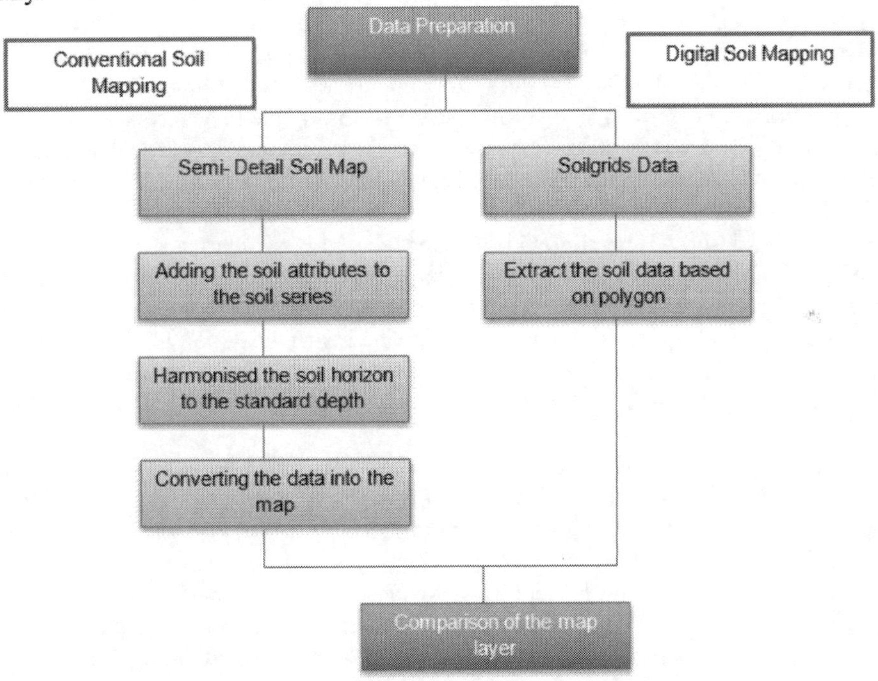


**Figure 5**. Process flow chart of the study

## 4. Conclusions

Conventional and digital soil mapping was compared in this study with the aim of developing a framework that assists in earth systems modelling in Malaysia. The comparison of the soil maps generally showed a significant difference for soil physical properties whereas no significant difference was reported for soil pH, a chemical property. Conventional soil data which were derived from observations have to be used in Earth Systems modelling and need to be integrated with digital soil maps such as SoilGrids. The developed framework in this study can be used to strengthen the accuracy of the available soil data in Earth system models related to climate change.

**References**

[1] Schils R, Kuikman P, Liski J, Oijen M van, Smith P, Webb J, Alm J, Somogyi Z, Akker J van den, Billett M, Emmett B, Evans C, Lindner M, Palosuo T, Bellamy P, Jandl R and Hiederer R 2008 *Review of Exciting Information on the Interrelations between Soil and Climate Change* (Wageningen: The Netherlands)

[2] Johnson-welch C, Alemu B, Msaki T P, Sengendo M, Kigutha H and Wolff A 2000 Improving Household Food Security: Institutions, Gender and Integrated Approaches *Broadening Access Strength. Input Mark. Syst.* 1–50

[3] Imeson A C and Lavee H 1998 Soil Erosion and Climate Change: the Transect Approach and the Influence of Scale *Geomorphology* **23** 219–27

[4] Dasgupta S, Hossain M M, Huq M and Wheeler D 2015 Climate Change and Soil Salinity: The Case of Coastal Bangladesh *Ambio* **44** 815–26

[5] Borrelli P, Robinson D A, Panagos P, Lugato E, Yang J E, Alewell C, Wuepper D, Montanarella L and Ballabio C 2020 Land Use and Climate Change Impacts on Global Soil Erosion by Water (2015-2070) *Proc. Natl. Acad. Sci. U. S. A.* **117** 21994–2001

[6] Sung C, Ishak C, Abdullah R, Othman R, Panhwar Q and Aziz M 2017 *Soil Properties (Physical, Chemical, Biological, Mechanical)*

[7] Wimalasiri E M, Jahanshiri E, Chimonyo V G P, Kuruppuarachchi N, Suhairi T A S T M, Azam-Ali S N and Gregory P J 2021 A Framework for the Development of Hemp (Cannabis sativa L.) as a Crop for the Future in Tropical Environments *Ind. Crops Prod.* **172** 1–16

[8] Besar N A, Suardi H, Phua M-H, James D, Mokhtar M Bin and Ahmed M F 2020 Carbon Stock and Sequestration Potential of an Agroforestry System in Sabah, Malaysia *Forests* **11** 1–16

[9] Rozainah M Z, Nazri M N, Sofawi A B, Hemati Z and Juliana W A 2018 Estimation of Carbon Pool in Soil, Above and Below Ground Vegetation at Different Types of Mangrove Forests in Peninsular Malaysia *Mar. Pollut. Bull.* **137** 237–45

[10] Hengl T, Jesus J M de, Heuvelink G B M, Gonzales M R, Kilibarda M, Blagotic A, Shangguan W, Wright M N, Geng X, Bauer-Marschallinger B, Guevera M A, Vargas R, MacMillan R A, Batjes N H, Leenaars J G B, Ribeiro E, Wheeler I, Mantel S and Kempen B 2017 SoilGrids250m: Global Gridded Soil Information based on Machine Learning *PLoS One* 1–40

[11] Jahanshiri E, Nizar N M M, Suhairi T A S T M, Gregory P J, Mohamed A S, Wimalasiri E M and Azam-Ali S N 2020 A Land Evaluation Framework for Agricultural Diversification *Sustainability* **12** 1–19

[12] Wimalasiri E M, Jahanshiri E, Suhairi T A S T M, Udayangani H, Mapa R B, Karunaratne A S, Vidhanarachchi L P and Azam-Ali S N 2020 Basic Soil Data Requirements for Process-Based Crop Models as a Basis for Crop Diversification *Sustain. 2020, Vol. 12, Page 7781* **12** 7781

[13] Moore A W, Russell J S and Ward W T 1972 Numerical Analysis of Soils: a Comparison of Three Soil Profile Models With Field Classification *J. Soil Sci.* **23** 193–209

[14] Tifafi M, Guenet B and Hatté C 2018 Large Differences in Global and Regional Total Soil Carbon Stock Estimates Based on SoilGrids, HWSD, and NCSCD: Intercomparison and Evaluation Based on Field Data From USA, England, Wales, and France *Global Biogeochem. Cycles* **32** 42–56

[15] De Sousa L M, Heuvelink G B M, Batjes N H and Kempen B 2018 SoilGrids: Using Big Data Solutions and Machine Learning Algorithms for Global Soil Mapping *Scientific Symposium FAIR Data Sciences for Green Life Sciences* pp 1–42

**Acknowledgement**

Authors acknowledge European Union's Horizon 2020 research and innovation programme, grant agreement No. 774234.

# Variability and trends of climate in east Wollega zone, Western Ethiopia

**Teka Bekuma[1]*, Girma Mamo[2] and Alemayehu Regassa[3]**
[1]Wollega University, Nekemte, Ethiopia
[2] Ethiopian Institute of Agricultural Research, Addis Ababa, Ethiopia
[3] Jimma University, Jimma, Ethiopia
*Corresponding author: tekabeka2017@gmail.com

**Abstract.** Climate change is a major concern for the environmental, social and economic sectors of our world. The purpose of the study was to examine climate variability and trends in East Wollega Zone, Western Ethiopia. Daily precipitation and temperature information for 37 years (1981-2017) were gathered from the Ethiopian National Meteorological Agency. Mann-Kendall trend test, Sen Slope, linear regression model, and coefficient of variation were used to analyze data. The result found that monthly, seasonal and annual rainfall observed an insignificant declining trend except for May, November, and spring and the change was only significant for August at a significance level of 0.05. Maximum and minimum temperature revealed upward warming trends on a monthly, seasonal and yearly basis. The outcome of the linear regression model indicated an insignificant relationship between monthly, seasonal, and annual rainfall and time at a confidence level of 0.05. A statistically significant relationship has been detected between annual maximum, minimum, and mean temperature and year at 0.05 significance level. With the exception of winter minimum temperature, the result found a statistically significant relationship for seasonal temperature. The variation of seasonal rainfall was moderate for summer (23%) and autumn (29%) while it was high for spring (33%) and winter (65%) and it was also moderate for annual (23%). In view of the fact that crop production is affected by rainfall and temperature variability, adaptation to its adverse effect is imperative. Cropping system decisions and adaptation plans in the area should better accommodate for fluctuations in rainfall and temperature.

**Keywords:** rainfall, temperature, Mann-Kendall (MK), trend, linear regression

## 1. Introduction

Climate change is among the greatest environmental, social, and economic issues locally, regionally, and globally. An agreement amongst researchers on human-induced warming of the planet has increased to 100% [1]. Climate change is an unavoidable reality that has already begun to happen [2]. Because of economic and population growth, human activities impacted the climate by increasing the amount of greenhouse gases in the atmosphere [3]. Agriculture and climate change are linked in two ways: agriculture cause climate change by emitting greenhouse gases, and climate change have a negative impact on agriculture in general [4].

Despite increases in certain crops, climate change is anticipated to have a negative effect on agriculture, jeopardizing global food security [5]. Extreme weather events, notably severe flooding and droughts, have adversely impacted crop productivity, widened yield gaps, and threatened agricultural sustainability in emerging and vulnerable parts of the world [6]. Because of their direct impact on agriculture, temperature and rainfall are the significant climate variables highly examined in climate studies [7]. Variations in rainfall patterns and soil erosion, both of which are critical for agricultural production, can be influenced by rising temperatures, which increase evapotranspiration and lower soil moisture [8].

Africa is the most susceptible continent on the planet, with frequent droughts, severe temperatures, and rains causing devastation on crop and food supplies [9]. Ethiopia is one of the most sensitive countries to climate change, and natural disasters such as drought and flooding occur on a regular basis [10], (Gashaw et al., 2017), and [12]]. Ethiopia has suffered climate change and fluctuation, as well as other environmental challenges [13]. Climate change and unpredictability disproportionately affect subsistence farmers, who endure unpredictable and geographically diverse impacts [14]. In Ethiopia, where rain-fed agriculture is the norm and exposure is worsened by widespread poverty and rapid population expansion droughts are the most devastating climatic hazard [15]. Though Ethiopia's influence of global warming has been minimal in recent decades, the temperature has risen at a rate of about 0.2°C per decade [16]. Rainfall was unpredictable in terms of both space and time [17]. From 1961 to 2015, mean temperature showed a considerable upward trend, whereas summer season rainfall showed a downward tendency from 1901 to 2015 [18].

Climate research is needed to quantify local effects and vulnerabilities, identify adaptation alternatives, and guide regional and national climate policies and initiatives [19]. The knowledge of climate variability and trends is very crucial for panning adaption decision and ensuring agricultural sustainability among small-scale farmers. Increased localized climate risks and extremes, including floods, changing rainfall patterns, and cropping season changes, are already affecting agriculture. Crop productivity in the area was negatively impacted by such catastrophic events and constant climate changes, posing a challenge to the country's efforts to achieve food security in a changing environment. In general, climate research in Ethiopia is restricted at the national level, and it is almost always missing at the local level. Climate studies must be locally relevant in order for their findings to be used. As a result, the purpose of this research was to fill this gap by examining climate variability and trends in Western Ethiopia's East Wollega Zone.

## 2. Materials and methods

### *2.1 Description of the study area*

The research was carried out in the Oromia National Regional State's East Wollega Zone in Western Ethiopia. It is one of the Oromia National Regional State's Zones, which includes 17 rural districts and 289 rural peasant associations. It is located 328 kilometers west of Ethiopia's capital, Addis Ababa. The zone's entire land area is around 14,102.5km$^2$, accounting for about 3.88% of the Oromia National Regional State's overall area.

#### *2.1.1 Location*

East Wollega Zone is found on Northing 8$^0$31'20"N to 10$^0$22'30"N and Easting 36$^o$06'00"E 37$^o$12'00"E. It is bordered on the North by Amhara National Regional State, on the South by Jimma zone, on the East by Horo Guduru Wollega and West Shewa zone, on the North-West by Benishangul Gumuz National Regional State, on the West by West Wollega zone, and on the South-West by Buno Bedelle zone. Figure 1 depicts a map of the research area.

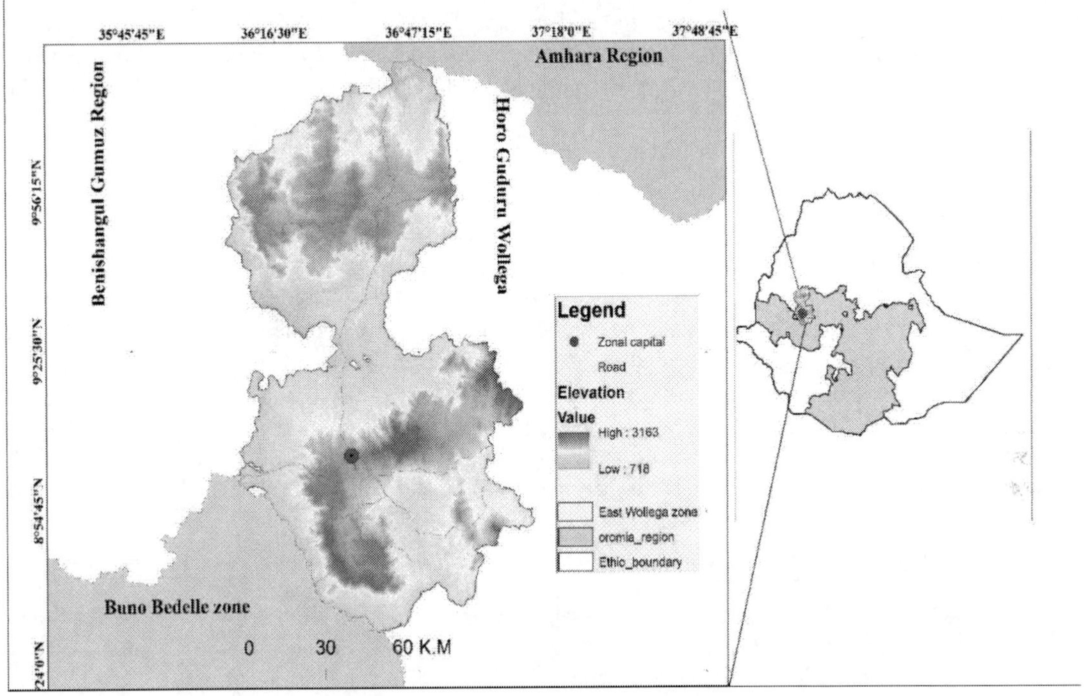


**Figure 1.** Location map of the research area

*2.1.2 Topography and climate*

The altitude of the zone ranges between 718 to 3163 meters above sea level. It consists primarily of low plateaus with isolated ranges, such as in the Jima Arjo district. The zone's climate is classified into three categories: highland (20.50%), midland (50.90%), and lowland (28.60%). The daily temperature ranges from 15 to 27°C [20]. Long summer rainy seasons (June to September), brief rainy seasons (March to April), and winter dry seasons alternate (December to February). The main rainy season in Ethiopia is from June to September, as is the case in much of the country's highlands [21].

*2.1.3 Farming system*

The dominant source of livelihoods in the study area is crop production and livestock rearing. Crop and livestock production are used for both domestic use and as a source of revenue. Maize, sorghum, teff, millet, wheat, and barley are the principal cereal crops farmed in the area. Crop production is an important livelihood activity for farmers and is mainly rain-fed subsistence agriculture. Cattle rearing rely on natural grasses and crop wastes that are retained in the traditional management approach [22]. The research areas soils have been severely eroded because of these environmentally unfriendly farming methods and excessive rainfall levels, resulting in nutrient loss, soil acidity, and overall land and natural resource degradation. As a result, most soil acidity-sensitive crops such as barley, bean, and wheat have been forced out of cultivation in some places [23].

*2.2 Data type and sources*

Historical time series climate data (temperature and rainfall) for 37 years (1981-2017) was collected from National Meteorology Agency (NMA). The study included both dependent and independent (explanatory) variables. In this case, the independent variable was time, while the dependent variables were temperature and rainfall.

*2.3 Analytical materials and methods*

*2.3.1 Mann-Kendall trend test*

Mann-Kendall (MK) test was employed for trend analysis. Kendall's tau, S statistics, Z statistics, and P-value were used to detect variation in rainfall and temperature. The P-values used to determine whether any apparent patterns are statistically significant or not. The decision was made based on level of significance ($\alpha$ =0.05) which compared against the p-value. There is no significant trend if the p-value is above the significance level ($\alpha$ =0.05); there is a significant trend if the p-value is below the significance threshold. The negative of MK Stat (S), Kendall's tau, and Z statistics value represents the declining trend while the positive value represents the increasing trend or change. Mann-Kendall discrimination is the most usually interest non-parametric experience for detecting climatic shifts and run the analysis [24]. The MK S Statistic is calculated as equation (1).

$$S = \sum_{i=1}^{n-1} \sum_{j=i+1}^{n} sign(\mathrm{Gj} - \mathrm{Gi}) \qquad (1)$$

Where $Gj$ and $Gi$ are the time series' successive data values in the years $j$ and $i$ ($j > i$) and n is the time series length. Positive S importance particularizes an increasing strike and negative importance depicts a falling trend in the data course and sign (Gj - G$i$) calculated as equation (2):

$$sign(\mathrm{Gj} - \mathrm{G}i) = \begin{cases} 1\, if\ (\mathrm{Gj} - \mathrm{Gi}) > 0 \\ 0\, if\ (\mathrm{Gj} - \mathrm{Gi}) = 0 \\ -1 if\ (\mathrm{Gj} - \mathrm{Gi}) < 0 \end{cases} \qquad (2)$$

The S-variance statistic's ($\sigma$2) is defined as equation (3).

$$\sigma^2 = \frac{n(n-1)(2n+5) - \sum t_i(i)(i-1)(2i+5)}{18} \qquad (3)$$

Where, $t_i$ represents the count of fasten to extent $i$ that have been completed. If the data sequence holds tied values, the numerator summation term is used. Equation (4) is to compute the standard test statistic (Zs):

$$Z_s = \begin{cases} \frac{s-1}{\sigma}\ for S > 0 \\ 0\ for\ S = 0 \\ \frac{s+1}{\sigma} for S < 0 \end{cases} \qquad (4)$$

The result of the test statistic trend's significance is measured in Zs. In reality, the null hypothesis, Ho, is tested using this test statistic. The null guess is rejected if | Zs| is bigger than Z$\alpha$/2, where $\alpha$ is the selected significance threshold, meaning that the trend is significant.

*2.3.2 The test of Sen's slope estimator*

Sen's slope estimator was used to estimate the average changes in a variable over time (increasing or decreasing) per period (monthly, seasonally, or annually) [25]. The slope of Sen is defined as equation (5):

$$Sen's\ Slope = Median\ \left(\frac{\mathrm{x}j - \mathrm{x}i}{\mathrm{j} - i} : i < \mathrm{j}\right) \qquad (5)$$

Where, xj and x*i* are data values at time j and *i* (*i* <j). A positive number of denotes an upward trend, whereas a negative value denotes a downward trend.

*2.3.3 Model of linear regression*
This model was fitted to historical rainfall and temperature data in order to describe a link between the dependent variables (rainfall and temperature) and the independent variable (time). It is a useful technique for examining the links between variables by linking one variable to a group of variables [26]. A linear regression equation defined as equation (6).

$$Y = c + dx \quad \ldots\ldots (6)$$

Where the response variable is y, the slope is d, the explanatory variable is x, and the intercept is c.

*2.3.4 Coefficient of variation*

The temporal variability of temperature and rainfall was calculated using the coefficient of variation (CV). The coefficient of alteration is a spread measure that expresses the amount of variance compared to the mean. A higher CV appraise directs greater variability, whereas a lower CV indicates less variability, and is calculated as equation (7).

$$CV = \frac{\sigma}{\mu} x100 \quad \ldots\ldots (7)$$

Where CV stands for coefficient of variation; σ is standard error and μ is mean of rainfall and temperature.

## 3. Discussion of the findings

*3.1. Rainfall patterns and trends*
The two-sided Mann-Kendall test was performed to examine whether there is a statistically significant monotonic increasing or decreasing trend in monthly, seasonal, and yearly rainfall data sequence, as shown in Table 1. The results demonstrated that, except for August, no significant trend could be detected at the 0.05 significance level on a monthly, seasonal, or annual basis. Except for May, November, and the Spring season, the S statistics, Z statistics, and Kendall's tau all exhibit negative values, indicating a decreasing trend in monthly, seasonal, and annual rainfall. The Sen's slope of a trend line displays a declining magnitude in rainfall per month, seasonal, and yearly, except for May, November, and spring, which have a positive slope. The findings of a research done in Northwest Ethiopia were similar, with rising temperatures and decreasing rainfall amounts [27] . Figure 2 and Figure 3 show Sen's slope for monthly and seasonal rainfall.

**Table 1.** Monthly, seasonal and yearly rainfall Sen's slope and Mann-trend Kendall's

| Period | MK Stat (S) | Kendall's tau | Z Statistic | P-value | α | Sen's Slope | Sig. at α =0.05 |
|---|---|---|---|---|---|---|---|
| January | -70 | -0.1051 | -0.9024 | 0.3668 | 0.05 | -0.171 | Insignificant |
| February | -82 | -0.1231 | -1.0594 | 0.2894 | 0.05 | -0.272 | Insignificant |
| March | -106 | -0.1592 | -1.3733 | 0.1697 | 0.05 | -0.574 | Insignificant |
| April | -144 | -0.2162 | -1.8703 | 0.0614 | 0.05 | -1.287 | Insignificant |
| May | 96 | 0.1441 | 1.2425 | 0.2141 | 0.05 | 1.857 | Insignificant |
| June | -34 | -0.0511 | -0.4316 | 0.6660 | 0.05 | -0.898 | Insignificant |
| July | -54 | -0.0811 | -0.6932 | 0.4882 | 0.05 | -0.941 | Insignificant |
| August | -212 | -0.3183 | -2.7596 | 0.0058 | 0.05 | -3.125 | **Significant** |
| September | -64 | -0.0961 | -0.8239 | 0.4100 | 0.05 | -1.256 | Insignificant |
| October | -16 | -0.0240 | -0.1962 | 0.8445 | 0.05 | -0.262 | Insignificant |

| | | | | | | | |
|---|---|---|---|---|---|---|---|
| November | 44 | 0.0660 | 0.5624 | 0.5738 | 0.05 | 0.206 | Insignificant |
| December | -44 | -0.0661 | -0.5624 | 0.5738 | 0.05 | -0.107 | Insignificant |
| Winter | -62 | -0.0930 | -0.7978 | 0.4250 | 0.05 | -0.470 | Insignificant |
| Spring | 4 | 0.0060 | 0.0392 | 0.9687 | 0.05 | 0.067 | Insignificant |
| Summer | -96 | -0.1441 | -1.2425 | 0.2141 | 0.05 | -4.414 | Insignificant |
| Autumn | -36 | -0.0541 | -0.4578 | 0.6471 | 0.05 | -1.140 | Insignificant |
| Annual | -94 | -0.1411 | -1.2163 | 0.2239 | 0.05 | -6.449 | Insignificant |

Note: P-value of less than or equal to 0.05 is significant, while one more than 0.05 is not

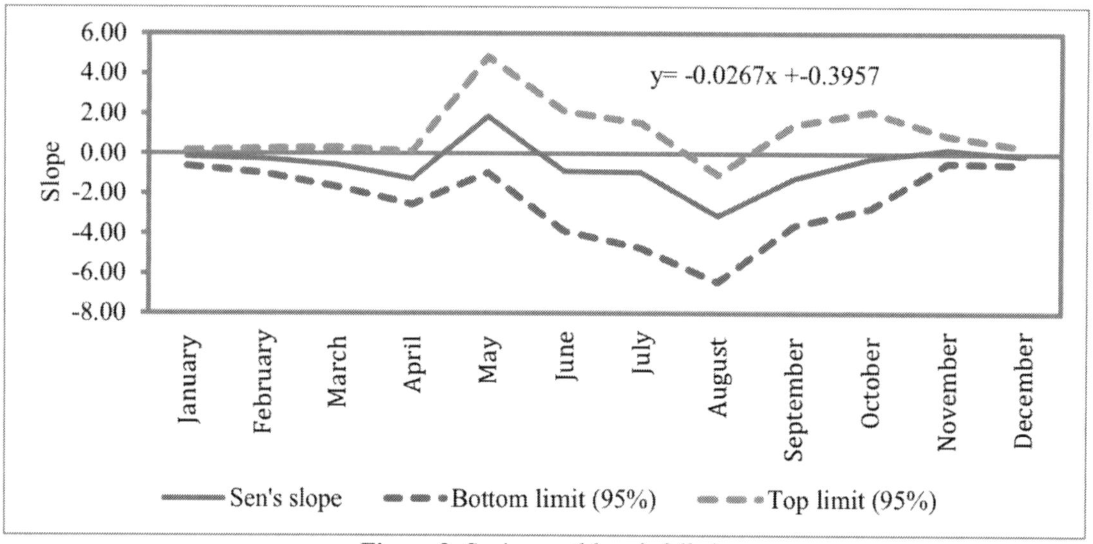


**Figure 2.** Sen's monthly rainfall slope

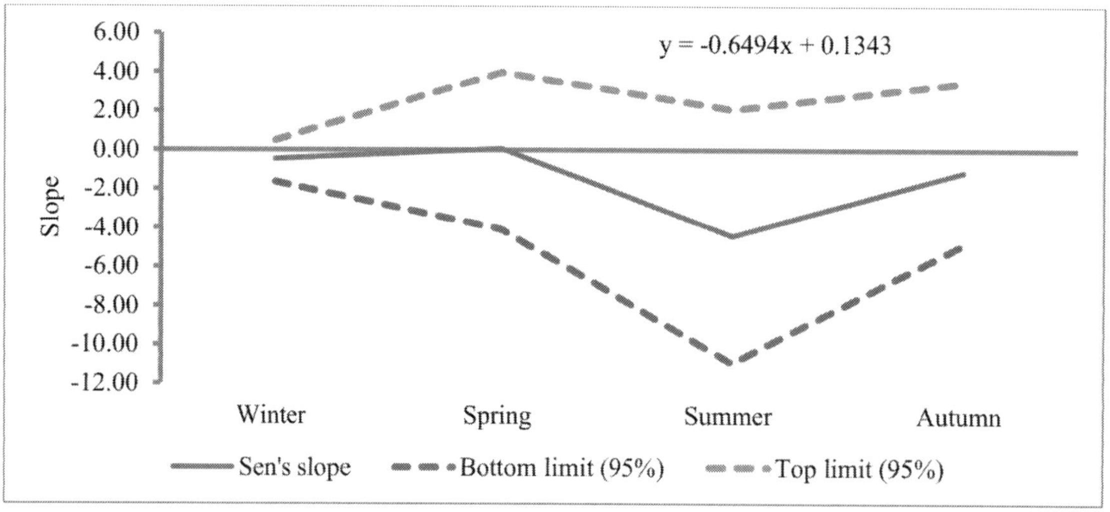


**Figure 3.** Sen's seasonal rainfall slope

### *3.2 Trend of temperature*

*3.2.1 Minimum monthly temperature trend.* Table 2 displays the monthly average minimum temperature for the period of 1981-2017, which demonstrates a warming trend. The Mann-Kendall test found a significant warming tendency for the months of April, May, July, August, September, and October at the 0.05 significance level. A notable trend has not been observed for the months of January, February, March, June, November, and December. The direction of change in minimum temperature for all months was increased, as evidenced by positive Mann-Kendall Stat (S), Kendall's tau, and Z Statistics. The presence of positive change per month was also demonstrated by Sen's slope. Sen's slope for monthly minimum temperature is shown in Figure 4.

**Table 2.** Mann-trend Kendall's test findings for monthly minimum temperatures

| Months | MK Stat (S) | Kendall's tau | Z Statistic | P-value | Sen's Slope | Sig. at α =0.05 |
|---|---|---|---|---|---|---|
| Jan | 94 | 0.1414 | 1.2165 | 0.2238 | 0.023 | Insignificant |
| Feb | 109 | 0.1640 | 1.4129 | 0.1577 | 0.020 | Insignificant |
| Mar | 97 | 0.1458 | 1.2557 | 0.2092 | 0.015 | Insignificant |
| Apr | 258 | 0.3880 | 3.3618 | 0.0008 | 0.040 | Significant |
| May | 186 | 0.2801 | 2.4206 | 0.0155 | 0.022 | Significant |
| Jun | 146 | 0.2202 | 1.8974 | 0.0578 | 0.011 | Insignificant |
| Jul | 181 | 0.2728 | 2.3552 | 0.0185 | 0.013 | Significant |
| Aug | 172 | 0.2594 | 2.2378 | 0.0252 | 0.008 | Significant |
| Sep | 312 | 0.4699 | 4.0689 | 0.00004 | 0.015 | Significant |
| Oct | 180 | 0.2707 | 2.3415 | 0.0192 | 0.033 | Significant |
| Nov | 107 | 0.1608 | 1.3865 | 0.1656 | 0.023 | Insignificant |
| Dec | 16 | 0.0240 | 0.1962 | 0.8445 | 0.005 | Insignificant |

Note: P-value of less than or equal to 0.05 is significant, while one more than 0.05 is not.

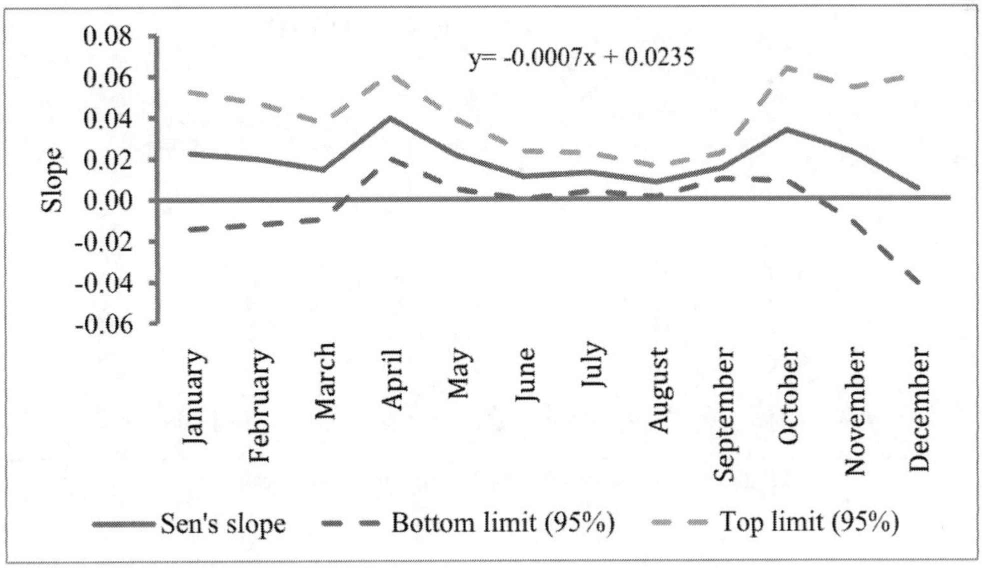


**Figure 4.** Sen's slope for monthly minimum temperature

*3.2.2 Trend of monthly maximum temperature.* The monthly maximum temperature for the period of 1981-2017 shows a warming trend, as shown in Table 3. The Mann-Kendall test has indicated a significant warming tendency at the 0.05 significance level for all months except January, May, November, and December. For all months, the positive Mann-Kendall Stat (S), Kendall's tau, and Z Statistics demonstrated an increasing trend in maximum temperature. The presence of positive change was also observed by Sen's slope for all months. Sen's slope for monthly maximum temperature is shown in Figure 5.

**Table 3.** Mann-trend Kendall's test findings for monthly maximum temperatures

| Months | MK Stat (S) | Kendall's tau | Z Statistic | P-value | Sen's Slope | Sig. at α =0.05 |
|---|---|---|---|---|---|---|
| January | 146 | 0.2195 | 1.8968 | 0.05786 | 0.036 | Insignificant |
| February | 229 | 0.3446 | 2.9828 | 0.00286 | 0.077 | Significant |
| March | 286 | 0.4294 | 3.7275 | 0.00019 | 0.077 | Significant |
| April | 244 | 0.3664 | 3.1782 | 0.00148 | 0.101 | Significant |
| May | 22 | 0.0330 | 0.2747 | 0.78360 | 0.010 | Insignificant |
| June | 177 | 0.2664 | 2.3025 | 0.02131 | 0.029 | Significant |
| July | 161 | 0.2419 | 2.0928 | 0.03637 | 0.017 | Significant |
| August | 152 | 0.2286 | 1.9752 | 0.04824 | 0.016 | Significant |
| September | 220 | 0.3308 | 2.8648 | 0.00417 | 0.028 | Significant |
| October | 187 | 0.2810 | 2.4329 | 0.01498 | 0.029 | Significant |
| November | 139 | 0.2092 | 1.8053 | 0.07102 | 0.013 | Insignificant |
| December | 83 | 0.1249 | 1.0727 | 0.28340 | 0.011 | Insignificant |

Note: P-value of less than or equal to 0.05 is significant, while one more than 0.05 is not

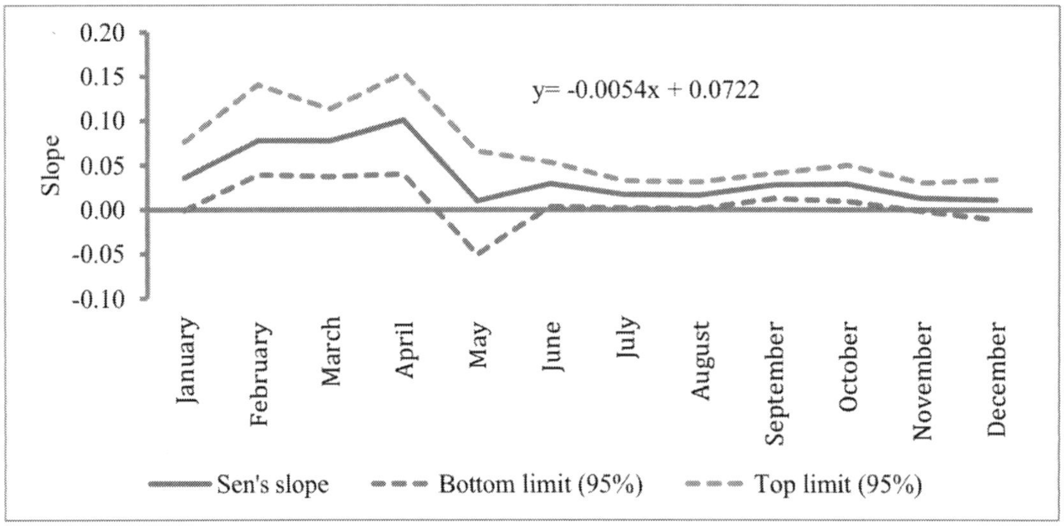


**Figure 5.** Sen's slope for monthly maximum temperature

*3.2.3 Seasonal and yearly temperature trends.* Table 4 displays the results of the Mann-Kendall trend test, which revealed an increasing warming trend for seasonal and yearly maximum and minimum temperatures. Except for the winter minimum temperature, the tendency of maximum and minimum temperatures on both a seasonal and annual basis demonstrated a statistically significant warming trend at the 0.05 significance level. The presence of seasonal and annual temperature increment is further confirmed by the Sen's slope. Seasonal and yearly maximum and minimum temperatures have increased throughout the period of 1981-2017, according to positive MK test Statistics (S), Kendall's tau, and Z Statistics. Sen's slope is shown in Figure 6 and Figure 7 for the minimum and maximum seasonal temperatures, respectively. A study of the upper Blue Nile River Basin found a similar trend in yearly minimum and maximum temperatures from 1981 to 2010 [28].

**Table 4.** Mann-Kendall trend test results for seasonal and yearly temperatures

| Variable | Seasons | MK Stat (S) | Kendall's tau | Z Statistic | P-value | Sen's Slope | Sig. at α =0.05 |
|---|---|---|---|---|---|---|---|
| Tmax | Winter | 253 | 0.3414 | 3.0484 | 0.00230 | 0.044 | Significant |
| | Spring | 264 | 0.3565 | 3.1817 | 0.00146 | 0.062 | Significant |
| | Summer | 182 | 0.2458 | 2.1897 | 0.02855 | 0.019 | Significant |
| | Autumn | 255 | 0.3448 | 3.0733 | 0.00212 | 0.020 | Significant |
| Tmin | Winter | 107 | 0.1444 | 1.2823 | 0.19970 | 0.013 | Insignificant |
| | Spring | 284 | 0.3835 | 3.4237 | 0.00062 | 0.025 | Significant |
| | Summer | 209 | 0.2824 | 2.5165 | 0.01185 | 0.010 | Significant |
| | Autumn | 281 | 0.3797 | 3.3876 | 0.00071 | 0.025 | Significant |
| Annual | Tmax | 262 | 0.3934 | 3.4136 | 0.00064 | 0.035 | Significant |
| | Tmin | 226 | 0.3393 | 2.9427 | 0.00325 | 0.018 | Significant |

Note: P-value of less than or equal to 0.05 is significant, while one more than 0.05 is not.

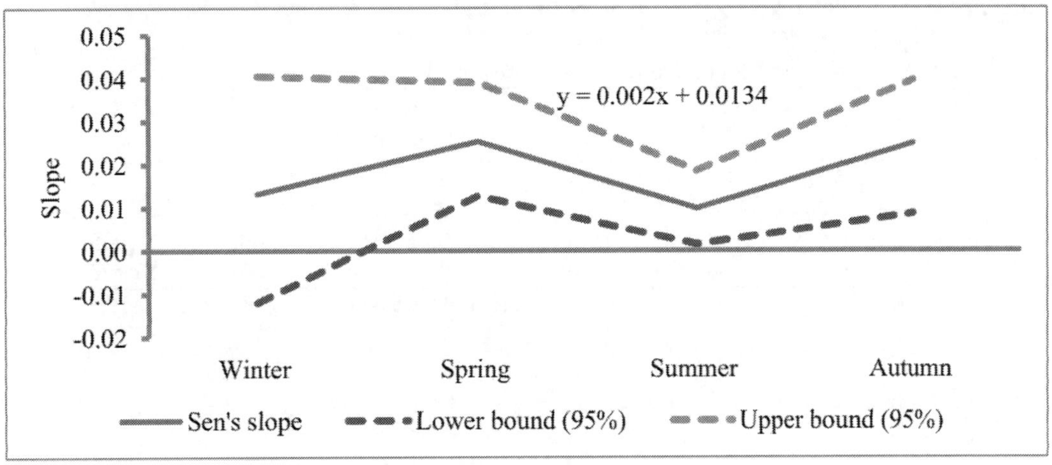


**Figure 6.** Sen's slope for seasonal minimum temperature

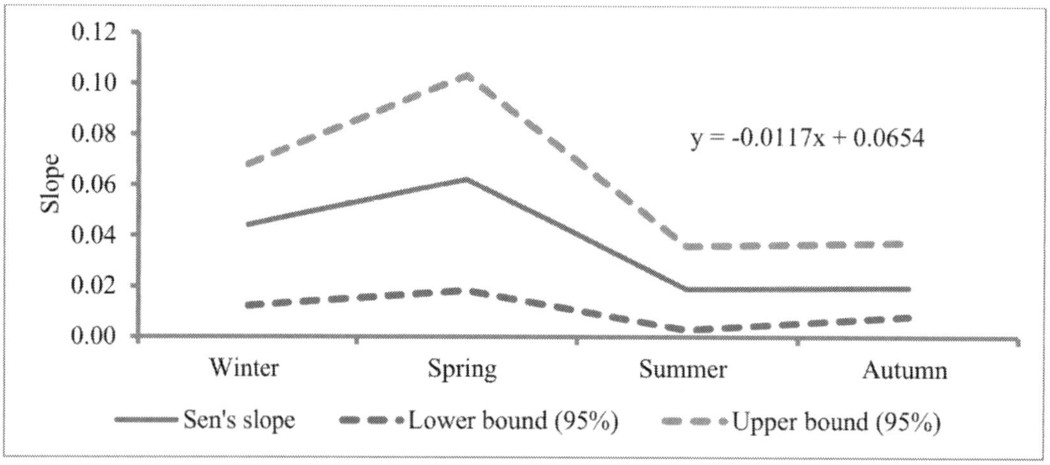


**Figure 7.** Sen's slope for seasonal maximum temperature

### *3.3 Summary of the fitted linear regression Model*

*3.3.1 Annual rainfall and temperature in relation to time.* The findings of the fitted linear model to analyze the relationship between rainfall, temperature, and time are shown in Table 5. At the 95% confidence level, the results showed that there was no significant association between annual rainfall and time. The fitted model's R-Squared score suggests that 4.52% of the variability in annual rainfall is explained by the fitted model, while the correlation coefficient of -0.21 indicates a weak relationship. At the 95% confidence level, there was a statistically significant association between annual maximum, minimum, and mean temperature and year. The R-squared data revealed that the variability in annual maximum, minimum, and mean temperatures was 21.30%, 20.34%, and 26.33%, respectively. The correlation coefficients for maximum, minimum, and mean temperature were 0.46, 0.45, and 0.51, respectively, indicating a moderate link between the variables. Figure 8 and Figure 9 show annual rainfall and temperature time series plots, respectively.

**Table 5.** Annual rainfall and temperature results from the fitted linear regression model

| Variable | Linear Regression Equation | $R^2$ (%) | correlation coefficient | P-Value | Relationship |
|---|---|---|---|---|---|
| Rainfall | y = −8.18x + 18179.1 | 4.52 | -0.21 | 0.2065 | Not significant |
| Tmax | y = 0.04x -56.9448 | 21.30 | 0.46 | 0.0040 | Significant |
| Tmin | y = 0.02x -21.5948 | 20.34 | 0.45 | 0.0051 | Significant |
| Tmean | y = 0.03x -40.4901 | 26.33 | 0.51 | 0.0012 | Significant |

Note: P-value of less than or equal to 0.05 is significant, while one more than 0.05 is not.

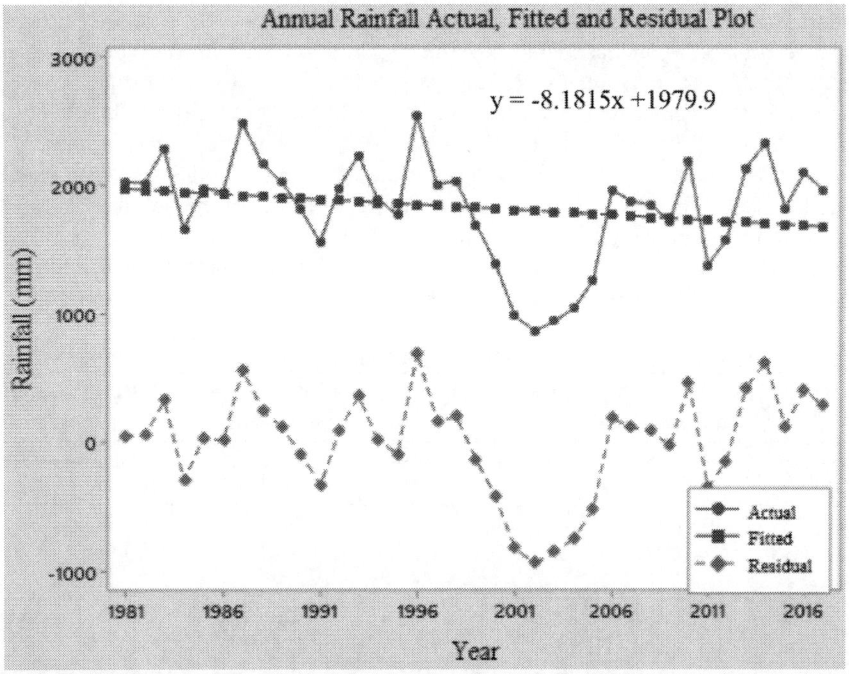


**Figure 8.** Annual rainfall actual, fitted and residual plot using linear regression model

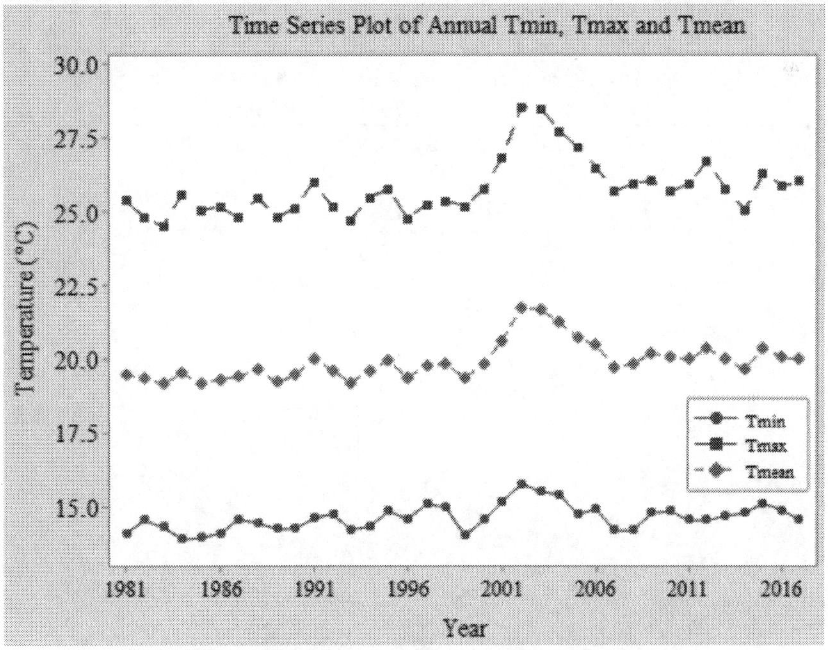


**Figure 9.** Yearly temperature time series plot

*3.3.2 Seasonal rainfall and temperature in relation to time.* The findings of the fitted linear regression model to explore the association between seasonal temperature and rainfall with time are shown in Table 6. Except for the winter minimum temperature, the results demonstrated a statistically significant link between seasonal minimum and maximum temperatures and time at the 95% confidence level. The fitted model exhibited 14.68%, 22.29%, 15.29% and 12.66% of variability in winter, spring, summer and autumn maximum temperature respectively. Winter, spring, summer, and autumn minimum temperatures varied by 2.22%, 26.63%, 18.48%, and 21.04%, respectively. At the 95% confidence level, there was no statistically significant association between winter, spring, summer, and autumn rainfall and time. The fitted model's R-squared statistics indicated that rainfall variability in winter, spring, summer, and autumn was 8.41%, 0.14%, 7.143%, and 0.79%, respectively. Figure 10 depicts a seasonal rainfall time series plot.

**Table 6.** Summary of linear regression model for seasonal rainfall and temperature

| Season | Variable | Linear Regression Equation | $R^2$ (%) | Correlation coefficient | P-Value | Relationship |
|---|---|---|---|---|---|---|
| Winter | Tmax | y= 0.05x -73.62 | 14.68 | 0.3832 | 0.0192 | Significant |
| | Tmin | y= 0.01x -9.51 | 2.22 | 0.1489 | 0.3791 | Insignificant |
| Spring | Tmax | y= 0.07x -103.72 | 22.29 | 0.4721 | 0.0032 | Significant |
| | Tmin | y= 0.03x -34.73 | 26.63 | 0.5160 | 0.0011 | Significant |
| Summer | Tmax | y = 0.02x -18.83 | 15.29 | 0.3911 | 0.0167 | Significant |
| | Tmin | y= 0.01x -7.18 | 18.48 | 0.4299 | 0.0079 | Significant |
| Autumn | Tmax | y= 0.03x -31.51 | 12.66 | 0.3558 | 0.0307 | Significant |
| | Tmin | y= 0.02x -35.16 | 21.04 | 0.4587 | 0.0043 | Significant |
| Winter | | y = - 1.02x + 2089.94 | 8.41 | -0.2900 | 0.0817 | Insignificant |
| Spring | Rainfall | y= - 0.40x +1149.4 | 0.14 | -0.0375 | 0.8258 | Insignificant |
| Summer | | y= - 5.74x + 12464.6 | 7.143 | -0.2673 | 0.1098 | Insignificant |
| Autumn | | y= - 1.02x + 2475.14 | 0.79 | -0.0889 | 0.6007 | Insignificant |

Note: P-value of less than or equal to 0.05 is significant, while one more than 0.05 is not.

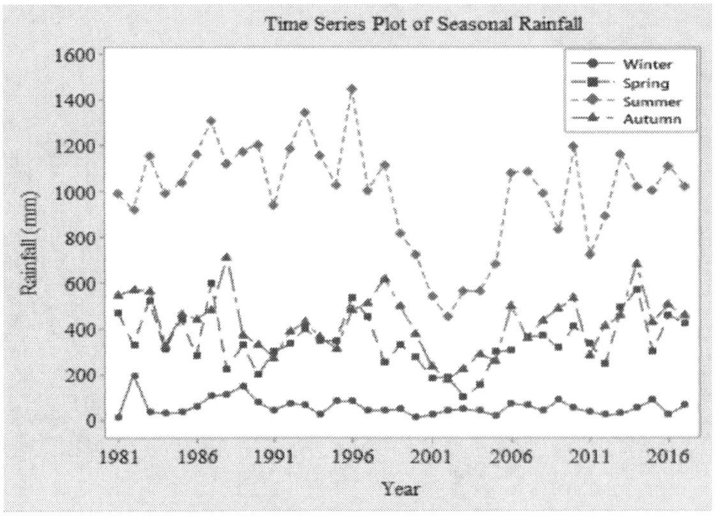


**Figure 10.** Time series plot of seasonal rainfall

**Table 7.** Coefficient of variation of rainfall and temperature

| Month | Rainfall (mm) | | | Tmax (°C) | | | Tmin (°C) | | |
|---|---|---|---|---|---|---|---|---|---|
| | Mean | SD | CV (%) | Mean | SD | CV (%) | Mean | SD | CV (%) |
| January | 17.45 | 21.30 | 122 | 27.15 | 1.57 | 6 | 13.03 | 1.15 | 9 |
| February | 22.49 | 20.53 | 91 | 29.08 | 1.87 | 6 | 14.62 | 1.12 | 8 |
| March | 56.18 | 36.72 | 65 | 30.30 | 1.56 | 5 | 16.17 | 0.65 | 4 |
| April | 90.83 | 41.44 | 46 | 29.67 | 1.94 | 7 | 16.58 | 0.78 | 5 |
| May | 199.70 | 87.36 | 44 | 27.28 | 2.19 | 8 | 16.56 | 0.63 | 4 |
| June | 296.39 | 95.91 | 32 | 23.83 | 0.79 | 3 | 15.70 | 0.36 | 2 |
| July | 342.36 | 96.66 | 28 | 22.00 | 0.68 | 3 | 14.85 | 0.35 | 2 |
| August | 353.61 | 92.28 | 26 | 22.15 | 0.49 | 2 | 14.94 | 0.24 | 2 |
| September | 277.61 | 74.52 | 27 | 23.05 | 0.63 | 3 | 15.01 | 0.28 | 2 |
| October | 116.06 | 75.72 | 65 | 24.31 | 1.00 | 4 | 13.46 | 0.94 | 7 |
| November | 32.88 | 22.66 | 69 | 24.83 | 1.17 | 5 | 12.20 | 1.11 | 9 |
| December | 18.70 | 19.86 | 106 | 25.57 | 1.41 | 5 | 11.99 | 1.51 | 13 |
| Winter | 58.63 | 37.93 | 65 | 27.27 | 1.43 | 5 | 13.21 | 0.83 | 6 |
| Spring | 346.71 | 116.05 | 33 | 29.09 | 1.52 | 5 | 16.44 | 0.54 | 3 |
| Summer | 992.36 | 232.43 | 23 | 22.66 | 0.57 | 3 | 15.17 | 0.28 | 2 |
| Autumn | 426.55 | 124.73 | 29 | 24.06 | 0.85 | 4 | 13.56 | 0.57 | 4 |
| Annual | 1824.25 | 416.60 | 23 | 25.77 | 0.97 | 4 | 14.59 | 0.43 | 3 |

Note: SD stands for Standard Deviation, and CV is for Coefficient of Variation

### *3.4 Rainfall and temperature variability*

Table 7 illustrates the coefficient of variation values based on [29] as (CV < 20) low variability, (20 < CV < 30) modest variability, and (CV > 30) significant variability in rainfall. Accordingly, the seasonal variation of rainfall was moderate for summer (23%) and autumn (29%) whereas it was high for spring (33%) and winter (65%), and yearly rainfall variation was moderate (23%). The temperature variation was classified as (steady < 5°C; modest 5 ≤ 10°C; and excessive, ≥10°C) according to [30]. Seasonal maximum temperature was stable for summer (3%) and autumn (4%) while it was moderate for winter and spring (5%). The seasonal minimum temperature was stable in the spring (3%), summer (2%), and autumn (4%), but moderate in the winter (6%). The annual maximum and minimum temperatures remained stable. Except for the severe variability in the minimum temperature in December, the temperature variation in the area was classified as stable and moderate. Monthly, seasonal, and yearly rainfall coefficients of variation are shown in Figure 11.

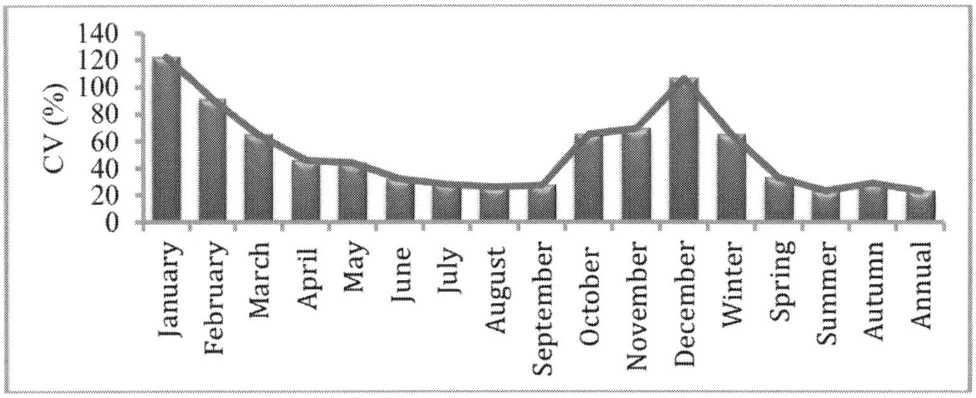


**Figure 11.** Coefficients of variance for monthly, seasonal, and yearly rainfall

## 4. Conclusion

Climate change is one of the world's most pressing environmental, social, and economic issues. The purpose of this study was to investigate the trend and variability of rainfall and temperature in the East Wollega Zone of Oromia National Regional State. The result of the study found that, except for May, November, and spring, the monthly, seasonal, and yearly rainfall all showed an insignificant declining trend. In the monthly, seasonal, and annual time scales, maximum and minimum temperatures revealed an increasing warming trend. At the 0.05 significance level, the results of the linear regression model revealed an insignificant connection between monthly, seasonal, and yearly rainfall and time. There was a statistically significant link between annual maximum, minimum, and mean temperature and year at 0.05 significance level. With the exception of the winter minimum temperature, a statistically significant connection between seasonal temperature and time was identified at the 0.05 significance level. The variance of monthly, seasonal, and annual minimum and maximum temperatures ranged from stable to moderate, with the exception of the December minimum temperature, which showed severe variability. Summer and autumn had a moderate seasonal fluctuation in rainfall, whereas spring and winter had a significant variation. The fluctuation in annual rainfall was moderate. Because crop production is influenced by rainfall and temperature variability, smallholder farmers must learn to adapt to their negative repercussions. It is better to consider this variability when deciding a cropping system and planning adaptation in the area. Despite the fact that the climate system is complicated and has many variables, the current study only considers rainfall and temperature. As a result, future research should take into account more climate parameters for more in-depth investigations in the area.

## References

[1] Powell J 2017 Scientists Reach 100% Consensus on Anthropogenic Global Warming *Bull. Sci. Technol. Soc.* **37** 183–4

[2] Adefisan A 2018 Climate Change Impact on Rainfall and Temperature Distributions Over West Africa from Three IPCC Scenarios *J Earth Sci Clim Chang.* **9** 476

[3] IPCC 2014 *Summary for Policymakers. In Climate Change 2014: Synthesis Report*

[4] Kaushik G and Sharma K C 2015 Climate Change and Rural Livelihoods-Adaptation and Vulnerability in Rajasthan *Glob. Nest J.* **17** 41–9

[5] Nelson G C, Rosegrant M W, Koo J, Robertson R, Sulser T, Zhu T, Ringler C, Msangi S, Palazzo A, Batka M, Magalhaes M, Valmonte-santos R, Ewing M and Lee D 2009 *Climate Change Impact*

*on Agriculture and Costs of Adaptation* (Washington, D.C: , IFPRI Food policy report,International Food Policy Research Institute)

[6] Onu F M and Ikehi M E 2018 Mitigation and Adaptation Strategies in the Face of Climate Change *J. Agric. Vet. Sci.* **9** 67–71

[7] Sayemuzzaman M, Jha M K, Mekonnen A and Schimmel K A 2014 Subseasonal Climate Variability for North Carolina, United States *Atmos. Res.* **145–146** 69–79

[8] Chijioke O B, Haile M and Waschkeit C 2011 *Implication of Climate Change on Crop Yield and Food Accessibility in Sub - Saharan Africa*

[9] Berhane A, Hadgu G, Worku W and Abrha B 2020 Trends in Extreme Temperature and Rainfall Indices in the Semi-arid Areas of Western Tigray, Ethiopia *Environ. Syst. Res.* **9** 1–20

[10] Echeverría D and Tornton A 2011 *Review of Current and Planned Adaptation Action : Southern Africa and Planned Adaptation Action : Southern Africa*

[11] Gashawl T, Mebrat W, Hagos D and Nigussie A 2017 Climate Change Adaptation and Mitigation Measures in Ethiopia *J. Biol. Agric. Healthc.* **4** 148–53

[12] Wudu Abiye ABEBAW 2020 Review on Climate Change Adaptation and Mitigation Practices in Ethiopian Agriculture Sector *Black Sea J. Agric.* **3** 340–3

[13] Zegeye H 2018 Climate Change in Ethiopia : Impacts , Mitigation and Adaptation *Int. J. Res. Environ. Stud.* **5** 18–35

[14] Morton J F 2007 The Impact of Climate Change on Smallholder and Subsistence Agriculture *Proc. Natl. Acad. Sci. U. S. A.* **104** 19680–5

[15] Koo J, Thurlow J, Eldidi H, Ringler C and Pinto A D E 2019 *Building Resilience to Climate Shocks in Ethiopia* (Washington, DC: International Food Policy Research Institute)

[16] Seyoum Y H 2015 *Perceptions, Vulnerability and Adaptation to Climate Change in Ethiopia : The Case of Smallholder Farmers in Sidama*

[17] Abegaz W B and Mekoya A 2020 Rainfall Variability and Trends over Central Ethiopia *Int. J. Environ. Sci. Nat. Resour.* **24** 215–27

[18] Elzopy K A, Chaturvedi A K, Chandran K M, Gopinath G, Naveena K and Surendran U 2021 Trend Analysis of Long-term Rainfall and Temperature Data for Ethiopia *South African Geogr. J.* **103** 381–94

[19] Georgis K 2015 Planning National Adaptation Responses to Climate Change *Clim. Chang. Agric. Food Secur.* 1–8

[20] Firamye G, Asamenew T and Mezene W 2015 Study on Prevalence of Poultry Coccidiosis in Nekemte town, East Wollega, Ethiopia *African J. Agric. Res.* **10** 328–33

[21] Fita T 2014 White Mango Scale, Aulacaspis tubercularis, Distribution and Severity Status in East and West Wollega Zones, Western Ethiopia *Sci. Technol. Arts Res. J.* **3** 1–10

[22] Abera D, Birhanu T and Negassa D 2014 Prevalence of Bovinetrypanosomiasis and Associated Risk Factors in East Wollega Zone, Western Ethiopia *Eur. J. Biol. Sci.* **6** 33–9

[23] Chimdi A 2015 Assessment of the Severity of Acid Saturations on Soils Collected from Cultivated Lands of East Wollega Zone, Ethiopia *Sci. Technol. Arts Res. J.* **3** 42–8

[24] Mann H B 1945 Non-Parametric Test Against Trend *Econometrica* **13** 245–59

[25] Sen P K 1968 Estimates of the Regression Coefficient Based on Kendall's Tau *J. Am. Stat. Assoc.* **63** 1379–89

[26] Verbeek M 2017 Using Linear Regression to Establish Empirical Relationships *IZA World Labor* **336** 1–11

[27] Teshome M 2017 Perceived Impact of Climate Change on Crop Yield Trend in Denbia Woreda of Amhara Region, Northwest Ethiopia *MOJ Ecol. Environ. Sci.* **2** 1–12

[28] Mengistu D, Bewket W and Lal R 2014 Recent Spatiotemporal Temperature and Rainfall Variability and Trends over the Upper Blue Nile River Basin, Ethiopia *Int. J. Climatol.* **34** 2278–92
[29] Hare F K 1983 Climate and Desertification *Revised analysis (WMO-UNDP) WCP-44* (Geneva, Switzerland) pp 5–20
[30] Wehrly K E, Wiley M J and Seelbach P W 2003 Classifying Regional Variation in Thermal Regime Based on Stream Fish Community Patterns *Trans. Am. Fish. Soc.* **132** 18–38

**Acknowledgements**
I would like to express my sincere gratitude to Wollega University for providing me with the chance to pursue my PhD in climate change mitigation and adaptation.

# Raster correlation method for vegetation density and surface temperature in Kupang City

**Arfita Rahmawati[1]* and Bella Theo Tomi Pamungkas[2]**

[1]Geography Education Departement, Nusa Cendana University, Jl. Laksda Adi Sucipto, Penfui, Indonesia, 85001
[2]Geography Education Departement, Nusa Cendana University, Jl. Laksda Adi Sucipto, Penfui, Indonesia, 85001
*Corresponding author: ararfita@gmail.com

**Abstract.** The phenomenon of urban heat is caused by global warming that occurred in Kupang City. This causes changes in vegetation density and microclimate especially surface temperature. The study aims to known determine the value of vegetation density and surface temperature in Kupang City in 2010, 2015, and 2021. The raster correlation method was used to study the changes in vegetation density and surface temperature in Kupang City. The research was using the data from Landsat 7 ETM+ satellite imagery in 2010, Landsat 8 OLI TIRS in 2015, and 2020. The changes in vegetation density in the Kupang City area could be seen from the results of processing the satellite image data obtained in 2010 was medium vegetation (the vegetated land cover was in the form of coconut plantation and mixed plantation), in 2015 and 2020 was dominated by low vegetation (grasses and bushes). The results of this study showed that in 2010 the dominant temperature class was 26$^0$C - 28$^0$C. However, the temperature class increased until 2015 and 2020 where the dominant temperature was 29$^0$C - 31$^0$C. The regression coefficient value is negative, the meaning is when the vegetation is denser, the surface temperature value will be lower.

**Keywords**: raster correlation, vegetation density, surface temperature.

## 1. Introduction

The phenomenon of urban heat due to global warming that occurs in almost all cities in Indonesia is a microclimate phenomenon that cannot be avoided. Microclimate issues are an important factor in achieves sustainability within cities [1]. The phenomenon of urban heat increases the ambient temperature significantly [2]. The green space can reduce the impact of rising surface temperatures, by reducing heat from surfaces temperature especially at night [3]. The Greenspace is certainly related to the level of vegetation density. Vegetation density has benefits in the course of the global carbon cycle and mitigation of climate change [4]. Vegetation with a large density can provide protection for people from rising surface temperatures on land [5]. When vegetation density continues to decrease as a result of uncontrolled population density will cause other environmental problems that are mutually sustainable.

Kupang City is the capital city of East Nusa Tenggara Province, which is always development. The recent growth and development of Kupang City is the impact of the rapid development promoted by the central government [6]. Kupang city is a semi-arid area with significant differences in duration of rain (3–4 months) and dry season (8–9 months) [7]. With a small duration of rainy months, this will have an impact on the urban heat phenomenon.

Various previous studies related to surface temperature and vegetation density by Sendi Al Mukmin with the title of analyzing the effect of land cover changes on the distribution of surface temperature and its relationship to urban heat islands with the results of high-temperature research found in the downtown area which then decreased towards the village [8]. Duncan in his research entitled turning down the heat: an enhanced understanding of the relationship between urban vegetation and surface temperature at the city scale stated that random forests showed vegetation types equal to 31.84% of the out-of-bag variance in summer, that an increase in large vegetation cover in a location has an impact on cooling [9]. This study aims to determine changes in vegetation density and surface temperature in the Kupang City in 2010, 2015 and 2020 and their correlation.

Based on the problems above, the study needs to identify the effect of changes in vegetation on surface temperature in Kupang City. it can be seeing the vegetation density changes from the previous few years and the extent to which the presence of vegetation affects the surface temperature of the area. As well as providing input and information to the Kupang City government and related institutions on how the vegetation density and surface temperature are presented in the form of a map between 2010, 2015 and 2020 so the local government makes attention to the environmental conditions of Kupang City in particular. Especially to provide important information of the existence of vegetation.

## 2. Materials and method

The data used in this study are Landsat 7 ETM+ satellite imagery in 2010 and Landsat 8 OLI/TIRS satellite imagery for 2015 and 2020. The method used the Normalized Difference Vegetation Index (NDVI) analysis technique to analyze vegetation density in Kupang City. Normalized Difference Vegetation Index (NDVI) was used to calculate the vegetation density value from digital number processing of brightness data from multiple channels satellite sensor data [10]. In NDVI analysis, the mathematical equations that can be used to calculate NDVI are as follows [11]:

$$NDVI = \frac{R_{NIR}-R_{RED}}{R_{NIR} + R_{RED}} \quad (1)$$

Where $R_{NIR}$ is near-infrared band and $R_{RED}$ is infrared band. Next used an analysis of the surface temperature algorithm used the mathematical formula. The Mono-window Brightness Temperature algorithm used to correct the atmospheric parameters may affect the data obtained by sensors in space as a result of absorption and scattering of electromagnetic radiation [12]. Mono-window Brightness Temperature algorithm to convert the digital number to get the surface temperature value. The following are the steps in analyzing surface temperature using the Mono-window Brightness Temperature method [13]:

$$L_{\lambda =} L_{\min(\lambda)} + \{L_{\max(\lambda)} - L_{\min(\lambda)} / Q_{max}\} \, x \, QDN \quad (2)$$

Description,
Lλ : radiant spectral
Lmax(λ) : maximum radiant spectral
Lmin(λ) : minimum radiant spectral
Qmax : maximum K value digital number
QDN : digital number

$$Tb = \frac{K_2}{In(\frac{K_1}{L\lambda}+1)} \quad (3)$$

Description,
Tb : satellite brightness temperature (°K)
K1 : maximum spectral radiance
K2 : absolute temperature calibration constant (°K)
Lλ : radiant spectral

$$T_{Celcius} = T_{Kelvin} - 273 \tag{4}$$

Description,
$T_{Celcius}$ : temperature ($^0$C)
$T_{Kelvin}$ : temperature ($^0$K)

Then the last step used a raster correlation with the independent variable the value of the vegetation density being the variable (x) and the dependent variable (y) the surface temperature. For raster correlation used simple regression until get correlation between two variables:

$$Y = a + bx \tag{5}$$

Description,
Y : dependent variable (surface temperature)
a : the price of Y when X = 0 (constant price)
b : the direction number of the regression coefficient
X : independent variable (vegetation density and land cover)

The research method is described in the form of a research flow chart as Figure 1 follows:

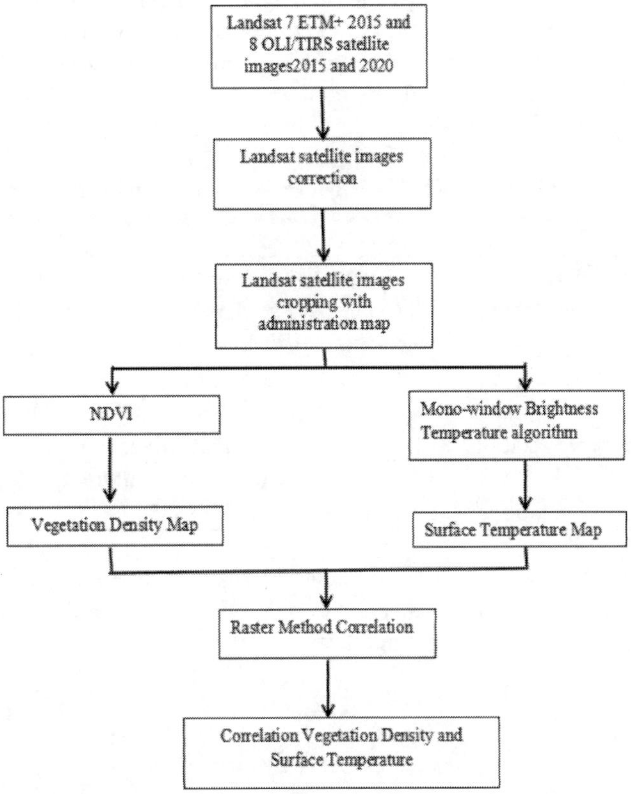


**Figure 1.** Flow diagram

## 3. Results and discussion

### *3.1. Vegetation density*

The vegetation density map is needed from the results of the NDVI transformation for that in the NDVI transformation. The year 2010 had a minimum value of -0.480461 and a maximum value of 0.8446 and 2015 had a minimum value of -0.16945 and a maximum value of 0.554608. In 2020 had a minimum value of -0.860565 and a maximum value is 0.81257 (Figure 2). This diversity of NDVI values need to be shareded or simplified into simple classes to form a pattern of the same or homogeneous NDVI values for interpretation process.

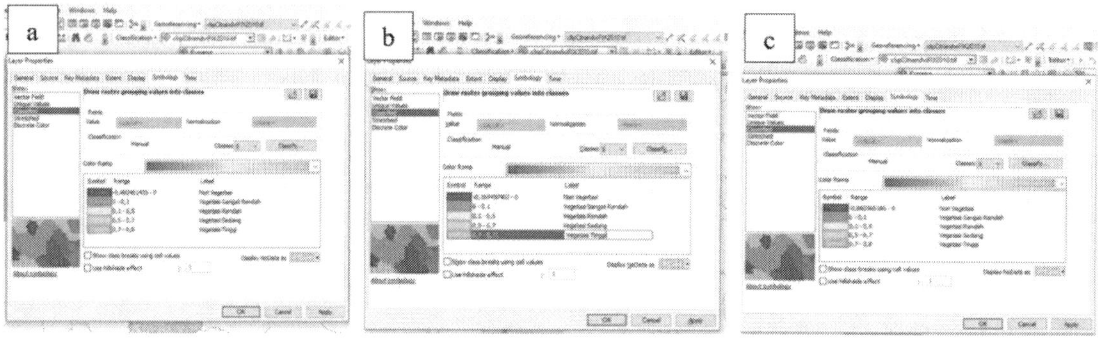

**Figure 2.** NDVI 2010 (a); NDVI 2015 (b); NDVI 2020 (c)

Based on the classification of the NDVI value, five classes of vegetation density are obtained where the five classes include non-vegetation classes marked with a vegetation index value (0 - 0.1), very low vegetation class with a value (0.1 - 0.5), low vegetation with a value (0.51 - 0.7), medium vegetation with a value (0.51 - 0.7) and high vegetation marked with a value (> 0.71). For NDVI classification of vegetation density based on Table 1 [14]:

**Table 1.** NDVI classification of vegetation density

| Class | Vegetation Density | Type |
|---|---|---|
| < 0 | Non-Vegetation | Water bodies like rivers, lakes, etc. |
| 0 – 0,1 | Lower Vegetation | Open land settlements covered with asphalt or paving or asphalt roads |
| 0,11 – 0,5 | Low Vegetation | Ground cover vegetation (such as grasses and bushes), dirt roads, empty fields without asphalt or paving. |
| 0,51 – 0,7 | Medium Vegetation | the vegetated land cover was in the form of coconut plantation and mixed plantation. |
| > 0,71 | High Vegetation | forested vegetation |

The results of this classification can create new information in the form of a layout map of the vegetation density of Kupang City which can describe the results of processing Landsat satellite imagery through NDVI transformation using the red and near-infrared band spectral values for Landsat 7 and 8.

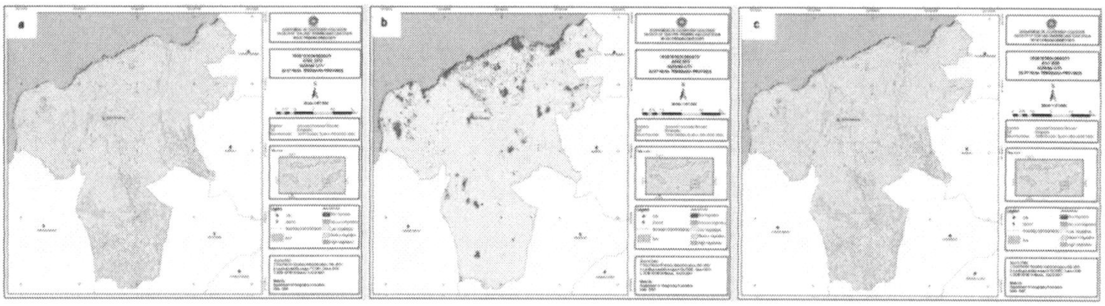

**Figure 3.** Vegetation density 2010 (a); vegetation density 2015 (b); vegetation density 2020 (c)

The vegetation density map can provide an overview of the vegetation density in Kupang City (Figure 3). The area of land that has vegetation according to the class that has been created, Table 1 will describe the results of the calculation of the area in hectares (Ha) of vegetation density class in each year of recording Landsat satellite imagery in 2010, 2015 and 2020:

**Table 2.** Area of vegetation density in Kupang City 2010, 2015, and 2020

| Year | Area (Ha) | | | | |
|---|---|---|---|---|---|
| | Non Vegetation | Lower Vegetation | Low Vegetation | Medium Vegetation | High Vegetation |
| 2010 | 53,19 | 95,49 | 3802,95 | 7391,25 | 4309,74 |
| 2015 | 145,08 | 731,52 | 14768,82 | 7,47 | - |
| 2020 | 132,21 | 210,15 | 11806,29 | 3235,68 | 268,47 |

Source: analysis result, 2021

Based on table 2 in 2010 the dominant vegetation is medium vegetation with area of 7391.25 Ha. The dominant vegetation density in 2015 is low vegetation class where the area is 14768.82 Ha. In 2020, the dominant vegetation density is the low vegetation class where the land area reaches 11806.29 Ha.

### *3.2. Surface temperature*

Utilization of Landsat satellite imagery used thermal channels had been devoted to utilization in mapping surface temperature and soil moisture. That can make it easier to analyze and detect temperatures on the earth's surface. Analysis in surface temperature in Kupang City used Landsat 7 ETM+ satellite imagery and Landsat 8 OLI TIRS satellite imagery which had a thermal channel to detect temperature over a wide area. These two Landsats have different thermal channels inside, the Landsat 7 ETM+ thermal channel in band 6 and on Landsat 8 OLI TIRS the thermal channel in bands 10 and 11.

Figure 4 shows the surface temperatures in 2010, 2015, and 2020, and Table 3 describes Figure 4. The surface temperature in 2010 was the lowest temperature of $17^0$C – $19^0$C and the highest temperature was $35^0$C – $37^0$C. In 2015 and 2020 was the lowest temperature $20^0$-$22^0$C and the high temperature was $35^0$C – $37^0$C. In 2010 the dominant surface temperature was $26^0$C - $28^0$C had a total area of 8215.65 Ha, but in 2015 the most dominant surface temperature was no longer at $26^0$C - $28^0$C but the most dominant was at $29^0$C - $31^0$C, with area of 9673.83 Ha. In 2020 the dominant temperature is still at $29^0$C - $31^0$C with the same area value as 2015 is 9673.83 Ha.

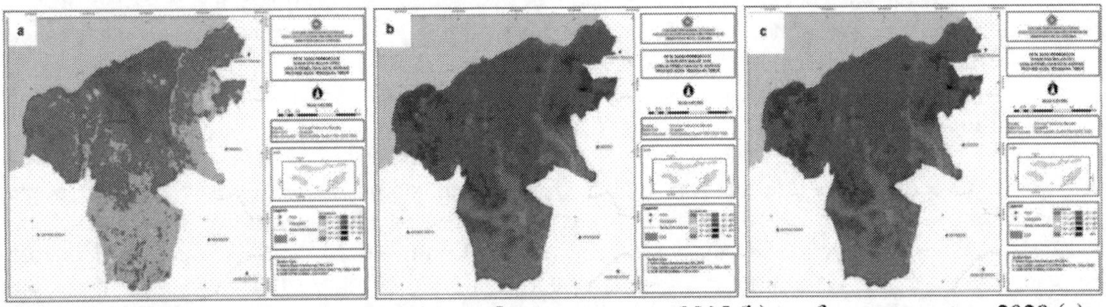

**Figure 4.** Surface temperature 2010 (a); surface temperature 2015 (b); surface temperature 2020 (c)

**Table 3.** Area of surface temperature in Kupang City 2010, 2015, and 2020

| Range of Surface Temperature ($^0$C) | Area (Ha) | | |
|---|---|---|---|
| | 2010 | 2015 | 2020 |
| 11 – 13 | 0 | 0 | 0 |
| 14 – 16 | 0 | 0 | 0 |
| 17 – 19 | 97.2 | 0 | 0 |
| 20 – 22 | 149.67 | 0.27 | 0.27 |
| 23 – 25 | 5218.83 | 359.1 | 359.1 |
| 26 – 28 | 8215.65 | 4178.25 | 4178.25 |
| 29 – 31 | 1828.44 | 9673.83 | 9673.83 |
| 32 – 34 | 230.22 | 1427.31 | 1427.31 |
| 35 – 37 | 0.36 | 14.13 | 14.13 |
| >38 | 0 | 0 | 0 |

Source: analysis result, 2021

*3.3. Raster correlation vegetation density and surface temperature in Kupang City*
Based on the results of NDVI processed in the form of vegetation density for the period in 2010, 2015, and 2020, the effect of changes in vegetation density on the results of surface temperature processing in the Kupang City area used a simple regression correlation. The result is as Figure 5 follows:

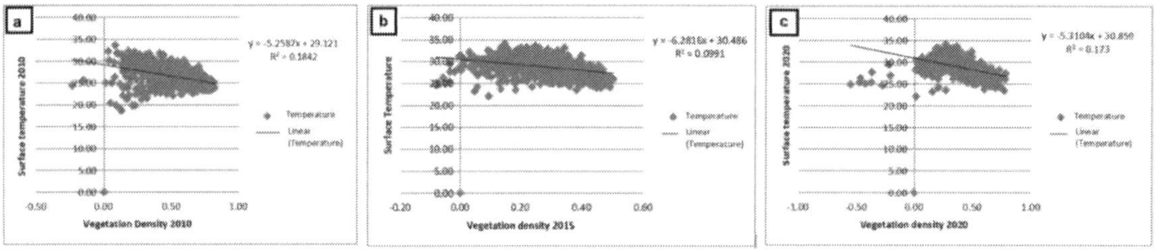


**Figure 5.** Correlation vegetation density and surface temperature 2010 (a); 2015 (b); 2020 (c)

In 2010 the equation y = -5.2587x + 29.121 generated with an effect of 18%. In 2015 the equation y = -6.2816x + 30.486 was generated with an effect of 9%. In 2020 the equation y = -5.3104x + 30.859 generated with an effect of 17%. From the three equations, the regression coefficient is negative, which means that the two parameters are inversely proportional.

## 4. Conclusions

The dominant vegetation density in 2010 was a medium vegetation class (the vegetated land cover was in the form of coconut plantation and mixed plantation) then the dominant vegetation class in 2015, the dominant class was a low vegetation class (grasses and bushes). 2020 in July the vegetation class that dominates is still a low vegetation class. The surface temperature in 2010 the dominant temperature class was 26$^0$C - 28$^0$C. However, the temperature class increased until 2015 and 2020 where the dominant temperature was 29$^0$C - 31$^0$C. From 2010, 2015, and 2020, the average value of the effect of vegetation density with surface temperature was 14.67%. Then the coefficient value is negative, meaning that the denser the vegetation, the lower the surface temperature. The continuous increase in surface temperature values by a decrease in vegetation density in Kupang City needs by government and community to realize the importance of high-density vegetation such as forested vegetation.

**References**

[1] Jin H, Cui P, Wong N H and Ignatius M 2018 Assessing the effects of urban morphology parameters on microclimate in Singapore to control the urban heat island effect *Sustainability (Switzerland)* **10**

[2] Li G, Zhang X, Mirzaei P A, Zhang J and Zhao Z 2018 Urban heat island effect of a typical valley city in China: Responds to the global warming and rapid urbanization *Sustainable Cities and Society* **38** 736–45

[3] Schettini E, Blanco I, Campiotti C A, Bibbiani C, Fantozzi F and Vox G 2016 Green control of microclimate in buildings *Agriculture and Agricultural Science Procedia* **8** 576–82

[4] Xu L, Shi Y, Fang H, Zhou G, Xu X, Zhou Y, Tao J, Ji B, Xu J, Li C and Chen L 2018 Vegetation carbon stocks driven by canopy density and forest age in subtropical forest ecosystems *Science of The Total Environment* **631–632** 619–26

[5] Nuruzzaman Md 2015 Urban heat island: causes, effects and mitigation measures - a review *International Journal of Environmental Monitoring and Analysis* **3** 67

[6] Liem Y, Mahagarmitha R R, Studi P, Arsitektur M, Arsitektur F, Wahidin J and No S 2019 Habitat perkotaan di Kota Kupang *ATRIUM* 31–40

[7] Nait C, Bisri M, Soetopo W, Prayogo T, Limantara L M and Krisnayanti D S 2021 Identification of climate change impact at Raknamo Dam, Kupang, Indonesia *IOP Conference Series: Earth and Environmental Science* **724** 012069

[8] Al Mukmin S A, Wijaya A and Sukmono A 2016 Analisis pengaruh perubahan tutupan lahan terhadap distribusi suhu permukaan dan keterkaitannya dengan fenomena Urban Heat Island *Jurnal Geodesi Undip* **5** 224–33

[9] Duncan J M A, Boruff B, Saunders A, Sun Q, Hurley J and Amati M 2019 Turning down the heat: An enhanced understanding of the relationship between urban vegetation and surface temperature at the city scale *Science of The Total Environment* **656** 118–28

[10] Philiani I 2016 Pemetaan vegetasi hutan mangrove menggunakan metode Normalized Difference Vegetation Index (NDVI) di Desa Arakan, Minahasa Selatan, Sulawesi Utara. SOIJST Vol. 1 (2):211-222 *Surya Octagon Interdisciplinary Journal of Technology* **1**

[11] Riko Y, Meha A I and Prasetyo S Y J 2019 Perubahan konversi lahan menggunakan NDVI, EVI, SAVI dan PCA pada Citra Landsat 8 (studi kasus : Kota Salatiga) *Indonesian Journal of Computing and Modeling* **1** 25–30

[12] Ndossi M I and Avdan U 2016 Application of open source coding technologies in the production of Land Surface Temperature (LST) maps from Landsat: A PyQGIS plugin *Remote Sensing* **8**

[13] Rongali G, Keshari A K, Gosain A K and Khosa R 2018 A mono-window algorithm for land surface temperature estimation from landsat 8 thermal infrared sensor data: A case study of the beas river basin, India *Pertanika Journal of Science and Technology* **26** 829–40

[14] Noviyanti I K and Roychansyah M S 2019 Analisis ketersediaan ruang terbuka hijau dengan NDVI menggunakan citra satelit Worldview 2 di Kota Yogyakarta *Majalah Ilmiah Globe* **21** 63

**Acknowledgement**

This study was funded by the Faculty of Teacher Training and Education, Nusa Cendana University.

# Decision-making in climate change adaptation through a cross-sectoral approach: review

**Aastha Lamichhane[1], Muhamad Khoiru Zaki[2], Emmanuel Okiria[3] and Keigo Noda[2]***

[1] Graduate School of Natural Science and Technology, Gifu University, Japan
[2] Faculty of Applied Biological Sciences, Gifu University, Japan
[3] The United Graduate School of Agricultural Sciences, Gifu University, Japan
*Corresponding author: anod@gifu-u.ac.jp

**Abstract**. Climate change adaptation (CCA) is an emerging topic in recent years since global temperature is continuing to rise and posing threats to natural biodiversity and human life. Its diverse nature requires efforts from a broad range of sectors to cope or adjust. This review paper aims to systematically study the cross-sectoral approaches in CCA decision making and develop probable strengths and shortcomings of those approaches. Literatures concerned with Multi-Criteria Decision-Making Method, or Multi-Criteria Analysis (MCA) are reviewed since it is considered as a holistic approach to integrate multiple sectors and combine monetary and non-monetary terms prevailing in sectoral and regional aspects, mainly in agriculture and water resource management. The literatures were searched through Scopus and PRISMA method was adopted to systematically refine the published articles based on our criteria. Out of 383 articles discovered, 139 were related to CCA out of which, 33 articles which applied MCA as their methodological approach were shortlisted for the core study. The result showed that MCA is extensively used in CCA decision making, prioritizing options, and formulating adaptation strategies at local and regional scale and considered as a flexible, transparent, and effective method because of an active engagement of stakeholders and experts' judgement. But its inability to address the underlying uncertainties of climate change scenario is one of the major drawbacks seen. Thus, many literatures suggest incorporating Sensitivity analysis, Dynamic Adaptive Pathways, Real Option Analysis, or Robust Decision-Making Analysis with MCA to overcome those deep uncertainties.

**Keywords** : climate change, MCA, water resource management, adaptation strategies

## 1. Introduction

In natural systems, Climate Change Adaptation (CCA) can be defined as the adjustment to the changing climate and its effects where anthropogenic influences may increase the coping capacity; whereas, in human systems, adaptation is the adjustment by humans with the foreseeable changing climate for minimizing damage or increasing benefits [1]. Adaptation in climate change is basically the adjustments at community or individual level to cope with the climatic-economic–social–political dynamics [2,3]. While, adaptation assessment refers to the estimation of the options that help to cope to the changing climate by evaluating the certain factors related like applicability, feasibility, cost-effectiveness and expected benefits [4]. Policy making in adaptation needs a holistic approach and incorporates wide range of sectors, sub sectors to formulate better decisions, where, ensuring sustainability of those formulated decisions is an important aspect [5].

A planned adaptation option requires the integration from both top and bottom level and the derived options should be locally acceptable and suitable [6,7]. It should always be context-specific and consider all the factors such as, regions, climate, environmental, social, and political conditions by incorporating both local stakeholders and experts in the related field. Such holistic plans help any community to cope with the consequences of climate change through risk minimization [8]. In such scenarios, concerned stakeholders are the ones to understand the social status and generate required remedies for their own area and hence, their involvement enhances the reliability of the adaptation option at the local level [9]. In order to respond to the long-term challenges of climate change, the decision makers should incorporate a broader sectors and experts from diverse fields to deal with the high uncertainty linked with it [1,10,11]. At the present context, Multi-criteria analysis (MCA) or Multi-criteria Decision-Making Method (MCDM) stands as a convenient and effective decision-making tool which can address a wide range of sectors, combine costs and benefits along with other qualitative options, and is a widely accepted method that considers a diverse range of actors to assess options against a set of criteria. Several countries have already formulated various adaptation tools and frameworks through MCA to aid in drafting adaptation plans and have been applied by different bilateral, multilateral, and non-governmental organizations [1]. Not only in the developed world, MCA is seemed to be applicable in Least Developed Countries (LDCs) as it has been successfully applied for evaluating farmer`s adaptation to climate variability due to drought [12], adaptation options for agriculture and flood risk reduction [4,5,13,14].

Multi-criteria analysis is a participatory, systematic and structured framework for assessing and prioritizing adaptation options under a broad scenario which allows normative judgement through the involvement of experts and stakeholders [5,10,15–18]. However, climate change is a dynamic term and the adaptive capacity could be weakened by the underlying uncertainties associated with it, whether that be socio-economic or climatic variables[19] and it is still a question that MCA can address those anomalies while formulating strategies or not. This review paper aims to study about the past applications of MCA in different fields across diverse regions of the world and analyze criteria required for developing adaptation strategies in different scenarios.

## 2. Methodology

Published journal articles were collected from scientific databases such as Scopus and Google Scholar. Preferred Reporting Items for Systematic Reviews and Meta-Analyses (PRISMA) method was used to analyze the articles extracted from Scopus to sort out the relevant studies. PRISMA method is a systematic technique used to screen the available articles through checklist and flow diagram [20]. PRISMA 2020 checklist and flow diagram (Figure 1) was used to perform the analysis through 27-item recommendation list on topics such as title, abstract, introduction, methods, results and discussion.Through a rigorous study and out of 383 research articles discovered, 139 papers were selected which solely analysed climate change adaptation options in different regions and out of those, 51 papers were found eligible for the review and 35 case studies were shortlisted which included MCA to formulate adaptation strategies.

The papers published between 2010 to 2021 (until June) were searched to understand the trends in climate change adaptation and were selected based on the following criteria:

a) papers those focused on implementation of actual adaptation practice rather than being limited to plans and policies.
b) consider both monetary and non-monetary adaptation options,
c) including the involvement of multiple sectors during the planning of adaptation policies,
d) adaptation planning and implementation through multi-criteria decision-making methods,

A systematic literature review was carried out by using PRISMA method and the articles were selected on the following processes: (1) Identification: The articles were identified in the scientific database; (2) Screening: All the articles were screened firstly by title and abstract, date of publishment, area; (3) Eligibility: Those papers which fell into our category of decision making in climate change adaptation focusing on water resource management, flood risk assessment/reduction, agricultural sector was selected; (4) Inclusion: Articles using MCA as at least one of their methods

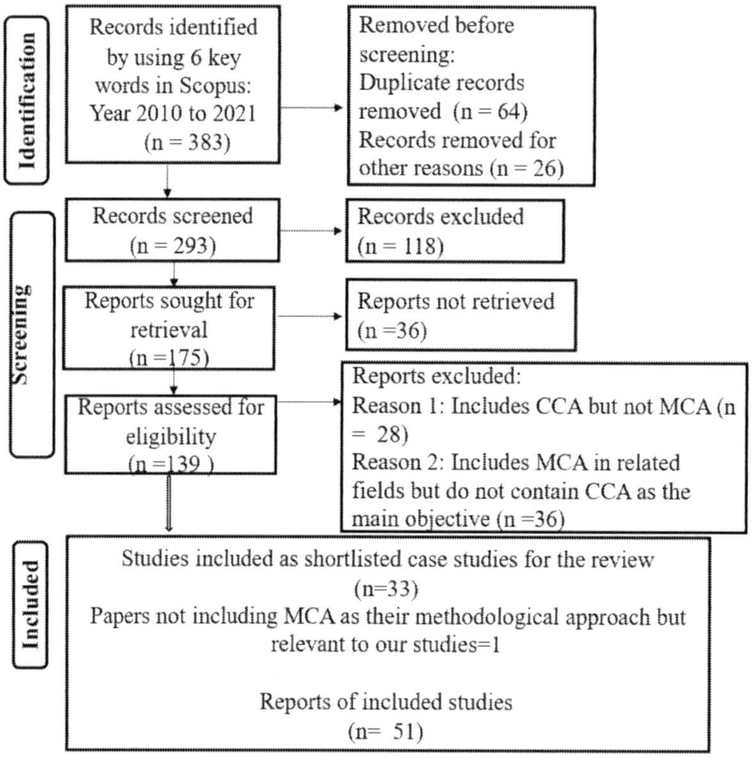


**Figure 1.** Flowchart of article selection through PRISMA method

## 3. Results and discussion

The literature review identified an exponential growth in the number of publications in the recent years. From 2010 to 2021 June, a total of 383 articles were displayed and we found that 139 of them met our research objectives of climate change adaptation and decision making Figure **2**. They were further screened for eligibility test through PRISMA method. About 128 research articles, 4 book chapter and 3 review paper. Multiple case studies of climate change adaptation between 2010 to 2021 came across in different parts of the world but those which met our research objectives, included MCA as their methodological approach for CCA fell under our eligibility criteria and thus, are shortlisted in this paper. Major area of focus of this review are: (1) CCA in agriculture; (2) Water resource management (WRM); and (3) Flood risk management.

Figure 2 shows an increasing trend in the number of publications from 2010 to 2016 with some fluctuations in 2017 and 2018. In 2019, highest (n=22) articles were published and that of 2021(until June), the published articles are 14 which is about 63 % of the whole year. According to our observations, extensive planning literatures have focused on IWRM flood control, drought management, followed by crop management in agricultural sector. A few papers cored their research into development of spatial planning in the urban area in order to diminish the flood loss which is a remarkable notation.

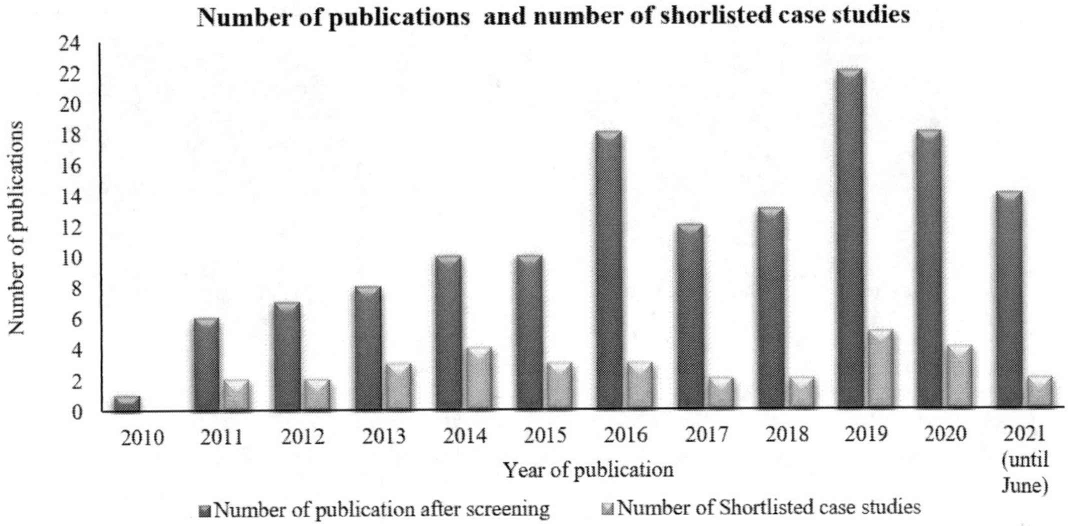


**Figure 2.** Distribution of 139 articles and shortlisted case studies from 2010 to 2021 (until June)

### *3.1 Categorical distribution of literature*

Figure 3 shows the articles found on the database "Scopus" based on the keywords and subject areas and shortlisted case studies based on continents and area of focus. Figure 3
(a) shows the research papers that were found by searching through 6 keywords: climate change, Multi-criteria Analysis, Climate Change Adaptation, Decision-making, Adaptation Management and Risk Assessment. The results showed that the papers with the keyword "Climate Change" are highest followed by "Multi-criteria Analysis". Figure 3
(b) shows the percentage of articles discovered through subject area: Environmental science, agricultural science, energy, social science, earth and planetary science, engineering where, relevant papers were found to be published within the sector "Environmental Science".

Figure 3(c) shows the shortlisted case studies based on continents and Figure 3(d) shows shortlisted case studies by area of focus. From Figure 3(c), it can be speculated that about an equal proportion of relevant research papers have been included from Asia and Europe (35% and 34% respectively); followed by Africa (16%) and Australia (9%). On a contrary, minimum number of studies were eligible from North America and South America for our research (2% and 3% respectively). Likewise, figure 3(d) represents the proportion of shortlisted case studies based on area of focus and it shows that agriculture and water resource management are the mostly researched sector in climate change adaptation (26% each), followed by flood management (24%) and 18% adopted MCA for agricultural water or community-based adaptation as "mixed" sectors. A few numbers (n=2, 6%) were selected from spatial planning which also incorporated flood risk management.

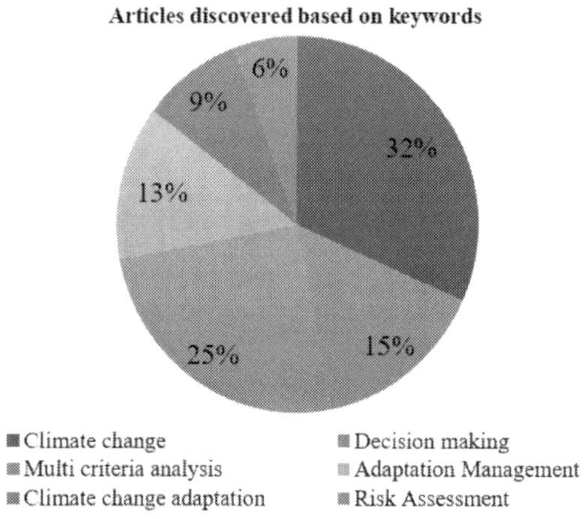


(a) Articles discovered based on keywords

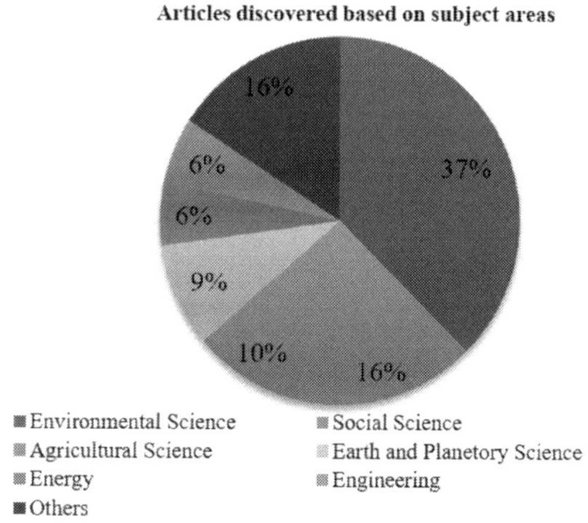


(b) Articles discovered based on subject areas

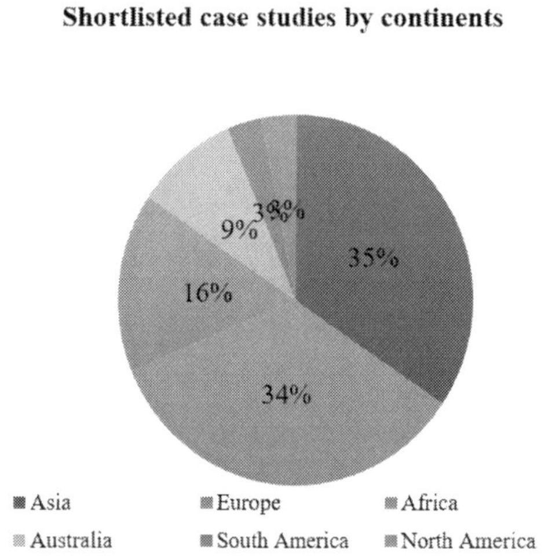


(c): Shortlisted case studies by continents

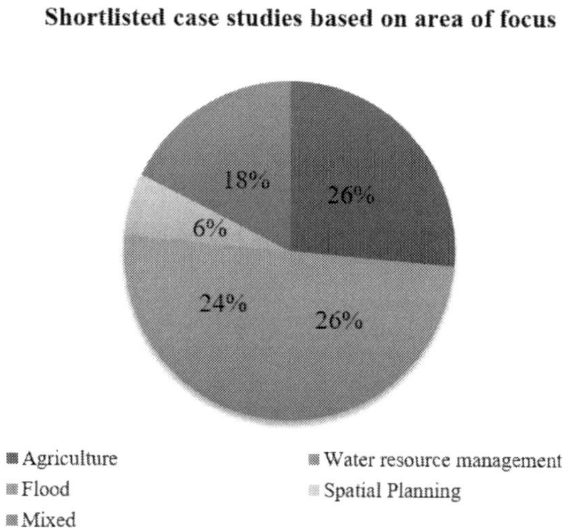


(d) Shortlisted case studies by area of focus

**Figure 3.** Categorical distribution of discovered and shortlisted articles

### *3.2 Shortlisted case studies based on methodological approach*

After screening the articles thoroughly through title and abstracts, eligible research articles were further analyzed based on methodological approaches. As there are multiple methods of conducting MCA, our team further differentiated them according to the methods applied. It was found that, 16 articles adopted MCA as their only methodological approach [4,5,12,18,21–32], and an article by Khadka et al., [33] applied a similar method called multi-variate analysis for community-based adaptation action plan. Another common method applied to derive alternatives was Analytic Hierarchy Process (AHP) [34], followed by its combined method with Technique for Order of Preference to Ideal Solution (TOPSIS) [35,36]. Several novelties in combination of MCA methods were encountered where, three articles combined AHP with Geographic Information Systemn (GIS) and Remote Sensing (RS) where Digital Elevation Model (DEM) images were used for flood risk assessment [19,37,38]. Likewise, Baláž et al., 2021 combined AHP and Delphi methods for decision making in water management [6]. Few articles [7,16] formulated new concept of the combination of top-down approaches with bottom-up approaches, for example, Water Evaluation and Planning System (WEAP) model with other methods of MCA like AHP, *VIekriterijumsko KOmpromisno Rangiranje* (VIKOR) and Fuzzy Ordered Weighted Averaging (FOWA). Similarly, Worku et al., [14] integrated MCA with Soil and Water Assessment Tool SWAT) and AHP for determining alternatives. Lawrence et al., [39] developed a hybrid process was developed to address deep uncertainties in CCA by combing Dynamic Adaptive Pathways (DAP) and Real Option Analysis (ROA). Likewise, Nigussie et al., [40] applied another method of MCA called Preference Ranking Organization Method for Enrichment of Evaluations (PROMETHEE II), Batisha [37] applied Fuzzy Decision Making Technique (DMT) and Varela-ortega et al., [41] integrated economic hydrologic modelling on agriculture and water sectors with MCA or land suitability projections for climate change adaptation, whereas, Cornwell et al., [42] adopted GIS based MCA with fuzzy membership technique.

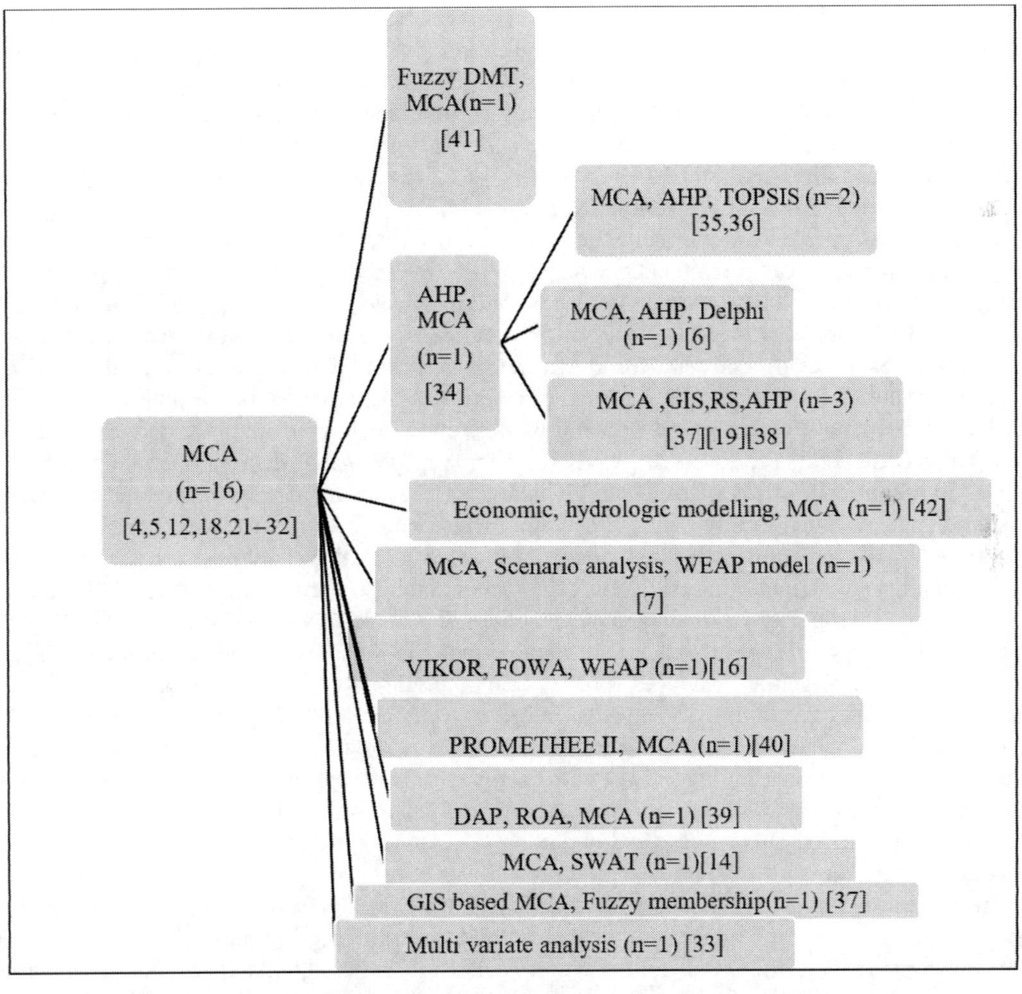


**Figure 4.** Shortlisted case studies based on methodological approach

### *3.3 MCA in decision-making*

Nine research articles have identified adaptation options in agricultural sector where the prioritization of adaptation options are based on MCA approach and besides that, two articles focused on agricultural water adaptation [16,43]. Various methods of MCA have been applied in agricultural sector like AHP, TOPSIS, Fuzzy and combination of those methods can also be seen in multiple research articles [35]. The application of MCA in the water management sector is enormously found in the literatures where it is mainly targeted for IWRM, flood risk assessment, risk analysis and managing other water related disaster like drought. Seventeen articles are shortlisted in this sector and out of them, seven articles solely focused on flood risk assessment and formulation of adaptation options to cope with it, whereas incorporation of spatial planning along with building flood resilience plan can also be found in Meng et al. [44].

The review paper revealed that the global focus has been slowly shifting towards adaptation options from the previous trends of impacts assessment. In recent years, MCA has been excessively used in the sectors of water resource management, agriculture and flood risks management [7,13,21]. MCA involves a participatory approach which is a bottom-up method that addresses vulnerability and risk management by considering adaptive capacity and adaptation measures, but it solely cannot ensure the applicability and certainty of the prioritized options. Cornwell et al., [32] suggest to define adaptation within basic three questions: what to adapt, how to adapt and who adapts and it must be inclusive of the following four qualities: effectiveness, efficiency, transparency (in terms of costs and benefits) and legitimacy [32]. To achieve this, multiple literatures found that simultaneous coordination between scientific groups, researchers and technical experts with the concerned stakeholders play a vital role in formulating relevant adaptation options rather than just involving DMs and stakeholders [7,18,21,39]. Most of the literatures have applied compensatory and mixed methods (assessing both qualitative and quantitative options) like AHP and MAT of MCA [6,45,46]; while few have totally relied on the quantitative data adopting weighted summation method [4,7,21].

In water management sector, a novel concept has been formulated by Bhave et al., [7] for a large scale planning for a broader region, or country level, by incorporating top-down approach like hydrological modelling, climate modelling and climate projections with the participatory bottom up approach like MCA for assuring better adaptation options. However, he suggests that there are some underlying drawbacks associated with top-down methods, first is it strongly relies on future climate projections and the second is uses complex scientific equations which might not be appropriate for local or small level adaptation planning and decision makers due to data unavailability. However, in some studies (Chen et al., [47]; Preston et al., [19]) incorporating geospatial data and adopting GIS based visualization of future climate scenarios with MCA found to be beneficial in increasing the applicability of adaptation options and also assess the exposure of different hazards and properties at risk. This could also support the local government in adaptation planning. The integration between these techniques gave more accurate results along with transparency and accountability. Further, Lee et al., [18] concluded that in decision making process, management practices and correlated factors like energy usage and efficiency should be considered to ensure sustainable planning and environmental sustainability. The selection of water management practices should focus on multiple performance indicators to obtain a clearer vision of tradeoffs and overall impacts. In order to explore multiple pathways, Lawrence et al., [39] concluded that by integrating participatory tool with several other analysis in the context of coastal adaptation, could enhance the ability to cope with changing risks over time and the quality of the options for the long term planning. MCA has also been successfully used in the planning of adaptation options for drought prone region as an integrated approach to address agricultural water availability in the watershed in future climate variabilities [12].

In the context of least developed countries, MCA is more efficient as it is capable to support real decision processes, identify complexities of the options, assess feasibility and different stakeholders interest can be incorporated into decision processes [5,13,21]. In decision making, proper estimation of exposure and vulnerability is an essential factor before the identification of adaptation strategies as the adaptive capacity is hindered by these two factors. Hence, prior to assessing adaptation strategies, climate risks assessment by evaluating factors such as exposure and vulnerability would assist decision makers to draft reliable adaptation strategies for future probabilities of natural disasters like flood [47,48]. However, Smit et al., [2] concluded that the participatory bottom-up approach integrates the economical-social-political structures into adaptation planning and their relationship reflects the adaptive capacity of any community. Also the application of certain frameworks like MCA4 Climate adopted by Miller and Belton, seemed to assure policy makers to systematize wide range of water related options for combating climate change and they recommend that incorporating sensitivity analysis highly addresses the future uncertainties and increases the robustness in policy making [5]. Some review articles (Abdullha et al., [10] and de Brito et al., [15]) discovered that AHP or AHP combined with other methods like TOPSIS, Simple Additive Weighting Method (SAW) is the most popular method adopted in flood risk mapping in the world which is believed to handle uncertainties linked with flood events and future criteria weights [10,15].

Similarly, Chen et al., [47] discovered that the integration of spatial data with Spatial Multi Criteria Decision Making (SMCDM) even for extreme weather conditions like heavy rainfall, cyclones etc. could result to a standardized process and improve efficiency, transparency, and accountability. But only the options with high feasibility are selected as high priorities and to uplift the gross benefits for sustainable water resource management and practices, the options with multiple performance indicators could be included [18]. Nevertheless, it is also evident that the options with high priority might not have high feasibility and might create conflicts while implementing the plan [30].

### *3.4 Dealing with uncertainties*

Climate change is a dynamic process, and the severity of the unseeable future situations create more complexities in developing adaptation strategies for the future. The level, extent, duration and dynamics associated with the biophysical and socio-economic processes are important factors to be considered while drafting adaptation plans [49]. MCA itself is unable to perform uncertainty analysis and hence, it is better to combine MCA with other analyzing tools in order to discover deep uncertainties and omit partial understandings [7,39], Correspondingly, in order to respond to the long term challenges of climate change, the decision makers should incorporate a broader sectors and experts from diverse fields to deal with the high uncertainty linked with it [1,10,11].

For a better formulation of adaptation strategies, it should be scenario driven, participatory, scientific, should take account of related uncertainty, economically viable and consider about possible maladaptation risks associated with any options [3]. That is why several literatures suggest that integration of technical, social, and institutional aspects could help to combat with the constantly evolving climate where, performing "Sensitivity Analysis" boost up decision making processes in climate sensitive sectors like agriculture, water management and flood risks [5,18,49]. Sensitivity analysis is a process used to assess the performance of prioritized adaptation options after scoring, weighting, and ranking by stakeholders and experts. If the options are found to be poor performing in the future variabilities, they are replaced with alternative options and assessed again. This process helps to formulate relevant adaptation option applicable for that region. It seemed to be helpful for combating the future climatic, socioeconomic, or political variable and its application leads to a better formulation of adaptation strategies, provide space for the allocation of alternative options if required, and ensures the ability of prioritized adaptation options in unseeable climate dynamics. This increases robustness, performance, suitability, reliability, and formulate most applicable adaptation options. Likewise, sensitivity analysis in multiple climate scenarios helps in smooth making of policy portfolios, support inclusive planning process with socially reliable and acceptable adaptation options [5,7,15,18]. Chen et al., [47] proved that such integrated tools can be used to analyze uncertainty by evaluating different indices like climate, hydrology, topography along with the assessment between sectors and stakeholders.

In parallel, Lawrence et al., [39] suggests that, in order to compensate the shortcomings of MCA in dealing with uncertainties, hybrid methods like combining MCA with DAP, ROA and Robust Decision Making Analysis (ROMA) would overcome the limitation of one another and design robust options and adaptation strategies [39]. The former method helps to formulate strategies under deep uncertainties in which plausible futures and transient scenarios are compared with objectives under uncertain global and regional changes, links short term targets and long term goals and assess their effectiveness [39,50]. Likewise, ROA is widely used in large-scale projects or public infrastructures and could be used to evaluate the options in accordance with future scenarios for robust and adaptive pathways in the non-stationarity of climate change [51]. The latter (ROMA) concerns about building better decisions rather that predictions and uses simulation models to perform robustly under various plausible future scenarios

[39,52]. It focuses on developing less vulnerable strategies for long term resilience and formulate no/low regret options, then identifies and characterizes high risks and vulnerabilities associated with the strategies [53]. These methods are known as data and resource intensive as they consist of multiple simulation models of multiple plausible future scenarios. But this problem can be tackled by developing simplified models and limiting the number of uncertainties. But the simplified models could be data deficit and may have the chance to overlook most vulnerable scenarios [51]. Thus, despite of the data intensiveness of such applications to DMs, the final shared decision formulated by such complex process has a greater reach towards the priority area and would be applicable in the real-life context.

### *3.5 Shortlisted case studies with novel methods relevant to our objective*

*3.5.1 Combined top-down and bottom-up approach using WEAP model.* The study conducted by Bhave et al., has been shortlisted as one of the relevant articles which seems to be powerful in addressing the underlying uncertainties of climate variabilities in decision-making. In this paper, a top-down approach and a bottom-up approach has been combined in developing climate-change adaptation options in the water resource management sector [7]. The bottom-up approach is MCA consisting of the involvement of concerned stakeholders for prioritization of the locally applicable options. The top-down approach is a hydrological model called WEAP model, which is used to simulate hydrological variabilities, direct water resource management alternatives. WEAP model was used to evaluate the performance of the derived alternatives in future streamflow. The result showed that this kind of novel concept led to a locally applicable and acceptable adaptation options along with the information about the hydrological behaviour of the future climate change. Decision making in adaptation at the regional scale incorporates a wide range of area, including its climatic and hydrological variabilities that might hinder the socio-economic aspects in the future climate scenarios. Hence it is important for any adaptation option to address those dynamics of future climate change so that the drafted plans will be suitable to address the vulnerability of local people. In that sense, the combination of top down and bottom-up approach carried out by the author has led to the formulation of stronger and locally acceptable adaptation options. However, this research lacks the information beyond climatic variabilities like deep social and economic factors which could have resulted to a clear vision.

*3.5.2 Coupling methods.* Other shortlisted articles are Preston et al. [19] and Chen Y et al. [47] where a novel concept of integrating MCA with spatial analysing tool like GIS and RS data have been applied. The results showed that, such coupling methods not only enhance the quality of adaptation strategy assessment, but also links spatial variability with decision making[47]. It has also demonstrated that the applicability and reliability of the adaptation strategies increased by combining RS data and GIS analysis with MCDA and AHP. Similar method was applied by Preston et al., [19] who found out that the consideration of geospatial data could compensate the subjective judgement of MCA, address uncertainties to some extent in the context of climatic, hydrologic projections and enhance decision support for adaptation planning. The spatial data analysis also aids in addressing underlying geographical uncertainties of the future, thus, leading to highly accountable, transparent, visualized, capable and efficient action plans for complex aspect like flood risk evaluation and reduction. are Hence, it might attract the concern of decision makers towards spatial variabilities as well, which is a significant aspect to adapt with climate change impacts.

## 4. Conclusions

This review paper has thoroughly analyzed the research papers which adopted multi-criteria decision-making analysis in the formulation of climate change adaptation options. One of the major objectives was to search for the papers cored in formulation of climate change adaptation strategies and decision-making tool bid by them. Another objective was to evaluate the applicability of multi-criteria analysis (MCA) in the sectors of water resource management, flood control, drought management, agriculture. MCA was chosen as a base method for the assessment because of its ability to involve multiple sectors

into a discrete module and address their problems through both quantitative and participatory processes. From the review, firstly, we identified the significance of MCA in the decision-making process in different regions of the world along with the institutional and statistical processes incorporated within it. Secondly, we figured out the transparency, clarity and problematic trade-offs experienced between decision makers and community.
From our review, following criteria were figured out which are essential points while formulating adaptation options:

- Policy making should be holistic and should incorporate wide range on sectors.
- Uncernainty is the biggest challenge in the future climate change adaptation, hence, it's an essential aspect to be analysed.
- The formulated adaptation plans will be reliable if it's has a purpose of long-term goal and is sustainable.
- Another important aspect is economic feasibility without which any plans cannot work properly and considering future market risk and policy interventions.
- Participatory approach combined with technical approaches like hydrolocal modelling could be a suitable option while drafting adaptation plans.
- As MCA itself doesn`t address uncertainties, pathways for alternative scenarios can be developed by combining it with other analysis like sensitivity analysis or real option analysis assess the possible effects on options in changing climate.
- Economic benefits and environmental sustainability always go hand on hand.
- For a broader region, geospatial data like GIS and Remotely sensed images could help in decision making.
- Vulnerability assessment and ecosystem consensus are crucial aspects in decision making for assessing and analyzing future risk of exposure and vulnerability.
- Technical assistance like experts and researchers play a vital role in decision-making process to make more scenario driven and reliable options.
- Compensability: The possibility of bad performing options could be compensated by good performing ones.
- Maladaptation: To reduce the risks of maladaptation, detailed analysis of future climate scenario could minimize maladaptation while dimplementing adaptation plans.

We also found few drawbacks linked with MCA in terms of generalizing all qualitative options into a same value (by standardization process) for assessment. Also, it is found that the underlying uncertainties of future changing variables like political situations, extreme events, natural disasters could affect the actual applications of the prioritized options. Hence, the most significant finding of this paper is that decision makings and planning inclusive of Sensitivity Analysis could identify robust options for future climate, social and political variables. Otherwise, future unexpected possibilities could interfere in policy interventions, which might result in economic and environmental clashes. Further, adaptive capacity depends upon the social infrastructure and economic condition of the people of that region, hence it varies with communities and countries where, developing countries might need assessment in extra depth.

## 5. Recommendation

As climate is dynamic and is always evolving, the adaptation measures planned today must be compatible with the effects of future scenarios. But if the adaptation strategies are focused only on the current climate scenarios, it could lead to maladaptation in the future conditions. This could not only affect the adaptation status but also cause huge social and economic damages. Also, future scenarios vastly differ with geographical conditions and political influences. Thus, detailed studies focusing on different vulnerable regions of the world is essential. If the assessment is done separately in developed and developing countries, that could enhance the essence of this paper and the review can be made more context-specific, reliable, and scenario-driven.

## 6. Limitations of this Study

This review paper generally covers the application of multi-criteria analysis methods in CCA decision making in various parts of the world. But there are still limitations with the paper selection. The articles selected here are from a single database. As a further study, meta-analyses could be conducted through other databases in a diverse way to expand the results broadly. Furthermore, the applicability of MCA might differ with the country's economic status, climate vulnerability and topographical situation as well. Thus, in the future, further consideration of such aspects could give profound results.

## 7. Challenges

The major challenge was in finding right papers regarding MCA in different countries which focused on Climate change Adaptation. As compared to other databases, Scopus showed high number of related articles, so we sticked with that. Another challenge was in terms of paper selection. As many papers considered climate vulnerability with adaptation, it was difficult to analyze the differences and sort out the relevant paper.

## References

[1] Field G B, Barros V, Stocker T F, Qin D, Dokken D J, Ebi K L, Mastrandrea M D, Mach K J, Plattner G K, Allen S K, Tignor M and Midgley P M 2012 *Managing the risks of extreme events and disasters to advance climate change adaptation* (Cambridge)

[2] Smit B and Wandel J 2006 Adaptation, adaptive capacity and vulnerability *Global Environmental Change* **16** 282–92

[3] IPPC 2014 *Climate change 2014: impacts, adaptation, and vulnerability* ed C B Field, V R Barros, D J Dokken, K J Mach, M D Mastrandrea, T E Bilir, M Chatterjee, K L Ebi, Y O Estrada, R C Genova, B Girma, E S Kissel, A N Levy, S MacCracken, P R Mastrandrea and L L White (Cambridge, United Kingdom and New York: Cambridge University Press)

[4] Haque A N, Grafakos S and Huijsman M 2012 Participatory integrated assessment of flood protection measures for climate adaptation in Dhaka: *Environment and Urbanization* **24** 197–213

[5] Miller K A and Belton V 2014 Water resource management and climate change adaptation: a holistic and multiple criteria perspective *Mitigation and Adaptation Strategies for Global Change* **19** 289–308

[6] Baláž V, Dokupilová D and Filčák R 2021 Participatory multi-criteria methods for adaptation to climate change *Mitigation and Adaptation Strategies for Global Change* **26** 15

[7] Bhave A G, Mishra A and Raghuwanshi N S 2014 A combined bottom-up and top-down approach for assessment of climate change adaptation options *Journal of Hydrology* **518** 150–61

[8] Smith J B, Klein R J T and Huq S 2003 *Climate change, adaptive capacity and development* (London: Imperial College Press)

[9] Gajanan Bhave A, Mishra A and Groot A 2013 Sub-basin scale characterization of climate change vulnerability, impacts and adaptation in an Indian River basin *Regional Environmental Change* **13** 1087–98

[10] Abdullah M F, Siraj S and Hodgett R E 2021 An overview of multi-criteria decision analysis (MCDA) application in managing water-related disaster events: analyzing 20 years of literature for flood and drought events *Water* **13** 1358

[11] Handmer J, Honda Y, Kundzewicz Z W, Arnell N, Benito G, Hatfield J, Mohamed I F, Peduzzi P, Wu S, Sherstyukov B, Takahashi K and Yan Z 2012 Changes in impacts of climate extremes: human systems and ecosystems *Managing the risks of extreme events and disasters to advance climate change adaptation* ed C B Field, V Barros, T F Stocker, D Qin, D J Dokken, K L Ebi, M D Mastrandrea, K J Mach, G K Plattner, S K Allen, M Tignor and P M Midgley (Cambridge, UK and New York, USA: Cambridge University Press) pp 231–90

[12] Benegas L, Jiménez F, Locatelli B, Faustino J and Campos M 2009 A methodological proposal for the evaluation of farmer's adaptation to climate variability, mainly due to drought in watersheds in Central America *Mitigation and Adaptation Strategies for Global Change* **14** 169–83

[13] Haque A N 2016 Application of multi-criteria analysis on climate adaptation asessment in the context of least developed countries *Journal of Multi-Criteria Decision Analysis* **23** 210–24

[14] Worku G, Teferi E, Bantider A and Dile Y T 2020 Prioritization of watershed management scenarios under climate change in the Jemma sub-basin of the Upper Blue Nile Basin, Ethiopia *Journal of Hydrology: Regional Studies* **31** 100714

[15] de Brito M M and Evers M 2016 Multi-criteria decision-making for flood risk management: A survey of the current state of the art *Natural Hazards and Earth System Sciences* **16** 1019–33

[16] Golfam P, Ashofteh P S and Loáiciga H A 2019 Evaluation of the VIKOR and FOWA multi-criteria decision making methods for climate-change adaptation of agricultural water supply *Water Resources Management* **33** 2867–84

[17] Department for Communities and Local Goverment 2009 *Multi-criteria analysis: a manual* (Wetherby, West Yorkshire: Communities and Local Government Publications)

[18] Lee M, Yu C Y, Chiang P C and Hou C H 2018 Water–energy nexus for multi-criteria decision making in water resource management: a case study of Choshui River Basin in Taiwan *Water* **10** 1740

[19] Preston B L, Maloney M, Thomsen D, Smith T, Mangoyana R and Conlon B 2013 *A multi-criteria analysis of coastal adaptation options for local government* (Sydney)

[20] Selçuk A A 2019 A guide for systematic reviews: PRISMA *Turkish Archives of Otorhinolaryngology* **57** 58

[21] Acharjee T K, Hellegers P, Ludwig F, van Halsema G, Mojid M A and van Scheltinga C T 2020 Prioritization of adaptation measures for improved agricultural water management in Northwest Bangladesh *Climatic Change* **163** 431–50

[22] Verkerk P J, Sánchez A, Libbrecht S, Broekman A, Bruggeman A, Daly-Hassen H, Giannakis E, Jebari S, Kok K, Klemenčič A K, Magjar M, de Arano I M, Robert N, Smolar-Žvanut N, Varela E and Zoumides C 2017 A participatory approach for adapting river basins to climate change *Water* **9** 958

[23] Ghosh S and Mistri B 2021 Assessing coastal vulnerability to environmental hazards of Indian Sundarban delta using multi-criteria decision-making approaches *Ocean & Coastal Management* **209** 105641

[24] Baker I, Peterson A, Brown G and McAlpine C 2012 Local government response to the impacts of climate change: An evaluation of local climate adaptation plans *Landscape and Urban Planning* **107** 127–36

[25] Porthin M, Rosqvist T, Perrels A and Molarius R 2013 Multi-criteria decision analysis in adaptation decision-making: a flood case study in Finland *Regional Environmental Change* **13** 1171–80

[26] Bojovic D, Bonzanigo L, Giupponi C and Maziotis A 2015 Online participation in climate change adaptation: A case study of agricultural adaptation measures in Northern Italy *Journal of Environmental Management* **157** 8–19

[27] Ceccato L, Giannini V and Giupponi C 2011 Participatory assessment of adaptation strategies to flood risk in the Upper Brahmaputra and Danube river basins *Environmental Science & Policy* **14** 1163–74

[28] Champalle C, Ford J D and Sherman M 2015 Prioritizing climate change adaptations in Canadian Arctic Communities *Sustainability* **7** 9268–92

[29] Kandari A M, Baja S, Ala A, Kaimuddin, Kasim S and Taufik Y 2019 Promoting sustainable agricultural management through spatio temporal optimization of food crop land based on pedo-agroclimate at Kalalasi region, Southeast Sulawesi, Indonesia *IOP Conference Series: Earth and Environmental Science* **383** 012005

[30] Füssel H M 2009 Ranking of national-level adaptation options. An editorial comment *Climatic Change* **95** 47–51

[31] Baills A, Grandjean G, Maspataud A, Ettinger S, Abad J, Dias N, Albris K, Hemmers J, Clegg G and Martucci C 2020 The ESPREssO action database: collecting and assessing measures for disaster risk reduction and climate change adaptation *International Journal of Disaster Risk Reduction* **48** 101599

[32] Cornwell E, Sposito V and Faggian R 2021 Agricultural adaptation mainstreaming and its study through a systemic adaptation assessment framework: a sub-alpine case-study *Journal of Rural Studies* **84** 22–30

[33] Khadka C, Aryal K P, Edwards-Jonášová M, Upadhyaya A, Dhungana N, Cudlin P and Vacik H 2018 Evaluating participatory techniques for adaptation to climate change: Nepal case study *Forest Policy and Economics* **97** 73–82

[34] Simane B and Zaitchik B F 2014 The sustainability of community-based adaptation projects in the Blue Nile Highlands of Ethiopia *Sustainability* **6** 4308–25

[35] Stričević R, Srdjević Z, Lipovac A, Prodanović S, Petrović-Obradović O, Ćosić M and Djurović N 2020 Synergy of experts' and farmers' responses in climate-change adaptation planning in Serbia *Ecological Indicators* **116** 106481

[36] Golfam P, Ashofteh P S, Rajaee T and Chu X 2019 Prioritization of water allocation for adaptation to climate change using multi-criteria decision making (MCDM) *Water Resources Management* **33** 3401–16

[37] Batisha A F 2015 Implementing fuzzy decision making technique in analyzing the Nile Delta resilience to climate change *Alexandria Engineering Journal* **54** 1043–56

[38] Chen Y, Liu R, Barrett D, Gao L, Zhou M, Renzullo L and Emelyanova I 2015 A spatial assessment framework for evaluating flood risk under extreme climates *Science of The Total Environment* **538** 512–23

[39] Lawrence J, Bell R and Stroombergen A 2019 A hybrid process to address uncertainty and changing climate risk in coastal areas using dynamic adaptive pathways planning, multi-criteria decision analysis & real options analysis: A New Zealand application *Sustainability* **11** 406

[40] Nigussie Y, van der Werf E, Zhu X, Simane B and van Ierland E C 2018 Evaluation of climate change adaptation alternatives for smallholder farmers in the Upper Blue-Nile Basin *Ecological Economics* **151** 142–50

[41] Varela-Ortega C, Blanco-Gutiérrez I, Esteve P, Bharwani S, Fronzek S and Downing T E 2016 How can irrigated agriculture adapt to climate change? Insights from the Guadiana Basin in Spain *Regional Environmental Change* **16** 59–70

[42] Cornwell E, Sposito V and Faggian R 2020 Land suitability projections for traditional sub-alpine cropping in the Australian Alps and Chilean Dry Andes. A combined biophysical and irrigation potential perspective *Applied geography* **121** 1–11

[43] Rotich S C and Mulungu D M M 2017 Adaptation to climate change impacts on crop water requirements in Kikafu catchment, Tanzania *Journal of Water and Climate Change* **8** 274–92

[44] Meng M, Dabrowski M and Stead D 2020 Enhancing flood resilience and climate adaptation: The State of the Art and New Directions for Spatial Planning *Sustainability* **12** 7864

[45] Aulong S, Chaudhuri B, Farnier L, Galab S, Guerrin J, Himanshu H and Prudhvikar Reddy P 2011 Are South Indian farmers adaptable to global change? A case in an Andhra Pradesh catchment basin *Regional Environmental Change* **12** 423–36

[46] Nambi A A, Bahinipati C S, Raghunath R and Nagendran R 2015 Farm household level adaptation metrics for agriculture and water sectors *International Journal of Climate Change Strategies and Management* **7** 27–40

[47] Chen Y, Liu R, Barrett D, Gao L, Zhou M, Renzullo L and Emelyanova I 2015 A spatial assessment framework for evaluating flood risk under extreme climates *Science of The Total Environment* **538** 512–23

[48] Jhong B C, Huang J and Tung C P 2019 Spatial assessment of climate risk for investigating climate adaptation strategies by evaluating spatial-temporal variability of extreme precipitation *Water Resources Management: An International Journal, Published for the European Water Resources Association (EWRA)* **33** 3377–400

[49] Howden S M, Soussana J F, Tubiello F N, Chhetri N, Dunlop M and Meinke H 2007 Adapting agriculture to climate change *Proceedings of the National Academy of Sciences* **104** 19691–6

[50] Haasnoot M, Kwakkel J H, Walker W E and ter Maat J 2013 Dynamic adaptive policy pathways: A method for crafting robust decisions for a deeply uncertain world *Global Environmental Change* **23** 485–98

[51] Ihm S H, Seo S B and Kim Y O 2019 Valuation of water resources infrastructure planning from climate change adaptation perspective using real option analysis *KSCE Journal of Civil Engineering* **23** 2794–802

[52] Marchau V A W J, Walker W E, Bloemen P J T M and Popper S W 2019 *Decision making under deep uncertainty* (Switzerland: Springer)

[53] Dittrich R, Wreford A, Topp C F E, Eory V and Moran D 2017 A guide towards climate change adaptation in the livestock sector: Adaptation options and the role of robust decision-making tools for their economic appraisal *Regional Environmental Change* **17** 1701–12

**Acknowledgement**

This work was supported by JST SATREPS (JPMJSA1502), RISTEX (JPMJRX20I8), SICORP (JPMJSC20E3), and JSPS KAKENHI (JP19H03069; 20KK0242), and ERCA(JPMEERF20202004), Japan.

# Diurnal rainfall cycle on small tropical island as seen by the 20 years half-hourly GPM-IMERG Satellite precipitation product

**I Putu Winsen Putra Arsana[1], Abd. Rahman As-syakur*[1], Ida Bagus Mandhara Brasika[1], I Wayan Andi Yuda[2] and Bagus Satya Mas[1]**

[1]Marine Science Department, Faculty of Marine and Fisheries, Udayana University, Bukit Jimbaran Campus, Bali, 80361 Indonesia
[2]Indonesian Agency for Meteorology, Climatology, and Geophysics (BMKG), Negara Office, Jalan Leli No. 9 BB Agung, Negara, Bali 82212, Indonesia
*Corresponding author: ar.assyakur@pplh.unud.ac.id

**Abstract**. The diurnal rainfall cycle is one of the dominant factors causing the high variability of convection activity in the Indonesian Maritime Continent (IMC) and vulnerable to climate change. Research on diurnal rainfall cycles in small tropical islands (areas less than 6150 km2), such as in the IMC is important because of the lack of understanding of its characteristics. The aims of current study are to investigate spatial characteristics of the diurnal rainfall cycle over Bali and their surrounding islands using half-hourly rainfall that estimated from the Integrated Multi-satellitE Retrievals for GPM (IMERG) for the period 2001−2020. Statistical scores employed to analyse the diurnal rainfall cycle characteristics is means analysis that done during year observation on long-term half-hourly and annually. The result of half-hourly IMERG rainfall estimation confirms for the first time that the rainfall in Bali contributes to the diurnal rainfall variability over the eastern part of Java Island. In Bali, rainfall start to initiate during 1200 local time (LT, +8 UTC) mostly began in high elevation located in the central part of the island, on the southern slopes of the mountain range. Rainfall over land continues to increase and slightly spread to the west until reaches its peak in 1400 LT. Rainfall began to depletion in the central of the island and at 1530 LT spread toward the western and merged with the existing rainfall in the east of Java Island. In the night (2200 LT), rainfall leaves Bali island towards the Bali strait and Bali sea.

**Keywords** : diurnal rainfall cycle; small island; IMERG; remote sensing

## 1. Introduction

Indonesia is a tropical region located at the equator and has maximum rainfall and sunlight. Solar activity and wind patterns affect rainfall patterns in Indonesia [1]. Rainfall over Indonesia is strongly influenced by monsoon climatic conditions which result in clear differences between wet and dry seasons [2]. The pattern and distribution of rainfall in the Indonesian Maritime Continent (IMC) is influenced by the diurnal cycle of rainfall [3].

Diurnal rainfall in the tropics, dominated by the circulation of land breezes and sea breezes [4][5]. Land breezes and sea breezes result in distinct characteristics of rain [6]. The diurnal rainfall is influenced by the convection process. The convection process in the afternoon and evening occurs on land and at night and in the morning dominantly occurs in the ocean [7]. In the Indonesian Maritime Continent (IMC) area, the shape of the island, the topographical conditions and the shape of the coastline cause the diurnal cycle of rainfall to be inhomogeneous for all regions [8].

Different topographic characteristics in some region will affect the air temperature which is related to the evaporation rate in making clouds and producing intensity of rain [9]. Characteristics of the diurnal rain cycle in mountain area will show a different pattern with lowland areas. Mountains are a physical barrier to the movement of rain. Different forms of straight and concave coastlines have different characteristics of diurnal rainfall [10]. On the shoreline with a concave shape will produce high rainfall in the ocean because, the shoreline with a concave shape has a convergence of onshore winds that is preponderant than the straight shoreline [11].

Many experts have tried to conduct some research about characteristics and mechanisms of the diurnal cycle of rainfall were carried out mainly on large islands in Indonesia such as Sumatra [12] [13], Kalimantan [14], Java [9] and Papua [14]. The present research aims to observe the distribution of rain on the island of Bali to determine the effect of large islands on rainfall patterns in small tropical islands.

## 2. Materials and method

### 2.1. Time and place

This research was conducted on the island of Bali by looking at the distribution of rainfall during 2001 until 2020. The location is chosen to see the distribution movement of rain from or to Bali as shown in Figure 1. Bali has a mountain range that divides it into two parts of the watershed that affects the north and south, has different rainfall as shown in Figure 2. The location was chosen as a representative the conditions of the one tropical complex small island in the Indonesian Maritime Continent (IMC)

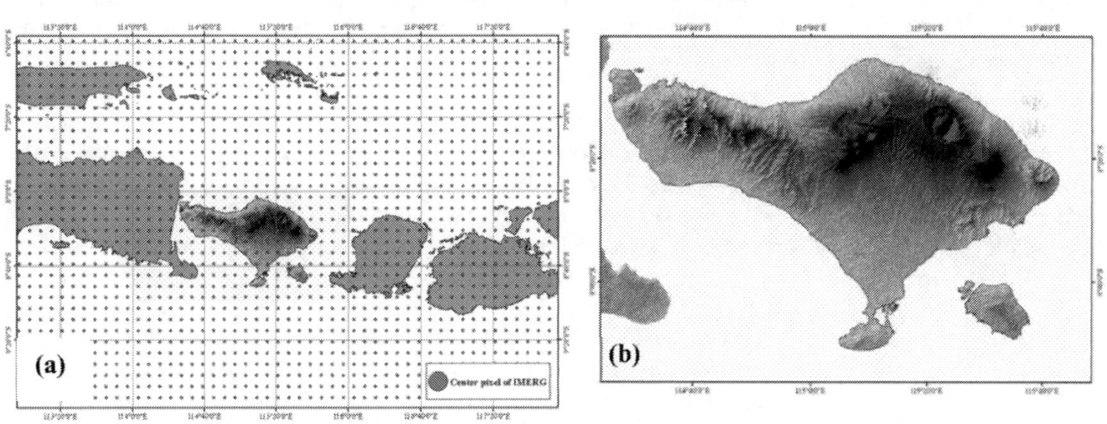


**Figure 1.** (a) Study area with pixel of IMERG, (b) Bali topography

### 2.2. Material

The data used is obtained by downloading from the https://disc.gsfc.nasa.gov site, which is in the form of diurnal rainfall data every half hourly which was downloaded for 20 years from 2001 to 2019. Half-hourly rainfall is estimated from the Integrated Multi-satellitE Retrievals for GPM (IMERG) version 6B. The downloaded data only contains the coordinates selected by coordinates information (Bali and their surrounding island) that has been determined through the Google Earth Application. The downloaded data is data with the '.nc4' extension format which will be converted in the '.txt' extension format. The data contains information on longitude, latitude, and precipitation.

*2.3. Methods*
Data that has been downloaded and collected will be processed. Processing is done by merging data files in the extension format '.nc4' using python software. In this stage, the value of precipitation every half hourly will be averaged, then the results obtained an average of 20 years of precipitation data. After processing, the result is a file with the extension format '.txt' which will then be processed using the ArcGIS application. The extension format is changed, to simplify processing in ArcGIS applications. The value of the '.txt' file is data that contains points and will be converted to raster data. After processing, a map of the distribution of diurnal rainfall will be obtained and the addition of shapefile data file for the island of Bali and its surroundings to indicate the distribution of the distribution occurs on the island.

*2.4. Data analysis*
Data were analyzed by averaging the data that had been downloaded and collected by half hourly. This analysis is used to see how the state of diurnal rainfall every half hour for 20 years on the island of Bali and it is surrounding. The calculations used in analyzing the data are for example in Equation (1):

$$X_{\text{hhmm}} = \frac{(hhmm_a + hhmm_b + hhmm_c)}{n} \quad (1)$$

where $X_{hhmm}$ is average rainfall of hours "hh" minutes "mm" (mm); $hhmm_a$ is rainfall data hour of "hh" minute "mm" date "a"; $hhmm_b$ (mm); is rainfall data hour of "hh" minute "mm" date "b" (mm); $hhmm_c$ is rainfall data hour of "hh" minute "mm" date "c" (mm); n is amount of data.

## 3. Results and discussion

*3.1. Annual rainfall distribution over 20 years*
The results of the analysis of diurnal rainfall for 20 years from 2001 to 2020 are shown in Figure 2. The island of Bali has unique topographic characteristics. The area is divided into 2 parts, the northern and southern parts are separated by mountains, as shown in Figure 2. Bali has a height variation of 0 – 3028 m above sea level. Topographic diversity can increase the variability of the temporal spatial level in an area [15].

The map shows that the range of rainfall values in Bali and its surroundings is 0.19 mm/hour to 0.304 mm/hour marked in dark blue with the lowest value and red color indicating the highest value. The value with the highest intensity on the island of Bali is located in the middle of the mountain range with the highest value being 0.28 mm/hour. According to Qian [9], the peak of rain in the mountains can be caused by valley winds that come and the convergence of sea breezes during the day. The value of rainfall falls in low land areas when moving away from mountainous areas. This can be caused by differences in air dynamics during the dry and wet seasons that enter the island of Bali at low altitudes and move vertically towards mountainous areas and produce higher spatial values of rainfall [15].

The value of rain in ocean can be determined from the size of a waters [10]. The waters of the Bali Strait and Lombok Strait have different rainfall distribution values. The waters of the Lombok strait tend to be blue which is indicates low values ranging from 0.13 mm/hour to 0.16 mm/hour. In the waters of the Bali strait, the distribution value shown has a greater value, ranging from 0.18 mm/hour to 0.21 mm/hour. Fujita et al. [10] explained that the wider a strait, the smaller the rainfall value because, there are no high mountains in each strait.

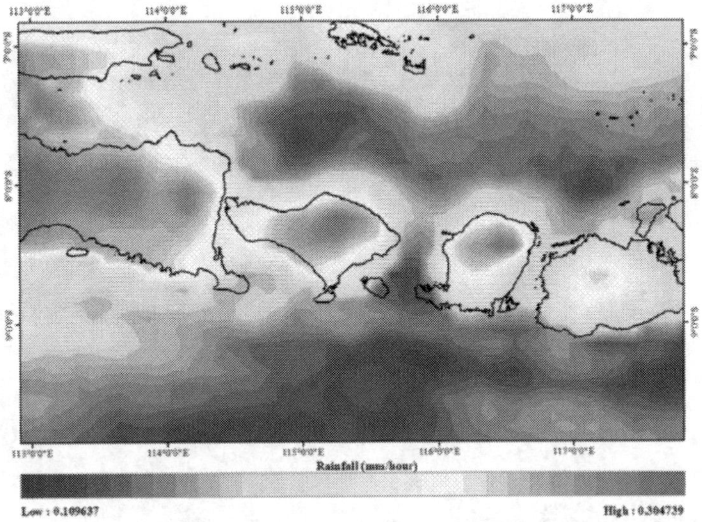


**Figure 2**. Annual rainfall distribution pattern of Bali and its surroundings (2001-2020)

### *3.2. Diurnal Cycle distribution rainfall*

Figure 3 shows the results of the variation of the diurnal rain distribution, the results of the average half hourly rainfall during 2001-2020. The map shows the distribution of rain in the morning from 00:00LT to 03:30LT (local time in Bali). The rainfall values displayed are 0.05 mm/hour to 1.0 mm/hour. Rain distribution tends to have a value of 0.05 mm/hour to 0.5 mm/hour.

The peak of rainfall at 00:00 LT to 00:30 LT on the island of Bali begins in the northern and southern areas with values ranging from 0.20 mm/hour. In the central mountainous area of the island of Bali, the results are drier with rainfall values of 0.13 mm/hour to 0.19 mm/hour. The surrounding islands have almost the same distribution pattern, tending to be drier than the seas.

At 01:00 LT to 01:30 LT the results show that the rain is moving towards the east and begins to spread in the northern waters and southern waters of Bali. Meanwhile, the movement of rain in the Bali Strait and Lombok Strait is not as wide as other waters. The Bali Strait is flanked by Java Island and the Bali Island, while the Lombok Strait is flanked by Bali Island and Sumbawa Island, both of which have in common, namely having a narrower area. Mainland Bali shows the area begins to dry up until 03:30 LT towards the north with rainfall values of 0.1 mm/hour to 0.2mm/hour.

Figure 3 (b) shows the distribution of rainfall that occurs from 04:00 LT to 07:30 LT, there is no clear peak of rainfall. In this period, rain only appears in the north and south with a small value of 0.09 mm/hour to 0.16 mm/hour. Rain is still dominant in the seas, but slowly away from the mainland towards the north and south.

The movement of rain from land to sea can be seen at 04:00 LT to 07:30 LT on the Java Island which directly affects the distribution of rain on Bali and Lombok which is also influenced by Sumbawa Island due to the presence of land winds resulting in convective movements. Based on the study of Alfahmi et al. [11], the dominant wind also affects the propagation and intensity of diurnal rainfall. Mori [12] explains that in his research scheme, the convergence of rainfall along the seas.

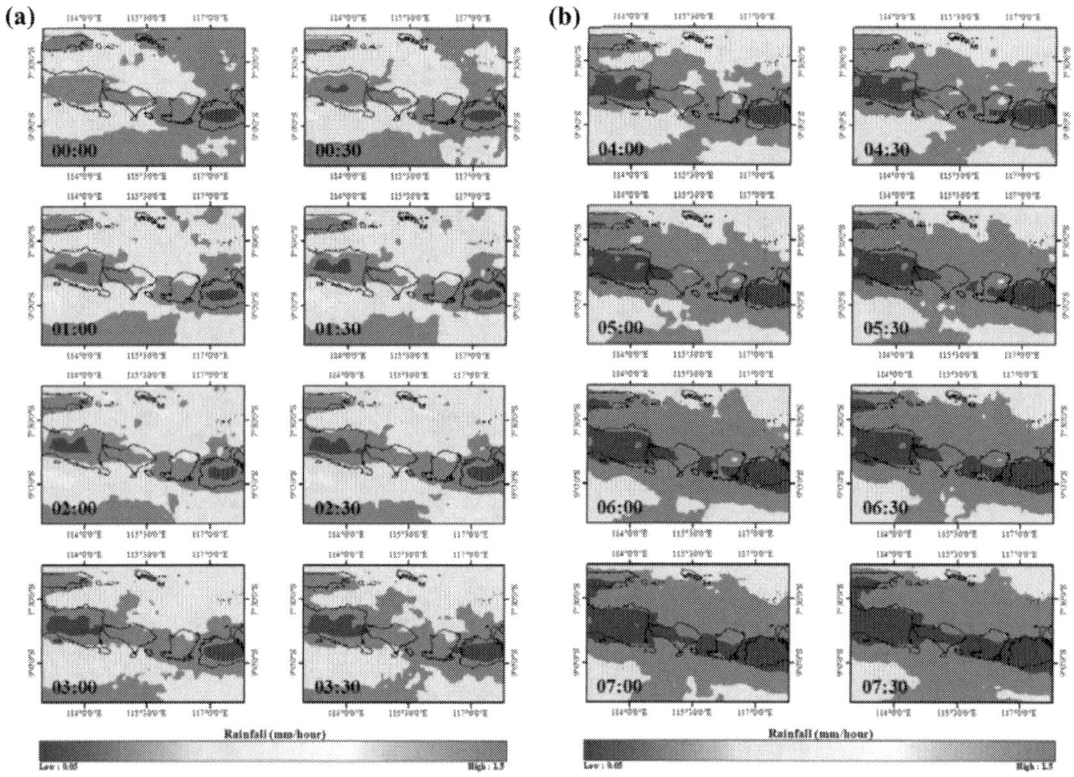


**Figure 3**. Half hourly rain distribution (a) 00:00 to 03:30 (LT, +8 UTC) (b) 04:00 to 07:30 (LT, +8 UTC)

Figure 4 (a) shows the distribution of rain from 08:00 LT to 11:30 LT on the island of Bali and its surroundings. It can be seen that the land has a small value ranging from 0.02 mm/hour to 0.09 mm/hour. The sea areas north and south of Bali Island show values ranging from 0.12 mm/hour to 0.19 mm/hour. The high intensity of rain in the sea area is higher than the land, indicating that during this period the land breeze still affects the distribution of rain.

At 11.30 LT (figure 4b) shows a new peak of rain intensity that occurred in the central area of the island of Bali which has a mountainous topography. The value of rainfall intensity in the area reaches 0.17 mm/hour to 0.24 mm/hour. The peak of rain intensity is also seen on Lombok Island, which has the same topography with rainfall intensity of 0.19 mm/hour to 0.22 mm/hour.

Figure 4 (b) shows the distribution of rainfall for the period from 12:00 LT to 15:30 LT. at 12:00 LT it rains on every island. This is due to the intense sea breeze activity during the day. The difference in the level of warming that is different on land and sea makes the movement of sea breezes stronger. The area where it rains almost has the same topography, namely mountainous topography. The appearance of rain on Bali Island is characterized by rainfall intensity of 0.28 m/hour to 0.37 mm/hour.

The distribution at 12:00 LT to 15:30 LT shows that the distribution of rainfall starts from the mountainous area and spreads throughout the land. The peak of rainfall occurs throughout the island at 14:00 LT with an intensity of 0.63 mm/hour to 0.94 mm/hour and begins to fade at 14:30 LT. According to Qian [16] the highlands tend to have a stronger onshore wind circulation than the lowlands which allows the distribution of rain to occur from mountainous areas to the whole land.

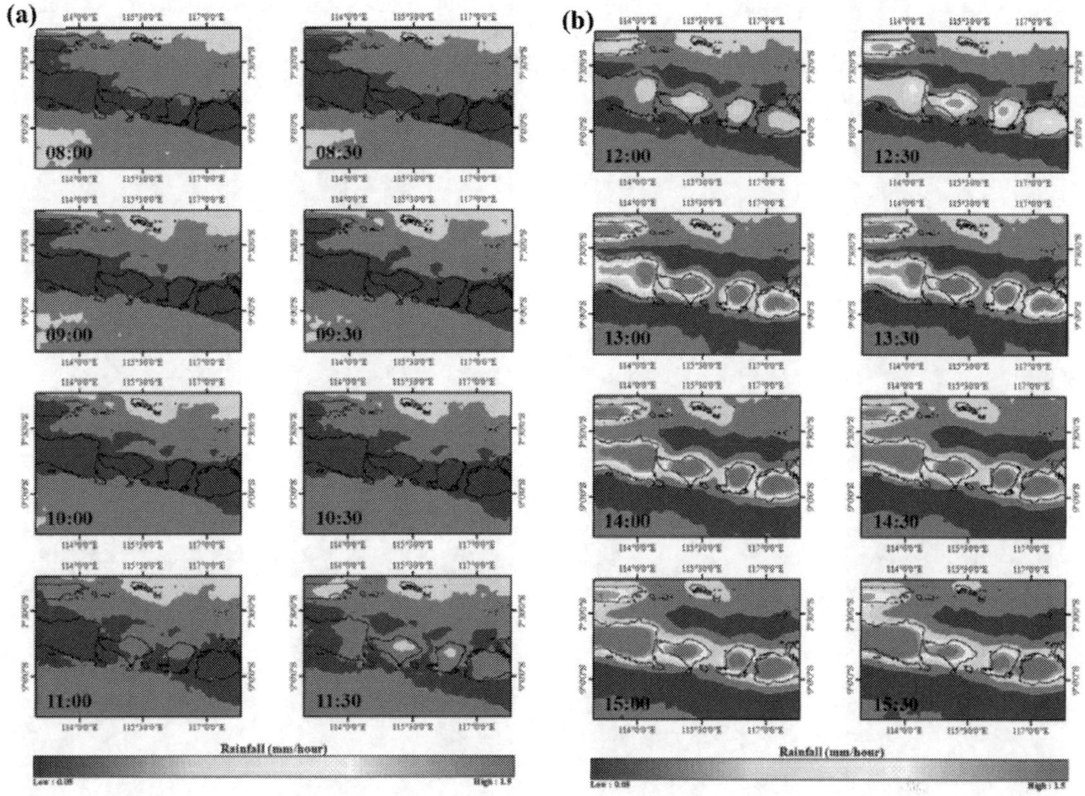


**Figure 4**. Half hourly rain distribution (a) 08:00 to 11:30 (LT, +8 UTC) (b) 12:00 to 15:30 (LT, +8 UTC)

The intensity of rain in the southwest Bali which has a concave coastline topography has a value of 0.53 mm/hour to 0.69 mm/hour which is greater than the eastern area of the island of Bali with a value of 0.27 mm/hour to 0.41 mm/hour. According to Alfahmi [11], the peak of rainfall above the concave coastline is located at the concave position which is assumed to have a smooth line. This area encourages the convergence of sea breezes and land breezes which have a direction perpendicular to the shoreline. The movement of rainfall on a concave coastline is slower than the propagation of rain from the top of the mountain to the ocean [11]. Meanwhile, the eastern area of Bali has an irregular coastline, creating shifting peaks of rain and having an irregular shape.

Figure 5 (a) shows the distribution of rainfall in the period 16:00 LT to 19:30 LT. Rain on the Bali Island moves slowly towards the Java Island. Meanwhile, the rain on Lombok Island moved towards Sumbawa Island. This result is not in line with research conducted by Alfahmi [11] in Cenderawasih Bay and Tolo Bay, in the afternoon the rain moves from the mountains to the ocean.

The movement of rain from Bali to Java can be influenced by moving land winds, and the size of the Bali Strait is not too wide compared to the Lombok Strait. There are mountains on every land that connect the rain so that the movement of rain in Bali moves towards the island of Java. The characteristics of the Alas Strait (between the islands of Lombok and Sumbawa) are also almost the same as the Bali Strait which causes rain to move from Lombok to Sumbawa.

At 19:30 LT to 20:30 LT, the mainland of Bali Island and the surrounding waters dries up and tends to have rainfall intensity values of 0.18 mm/hour to 0.28 mm/hour. The mainland of Bali dries up again in the north and has a light intensity value of 0.17 mm/hour to 0.27 mm/hour. At night, the ocean shows a higher rainfall value than in the afternoon. The movement of rain from land to sea is influenced by land winds which cause peaks of rain in the ocean. The peak of rain that occurs in the ocean occurs until the early hours of the morning (Figure 3a).

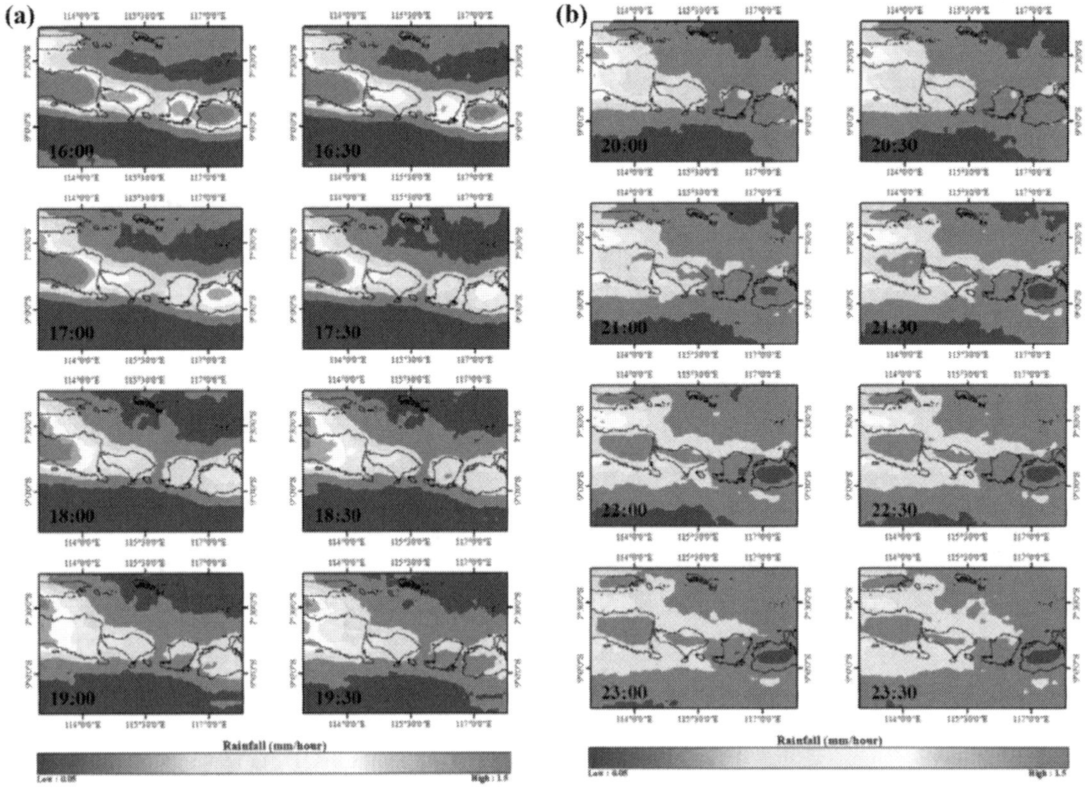


**Figure 5**. Half hourly rain distribution (a) 16:00 to 19:30 (LT, +8 UTC) (b) 20:00 to 23:30 (LT, +8 UTC)

## 4. Conclusion

In this study, we found that based on a half hour diurnal rainfall distribution on the island of Bali and its surroundings for 20 years (2001 – 2020) it showed that the peak of rainfall on the mainland occurred during the day (12:00 LT – 17:00 LT). Rainfall began at sea occurs from midnight to early morning (00:00 LT – 04:00 LT). Several factors affect the distribution of diurnal rainfall on the island of Bali. Land breezes and sea breezes affect the duration of rain that occurs on land and sea. Mountain topography, coastline, sea area affects the value of the diurnal rainfall distribution on the island of Bali.

## References

[1] Sinambela W, Dani T, Rusnadi I E, Tejo J, Matahari N P and Antariksa D 2008 Pengaruh aktivitas matahari pada curah hujan di Indonesia *J. Sains Dirgant.* **5** 149–68
[2] Susilokarti D, Arif S S, Susanto S and Sutiarso L 2017 Identifikasi perubahan iklim berdasarkan data curah hujan di wilayah selatan Jatiluhur Kabupaten Subang, Jawa Barat *agriTECH* **35** 98–105
[3] Yamanaka M D 2016 Physical climatology of Indonesian maritime continent: An outline to comprehend observational studies *Atmos. Res.* **178–179** 231–59

[4] Qian J H 2008 Why Precipitation Is Mostly Concentrated over Islands in the Maritime Continent *J. Atmos. Sci.* **65** 1428–41
[5] Crosman E T and Horel J D 2010 Sea and Lake Breezes: A Review of Numerical Studies *Boundary-Layer Meteorol. 2010 1371* **137** 1–29
[6] Wu Q, Ruan Z, Chen D and Lian T 2015 Diurnal variations of tropical cyclone precipitation in the inner and outer rainbands *J. Geophys. Res. Atmos.* **120** 1–11
[7] Yang G-Y and Slingo J 2001 The diurnal cycle in the tropics *Mon. Weather Rev.* **129** 784–801
[8] Ohsawa T, Ueda H, Hayashi T, Watanabe A and Matsumoto J 2001 Diurnal Variations of Convective Activity and Rainfall in Tropical Asia *J. Meteorol. Soc. Japan. Ser. II* **79** 333–52
[9] Qian J H, Robertson A W and Moron V 2010 Interactions among ENSO, the Monsoon, and Diurnal Cycle in Rainfall Variability over Java, Indonesia *J. Atmos. Sci.* **67** 3509–24
[10] Fujita M, Kimura F and Yoshizaki M 2010 Morning Precipitation Peak over the Strait of Malacca under a Calm Condition *Mon. Weather Rev.* **138** 1474–86
[11] Alfahmi F, Boer R, Hidayat R, Perdinan and Sopaheluwakan A 2019 The Impact of Concave Coastline on Rainfall Offshore Distribution over Indonesian Maritime Continent *Sci. World J.* **2019** 6839012
[12] Mori S, Jun-Ichi H, Tauhid Y I, Yamanaka M D, Okamoto N, Murata F, Sakurai N, Hashiguchi H and Sribimawati T 2004 Diurnal Land–Sea Rainfall Peak Migration over Sumatera Island, Indonesian Maritime Continent, Observed by TRMM Satellite and Intensive Rawinsonde Soundings *Mon. Weather Rev.* **132** 2021–2039
[13] As-syakur A R, Imaoka K, Ogawara K, Yamanaka M D, Tanaka T, Kashino Y, Nuarsa I W and Osawa T 2019 Analysis of spatial and seasonal differences in the diurnal rainfall cycle over sumatera revealed by 17-year TRMM 3B42 dataset *Sci. Online Lett. Atmos.* **15** 216–21
[14] Ichikawa H and Yasunari T 2008 Intraseasonal Variability in Diurnal Rainfall over New Guinea and the Surrounding Oceans during Austral Summer *J. Clim.* **21** 2852–68
[15] Rahmawati N and Lubczynski M W 2018 Validation of satellite daily rainfall estimates in complex terrain of Bali Island, Indonesia *Theor. Appl. Climatol.* **134** 513–32
[16] Qian J H, Robertson A W and Moron V 2013 Diurnal Cycle in Different Weather Regimes and Rainfall Variability over Borneo Associated with ENSO *J. Clim.* **26** 1772–90

**Acknowledgement**
The authors are thankful to GPM - IMERG data providers for making it available to carry on this study (https://pmm.nasa.gov/gpm).

# Impact of climate change adaptation on paddy yield in dry zone Sri Lanka: A case study using agricultural production systems simulator (APSIM) model

**Eranga M Wimalasiri *, A D Ampitiyawatta, P K Dissanayake[1] and A S Karunaratne**

Department of Export Agriculture, Faculty of Agricultural Sciences, Sabaragamuwa University of Sri Lanka, Belihuloya, 70140, Sri Lanka
*Corresponding author: eranga@agri.sab.ac.lk

**Abstract**. Climate change in terms of increased temperature and variation of rainfall patterns is evident in Sri Lanka. Being a country with an agro-based economy, the impact of the changes of the climate negatively affects crop production and a possesses threat to food security. Using a crop modelling approach, the impact of climate change adaptation on paddy, which is the major crop in the country was assessed in Thanamalwila (6.43°N, 81.09°E), Sri Lanka. The calibrated and validated Agricultural Production Systems Simulator (APSIM) model was used for yield simulations. Bias-corrected future (2040-2069) climate data were obtained from 3 general circulation models; BNU-ESM, CSIRO-Mk3-6-0, and GFDL-ESM2M under the RCP4.5 scenario. Short season (BG300) and long season (BG358) cultivar, late and early planting (both by two weeks) were used as adaptation methods. Average paddy yield under baseline (1980-2009) climate for BG300 and BG358 were 3711 and 4524 Kg ha$^{-1}$, respectively. In BG300, the average future yield decreased in all 3 GCMs while in BG358 (long season), the average yield increased under 2 GCMs except CSIRO-Mk3-6-0. On average, delayed planting by 14 days as an adaptation measure increased the yield of BG300 (2.43%), the short season cultivar while in BG358, the average yield decreased by 3.55%. Except for CSIRO-Mk3-6-0 (in BG358), paddy yield in both cultivars decreased on early planting. This simple crop modelling exercise has shown the possibility of selection of cultivar and planting dates as climate change adaptation strategies. Further, field experiments and detailed crop model simulations are essential to validate the findings.

**Keywords**: APSIM, climate change adaptation, crop modelling, food security, rainfed agriculture

## 1. Introduction

Climate change is one of the major constraints to crop production in many parts of the world including Sri Lanka. According to previous studies, climate change is evident in Sri Lanka in terms of rainfall variability, increased temperature, and a higher frequency of extreme climatic events [1] [2]. With the climate changes, the yields of key crops in tropical regions such as paddy and maize are expected to decline [3] [4]. Due to climate change, it is expected that paddy yield will be decreased by 27.5% in China for the 2041-2070 period [5]. Sri Lanka is a tropical country on which rainfed agriculture is prominent [6] and around 34% of the total cultivated area of the country is under rice cultivation. Therefore, being a country with an agricultural-based economy, the impact of the changes of the climate negatively affects crop production and a possesses threat to the food security in Sri Lanka.

Different approaches that include empirical relationships and crop models have been used to study the impact of climate change and climate adaptation worldwide. Out of them, crop models act as one of the fast and reliable tools to assess the climate change impact and adaptation strategies under diverse agroecological conditions and crop varieties [7]. The Agricultural Production Systems Simulator (APSIM) is one of the crop modelling platforms to simulate crop production systems under climatic and management options [8]. Previously, it has been successfully used to simulate rice production in Sri Lanka [9] [10] [11].

Adaptation options are one of the best solutions to minimize crop yield losses due to climate change [1]. Farmers in Sri Lanka have their own adaptation measures to avoid unfavourable weather conditions and yield losses [1] [2]. However, they were not studied in detail using crop models. Therefore, the objective of this study is to assess the impact of climate change adaptation on paddy using the crop modelling approach. Paddy yield will be simulated using the Agricultural Production Systems Simulator (APSIM) model.

## 2. Materials and method

### 2.1. *Study area*

Thanamalwila (6.43°N, 81.09°E) which is located in the dry zone of Sri Lanka was selected as the study location. The annual average rainfall of the area is less than 1500 mm. A bi-modal pattern of rainfall can be observed in the area on which two peaks in March-April and October-November lead to the onset of two growing seasons.

### 2.2. *Climate data*

Observed daily minimum and maximum temperature and rainfall data for the 1980-2009 period were collected from the Department of Meteorology Sri Lanka. Due to the unavailability of observed solar radiation, the data were obtained from the NASA POWER database [12]. Data were quality checked for errors and missing data.

Bias-corrected daily minimum and maximum temperature, rainfall, and solar radiation for mid-century climate (2040-2069 period) were obtained for Representative Concentration Pathway (RCP) 4.5 scenario. Data were downscaled for 3 General Circulation Models (GCMs) as; BNU-ESM (GCM1) of College of Global Change and Earth System Science, Beijing Normal University, China, CSIRO-Mk3-6-0 (GCM2) of Commonwealth Scientific and Industrial Research Organization, and the Queensland Climate Change Centre of Excellence, Australia and GFDL-ESM2M (GCM3) of Geophysical Fluid Dynamics Laboratory, USA. The data were downscaled from the Consultative Group for International Agricultural Research (CGIAR) Research Program on Climate Change, Agriculture and Food Security (CCAFS) database (http://ccafs-climate.org/).

### 2.3. *Model description*

The Oryza module of Agricultural Production Systems Simulator (APSIM - Version 7.10) [8] was used to simulate paddy growth and development. Being a process-based crop model, APSIM can simulate several crops, climate, and soil-related processes. The model requires daily rainfall, minimum and maximum temperatures, and solar radiation as climate data.

### 2.4. *Climate change adaptation options*

Two types of adaptation options were selected as cultivar and planting date. BG300, a short duration (3 months) cultivar, and BG358, a long duration (3 and half months) cultivar were selected. These two cultivars are prominent cultivars across Sri Lanka. Two planting dates were used as; late planting (2 weeks delayed planting) and early planting (2 weeks earlier planting).

### 2.5. *Yield simulation*

Pre-calibrated and validated APSIM Oryza model was used in this study [10]. Published genetic coefficients of paddy cultivars were used in simulations. All other parameters of the APSIM Oryza model were set as default. The planting date was set as 10th November. All other crop management data (plant densities, fertilizer applications, etc.) were collected from Wimalasiri et al. [9]. Paddy yield was simulated for both past/ baseline (1980-2009) and mid-century (2040-2069) climates.

## 3. Results and discussion

### 3.1. *Past and future climate*

The observed annual average of maximum (Tmax) and minimum (Tmin) temperatures of Thanamalwila were 32.1 °C and 23.3 °C, respectively. The average annual rainfall for the baseline period was 1121 mm. In all 3 GCMs considered, both Tmax and Tmin increased during the 2040-2069 period on which the highest Tmax and Tmin increments were reported from GFDL-ESM2M (1.3 °C) and CSIRO-Mk3-6-0 (1.7 °C) respectively. The average annual rainfall increased by 129 mm in BNU-ESM and by 507 mm in CSIRO-Mk3-6-0 while decreasing by 76 mm in GFDL-ESM2M. The variation of rainfall and minimum and maximum temperatures in both past and future climates are shown in Figure 1.

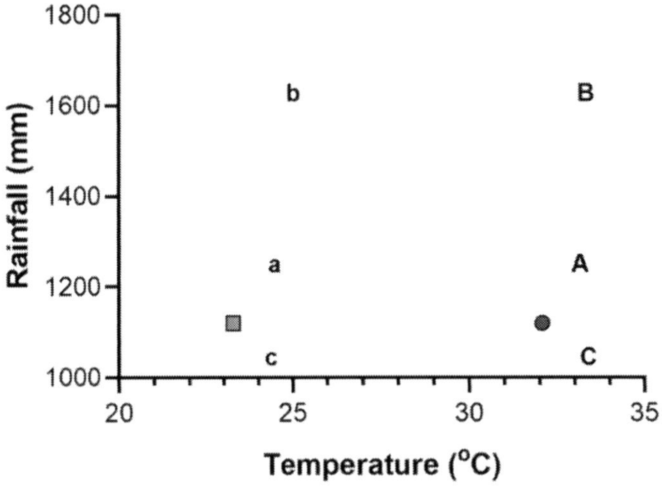


**Figure 1.** Variation of minimum and maximum temperatures and rainfall during 1980-2009 (symbols) and 2040-2069 (letters) period. In symbols, the square represents minimum temperature while the circle represents maximum temperature. Letters denote 3 GCMs while lower and uppercase letters represent minimum and maximum temperatures

### 3.2. *Paddy yield and yield change in future climate*

The average paddy yield for both past and future climates is presented in Table 1. Comparatively, a higher yield was reported from the long duration cultivar (BG358) than the short duration cultivar (BG300).

**Table 1.** Average paddy yield of two cultivars during past and future climates

| Climate | Yield (kg ha$^{-1}$) | |
|---|---|---|
| | **BG300** | **BG358** |
| Past climate | 3711±255 | 4524±327 |
| Future climate - BNU-ESM | 3646±218 | 4617±323 |
| Future climate - CSIRO-Mk3-6-0 | 3467±245 | 4455±361 |
| Future climate - GFDL-ESM2M | 3668±244 | 4575±335 |

When compared with the past yield, the average paddy yield decreased in all 3 GCMs in BG 300 with the highest reduction of 6.6% in CSIRO-Mk3-6-0. The average yield reduction of BG300 in BNU-ESM and GCM2 were 1.8 and 1.1% respectively. In the long duration cultivar (BG358), the paddy yield decreased by 1.5% in CSIRO-Mk3-6-0 while increased in 2 other GCMs by 2.1% (BNU-ESM) and 1.1% (GFDL-ESM2M). The results of this study agree with the previous findings on projected paddy yield for mid-century (2040–2070) in the Intermediate zone, Sri Lanka using other GCMs [10].

In contrast, some other studies in dry zone Sri Lanka have shown that the yield of BG300 cultivar is expected to increase in the year 2050 compared to the year 2017 [11]. The impact of climate change on crop yield depends on various factors that include location, season, and management practices. Therefore, detailed studies are needed on the performance of BG300 and BG358 in other agroecological zones in Sri Lanka.

### *3.3. Impact of adaptation options*

The paddy yield change under adaptation measures is shown in Figure 2. The impact of late and early plating on the baseline climate was initially studied. According to the model simulated paddy yield, the late planting increases the yield of short duration cultivar (BG300) by 1.92% while decreases in the long duration cultivar (BG358) by 1.43% in the current climate. Early planting by 14 days decreased the yield in both cultivars (by 0.64 and 0.56% respectively).

However, in the future climate, the delayed planting by 14 days increased the yield in all 3 GCMs for short season cultivar on which the values ranged from 1.71% (GFDL-ESM2M) to 3.09% (CSIRO-Mk3-6-0). In contrast, the yield of long duration cultivar (BG358) decreased on delayed planting in all 3 GCMs with the highest reduction of 4.51% in CSIRO-Mk3-6-0. In early planting by 14 days, the yield decreased in all the GCMs for short duration cultivar. However, early planting increased yield in CSIRO-Mk3-6-0 (0.11%) while decreasing in two other GCMs in the long duration cultivar.

Two adaptation strategies (cultivar selection and change of planting date) used in this study showed different sensitivities under different future climate scenarios. However, detailed field experiments and yield simulations using crops models are needed to validate the findings. Other than these options, farmers in Sri Lanka follow different climate change adaptation strategies that can be categorized into land management, crop management, and irrigation management [1]. The sensitivity of other climate change adaptation strategies can be assessed using the crop modelling approach.

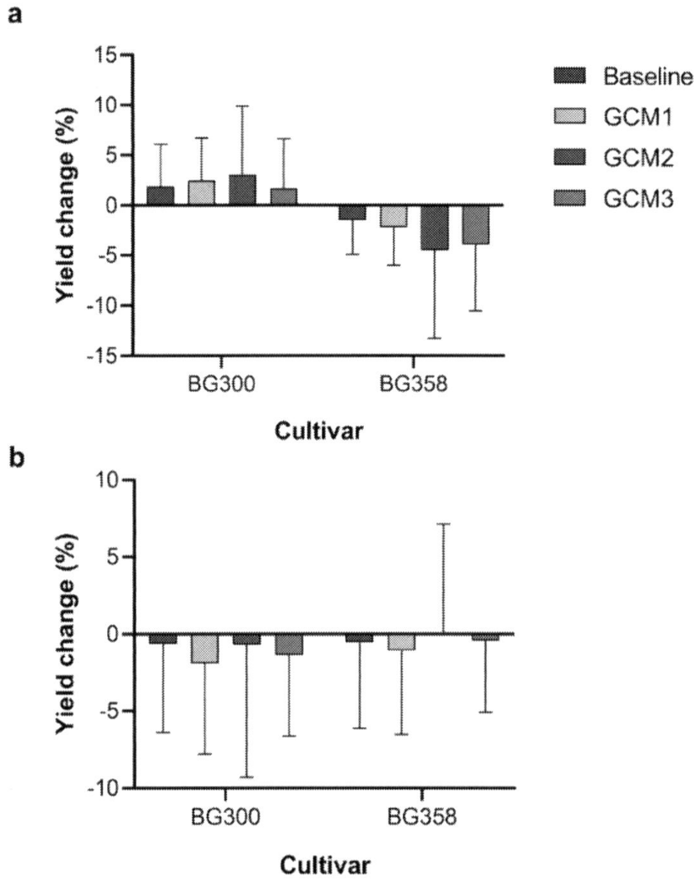


**Figure 2.** Variation of paddy yield change of baseline (1980-2009) and future (2040-2069) climate under late planting (a) and early planting (b)

## 4. Conclusions

Both minimum and maximum temperatures and rainfall (average) are expected to increase in Thanamalwila, Sri Lanka during the 2040-2069 period. The Agricultural Production Systems Simulator (APSIM) model has been successfully used to study the climate change adaption strategies in Sri Lanka. According to the model output, the short-duration rice cultivar (BG300) is at high risk than the long duration cultivar (BG358) in the mid-century (2040-2069) climate. This simple crop modelling exercise has shown that selection of cultivar and planting dates are possible climate change adaptation strategies in the dry zone, Sri Lanka.

## References

[1] Esham M and Garforth C 2013 Agricultural adaptation to climate change: insights from a farming community in Sri Lanka *Mitig. Adapt. Strateg. Glob. Chang.* **18** 535–49
[2] De Costa W A J M 2008 Climate change in Sri Lanka: Myth or reality? Evidence from long-term meteorological data *J. Natl. Sci. Found. Sri Lanka* **36** 63–88
[3] Lobell D B and Field C B 2007 Global scale climate–crop yield relationships and the impacts of recent warming *Environ. Res. Lett.* **2** 014002
[4] Malaviarachchi M A P W K, Costa W A J M De, Fonseka R M, Kumara J B D A P, Abhayapala

K M R D and Suriyagoda L D B 2015 Response of Maize (Zea mays L.) to a Temperature Gradient Representing Long-Term Climate Change Under Different Soil Management Systems *Trop. Agric. Res.* **25** 327

[5] Li Y, Wu W, Ge Q, Zhou Y and Xu C 2016 Simulating climate change impacts and adaptive measures for rice cultivation in hunan province, China *J. Appl. Meteorol. Climatol.* **55** 1359–76

[6] Wimalasiri E M, Ashfold M J, Walker S, Nissanka S P and Karunaratne A S 2017 The relationship between rainfall characteristics and Proso millet (*Panicum miliaceum* L.) cultivation in low country dry zone, Sri Lanka *Trop. Agric. Res. Ext.* **20** 32

[7] Ruane A C, Mcdermid S, Rosenzweig C, Baigorria G A, Jones J W, Romero C C and Dewayne Cecil L 2014 Carbon-temperature-water change analysis for peanut production under climate change: a prototype for the AgMIP coordinated climate-crop modeling project (C3MP) *Glob. Chang. Biol.* **20** 394–407

[8] Holzworth D P, Huth N I, deVoil P G, Zurcher E J, Herrmann N I, McLean G, Chenu K, van Oosterom E J, Snow V, Murphy C, Moore A D, Brown H, Whish J P M, Verrall S, Fainges J, Bell L W, Peake A S, Poulton P L, Hochman Z, Thorburn P J, Gaydon D S, Dalgliesh N P, Rodriguez D, Cox H, Chapman S, Doherty A, Teixeira E, Sharp J, Cichota R, Vogeler I, Li F Y, Wang E, Hammer G L, Robertson M J, Dimes J P, Whitbread A M, Hunt J, van Rees H, McClelland T, Carberry P S, Hargreaves J N G, MacLeod N, McDonald C, Harsdorf J, Wedgwood S and Keating B A 2014 APSIM - Evolution towards a new generation of agricultural systems simulation *Environ. Model. Softw.* **62** 327–50

[9] Wimalasiri E M, Jahanshiri E, Suhairi T A S T M, Udayangani H, Mapa R B, Karunaratne A S, Vidhanarachchi L P and Azam-Ali S N 2020 Basic Soil Data Requirements for Process-Based Crop Models as a Basis for Crop Diversification *Sustain. 2020, Vol. 12, Page 7781* **12** 7781

[10] Zubair L, Nissanka S P, Weerakoon W M W, Herath D I, Karunaratne A S, Prabodha A S M, Agalawatte M B, Herath R M, Yahiya S Z, Punyawardhene B V R, Vishwanathan J, Delpitiya P, Wijekoon A E N, Gunaratna J, Chandrasekara S S K, Wickramagamage P, Weerasinghe K D N, Navaratne C M, Perera R S, Gunesekara A I, Kumara G M P, Wallach D, Valdivia R O and McDermid S 2015 Climate Change Impacts on Rice Farming Systems in Northwestern Sri Lanka 315–52

[11] Amarasingha R K, Suriyagoda L D B, Marambe B, Galagedara L W and Punyawardena R 2018 Impact of climate change on rice yield in Sri Lanka: a crop modelling approach using Agriculture Production System Simulator (APSIM) *Sri Lanka J. Food Agric.* **4** 21

[12] Zhang T, Stackhouse P W, Chandler W S, Hoell J M, Westberg D and Whitlock C H 2008 A Global Perspective on Renewable Energy Resources: Nasa's Prediction of Worldwide Energy Resources (Power) Project *ISES Sol. World Congr. 2007, ISES 2007* **4** 2636–40

# The "jamban systems" for water conservation in the face of climate change and their benefit to local people

**Kristiyanto[1,2]*, Mashudi Alamsyah[1] and Mertua Ferry Siburian[1]**

[1]Indraprasta PGRI University,TB. Simatupang, Jl. Nangka Raya No.58 C, Jagakarsa, South Jakarta, 12530, Indonesia
[2]State Islamic University Syarif Hidayatullah Jakarta. Jl. Ir. H. Djuanda No. 95, Ciputat, South Tangerang 15412, Banten, Indonesia
Corresponding author: kristiyanto94@gmail.com

**Abstract.** Drought occurrence is a global problem that has an impact on both ecological systems and human beings, so some of the local people, chiefly in rural areas, have tried to create water harvesting using traditional techniques, one of which is the "small reservoir" type. Those types of water conservation and water availability on a small scale, alongside it, have become one strategy for coping with drought, which is caused by climate change impacts, alongside it is part of local economic development. Hence, this research aimed to explore and identify the "Jamban system" functions of local people in their daily lives, so the direct observation and in-depth individual interviews methods, thence the analytical contents of the field data that were used to identify and analyze it, alongside its ethnographic approach. The result of this research, qualitatively, shows that "Jamban systems" have many functions, which are used in aquaculture, the aqua-plants, the water availability or stock, and they also cope with climate change impacts. These were part of the multifunction that is used by the local people to meet their daily demand for family-scale water conservation, alongside the water conservation paradigm, which would be social resilience system shaped.

**Keywords:** ecological systems, human being, traditional technique, water conservation

## 1. Introduction

In urban and rural areas, watersheds (water crises) in meeting the water demand have been caused by both climate change impacts [1] and land use change impacts [2,3], so the season has changed (rainy to dry season). The drought is a global problem that impacts human beings and ecosystems, so indirectly resulting in the loss of welfare of local people, alongside land degradation and desertification phenomena [4–6]. As a result, it is critical to reducing response time when dealing with environmental hazards. The Jamban systems type developed is the one traditional technique that has been developed by the local people to obtain water availability (stock), [7] which has been considered to be a part of the "water harvesting methods" part of which many advantages for human beings' welfare goals, alongside water conservation and sustainable development.

Hence, the "Jamban systems" is a part of one solution or strategy that is used to cope with drought occurrences. Alongside it would be developing aquaculture and aqua-plants [2,3], and according to [8], traditional water management has considered the water supply to human beings sustainable. The development of "Jamban systems" is a part of the response to the watershed in which a rural area was attacked violently during the dry season (drought). These systems were able to cope with it [9–11], alongside the local people's increased wellbeing through both aquaculture and aqua-plant development.

Hence, developing water harvesting methods to increase water availability in rural areas has become a priority. Particularly for local people in plateau areas, where both human activities (anthropogenic) [12,13], as well as population growth and development, are required for their livelihood, such as living space, green areas, and so on [14]. Therefore, it is pressured by land-use change activated in rural areas (land conservation to land economic changes), which could be one of the problem symptoms of ecological disturbance that impacts the water cycle (hydrology) impacts the watershed (water resources), and it's slowly coming to extremes of drought occurrence. Hence, according to [15], human beings are confronted with threats to water availability (water supply) in their daily lives. Therefore, it is necessary to implement water management based on a water conservation paradigm, to supply sustainable water to human beings kindly and sustainably [16]. Hence, this research will be exploring Jamban systems functions and their contributions to the local people, which will be utilized to save traditional water harvesting systems and the water supply that are used to support their lives in a kind and sustainable way.

## 2. Materials and methods

In general, Cikalong village is a rural area type and it's geographically at least 100 masl to 500 masl, alongside it's a plateau type and it's a conservation area or protected area development [2,3], which could be shown in Figure 1. As a result, the local people can be autonomous in their need to stay alive and sustainably, but on the other hand, the local people have faced the watershed of the drought occurrence in each of the dry seasons every year, which leads to extreme phenomena. to the description of field problems above, some of the sites serve as samples of representative research, which is at least 1-5 sites of research objects that are used to collect data.

They both use direct observation and in-depth individual interview methods. It is based on a purposive random sampling approach based on the elevation of gradients [17] and its ecosystem types or characteristics, which is the distance between one of the research objects and the others, which is considered to be representative in taking field data that was documented, analyzed, and discussed comprehensively. These data for analysis by the ethnography approach [18] are based on direct observation and in-depth individual interviews with the results showing that the perceptions or interpretations of the local people are both the functions of "Jamban systems" and the various "Jamban systems" types developed in their daily lives.

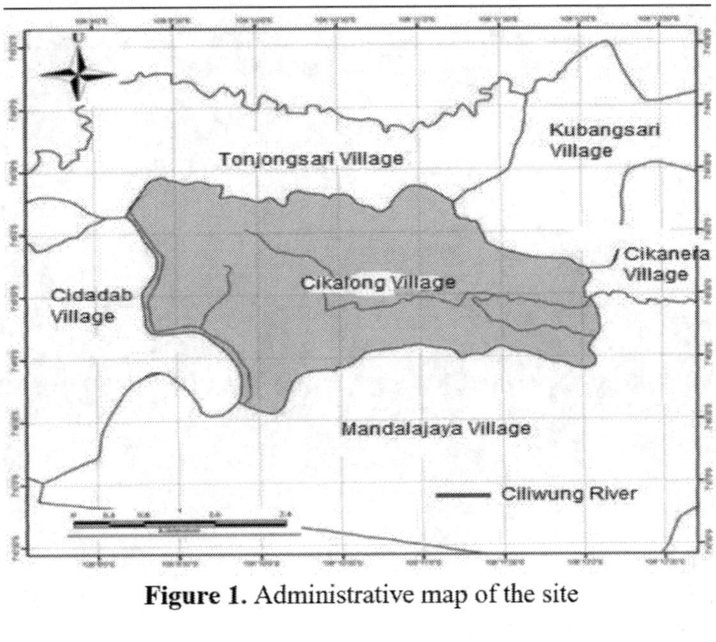


**Figure 1.** Administrative map of the site

Inventorying and analyzing the functions of Jamban systems and their contributions to local people has become a research topic, as has to determine and calculate the utilization of Jamban systems in their daily lives at this time, which were used for aquaculture and aqua-plant type development. Hence, discussing critically Jamban systems functions in a rural area now is pivotal and it has been considered to cope with the drought occurrence kindly alongside how the local people have developed Jamban systems as an alternative solution to a secondary need for development, which is all part of the research question that this research explores.

## 3. Results and discussion

The "Jamban systems" has developed in rural areas, where is part of the response to drought occurrence, alongside it used to the human of waste management, so the local people tried to use in developing it, therefore the each of dwelling in rural areas as Cikalong areas has used to traditional management of type development, which was as the systems to water closet functions, alongside as the water conservation type or methods developed [2,3], utilized as plants species to grow and developed (aqua-plants methods) (Figure 2.A), aquaculture methods (Figure 2.B), and the traditional technique of type development the others, which it leads to water conservation to multifunction of Jamban systems, and indirectly that the Jamban systems type is the one way to response or alternative solution, and it would be the welfare of local people, kindly, adaptively, and dynamically.

Refers to Kristiyanto [2,3] and Figure 2, showing that Jamban systems have become one solution or strategy for managing water conservation (water storage) at the local scale, alongside creating social resilience system goals [19]. This is one of the targets in creating sustainable or rural development [9] and that the Jamban system is the same as water harvesting methods, which are part of the response or adaptation type to climate change impacts [2,3]. Many benefits would be obtained in the use of Jamban systems as one of the methods in the surviving process, mainly in watershed management to meet the demand on a local scale (family). Therefore, Figure 2 describes how local people tried to survive in a watershed by saving traditional water harvesting systems [methods], alongside creating a local economy and ensuring the local people were able to utilize them [2,3]. Hence, some plant species grew and developed kindly and each of the research objects (1–5), with the vegetable being more dominant than the other plant species (Table 1), indirectly creating a food security paradigm or goals in rural areas on a local scale.

**Figure 2.** (A) The rural plants (*Aqua-plants)* and (B) It used to *aquaculture* development (varieties of fish, especially *Cyprinus carpio*, *Osphronemus goramy*, and forth) on jamban systems, which were considered to local economic created (welfare).

**Table 1.** Plant species grow and develop on "Jamban Systems," and they are used to

| No. | Local name | Family | Scientific name | Used to |
|---|---|---|---|---|
| 1 | Tomat | *Solanaceae* | *Solanum lycopersicum* | Vegetable |
| 2 | Cabe | *Solanaceae* | *Capsicum frutescens* | Vegetable |
| 3 | Seledri | *Apiaceae* | *Apium graveolens* | Vegetable |
| 4 | Daun Bawang | *Amaryllidaceae* | *Allium fistulosum* | Vegetable |

Source: Data primary of field observation results, 2020-2021

### *3.1. Jamban systems as ecological functions for the wellbeing of local people*

The Jamban system is the one way done by local people to manifest food security or water security [2,3] at the same time, to create sustainable water management. The complexity of the Jamban system as used to ecological function in rural areas is dynamic, which would be shown to prove that it can improve the welfare of local people by utilizing the Jamban system optimally and sustainably. Those activities could be identified as how local people can create ecological nuance by developing Jamban systems kindly.

The Jamban system, according to [20], is one of the solutions or strategies in water management that is based on local ecological knowledge perspective, which recycles and reuses was developed by local people in their surroundings, mainly in rural areas, and [21] also affixes that the Jamban system is part of indigenous or local knowledge developed and it's part of the traditional conservation technique type [22]. Many of the advantages that local people obtained from the Jamban system [2,3], which were automatically created to create an ecological system (ecosystems) that was more stable and sustainably, so it led to the carrying capacity of local people to obtain welfare. Creating a life cycle in Cikalong areas was one type of activity, where the local people could adapt to droughts that caused climate change impacts [23], and it's more caused by human activities in changing land use (land-use change) from conservation areas to economic areas massively and destructively (Figure 3).

**Figure 3.** (A) Land use change dynamic to economic areas' expansion (Farmland), (B) developing dwelling communities' expansion in rural areas.

Indirectly, the local people have changed land use massively and destructively, which has caused an unbalanced ecological system (ecological disturbance), and directly, it has changed the environment susceptible to hazard, mainly the loss of water cycle process (hydrological), in which drought is one phenomenon occurrence. For example, Figures A and B show how local people have utilized land-use change massively as it relates to changing farmland areas, agricultural land, and housing (conservation areas to economic areas), resulting in a loss of ecological potential or functions in rural areas, such as in Cikalong village, Tasikmalaya, West Java. This matter is part of the local activities accumulated in their surroundings that indirectly degrade water quality [24], one of which is defecating (fecal). It is based on field data, mainly in the Cikalong village areas, where almost all the local people have been utilizing the Jamban system (ponds) as the end of the human waste (defecation process), so it would be caused by diarrhea phenomena, mainly in children [25–27].

Figure 2 shows the traditional human waste developed by various types of type (B), which was created to have biotic aspects developed alongside creating ecological nuance, mainly in the food web dimension and its impacts on another type [28,29], in which it was created to have a life cycle by kindly, sustainably, and it's sustainable for social resilience system goals and it's part of the economic source for the local people in their daily lives.

### *3.2. The jamban systems as local economic sources of local development*

Each of the dwellings tried to make use of the Jamban systems as local economic sources. See Figure 1. On the other hand, the Jamban system includes harvesting of water functions, which is part of adapting to the drought phenomenon, so when in the dry season, the local people are prepared to be confronted with adaptability, kindly, and dynamically, because of the water stock (availability) in their surroundings (Jamban systems). Hence, many of the advantages that are attained by the local people with the Jamban systems, which is the minimum number of able to fish or plant species harvested (Figure 2), so ecologically and economically, each of the local people would be surviving on their natural potential with their local ecological knowledge, dynamically [22]. The Jamban systems are part of their aliveness. One was considered as an alternative solution to coping with field problems, which the local people were able to create and use strategies for confronting.

Economically, the local people tried to develop livestock. For example, there were goats, cows, chickens, and so forth. It's also an alternative solution to decrease local poverty by exploring natural resource potential kindly and dynamically based on local ecological knowledge perspective [29]. Figure 4 is one of the field data obtained based on direct observation, which shows that the local people try to utilize local natural potential by optimally using various livestock in their daily lives. Formerly and until now, urbanization has become a solution to welfare more quickly than as farmers, so many of the local people have shifted to becoming workers, traders, and so forth. It is part of one strategy to improve the economy and their lives (welfare).

**Figure 4.** (A) the chicken, which is used for plows in farmland, and (B) the buffalo, a livestock type that evolved in rural areas

## 4. Conclusion

On a local level, the Jamban system is the only means to deal with climate change. Water security is also important, as it contributes to the well-being of local residents in their daily life, even during the dry season. The jamban system, on the other hand, is an alternate solution to water availability, as well as a part of the adaptation and mitigation strategy to droughts induced by climate change impacts. Effectively, the Jamban system has brought water security and sustainability, and it is part of water conservation, so it has considered drought kindly. It could be developed to include aqua-plants and aquaculture methods as local economic and food security on a family scale.

**References**

[1] Bohle H G, Downing T E and Watts M J 1994 Climate change and social vulnerability: Toward a sociology and geography of food insecurity *Glob. Environ. Chang.* **4** 37–48
[2] Kristiyanto 2021 The "jamban systems" as coping to both of land use and climate change impacts in rural area *IOP Conf. Ser. Earth Environ. Sci.* **724** 012072
[3] Kristiyanto 2020 Traditional development of water conservation based on local knowledge: coping with climate change impacts in rural areas *IOP Conf. Ser. Earth Environ. Sci.* **423** 012067
[4] Andrew W 1993 Desertification as a global environmental issue *GeoJurnal* **31** 11–4
[5] Le Houérou H N 1996 Climate change, drought and desertification *J. Arid Environ.* **34** 133–85
[6] Abdi O A, Glover E K and Luukkanen O 2013 Causes and Impacts of Land Degradation and Desertification: Case Study of the Sudan *Int. J. Agric. For.* **3** 40–51
[7] Behailu B M, Pietilä P E and Katko T S 2016 Indigenous Practices of Water Management for Sustainable Services: Case of Borana and Konso, Ethiopia *SAGE J.* **6**
[8] Sharma N and Kanwar P 2009 Indigenous water conservation systems—A rich tradition of rural Himachal Pradesh *Indian J. Tradit. Knowl.* **8** 510–3
[9] Sachin S, Manohar, Rambirendra, Hemant, Shyambirendra and Takar 2018 A Review on Sustainable Development of Rainwater Harvesting System in Urban and Rural Areas *Int. Res. J. Eng. Technol.* **5** 511–4
[10] Loveson V J and Misra D D 2004 Rainwater harvesting technologies for mining rural areas **1** 66–70
[11] Dharmarao S S 2016 Water harvesting – a solution to drought and falling level of ground water *International Conference on Emerging Research trends in Applied Engineering and Technology*
[12] Helmers A, Platek A, Ponte M, Secen N and Cottenie K 2016 The impacts of anthropogenic disturbance on plant species richness in the freshwater lakes of Algonquin Provincial Park *SURG J.* **9** 5–13
[13] Flandroy L, Poutahidis T, Berg G, Clarke G, Dao M C, Decaestecker E, Furman E, Haahtela T, Massart S, Plovier H, Sanz Y and Rook G 2018 The impact of human activities and lifestyles on the interlinked microbiota and health of humans and of ecosystems *Sci. Total Environ.* **627** 1018–38
[14] Bergesen J D, Suh S, Baynes T M and Musango J K 2017 Environmental and natural resource implications of sustainable urban infrastructure systems *Environ. Res. Lett.* **12** 125009
[15] Fulazzaky M A 2014 Challenges of Integrated Water Resources Management in Indonesia *Water* **6** 2000–20
[16] Tuan L, Wyseure G and Viet L H 2004 Sustainable water management for rural development in the Mekong river delta, Vietnam *The second International Symposium on Southeast Water Environment* (Hanoi, Vietnam) pp 1–7
[17] Jin X, Zhang Y, Schaepman M, Clevers J and Su Z 2008 Impact of elevation and aspect on the spatial distribution of vegetation in the Qilian mountain area with remote sensing data *Int. Arch. Photogramm. Remote Sens. Spat. Inf. Sci.* **37** 1385–90
[18] Pelto G H and Armar-Klemesu M 2015 Identifying interventions to help rural Kenyan mothers cope with food insecurity: results of a focused ethnographic study *Matern. Child Nutr.* **11 Suppl 3** 21–38
[19] Folke C 2006 Resilience: The emergence of a perspective for social–ecological systems analyses *Glob. Environ. Chang.* **16** 253–67
[20] Beekman G B 1998 Water Conservation, Recycling and Reuse *Int. J. Water Resour. Dev.* **14** 353–64

[21] Cheserek G 2005 Indigenous knowledge in water and watershed management: "Marakwet" conservation strategies and techniques *FWU: Topics of Integrated Watershed Management* pp 25–33

[22] Clarke W C 1990 Learning from the Past: Traditional Knowledge and Sustainable Development on JSTOR *Contemp. Pac.* **2** 233–53

[23] Zhiña D, Montenegro M, Montalván L, Mendoza D, Contreras J, Campozano L and Avilés A 2019 Climate Change Influences of Temporal and Spatial Drought Variation in the Andean High Mountain Basin *Atmosphere (Basel).* **10** 558

[24] Govender T, Barnes J M and Pieper C H 2011 The impact of densification by means of informal shacks in the backyards of low-cost houses on the environment and service delivery in cape town, South Africa *Environ. Health Insights* **5** 23–52

[25] Cui P, Li J, Liu N and Duan Z jun 2018 Incidence of acute diarrheal illness in Chinese communities: A meta-analysis *BMC Gastroenterol.* **18** 144

[26] Pires M M, Prado P I and Guimarães P R 2011 Do Food Web Models Reproduce the Structure of Mutualistic Networks? *PLoS One* **6** e27280

[27] da Paz M G A, de Almeida M F and Gunther W M R 2012 Diarrhea in children and sanitation and housing conditions in periurban areas in the city of Guarulhos, SP - PubMed *Rev Bras Epidemiol* **15** 188–97

[28] Zhang L, Hartvig M, Knudsen K and Andersen K H 2014 Size-based predictions of food web patterns *Theor. Ecol.* **7** 23–33

[29] Woodley E 1991 Indigenous ecological knowledge systems and development *Agric. Hum. Values 1991 81* **8** 173–8

# Impact of climate change on the agriculture sector and its adaptation strategies

**Lilyk Eka Suranny*, Evi Gravitiani and Mugi Rahardjo**

Faculty of Economy and Business, Universitas Sebelas Maret, Indonesia
*Corresponding author: lilykekasuranny@student.uns.ac.id

**Abstract.** Climate change in Indonesia, including erratic rainfall intensity, extreme temperatures, El Nino and La Nina events, can cause farmers not to implement agricultural efforts optimally. Multiplier effect caused is the risk of decreased agricultural production and even crop failure that causes losses for farmers. The impact can broadly reduce farmers' welfare, food security and increase poverty, for which most Indonesians livelihood as farmers. The negative consequences of climate change require efforts to minimize the impact and anticipate through adaptation strategies. These research aims are to analyze the impact of climate change on the agriculture sector and develop an adaptation strategy to address the impact of climate change on the agricultural sector. This method uses review literature related to the impact of climate change on agricultural products and adaptation strategies to climate change in the agricultural sector. The impact of climate change caused a decline in agricultural production and even crop failure. The main actors of climate change that affect the agricultural sector include rain patterns, extreme climate events (floods and droughts), and increased air temperatures. Recommendations for adaptation strategies are using superior plants, adjusting planting time and pattern, water management technology (rain harvesting and irrigation), and biotic environmental strategies.

**Keywords:** climate change, agriculture, strategy, adaptation

## 1. Introduction

Indonesia is an agrarian country, where the agricultural sector is the livelihood of most of the population in Indonesia. This tropical climate condition makes most of Indonesia's population farmers. The agriculture sector is very vulnerable to changes in climate. This is because climate change affects planting patterns, production, and crop bags' cauldrons [1]. Climate change is a condition in which a change in world climate patterns results in extreme weather, erratic temperatures, and fluctuating rainfall.

Studies related to the impact of climate change on the agricultural sector state that reduced intensity of rain is the leading cause of the decline in farmers' harvests on dry land in Dharmaputri, India [2,3]. Climate variations such as prolonged droughts have a high impact on dryland crop yields. In Southeast Asia, vulnerability due to climate change causes a decrease in the production of grain and corn crops by about 10% [4,5]. While in Indonesia, cases due to climate change can cause a shortage of groundwater supply, lower soil moisture, decreased fertility and increased evaporation and rainfall [5]. Multiplier effect caused is the risk of decreased agricultural production and even crop failure that causes losses for farmers. The impact can broadly reduce farmers' well-being, food security and increase poverty. But in reality, agricultural production is becoming so low that even farmers are experiencing crop failure, which is one of the causes of climate change.

Understanding farmers regarding climate change is critical to anticipate and minimize climate change risk through mitigation and adaptation actions by adjusting to agricultural cultivation techniques [6,7]. Farmers usually take mitigation measures against climate change based on hereditary experience and knowledge carried out in their agricultural business activities. Farmer seeks to combine mitigation strategies with the experience they gain. From this, farmers will try to choose the most appropriate mitigation measures to minimize the risks due to climate change [7,8].

Research related to the impact of climate change on agricultural production has been widely done. Meanwhile, farmers' understanding of the impact of climate change is still lacking, so early anticipation of the impacts caused by climate change is still minimal. As a result, there is a decrease in agricultural production and even crop failure. Therefore, there needs to be integrated research and assessment between the impacts of climate change and adaptation strategies that are recommended primarily for farmers. This article aims to outline the main factors causing climate change, the impact of climate change, and adaptation strategies for farmers in dealing with climate change.

## 2. Research methods

This qualitative research is a literature review on relevant research on the impact of climate change on the agricultural sector. The data used in this study comes from the results of research that has been conducted and published in national and international journals. Review literature is a comprehensive overview of knowledge or research that others have previously done on a specific topic and conveys to readers the topic for further research ideas. The activity starts with searching for data sources according to the topic to be researched by reading the appropriate article title and reading the abstract. If the title and abstract match the required reference criteria, then read the entire article next analyzed. Articles that fit the criteria required by the researcher will be taken as a reference source. The method of analysis of the article's content used as a reference uses the analysis of journal content.

## 3. Results and discussions

### *3.1 Impact of climate change on agricultural production*

Research on the impact of climate change on agricultural production is a critical issue to achieve the three main goals of the Sustainable *Development Goals* (SDGs), namely: Alleviating all forms of poverty, ending hunger, achieving food security, and improving nutrition, and supporting sustainable agriculture, and ensuring healthy and supporting lives. Well-being for all ages [9]. When the impacts of climate change can be minimized through adaptation strategies, efforts to achieve the SDGs goals will be more optimal.

Based on a review of the literature submitted, the impact of climate change is vulnerable to affect the agricultural sector, leading to a decrease in agricultural production. Rice production in Indonesia decreased by about 25% due to climate change in 2014 [10,11]. Agriculture in Yogyakarta, Indonesia, is severely affected by the occurrence of climate change [11]. Climate change events include floods and droughts that can damage agricultural land and cause crop pest attacks [11–13]. Studies on El Nino, La Nina, and food supply in Java show that El Nino climate anomalies negatively affect rice production in Java Island. The barrage of events due to climate change is the leading cause of production failures in the agricultural sector [14].

Climate change can affect the social and economic sustainability of the agricultural sector, either directly or indirectly. The incidence of crop failure, causing low agricultural productivity with high production costs due to climate change, impacts the decline in farmers' incomes. Another impact is increasing the number of unemployed [11,15,16]. Food scarcity often occurs in several countries, including Indonesia, is caused by crop failure. Crop failure in food crops is due mainly to environmental factors, and one of the most influential is climate change [17].

Climate change can have an impact on the productivity of corn. The growth and development of the plant are determined mainly by elements of climate, such as air temperature. Air temperature affects the activity of plant life. Its effects include proses photosynthesis respiration, respiration, rationing, hancollision, pollination, and fertilization. One of the efforts of the adaptation to deal with the effects of climate change is to establish planting patterns and planting calendars based on consideration of climate policy [18,19]. In addition, the impact caused by climate change there is a rise and decrease in temperature, the instability of rain yang down, and the tidal events of seawater that are erratic. Climate change that occurs affects the quality and quantity of corn commodity corn. Rainfall is one climate element that supports water preparations, especially in a rainy and dry land. Rainfall that exceeds the limit causes an increase in the volume of water at ground level, thus affecting plant growth. Excessive rainfall can affect plant productivity resulting in impaired plant growth [19,20].

The three main factors associated with global climate change, which impact the agricultural sector, are 1) changes in rain patterns, 2) increasing extreme climate events (floods and droughts), and 3) discolouring of air temperatures [21,22]. There is a fairy one from 1879 to 2006; there has been a decrease in rainfall and changes in rainfall patterns in Tasikmalaya, West Java [18]. This is due to the occurrence of shifts at the beginning of the season and planting period that impact the decrease in the potential of one period of rice planting period (Figure 1). It was further conveyed that changes in rainfall patterns can reduce the availability of water in reservoirs, especially in Java (Jatiluhur reservoir in West Java and Gajah Mungkur and Kedung Ombo dams in Central Java [22].

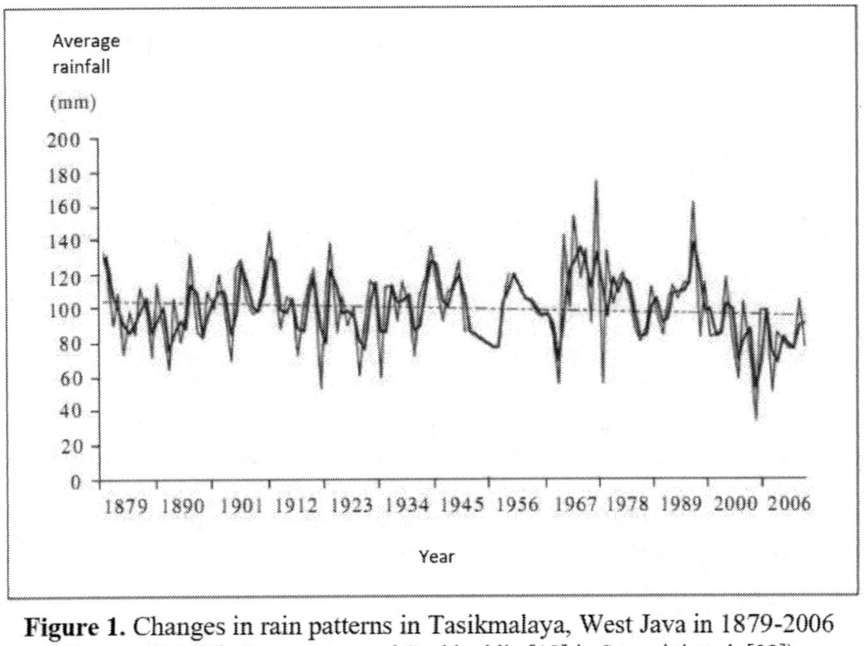


**Figure 1.** Changes in rain patterns in Tasikmalaya, West Java in 1879-2006
(Source: [18,22]. Runtunuwu and Syahbuddin [18] in Surmaini et al, [22])

The results of a global analysis of the climate change index, which is an index that rates future climate deviations compared to current ones, indicate that the value of climate deviations in Indonesia will increase in the future by 7 and 8 [22,23]. The value indicates that Indonesia will experience an increase in the frequency of extreme climate events such as floods and droughts in the future. The impact of flood and drought events on the destruction of rice paddy crops from 1989-2007 shows a significant upward trend (Figure 2). Based on the picture, the damage to rice crops due to the highest flood occurred in 1991, which impacted more than 60 thousand hectares.

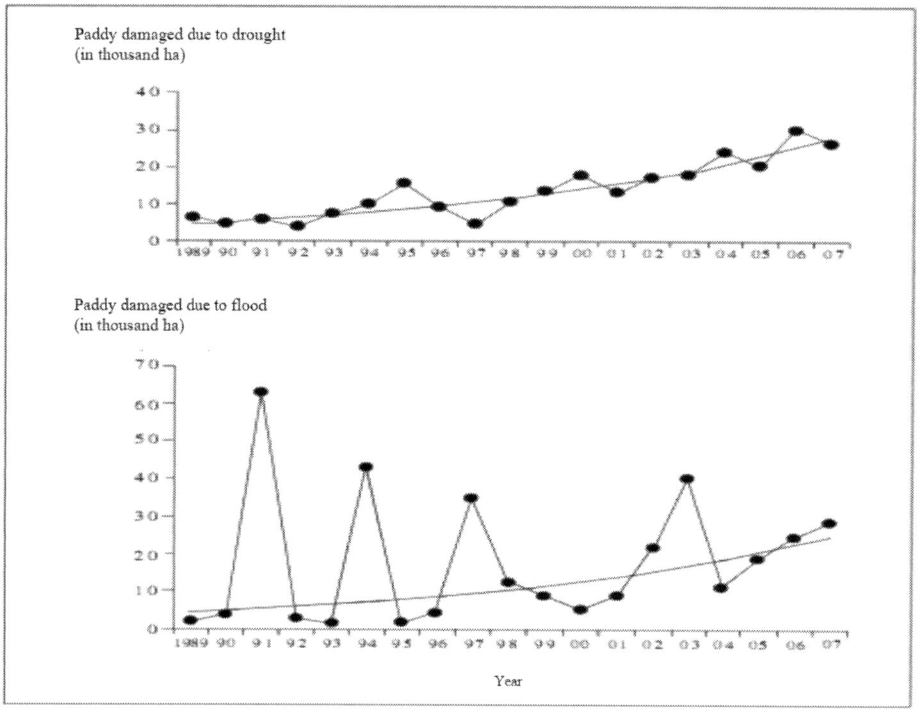


**Figure 2.** Extensive damage to rice crops due to floods and droughts (data processed from Ditlin Food Crops) Source: [22,23]

The increase in air temperature causes an increase in transpiration that results in decreased productivity of food crops, accelerates the ripening of fruit/seed, increases water consumption, decreases the quality of crop yields, and encourages the development of pests that grow crops [22,24]. Research related to the impact of rising air temperature on the decline in crop production, that the increase in air temperature up to 2°C in low-lying areas can reduce production by 40%, while in temperate and highland plains cause a decrease in the production of about 20% [22]. Furthermore, the temperature increase of at least 1°C will reduce the yield of rice crops by 10% [25]. A temperature rise of 1°C can decrease crop production by 5-7% [26]. The decline in crop production is caused by more periods of plant growth, the decline in respiration, and the formation of zinc [22,26].

### *3.2 Farmers' adaptation to climate change*

To deal with climate change, farmers must have strategies in the form of adaptation technology. Adaptation strategies that can be recommended to farmers to deal with climate change, including (1) adjustment of planting time and pattern, (2) the use of superior varieties resistant to drought, immersion, and salinity, (3) rain harvesting technology, (4) irrigation technology, (5) biotic environmental strategies. The strategy can be spelt out as follows:

*3.2.1 Adjustment of planting time and pattern.* Time adjustment and planting patterns feed one of the efforts to reduce the impact of climate change due to seasonal shifts and changes in rainfall patterns. Ministry of Agriculture has published a calendar map atlas of Planting Java Island as a guide for farmers in applying planting patterns. The planting calendar map is compiled to guide the planting patterns of mainly rice crop commodities, based on climate and water resources [24]. On the planting calendar map, three scenarios of climate events are created, namely wet year (TB), normal year (TN), and dry year (kindergarten). Sealing was in the planting calendar map delivered guidance on prediction climate so that the planting pattern can be adjusted to the conditions of climate and water resources.

*3.2.2 Use of superior varieties of plants.* The ministry of agriculture has released superior varieties of crops in anticipation of various climatic conditions. Select varieties of rice resistant to soaking, including Ciherang, Inpara 4, and Inpara 5 varieties. Superior rice varieties are resistant to drought, including Inpago 5 (potentially producing 6 tons/ha), Inpari 10 varieties (potentially producing 7 tons/ha). Increased rainfall can affect the growth of corn crops, especially in the early vegetative stages. The Ministry of Agriculture released drought-resistant corn varieties Bima 4 for hybrids and Lamuru types for composite types, with production potentials of 11.7 tons/ha and 7.6 tons//ha, respectively. Recommendations of soybean varieties resistant to puddles are Grobogan and Kawi varieties with yield potential of 3.4 tons/ha and 2.8 tons/ha, respectively. Soybean varieties are resistant to drought, namely Agromulyo varieties (planting age 82 days and production potential of 3.1 tons/ha) and Grobogan Varieties (planting age of 76 days and production potential of 3.4 tons/ha).

*3.2.3 Rain harvest technology.* Rain harvesting technology is one of the alternatives to the water dissing technology with the principle of retaining excess water in the rainy season and utilizing it again in the dry season to plant crops. Examples of utilization of rain harvest technology are the manufacture of convex and trench dams. Embung serves as a place of water catchment used to increase groundwater storage capacity. While trench dam technology is a technology to collect or stem the flow of water in a ditch (drainage *network),* which aims to hold surface water. The technology is used to anticipate the occurrence of drought by accommodating abundant water, overcoming the dangers of flooding and erosion, and increasing the development of agricultural businesses on dry land or dry conditions [27].

*3.2.4 Irrigation technology.* The Ministry of Agriculture conveyed guidance related to irrigation ecology that is recommended to overcome water checks in crops, including Rentengwells, capillary irrigation, drip irrigation, *macak-macak* irrigation, rotating irrigation, and eagle-brush irrigation. The application of various irrigation techniques can be adjusted to the needs of the topographical conditions of agricultural land. The aim is to meet the needs of water planting in conditions of limited water availability and increase the value of water use.

*3.3 Biotic environmental strategies*

Until now, pests are still an obstacle for farmers. Climate change has implications for increased intensity of disturbance of plant-disrupting organisms [24,27] and disease in plants resulting from increased temperature and humidity. For example, the development of brown stem pests (WBC) and tungro in rice, *bulai* disease in corn crops, soybean crop pests are *grapyak* caterpillars, which are pests of peas and *kutu kebul (Bemisiatabaci)* that can cause production loss of up to80% even crop failure [27]

Plant disrupting organisms (OPT) is an essential part of plant protection to maintain productivity. Pest control is naturally done by manipulating natural enemies of plant pests to be more environmentally friendly. This is done as a first choice before using chemicals such as insecticides. In principle, the natural enemy will continually develop following the development of pests. As long as the natural enemy can suppress pests, then control with chemicals is unnecessary because the balance of biology has been achieved. But when the development of natural enemies can no longer keep up with pests, meaning that biological balance is not achieved, other control tactics are needed, including the use of chemicals.

Brown stem control technology (WBC) uses a control threshold based on manipulating natural enemies in plant pests. This is done to reduce the use of insecticides and increase farmers' incomes [28]. Brown stem control technology (WBC) begins with monitoring plants to determine the economic threshold of natural enemy corrections using formulas [28]. The recommended insecticides can be used when the specified corrected economic threshold has been exceeded. The economic threshold is not a fixed price but will fluctuate in the price of grain and pesticides. When the price of grain increases, the economic threshold will fall and vice versa, but when the price of insecticides rises, the economic threshold will rise and vice versa.

## 4. Conclusion

The impact of climate change that has occurred in Indonesia, such as erratic rainfall intensity, extreme temperatures, El Nino and La Nina, causes farmers to be unable to implement optimal planting patterns. The agricultural business risk becomes higher, resulting in a decrease in agricultural production and even crop failure. To overcome these problems, the capacity of farmers' adaptation to climate change must be increased. Through this step, the vulnerability of farmers to the conditions of climate change impacts can be overcome. The main factors causing climate change that affect the agricultural sector include rain patterns, extreme climate events (floods and droughts), and increased air temperatures. Recommendations of adaptation strategies that can be done are using superior varieties of plants; adjusting planting time and pattern; rain harvest technology; irrigation technology, and biotic environmental strategies.

## References

[1] Nurdin N 2011 *Antisipasi Perubahan Iklim untuk Keberlanjutan Ketahanan Pangan* (Sulawesi Utara)

[2] Angles S, Chinnadurai M and Sundar A 2011 Awareness on Impact of Climate Change on Dryland Farmers *Indian J. Agric. Econ.* **66** 365–72

[3] Hidayati I N and Suryanto S 2015 Pengaruh Perubahan Iklim terhadap Produksi Pertanian dan Strategi Adaptasi pada Lahan Rawan Kekeringan *J. Ekon. Stud. Pembangunan.* **16** 42–52

[4] Boko M, Niang I, Nyong A, Vogel A, Githeko, Andrew, Medany M, Osman-Elasha B, Tabo R, Yanda and Pius Z 2007 *Africa Climate Change 2007: Impacts, Adaptation and Vulnerability: Contribution of Working Group II to the Fourth Assessment Report of the Intergovernmental Panel on Climate Change* (Tanzania: University of Dar-es-salaam)

[5] Sumaryanto 2016 Strategi Peningkatan Kapasitas Adaptasi Petani Tanaman Pangan Menghadapi Perubahan Iklim *Forum Penelit. Agro Ekon.* **30** 73

[6] Speranza C I 2010 *Resilient Adaptation to Climate Change in African Agriculture* (German: Deutsches Institut für Entwicklungspolitik (DIE))

[7] Rasmikayati E, Saefudin B R, Rochdiani D and Natawidjaja R S 2020 Dinamika Respon Mitigasi Petani Padi di Jawa Barat dalam Menghadapi Dampak Perubahan Iklim serta Kaitannya dengan Pendapatan Usaha Tani *J. Wil. dan Lingkung.* **8** 247–60

[8] Kurniawati F 2012 *Pengetahuan dan Adaptasi Petani Sayuran terhadap Perubahan Iklim (Studi Kasus: Desa Cibodas, Kecamatan Lembang, Kabupaten Bandung Barat)* (Bandung: Universitas Padjajaran)

[9] Barokatuminalloh, Purwaningsih Y, Mulyaningsih T and Daerobi A 2021 Perception and Adaptation Capacity of Farmers to Climate Change *Ann. Rom. Soc. Cell Biol.* **25** 5373–82

[10] Fadhliani Z 2016 *The Impact of Crop Insurance on Indonesian Rice Production*

[11] Saptutyningsih E, Diswandi D and Jaung W 2020 Does Social Capital Matter in Climate Change Adaptation? A lesson from Agricultural Sector in Yogyakarta, Indonesia *Land use policy* **95** 1–5

[12] Baehaki S E and Widiarta N 2011 *Hama wereng dan cara pengendaliannya pada tanaman padi* (Balai Besar Penelitian Tanaman Padi)

[13] Syahu R 2007 *Analisis Tingkat Serangan Wereng Batang Coklat (Nilaparvata lugens Stal.) Berdasarkan Faktor Iklim (Studi Kasus: 10 Kabupaten Endemik Di Provinsi Jawa Barat)* (Bogor: IPB University)

[14] Utami A W, Jamhari and Hardyastuti S 2011 El Nino, La Nina, dan Penawaran Pangan *J. Ekon. Pembang.* **12** 257–71

[15] Mohd E T, Chamhuri S, Rafiqul I M and Basri T 2011 Impact of Agricultural Supports for Climate Change Adaptation: a Farm Level Assessment *Am. J. Environ. Sci.* **7** 178–82

[16] Siwar C, Alam M, Murad W and Al-Amin A Q 2009 A Review of the Linkages between Climate Change, Agricultural Sustainability and Poverty in Malaysia *Int. Rev. Bus. Res. Pap.* **5** 309–21

[17] Sumarlin S, Karimuna L and Syaf H 2018 Pengaruh Faktor Iklim terhadap Pertumbuhan dan

Produksi Tanaman Jagung (Zea mays L.) *Berk. Penelit. Agron.* **6** 17–24
[18] Runtunuwu E and Syahbuddin H 2007 Perubahan Pola Curah Hujan dan Dampaknya terhadap Potensi Periode Masa Tanam *Tanah dan Iklim* **26** 1–12
[19] Herlina N and Prasetyorini A 2020 Effect of Climate change on planting season and productivity of maize (Zea mays L.) in Malang Regency *J. Ilmu Pertan. Indones.* **25** 118–28
[20] Mardawilis E R 2016 Pengaruh Curah Hujan terhadap Produksi Tanaman Pangan Kabupaten Kampar Provinsi Riau *Prosiding Seminar Nasional Lahan Suboptimal* pp 281–9
[21] Salinger M J 2005 Climate Variability and Change: Past, Present and Future- An overview *Increasing Clim. Var. Chang.* **70** 9–29
[22] Surmaini E, Runtunuwu E and Las I 2011 Upaya Sektor Pertanian dalam Menghadapi Perubahan Iklim *J. Litbang Pertan.* **30** 1–7
[23] Baettig M B, Wild M and Imboden D M 2007 A Climate Change Index: Where Climate Change may be Most Prominent in the 21st Century *Geophys. Res. Lett.* **34** 1–6
[24] Las I 2008 Menyiasati Fenomena Anomali Iklim bagi Pemantapan Produksi Padi Nasional pada Era Revolusi Hijau Lestari *Pengemb. Inov. Pertan.* **1** 83–104
[25] Peng S, Huang J, Sheehy J E, Laza R C, Visperas R M, Zhong X, Centeno G S, Khush G S and Cassman K G 2004 Rice Yields Decline with Higher Night Temperature from Global Warming *Proceedings of the National Academy of Sciences of the United States of America* vol 101 pp 9971–5
[26] Matthews R B, Kropff M J, Horie T and Bachelet D 1997 Simulating the Impact of Climate Change on Rice Production in Asia and Evaluating Options for Adaptation *Agric. Syst.* **54** 399–425
[27] Pertanian K 2011 *Pedoman Umum Adaptasi Perubahan Iklim Sektor Pertanian* (Jakarta: Badan Penelitian dan Pengembangan Pertanian, Kementerian Pertanian)
[28] Baehaki S E 2009 Strategi Pengendalian Hama Terpadu Tanaman Padi dalam Perspektif Praktek Pertanian yang Baik (Good Agricultural Practices) *Pengemb. Inov. Pertan.* **1** 65–78

**Acknowledgement**
Thank you to all those who have helped prepare this article, especially to the Magister of Economics and Development Studies at Sebelas Maret University.

IOP Conf. Series: Earth and Environmental Science 1016 (2022) 012039 doi:10.1088/1755-1315/1016/1/012039

# Strategy to improve regional resilience against flood disasters in the North Coast of Central Java

**Mohammad Muktiali[1]* and Rukuh Setiadi[1]**

[1]Department of Urban and Regional Planning, Diponegoro University, INDONESIA
*Corresponding author: alimukti890@gmail.com

**Abstract**. The impact of climate change has caused flooding in the North Coast of Central Java Province. The level of flood hazard in 14 urban/municipalities on the North Coast of Central Java is all in the high category. The potential population exposed to flooding in this area reaches around 11,589,000 people, where a third are vulnerable groups (the poor). The impact of the flood disaster has caused environmental damage as well as physical and economic losses. This article aims to examine strategies to increase the resilience of the North Coast of Central Java using the Socio-Ecological System (SES) approach, where there is a reciprocal relationship between aspects of nature/environment (Natural System) and socio-economic (Socio-Economic Systems). Given that vulnerability (V) in the Pantura area of Central Java is dominated by many poor people and vulnerable groups, the strategy that needs to be done to reduce vulnerability is through poverty alleviation programs. To reduce the level of exposure (E) in the form of large physical and economic losses, it is necessary to provide the urban infrastructure that ensures the sustainability of the socio-economic activity system can take place properly. In addition, it is also necessary to control land use change because this has reduced the capacity and storage capacity of water, as well as prevented flooding. Finally, to overcome environmental damage, it is necessary to carry out an environmental conservation program on the North Coast of Central Java in a comprehensive and sustainable manner.

**Keywords** : flood disasters, regional resilience, Socio-Ecological Systems (SES), coastal areas.

## 1. Introduction

The concept of resilience continues to be studied through the perspectives of various multidisciplinary sciences; and this has practical implications for aspects of planning, development and urban/regional restructuring [1–3]. The concept of regional/city resilience can be used as a guide for policymakers in dealing with increasingly complex regional/city dynamics with various systems that are interconnected with each other such as economic, social, ecological-environmental and environmental systems. other systems [4]. Resilience is needed to meet the challenges of sustainable development [5,6]. The same thing was stated by ICLEI [7] which states that cities designed and developed using best practices of sustainability and resilience, can support or even increase the capacity of the city's ecosystem both internally and externally. Regional and urban systems have grown dynamically exponentially with various unstable and non-linear elements and relationships that produce transformative characters. The interrelationships between natural and social processes in which humans, nature and culture engage in transformative and evolutionary dynamics characterize urban areas as socio-ecological systems [8,9]. For this reason, according to Li T et al. [10] who studied Social-Ecological Resilience (SER) in the context of Assessment-Mechanism-Management, suggested the need for a regional/city resilience study that comprehensively integrates the Nature-Social-Economy dimensions through various multidisciplinary approaches. Through these various approaches, researchers and policy makers are expected to gain a holistic and comprehensive understanding of regional/city resilience so that they are able to respond to changes in resources and a very dynamic environment.

According to data from the National Disaster Management Agency, disasters almost always affect the lives of Indonesian people. Not only casualties but also economic damage and losses which ultimately undermine the welfare and economy of the community. During January 2014, the disaster had caused inflation of 1.07%. Based on several estimates and studies of the economic impact of the disaster, it is estimated that flood and landslide damage in 16 regencies/cities in Central Java amounted to Rp. 2.01 trillion, while losses and flood damage in the Pantura of Java amounted to Rp. 6 trillion. Meanwhile, the tidal flood that occurred in Semarang City caused considerable economic losses, especially in key sectors such as irrigation, plantations, buildings, settlements, and rice fields. In 2010, the residential sector suffered a loss of 1.4 trillion rupiah. In the pessimistic scenario without any efforts to deal with it, the housing sector has the potential for losses of up to 2.6 trillion rupiahs in 2030, while in the optimistic scenario with adaptation efforts and increasing resilience to climate change, the settlement sector can reduce potential losses and reach 2. 1 trillion rupiahs [3].

Resilience is the ability of a system to predict and cope with shocks by utilizing available resources and opportunities. Meanwhile, the notion of regional resilience is the ability to anticipate, prepare, respond to changes and recovery caused by shocks [11]. It is closely related to governance, ecosystem services, green infrastructure, risk reduction, sustainability, complexity, and vulnerability of regional/urban systems [12,13]. A shock is an event that occurs quickly and suddenly and has the potential to result in loss of life and/or major damage to city assets. In this case, what is meant by shocks can be in the form of natural disasters, non-natural disasters and social disasters (according to Law Number 24 of 2017). Furthermore, Modica and Zoboli in Marin et al. [14] developed a framework for regional resilience to natural disasters based on a socio-ecological approach. The socio-economic system and its factors in local communities have strong links and even blend with nature/environment. This is because human activities are highly dependent on natural/environmental systems and are mutually influencing human life. The relationship between natural/environmental and socio-economic aspects in the framework of resilience to natural disasters can be seen in Figure 1 below. In the figure, the analyzed factors include 3 main things, namely: (i) factors that are only related to natural/physical aspects; (ii) factors that only affect link to the socio-economic system; (iii) factors originating from the interaction between natural systems and socio-economic systems. In general, the hazard aspect is directly related to the natural system, and is also influenced indirectly by the socio-economic system (Socio-Economic System) through anthropogenic contributions to global warming which results in natural disasters. On the other hand, exposure, vulnerability, and resilience are related to the socio-economic characteristics of the area being analyzed. Finally, the risk is related between natural systems and socioeconomic systems. According to Marin et al. [14] a socio-ecological-based regional resilience model (SES) was carried out by combining all components as follows:

$$R\,ij = f(\,H\,ij\,,E\,i\,,V\,i\,,Res\,i\,) \tag{1}$$

Where Hij is the level of natural disaster hazard (hazard) in region i and hazard j, Ei is socio-economic exposure (exposure), Vi is socioeconomic vulnerability (vulnerability), and Res i is socioeconomic resilience (resilience). Hazard in this case is a natural event that causes socio-ecological changes to occur in an area or community. Exposure is defined as all assets and people that can be harmed by a natural disaster; includes physical and socio-economic components. Vulnerability is defined as all the inherent characteristics of the exposed object/area that creates a potential hazard.

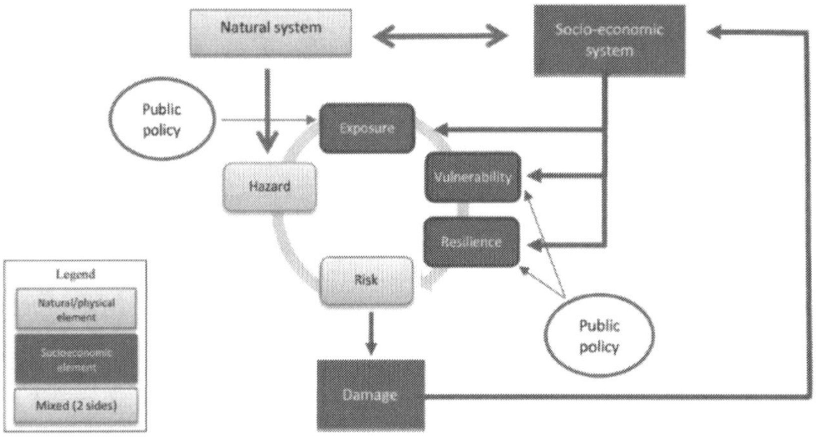


**Figure 1.** Framework for regional resilience to natural disasters based on the socio-ecological framework. Source: Marin et.al. (2021) [14]

Resilience is the ability of the system to recover or adapt so that the impact of disturbances affecting the system is reduced. Disaster risk (risk), according to ex-ante and restrictive interpretations, is the result of the interaction between the hazards of a natural event and the elements exposed to hazards and their vulnerabilities. From the equation above, it can be said that if at least one of the variables is very low which means the level of resilience is high, then the level of natural disaster hazard does not pose too much risk. Similarly, if there is no exposure (E) then the level of risk (R) will be relatively low, regardless of the level of hazard (H), vulnerability (V) and resilience (Res). Finally, as long as the vulnerability (V) and resilience (Res) measurements are carried out correctly, the risk level is low with a high level of resilience and/or low vulnerability, even though the level of hazard and exposure is high. This is because the area will not be much affected by natural disasters because the area is not prone to disasters (low area vulnerability) and/or can recover quickly from possible damage (high regional resilience).

The consequences of the complex dynamics of the human-centered natural and man-made environment have resulted in various theories and concepts of partial and sectoral resilience, for example, climate resilience, disaster resilience, socio-ecological resilience. On the other hand, because the socio-ecological developments that occur in regions/cities are very dynamic, the concepts and theories of regional/city resilience have also continued to develop over the last three decades and have made the relationship between the variables and determinants of regional/city resilience also change dynamically. This also has practical implications that cause the management of urban area resilience to be ineffective and holistic [2]. For this reason, so that the theory and concept of regional/city resilience are not only partial, and to become a formal theory, this research will be carried out with a comprehensive rational approach through analysis and approach to each element as a whole that is interrelated and relates to each other holistically. This article will examine the Social-Ecological-Systems (SES) approach in increasing the resilience of the northern coastal areas of Central Java against floods. The main question to be answered is "How is the resilience of the northern coastal areas of Central Java to floods in terms of the perspective of Social-Ecological Systems (SES)? What are the strategies and policies that need to be carried out by districts/cities in the northern coastal areas of Central Java in order to increase the resilience of their regions to flood disasters?

## 2. Materials and method

### 2.1. Materials

The study area in this study covers several districts/cities in the North Coast (Pantura) area of Central Java Province. The selection of this study area is based on the specific socio-ecological characteristics of the area which is a coastal area. The 14 regencies that became the research area included the cities of Tegal, Pekalongan, Semarang and the regencies of Rembang, Batang, Kudus, Pekalongan, Kendal, Demak, Jepara and Brebes (see Figure 2).

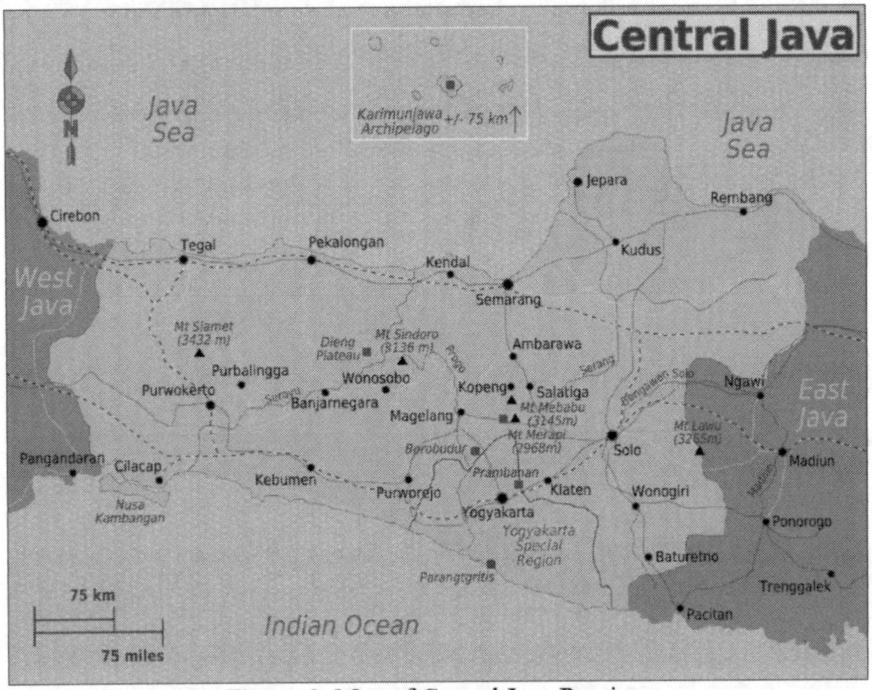


**Figure 2.** Map of Central Java Province

### 2.2. Methods

The study approach that will be used in this research is a triangulation approach (mixed methods). The basis of consideration used is because this method has the advantage of reducing bias that may arise in a research method. Case studies can combine quantitative and qualitative approaches (mixed methods). The quantitative approach method is a study approach that is measurable and supports qualitative research. The qualitative approach method accommodates verbal descriptions in place of numbers, or combines statistical processing with verbal processing with a fixed quantitative mindset. The data used in this study are in the form of data on the area of floods, physical and economic losses and the number of residents affected by floods. The data is sourced from the Central Statistics Agency (BPS) and the National Regional Disaster Management Agency (BNPBD) of Central Java Province and regencies/cities in the Pantura region of Central Java. For each variable, a scoring was carried out to determine the level of high, medium and low categories. More details can be seen in Table 1.

**Table 1.** Research variables and indicators

| Variables | Indicator |
|---|---|
| Flood hazard level (H) | Area affected by flood (ha) |
| Exposure level (E) | Physical and economic losses due to flooding (rupiahs) |
| Vulnerability level (V) | Total population of vulnerable groups affected by flooding (people) |

## 3. Results and discussion

This initial section describes the socio-ecological aspects related to the flood disaster that occurred in 14 districts/cities in the Pantura region of Central Java.

IOP Conf. Series: Earth and Environmental Science 1016 (2022) 012039 doi:10.1088/1755-1315/1016/1/012039

### *3.1. Potential flood threat (H) in the North Coast Region of Central Java*

The broad potential for flood hazard in the Pantura area of Central Java, which covers 14 city districts, is all categorized as high (severe) (Table 2). The area with the largest flood hazard is in Pati Regency which reaches 116,229 hectares, followed by Brebes which reaches 102,882 hectares.

**Table 2.** Potential areas of flood hazards in the Pantura Region of Central Java

| No | Municipalities/City | Hazard | |
|---|---|---|---|
| | | Area (Ha) | Category |
| 1 | Rembang | 70,435 | High |
| 2 | Pati | 116,229 | High |
| 3 | Kudus | 31,952 | High |
| 4 | Jepara | 62,648 | High |
| 5 | Demak | 90,012 | High |
| 6 | Kendal | 47,487 | High |
| 7 | Batang | 29,400 | High |
| 8 | Kab Pekalongan | 37,185 | High |
| 9 | Pemalang | 60,177 | High |
| 10 | Kab Tegal | 61,975 | High |
| 11 | Brebes | 102,882 | High |
| 12 | Kota Semarang | 24,423 | High |
| 13 | Kota Pekalongan | 83,700 | High |
| 14 | Kota Tegal | 3,965 | High |

Source : BNPB (2015)[15], processed

**Table 3.** Vulnerable population groups affected by flood disasters in Pantura Central Java

| No | Region | Potential of Exposed Population (Soul) | | | | |
|---|---|---|---|---|---|---|
| | | Total population | People Group | | | Category |
| | | | Vulnerable Group Rentan | Poor Resident | Disabled Resident | |
| 1 | Rembang | 536,678 | 78,000 | 196,410 | 2,571 | High |
| 2 | Pati | 1,076,574 | 172,285 | 310,693 | 6,900 | High |
| 3 | Kudus | 755,160 | 101,054 | 117,062 | 2,165 | High |
| 4 | Jepara | 983,629 | 150,222 | 214,626 | 4,186 | High |
| 5 | Demak | 1,113,059 | 156,031 | 366,487 | 4,328 | High |
| 6 | Kendal | 709,200 | 106,319 | 171,686 | 3,020 | High |
| 7 | Batang | 401,528 | 58,645 | 106,070 | 1,912 | High |
| 8 | Pekalongan | 713,460 | 106,141 | 243,974 | 3,201 | High |
| 9 | Pemalang | 1,014,323 | 157,159 | 344,217 | 3,464 | High |
| 10 | Tegal | 1,225,958 | 189,948 | 373,095 | 5,635 | High |
| 11 | Brebes | 1,483,429 | 223,601 | 631,183 | 6,445 | High |
| 12 | Kota Semarang | 1,279,821 | 164,230 | 113,595 | 1,211 | High |
| 13 | Kota Pekalongan | 296,353 | 39,161 | 37,061 | 740 | High |
| 14 | Kota Tegal | 246,091 | 34,153 | 45,815 | 1,070 | High |

Source : BNPB (2015)[15], processed

### *3.2. Vulnerable population groups (V) exposed to floods in the Pantura Region of Central Java*

The potential population exposed to flooding in the Pantura area of Central Java, which includes 14 urban districts, is in the high category (vulnerable). The most exposed area is in Brebes Regency which reaches 1,483,429 people, followed by Semarang City which reaches 1,279,821 people. Likewise, for the vulnerable group which includes the vulnerable age group, the poor and the disabled are mostly in Brebes Regency. More details can be seen in Table 3.

### *3.3. Exposure (E): potential flood losses in the Pantura Region of Central Java*

The potential for flood losses in the Pantura area of Central Java, which includes 14 urban districts, is mostly high category. The area with the highest loss rate was in Semarang City which reached a total loss of 8,258.3 billion, followed by Demak Regency which reached 7,113.3 billion. More details can be seen in Table 4 below.

**Table 4.** Potential flood losses in the Pantura of Central Java

| No | Region | Potential Losses (Billions) | | | Category |
|---|---|---|---|---|---|
| | | Physical Loss | Economic Loss | Total Loss | |
| 1 | Rembang | 747.6 | 604.2 | 1,351.8 | High |
| 2 | Pati | 2,714.1 | 1,897.4 | 4,611.5 | High |
| 3 | Kudus | 1,519.8 | 678.4 | 2,198.2 | High |
| 4 | Jepara | 1,477.1 | 909.3 | 2,386.4 | High |
| 5 | Demak | 3,585.5 | 3,527.8 | 7,113.3 | High |
| 6 | Kendal | 1,583.9 | 652.2 | 2,236.0 | High |
| 7 | Batang | 847.5 | 260.4 | 1,107.9 | High |
| 8 | Pekalongan | 1,655.1 | 518.7 | 2,173.8 | High |
| 9 | Pemalang | 2,603.6 | 1,290.9 | 3,894.4 | High |
| 10 | Tegal | 2,624.0 | 1,164.7 | 3,788.8 | High |
| 11 | Brebes | 3,360.2 | 2,487.7 | 5,847.9 | High |
| 12 | Kota Semarang | 8,242.4 | 16.0 | 8,258.3 | Medium |
| 13 | Kota Pekalongan | 1,425.4 | 25.6 | 1,451.0 | High |
| 14 | Kota Tegal | 3,194.2 | 44.9 | 3,239.1 | High |

Source : BNPB (2015)[15], processed.

### *3.4. Potential environmental damage due to Floods in Pantura Central Java*

The potential for environmental damage due to flooding in the Pantura area of Central Java, which covers 14 regencies and cities, is mostly in moderate category. The area with the highest environmental damage is Pati Regency, which covers an area of 42 hectares, followed by Kendal Regency which reaches 39 hectares. More details can be seen in Table 5.

### *3.5. Issues and problems of resilience against flood disasters in the North Coast Region of Central Java*

To examine problems and solutions, a socio-ecological framework is used; where there is a reciprocal relationship between aspects of nature/environment (Natural System) and socio-economic (Socio-Economic Systems). By referring to the equation: $R = H \times E \times V \times (1\text{-}Res)$, then the analysis of the problems and solutions for the resilience of the Pantura region against flooding can be described as Figure 3.

**Table 5.** Potential environmental damage due to floods in Pantura Central Java

| No | Region | Potential Environmental Damage | |
|---|---|---|---|
| | | Area (Ha) | Category |
| 1 | Rembang | 28 | High |
| 2 | Pati | 42 | Medium |
| 3 | Kudus | 18 | Medium |
| 4 | Jepara | 31 | High |
| 5 | Demak | 28 | Medium |
| 6 | Kendal | 39 | High |
| 7 | Batang | 29 | Low |
| 8 | Pekalongan | 32 | High |
| 9 | Pemalang | 28 | Medium |
| 10 | Tegal | 35 | Medium |
| 11 | Brebes | 34 | Medium |
| 12 | Kota Semarang | 32 | Medium |
| 13 | Kota Pekalongan | 8 | High |
| 14 | Kota Tegal | 8 | Medium |

Source: BNPB (2015)[15], processed

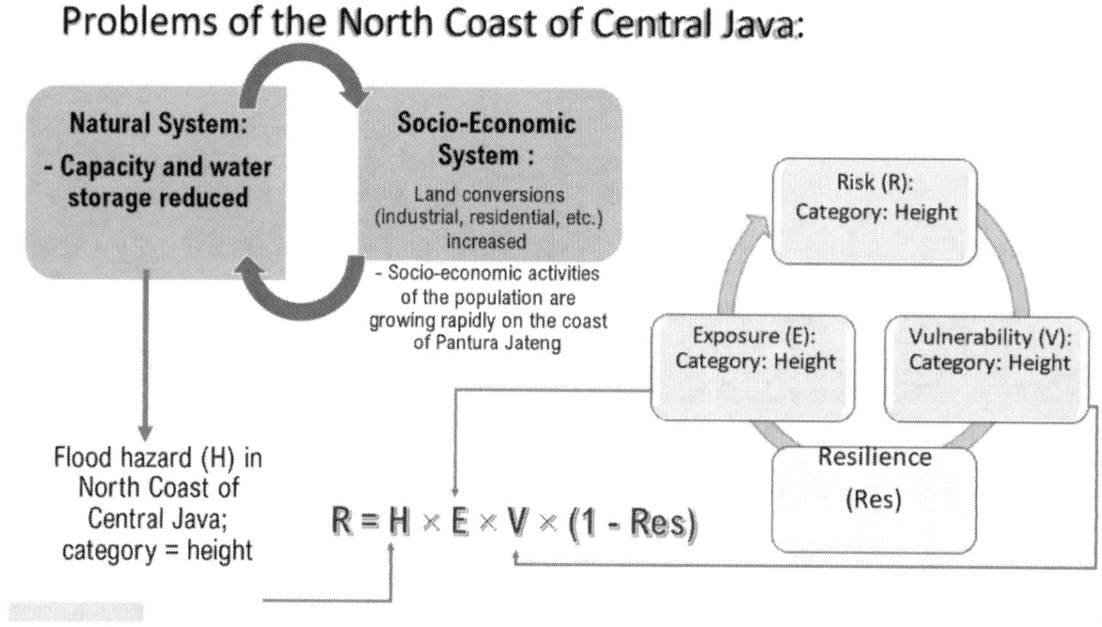


**Figure 3.** Issues and problems of resilience against flood disasters in the North Coast Region of Central Java. Source : Analysis result, 2021

Based on the scheme where hazard (H) is the level of flood disaster hazard (hazard) in 14 districts/cities in the Central Java region in the high category. In a socio-ecological perspective; The high level of flood hazard in the Pantura area of Central Java is caused by a causal relationship between the socio-economic system and the natural/environmental system. In this case, the increase in land conversion into industrial and residential land (Socio-Economy System) results in reduced water capacity and storage capacity (Natural System) causing flooding in the area. Groundwater storage is highly correlated with land use change and land restoration and is an important variable in water resource management and environmental restoration [16]. Research results in northwest China also show that climate change and land use change have reduced blue and green water resources [17]. The results of the study of the researches that have been carried out also prove that land use changes actually affect regional hydrological ecosystem servicesthrough changes in the physical character of the earth's surface and water cycle processes [18].

*3.6. Strategies to increase resilience to floods in the North Coast of Central Java*

In the North Coast (Pantura) of Central Java, it is indicated that the level of socio-economic vulnerability (vulnerability-V) and the level of socio-ecological exposure (exposure - E) are relatively high, resulting in a high level of disaster risk (risk - R). Thus, referring to the equation R = H × E × V × (1-Res), then to reduce the risk (R) is done by increasing resilience (resilience - Res). Resilience (Res) can be increased by reducing components V, E and R [14]. The results of the research in the district and city of Padang also show the same result that resilience can be increased through reducing vulnerability [19]. To reduce vulnerability (V) it is carried out with poverty alleviation programs for the poor in the Pantura region of Central Java. The programs provided include physical programs such as infrastructure development, social programs such as social safety nets and economic programs such as cash transfers. With these programs, the level of vulnerability can be reduced and resilience can be increased. To increase resilience, a comprehensive synergy and coordination between sectors is needed to make it more effective, efficient and sustainable [4]. To reduce the level of exposure (E) it can be done by carrying out environmental conservation, for example conservation of mangrove forests on the coast of the North Coast of Central Java. With this conservation will be able to reduce the level of community exposure to floods so that resilience will increase. Meanwhile, in relation to reducing the threat of flooding, it can be done by increasing the water storage capacity in the Pantura area. For this reason, it is necessary to control land conversion, especially into industry and housing so that land cover can be maintained so that water capacity and storage capacity increases and flood disasters can be reduced. More details can be seen in the following figure 4.

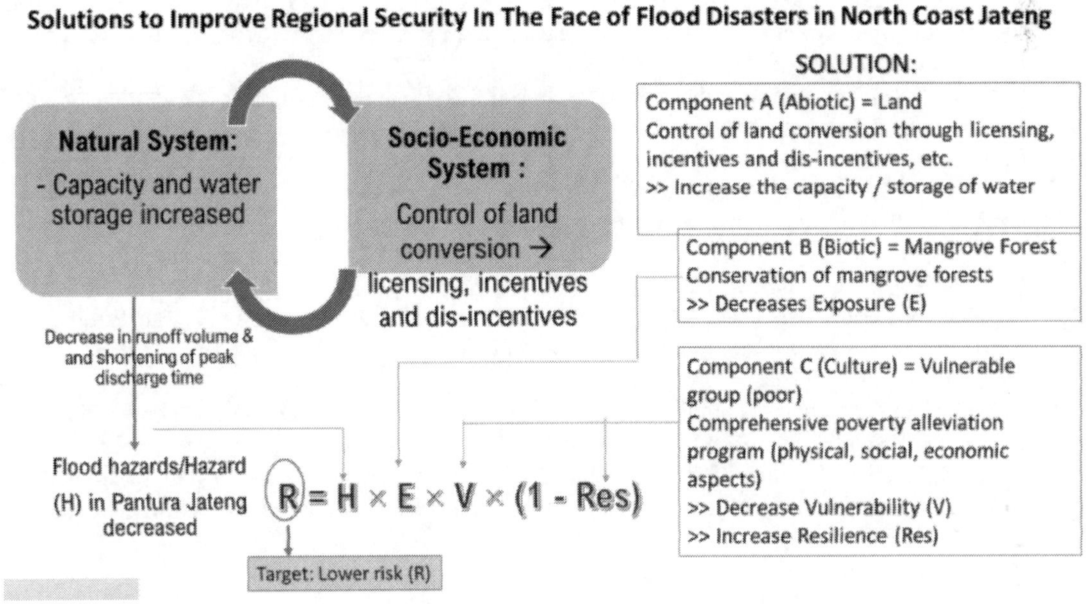


**Figure 4.** Strategies to increase resilience to flood disasters in the North Coast Region of Central Java. Source: Analysis results, 2021.

## 4. Conclusions

Given that vulnerability (V) in the Pantura area of Central Java is dominated by many poor people and vulnerable groups, the solution that needs to be done to reduce vulnerability is through poverty alleviation programs. To reduce the high level of socio-ecological exposure (E) in the form of large physical and economic losses, it is necessary to provide the urban infrastructure that ensures the sustainability of the socio-economic activity system can take place properly. In addition, it is also necessary to control the conversion of land into industry and housing because it causes the water capacity and storage capacity to decrease, causing flooding. Finally, to overcome environmental damage, it is necessary to carry out a comprehensive and sustainable environmental conservation program in the Pantura area of Central Java.

## References

[1] Peng C, Yuan M, Gu C, Peng Z and Ming T 2017 A review of the theory and practice of regional resilience *Sustainable Cities and Society* **29** 86–96

[2] Wang Y C, Moya Guerola M, Lin Y C, Hsieh Y P, Strong C, Tsai M C and Lin C Y 2019 Effects of childhood adversity and resilience on Taiwanese youth health behaviors *Pediatrics and Neonatology* **60** 368–76

[3] Walker J 2020 Genealogies of resilience: From conservation to disaster adaptation *More Heat than Life: The Tangled Roots of Ecology, Energy, and Economics* (Singapore: Palgrave Macmillan) pp 311–39

[4] Huck A, Monstadt J and Driessen P 2020 Mainstreaming resilience in urban policy making? Insights from Christchurch and Rotterdam *Geoforum* **117** 194–205

[5] Chelleri L and Olazabal M 2012 *Multidisciplinary perspectives on urban resilience* (Bilbao: Basque Centre for Climate Change)

[6] Elmqvist T, Fragkias M, Goodness J, Güneralp B, Marcotullio P J, McDonald R I, Parnell S, Schewenius M, Sendstad M, Seto K C and Wilkinson C 2013 *Urbanization, biodiversity and ecosystem services: Challenges and opportunities: A global assessment* (Dordrecht: Springer)

[7] ICLEI 2013 ICLEI—Local governments for sustainability

[8] Botequilha-Leitão A and Díaz-Varela E R 2020 Performance Based Planning of complex urban social-ecological systems: The quest for sustainability through the promotion of resilience *Sustainable Cities and Society* **56** 102089

[9] Suhelmi I R, Fahrudin A and Triwibowo F H 2014 Potential economic losses due to tidal inundation and flood at Semarang City *Forum Geografi* **28** 113–8

[10] Li T, Dong Y and Liu Z 2020 A review of social-ecological system resilience: Mechanism, assessment and management *Science of The Total Environment* **723** 138113

[11] de Roo G 2018 Ordering principles in a dynamic world of change – On social complexity, transformation and the conditions for balancing purposeful interventions and spontaneous change *Progress in Planning* **125** 1–32

[12] Ernstson H, Leeuw S E V der, Redman C L, Meffert D J, Davis G, Alfsen C and Elmqvist T 2010 Urban transitions: On urban resilience and human-dominated ecosystems *AMBIO* **39** 531–45

[13] Wilkinson C 2011 Social-ecological resilience: Insights and issues for planning theory: *Planning Theory* **11** 148–69

[14] Marin G, Modica M, Paleari S and Zoboli R 2021 Assessing disaster risk by integrating natural and socio-economic dimensions: A decision-support tool *Socio-Economic Planning Sciences* **77** 101032

[15] BNPB 2015 *Kajian Bencana* (Jakarta)

[16] Zhao Y, Wang Y, Sun H, Lin H, Jin Z, He M, Yu Y, Zhou W and An Z 2019 Intensive land restoration profoundly alters the spatial and seasonal patterns of deep soil water storage at watershed scales *Agriculture, Ecosystems & Environment* **280** 129–41

IOP Conf. Series: Earth and Environmental Science 1016 (2022) 012039 doi:10.1088/1755-1315/1016/1/012039

[17] Zhao A, Zhu X, Liu X, Pan Y and Zuo D 2016 Impacts of land use change and climate variability on green and blue water resources in the Weihe River Basin of northwest China *CATENA* **137** 318–27
[18] Jin G, Wang P, Zhao T, Bai Y, Zhao C and Chen D 2015 Reviews on land use change induced effects on regional hydrological ecosystem services for integrated water resources management *Physics and Chemistry of the Earth* **89–90** 33–9
[19] Kusumastuti R D, Viverita, Husodo Z A, Suardi L and Danarsari D N 2014 Developing a resilience index towards natural disasters in Indonesia *International Journal of Disaster Risk Reduction* **10** 327–40

# Adaptation of red rice farmers to long drought in Ponjong Districts Gunung Kidul Regency, Yogyakarta Indonesia

**Lestari Rahayu*, Zuhud Rozaki , Indardi and Ari Isdiana**

Department of Agribusiness, Universitas Muhammadiyah Yogyakarta, Jl. Brawijaya, Geblagan, Tamantirto, Kasihan, Bantul, Daerah Istimewa Yogyakarta 55183, Indonesia
*Corresponding author: lestari@umy.ac.id

**Abstract**. Adaptation of red rice farmers is an action taken by farmers to change crop cultivation during a prolonged drought, that cause by climate change. Adaptation consists of changes in planting patterns and crop cultivation. This study aims to describe the adaptation of red rice farmers and know the profile that influences the adaptation of red rice farmers to prolonged droughts. The study was conducted in three zones, including the northern zone (Umbulrejo Village), the central zone (Sumbergiri Village), and the southern zone (Sidorejo Village). This study was conducted in Ponjong Subdistrict purposively, on sampling conducted systematically random sampling with 200 respondents. The data were analyze using descriptive analysis, score analysis, and Spearman Rank correlation. The results showed that adaptations applied during prolonged droughts in all three zones were classified as "moderate." The factors that affected adaptation in all three zones include age, number of family members, farming experience, income, land area, and agricultural ownership. In contrast, gender and education do not affect adaptation.

**Keywords**: adaptation, cultivation, prolonged drought, red rice farming

## 1. Introduction

Red rice *gogo* rice is a commodity that has shown its existence and developed well in Gunungkidul. Gogo rice can grow well on dry land and provides benefits for farming communities that do not have rice fields. Segreng local red rice *gogo* rice is a type of red rice known in Gunungkidul. Segreng Handayani is a local rice variety native to Gunungkidul that has been released nationally since 2009. Because it includes a type of *gogo* rice that does not require high irrigation intensity, Segreng Handayani is suitable for growing in the Gunung kidul area, which is mainly dominated by marginal lands with low rainfall.

In Indonesia the improvement of red rice varieties has not received adequate attention [1]. In terms of price, red rice can compete with white rice. Red rice is a food that has high healthy [2] Red rice has a much higher nutritional content compared to white rice varieties [1]. This results in the selling price of red rice in the market is higher compared to white rice. Red rice contains carbohydrates, fats, proteins, fiber, and minerals; in addition, it also contains anthocyanins [3] Anthocyanins are phenolic compounds that are useful as antioxidants [2,4,5].

Gunungkidul Regency is the largest area in Yogyakarta Province, reaching 46.63% of the entire Yogyakarta region. Gunungkidul is a region classified as having dry land composed of limestone rocks [6] Gunungkidul regency most of its agricultural land is dry land (moorland) that needs development. Dryland conditions in Gunungkidul are suitable for growing red rice. The local variety developed is the

Segreng Handayani Variety. Ponjong district is an area that develops red rice. It is divided into three zones: the northern zone with hilly land and latosol land type, the centre zone with hilly land and gramosol soil type, and the southern zone with flat land and gramosol soil type [7].

Efforts to increase rice production in various regions are generally on land with a sound irrigation system. Irrigation is an essential component in improving agricultural yields and farmers' well-being [6]. However, one of the consequences of the decline in rice production is due to climate change. Farmers aware of climate change and its impact on food crop production have been able to develop livelihood strategies, as well as the adaptations they are doing in a way that can continuously be done to overcome the impact of uncertain climate change on food crop production [8]. There is a close relationship between climate change and agricultural production [9].

The impact of extreme climate change in the form of drought will cause crop failure. The dry season generally occurs for five months, namely in April - September. However, in the research area, the dry season occurs for seven months. This can be said to be a long drought. A prolonged drought is a dry season condition beyond its standard limit. Reduced rain intensity is a major contributing factor to the decline in crop yields [10]. According to the data, rainfall that occurred in 2009-2020 in the research area experienced fluctuating and every four years decreased so that the dry season became longer. This is likely to happen in 2022 after 2018 experienced a long drought (see Figure 1).

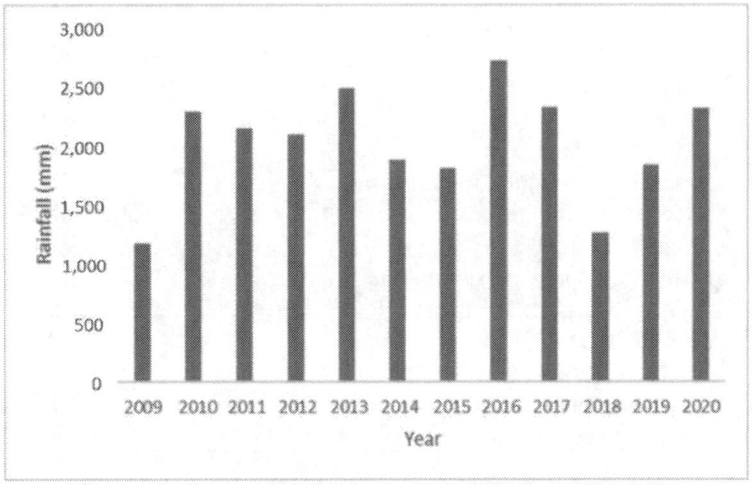


**Figure 1.** Gunungkidul District Rainfall Data 2009-2020

Productivity of red rice from 2015-2020 experienced fluctuations. In 2015 productivity amounted to 4.41 tons per ha, further experiencing productivity increases in 2016 of 5.15 tons per ha and two productivities of 6.01 tons per ha. However, in 2018 experienced a drought with a total rainfall a year of 1266 mm, the productivity of red rice decreased by 5.65 tons. Red rice productivity is influenced by rainfall conditions classified as relatively dry, namely Q < 1,670 mm. Rice productivity generally reaches 6 tons per hectare; even when its productivity reaches the optimal point, the yield reaches 8 tons per hectare calculated in its varieties and cultivation techniques [11].

The decline in red rice production is a big problem for farmers. Red rice farmers in Ponjong District are divided into three zones based on land type and physical condition. The southern zone (hilly and limestone soil, central zone (flatter land), northern zone (hilly and the soil is not limestone, there is a source of water) [12]. Different profile conditions will affect the characteristics of farmers, and differences in physical conditions on each land will affect climate change along with low rainfall so that there is a long dry season. It results in changes in income, accompanied by decreased rice productivity. So, to overcome this, farmers need to take adaptation actions. Adaptation is an adjustment to the environment. [13, 14]. Adapting during a prolonged drought aims to help farmers not experience a decline in crop yields and incomes. Adaptation actions that can be done include the adaptation of farmers to the process of crop cultivation. Adaptation is a way to organize the layout and order of planting over a period. Things that are done include applying the intercropping system, changing the planting time, soil arrangement, moving to *palawija,* planting the beginning of the rainy season, not planting at all, seed preparation, changing location, changing harvest time, increasing land area, using *genjah* varieties. [15]. This study aims to describe the adaptation of red rice farmers to the existence of prolonged droughts in Gunungkidul and analyses the relationship of farmers' profiles with the adaptation of red rice farmers to the existence of prolonged droughts in Gunungkidul.

## 2. Materials and method

This research was conducted in Ponjong Subdistrict because the location is a center of food crops, especially red rice. Location selection is also based on the region's potential vulnerability to climate change. The method of determining the sample used is systematic random sampling, and the method of estimating the proportion of samples is used to find out the number of farmers who are used as respondents Obtained the number of samples as many as 200 farmers from 399 farmers with a bound of error of 5%. This value indicates that this study will not make errors at a rate of more than 5%.

**Table 1**. Adaptation of red rice farmers to the process of crop cultivation

| **No.** | **Planting Pattern Adaptation** | **Score range** | | | |
|---|---|---|---|---|---|
| | | 1 | 2 | 3 | 4 |
| 1. | Implementation of Intercropping system | ND | RD | D | AD |
| 2. | Change planting time | ND | RD | D | AD |
| 3. | Decide in processing the soil | ND | RD | D | AD |
| 4. | Reduction of arable land area | ND | RD | D | AD |
| 5. | Moving from rice production to *palawija* | ND | RD | D | AD |
| 6. | Planting rice at the beginning of a rain | ND | RD | D | AD |
| 7. | Not planting plants at all | ND | RD | D | AD |
| 8. | Better and thorough preparation of seeds or seeds | ND | RD | D | AD |
| 9. | Moving from a location is at high risk to the impact of a prolonged drought. | ND | RD | D | AD |
| 10. | Decide or change harvest times | ND | RD | D | AD |
| 11. | Increase in arable land area | ND | RD | D | AD |
| 12. | Use of varieties of *genjah* | ND | RD | D | AD |

Measurements to find out farmers adapt to changes in crop cultivation are divided into four benchmarks, namely not doing (ND), meaning that farmers do not adapt during the long dry season, rarely doing (RD), meaning farmers only do one or two times and tend not to adapt, do (D) means farmers do it at a specific time, and always does (AD) means farmers implement adaptation strategies that are carried out regularly (Table 1).

Adaptation criteria are not responsive in a long dry season if farmers do not take any action at all. Unresponsive adaptation is the action of farmers who do not respond to an event, only know, and does not always apply. The moderate score criterion is the action of farmers in responding to events but not constantly adapting. Adaptation, the responsive criterion, is the farmer's action in response to an event that the farmer eventually adapts but with specific considerations. Very responsive criteria are the actions of farmers in responding to events by adapting without consideration.

Spearman rank correlation analysis is used to determine the relationship of factors that affect the level of adaptation of farmers in the long dry season.

$$r_s = 1 - \frac{6\sum {d_i}^2}{n(n^2-1)} \quad (1)$$

Where rs is spearman correlation coefficient, di is difference in score between two variables, and n is number of data or samples.

## 3. Results and discussion

### 3.1. *Adaptation of farmers to the process of cultivating crops*

Adaptation of Red Rice Farmers to plant cultivation is an activity to regulate the layout and planting arrangements, including the tillage period and the period of not being planted within a certain period [14]. This adaptation is a farmer's strategy due to the long dry season to get maximum production. The adaptation strategy consists of implementing an intercropping system that aims to obtain diverse harvests. Changing the planting time is the retreat of the planting process according to natural conditions [13]. Land management regulation is improving the treatment carried out on the land to be planted. Reducing land area is an adaptation of adaptation to low water conditions, switching to corn, cassava is an adaptation action because these plants do not need much water, planting at the beginning of the rainy season is a cultivation action carried out in the first rainy season, not planting at all is farmers' way to avoid crop failure, good seed preparation is an action to produce quality crops, changing planting locations is an act of moving the location from the moor to the yard, setting harvest time is the farmer's action in harvesting at the right time, increasing the cultivation area is a good action. It is done to produce a better crop. Adaptation to short-lived and drought-resistant plant varieties is an action by farmers so that farmers do not wait long to get production [16].

Farmers adapt to climate change by shifting planting times, changing crop variation, changing planting patterns, changing planting places and locations, based on their experience of climate change that takes place gradually Table 2. The southern zone is included in the unresponsive criteria, meaning farmers do not adapt to planting crops during the long dry season. The thing that causes the southern zone not to implement intercropping because the land owned by the area is a type of black soil that is fertile very good if the land is planted with red rice and not planted in a ride so that the results of red rice can be maximized. It is different from the central zone and the northern zone. It is red soil, so it is very suitable for a ride system. with corn and cassava [12].

**Table 2.** Adaptation of farmers to the cultivation process in all three zones

| Cultivation process | Southern Zone | | Central Zone | | Northern Zone | |
|---|---|---|---|---|---|---|
| | Score | Criterion | Score | Criterion | Score | Criterion |
| Intercropping Planting | 1.54 | Unresponsive | 3.43 | Very Responsive | 3.65 | Very Responsive |
| Change planting time | 2.95 | Responsive | 3.18 | Responsive | 3.14 | Responsive |
| Tilling the Land | 2.83 | Responsive | 3.29 | Responsive | 2.98 | Responsive |
| Reduction of Land Area | 1.95 | Lack of Responsiveness | 2.39 | Moderate | 2.14 | Lack of Responsiveness |
| Moving to *Palawija* | 2.72 | Moderate | 2.36 | Moderate | 2.14 | Lack of Responsiveness |
| Planting the Beginning of the Rainy Season | 3.23 | Responsive | 3.34 | Responsive | 3.05 | Responsive |
| Not Planting | 1.74 | Lack of Responsiveness | 1.64 | Lack of Responsiveness | 2.23 | Moderate |
| Seed Selection | 2.56 | Moderate | 3.02 | Responsive | 2.95 | Responsive |
| Switch Planting Locations | 1.77 | Lack of Responsiveness | 3.43 | Very Responsive | 3.52 | Very Responsive |
| Changing Harvest Time | 2.36 | Moderate | 2.61 | Moderate | 2.02 | Lack of Responsiveness |
| Land Area Addition | 1.81 | Lack of Responsiveness | 2.27 | Moderate | 2.03 | Lack of Responsiveness |
| Use of *Genjah* Varieties | 3.55 | Very Responsive | 2.34 | Moderate | 2.27 | Moderate |
| **Number of Scores** | **29.01 Moderate** | | **33.3 Moderate** | | **32.12 Moderate** | |

Adaptation in changing planting time in the three zones is included in the responsive criteria. The prolonged drought caused farmers to adapt to changes in planting time to adjust to the fall of the first rainy season. The purpose of farmers changing the planting time is so that the plants do not lack water so that farmers do not need much money when the cultivation process is carried out.

A cultivation process in deciding to cultivate the soil in all three zones belongs to the responsive criteria. It means farmers in all three zones believe that with suitable soil processing arrangements by increasing manure. Will produce a harvest that produces mal products with good quality. Reducing land area during a long dry season is an adaptation that is classified as less responsive. It is because farmers in the southern and northern zones think that if there is a reduction in arable area, it is the same as the harvest produced.

Adaptation of the cultivation process from rice farming to *palawija* farming in the northern zone is classified as less responsive. Farmers in the northern zone are less motivated to adapt by replacing rice plants with secondary crops during the long dry season. It is because farmers believe that the Segreng Handayani variety of red rice is resistant to dryland conditions. Adaptation by planting other than rice is that corn can be an advantage for farmers. Planting secondary crops does not require much water, and the capital spent by farmers is also not too large, compared to red rice, so for the cultivation process.

The cultivation of plants is very dependent on the flow of sufficient water. Adapting planting at the beginning of the rainy season is one way that farmers can do. Based on the data, the three zones fall under the criteria of responsiveness in adapting planting at the beginning of the rainy season. The transfer of cultivation sites from drought-prone land to yard land is a solution to reduce crop failure during the long dry season. Planting in paddy fields makes it easier for farmers to process irrigation.

The irrigation is obtained from the water of the drinking water company owned by each household. Farmers prefer to grow red rice with the area of land owned without increasing the land area. Farmers do not want to rent other people's land to be planted with red rice because it will increase capital while farmers' capital is minimal.

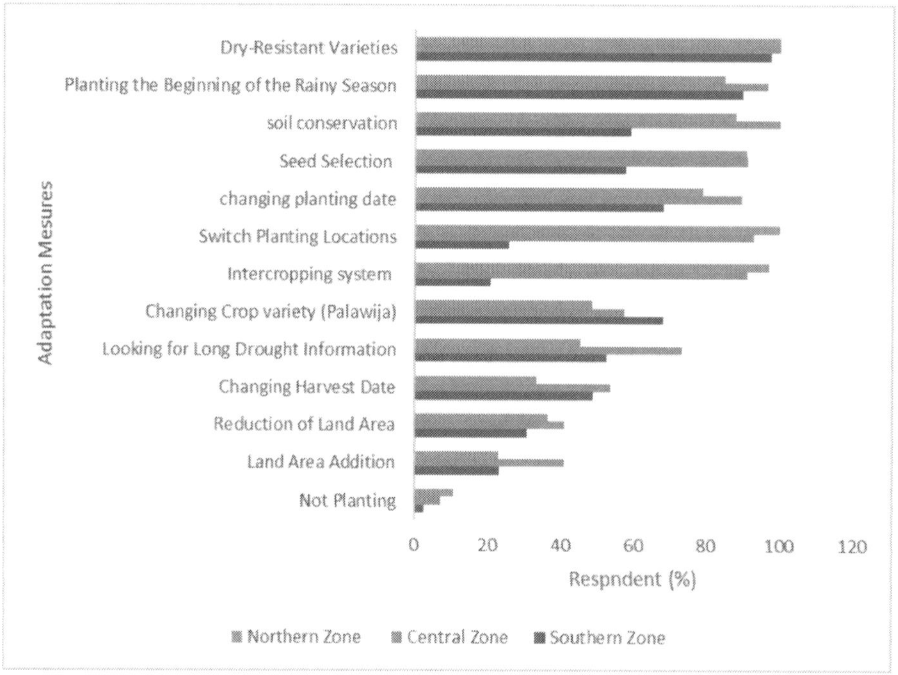


**Figure 2.** Adaptation measures adopted by farmers across three study areas in Gunungkidul Regency

The implementation of adaptation in the long dry season by farmers varies in the three zones of the Gunungkidul Regency (Figure 2). The main adaptation effort made by farmers is the use of dry-resistant rice varieties. The central and northern zones use 100% red rice varieties that are resistant to the dry season. The main reasons for using dry-resistant red rice varieties, planting date, soil conservation, and selecting superior seeds become more agriculturally dependent on rain and groundwater for plant cultivation in Gunungkidul Regency [17].

### *3.2. Farmer Profiles Affecting Adaptation*

The profiles of farmers that are considered to affect the occurrence of an adaptation include gender, age, education level, number of family members, farming experience, income, land area, and land ownership [18]. The farmer profile can affect the adaptation to changes in the crop cultivation process.

**Table 3.** Correlation Rank Spearman farmers profiles influence adaptation

| factors | southern zone | | Central zone | | northern zone | |
|---|---|---|---|---|---|---|
| | Rs | Sig. | Rs | Sig. | Rs | Sig. |
| X1 | -0.331 | 0.003** | -0.244 | 0.069 | -0.41 | 0.001** |
| X2 | -0.063 | 0.582 | -0.117 | 0.39 | -0.98 | 0.434 |
| X3 | 0.073 | 0.526 | 0.163 | 0.231 | 0.108 | 0.389 |
| X4 | 0.345 | 0.002** | 0.2 | 0.139 | 0.38 | 0.763 |
| X5 | -0.05 | 0.662 | -0.205 | 0.129 | -0.254 | 0.04* |
| X6 | 0.284 | 0.012* | 0.045 | 0.743 | 0.149 | 0.234 |
| X7 | 0.311 | 0.006** | -0.79 | 0.565 | -0.301 | 0.014* |
| X8 | 0.307 | 0.006** | 0.107 | 0.432 | -0.003 | 0.978 |

Where Rs is *spearman Rank* Correlation, X1 is age, X2 is gender, X3 is education , X4 is number of family members, X5 is farming experience, X6 is revenue, X7 is area, and X8 is tenancy status.

Based on Table 3, factors that have a significant relationship in the southern zone to farmers' adaptation include age, number of family members, income, land area, and agricultural ownership but gender and education do not affect adaptation.

**Age.** The value of coefficient correlation is negative, which means that the older the age of farmers will reduce the ability and memory of farmers to do something related to the world of agriculture. Old age factors also affect the acceptance of adaptation that is not maximal. According to [19] the higher the age of farmers, the more inefficient the farm. This happens because most farmers are elderly, including the category of unproductive anymore. The ability to manage farmers has dropped and impacted the decline in the technical efficiency of rice farming businesses.

**Many family members.** The correlation coefficient of the number of family members is a positive or unidirectional correlation coefficient, which means that more and more members in the family significantly affect adaptation with the number of family members who must be met, farmers must try to increase the productivity of new red rice to increase income so that basic needs and secondary needs can be met.

**Farming experience** The value of the correlation coefficient is (rs = -0.254). The value of farming experience is related to the adaptation of changes in working patterns, i.e., in the low category. The value of the correlation coefficient that is negative indicates the lower the farmer in adapting work patterns.

**Income.** Income variables have a significant association with the adaptation of red rice farmers. The correlation coefficient is positive or unidirectional, which means that many farmers' incomes affect changes in red rice cultivation run by farmers. If only relying on the results of red rice, income obtained by farmers can only be for basic needs.

**Land.** The value of the correlation coefficient is positive, which means that the wider the land owned by farmers, this will affect agricultural cultivation patterns. Farmers who have large land tend to choose to hunt their land to other farmers. With large land, farmers feel unable if they only work their land individually. By hunting other farmers to work on their land, it is felt that it will be more effective so that rice cultivation can be carried out on time.

**Tenancy Status.** The value of the correlation s a positive coefficient. This means that the higher the ownership of agriculture that has the status of leased land or owns and rent will affect the adaptation of working patterns. With rental land ownership, farmers must spend much money to pay the cost of the rent and the land itself and rent. With much need to cultivate rice, farmers change working patterns, namely by finding a side job to increase family income. Including cultivation needs in buying production facilities and other equipment [16].

## 4. Conclusions
Red rice farmers in the long drought adapt the plant cultivation process to the moderate category, meaning that some farmers in the three zones apply adaptation and some do not. Adaptations that are always carried out by farmers in the long dry season are using dry resistant rice varieties. Farmers' profiles that influence adaptation to prolonged droughts have a significant relationship in the three zones: age, farming experience, land area, number of family members, income, and agricultural ownership. Gender and education do not affect adaptation.

## References
[1] Arifuddin A, Wangiyana W and Aryana I G P M 2019 Effect of Relay-Planting Several Legume Species at Various Ages of Rice on Growth and Yield of Red Rice Grown Together with Legume Crops under Aerobic Irrigation System *Int. J. Environ. Agric. Biotechnol.* **4** 1704–10
[2] Suliartini N W S, Kuswanto K, Basuki N and Soegianto A 2016 Superior lines candidates evaluation of two local red rice Southeast Sulawesi cultivars (Indonesia) derived from gamma rays irradiation techniques *Int. J. Plant Biol.* **7** 64–7
[3] Shinta S, Indriyani S and Arisoesilaningsih E 2014 Morphological Variation of Six Pigmented Rice Local Varieties Grown in Organic Rice Field at Sengguruh Village, Kepanjen Subdistrict, Malang District *J. Trop. Life Sci.* **4** 149–60
[4] Anggraini T, Novelina, Limber U and Amelia R 2015 Antioxidant activities of some red, black and white rice cultivar from west sumatra, Indonesia *Pakistan J. Nutr.* **14** 112–7
[5] Gong E S, Liu C, Li B, Zhou W, Chen H, Li T, Wu J, Zeng Z, Wang Y, Si X, Lang Y, Zhang Y, Zhang W, Zhang G, Luo S and Liu R H 2020 Phytochemical profiles of rice and their cellular antioxidant activity against ABAP induced oxidative stress in human hepatocellular carcinoma HepG2 cells *Food Chem.* **318**
[6] Eiche E, Hochschild M, Haryono E and Neumann T 2016 Characterization of recharge and flow behaviour of different water sources in Gunung Kidul and its impact on water quality based on hydrochemical and physico-chemical monitoring *Appl. Water Sci.* **6** 293–307
[7] Widyastuti M, Riyanto I A, Naufal M, Ramadhan F and Rahmawati N 2019 Catchment Area Analysis of Guntur Karst Spring Gunung Kidul Regency, Java, Indonesia *IOP Conference Series: Earth and Environmental Science* vol 256
[8] Ayanwuyi, Kuponiyi E, Ogunlade F A and Jo O 2010 Farmers perception of impact of climate changes on food crop production in Ogbomosho Agricultural Zone of Oyo State , Nigeria *Glob. J. Hum. Soc. Sci.* **10** 33–40
[9] Winarto Y T, Stigter K, Dwisatrio B, Nurhaga M and Bowolaksono A 2013 Agrometeorological learning increasing farmers' knowledge in coping with climate change and unusual risks *Southeast Asian Stud.* **2** 323–49
[10] Angles S, Chinnadurai M and Sundar A 2011 Awareness on impact of climate change on dryland agriculture and coping mechanisms of dryland farmers *Indian J. Agric. Econ.* **66** 365–72
[11] Junaidi, Basri H and Sugianto 2019 Identifikasi Perubahan Unsur-Unsur Iklim Terhadap Produktifitas Padi Sawah Di Kabupaten Pidie Jaya *J. Agroristek* **2** 52–63
[12] Istiyanti E, Ayu V.C N and Rahayu L 2018 Development of Cassava Farming in Calcareous Land in Gunungkidul Regency Special Region of Yogyakarta *Proceedings of the International Conference on Food, Agriculture and Natural Resources (FANRes 2018)* vol 172 pp 31–5
[13] Ndamani F and Watanabe T 2015 Farmers' perceptions about adaptation practices to climate

change and barriers to adaptation: A micro-level study in Ghana *Water* **7** 4593–604
[14] Owombo P T, Koledoye G F, Ogunjimi S I, Akinola A, Deji O F and Bolarinwa O 2014 Farmers adaptation to climate change in Ondo State, Nigeria: A gender analysis *J. Geogr. Reg. Plan.* **7** 30–5
[15] Ding Y, Wang W, Zhuang Q and Luo Y 2020 Adaptation of paddy rice in China to climate change: The effects of shifting sowing date on yield and irrigation water requirement *Agric. Water Manag.* **228**
[16] Abid M, Scheffran J, Schneider U A and Ashfaq M 2015 Farmers' perceptions of and adaptation strategies to climate change and their determinants: The case of Punjab province, Pakistan *Earth Syst. Dyn.* **6** 225–43
[17] Westengen O T and Berg T 2014 Crop adaptation to climate change in SSA: The role of genetic resources and seed systems *Agric. Food Secur.* **3** 327–43
[18] Brown P R, Tuan V Van, Nhan D K, Dung L C and Ward J 2018 Influence of livelihoods on climate change adaptation for smallholder farmers in the Mekong Delta Vietnam *Int. J. Agric. Sustain.* **16** 255–71
[19] Triyono T, Handoyo Mulyo J, Masyhuri M and Jamhari J 2016 Pengaruh Karakteristik Struktural dan Manajerial Terhadap Efisiensi Usahatani Padi di Kabupaten Sleman *Agrar. J. Agribus. Rural Dev. Res.* **2** 1–8

# Farmers' perception and objective validation of the impact of climate change on farming activities in Northeast Thailand

**Miki Nodera[1], Muhamad Khoiru Zaki[2], Mallika Srisutham[3], and Koshi Yoshida[4], and Keigo Noda[2*]**

[1]Graduate School of Natural Science and Technology, Gifu University, Japan
[2]Faculy of Applied Biological Sciences, Gifu University, Japan
[3]Faculty of Soil Science and Environment, Khon Kaen University, Thailand
[4]Graduate School of Frontier Sciences, The University of Tokyo, Japan
*Corresponding author: anod@gifu-u.ac.jp

**Abstract**. In general, there is a gap between people's subjective and objective assessment of risk and resilience and objective resilience is larger than a subjective one. The aim of this paper is to analyze farmers' perception of climate change comparing with objective data. We conducted the questionnaire survey in Khon Kaen Province in Northeast Thailand and clarified the farmers' profile, agricultural management, and perception of climate change. In addition, comparing to questionnaire result and objective data, we discussed how much the actual farmers' perception of climate change is different from the objective data. As a result, the farmers perceptions to the rainfall amount in 10 years and trend of objective data were not consistent. This gap suggests that farmers perceived changes in short-term rainfall as climate change rather than changes in long-term rainfall, and that farmers may perceive their production as decreasing due to the increased amount lost due to the expansion of paddy fields into unsuitable areas.

**Keywords** : farmers' perception, rainfall, rice production, questionnaire survey, Thailand

## 1. Introduction

Monsoon Asia has about 52% of the total population of the world [1], and future population is anticipated to increase, accompanied by natural disaster risks. According to the Fifth Assessment Report of Intergovernmental Panel on Climate Change (IPCC AR5), the amounts of streamflow in some rivers are projected to increase with high confidence. And the intensity of extreme floods with the same frequency would be stronger [2]. There are two countermeasures to the climate change; one is emission reduction (mitigation) to suppress global warming, and the other is adaptation measures to the adverse effects of global warming. Agriculture is a coupled human-natural system so the definition of adaptation in agriculture includes those of the two systems. Emerging adaptations in agriculture to climate change are diverse. For example, switching crop types, moving agricultural areas to other areas, and changing agronomic practices in response to observed climate change [3].

Northeast Thailand is located in the range of 1-18°N and 101°-105°E and has an area and population approximately one-third of Thailand [4]. The rainy season is from May to October, and rice production is mainly carried out during this season. In Northeast Thailand soil has low fertility and low water retention capacity because soil is mostly covered with sandy soil. Salt accumulation to the soil surface eluted from underground rock salt layer is a serious problem in various places. The irrigation rate (ratio of irrigated area to farmland area) is the lowest in the country at 10.6% and the most farmlands are rainfed. This is because nearly 80% of the Northeast Thailand is gentle relief plains, and there are few suitable site for large-scale dam construction where large capacity of the water storage is expected [5]. In this region, some adaptation measures have already taken. To tackle with projected climate-change impacts, the Thai government's Office of Natural Resource and Environmental Policy and Planning (ONEP) formulated the Climate Change Master Plan in 2015 and a National Adaptation Plan (NAP) in 2019. Adaptation measures include weather-index insurance and King Bhumibol's New Theory on Land and Water Management. Weather-index insurance is an insurance to paid when the rainfall amount at the standard observation point falls below the index value [6]. King Bhumibol's New Theory emphasizes a proper land and water management for the optimum benefits [7].

The relationship between perceived risk and behavioral adaptation depends on factors such as first-hand experience, access to resources needed to undertake protective actions, perception of the efficacy of those actions and personal responsibility; especially, first-hand experience of extreme events or stronger negative emotional experiences increase behavioral adaptation [8]. There is a gap between people's subjective and objective assessment of risk and resilience; objective resilience is larger than subjective one [9].

Shretha et al. [10] and Bastakoti et al. [11] found that 90% or more farmers perceived climate change and tried to adapt to climate change. Manandhar et al. [12] also found that 45% of household recognized climate change. In addition, they said that access to information on climate change is an important factor that influences the likelihood of local people taking adaptation measures. Arunrat et al. [13] found that there were farmers who would not use any adaptation strategies because of lack of information about impacts of climate change and limited knowledge about adaptation measures. These studies adopted questionnaire survey, but they did not discuss the reasons why farmers perceived climate change based on the objective data such as rainfall data or production data. Therefore, the aim of this paper is to analyze farmers' perception of climate change comparing with objective data. We conducted the questionnaire survey in Khon Kaen Province in Northeast Thailand and clarified the farmers' profile, agricultural management and perception of climate change. In addition, comparing the questionnaire result with objective data, we discussed how much the actual farmers' perceptions of climate change are different from the objective data.

## 2. Materials and method

### *2.1. Questionnaire survey*

Three villages were selected to compare the different availability of agricultural resources in terms of salt damage and irrigation water in Khon Kaen province: Ban Du Yai, Ban Kao Noi, and Ban Du Noi (Figure 1). The number of samples collected in Ban Du Yai, Ban Kao Noi and Ban Kao Noi were 24, 21 and 27, respectively. The survey in Ban Du Yai was conducted on 26 August 2018, where farmers were suffering from salt damage. The surveys in Ban Du Yai and Ban Kao Noi were conducted on 6 and 7 October 2018, respectively. In Ban Du Yai, 67% of the farmers used pond water for irrigation, while the farmers were suffering from water shortage. In Ban Kao Noi, 62% of the farmers used pond water and 82% of the farmers used ground water for irrigation. A total of 72 samples were collected for analysis. These villages are located in a few kilo meters south of Khon Kaen University. And each village is located in a few kilo meters.

Questionnaires to the local farmers were conducted in face-to-face method by the students of Khon Kaen University and the collection rate was 100%. The items of the questionnaire were divided into three categories: farmers' recognition of seasons, farmers' perception of climate change, and farm management. The questions about farmers' recognition of seasons are composed of "how did you recognize the rainy season start/end", and "how did you decide when you start to plant/harvest rice". The questions about farmers' perception of climate change are composed of "why do you think the rainfall has changed in these 10 years", and "if you think the rainfall has changed in these 10 years please let us know the effect on your farming". The questions about farm management are composed of "agricultural income/cultivation area is stable or not", and "why do you think cultivation area is not stable". A cross-analysis was conducted by categorizing farmers who answered that they used/did not use weather news when they judged rainy season start, and by categorizing farmers by the reasons why they perceived rainfall change.

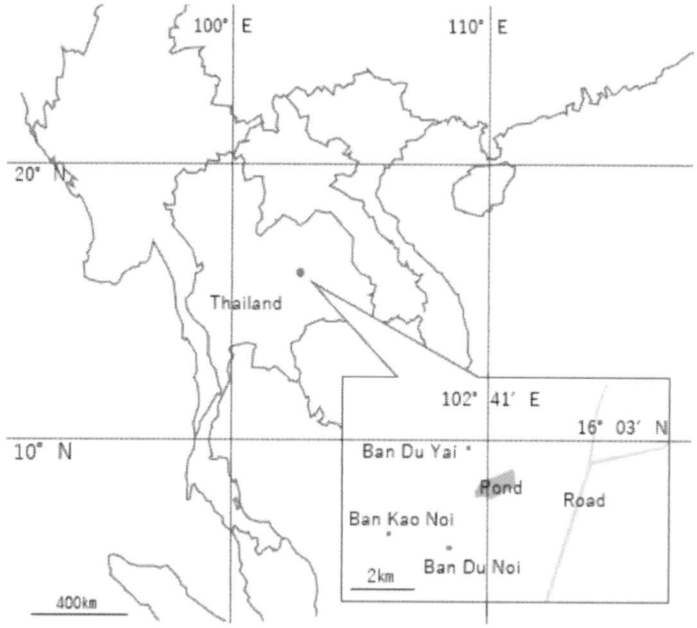


**Figure 1.** Map of survey area

### *2.2. Trend analysis of rainfall and agriculture data*

We analyzed objective data of rainfall monitoring and agricultural statistics in order to validate the results of questionnaire survey. The rainfall data monitored by Thai Meteorological Department (TMD) was adopted: daily rainfall in 10 km grid from 1981 to 2017. The grid data corresponding to the villages was used for the analysis. The agricultural statistics data from Office of Agricultural Economics (OAE) was adopted; we extracted planting area, harvested area and production of rice in the rainy season (major rice) from 1981 to 2017 in Khon Kaen province. The lost area is planting area minus harvested area, and harvested area rate is harvested area rate divided by planting area. The Mann-Kendall test [14], a non-parametric method, was applied to the trend analysis for rainfall and production.

## 3. Results

### *3.1. Result of questionnaire survey*

Figure 2 shows the result of questionnaire in all farmers. The question about farmers' recognition of seasons shows 70.8% of farmers judged the onset of the rainy season by using the weather forecast. And 66.7% of farmers judged the offset of rainy season by using the weather forecast. In addition, the question about timing of planting rice shows 63.9% of farmers planted rice when they thought rainy season started. On the other hand, the question about the timing of harvesting rice had incorrect choices so many farmers selected other. Other includes some answers such as "same as last year" or "rainy season end". The question about farm management shows 24% of farmers answered agricultural income was unstable, while 32% of farmers answered that cultivation area was not stable. In addition, many farmers thought that cultivation area was not stable because of water shortage. Some farmers thought that cultivation area was stable because they had enough water.

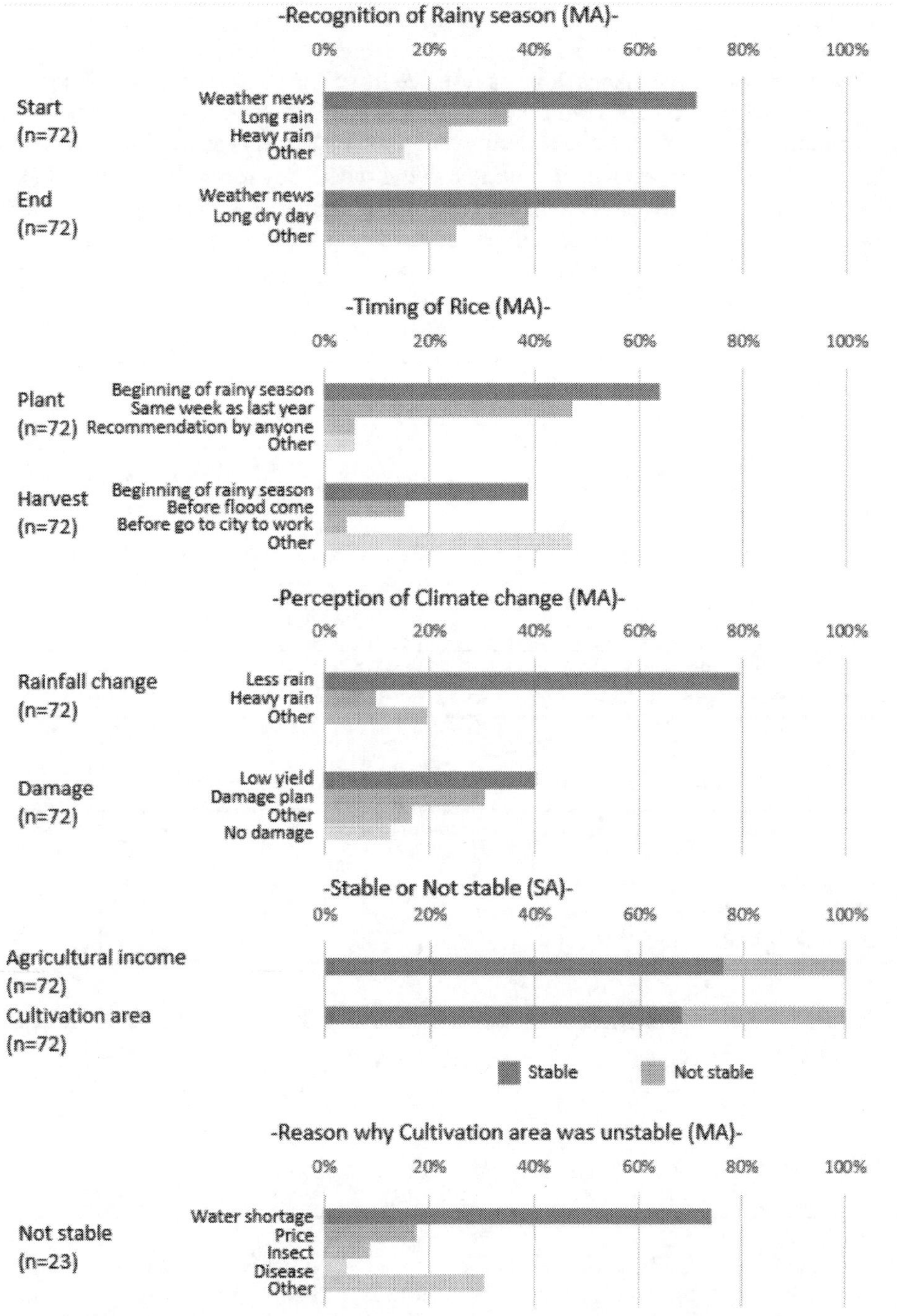


**Figure 2.** Questionnaire result of all farmers

The question about farmers' perception of climate change shows 79.2% of farmers thought that rainfall has changed because of less rain. And because of climate change 40.3% of farmers thought yields have decreased, and 30.6% of farmers thought rice was suffering. On the other hand, 2.9% of farmers thought there is no impact. Table 1a shows the result of the cross analysis of questionnaire. In terms of farmers perception of climate change, 81% of farmers who do not use weather news when they recognize seasons answered "Less rain", while 78.4% of farmers who use weather news answered " Less rain" and 13.7% answered "Other" (Table 1).

In terms of cultivation area stability, Table 2 presented that 57.1% of farmers who do not use weather news answered "Stable", while 72.5% of farmers who use weather news answered "Stable". In addition, the reason why farmers have an unstable in cultivation area is majorly caused by water shortage as shown in Table 3. 79% of farmers who use weather news when they recognize seasons and plant rice when rainy season start answered cultivation area was stable. On the other hand, 64.3% of farmers who do not use weather news when they recognize seasons and plant rice when rainy season start answered cultivation area was stable as shown in Table 4.

**Table 1.** Farmer`s perception about climate change

| Perception of climate change – rainfall change - (MA) | Less rain (%) | Heavy rain (%) | Other (%) |
|---|---|---|---|
| Don't use weather news (n=21) | 81.0 | 14.3 | 4.8 |
| Use weather news (n=51) | 78.4 | 7.8 | 13.7 |

**Table 2.** Farmer`s perception about cultivation area stability on climate change

| Stable or Not stable - Cultivation area - (SA) | Stable (%) | Not stable (%) |
|---|---|---|
| Don't use weather news (n=21) | 57.1 | 42.9 |
| Use weather news (n=51) | 72.5 | 27.5 |

**Table 3.** The reasons why cultivation area was unstable

| Reasons why cultivation area was unstable (MA) | Water shortage (%) | Other (%) |
|---|---|---|
| Don't use weather news (n=9) | 77.8 | 22.2 |
| Use weather news (n=14) | 71.4 | 28.6 |

**Table 4.** Cross analysis between recognize seasons and plant rice on climate change

| Stable or Not stable - Cultivation area - (SA) | | Stable (%) | Not Stable (%) |
|---|---|---|---|
| Recognize seasons×Plant rice | Non-weather×Rainy season start (n=14) | 64.3 | 35.7 |
| | Weather news ×Rainy season start (n=29) | 79.3 | 20.7 |

Note:
Don't use weather news: Those who do not use weather news when farmers recognize seasons
Use weather news: Those who use weather news when farmers recognize seasons
The dominator is the number of people who chose "Don't use weather forecast" or "Use weather news".

### *3.2. Trend of objective data*

Table 5 shows the result of objective data analysis. Annual rainfall and monthly rainfall were analyzed using rainfall data monitored by TMD from 1981 to 2017. Average annual rainfall was 1236.9±218.4 mm. Average rainfall in rainy season was 1047.5±193.3 mm. Average rainfall in dry season was 189.5±77.7mm. The results of the Mann-Kendall test showed a significant increase trend in annual rainfall and rainfall in rainy season from 1981 to 2017.

Planting area, harvested area, production, yield, lost area, and harvested area rate were analyzed from 1981 to 2017. Planting area, harvested area, production and yield was 333059.8±51037.3 ha, 301459.5±48648.9 ha, 566777.4±148656.3 t and 1856.4±265.4 kg/ha and lost area was 31600.3±20581.6 ha. Harvested area rate was 90.5±6.1%. Mann-Kendall test results showed a significant increase trend in planting area, harvested area, production, yield and lost area from 1981 to 2017. On the other hand, there was a significant decrease in the harvested area rate.

**Table 5.** The result of objective data analysis from 1981 to 2017

| Name | Average | SD | Mann-Kendall test | |
|---|---|---|---|---|
| | | | Sen's slope | Z-Value |
| Annual rainfall [mm] | 1236 | 218 | 8.110 | 3.28** |
| Rainy season rainfall [mm] | 1047 | 193 | 7.526 | 2.76** |
| Dry season rainfall [mm] | 189 | 77 | 1.175 | 1.03 |
| Planting area [ha] | 333059 | 51037 | 4057.374 | 5.43*** |
| Harvested area [ha] | 301459 | 48648 | 3061.36 | 3.96*** |
| Production [t] | 566777 | 148656 | 11862.123 | 5.64*** |
| Yield [kg/ha] | 1856 | 265 | 18.049 | 5.43*** |
| Lost area [ha] | 31600 | 20581 | 884.669 | 2.81** |
| Harvested area rate [%] | 90 | 6 | -0.002 | -1.97* |

Note:
Sen's slope: The amount of change per year
Z-value: Standardized Test Statistic
* : When $|Z| > 1.96$, the null hypothesis is rejected at significance level 0.05.
** : When $|Z| > 2.576$, the null hypothesis is rejected at significance level 0.01.
*** : When $|Z| > 3.292$, the null hypothesis is rejected at significance level 0.001.

## 4. Discussion

In terms of farmers' perception of the seasons, the results of the questionnaire analysis (Figure 2) show that many farmers use weather news as means of recognizing the seasons. Furthermore, in terms of farmers' perception of the climate change, the results of the cross-analysis (Tables 1 to 4) shows that there is a difference in perception between those who use the weather news and those who do not use it. Those who use weather news were less likely to think rainfall change due to "Less rain" and more likely to think because of "Other" than those who do not use weather news. On the other hand, the result of the rainfall trend showed that rainfall has increasing trend. These results suggest that the perceptions of farmers who use weather news are more correct than farmers who so not use it. In addition, most farmers who use weather news when they recognize seasons and plant rice when the rainy season starts answered cultivation area was stable, which indicates that the obtained information from the weather news is useful for farmers when they are farming. In other words, obtaining information about the onset of the rainy season and other aspects of cultivation from familiar media, such as weather news, is important for farmers to take adaptation actions. This is because access to information on climate change is important for adaptation practice [8,12,13].

The results of the questionnaire analysis showed that farmers thought that rainfall has decreased and that rainfall change have affected production (Figure 2). However, the results of the objective data analysis showed that the rainfall and planting area, harvested area, production and yield were on the rise (Table 5). These gaps suggest that farmers' perceptions of climate change did not match with the results of objective data. Difference in perception of rainfall change is thought due to that farmers perceived climate change as short-term changes by rainfall patterns, such as the dry spell or the occasional serious drought memory, rather than the long-time change in rainfall [10]. Difference in perception of production damage by climate change implies that cultivation area was expanded to the land where productivity was unstable or the water availability was low, because the lost area has increased and the harvested area rate has decreased. Wilson et al. [8] said that the relationship between perceived risk and behavioural adaptation depends on factors such as first-hand experience. As the area harvested rate has the decreasing trend, farmers experienced lost area as damage every year and may perceive an increase in lost area as a risk.

## 5. Conclusion

In this study, the results of the questionnaire survey were analyzed on the basis of objective data. Farmers' perceptions of climate change derived from the questionnaire survey did not agree with the results of the objective data analysis. It was suggested that the reason for the discrepancy may be that farmers perceived changes in short-term rainfall as climate change rather than changes in long-term rainfall, and that farmers may perceive their production as decreasing due to the increased amount lost due to the expansion of paddy fields into unsuitable areas. However, the scope of this study was narrow and the analysis was not conducted using a land use map. Therefore, future analyses using a broader area of data and land use maps will be necessary.

## References

[1] FAO 2015 The State of Food and Agriculture 2015 *https://www.fao.org/publications/sofa/2015/en/*

[2] Tachikawa Y, Hanittinan P, Yorozu K, Ichikawa Y and Kim S 2017 D4PDF o mochiita indoshina hantou ni okeru kasenryuryo yosoku (Forecasting River Discharge Rate in Indochina Peninsula Using d4PDF) *Theme D: Precise Impact Assessments on Climate Change, Program for Risk Information on Climate Change, Report for Fiscal year 2018* (Japan) pp 72–3

[3] UNEP - UN Environment DTU Partnership 2017 Adaptation Gap report 2017 *https://www.unep.org/resources/adaptation-gap-report-2017*

[4] Takagaki M 1987 Climate and Agriculture in Northeastern Thailand *J. Agric. Meteorol.* **42** 385–9

[5] JICA 2010 JICA Report *https://openjicareport.jica.go.jp/617/617_122.html*

[6] Yoshida K, Homma K, Maki M, Noda K, Shirakawa H, Oki K, Sritumboon S and Srisutham M 2017 Possibility of weather index insurance as a climate change adaptation measure in northeastern Thailand *JSCE Proc. G* **73** I_377-I_383

[7] Kositgittiwong D and Ekkawatpanit C 2016 A Study on Integrated Water Resources Management and King Bhumibol's New Theory on Land and Water Management in Wattananakorn area, Sa Kaeo Province, Thailand *Glob. J. Eng. Technol. Rev.* **1** 69–77

[8] Wilson R S, Herziger A, Hamilton M and Brooks J S 2020 From incremental to transformative adaptation in individual responses to climate-exacerbated hazards *Nat. Clim. Chang. 2020 103* **10** 200–8

[9] Agrawal N, Elliott M and Simonovic S P 2020 Risk and Resilience: A Case of Perception versus Reality in Flood Management *Water* **12** 1254

[10] Shrestha R P, Chaweewan N and Arunyawat S 2017 Adaptation to Climate Change by Rural Ethnic Communities of Northern Thailand *Climate* **5** 57

[11] Bastakoti R C, Gupta J, Babel M S and van Dijk M P 2013 Climate risks and adaptation strategies in the Lower Mekong River basin *Reg. Environ. Chang.* **14** 207–19

[12] Manandhar S, Pratoomchai W, Ono K, Kazama S and Komori D 2015 Local people's perceptions of climate change and related hazards in mountainous areas of northern Thailand *Int. J. Disaster Risk Reduct.* **11** 47–59

[13] Arunrat N, Wang C, Pumijumnong N, Sereenonchai S and Cai W 2017 Farmers' intention and decision to adapt to climate change: A case study in the Yom and Nan basins, Phichit province of Thailand *J. Clean. Prod.* **143** 672–85

[14] Nishioka M and Takara K 2004 Hydrological time series trend fluctuation by Mann-Kendall test *J. Soc. Hydrol. Water Resour.* **17** 343–53

**Acknowledgement**
This work was supported by JST SATREPS (JPMJSA1502) and JSPS KAKENHI (JP19H03069; JP20J10806), Japan.

# Carbon dioxide ($CO_2$) gas emission based on land use types of peatland in Kinali, West Pasaman Regency

**Septia Dwi Syahputri[1], Hermansah[2] and Yulnafatmawita*[3]**

[1]Soil Science Department Andalas University, Kampus Unand Limau Manis Padang, Indonesia, 25614
[2]Soil Chemical Laboratory, Andalas University, Kampus Unand Limau Manis Padang, Indonesia, 25614
[3]Soil Physical Laboratory, Andalas University, Kampus Unand Limau Manis Padang, Indonesia, 25614
*Corresponding author: yulnafatmawita@agr.unand.ac.id

**Abstract.** Carbon dioxide gas which is one of the greenhouse gases stated at Kyoto Protokol in 1997 can come from peatland. Conversion of peatland into farming land can accelerate the decomposition process of OM which will emit carbon dioxide gas ($CO_2$) into the atmosphere. This study was aimed to measure $CO_2$ emission from several types of peatland in Kinali District, West Pasaman Regency, West Sumatra Province. The research was conducted using a survey method based on the land use type (purposive sampling). The measurement of $CO_2$ was conducted in the field using an infrared non-dispersive sensor (NDIR) found in the Carbon Dioxide ($CO_2$) Monitor tool. Soil properties analyzed were organic-C, total N, and C/N ratio. The results showed that the different use in peatland has no significant effect on the value of $CO_2$ emission. Carbon dioxide emission has a positive correlation ($r = 0.68$) with soil organic carbon content. The highest average of $CO_2$ emission was 0.92 $g/m^2/d$ (under oil palm plantation) from soil surface and 1.56 $g/m^2/d$ (under seasonal cropland) from the surface of soil on 30 cm depth. The lowest $CO_2$ emission was 0.68 $g/m^2/d$ (under mixed garden) from both soil surfaces.

**Keywords:** carbon dioxide ($CO_2$), emission, land use, peatland, soil organic matter

## 1. Introduction

Carbon dioxide is one of the greenhouse gases being highly considered as a contributor to global warming. This gas ($CO_2$) was mainly derived from fossil fuel oxidation such as from industries, transportation, as well as electricity, and it can also come from agriculture activities. Land-use change from forest to agricultural land, especially for intensive cultivation, will emit $CO_2$ to the atmosphere. Tillage could accelerate $CO_2$ emission from mineral soils as organic matter (OM) is oxidized due to improving soil aeration of the soil [1].

Peatland is one of marginal soil, besides mineral soils, which is being exploited for agriculture purposes. Land-use change from natural peatland into farming land has changed the characteristics of the peatland. The first action being conducted in managing peatland for farming is by changing the water table to meet the need of specific crops. Different land use will cause the different depths of the water table. It means that it will differ the depth of peatland being unsaturated. This will cause different processes of organic matter (OM) decomposition and finally will cause different gas to be emitted to the atmosphere.

Under saturated conditions, as in natural ones, the type of gas being emitted from peatland is methane ($CH_4$). Methane contributed to 10% of the total GHG causing global warming in the US in 2019. It was only one-eighth of $CO_2$ emission. However, it has powered 25 times higher than $CO_2$ gas [2]. It contributed approximately 0.5 °C of warming since pre-industrial times, it was number two after $CO_2$.

Carbon dioxide is a type of gas being produced under aerobic conditions. High organic matter content in peatland will cause high $CO_2$ emissions as it is being dried or as the water table is lower. However, cultivation could also add some OM to the soil. As reported by Maswar *et al.,* [3] the decomposition of the palm leaves on the soil surface of oil plantation could return carbon to the soil for approximately 9.39 t C/ha - 12,23 t C/ha each year.

Some factors are affecting the amount of $CO_2$ emission from peatland besides the water table depth. Air temperature, humidity, soil water content, soil OC, C/N ratio are among the factors affecting peat oxidation [1]. Carbon dioxide gas varies between 300-600 ppm depending on place and time. The critical value of $CO_2$ gas in the air was 5000 ppm based on SNI 19-0232-2005 [4]. If it is > 5000 ppm it will be toxic to human health. The concentration of $CO_2$ gas in the Indonesian atmosphere tended to increase by time from 373 to 383 between 2004 and 2010. If it reaches 550 ppm, the air temperature will increase by 3°C [5].

Land-use change (LUC) from the original into the farming ecosystem has globally increased $CO_2$ emission by more than 20% in 18 years (1990-2008). Based on Maswar *et al.* [3] $CO_2$ emission could reach 632 T/year in Indonesia. This emission was due to some factors, mostly OM decomposition after deforestation, drained peatland, and loss of vegetation being able to absorb atmospheric $CO_2$.

Carbon dioxide emission in Sumatra and Borneo reached 2 billion tons from land-use change and burning of peatland. As reported [5] from Wetland International, a world organization in wetland management and sustainability. Even though peatland burning can be rehabilitated, draining peatland for agriculture will always emit $CO_2$ to the atmosphere.

Husnain *et al.* [6] found that $CO_2$ emission from peatland under secondary forest, oil palm, acacia, and rubber plantation was 61±25, 66±25, 59±19, 52±17 Mg/ha/y, respectively from a secondary forest, and oil palm plantation, acacia, and rubber plantation, while under bare land ranged between 24-67 Mg $CO_2$/ha/y. Nusantara *et al.* [7] reported that there was a difference in $CO_2$ emission between the wet and dry seasons. $CO_2$ gas emission from oil palm plantations and secondary forest increased under wet season and from corn plantation increased under dry season. Then, peatland burning for 2.79 million ha in the year 2006 emitted 98.38-180.38 million tons of $CO_2$ to the atmosphere [8].

Peat soil in Sumatra reached 5.85 million ha or approximately 43.56% of total peatland in Indonesia [9]. From that area, approximately 100.87 Ha in West Sumatra [10], and about 12.045 ha [11] was found in Kinali, West Pasaman Regency. Land-use change from natural to agricultural land also happened in Kinali. Since the last few decades, the natural peatland there had been converted into agricultural lands. This will affect the management given to the soil and finally will affect the properties of the soil, especially the depth of the water table. Based on the condition above, this research was conducted to measure the amount of $CO_2$ emitted from the peatland under several land-use types in Kinali.

## 2. Materials and method

### *2.1. Materials*

This research took part in Kinali, West Pasaman Regency, West Sumatra Province, Indonesia and in Soil Laboratory Universitas Andalas Padang from January to March 2021. The geographical position is between 00° 03' N – 00° 11' S and between 99°45'–99°03' E [12]. It is approximately 160 km from the capital of West Sumatra. The total area of Kinali is 482.64 Km$^2$, having two subdistricts, Kinali dan Katiagan. The peatland (Histosols) area was about 15,351.67 ha (31.81%) in Kinali. The Histosols belonged to the great group Toposaprist having <2 m depth (6,507.32 ha), Troposaprist > 2 m depth (8,844.39 ha).

### 2.2. Methods

This research used a survey method, in which the soil was sampled purposively based on the land use types. There were 5 main types of land use (oil palm plantation, bushland, mixed garden, open land, and seasonal cropland) found in the research site. At each type of land use, three locations were chosen for $CO_2$ emission. The $CO_2$ emission was measured under natural conditions using an infrared non-dispersive sensor (NDIR) found in the Carbon Dioxide ($CO_2$) Monitor Tool. $CO_2$ emission was measured at 2 soil depths, the first was at the soil surface and the second was at the surface of the soil on the 30 cm depth. The second was conducted by removing soil from 30 cm depth. Then, the $CO_2$ emission was directly measured. Besides $CO_2$ gas emission, the air temperature, as well as relative humidity, was also measured by using this NDIR tool.

Soil samples were also taken from each type of land use at 0-30 cm and 30-60 cm for soil organic-C (using wet oxidation method) and soil moisture content (using the gravimetric method) analyses. Other soil properties (bulk density, total soil porosity, total-N, fiber content, ash content, and depth of water table) were derived from data resulting from Yulnafatmawita *et al.* [13].

Based on data from Climatology Station [14] from 2013-2019 in Padang Pariaman, Kinali received an annual rainfall of 3,534.71 mm on average. The lowest monthly rainfall (155 mm) was found in July and the highest was in October (527 mm) each year. According to Schmidt and Ferguson [15], by comparing the number between dry and wet months (Q value), the climate in the research area was classified into A-type (Q=0.013) or very wet. Data collected were statistically analyzed using JMP as well as some correlation between the parameters using simple linear correlation. The data are presented in form of Graphs, Tables, and correlations.

## 3. Results and discussion

### 3.1. Land-use types

According to local people in Kinali, the peatland for some types of use was originally peatland forest. The forest was cleared for oil palm plantation and some were abandoned (bushland). Then, the plantation was converted again into other types of land use, some were inter-cropped with perennial crops (mix garden), some was converted into corn cultivated land (seasonal cropland), and some was being open without vegetation.

*3.1.1 Oil palm plantation.* Oil palm plantation, 77.43% of peatland in Kinali, was under local society management which was conventionally managed. The plantation was fertilized using synthetic fertilizers, some using manure, and some was added nothing. The depth of the water table at the oil palm plantations ranged between 37 to 39 cm.

*3.1.2 Mix garden.* This mixed garden was originally an oil palm plantation that was not properly maintained. Therefore, the crops could not grow well, and then it was replaced with some other crops. The water table of this land was ± 67 cm in depth.

*3.1.3. Bushland.* Bushland was originally forest land then it was abandoned after being cleared. Therefore, there is some bush growing well on the land surface. The water table depth was between 42 and 59 cm.

*3.1.4. Seasonal cropland.* In the last two years, the oil palm plantation was converted into corn (*Zea mays* L.) land. This water table depth was between 35 and 50 cm.

*3.1.5. Open land.* The open land was an oil palm plantation before being replanted. The depth of the water table was between 41and 47 cm.

### 3.2. *Soil organic carbon and soil moisture content*

Based on Table 1, oil palm plantation showed the highest soil organic carbon content (38.19%) at 0-30 cm soil depth, but it was slightly below seasonal cropland even though both were not significantly different. It was 246% of that under mix garden or 236% of that under open land at 0-30 cm soil depth, and 254% of that under mix garden at 30-60 cm soil depth. It could be affected by the water table of the soil. As stated that oil palm plantation had the shallowest water table (38 cm) among the land use types, and the deepest was under mix garden (66.33 cm) [16]. The water table will affect the moisture content of the soil. Soil having shallow water table will have higher soil moisture content than that under deep one. Water table close to soil surface would cause less drained soil depth which could be attacked by microorganisms. Therefore, it was less OC decomposition meaning higher SOC content.

**Table 1.** Soil organic carbon and soil moisture content of peatland on soil surface 0-30 cm and 30-60 cm at five different land-use types

| Land-use | SOC | | Soil Moisture | |
|---|---|---|---|---|
| | 0-30 cm | 30-60 cm | 0-30 cm | 30-60 cm |
| | ..................%.................. | | ...............%.............. | |
| Oil Palm Plantation | 38.19a | 43.20a | 101.95a | 113.44ab |
| Seasonal Crop Land | 31.55b | 48.02a | 75.92b | 119.33a |
| Bush Land | 22.66c | 25.53b | 72.23b | 68.42bc |
| Open Land | 16.17d | 29.23b | 41.64c | 73.49bc |
| Mix Garden | 15.53d | 16.98c | 40.27c | 52.80c |

Note: Data followed by small letter on the same column was significantly different based on *Student's t* at 0.05

Besides due to the soil moisture content as affected by the depth of the water table, the amount of SOC content of the soil was also determined by the management given to the soil and the maintenance. Peatland for oil palm plantation was professionally managed especially the water table depth, and the land was not cultivated. While, on the other hand, the water table of the peat soil under mix garden was not well-managed and maintained, the land was also some cultivated for annual crops. Therefore, the amount of the SOC decreased by time under mix garden.

### 3.3. *Air temperature and humidity*

Air temperature (T) and relative humidity (rH) during $CO_2$ emission measurement was significantly different among the types of land use (Table 2). Air temperature under oil palm plantations was found the highest, it was 13.5% higher than that under open land (the lowest). Based on Lu *et al* [16] the temperature in the location was classified as higher (30-45°C). Relative air humidity was not different among the land use being measured during $CO_2$ accounting. The air humidity was average for 84.66% which belongs to humid [17].

Soil moisture was significantly different among the land use types. The highest soil moisture was found under oil palm plantations (101.95%) which were 2.45% of that under open land. High soil moisture content under oil palm plantation was due to high SOC content under that soil (Table 1).

**Table 2.** Air temperature and humidity on peatland at different types of land use in Kinali

| Land-use | Air Temp. (°C) | Humidity (%) |
|---|---|---|
| Oil Palm Plantation | 36.77a | 82.65a |
| Mix Garden | 36.40ab | 85.52a |
| Bush Land | 34.41bc | 87.22a |
| Seasonal Crop Land | 33.67c | 83.46a |
| Open Land | 32.41c | 84.45a |

Note: Data followed by small letter on the same column awas significantly different based on *Student's t* at 0.05

### *3.4. $CO_2$ emission*

Based on Figure 1 and the data presented on Table 3, it was shown that there was no significant difference of $CO_2$ emission from peat soil among the types of land use either from soil surface or at the surface of soil on the 30 cm depth. This was due to the effect of some factors, among of which was relative humidity that was not different among the types of land use. Even though the air temperature (Table 2) was different among the land use, the $CO_2$ emission from the land use types was not different. If compared to the amount of $CO_2$ emission allowed [4], the amount of $CO_2$ emission from the any types of peat land was still below the threshold, it was only 35-48% of that at soil surface, and 35-80% of that at surface of soil on the 30 cm soil depth.

The amount of $CO_2$ emission from lower soil surface (on 30 cm depth) was higher than those from soil surface. It increased by 1-87% on the 30 cm soil depth compared to $CO_2$ emission on the soil surface. Higher $CO_2$ emission at the lower soil depth was due to. Then the emission of $CO_2$ from peat land followed the tendency of soil organic carbon and moisture content. High SOC in the soil caused higher amount of carbon degraded and finally emitted $CO_2$ to the atmosphere. This was in line with a result reported [18] that the more organic matter added to the soil, the higher the amount of $CO_2$ emitted from the soils. The highest soil moisture content was associated with high OC accumulated in the soil. Therefore, high soil moisture content increased $CO_2$ emission from the land. The tendency of highest $CO_2$ emission was found under oil palm plantation.

$CO_2$ respiration from peat soil highly affected by the water table depth [19–21]. Boonman *et al.* [21] reported that increasing ditch water levels decreased rates of peat respiration. Peat land having shallower water table (40 cm) emitted higher $CO_2$ than that under deeper water table (80 cm) [22]. However, soil moisture variation did not always significantly affected $CO_2$ emission in the humified peat soils [23].

Between the two soil depths being measured, $CO_2$ emission from lower (30 cm) depth gave higher values but they were not significantly different. This could be affected by the water table of the soil which was less than 100 cm for all types of land use. Tendency of increasing $CO_2$ emission from lower soil depth was supported to by higher SOC content (Table 1) in the depth. High OC content means more materials can be turned into $CO_2$ through enzymatic oxidation by microorganisms.

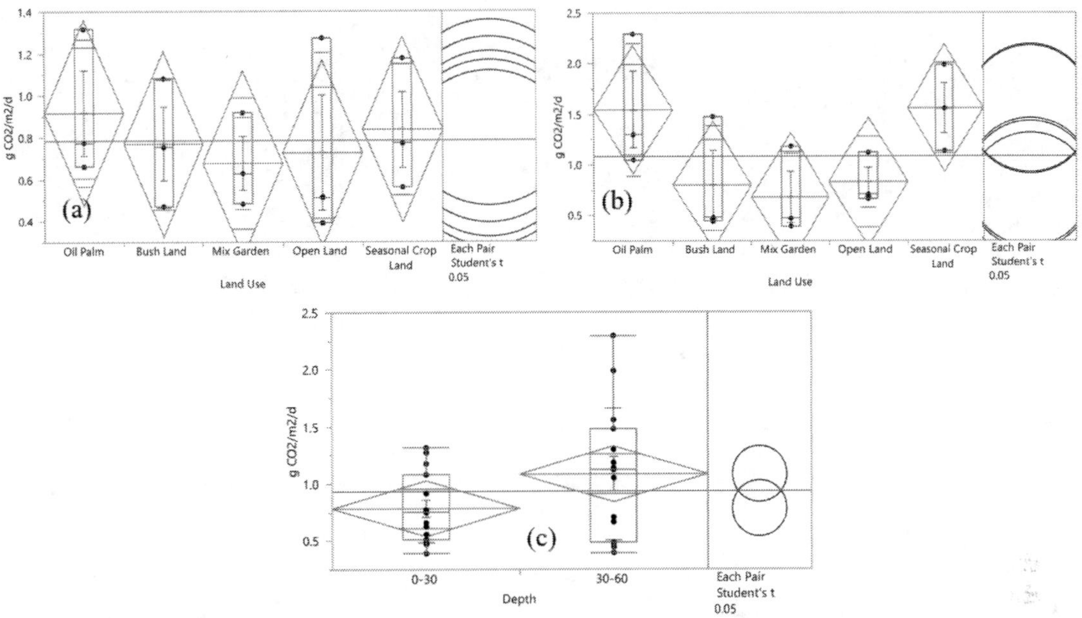


**Figure 1.** $CO_2$ emission from peat land under different types of land use at 0-30 cm (a) and 30-60 cm (b) and comparison between (c) the soil depth

**Table 3.** Soil depth of peatland on soil surface 0-30 cm and 30-60 cm at five different land-use types

| Land-use | Soil depth | |
|---|---|---|
| | 0 cm | 30 cm |
| | ………………g $CO_2/m^2/d$……………… | |
| Oil Palm | 0.92 a | 1.55 a |
| Seasonal Crop Land | 0.84 a | 1.56 a |
| Bush Land | 0.77 a | 0.80 a |
| Open Land | 0.73 a | 0.83 a |
| Mix Garden | 0.68 a | 0.68 a |
| **Average** | **0.79 a** | **1.08 a** |

Note: Data followed by small letter on the same column was significantly different based on *Student's t* at 0.05

### *3.5. Correlation between $CO_2$ emission and soil properties*

The correlation between $CO_2$ emission from peatland and the soil properties was presented in Table 4. Based on linear regression analyses, in general, $CO_2$ emission from soil surface did not significantly correlate ($R^2$<0.5) to the soil physical and chemical properties at the 0-30 cm soil depth. However, there were high correlations between some of soil properties to the $CO_2$ emission from 30 cm soil depth. Among the soil properties, soil BD and total-N highly affected $CO_2$ emission, then followed by SOM, fiber, and ash content, as well as soil moisture.

**Table 4.** Correlation between soil properties and $CO_2$ emission from a peatland in Kinali, Pasaman

| Soil Properties | Equation | $R^2$ | r |
|---|---|---|---|
| **At 0-30 cm Soil Depth** | | | |
| SOM | Y=20.88X + 1138.5 | 0.1868 | 0.43 |
| Temperature | Y=-1.2837+2076.4 | 0.0001 | 0.00 |
| Air humidity | Y=52.33X - 2398.2 | 0.0710 | 0.27 |
| Soil moisture | Y=7.5385X + 1531.3 | 0.0657 | 0.26 |
| Fiber content | Y=32.73X + 1612.9 | 0.2054 | 0.45 |
| Ash content | Y=-20.876X + 3226.1 | 0.1868 | -0.43 |
| Water table | Y= -12.037X + 2615.2 | 0.0286 | -0.17 |
| C/N ratio | Y=-45.846X + 3363.7 | 0.1328 | -0.36 |
| Total-N | Y=817.93X + 1281 | 0.2225 | 0.47 |
| BD | Y=-3391X + 3691.3 | 0.2634 | -0.51 |
| **At 30-60 cm Soil Depth** | | | |
| SOM | Y=50.022X - 3.253 | 0.4889 | 0.70 |
| Temperature | Y=754.18X + 26609 | 0.1797 | 0.42 |
| Air Humidity | Y=-77.315X + 9660.8 | 0.0491 | 0.22 |
| Soil Moisture | Y=25.926X + 591.02 | 0.3572 | 0.60 |
| Fiber content | Y=78.927X + 1371.2 | 0.5534 | 0.74 |
| Ash Content | Y=-50.022X + 49999 | 0.4889 | -0.70 |
| Water Table | Y=-64.054X + 5912.1 | 0.2240 | -0.47 |
| C/N ratio | Y=-42.482X + 4040.9 | 0.1227 | -0.35 |
| Total-N | Y=2345.2X -57.117 | 0.6506 | 0.81 |
| BD | Y=-7418.6X + 6168.5 | 0.6814 | -0.83 |

Soil BD, ash content, water table, and C/N ratio negatively correlated to the $CO_2$ emission from peatland. The higher the soil BD, the more compacted the soil, causing the more difficult the OM to be enzymatically decomposed. Then, the higher the soil BD means the lower the probability of SOM to be decomposed by microorganisms. The ash content as an indicator of peat maturity negatively affected $CO_2$ emission from peatland. The higher the ash content was the higher the maturity level of the peat soil, the less the SOC content in it. Therefore, as the ash content increases, the amount of OM going to be decomposed decreases, contributing less $CO_2$ emitted to the atmosphere.

Unlike ash, fiber content was positively correlated to the amount of $CO_2$ emission from peatland. The high fiber content of peat soils means there is still more OM left in the soil which can be decomposed and produced $CO_2$ that can be emitted to the atmosphere. Like the fiber, soil moisture content positively affected $CO_2$ emission. High soil moisture content is associated with high OM content in the soils because OM could retain more amount of water compared to its weight. Therefore, the higher the soil moisture, was the more $CO_2$ can be produced.

Total nitrogen (N) positively linearly influenced $CO_2$ emission from peat soil to the atmosphere. High N content will increase $CO_2$ emission from peatland. This could be found to be true since the higher the N content of the OM was the more preferable the materials be attacked by organisms. This was due to the need for N by organisms to proliferate themselves before attacking other OM in the soil. As the amount increases, the microbes will faster degrade the OM, contributing more $CO_2$ to the atmosphere.

The deeper the soil water table, the more the amount of SOM being available to be decomposed by soil organisms, as a consequence the more the amount of $CO_2$ will be emitted to the atmosphere. This was due to the effect of draining the peatland, increasing oxidation zone as well as activity of decomposing microbes. Since the water table of the peatland under all types of land use being measured was ≤ 60 cm [13] the effect on $CO_2$ emission was not significant.

## 4. Conclusion

Based on data resulting from this research, it could be concluded that $CO_2$ emission from a peatland in Kinali was < the critical level stated by SNI (5000 ppm). Their highest $CO_2$ emission was found under oil palm plantation (0.92 g $CO_2/m^2/d$ at soil surface and 1.55 g $CO_2/m^2/d$ at the depth of 30 cm). Then, it was followed by seasonal cropland (0.84 and 1.56 g $CO_2/m^2/d$), bushland (0.77 and 0.80 g $CO_2/m^2/d$), open land (0.73-0.83 g $CO_2/m^2/d$), and mix garden (0.68 g $CO_2/m^2/d$, respectively for both depths. $CO_2$ emission negatively correlated to BD (r=-0.83) and ash content (r=-0.70), but positively correlated to total-N (r=0.81), fiber content (0.74), OC content (r=0.70), and moisture content (r=0.60) of the soil.

## Reference

[1] Yulnafatmawita 2005 *Fractionation of soils based on bonding energy and aggregate size : a method for studing the effect of structural hierarchy on degradation process* (Australia)
[2] EPA 2021 Overview of Greenhouse Gases *United States Environmental Protection Agency*
[3] Maswar, Haridjaja O, Sabiham S and van Noordwijk M 2011 Kehilangan karbon pada berbagai tipe penggunaan lahan gambut tropika yang didrainase (carbon loss from several landuse type on tropical peat lands drained) *Jurnal Tanah dan Iklim* **34** 13–25
[4] SNI 2005 *Nilai Ambang Batas (NAB) zat kimia di udara tempat kerja* (Jakarta)
[5] Samiaji T 2011 Gas CO2 di wilayah Indonesia *Berita Dirgantara* **12** 68–75
[6] Husnain, Wigena I G P, Dariah A, Marwanto S, Setyanto P and Agus F 2014 CO2 emissions from tropical drained peat in Sumatra, Indonesia *Mitigation and Adaptation Strategies for Global Change 2014 19:6* **19** 845–62
[7] Nusantara R W, Sudarmadji, Djohan T S and Haryono E 2015 Emisi CO2 tanah akibat alih fungsi lahan hutan rawa gambut di Kalimantan Barat (Soil emissions of CO2 due to land use change of peat swamp forest at West Kalimantan) *Jurnal Manusia dan Lingkungan* **21** 268–76
[8] Dariah A and Maswar 2016 Isu lingkungan gambut tropika Indonesia *Lahan gambut Indonesia* (Bogor: Badan Penelitian dan Pengembangan Pertanian ) pp 101–29
[9] Anda M, Ritung S, Suryani E, Sukarman, Hikmat M, Yatno E, Mulyani A, Subandiono R E, Suratman and Husnain 2021 Revisiting tropical peatlands in Indonesia: Semi-detailed mapping, extent and depth distribution assessment *Geoderma* **402** 115235
[10] Wahyunto, Nugroho K and Agus F 2016 Perkembangan pemetaan dan distribusi lahan gambut di Indonesia *Lahan Gambut Indonesia* (Bogor: Badan Penelitian dan Pengembangan Pertanian) pp 33–60
[11] Mawardi E 2011 *Kajian potensi sumberdaya lahan dan realisasi pengembangan usahatani jagung di Sumatera Barat* (Sumatera Barat)
[12] BPS Kab. Pasaman Barat 2020 *Kabupaten Pasaman Barat Dalam Angka 2020* (Pasaman Barat: Badan Pusat Statistik Kabupaten Pasaman Barat)
[13] Yulnafatmawita, Syahputri S and Hermansah 2021 Degree of peatland maturity at different land use types in Kinali, West Sumatra Indonesia *International Seminar on Tropical Peatland ISTP 2021* (Bogor: Perkumpulan Masyarakat Gambut Indonesia)
[14] BMKG Sta. Klim. Kelas II Padang Pariaman *Badan Meterologi, Klimatologi, dan Geofisika*
[15] Schmidt F H and Ferguson J H A 1951 *Rainfall types based on wet and dry period ratios for Indonesia with Western New Guinee* (Jakarta: Kementerian Perhubungan, Djawatan Meteorologi dan Geofisik)
[16] Lu J, Zhang X, Li P and Zhu Y 2015 Case-based FCTF reasoning system *Applied Sciences* **5** 825–39
[17] Ray R L, Griffin R W, Fares A, Elhassan A, Awal R, Woldesenbet S and Risch E 2020 Soil CO2 emission in response to organic amendments, temperature, and rainfall *Scientific Reports* **10** 5849
[18] Rumbang N 2015 A study of carbon dioxide emission in different types of peatland use in Kalimantan *Ilmu Pertanian (Agricultural Science)* **18** 9–17

[19] Astiani D, Burhanuddin, Taherdjadeh M and Curran L M 2016 Effects of water table level on soil CO2 respiration in West Kalimantan forested and bare peatland: An experimental stage *Nusantara Bioscience* **8** 201–6
[20] Husnain, Sipahutar I A, Purnomo J, Widyanto H and Nurhayati 2017 CO2 emissions from tropical peat soil affected by fertilization *Journal of Tropical Soils* **22** 1–9
[21] Boonman J, Hefting M M, van Huissteden C J A, van den Berg M, van Huissteden J, Erken G, Melman R and van der Velde Y 2021 Cutting peatland CO2 emissions with rewetting measures *Biogeosciences Discuss* **bg-2021-276**
[22] Kechavarzi C, Dawson Q, Bartlett M and Leeds-Harrison P B 2010 The role of soil moisture, temperature and nutrient amendment on CO2 efflux from agricultural peat soil microcosms *Geoderma* **154** 203–10
[23] Berglund Ö and Berglund K 2011 Influence of water table level and soil properties on emissions of greenhouse gases from cultivated peat soil *Soil Biology and Biochemistry* **43** 923–31

**Acknowledgement**
Authors would like to thanks all participants helping in data collecting in the field and analyzing soil samples in the soil laboratory Andalas University, Padang Indonesia.

# Factors influencing carbon emissions disclosures in high profile companies: some Indonesian evidence

**S R Ika[1], Yuliani[1], A Okfitasari[2] and A K Widagdo[3*]**

[1]Department of Accounting, Janabadra University, Jl Tentara Rakyat Mataram 55-57 Yogyakarta, 55231, Indonesia
[2]Department of Accounting, Duta Bangsa University, Jl. Bhayangkara 55, Tipes, Solo, 57154, Indonesia
[3]Faculty of Economics and Business, Sebelas Maret University, Jl Ir Sutami 36 A Solo, 57771, Indonesia
[*]Corresponding author: widagdo1998@staff.uns.ac.id

**Abstract.** The purpose of this article was to examine the influence of ownership structure, environmental performance, financial performance, and company size on carbon emissions reporting of high profile companies in Indonesia. Ownership structure was measured by foreign ownership, institutional ownership, and concentrated ownership, while environmental performance was measured by the Company Performance Rating Program in Environmental Management released by the Ministry of Environment (PROPER) and ISO 14001 certification. The disclosure of carbon emissions was assessed by an index drawn from the carbon disclosures project (CDP). High profiles companies were chosen as the sample since the companies create carbon emissions in their operation which contributes to global climate change. Utilizing 102 companies in 2019 as a sample, results of multivariate regression suggest that institutional ownership and size positively influence the level of carbon emission disclosure. These results indicate that larger companies and companies owned by an institution have a higher tendency to report carbon emissions either in the firm's annual report or in the sustainability reporting. The findings motivated the listed companies in reporting their carbon emission which is voluntary in nature. Such findings may give an understanding to the capital market regulatory body in releasing regulations that encourage listed companies to report their carbon emission.

**Keywords :** carbon emission, company regulation, ISO 14001

## 1. Introduction

The problem of climate change has pushed companies to become more responsible in operating their business. The main cause of climate change that many countries have to deal with is carbon emissions. Corporations provide a contribution to a substantial share of greenhouse emissions. As stakeholders are now very concerned with environmental matters, companies are mindful that their environmental accountability practices are valuable facts for stakeholders, including carbon pollution information. Disclosure of carbon emissions remains voluntary in nature by the Indonesia Stock Exchange (IDX). Given that environmental information concerns stakeholders, particularly customers, the company's governance process is assumed to ensure that more information about carbon emissions is needed to be disclosed in the company annual report or the sustainability reporting.

The purpose of this article is to investigate factors that may influence the extent of carbon emissions disclosures. As this type of information is voluntary, it is urgent to look at factors that affect the extent of disclosure. The factors that will be examined are ownership structure, environmental performance, leverage, and company size.

## 2. Literature review and hypotheses

### *2.1. Previous studies*
There are studies in Indonesia that examined the determinants of carbon emissions disclosures such as Putri and Yanto [1], Hermawan et al. [2], Budiharta and Kacaribu [3], and Chariri et al. [4]. The difference between the present study and previous literature is the sample and time frame (high profile companies with more recent data) and the examined variable, i.e. environmental performance that captures both ISO certification on the environment and PROPER, the environmental performance rating for companies assessed by the Ministry of Environment.

### *2.2. Hypotheses development*

*2.2.1. Foreign ownership.* The extent to which an organization discloses its environmental information could be affected by both the environmental culture of the nation in which it operates and the situation in the parent country [5]. According to legitimacy theory, multinational corporations share more details about their social and environmental activities than domestic corporations do [6]. Barako et al. [7] and Haniffa and Cooke [8] established a strong and favourable relationship between international ownership percentage and the amount of voluntary reporting. They assert that there is a greater need for transparency as a way of monitoring the management activities of foreign owners' subsidiaries. Furthermore, foreign investors may have a substantial impact on how they handle corporate financial statements to comply with foreign reporting standards [7]. We expect that companies with the international stock owner would share more knowledge about climate change-related problems than companies with domestic share owners. Hence, this study will test the following hypothesis.

H1: Foreign ownership positively influences the level of carbon emissions disclosure.

*2.2.2. Institutional ownership.* According to Choi et al. and Akbas et al. [9], [10] the increased institutional ownership would increase the company's monitoring and therefore expose more of the company's climate change disclosures. Institutional ownership has a considerable power to encourage the management to run the business in a sustainable manner which thereby increases the carbon emission disclosure level. Borghei-Ghomi and Leung [11] and Akbaş and Canikli [10] found a consistent positive association between institutional ownership and voluntary disclosure on carbon emission. Therefore, we predict the following hypothesis.

H2: institutional ownership positively influences the level of carbon emissions disclosure.

*2.2.3. Concentrated ownership.* Previous literature has identified ownership concentration, defined as the ownership of a company's capital stock by a small number of shareholders, as a determinant in voluntary disclosure [11]. Thus, when companies with a lower ownership concentration have a larger shareholder base, they will face increased pressure from shareholders to reduce the information asymmetry between them by disclosing as much as possible the company information [12]. In contrast, in highly concentrated ownership companies, public exposure is less so that social and environmental activities of the companies are also lesser. Gonzalez-Gonzalez and Ramirez [12] found a negative relationship between concentrated ownership and climate change disclosure level. Therefore, we expect the following hypothesis.

H3: Concentrated ownership negatively influences the level of carbon emissions disclosure.

*2.2.4. Environmental performance.* According to legitimacy theory, business practices are considered legitimate if they conform to societal standards, traditions, and beliefs [7] As a result, businesses must demonstrate their concern for social and environmental concerns to earn society's legitimacy. Carbon emission disclosure can be an effective way for businesses to demonstrate their commitment to environmental issues, especially the impact of carbon emissions. When businesses adopt environmental management programs, the level of carbon emission disclosure is perceived to rise. ISO certifications in the environmental sector, specifically ISO 14001, and The Ministry of Environment environmental performance evaluation called PROPER rating can be viewed as critical indicators that businesses have implemented environmental management systems. This means that firms certified to ISO 14001 will report more details about their carbon emissions than firms not certified to ISO 14001 will. Accordingly, companies that receive a good rating in environmental performance will report more on carbon emission of business operation than companies that receive a bad rating. Therefore, this study expects the following hypothesis.

H4: Environmental performance positively influences the level of carbon emissions disclosure.

*2.2.5. Leverage.* Pahuja [13] states that if an organization wishes to borrow from international donors, it must provide a report demonstrating the environmental consequences of their activities, the environment-damaging possibilities, and the company must identify potential environmental threats and establish adequate plans to deal with them. This is consistent with expectations that businesses with a higher leverage ratio have a greater inclination to provide environmental information than those with less debt. Thus, we will test the following hypothesis.

H5: Leverage ratio positively influences the level of carbon emissions disclosure.

*2.2.6. Company size.* Previous studies have identified a positive relationship between firm size and the extent of social and environmental reporting [2], [5], [14]. This favourable relationship is based on the notion that bigger firms, on average, are involved with a greater variety of industries and operate on a global scale. Additionally, bigger organizations must appease a greater range of customers who may be involved in the company's environmental management and policies. As a result, these businesses face greater social and administrative pressure to reveal environmental information than smaller businesses [5], [9]. Additionally, the environmental transparency process is expensive, and bigger firms are more likely to be able to invest time in preparing and disclosing environmental information than mid- to small-sized businesses [5]. Therefore, this study considers the following hypothesis.

H6: Company size positively influences the level of carbon emissions disclosure.

## 3. Research method

### *3.1. Sample selection and research data*

The present study investigates the level of carbon emissions disclosures of high profile companies listed on IDX in 2019. The identification of high profile companies refers to Rupley et al. [15] and Choi et al. [9] regarding companies working in emissions-intensive industries. Therefore, this study takes companies that are included in the sector of basic industry, energy, consumption (primary and non-primary), and transportation as classified under the IDX industry classification. There were 166 companies in these industries which were then selected as a sample based on the following criteria. First, the company annual report or sustainability reporting is available. Second, the company has a carbon emissions disclosure and a policy to reduce carbon emissions. Third, the company has ISO 14001 certification and or PROPER environment rating. These criteria produce 102 companies as the final sample for the study. The year 2019 was used because it was the latest data when this study began.

Data related to ownership structure, financial performance, environmental performance are available in the company annual report. While data related to carbon emissions disclosure are available either in the company annual report or in the sustainability reporting.

### *3.2. Carbon emission disclosure measurement*

This research utilizes an index based on Choi et al. [9] to assess the carbon emission disclosure level. The use of the index is consistent with previous studies in carbon emission disclosure such as Putri and

Yanto [1], Hermawan and Gunardi [16], Budiharta and Kacaribu [3], and Kalu et al. [17]. The index consists of 18 items which cover 5 carbon emissions disclosures theme namely Climate change: risks and opportunities, Greenhouse gas emissions accounting, Energy consumption accounting, Greenhouse gas reduction and cost, and Carbon emission accountability. Since the index is unweighted, the presence of a statement relates to the item in the index in the company carbon emission reporting is given a score of one; and zero otherwise. The carbon emission disclosure level of a company (CED) is the sum of the score of the index divided by 18, the maximum possible score of a company could achieve.

### *3.3. The measurement of independent variables*

The present study examines six independent variables that include three variables in ownership structure, environmental performance variable, leverage, and size of the company. Foreign ownership (FOR) was identified by the stock owned by foreign investors divided by outstanding stock. Institutional ownership (INS) was quantified by the stock owned by the institution divided by outstanding stock. Block holder ownership or ownership concentration (OC) was assessed by the sum of stock owned by substantial shareholders, i.e. shareholders or investors holding 5% of stock or more.

Environmental performance (EP) was assessed by ISO certification on environment received by the company, called ISO 14001 and or PROPER, the environmental performance rating for companies released by the Ministry of Environment. PROPER rating is based on colour, ranging from the lowest, black to the highest, gold. Black and red are a colour indicator that are below the standard. While blue means the company has made environmental management efforts, which are required according to the prevailing laws and regulations. Green and gold are rating indicators that are beyond the standard. While green means the company is beyond compliance, gold means the company has consistently demonstrated environmental excellence in its production and service processes and conducts business in an ethical and socially responsible manner. In this research, none of our samples get a black or red rating. Therefore, we take the blue colour as the baseline indicator. To measure environmental performance, this research employs a nominal variable. If a company gets blue rating, it receives a score of zero. If a company get a green color it receives a score of one. If a company has ISO 14001 certification it gets a score of two, and if a company achieves a gold rating, it receives a score of three. Some companies have a green ratings as well as ISO 14001 certification. For the scoring purposes, we give whichever is the highest. Financial performance was assessed by leverage ratio (LEV), i.e. total debt to the total asset. Meanwhile, this study utilizes the natural logarithm of a total assets to measure company size (SIZE).

### *3.4. Hypotheses testing*

To examine the influence of independent variables as mentioned above on carbon emissions disclosure, this study conducts multivariate regression analysis. The regression model is as follows. This study also assesses the classical assumption test for the validity of the regression results

$$CED = \alpha + \beta1FOR + \beta2INS + \beta3OC + \beta4EP + \beta5LEV + \beta6SIZE + e \quad (1)$$

## 4. Result and discussion

### *4.1. Descriptive statistic*

Table 1 presents the descriptive statistic of all variables. The minimum score of carbon emission disclosures is 4 represents 22% of total carbon emission disclosure in the index. The average score for the disclosure is 12.97 which counts for about 72 % of the index. The number of carbon emission disclosure in high profile industries is relatively high as compared to mining industry 24% [1] and non-financial companies 34% [3]. Probably, the time frame and the sample make the results arethe different.

As presented in the table, the mean score for foreign ownership and institutional ownership is 32.19% and 49.81% respectively. While block holder ownership is 75.27%. The result suggests that share ownership in IDX is highly concentrated, the share owned by the public is minimum. The finding is consistent with Brown [18] which states that ownership concentration in Indonesia is high and the shares remain in the hand of the founding family. Regarding environmental performance, the blue rating is achieved by the highest number of samples, 40 companies representing 39.2% of the sample. Thirty-five companies have ISO 14001 certification. While the number of samples who get a green and gold ratings in PROPER is 25 and 2 companies respectively. The average value of leverage and company size is 3.13 and 29.71 respectively.

**Table 1.** Descriptive statistics

| | Minimum | Maximum | Mean | Standard Deviation |
|---|---|---|---|---|
| CED | 4 (22%) | 18 (100%) | 12.97 (72%) | 3.25 |
| FOR | 0.00 | 99.77 | 32.19 | 31.03 |
| INS | 0.00 | 98.31 | 49.81 | 31.98 |
| OC | 22.67 | 100 | 75.27 | 16.63 |
| LEV | 1.72 | 23.91 | 3.13 | 3.03 |
| SIZE | 26.98 | 33.49 | 29.71 | 5.251 |
| EP | 0 | 3 | 0.99 | 1.55 |
| Frequency of EP | 0 = 40 | 39.2% | | |
| | 1 = 25 | 24.5% | | |
| | 2 = 35 | 34.3% | | |
| | 3 = 2 | 2 | | |
| N | 102 | 102 | 102 | 102 |

*4.2. Results of hypotheses testing*

To investigate the influence of ownership structure, environmental performance, leverage, and company size on the carbon emissions disclosure level of the sample company, this study utilizes multivariate regression. The results of the test are portrayed in Table 2.

**Table 2.** Result of multivariate regression analysis

| | B | T | Sig | Tolerance | VIF |
|---|---|---|---|---|---|
| Constant | -9.935 | -1.394 | 0.166 | | |
| FOR | 0.016 | 1.214 | 0.228 | 0.551 | 1.814 |
| INS | 0.021 | 1.702 | 0.032** | 0.563 | 1.777 |
| OC | -0.011 | -0.570 | 0.570 | 0.803 | 1.245 |
| EP | 0.394 | 1.180 | 0.241 | 0.975 | 1.026 |
| LEV | -0.042 | -0.434 | 0.666 | 0.967 | 1.034 |
| SIZE | 0.736 | 3.365 | 0.001*** | 0.776 | 1.288 |
| F Value | | | 3.965 | | |
| F sig | | | 0.001*** | | |
| Adjusted $R^2$ | | | 0.150 | | |
| Normality test Sig | | | 0.589 | | |

Note: *** refers significant at 1%; ** refers significant at 5%.

Based on Table 2, institutional ownership positively influences the carbon emission disclosure level, significant at 5% level. It suggests that institutional investors require environmental friendly business operation that in turn increases the level of carbon emission disclosure of the company. The results are consistent with the study in Australia [9] and Turkey [10] which found that institutional ownership are likely to improve carbon emissions disclosure. The results however are not similar to studies in Indonesia [2] and [1]. Probably, the difference in the sample and period of observation contribute to the inconsistent finding.

Table 2 assert that company size also has a positive relationship with the carbon emissions disclosure level. It indicates that the larger the asset the higher the level of carbon emission disclosures. The result is consistent with [3], [9] which found that companies have bigger asset tend to produce more carbon emission disclosure in their annual report or sustainability reporting.

Table 2 also documents that foreign ownership, concentrated ownership, environmental performance, and leverage do not have any impact on the level of carbon emissions disclosure in high profile companies in IDX. The result that leverage does no influence on carbon emission reporting is consistent with [10].

The value of adjusted $R^2$ in Table 2 is 0.15 suggesting that ownership concentration, environmental performance, leverage, and company size may predict around 15% of the variability of the factors influencing carbon emission disclosures. The remaining 85% represents the variables that are not investigated in our regression model. In relation to the classical assumption test, the regression model passes the normality assumption assessment, as indicated by the normality test Kolmogorov Smirnov sig value that 0.589, exceed the cutting point of 0.05. The regression model also free from multicollinearity since the VIF (variance inflation factor) and Tolerance less than ten and more than 0.1 respectively.

## 5. Conclusion and limitation

The examination of the factors influencing the carbon emission disclosure level of high profile companies in Indonesia reveals that institutional ownership and size are positively associated with the disclosure level. Carbon emissions produced by the companies contribute to the global climate change. The results indicates that institutional investors encourages the management to report more on carbon emission disclosures in the annual report or sustainability reporting. The results also imply that companies whose resources are greater as indicated by the total assets will have more incentives to conduct sustainable business by reporting environmental preservation activities in carbon emissions disclosures.

This research has limitation that may be improved in the next research. The adjusted $R^2$ is relatively low (15%) which can be increased by adding more independent variables. There are some factors such as corporate governance characteristics, financial ratios, and firm value that can be explored in explaining the motivation of carbon emissions reporting.

## References

[1] Halimah N P and Yanto H 2018 Determinant of Carbon Emission Disclosure at Mining Companies Listed in Indonesia Stock Exchange *KnE Soc. Sci.* **3** 127

[2] Hermawan A, Aisyah I S, Gunardi A and Putri W Y 2018 Going green: Determinants of carbon emission disclosure in manufacturing companies in Indonesia *Int. J. Energy Econ. Policy* **8** 55–61

[3] Budiharta P and Kacaribu H E P B 2020 The Influence of Board of Directors, Managerial Ownership, and Audit Committee on Carbon Emission Disclosure: A Study of Non-Financial Companies Listed on BEI *Rev. Integr. Bus. Econ. Res.* **9** 75–87

[4] Chariri A, Januarti I and Yuyetta E N A 2018 Audit Committee Characteristics and Carbon Emission Disclosure *E3S Web Conf.* **73** 1–5

[5] Monteiro S M da S and Aibar-Guzmán B 2010 Determinants of environmental disclosure in the annual reports of large companies operating in Portugal *Corp. Soc. Responsib. Environ. Manag.*

**17** 185–204
[6] Kiliç M 2016 Online corporate social responsibility (CSR) disclosure in the banking industry: Evidence from Turkey *Int. J. Bank Mark.* **34** 550–69
[7] Barako D G, Hancock P and Izan H Y 2006 Corporate Disclosure by Kenyan Companies *Corp. Gov.* **14** 107–25
[8] Haniffa R M and Cooke T E 2002 Culture, corporate governance and disclosure in Malaysian corporations *Abacus* **38** 317–49
[9] Choi B B, Lee D and Psaros J 2013 An analysis of Australian company carbon emission disclosures *Pacific Account. Rev.* **25** 58–79
[10] Akbaş H E and Canikli S 2019 Determinants of voluntary greenhouse gas emission disclosure: An empirical investigation on Turkish firms *Sustain.* **11** 107
[11] Borghei-Ghomi Z and Leung P 2013 An Empirical Analysis of the Determinants of Greenhouse Gas Voluntary Disclosure in Australia *Account. Financ. Res.* **2**
[12] Gonzalez-Gonzalez J M and Ramirez C Z 2016 Voluntary carbon disclosure by Spanish companies: an empirical analysis *Int. J. Clim. Chang. Strateg. Manag.* **8** 57–79
[13] Pahuja S 2009 Relationship between environmental disclosures and corporate characteristics: A study of large manufacturing companies in India *Soc. Responsib. J.* **5** 227–44
[14] Ika S R, Nugroho J P, Achmad N and Widagdo A K 2021 The impact of corporate governance on environmental reporting : Evidence from the Indonesian manufacturing industry *IOP Conf. Ser. Earth Environ. Sci.* **739** 1–7
[15] Rupley K H, Brown D and Marshall R S 2012 Governance, media and the quality of environmental disclosure *J. Account. Public Policy* **31** 610–40
[16] Hermawan A and Gunardi A 2019 Motivation for disclosure of corporate social responsibility: Evidence from banking industry in Indonesia *Entrep. Sustain. Issues* **6** 1297–306
[17] Kalu J U, Buang A and Aliagha G U 2016 Determinants of voluntary carbon disclosure in the corporate real estate sector of Malaysia *J. Environ. Manage.* **182** 519–24
[18] Brown R A 2006 *The Rise of the Corporate Economy in Southeast Asia* (London and New York: Routledge)

# The readiness of local regulations in supporting low carbon development in Central Java Indonesia

**Rachman Djamal, Alfina Handayani* and Tri Risandewi**

Regional Planning, Research and Development Agency of Central Java (Bappeda Provinsi Jawa Tengah), Jl. Pemuda No 127-133 Semarang, Central Java, Indonesia
*Corresponding author: alfinahandayani16@gmail.com

**Abstract**. During the last two decades, Indonesia has experienced rapid economic progress and a reduction in poverty rates. On the other hand, the impact of economic growth causes many problems to the environment. Therefore, the Central Government developed the concept of low-carbon development. Low Carbon Development is a new development platform that aims to sustain economic and social growth through low greenhouse gas (GHG) emission development activities and minimize natural resource exploitation. This study aims to determine what extent the regulation readiness of Central Java Province on supporting low carbon development. The approach used in this research is descriptive qualitative. Three administrative regions (Semarang City, Sukoharjo Regency, and Cilacap Regency) were empirical observation areas. FGDs were carried out in the three regions to explore various information on programs related to LCD readiness and its implementation. The results indicate that local governments already have regulations and policies that support low-carbon development. However, they still often encounter many problems, such as limited authority and resources of local government, the dualism of interest, policy implementation was still in the planning stage, and there is no synergy between regional agencies.

**Keywords** : low carbon development, Central Java, regulation readiness

## 1. Introduction

Over the last two decades, Indonesia has experienced rapid economic progress and poverty reduction. In 2000-2019, the growth rate of Indonesia's Gross Domestic Product was between 5-6% per year [1] [2][3]. On the other hand, the impact of economic development has also resulted in many environmental problems in Indonesia [1][4][5]. In addition, development also contributes to climate change [6]. The results of the calculation of the national greenhouse gas (GHG) inventory show that the level of GHG emissions in 2019 was 1,866,552 Gg CO2e, an increase of 680,324 Gg CO2e compared to the emission level in 2000 [7].

As a part of Indonesia's commitment to climate change mitigation, the Ministry of Central Planning (Bappenas) had formulated a policy of "Low Carbon Development Planning (LCDP)" this initiative objection is to achieve inclusive economic growth through equitable development and poverty alleviation efforts, at the same time, maintaining both environmental quality and the availability of natural resources. Low Carbon Development supports the achievement of the 13th Sustainable Development Goals (SDGs), which is about Climate Action.

Low Carbon Development (LCD) is a new development platform that aims to maintain economic and social growth through development activities with low GHG emissions and reducing the exploitation of natural resources [1]. The outcome of low-carbon development can be measured by 1) Intensity of emission/carbon and 2) Reduction of GHG emission/absorption [1] .

The basic principles of LCD are: first, implementing evidence-based policies and applying an integrated systems-based approach. Second, placing carrying capacity (including GHG emissions) is a significant part of preparing and planning development targets. Third, emphasizing trade-off policy analysis to balance economic and social development goals with environmental management objectives. Fourth, applying HITS (Holistic, Integrated, Thematic, Spatial) principles, and fifth, active involvement of development planners with environmental stakeholders.

The Central Government's commitment to low carbon development started in 2020 or existed in the 2020-2024 RPJMN. Meanwhile, the Central Java Provincial government has formulated and started implementing its development agenda as stated in the 2018-2023 RPJMD. To encourage the transition and to support the achievement of the national targets in the context of LCD, the study was needed to find out how the regulations were applied by the local government.

Previous research showed that local governments already had the regulations and policies to reduce the impact of climate change. Meanwhile, a strong commitment was needed [8]. Research conducted by Qodriyatun [9] regarding regional readiness in implementing central government policies suggests that coordination between the central and local governments was urgently needed in every activity plan accompanied by regulatory support. Furthermore, even in developed countries like Australia, local governments were not effectively planning for climate impacts [10]. In addition, the lack of insight and mindset about the importance of sustainable development is also one of the obstacles in implementing policies that support the LCD [11]. However, there was limited information related to regional readiness for low-carbon development.

Therefore, this study aims to identify the extent to which regulations are prepared by the Central Java Provincial Government, especially in the three study areas, namely Semarang City, Sukoharjo Regency, and Cilacap Regency supporting low-carbon development. The readiness of regional regulations, in this case, is reviewed in a broader scope, not only limited to legal instruments but also concerning strategies and programs for the development of the Central Java Province as well as the problems faced.

## 2. Method

This research was carried out from August to November 2019 in 3 areas that were Semarang City, Sukoharjo Regency, and Cilacap Regency. Those were selected purposively referring to the index map of climate change vulnerability of Central Java Province (KLHK, 2015). In addition, the location selection refers to other pieces of literature from previous research, BPS data, planning documents for the district/city of the research location.

This research is qualitative descriptive research, using primary data. Data were collected through two steps: pre-survey and Focus Group Discussions (FGDs). The pre-survey was to obtain initial information from key informants, to identify the problems in implementing the climate change program. The results of the preliminary survey were research design and research instruments. The next stage after the pre-survey was FGD. The objective of the FGD was to deeply identify the implementation of climate change actions in terms of planning, program, strategy. FGD involved various stakeholders, such as local government agencies, universities, state-owned enterprises, and NGOs. The collected data were edited, summarized, and analyzed to achieve the objective of the study.

## 3. Results and discussion

### *3.1. Research site profile*

The city of Semarang, which is the capital of Central Java, is one area that contributes to greenhouse gas emissions, especially in the transportation and energy sectors. It contributed 4,011,692.22 tons of $CO_2e$ carbon emission in 2018, which was the most contribution obtained from the energy and transportation sector of 3,270,166, 57 tons of CO2e (82.07%). Besides transportation, the existence of industrial areas contributes to climate change, they are scattered in several places. Furthermore, the rapid development of the city influences the increased occupancy rate. It also has the potential to contribute to greenhouse gas emissions from the waste sector.

Sukoharjo Regency is one of the most emission contributors from the industrial sector in Central Java Province. The total carbon emission of Sukoharjo Regency in 2018 was 5,201.23 CO2e. The energy and agriculture sectors contributed 4,528.42 CO2e (87.3%), and 479.94 CO2e (8.99%). This district was first familiar with the issue of climate change and has developed policies related to climate change.

Cilacap Regency is the district with the highest emitter (30%) in Central Java Province. Its carbon emission contribution is 21,400,706.76 tons of CO2e, the largest of which comes from the Industrial Processes and Product Use (IPPU) sector. In the context of low-carbon development, Cilacap Regency has the potential to reduce emission/carbon intensity.

The Central Java Provincial Government already has a GHG regional action plan [12], but the District/City has not prepared RAD-GRK because it is not mandatory, and there are no specific guidelines in the Regional Strategic Plan. In addition, it was still waiting for a review of the Provincial action plan-greenhouse gas RAD-GRK as a guide for local governments. In the transition process from RAD-GRK to LCDP, the three research sites had never received assistance in the preparation of action plans for low-carbon development areas.

### *3.2. Regional Regulation Readiness Condition*

Regional readiness conditions in supporting low-carbon development regarding conformity aspect with planning documents, regulations, policies (programs/activities), and the obstacles faced. Collecting data using FGDs were then analyzed and shown in Table 1.

**Tabel 1**. Profile of manucipality/ regencies on readiness in supporting low carbon development in central java

| Municipalities/ Regencies | Readiness Condition | | | |
|---|---|---|---|---|
| | Conformity aspect with planning documents | Regulations | Policies (programs/ activities) | Obstacles |
| Semarang Municipality | suitable | not optimal | support | - Limited authority of local government,<br>- Dualism of interest |
| Sukoharjo Regency | suitable | not optimal | support | - Limited human resources<br>- Policy implementation was still in the planning stage, Limited emission data<br>- There was no synergy between local government organizations |
| Cilacap Regency | suitable | not optimal | support | - Not optimal integration of Low Carbon Development Planning Guidelines into Planning documents<br>- Nomenclature<br>- There was no preliminary study in determining the national strategic project |

Source: Primary data, 2019

*3.2.1. Readiness of the Semarang Regency Government.* The readiness of the Semarang City Government in supporting low-carbon development is already implied in the vision contained in the 2005-2025 long-term development planning document (RPJPD), which is to realize "Semarang Kota Metropolitan yang Religius, Tertib, dan Berbudaya" and also clarified in the vision of the Semarang City RPJMD 2016-2021, namely: "Semarang Kota Perdagangan dan Jasa yang Hebat Menuju Masyarakat semakin sejahtera".This vision was followed up with the 3rd mission "Mewujudkan kota metropolitan yang dinamis dan berwawasan lingkungan" where development was prioritized on optimizing the use of spatial planning and increasing the development of regional infrastructure that is planned, aligned, harmonious, balanced, and equitable while still taking into account the concept of environmentally sound and sustainable development.

Several regulations that have been set by the Semarang City Government in supporting low-carbon development include Regional Regulation no. 13 of 2006 concerning Environmental Control, Regional Regulation no. 14 of 2011 concerning the Spatial and Regional Planning of the City of Semarang in 2011-2031, Regional Regulation no. 11 of 2017 concerning Amendments to the Regional Regulation of Semarang City No. 6 of 2016 concerning the RPJMD of Semarang City for 2016-2021, and also the Regulation of the Mayor of Semarang No. 24 of 2019 concerning Green Buildings. Some of these regulations are a form of commitment from the Semarang City Government to support the reduction of greenhouse gas emissions and low-carbon development.

The Semarang City Government's policies in supporting Low Carbon Development are followed up with the implementation of activity programs that have been listed in the Semarang City RPJMD 2016-2021 document, namely:

1. Water Resources Development and Maintenance Program;
2. Flood Control Program;
3. Road and Bridge Construction and Maintenance Program;
4. Traffic Control and Security Program;
5. Transportation Service Improvement Program;
6. Program to Improve Quality and Access to Information on natural resources and the environment;
7. Waste Management Performance Development Program;
8. Program for Capacity Building for climate change mitigation and adaptation;
9. Residential environment facilities and infrastructure program;
10. Green Open Space management program;
11. Program to improve the quality and reach of wastewater;
12. Drinking water supply and management program;
13. Disaster prevention and preparedness programs

In addition to these programs, several activities that have been and will be implemented by the Semarang City Government in supporting the reduction of greenhouse gas emissions and low-carbon development include the new renewable energy sector: development of waste processing into electrical energy in Jatibarang, power plant micro-hydro in Banyumanik, development of solar power plants in transportation sector: use of bus fleets 70% gas fuel and 30% fuel oil, smart driving, parking lot arrangement, intelligent transport system development, agriculture sector: low emission rice cultivation, biogas development , use of organic fertilizers and urban greening. The obstacles faced by the Semarang City Government in low-carbon development and reducing greenhouse gas emissions are the limited authority possessed by the local government so that there are several sectors such as energy and mining which are the authority of the provincial government, the dualism of interests where the region requires an industrial area because able to absorb labor but on the other hand the existence of an industry that is not pro-environment causes environmental degradation to be higher.

*3.2.2 Readiness of the Sukoharjo Regency Government.* The readiness of the Sukoharjo Regency Government in implementing Low Carbon Development can be seen in the planning documents of the District Long Term Development Plan (RPJPD). Sukoharjo 2005-2025 and the Medium Term

Development Plan (RPJMD) of Sokoharjo Regency for 2016-2021 as well as several regulations, both regional regulations and the Regent's Circular on the environment. In the Sukoharjo Regency RPJPD document, to realize the Vision of "-*Sukoharjo yang Maju, Adil, dan Makmur* ", one of the missions carried out is to create sustainable development by preserving the environment. The main target to be achieved by the Sukoharjo Regency Government is to realize optimal management of natural resources and the environment while maintaining the preservation of functions in supporting life with the direction of development to realize sustainable and competitive natural resources and the environment. In addition, the medium-term planning document of the Sukoharjo Regency RPJMD 2016-2021 also includes a mission to strengthen regional economic independence by mobilizing regional superior sectors by realizing natural resource conservation, the environment and disaster management.

Related to regulations that support low-carbon development, the Sukoharjo Regency Government has made Regional Regulation No. 660.1/1396 of 2017 concerning the Appeal for the Establishment of Climate Villages in sub-districts throughout Sukoharjo Regency. In addition, the response of the Sukoharjo Government in dealing with climate change deserves appreciation, namely carrying out an inventory of greenhouse gases from the energy sector, Industrial Processes and Product Use, agriculture, forestry and waste. This action was followed up with the formation of a Working Group Team on Greenhouse Gases based on the Decree of the Regent of Sukoharjo and in 2018 carried out an inventory of GHG emissions using the SIGNSMART application of the Ministry of Environment and Forestry.

In the Sukoharjo Regency RPJMD document, there are several programs and activities that support climate change management and low-carbon development, including (DLH Sukoharjo Regency, 2019):

1. Climate Village Program (Proklim);
2. Reforestation Program (procurement of seeds and distribution);
3. Waste Reduction Program (restriction of waste generation, waste utilization and recycling);
4. Waste Handling Program (use of waste as raw material for 3 Reduce, Recycle, Reuse Waste Disposal Sites, Integrated Waste Disposal Sites, and compost houses as well as energy sources for biodigesters and thermal processes);
5. Natural Resources and Soil Conservation Program (procurement of infiltration wells, biopori, soil damage status of biomass production in each sub-district);
6. Adiwiyata School Program.

The Climate Village Program is a national scope program managed by the Ministry of Environment and Forestry to increase the involvement of the community and other stakeholders to strengthen adaptation capacity to the impacts of climate change and the reduction of greenhouse gases. The program for the establishment of the Climate Village of Sukoharjo Regency which is in line with the central government's policy in tackling climate change and is also able to provide benefits both from an economic perspective (adequacy of clean water Rp. 14,200/day, food security Rp. 8,013/day, waste management Rp. 2,450/day ) and able to save energy up to Rp. 1,820/day or an average of Rp. 16,763/day or Rp. 502,890/month (based on the calculation of the Environmental Service of Sukoharjo Regency, 2019).

The valuable benefits, that they cannot be counted by the community, were health, brotherhood, kinship, andas well as environmental security. Even the Regent of Sukoharjo also issued Circular No. 660.1/1396 regarding the establishment of a Climate Village in the sub-districts of the Sukoharjo Regency. In 2018, there were 22 climate villages formed in Sukoharjo, and one of them even received an award from the Ministry of Environment and Forestry as the National Main Proklim for Climate Village, Ngadirejo Village, Kartasura District. In 2019, the District Government. Sukoharjo has formed 31 climate villages and is the largest in Central Java Province.

In addition to the Climate Village program, the Government of Sukoharjo also has several programs related to low-carbon development, namely (Bappelbangda Kab. Sukoharjo, 2019):

1. Increased Application of Livestock Technology
2. Residential Healthy Environment
3. Protection Conservation of natural resources
4. Improving the Quality and Access to Information on natural resources and the environment
5. Improved Pollution Control
6. Green Open Space Management
7. Forest and Land Rehabilitation
8. Control of Environmental Pollution and Destruction

The implementation of regional policies that support LCDs includes the preparation of Regional Action Plans for Greenhouse Gases, preparation of Strategic Environmental Assessment documents in the Regional Spatial Plan and RPJMD, preparation of Regional Environmental Management Performance Information, Regional Natural Resources Accounts, and Regional Environmental Status. Through several implementations of these policies, it is hoped that they will be able to provide data support, information systems, and planning documents. Constraints faced include limited human resources, no support for regulations from the Central Government, limited data availability that makes it difficult for regional apparatus organizations to fill out the GHG emission application system of the Ministry of Environment and Forestry, and there is no synergy between regional organizations. related to low carbon development.

*3.2.3. Readiness of the Cilacap Regency Government.* The readiness of the Cilacap Regency Government in implementing Low Carbon Development can be seen in the Regional Long Term Development planning document (RPJPD) Kab. Cilacap 2005-2025 and Cilacap Regency Medium Term Development Plan (RPJMD) 2017-2022. Implicitly in the vision of the RPJPD "Cilacap yang Berbudaya dan Sejahtera" has supported the concept of the LCD. Regarding the vision of the Cilacap Regency RPJMD, namely More Prosperous Equitable " Bangga Mbangun Desa", especially in the 5th mission, namely Developing and building regional infrastructure by paying attention to environmental aspects in the sustainable use of natural resources. The targets to be achieved include increasing disaster preparedness, increasing the growth of leading sectors, and realizing a quality living environment.

The readiness of the Cilacap Regency Government in implementing Low Carbon Development can be seen in the Regional Long Term Development planning document (RPJPD) Kab. Cilacap 2005-2025 and Cilacap Regency Medium Term Development Plan (RPJMD) 2017-2022. Implicitly in the vision of the RPJPD "*Cilacap yang Berbudaya dan Sejahtera*"" has supported the concept of the LCD. Paying attention to the vision of the Cilacap Regency RPJMD, namely More Prosperous Equitable "Proud to Build Villages", especially in the 5th mission, namely Developing and building regional infrastructure by paying attention to environmental aspects in the sustainable use of natural resources. The targets to be achieved include increasing disaster preparedness, increasing the growth of leading sectors, and realizing a quality living environment development that contributes to Greenhouse Gases.

The preparation of the Cilacap Regency RPJMD has also taken into account environmental analysis with Strategic Environmental Study (SEA) and integrated the issue of climate change. Support for the LCD can be seen in several regulatory products, such as the Cilacap Regent's Regulation Number 256 of 2018 concerning the 2018-2022 SDGs regional action plan.

1. The policies of the Cilacap Regency Government in supporting Low Carbon Development are followed up with the implementation of activity programs that have been listed in the Semarang Regency RPJMD document for 2018-2022, namely:
2. Natural resource protection and conservation program
3. Environmental management
4. Community capacity building in environmental management

5. Control of pollution and environmental damage
6. Supervision and determination of environmental law
7. Waste management performance development

Based on the results of discussions in the FGD, the Cilacap Regency Government strongly supports the concept of low carbon development and has carried out several collaborations, including with Pertamina to implement more sustainable development. However, there are still some obstacles that are faced, namely (1) The integration of Low Carbon Development Planning Guidelines into the Development Planning document is not yet optimal (needs a 'forced' policy). (2) The nomenclature, Codification of Programs, in the development planning guidelines (Permedagri No. 13 of 2006 and Permendagri No. 90 of 2019) have not been fully 'optimized' for local government agencies to be used as guidelines for the preparation of Strategic Plans and Work Plans (Cross-Sector Affairs), (3) Determination of PSN (Central Government Policy) needs to be preceded by a study of the carrying capacity and capacity of the locus of activity, especially in industrial development that contributes to Greenhouse Gases.

## 4. Conclusions

The Semarang City Government, Sukoharjo Regency and Cilacap Regency have conformed the concept of low carbon development with regional long- and medium-term planning documents. The readiness of regulations related to low-carbon development is manifested in various policies, such as Regional Regulations on waste management, green buildings, environmental control, circulars calling for the establishment of climate villages in sub-districts. Constraints faced by local governments include limited local government authority, dualism of interests, limited human resources, policy implementation is still in the planning stage, limited emission data, and there is no synergy between local government agencies. Recommendations that must be made by local governments are to increase commitment and understanding in the implementation of the LCD and to develop guidelines as a basis for preparing regional action plans in districts/cities.

## References

[1] Kementrian PPN/Bappenas 2019 *Low Carbon Development: A Paradigm Shift Towards a Green Economy in Indonesia*
[2] Soleh A 2015 Pertumbuhan Ekonomi Dan Kemiskinan Di Indonesia *EKOMBIS Rev. J. Ilm. Ekon. dan Bisnis* **2** 197–209
[3] Central Bureau of Statistics 2020 Pendapatan Nasional (National Income of Indonesia) *Buku Publ. Stat.* **1** 1–186
[4] Indarto K D and Rahayu S 2015 Dampak Pembangunan Perumahan Terhadap Kondisi Lingkungan, Sosial Dan Ekonomi Masyarakat Sekitar Di Kelurahan Sambiroto, Kecamatan Tembalang *Tek. PWK (Perencanaan Wil. Kota)* **4** 428–39
[5] Putri S F 2020 Hubungan Pembangunan Ekonomi Terhadap Kualitas Lingkungan Hidup Di Provinsi Jawa Timur *J. Din. Ekon. Pembang.* **2** 58
[6] Gomez-Echeverri L 2018 Climate and development: Enhancing impact through stronger linkages in the implementation of the Paris Agreement and the Sustainable Development Goals (SDGs) *Philos. Trans. R. Soc. A Math. Phys. Eng. Sci.* **376**
[7] Kementerian LHK 2021 Lapoaran Inventarisasi Gas RUmah KAca (GRK) dan Monitoring, Pelaporan, Verifikasi (MPV) **6** 951–2
[8] Handayani A 2021 Local government response to the impacts of climate change: A review of climate change strategies in Pekalongan, Central Java, Indonesia *IOP Conf. Ser. Earth Environ. Sci.* **724**
[9] Qodriyatun S N 2017 Kesiapan Pemerintah Daerah Provinsi Riau Dan Sumatera Selatan Dalam Pelaksanaan Kebijakan Restorasi Gambut *Aspirasi* **8** 113–32
[10] Baker I, Peterson A, Brown G and McAlpine C 2012 Local government response to the impacts of climate change: An evaluation of local climate adaptation plans *Landsc. Urban*

*Plan.* **107** 127–36
[11] Tay R and Rusmiwari S 2019 Implementasi Kebijakan Pembangunan Berkelanjutan *J. Ilmu Sos. dan IlmuPolitik* **8** 217–22
[12] GIZ 2013 Kajian kesiapan Pelaksanaan RAD-GRK Propinsi Jawa Tengah 246

**Acknowledgement**
We would like to thank Dr. Rukuh Setiadi from Diponegoro University and all who had been involved and assisted in this research.

# The legal framework for climate change on power generating policy in Indonesia

**Maskun[1]*, Marthen Napang[1], Sri Susyanti Nur[2], Achmad[2], Naswar[2], A Ilmar[2], and H Assidiq[3]**

[1] International Law Department, Faculty of Law, Hasanuddin University, Jl. Perintis Kemerdekaan km 10, Makassar, South Sulawesi 90245 Indonesia
[2]Civil Law Department, Faculty of Law, Hasanuddin University, Jl. Perintis Kemerdekaan km 10, Makassar, South Sulawesi 90245 Indonesia
[3]Assistant Researchers, Faculty of Law, Hasanuddin University, Jl. Perintis Kemerdekaan km 10, Makassar, South Sulawesi 90245 Indonesia
*Corresponding author: maskunmaskun31@gmail.com

**Abstract.** The energy sector is the second largest contributor to greenhouse gas emissions after agriculture and forestry. The issue of increasing greenhouse gas emissions resulting from the 35,000-Megawatt power plant program is undoubtedly related to Indonesia's commitment to the 2015 Paris Agreement. The purpose of this study is to identify the legal framework of the Indonesian government's policy towards the 35,000-Megawatt power generation program. The method used is normative legal research conducted by reviewing various power generating-related legal instruments and principles in Indonesia. The results show that constructing a power plant based on coal will affect Indonesia's commitments in the 2015 Paris Agreement. Indonesia has not yet passed the draft Renewable Energy Law; therefore, it is necessary to harmonize some Indonesian legal instruments to deal with this issue. In connection to climate change, coal is one of the significant contributors to climate change. The National Electricity Company (PLN) and IPP power plants need to increase electricity growth by 8.6%, increasing the number of power plants to 80,538 MW from 48,000 MW. In this case, the need for coal fuel in power plants is 60%, the rest comes from new renewable energy (water, geothermal, and solar) by 5%, and the remaining 45% comes from other premier fuels (diesel and petroleum).

**Keywords:** Indonesian power generating policy, climate change, legal framework

## 1. Introduction

The environment is an essential factor in human life and all living things in this world. A series of revolutions have occurred in the political, economic, and social fields in several countries, none of which can significantly change human values and behaviour in carrying out life as big as the contribution of the environmental revolution. Environmental issues became a global issue and began to be noticed by activists of international relations after the cold war.

One of the environmental issues that have a significant influence on all components of life and living systems for many people today is the phenomenon of climate change. Climate change is present as a phenomenon of environmental damage that impacts almost every field of life that threatens the existence of human life, both at the local, national, and global levels [1]. Climate change occurs due to the process of global warming, namely the increase in the average temperature of the earth's surface temperature due to the accumulation of heat retained in the atmosphere as a result of the greenhouse effect [2].

The Intergovernmental Panel on Climate Change (IPCC) reports that climate change is already happening. Earth's temperature has increased by about 0.8 degrees Celsius over the last century [3]. In the previous three decades, the conditions have been warmer than in the last decade. The increase in global CO2 has increased significantly from year to year. The rate of growth in CO2 concentration in 1960 was around 0.7 ppm/year and improved to 2.38 ppm/year in 2014 [4].

It is recorded that at least an increase of 80% from the initial amount in 1970 was 21 Gigatons to 38 Gigatons in early 2004. This increase was accounted for by contributing 77% of the total greenhouse gas emissions produced by humans in 2004 as a whole. The increase in the period 1995 to 2004 was recorded at the highest intensity compared to 1970-2014 [5].

The IPCC report states that increases in global welfare primarily drive greenhouse gas (GHG) emissions. Compared to the pre-industrial era, the average temperature will increase by 3-5 degrees Celsius by the end of this century. Significant increases came from the energy use, transportation, and industrial sectors. Meanwhile, other contributors in a smaller amount came from households, the forestry sector, and the plantation sector [5].

Rob Jackson had issued a new report on carbon dioxide emissions which reached a record high in 2019. According to Jackson, 37 billion tons of carbon dioxide was triggered by the great demand for oil and natural gas. It means that global carbon emissions rose 0.6 percent from the 2018 data, which was 2.1 percent. Jackson and other Global Carbon Project researchers warn that carbon emissions will continue to increase over the next decade unless policies on energy, transportation and industry change significantly [6].

The energy sector is the second-largest contributor to GHG emissions (40%) after agriculture and forestry (45%). Ministry of Environment and Forestry Smart Data Sign showing power plant-sourced mainly from coal accounted for a third (200 million tons of CO2) of total energy sector emissions in 2014 [6].

The end of 2015 became a momentum for global action in fighting climate change through the Conference of the Parties (COP) 21 in Paris. In his speech at the meeting, President Joko Widodo stated that as one of the largest forest-owning countries that are the lungs of the world, Indonesia has chosen to be part of the solution. Indonesia is committed to reducing GHG emissions by 29% in 2030 under a business-as-usual scheme, and 41% receive international financial support [2]. In the same year, the government has planned to increase the electricity supply by 35,000 MW in 2019 as a promise for developing the Nawa Cita under the Jokowi-JK administration [7]. Meet national electricity needs with an initial plan of 60% in 2019.

According to Greenpeace data, 60% or around 22,000 MW will be generated from coal-fired power plants (PLTU). Every 2,000 MW of coal-fired power plants will produce an average of 10.8 million tons of carbon emissions per year, in the range of 2019 to 2030 if all coal-fired power plants operate in 2019 under the government's target, this power plant will produce carbon emissions of 1.3 Giga Tons which of course will cause more increasing concentration of greenhouse gas emissions [8].

There is a contradiction between the government's commitment to reduce GHG emissions and the plan development action in the energy sector, namely the construction of a 35,000-Megawatt power plant, mainly using coal as fuel. With the tendency to increase GHG emissions plus the government's plan to use coal-fired energy sources on a large scale, GHG emissions will double, which is undoubtedly a significant contributor to climate change. The existence of different circumstances that tend in the opposite direction raises the question of the climate change legal framework for the Indonesian government's 35,000-Megawatt power generation program.

## 2. Research methods

This research is normative-empirical, using data collection techniques, namely library research, by studying, exploring, and analyzing many legal materials. The types of data are primary data, namely direct data obtained in the field and secondary data, namely data obtained from the results of library studies. The data were analyzed and presented qualitatively to analyze data obtained from field and library studies.

## 3. Results and discussion

### *3.1. The 35,000-megawatt power generation program*

The development of electricity facilities in Indonesia is stated in the Business Plan for the Provision of Electricity (RUPTL) 2016 to 2025, prepared by the Ministry of Energy and Mineral Resources (ESDM) of the Republic of Indonesia made PT. The State Electricity Company (PLN) Persero is the main motor for implementing electricity supply activities in Indonesia. The government's plan to encourage electricity sufficiency is through the 35,000-megawatt (35 gigawatts) program until 2019.

Ownership of power plant development projects planned in the RUPTL is adjusted to PLN's funding capacity as a corporation. Considering the substantial investment needs in the electricity sector, PLN is deemed unable to build all the necessities for new power plants unilaterally. In dealing with this, the government divides to provide opportunities for PLN / IPP to participate in the construction of some power plant projects, or by non-IPP third parties with a business model certain activity such as power wheeling, as well as excess power cooperation, determination of separate business areas [9].

Until 2015, the installed capacity of PLN and IPP power plants in Indonesia was 48,000 MW consisting of 44,824 MW in the Java-Bali system and 10,091 MW in the Sumatra electricity system and 4,150 MW in Eastern Indonesia. When considering the rental power plant of 3,703 MW, the installed capacity of the generator will be 51,348 MW. With an estimated increase in the economy in 2016-2025 that affects the demand electrification, the government plans to increase electricity growth by 8.6%, increasing the number of power plants to 80,538 MW. One of the Ministry of Energy and Mineral Resources programs is to meet the 35,000 MW Power Generation program planned to take place in 2015-2019 [9]. The separation of 35,000 MW Power Generation program can be seen in Table 1, as follows:

**Table 1.** The separation of 35,000 MW Power Generation program

| Power Plant | IPP | PLN | Total |
|---|---|---|---|
| PLATE | 17,598 | 2,215 | 19,813 |
| PLTA/PLTM | 582 | 1,389 | 1,791 |
| PLTG/MG/GU | 6,123 | 6,785 | 12,908 |
| PLTP | 555 | 170 | 725 |
| PLT Bayu | 180 | - | 180 |
| PLT Biomass | 30 | - | 30 |
| **Total** | **25,068** | **10,559** | **35,627** |

Based on data from the Directorate General of Electricity Report on the 35,000 MW and 7,000 MW power plant development program, it is stated that of the total planned 35,000 MW program, the portion of the use of coal fuel in power plants is 60%, 5 % coming from new and renewable energy (EBT - water, geothermal, and solar), and 45% coming from other premier energy such as diesel and petroleum [10].

In the 2016-2025 RUPTL, several sections explain the consequences of emissions generated when all power plants (including the 35,000MW program) have been completed and started operating. CO2 emissions will be generated if Indonesia's electricity production is carried out by fuel mix increase almost two times from 211 million tons in 2016 to 395 million tons in 2025. Of the 395 million tons of emissions, 317 tons (80%) are burning coal [9].

Indonesia's average grid emission factor in 2016 is 0.851 kgCO2/kWh, which will increase to 0.871 kg/kWh in 2022 due to the operation of many coal-fired power plants. It is also due to the delay in PLTP and PLTA projects and the reduced gas supply for power plants. However, according to the government in the RUPTL, at a later stage, the emission rate will decrease to 0.749 kgCO2/kWh in 2025 due to the positive contribution from the use of gas, geothermal, water and other EBT sources.

The RUPTL document also explains that if there is no aggressive addition of NRE, the emission factor will increase by 0.851 KgCO2/kWh in 2016 to 0.960 KgCO2/kWh in 2025. However, with aggressive NRE development, the CO2 emission factor will decrease to 0.749KgCO2/kWh in 2025, RUPTL (20% NRE), and 0.714 KgCO2/kWh if the 25% EBT target is met.

### *3.2. Indonesia's commitment to the nationally determined contribution (NDC) Paris agreement 2015*

In Indonesia's first NDC document, the government has set ambitious goals regarding consumption from sustainable production related to food, water and energy. According to the government of Indonesia, this goal will be achieved through empowerment and capacity building, improving essential health and education services, technological innovation, and sustainable natural resource management with good governance principles.

The Second National Communication Indonesia document in 2010 recorded that Indonesia's GHG emissions were estimated at 1.8 GtCO2e in 2005. It is an increase of 0.4 GtCO2e compared to 2005. Based on the First Biennial Update Report (BUR) document) submitted to the UNFCCC in January 2016, Indonesia's national GHG emissions were 1,453 GtCO2e in 2012, which shows an increase of 0.452 GtCO2e from 2000 [11].

Furthermore, the INDC divides the commitment to reduce GHG emission levels into 2, namely Unconditional and Conditional Reduction. Indonesia voluntarily commits to reduce GHG emissions on its own by 26% compared to the BAU scenario in 2020. This commitment is necessary for moving towards a more ambitious commitment to reduce GHG emissions by 2030. The BAU scenario is projected to be 2.869 GtCO2e in 2030, which is an update of the BAU scenario in the INDC due to conditions the latest from the development of energy policy, especially coal plants. In addition, Indonesia is to increase its contribution to reduce GHG emissions by up to 41% by 2030, depending on the availability of international support in the form of funding, transfer, and technology development well as capacity building.

The NDC document explains that the emission projection for the energy sector in the BAU sector is 1,669 MTon CO2e in 2030, while in the CM1 scenario, the emission projection is recorded by the government at 1,355 MTon CO2e and 1,271 MTonCO2e in CM2 scenarios requiring international support.

### *3.3. Analysis of the 35,000-megawatt power generation program on Indonesia's commitment in the 2015 Paris agreement*

As previously explained, implementing the 35,000-megawatt power plant program will increase GHG emissions because the percentage of coal fuel use is more significant than EBT-based fuels, which are environmentally friendly. Suppose it is associated with several principles in international environmental law. In that case, the use of not ecological friendly fuels seems likely to be inconsistent with the spirit of ecological conservation. Therefore, the principle of sustainable development is required for a state to meet the current generation without compromising their ability to meet their needs.

GHG emissions resulting from the use of coal fuel can affect climate conditions and become a significant contributor to climate change in the long term. Even in the short time when All coal-based power plants began operating in 2019 according to the government's target, they can produce air pollution, which significantly affects the quality of the air inhaled by the community, especially those around coal-fired power plants. It is undeniable that the potential health problems that will arise due to air pollution will be hazardous. Greenpeace estimates that premature death tolls could rise to 15,700 people/year in Indonesia due to the increased risk of chronic disease in adults and acute respiratory infections in children [12].

In the context of other environmental law principles, such as the principle of early prevention measures and the potential dangers of using coal fuel, the government should take preventive measures to minimize or avoid the harmful effects. The government actions can divert the fuel mix used by prioritizing the use of NRE, which has enormous potential in Indonesia and reduces the use of coal.

The electricity sector can actually reduce emissions of 260 MtCO2e. Around 225 MtCO2e is obtained from increasing the use of EBT sources such as optimizing geothermal energy and utilizing biomass for electricity generation. An additional 47 MtCO2e of emission reductions can be obtained through demand-side management actions, which can reduce the level of electricity demand [13].

Even though the development of NRE requires a large amount of funding, the financing of GHG emission reductions in the sector power generation varies from USD10 to USD40 per tCO2e. In addition, the previously planned geothermal capacity of 6 Gigawatts in 2020 costs USD27 per tCO2e utilization of biomass waste originating from palm oil management waste and agricultural waste, requires a charge of USD45 per tCO2e [13].

The previous explanation can also analyze whether the 35,000 MW power plant, which is predicted to result in a significant increase in GHG emissions, can affect Indonesia's commitment to reduce GHG emissions as stated in the first NDC document in 2015 Paris Agreement. According to the head of the legal section of the secretariat general of climate change control, Indonesia's first NDC document could only take effect after 2020. Therefore, before 2020, the Kyoto Protocol regime would still apply. As it is known, Law No. 16 of 2016 concerning Ratification of the Paris Agreement 2015 is a legal regulation for Indonesia to support GHG emission reduction.

Those efforts of global countries in tackling climate change are one form of an increase in the earth's temperature caused by GHG emissions that have been going on for a long time. They are formulated in a model of emission reduction commitments by countries globally under the UNFCCC. After the UNFCCC comes into force, the world is entering a new era of GHG emission reduction commitments to mitigate climate change.

According to expert Bodansky, many parties question the type of commitment to be made by the states. Major disagreements over the type or variety of commitments to be made usually arise from discussing the binding power of a commitment. According to him, there are at least 4 (four) types of commitment itself in its implementation [13].

First is a commitment to have no definite binding force. The debate often arises when a country wants to commit the consideration of whether the commitment will be politically or legally binding. In this form, the commitments are usually made only by prioritizing recommendations without any obligations. In the text of international environmental agreements, this form uses "should" rather than "shall", applied in the UNFCCC document. The series of commitments can only describe serious attention and presence awareness of changing climatic conditions from time to time [13].

Second is a form of one-way commitment, which Bodansky refers to as "No Lose" commitment. This commitment model has similarities with the first commitment model, and it means that the parties participating in the commitment can benefit if the commitment is implemented. The third is a commitment to legally binding for the parties participating in it. This commitment is usually indicated by the use of the word "shall" in environmental agreements international, and it can be found in the Kyoto Protocol commitments. Fourth is a commitment to be implemented, which relies on complete readiness. This commitment usually includes a specific rule enforcement system, ranging from possible violations to mechanisms dispute resolution in it precisely [13].

The 2015 Paris Agreement was a commitment of each country to determine the emission reduction figures they can implement. Unlike its predecessor, the Kyoto Protocol provides a more definite benchmark for the targets to be achieved from efforts to reduce global emissions. However, the Paris Agreement regime provides an opportunity for all developing and developed countries to contribute to reducing GHG emissions, which is different from the Kyoto Protocol regime, which only involves developed countries as actors to reduce global warming emissions.

The 2015 Paris Agreement is also the compatibility between the reduction commitments conveyed and the realization on the ground. However, it is still unclear if sanctions mechanisms will be given to countries that cannot fulfil their GHG emission reduction commitments. It only contains provisions regarding the settlement of disputes. As it is known that the settlement of disputes in the UNFCCC only refers to the different interpretations between two countries and implementation problems of the convention. Therefore, if Indonesia can not fulfil its commitment to reduce GHG emissions by 29% with its efforts and 41% by international support, there is no sanction can be applied to Indonesia

## 4. Conclusion

The increase in GHG emissions resulting from the 35,000-megawatts power generation program will influence Indonesia's commitment to the 2015 Paris Agreement. Even though the emission calculations from the program have been included, Indonesia will have difficulty achieving the commitments that have been stated. The flexible nature of commitments can be used as an opportunity for Indonesia to review the commitments contained in the NDC. The unclear sanctions system for non-fulfilment of commitments by a country in the Paris Agreement also affects the final result of efforts to reduce GHG emissions to control climate change.

## References

[1] Kementerian Lingkungan Hidup dan Kehutanan 2021 Mengenai perubahan iklim *http://ditjenppi.menlhk.go.id/kcpi/index.php/info-iklim/perubahan-iklim*
[2] KLKH 2020 *Laporan inventarisasi gas rumah kaca (GRK) dan monitoring, pelaporan dan verifikasi (MPV)* (Jakarta, Indonesia: Kementerian Lingkungan Hidup dan Kehutanan)
[3] IPCC 2021 About the IPCC https://www.ipcc.ch/about/
[4] Martono and Komala N 2018 Kondisi konsentrasi karbon dioksida di Bukittinggi selama kejadian el Nino 2015 *J. Kim. dan Pendidik. Kim.* **3** 118–25
[5] Jaworowski Z 2007 *CO2: The greatest scientific scandal of our time* (Washington, USA: EIRScience)
[6] CNN 2019 Emisi karbon dioksida global capai rekor tertinggi tahun 2019 *https://www.cnnindonesia.com/teknologi/20191205191747-199-454565/emisi-karbon-dioksida-global-capai-rekor-tertinggi-tahun-2019*
[7] Bappenas 2014 *Rancangan pembangunan jangka menengah nasional (RPJMN) 2015-2019. Buku I Agenda Pembangunan Nasional* (Jakarta, Indonesia: Bappenas RI)
[8] Saturi S and Nugraha I 2015 Indonesia targetkan penurunan emisi karbon 29% pada 2030 *http://www.mongabay.co.id/2015/09/02/indonesia-targetkan-penurunan-emisi-karbon-29-pada-2030*
[9] PT. Perusahaan Listrik Negara (Persero) 2016 *Pengesahan rencana usaha penyediaan tenaga listrik PT Perusahaan Listrik Negara (Persero) Tahun 2016 s.d. 2025* (Jakarta, Indonesia: PT PLN (Persero))
[10] Direktorat Jenderal Ketenagalistrikan 2017 *Laporan perkembangan program pembangkit listrik 35.000 MW & 7.000 MW* (Jakarta, Indonesia: Kementerian ESDM RI)
[11] KLHK 2015 *Indonesia first biennial report (BUR). Under the United Nation Framework Convention on Climate Change* (Jakarta, Indonesia: KLHK)
[12] Greenpeace 2015 *Ancaman maut PLTU Batubara* (Jakarta, Indonesia: Greenpeace)
[13] Bram D 2016 *Hukum perubahan iklim* (Malang, Indonesia: Setara Press)

IOP Conf. Series: Earth and Environmental Science 1016 (2022) 012046 doi:10.1088/1755-1315/1016/1/012046

# Coal combustion emission and corruption in ASEAN: Does government integrity plays moderation role?

**Rafiq Azzam Al Afif*[1], Ahmad Daerobi[2], Bhimo Rizky Samudro[2] and Evi Gravitiani[2]**

[1]Master Program of Economic Development, Sebelas Maret University, Indonesia.
[2]Departement of Economic Development, Faculty of Economics and Business, Sebelas Maret University, Indonesia
*Corresponding author: rafiqazzam@student.uns.ac.id

**Abstract**. Coal energy is a cheap and abundant source of electricity in ASEAN. Despite this, the reality of $CO_2$ emissions from coal gets a red report card on the issue of climate change. In terms of the institutional economy, corruption is the main enemy in the issue of environmental pollution ; the cleaner government agencies, the lower the environmental pollution. In particular, the research aims to find out the moderating effect of government integrity concerning corruption and $CO_2$ emissions from coal in ASEAN. But to help construct the analysis, the paper also previously looked into the relationship between corruption and $CO_2$ emissions from coal and found out whether economic development played a mediating effect in that relationship. Using ECM panel analysis with a sample of 9 ASEAN member states, the paper found that: (1) In the long and short term, the lower level of corruption will reduce $CO_2$ emissions from coal. (2) FDI plays a mediating effect between corruption and $CO_2$ emissions from coal. The scheme is that the lower the level of corruption, the FDI will decrease and reduce $CO_2$ emissions. (3) Government integrity strengthens the relationship between corruption and $CO_2$ emissions from coal, meaning that the more honest and non-corrupt the government will have an effect that can strengthen the relationship between corruption and $CO_2$ emissions from coal. This study recommends that the government in ASEAN member states should be keeping public trust because it will decrease the level of corruption and can save climate change in the future.

**Keywords** : CO2, Climate Change, Coal, Corruption, and Government Integrity

**JEL classification** : Q4 ; Q5 ; K2 ; K4 : K23

## 1. Introduction

ASEAN leaders at the 36th ASEAN Minister on Energy Meeting in Singapore agreed that coal in their countries is abundant and has low power generation costs. The growth of coal energy power generation from 2000 until 2016 in ASEAN is increasing rapidly, so that some ASEAN member countries such as Indonesia, Malaysia, Philippine, and Cambodia the power generation that using coal energy are more than 50% from total energy used. Meanwhile the Thailand Government has succeeded in conducting an electrification program in his country and Vietnam received appreciation from the World Bank for its success in implementing electrification programs in rural areas up to 99%[1].

In the other sector electrification program used coal energy in ASEAN has negatively impact on increase $CO_2$ emissions, which is coal combustion had contributed up to 40% of the total $CO_2$ emissions in the world [2]. According to McGlade and Ekins [3], if 40 years (since 2010) coal is not used, then the earth's temperature target of < 2 Celcius will be achieved and if it still used will be giving bad impact on climate change. The last 20 years, extreme weather events given a bad impact on the human resources and the economic sector in Myanmar, Vietnam, Thailand, and Philipines_[4]. Moreover, the climate change in the future of ASEAN has an impact on geographical and social-economic conditions, such as rising sea levels, extreme weather, deaths, and economic losses_[5]. In many developing countries that have abundant natural resources, corruption is an institutional social factor that is considered the main cause of environmental degradation [6]. Corruption will distort resource allocation, reduce the productivity of public spending, lower investment, and slow economic growth. From the institutional sectors, the stakeholders misappropriating the power to him and his friends, which is corrupt behavior is either case [7,8].

This research refers to Liu and Dong [9] about mediating and moderating variables in the relationship between corruption and environmental degradation. The novelty from this research is uses FDI as a mediator variable and Government Integrity Index as a moderation variable and then uses IPAT model as a variable control and for capturing the level of corruption used Corruption Perception Index from International Transparency. This study has 3 purposes, the first is to find out the correlation between corruption and $CO_2$ emissions from coal combustion. The second is to find out whether economic development variables have a mediating effect on the relationship between corruption and coal $CO_2$ emissions. The last is to find out whether Government Integrity Index plays a moderation role in the correlation between corruption and $CO_2$ emissions from coal combustion.

## 2. Materials and method

### *2.1.* Materials

The data in this research were obtained from International Energy Agency (IEA), World Bank Database (WBD), Heritage Foundation (HF), and Transparency International (TI). The paper use panel data with all ASEAN member countries from 2009 to 2018 (Table 1), because of data availability.

**Table 1.** Data Information

| Data Name | Symbol | Description | Source | Unit |
|---|---|---|---|---|
| Coal CO2 | I | Carbon dioxide emissions from coal combustion | IEA | Mt |
| Corruption Perception | CPI | Corruption Perception Index | TI | % |
| Economic Development | FDI | Foreign Direct Investment inflows in USD | WBD | Billion |
| Gross Domestic Product | GDP | Real GDP or constant price 2010 in USD | WBD | Million |
| Government Institution Performance | GI | Government Integrity Index | HF | % |
| Population | P | The total number of people or residents in an area | WBD | Million people |
| Total Energy Supply by coal | TES | This data is made up of production + imports - exports - international marine bunkers - international aviation bunkers ± stock changes | IEA | Terajoules (TJ) |
| Total Final Energy Consumption by coal | TFC | The sum of final consumption by coal resources | IEA | Terajoules (TJ) |

The paper uses the IPAT decomposition model from Kaya identity based on Chontanawat research [10]. The main variables in the model are CO2 from coal combustion, population, energy intensity, energy mix, emission factor, and GDP per capita. The following is Kaya Identity equation based on chantonawat research:

$$I_i = P_i \times \frac{GDP_i}{P_i} \times \frac{TES_i}{GDP_i} \times \frac{GDP_i}{P_i} \times \frac{TFC_i}{TES_i} \times \frac{I_i}{TFC_i} = P_i \times A_i \times E_i \times F_i \times C_i \quad (1)$$

Where:

| | |
|---|---|
| A | GDP per capita constant in 2010 USD (thousand USD) |
| E | Energy intensity (TJ per million USD) |
| F | Energy mix (TJ) |
| C | Emision Factor (Mt per TJ) |
| i | refer to the ASEAN member countries; Indonesia, Malaysia, Singapore, Philipines, Myanmar, Brunei, Cambodia, Thailand, Laos, and Vietnam. |

To find out the first purpose used Corruption Perception Index as a variable that represents control of corruption. For the second purpose, the paper uses foreign direct investment inflow to represent economic development. For the third purpose, the paper uses the Government Integrity Index from Heritage Foundation to represent the institutional performance. Government Integrity Index is composed of five-factor, that is irregular payments and bribes, transparency of government policymaking, absence of corruption, perceptions of corruption, and governmental and civil service transparency. The score of this index is derived from Corruption Perception Index from Transparency International, where 0 means that country is perceived as having lowly government integrity and 100 means that highly government integrity. The research flow chart is presented in Figure 1.

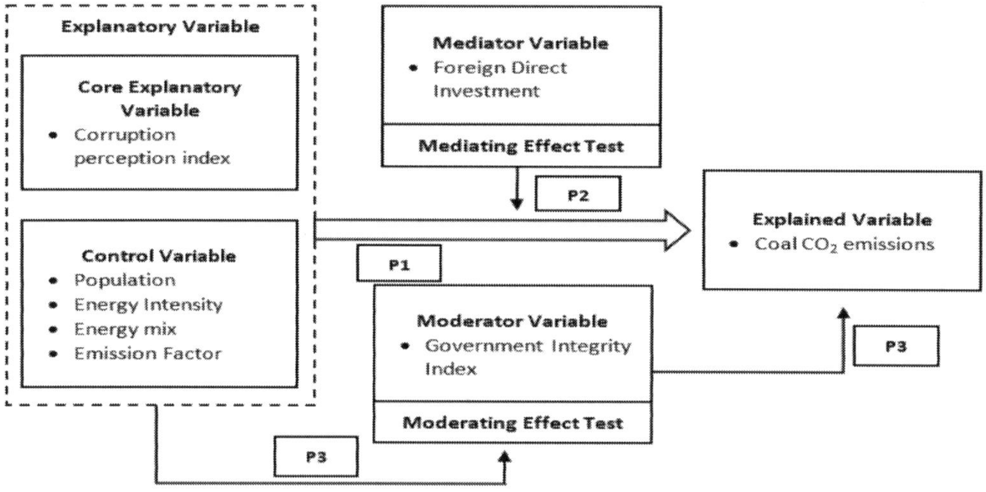


**Figure 1**. Research flow chat
(Source: Liu and Dong [9])

### *2.2. Methods*

Panel ECM analysis with the Ordinary Least Square technique is used. The paper using Philip Peron – Fisher (PP-Fisher) method for stationarity test and the cointegrity test using Levin, Lin & Chu method. Meanwhile, The Chow and The Hausman Test are used to determining the best model. According to Bagi in Iqbal [11], panel data analysis only use two classical assumption tests, that is multicollinearity and heteroscedasticity test. Even so in this research, where Pearson Correlation Test and Glejser Test are used.

IOP Conf. Series: Earth and Environmental Science 1016 (2022) 012046 doi:10.1088/1755-1315/1016/1/012046

### 2.3. *Subtitles*

$$lnI_{it} = \alpha_1 CPI_{it} + \theta_1 lnP_{it} + \theta_2 A_{it} + \theta_3 E_{it} + \theta_4 F_{it} + \theta_5 C_{it} + \mu_{it} + \varepsilon_{it} \quad (2)$$

Eq. (2) is used to find out the first purpose. Where if the value of $\alpha_1$ is less than zero and the prob-value is less than 0.1, so indicates that high control of corruption will decrease the coal CO2 emission. As for the short-term equation from eq. (2) or the ECM model is formulated as follows:

$$D(lnI)_{it} = \alpha_1 D(CPI)_{it} + \theta_1 D(lnP)_{it} + \theta_2 D(A)_{it} + \theta_3 D(E)_{it} + \theta_4 D(F)_{it} + \theta_5 D(C)_{it} + ECT + \mu_{it} + \varepsilon_{it} \quad (3)$$

If the value of ECT is negative and the prob-value is less than 0.1, so indicates there is a correlation between the short-term and the long-term model.

$$FDI_{it} = \alpha_2 CPI_{it} + \theta_1 lnP_{it} + \theta_2 A_{it} + \theta_3 E_{it} + \theta_4 F_{it} + \theta_5 C_{it} + \mu_{it} + \varepsilon_{it} \quad (4)$$

$$lnI_{it} = \alpha_3 CPI_{it} + \beta_1 FDI_{it} + \theta_1 lnP_{it} + \theta_2 A_{it} + \theta_3 E_{it} + \theta_4 F_{it} + \theta_5 C_{it} + \mu_{it} + \varepsilon_{it} \quad (5)$$

Eq. (4) and (5) are used to find out the second purpose and to analyze the mediating effect of foreign direct investment. Where if the prob-value of $\alpha_2$, $\beta_1$, and $\alpha_3$ is less than 0.1, so indicates that mediating effect of foreign direct investment is partially significant.

$$I_{it} = \alpha_4 CPI_{it} + \beta_2 lnGI_{it} + \gamma_1 CPI_{it} \times lnGI_{it} + \theta_1 lnP_{it} + \theta_2 A_{it} + +\mu_{it} + \varepsilon_{it}\theta_3 E_{it} + \theta_4 F_{it} + \theta_5 C_{it} + \mu_{it} + \varepsilon_{it} \quad (6)$$

Eq. (5) is used to find out the third purpose and to analyze the mediator effect of government integrity. Where if the prob-value of $\gamma_1$ is less than 0.1, so indicates that government integrity has an impact on the influence of the corruption control on coal CO2 emission, and the value of $\gamma_1$ reflects the extent of the moderating effect.

## 3. Result and discussion

### 3.1. *Model selection*

The estimates of all equation using Eviews 10, result that eq.2, 4, 5, and 6 choose Fixed Effect Model (FEM) as the best model and the best model eq.3 is Common Effect Model (CEM). The result is presented in Table 2.

**Table 2.** Model selection

| Test Name | Prob - Chi-Square | | | | |
|---|---|---|---|---|---|
| | Eq.(2) | Eq.(3) | Eq.(4) | Eq.(5) | Eq.(6) |
| Chow-Test | 0.000* | 0.228 | 0.000* | 0.000* | 0.000* |
| Hausman-Test | 0.000* | 0.261 | 0.000* | 0.000* | 0.000* |

Note: * significant at 0.01; ** significant at 0.05; *** significant at 0.1

### 3.2. *Classic assumption test*

The result of the Pearson Correlation Test shows that eq.5 and 6 are not free from the multicollinearity assumption. Meanwhile, for Glejser Test shows that all equations are free from the heteroscedasticity assumption. The result is presented in Table 3 and Table 4.

**Table 3.** Multicollinearity test

| Independent Variable Name | Pearson Correlation Value | | | | |
|---|---|---|---|---|---|
| | Eq.(2) | Eq.(3) | Eq.(4) | Eq.(5) | Eq.(6) |
| | *CPI Variable* | | | | |
| Log(P) | -0.403 | -0.057 | -0.403 | -0.403 | -0.403 |
| A | -0.105 | -0.116 | -0.105 | -0.105 | -0.105 |
| E | -0.115 | -0.130 | -0.115 | -0.115 | -0.115 |
| F | -0.192 | 0.090 | -0.192 | -0.192 | -0.192 |
| C | -0.039 | 0.154 | -0.039 | -0.039 | -0.039 |
| Log(GI) | - | - | - | - | 0.931* |
| CPI*GI | - | - | - | - | 0.997* |
| FDI | - | - | - | -0.837* | - |

Note: * More than 0.8

**Table 4.** Heteroskedasticity test

| Independent Variable Name | Prob. Value | | | | |
|---|---|---|---|---|---|
| | Eq.(2) | Eq.(3) | Eq.(4) | Eq.(5) | Eq.(6) |
| CPI | 0.115 | 1.000 | 0.991 | 0.583 | 0.828 |
| Log(P) | 0.629 | 1.000 | 0.791 | 0.940 | 0.581 |
| A | 0.889 | 1.000 | 0.988 | 0.983 | 0.879 |
| E | 0.955 | 1.000 | 0.778 | 0.916 | 0.763 |
| F | 0.938 | 1.000 | 0.598 | 0.757 | 0.807 |
| C | 0.950 | 1.000 | 0.888 | 0.726 | 0.664 |
| ECT(-1) | - | 1.000 | - | - | - |
| Log(GI) | - | - | - | - | 0.447 |
| CPI*GI | - | - | - | - | 0.863 |
| FDI | - | - | - | 0.618 | - |

Note: * significant at 0.01; ** significant at 0.05; *** significant at 0.1

### *3.3. Stationery and cointegrity test*

The result of the stationery test using PP-Fisher method for eq.2 shows that all variables are stationer at the 1st difference. As for the result of the cointegrity test using Levin, Lin & Chu method shows that residual is stationer at the level. It is indicated that between long and short term model has been correlated or balanced. The result of the test is presented in Table 5.

**Table 5.** Stationery and cointegrity test

| Variable Name | Prob. Value | |
|---|---|---|
| | *Stationery Test* | *Cointegrity Test* |
| Log(I) | 0.000* | - |
| CPI | 0.000* | - |
| Log(P) | 0.000* | - |
| A | 0.000* | - |
| E | 0.000* | - |
| F | 0.000* | - |
| C | 0.000* | - |
| Residual | - | 0.0003 |

Note: * significant at 0.01; ** significant at 0.05; *** significant at 0.1

### 3.4. *Estimates result*

**Table 6.** Estimates result

| Variable Name | P1 Coefficient | P1 Coeff. Eq.(3) | P2 Coeff. Eq.(4) | P2 Coeff. Eq.(5) | P3 Coefficient |
|---|---|---|---|---|---|
| CPI | -0.109* | -0.040* | -0.939* | -0.085* | -6.268* |
| LnP | 21.941* | 19.052** | 178.201* | 17.243* | 220.129* |
| A | -2.130 | 2.960 | -1.150 | -1.830 | 7.950* |
| E | -7.500 | -1.060 | 3.820 | -8.510 | -2.650 |
| F | -0.017* | -0.013* | 0.141* | -0.021* | 0.167* |
| C | 13.751* | 11.740* | -46.806 | 14.985* | -140.162 |
| FDI | - | - | - | 0.026* | - |
| LnGI | - | - | - | - | -23.502** |
| CPI*GI | - | - | - | - | 1.755* |
| ECT(-1) | - | -0.459* | - | - | - |

Note: * significant at 0.01; ** significant at 0.05; *** significant at 0.1; P = purpose

After fulfilling the requirements for the ECM panel regression, this section will discuss the result of the estimates equation. The result of the estimates equation is presented in Table 6.

*3.4.1. The first purpose.* Based on P1 in Table 6, in the long and short term shows that CPI are significant at level 0.01, which means in the long and short term control of corruption will significantly reduce coal CO2 pollution in the ASEAN region. The result has similarities with Liu and Dong [9] and Zhang et.al [12]. In the long and short term CPI has a negative effect on coal CO2 pollution. The value of correlation between CPI and coal CO2 pollution is minus 0.109 and in short term is 0.040. It means in the long term if the CPI score increase by 1%, then the coal CO2 pollution will decrease by 0.109% and in the short term, CO2 pollution will decrease by 0.040%.

*3.4.2. The second purpose.* Based on Table 6, the prob-value of CPI ($\alpha_2$) in coefficient Eq.4 and FDI ($\beta_1$), and CPI ($\alpha_3$) in coefficient Eq.5 are significant at level 0.01 or less than 0.1. It means that the level of economic development role as a mediator variable in the relationship between corruption and coal CO2 pollution. CPI has a negative effect on FDI and FDI has a positive effect on coal CO2 pollution. It means, directly, if the control of corruption in ASEAN is effective, it will decrease FDI inflows in this region, and indirectly FDI as a role mediator in the relationship. The scheme is, when the control of corruption in ASEAN success, it will decrease FDI inflows, and a decrease in FDI inflows will reduce coal CO2 pollution. The scheme of the correlation is presented in figure 2.

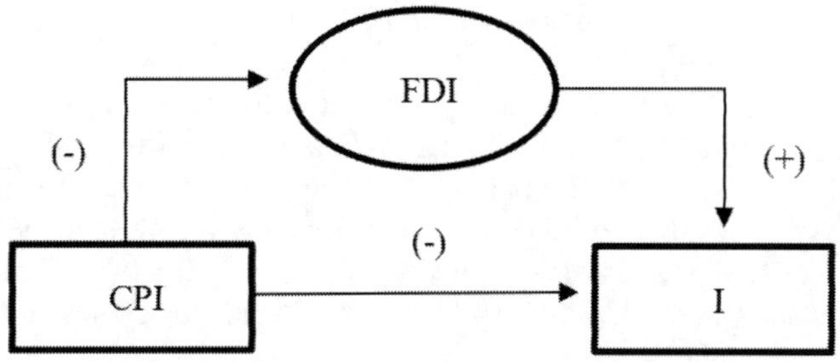


**Figure 2.** The scheme of mediator variable

Note: (-) = negative effect; (+) = positive effect

*3.4.3. The third purpose.* Based on Table 6, the prob-value of CPI*GI ($\gamma_1$) is significant at level 0.01 or less than 0.1. This means Government Integrity plays a moderating role in the relationship between corruption and coal CO2 pollution. The moderation coefficient that is roled by GI is positive effect, it is indicated that Government Integrity strengthens the relationship between CPI and coal CO2 pollution. It means, the more the government in ASEAN has integrity, it is can reduce coal CO2 pollution in the region. The scheme of this relationship is presented in figure 3.

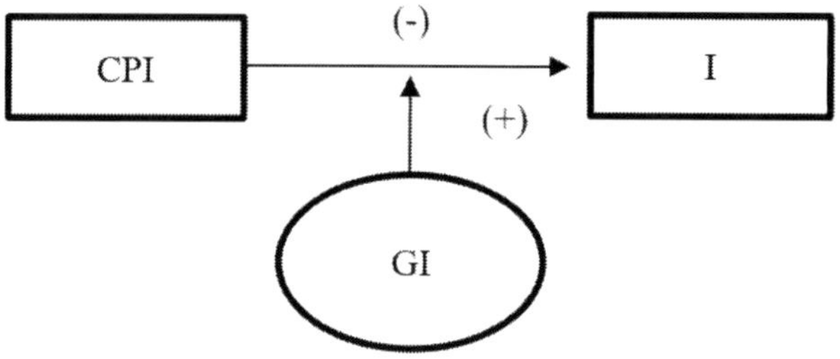


**Figure 3.** The scheme of Moderation Variable

Note: (-) = negative effect; (+) = positive effect

### *3.5. Discussion*

The result of P1 has similarities with Liu and Dong [9], Zhang et.al [12], Pellegrini and Gerlagh [13], Ozturk and Al-Mulali [14]. According to Liu and Dong [9], more non-corrupt governments will decrease the local haze pollution. It means, that control of corruption indirectly reduce CO2 emissions from haze pollution. Moreover, Zhang et.al [12] argue that in the countries with lower emissions, curbing corruption would have a more significant effect, but for a country with higher emissions not significant. According to Pallegrini and Gerlagh [13] curbing corruption in the level of countries will directly affect the implementation of environmental policies and the quality of civil servants. More, good environmental control will force companies to choose clean production methods. In addition, good quality civil servants will increase awareness of environmental laws.

Based on figure 2, that CPI has a negative effect on FDI inflows. The result has similarities with Jalil et.al [15], Qureshi et.al [16] and Leal et.al [17] . According to Jalil et.al [15], that in countries with bad institutional systems (for studies in Asia and Africa), bribery reduces the impact of institutional inefficiency, thereby providing an avenue for FDI. Moreover, Leal et.al [17] argue that corruption can accelerate bureaucratic processes and take on the role of business facilitator. There are indications that investors prefer countries with loose environmental regulations. More specifically, Qureshi et.al [16] in their research found that controlling corruption has a negative effect on FDI. It means that higher corruption will attract incoming foreign investment. To get higher profits, multinational companies tend to shift polluting labor-intensive industries to developing countries, thereby exacerbating environmental pollution in the host country [9].

The result of the third purpose shows that Government Integrity plays a moderation role between corruption and coal CO2. The Government Integrity Index is about honesty, trust, responsibility, and transparency from The Government in a country. The more honest, trusty, responsible, and transparent a Government, will have an indirect effect on reducing coal CO2 pollutant, especially in ASEAN. The lower government integrity shows that the lower control of corruption in a country is because corruption can hinder democracy and the rule of law [7]. Machiavelli's famous statement (in Leitao) which reads "*Favori agli amici, nemici della legge*" which means help for friends who are enemies of the law, is an illustration of abuse of office in the institutional sector which is often encountered in various corruption cases. This is similar to the phenomenon of buying and selling positions and regulations for private interests between stakeholders and the private sector.

## 4. Conclusion

Coal energy in ASEAN is abundant and has low power generation costs, but on the other side, coal combustion had contributed to 40% of the total CO2 emissions in the world and has a bad impact on climate change. This research has 3 purposes, to find out the correlation between corruption and CO2 emissions from coal combustion, to find out whether economic development variables have a mediating effect on the relationship between corruption and coal CO2 emissions, and to find out whether Government Integrity Index plays a moderation role in the correlation between corruption and CO2 emissions from coal combustion. The result of this research are; the first, in the long and short term control of corruption has a negative effect on coal CO2 emissions; the second, the level of economic development role as a mediator in the relationship between corruption, an increase of control of corruption will reduce FDI inflows, and the next, given effect on reducing coal CO2 emissions in ASEAN. Third, Government integrity plays a moderating role in the relationship between corruption and coal, exactly strengthening the relationship between control of corruption and CO2 emissions. The research recommends that the government in ASEAN member states should be keeping public trust, responsible in work, keep honesty in work, and applicated good governance because it will decrease the level of corruption and can save climate change in the future. For the next study, the paper suggests using another mediating, moderation, and control variable to analyze corruption and coal CO2 emissions or fixed with the model but change your location especially in the upper-middle or high-income country.

## References

[1] Yasuo O, Fumiko Y and Phoumin H 2019 Study on The Biomass and Coal Co-Combustion in the ASEAN Region *ERIA Research Project Report* (Jakarta: Economic Research Institute For Asean and East Asia).

[2] Edwards G A S 2019 Coal and climate change *Wiley Interdiscip. Rev. Clim. Chang.* **10** 1–16

[3] McGlade C and Ekins P 2015 The geographical distribution of fossil fuels unused when limiting global warming to 2°C *Nature* **517** 187–90.

[4] Eckstein D, Hutfils M-L and Winges M 2019 *briefing paper, global climate risk index 2019, Who Suffers Most From Extreme Weather Events? Weather-related Loss Events in 2017 and 1998 to 2017.*

[5] Overland I, Sagbakken H F, Chan H-Y, Merdekawati M, Suryadi B, Utama N A and Vakulchuk R 2021 The ASEAN climate and energy paradox *Energy Clim. Chang.* **2** 100019.

[6] Damania R 2002 Environmental controls with corrupt bureaucrats *Environ. Dev. Econ.* **7** 407–27.

[7] Leitao A 2016 Corruption and the environment *Socialomics* **5** 8–12

[8] Imperatives S 1987 Report of the World Commission on Environment and Development : Our Common Future

[9] Liu Y and Dong F 2021 Haze pollution and corruption : A perspective of mediating and moderating roles *J. Clean. Prod.* **279** 123550.

[10] Chontanawat J 2018 Decomposition analysis of of CO Decomposition analysis emission in in ASEAN : ASEAN : an an extended extended IPAT model. *Energy Procedia* **153** 186–90.

[11] Iqbal M 2015 Regresi Data Panel ( 2 ) " Tahap Analisis " *Blog Dosen Perbanas* pp 1–7

[12] Zhang Y J, Jin Y L, Chevallier J and Shen B 2016 The effect of corruption on carbon dioxide emissions in APEC countries: A panel quantile regression analysis *Technol. Forecast. Soc. Change* **112** 220–7.

[13] Pellegrini L and Gerlagh R 2006 An empirical contribution to the debate *Environ. Dev.* **15** 332–54

[14] Ozturk I and Al-Mulali U 2015 Investigating the validity of the environmental Kuznets curve hypothesis in Cambodia *Ecol. Indic.* **57** 324–30.

[15] Jalil A, Qureshi A and Feridun M 2016 Is corruption good or bad for FDI? Empirical evidence from Asia, Africa and Latin America *Panoeconomicus* **63** 259–71.

[16] Qureshi F, Qureshi S, Vinh Vo X and Junejo I 2021 Revisiting the nexus among foreign direct

investment, corruption and growth in developing and developed markets *Borsa Istanbul Rev.* **21** 80–91.
[17] Leal P H, Caetano R V and Marques A C 2021 Is the relocation of polluting industries prompted by fdi flow and stock, globalisation, corruption and regulation? *Int. J. Environ. Res. Public Health* **18** 1–29.

# The impact of audit committee characteristics, financial performance, and listing age on greenhouse gas emission disclosures of highly emitted industry in Indonesia

**A K Widagdo[1], B A Rahanyamtel [2] and S R Ika[2]***

[1] Faculty of Economics and Business, Sebelas Maret University, Jl Ir Sutami 36 A Solo, 57771, Indonesia
[2] Department of Accounting, Janabadra University, Jl Tentara Rakyat Mataram 55-57 Yogyakarta, 55231, Indonesia
[3]Corresponding author: ika@janabadra.ac.id

**Abstract.** Greenhouse gas or carbon emissions produced by manufacturing operations and other highly emitted industries are causes of global warming. Therefore, either in the sustainability reporting or in the corporate social and environmental reporting section, as stated in the annual report, the company usually discloses its activities related to carbon emission handling for sustainable business. In Indonesia, however, the extent of carbon emission disclosures is voluntary. The objective of this study was to investigate the effect of audit committee characteristics, financial performance, and listing age on carbon emissions reporting of highly emitted companies in Indonesia. Audit committee characteristics were measured by the number of audit committee members and the number of audit committee meetings, while Altman financial distress model measured financial performance. A checklist based on the Carbon Disclosure Project (CDP) evaluated the greenhouse gas emissions disclosures. This study uses 99 companies of highly emitted industry listed on the Indonesia Stock Exchange. Results of multiple regression analysis indicate that the number of audit committee meetings positively affects the greenhouse gas emissions report. The result suggests that the more active the audit committee in the company in conducting meetings, the higher the company's incentives to disclose carbon emission in the company's annual report or the sustainability reporting. The study provides insight into the regulation released by the capital market authority agency regarding strengthening factors that may influence listed companies to report their carbon emission.

**Keywords** : GHG, financial distress model, carbon disclosure project, carbon emission

## 1. Introduction

Companies have been forced to become more environmentally friendly due to the threat posed by climate change. In light of the threat, it is appropriate to urge widespread mitigation efforts. Corporate carbon emissions must be measured, disclosed, and reduced as part of the global carbon reduction effort, an essential component of the effort [1]. The purpose of greenhouse gas disclosure is to raise public understanding of the extent to which emissions are occurring, leading to more environmentally sustainable policymaking [2].

Previous literature confirms that the corporation's environmental accountability policies, such as disclosing data on carbon emissions, are useful to stakeholders [2–4]. The Indonesia Stock Exchange (IDX) does not require companies to disclose their carbon emissions. Previous studies assert that the company's governance structure ensures that additional details regarding carbon emissions are included either in an annual report or in the company's sustainability reporting [5–9].

The present research aims to test the determinants of the level of GHG emission disclosure. The determinants are audit committee characteristics, financial performance, and listing age. In conjunction with the Indonesia Government's effort in reducing carbon emissions, it is noteworthy to explore the factors that may encourage the highly emitted industry to disclose their carbon emissions either in annual reports or sustainability reporting. The extent of GHG emission disclosures is voluntary in Indonesia.

## 2. Literature review and hypotheses

### *2.1. Previous studies*

Previous researches have examined factors associated with the extent of GHG emission reporting both in Indonesia [3,5,6,10] and in the international arena, such as in Australia [2], in Spanish [11], in Italy [4], in the Netherlands [12], and in Nordics country [7]. Among those studies, few studies discuss the involvement of audit committees on the extent of GHG emission disclosures, especially in Indonesia. If any, the proxy used for the audit committee variable is limited to the number of audit committee members [6,13]. In contrast, the number of audit committee members is only one aspect of measuring the audit committee's effectiveness [14]. Chariri et al. [7] tested several proxies of audit committee effectiveness such as the number of members, the number of meetings, independence, but his study took a sample of Nordic countries. Therefore, this study attempt to test a proxy other than the number of audit committee members to measure its effectiveness, namely audit committee meetings concerning GHG emission disclosures in Indonesia.

### *2.2. Hypotheses development*

*2.2.1. Audit committee characteristics.* According to agency theory [15], issues can occur in the relationship between a principal (shareholder) and an agent (business management) as a result of the separation of ownership and management. As a result, agency theory strives to eliminate information asymmetries by including monitors and building systems that can protect shareholders against conflicts of interest on the side of management [16]. The audit committee is a committee formed by and responsible to the Board of Commissioners in helping carry out the duties and functions of the Board of Commissioners [17]. The audit committee's role may be used to resolve agency issues through its supervision responsibilities for financial reporting quality.

The number of audit committee members represents the resources of an audit committee in performing its oversight duties [18]. Therefore, the higher the number of audit committee members, the more influential the supervision and control of a business, including corporate social and environmental reporting [8]. Akhirah and Kiswanto [13] find a positive relationship between the number of audit committee members and the level of GHG emission disclosures.

The number of audit committee meetings represents the diligence of an audit committee in managing its supervision duties [18]. Therefore, the audit committee's ability to supervise management's financial reporting and information disclosure is dependent on how often members attend meetings. Audit committee members who attend regular meetings are more successful in monitoring tasks [9]. The frequency of audit committee meetings also allows members to examine the company's accounting standards, disclosures, and estimations. The more engaged the audit committee members are in regular meetings, the more possibilities they can debate and evaluate corporate financial reporting processes [19,20]. According to Chariri et al. [7], the more frequently audit committee members attend regular meetings, the higher the amount of GHG emission disclosure. Hence, this research will investigate the following hypotheses.

H1: The number of audit committee members positively affects the extent of GHG emission disclosure.

H2: The number of audit committee meetings positively affects the extent of GHG emissions disclosure.

*2.2.2. Financial performance.* Corporations that perform well financially have the financial capability to make environmental decisions, namely disclosure of environmental activities [21]. Hence, it is expected that a company whose financial condition is healthy is likely to conduct more activities in carbon emissions mitigation, increasing the GHG emissions disclosure. In this research, the Altman Z-score prediction of bankruptcy [22] is used as a proxy for financial performance. The assessment of Z-score may distinguish those who are financially healthy, in a grey area that needs more attention in managing the business, and financially distressed. Previous studies, to the best of authors' knowledge, have not yet examined the use of a set of ratios instead of a single ratio like return on asset, return on equity, leverage [4,5,10,12,23,24]. Using high leverage as a term for financial distress, Choi et al. [2] posit that GHG emissions disclosure tends to be lower in financially distressed firms. Hence, this research states the following hypothesis.

H3: Firms with healthy financial performance tend to increase GHG emissions disclosures.

*2.2.3. Company age.* Company age is essential in calculating carbon emissions since it indicates stakeholder strength, strategic approach, and financial success [25]. Environmental disclosure and policy-determining environmental conservation initiatives can boost a company's added value. Mature firms are believed to have more experience and better understand their constituents' informational demands [10]. Older businesses may demonstrate that a business may thrive in the face of a range of stakeholder expectations and government requirements. By publishing carbon emissions, the firm expects to have a beneficial influence and establish a favorable reputation that will enable it to function correctly. Previous studies have demonstrated that a company's age influences its carbon disclosure [23,24]. Therefore, this study proposes the following hypothesis.

H4: Company age positively affects the extent of GHG emissions disclosure.

*2.2.4 Profitability.* According to Haniffa and Coke [26], management's capacity to generate profit must complement their ability to fulfill their social duties. Through corporate social and environmental disclosure, a business conveys that it is interested in more than just profit and the social context. Businesses with a high profit margin are more concerned with the environment [24,27] and more capable of performing social disclosure than businesses with a low profit margin [28]. Some studies [2,3,24] discovered that businesses with high profitability were able to finance the additional time and material resources necessary for improved voluntary reporting and disclosure of GHG to overcome external pressures. However, some studies fail to prove that profitability relates to the extent of GHG emissions disclosure [4,5,12]. Thus, this study will investigate the ensuing hypothesis.

H5: Profitability positively affects the extent of GHG emissions reporting.

*2.2.5 Corporate size.* Earlier research confirms the inverse relationship between corporate size and the extent of GHG emission disclosures [2,4,5,11]. This beneficial association is based on the assumption that larger organizations engage in a broader range of sectors and operate globally. Additionally, larger firms must accommodate a broader spectrum of consumers who may be active in the environmental management and policies of the corporation. As a result, big enterprises are subjected to more societal and administrative pressure to release environmental information than smaller businesses [2,11,29]. Additionally, environmental transparency is complex and expensive, and larger enterprises are more likely to devote time to producing and publishing environmental information than smaller businesses [29]. As a result, the following hypothesis is tested in this study.

H6: Corporate size positively affects the extent of GHG emission disclosures.

## 3. Research method

### *3.1. Population, sampling, and data resources*
The population of this study is companies in the highly emitted industry listed on IDX in 2020. The year 2020 was the newest data available at the start of this study. Besides, the year is the first financial data year after the covid-19 pandemic spreading in Indonesia. Therefore, this research may give an understanding of the GHG emissions disclosure during the pandemic time. The highly emitted industry is based on previous studies' industry classification such as manufacture, energy, transportation [2,30]. The population of the study is 166 companies. The sample selection procedures are as follows. First, the company has an annual report or sustainability reporting publicly available. Second, it has a GHG emissions disclosure policy in place and a policy to minimize its GHG emissions. Third, data related to the examined variables are complete. There are 56 companies out of 166 with incomplete data, or the annual reports are unavailable. Eleven companies are excluded due to outliers. This study gets 99 firms as the final sample based on these procedures.

Data associated with audit committee characteristics, financial performance, listing age, and control variables (profitability and size of the company) are obtained from the annual report. Meanwhile, data were obtained from the corporate annual report or sustainability reporting in calculating GHG emissions disclosures.

### *3.2. The assessment of Greenhouse gas emissions disclosures*
This study examines the GHG emission disclosure level by utilizing a checklist series developed by Choi et al. [2]. The checklist was based on information supplied by the CDP (Carbon Disclosure Project) and has 18 items that address five GHG emissions disclosure key points: climate change: dangers and possibilities, carbon emissions reporting, energy consumption reporting, greenhouse gas reduction and cost, and carbon emission responsibility. The checklist evaluates the existence of a statement in a company's GHG emissions reporting. If an item in the checklist is disclosed, it obtains a score of one; otherwise, it receives a score of zero. The sum of the checklist score is then divided by 18. The number refers to the level of GHH emission disclosure of a business. The application of Choi et al. 's [2] content analysis in measuring GHG emissions reporting follows previous researches [1,3,5,6,31].

### *3.3. Independent variables*
The independent variables of this research are audit committee characteristics, financial performance, and company listing age, while profitability and size are the control variables. Audit committee characteristics include the number of audit committee members (ACM) and the number of audit committee meetings within a year (ACMEET). Data relates to ACM and ACMEET are available in the annual report, specifically in the audit committee report section.

This study utilizes the Altman Z-score financial distress prediction model to appraise the highly emitted industry's financial performance (FP). When it comes to predicting the potential bankruptcy of a firm, Altman's Z is one of the most well-known statistically-based prediction models [32]. The following is how the equitation is presented.

$$Z = 1.2X1 + 1.4X2 + 3.3X3 + 0.6X4 + 1.0X5 \quad (1)$$

Where

X1 = working capital divided by total asset

X2= retained earnings divided by total asset

X3= earnings before interest and taxes to total asset

X4= market capitalization divided by the book value of debt

X5= sales divided by total asset

Following Altman [22], a higher Z-score indicates a more favorable financial performance for the firm. The Z-score is divided into three categories, which are as follows.

If Z is above 2.99, the company is in good financial condition and not in distress.

If Z is between 1.81 and 2.99, the company's financial condition is in the grey area.

If Z is below 1.81, the company is in bad financial condition and included in the distressed condition.

The company's listing age (AGE) is when the company was first listed in IDX until December 31, 2020. The data is available in the annual report. Profitability was calculated by the return on asset (ROA), while firm size (SIZE) was assessed by the total asset transformed to the natural logarithm.

### *3.4. Regression model*

This research applies multiple regression analysis to test the effect of audit committee characteristics, financial performance, listing age, and control variables on GHG disclosures. The regression model is presented as follows. This research also conducts normality and multicollinearity test of classical assumption tests to validate the regression results.

$$GHG = \alpha + \beta 1AGE + \beta 2FP + \beta 3ACM + \beta 4ACMEET + \beta 5ROA + \beta 6SIZE + e \qquad (2)$$

## 4. Result and discussion

### *4.1. Descriptive statistic*

Data characteristics of all variables used in this study are presented in Table 1. The lowest score of GHG emissions disclosures is 4 out of 18, referring to 22% of the total disclosures. The highest score is 18, representing 100% of the total disclosure in the index. The mean score of GHG disclosures is about 72%. The score is better than those in manufacturing companies 18.7% [33], in non-financial firms, 34% [6], and in the mining industry, 24% [31]. The highest score of GHG was achieved by four companies in the mining industry, namely PT Vale Indonesia, PT. Medco Energi Internasional, PT. Adaro Energy, and PT. Bukit Asam. The difference is probably due to the period and industry sample.

As shown in Table 1, the listing age of the sample companies is 7189 days on average, equal to 19 years and eight months. The Altman Z score of financial performance is 2.41 on average, indicating the sample companies are in a grey performance area since the value is below 2.99. Surprisingly, during the pandemic year, a company achieved a Z score of 18.81, indicating the company had an excellent financial performance. The mean score of SIZE and ROA are 29.71 and 0.04, respectively. The ROA mean score indicates that the sample company has achieved a profit of 4 percent during the pandemic.

Table 1 presents the average number of audit committee members and audit committee meetings, 3 and 8 respectively. These numbers comply with Financial Service Authority Regulation [17], stating that the audit committee members are at least three persons. The audit committee conducts regular meetings at least once in three months.

**Table 1.** Descriptive statistics

| | Minimum | Maximum | Mean | Standard Deviation |
|---|---|---|---|---|
| GHG | 4 (22%) | 18 (100%) | 12.90 (72%) | 3.35 |
| AGE | 767 | 14704 | 7189.4 | 3924 |
| FP | -5.75 | 18.81 | 2.41 | 3.11 |
| ACM | 3 | 5 | 3.27 | 0.5 |
| ACMEET | 2 | 57 | 8.31 | 7.91 |
| SIZE | 25.22 | 29.68 | 29.71 | 1.72 |
| ROA | -1.01 | 1.55 | 0.04 | 0.23 |
| N | 99 | 99 | 99 | 99 |

### *4.2. Results of multiple regression*

Multiple regression analysis is used to assess the impact of AGE, FP, audit committee characteristics, ROA, and SIZE on GHG. Table 2 shows the results of the analysis. As presented in the table, ACMEET is the only significant variable in explaining the extent of GHG. The number of audit committee meetings is positively associated with the level of greenhouse gas emissions disclosure. The result suggests that the more active the audit committee conducts a meeting during the year, the higher the level of the GHG reporting of the company. The result is consistent with previous studies that state that an audit committee's meeting frequency increases the quality of financial reporting [7,34]. Meanwhile, the number of audit committee members does not affect the level of GHG disclosures. This finding contradicts Chariri et al. [7], who find that the number of audit committee members positively affects carbon emissions disclosures in Nordic companies.

**Table 2.** Result of multiple regression analysis

| | B | t | Sig | Tolerance | VIF |
|---|---|---|---|---|---|
| Constant | 2.437 | 0.419 | 0.676 | | |
| AGE | 0.00005 | 0.667 | 0.507 | 0.963 | 1.038 |
| FP | 0.075 | 0.675 | 0.501 | 0.889 | 1.125 |
| ACM | 0.116 | 0.174 | 0.862 | 0.739 | 1.353 |
| ACMEET | 0.113 | 2.389 | 0.019** | 0.760 | 1.315 |
| ROA | 0.386 | 0.259 | 0.796 | 0.909 | 1.100 |
| SIZE | 0.288 | 1.390 | 0.168 | 0.834 | 1.199 |
| F Value | | | 2.277 | | |
| F sig | | | 0.043** | | |
| Adjusted $R^2$ | | | 0.072 | | |
| Kolmogrov Smirnov Sig | | | 0.200 | | |

Note: ** implies significant at 5%

Table 2 also shows that AGE and FP do not impact GHG. The results indicate that listing age and financial performance does not affect greenhouse gas emissions reporting. ROA and SIZE do not have any impact either on GHG. The results denote that profitability and the company's asset do not influence the level of greenhouse gas emissions disclosures. The results are not similar with Borghei-Ghomi and Leung [23], who find that company size and company age positively affect carbon emissions disclosure in Australia. The results are also contradictory with Hapsoro and Ambarwati [24], examining the impact of company age and profitability in the oil, gas, and coal industry in international companies under the OSIRIS database. The country setting, sample, and time frame are probably the reasons for inconsistencies.

The adjusted $R^2$, as presented in Table 2 is 0.072. The number suggests that the predicting power of AGE, FP, ACM, ACMEET, ROA, and SIZE on GHG variability is about 7.2 %. The other 92.8% are other variables not examined in the regression model. The regression model in this study passes the normality of the classical assumptions test as the value of the Kolmogorov Smirnov sig is 0.200, far above 0.05. The model also passes the multicollinearity of classical assumptions test because the value of VIF is lower than 10 and the Tolerance is higher than 0.1.

## 5. Conclusion and directions for the subsequent research

This article investigates the impact of audit committee characteristics, financial performance, and listing age on the extent of greenhouse gas emissions disclosure. The disclosures are closely related to the activities done by the company in adapting and mitigating the consequence effect of carbon emissions from company operations to global warming. The results indicate that the number of meetings held by the audit committee stimulates the company's management decision to produce a higher level of greenhouse gas emissions disclosures.

There are several limitations to this research that may be overcome in a subsequent study. First, the adjusted $R^2$ is only 7.2 %, which could be improved by adding new independent variables. The involvement of audit committee expertise and other corporate governance factors such as independent commissioners may be investigated further as the determinants factors of greenhouse gas emissions reporting. Second, this study explores the level of disclosures in a period during the covid-19 pandemic. The subsequent research may examine the difference in greenhouse emissions disclosures level between pre and post covid-19.

**References**

[1] Kalu J U, Buang A and Aliagha G U 2016 Determinants of voluntary carbon disclosure in the corporate real estate sector of Malaysia *J. Environ. Manage.* **182** 519–24
[2] Choi B B, Lee D and Psaros J 2013 An analysis of Australian company carbon emission disclosures *Pacific Account. Rev.* **25** 58–79
[3] Hermawan A, Aisyah I S, Gunardi A and Putri W Y 2018 Going green: Determinants of carbon emission disclosure in manufacturing companies in Indonesia *Int. J. Energy Econ. Policy* **8** 55–61
[4] Jaggi B, Allini A, Macchioni R and Zagaria C 2018 The Factors Motivating Voluntary Disclosure of Carbon Information: Evidence Based on Italian Listed Companies *Organ. Environ.* **31** 178–202
[5] Nasih M, Harymawan I, Paramitasari Y I and Handayani A 2019 Carbon emissions, firm size, and corporate governance structure: Evidence from the mining and agricultural industries in Indonesia *Sustain.* **11**
[6] Budiharta P and Kacaribu H E P B 2020 The Influence of Board of Directors, Managerial Ownership, and Audit Committee on Carbon Emission Disclosure: A Study of Non-Financial Companies Listed on BEI *Rev. Integr. Bus. Econ. Res.* **9** 75–87
[7] Chariri A, Januarti I and Yuyetta E N A 2018 Audit Committee Characteristics and Carbon Emission Disclosure *E3S Web Conf.* **73** 1–5
[8] Alyousef L and Alsughayer S 2021 The relationship between corporate governance and voluntary disclosure: The role of boards of directors and audit committees *Univers. J. Account. Financ.* **9** 678–92
[9] Karamanou I and Vafeas N 2005 The association between corporate boards, audit committees, and management earnings forecasts: An empirical analysis *J. Account. Res.* **43** 453–86
[10] Herawaty V and Veronica Pernanda P 2020 Carbon emission disclosure, competition and company's characteristics *Int. J. Sci. Technol. Res.* **9** 1714–9
[11] Gonzalez-Gonzalez J M and Ramirez C Z 2016 Voluntary carbon disclosure by Spanish companies: an empirical analysis *Int. J. Clim. Chang. Strateg. Manag.* **8** 57–79
[12] van de Burgwal D and Vieira R J O 2014 Environmental disclosure determinants in Dutch listed companies *Rev. Contab. Finanças - USP* **25** 60–78
[13] Akhiroh T and Kiswanto 2016 The Determinant Of Carbon Emission Disclosures *Account. Anal. J.* **5** 326–36
[14] Ika S R and Mohd Ghazali N A 2012 Audit committee effectiveness and timeliness of reporting: Indonesian evidence *Manag. Audit. J.* **27**
[15] Jensen M C and Meckling W H 1976 Theory of the firm: Managerial behavior, agency costs and ownership structure *J. financ. econ.* **3** 305–60
[16] Fama E F and Jensen M C 1983 Separation of Ownership and Control *J. Law Econ.* **26** 301–25
[17] Financial Services Authority 2015 *Financial Services Authority Regulation Number 55/POJK.04/2015 dated December 23, 2015 regarding the Establishment and Guidelines for the Work Implementation of the Audit Committee*
[18] DeZoort F T, Hermanson D R, Archambeault D S and Reed S A 2002 Audit committee effectiveness: A synthesis of the audit committee literature *J. Account. Lit. Vol. 21, 2002, pp. 38-75* **21** 38–75

[19] Hundal S 2013 Independence, Expertise and Experience of Audit Committees: Some Aspects of Indian Corporate Sector *Am. Int. J. Soc. Sci.* **2** 58–75
[20] Vafeas N 2005 Audit Committees, Boards, and the Quality of Reported Earnings* *Contemp. Account. Res.* **22** 1093–122
[21] Luo L, Tang Q and Yi-cheniLan 2013 Comparison of Propensity for Carbon Disclosure between Developing and Developed Countries *Account. Res. J.* **26** 6–34
[22] Altman E I 1968 Financial Ratios, Discriminant Analysis and the Prediction of Corporate Bankruptcy *J. Finance* **23** 589–609
[23] Borghei-Ghomi Z and Leung P 2013 An Empirical Analysis of the Determinants of Greenhouse Gas Voluntary Disclosure in Australia *Account. Financ. Res.* **2**
[24] Hapsoro D and Ambarwati A 2018 Antecedents and Consequences of Carbon Emissions' Disclosure: Case Study of Oil, Gas and Coal Companies in Non-Annex 1 Member Countries *J. Indones. Econ. Bus.* **33** 99
[25] Roberts R W 1992 Determinants of Corporate Social Responsibility Disclosure an Application of Stakeholder Theory *Accounting, Organ. Soc.* **17** 595–612
[26] Haniffa R M and Cooke T E 2005 The impact of culture and governance on corporate social reporting *J. Account. Public Policy* **24** 391–430
[27] Hermawan A and Gunardi A 2019 Motivation for disclosure of corporate social responsibility: Evidence from banking industry in Indonesia *Entrep. Sustain. Issues* **6** 1297–306
[28] Gunawan J 2015 Corporate social disclosures in Indonesia: Stakeholders' influence and motivation *Soc. Responsib. J.* **11** 535–52
[29] Monteiro S M da S and Aibar-Guzmán B 2010 Determinants of environmental disclosure in the annual reports of large companies operating in Portugal *Corp. Soc. Responsib. Environ. Manag.* **17** 185–204
[30] Rupley K H, Brown D and Marshall R S 2012 Governance, media and the quality of environmental disclosure *J. Account. Public Policy* **31** 610–40
[31] Halimah N P and Yanto H 2018 Determinant of Carbon Emission Disclosure at Mining Companies Listed in Indonesia Stock Exchange *KnE Soc. Sci.* **3** 127
[32] Altman E I 1983 *Corporate financial distress and bankruptcy* (New York: John Wiley & Sons, Inc)
[33] Solikhah B, Wahyuningrum I F S, Yulianto A, Sarwono E and Widiatami A K 2021 Carbon emission report: A review based on environmental performance, company age and corporate governance *IOP Conf. Ser. Earth Environ. Sci.* **623**
[34] Ahmed Haji A 2015 The role of audit committee attributes in intellectual capital disclosures: Evidence from Malaysia *Manag. Audit. J.* **30** 756–84

# Mapping land use and surface water quality for urban clean water resource

**Erma Fitria Rini*[1,2], Paramita Rahayu[1,2] and Hakimatul Mukaromah[1,2]**

[1]Urban and Regional Planning Program, Engineering Faculty, Universitas Sebelas Maret, Jl. Ir. Sutami 36 A, Kentingan, Surakarta, 57126, Indonesia
[2]Regional Development Information Center, Institute of Research and Community Service, Universitas Sebelas Maret, Jl. Ir. Sutami 36 A, Kentingan, Surakarta, 57126, Indonesia
*Corresponding author: erma@staff.uns.ac.id

**Abstract**. The population growth and water needs, cumulated with climate change will aggravate the global water scarcity crisis. Surakarta city initiated to use their three rivers across the city as one of the surface water resources to ensure water availability and preserve the environment. This paper aims to give an understanding of the riverbanks' land use characteristics and river water quality in Surakarta's three major rivers. The land use data was gathered from satellite images, while the river water quality was gathered from secondary data. Spatial grouping analysis by geographic information system was used to examine the characteristic based on land use and water quality in three major rivers across Surakarta. The results show that there are some different typologies of land use that affect water quality. Therefore, the government needs to improve the land use management to improve the surface water capacity to become one of the urban clean water resources.

**Keywords** : land use; mapping; surface water; water resource

## 1. Introduction

The world's population living in urban areas is predicted to increase from 55% in 2018 to 60% in 2030 [1]. Rapid urbanization has triggered the government to meet various needs of the community, especially basic needs such as shelter, water and food sources. Natural resources to fulfil that needs, including water, have been affected by climate change. Some regions are vulnerable to projected climate change with regard to water availability [2]. The population growth and water needs, cumulated with climate change will aggravate the global water scarcity crisis. Water scarcity is likely to occur in regions where populations and water consumption are rising fast [2,3].

In urban areas, the majority of communities are facilitated by networked piped water to meet water needs so that people's access to water becomes easier. Urban households using networked water spend more money on water and consume more quantity of water [4]. In urbanized cities, water consumption tends to be higher as well as the community become wealthier [5]. Unfortunately, the needs and ease of access to water are not equal to the availability of abundant sources of clean water. With increasing population, rising living standards, and limited resources to meet human needs, efficiency in resource utilization becomes important. Cities take the lead in making comprehensive planning for water resource efficiency due to rapid urbanization in cities. Water resource efficiency is instrumental in increasing efficiency in water management for green and resilient urban development. In coping with challenges in urbanization, infrastructure is one of the key areas that Asian cities should focus on [5]. However, the development of water management infrastructure needs to be based on the study of the characteristics and problems of alternative sources of water and the things that affect it.

One alternative source of water is surface water, including river water. Rivers play many important roles in ecosystems and society, including as one of the water resources [6,7]. To make river water a source of water, an assessment of the river and its surrounding areas needs to be carried out. It is important to identify land use in its surrounding area because it will affect the water entering the river. Changes in land use exacerbated by climate change will affect the quantity and quality of water resources [8]. The land-use type like housing, government facilities, and industries within river basins causes increased waste and polluted rivers [9]. Changes in land use can also affect water quality and water availability in the river [10]. This study identifies land use characteristics in the river basin and river water quality then classifies the river basin according to these two characteristics. This grouping is expected to recommend for decision-makers to make development policies in the context of using surface water as a source of urban water.

## 2. Methods

This research uses land use and river water quality data in 2018 as spatial and quantitative data. Land use information was gathered from satellite imagery of 2018, while the river water quality was gathered from secondary data from environmental services of Surakarta City. Surface water quality gathered from river water quality of Pepe River, Anyar River, and Premulung River as three main rivers in Surakarta. While the land use covered 500 meters area from each river.

Spatial grouping analysis by geographic information system was used to examine the characteristic of land use and water quality in three major rivers across Surakarta. Grouping analysis is based on 15 variables (two variables of land use and 13 variables of river water quality): land-use type, land area, amount of TSS, TDS, pH, BOD, DO, Phosphate, Nitrate, Nitrite, Chlorine, Sulphur, Total Coli, COD, Detergent.

## 3. Results and discussion

The Surakarta City Government through environmental services always monitors the quality of river water that passes through the city every year. River quality monitoring is carried out not only to monitor its quality but also to determine environmental priority issues in the city. Pollutant content in Premulung, Anyar, and Pepe River from ten sample points in 2018 is shown in Table 1.

**Table 1.** River water quality of Surakarta on 2018 [11]

| River segments | TSS (mg/L) | TDS (mg/L) | pH | BOD (mg/L) | DO (mg/L) | Phosphate (mg/L) | Nitrate (mg/L) | Nitrite (mg/L) | Chlorine (mg/L) | Sulphur (mg/L) | Total Coli/100ml | COD (mg/L) | Detergent (ug/L) |
|---|---|---|---|---|---|---|---|---|---|---|---|---|---|
| Premulung downstream | 9 | 513 | 7.22 | 16.0 | 0.9 | 1.79 | 1.1 | 0.002 | 0.11 | 0.002 | 86,500 | 31.70 | 0.002 |
| Premulung upstream | 3 | 365 | 7.50 | 3.4 | 1.9 | 1.29 | 0.7 | 0.002 | 0.06 | 0.015 | 62,000 | 13.50 | 0.015 |
| Premulung midstream 1 | 6 | 368 | 7.48 | 6.0 | 1.3 | 1.16 | 0.7 | 0.024 | 0.13 | 0.003 | 91,100 | 18.20 | 0.003 |
| Premulung midstream 2 | 23 | 559 | 7.58 | 24.0 | 1.0 | 0.72 | 1.6 | 0.006 | 0.14 | 0.017 | 96,500 | 70.40 | 0.017 |
| Anyar downstream | 33 | 291 | 7.62 | 3.4 | 1.4 | 0.31 | 1.1 | 0.167 | 0.11 | 0.020 | 7,900 | 25.50 | 54.60 |
| Anyar upstream | 2 | 228 | 7.53 | 3.4 | 1.3 | 0.23 | 1.8 | 0.016 | 0.80 | 0.012 | 5,200 | 26.34 | 72.60 |
| Anyar midstream | 52 | 270 | 7.64 | 5.6 | 1.3 | 0.33 | 1.1 | 0.042 | 0.10 | 0.038 | 59,800 | 20.90 | 76.50 |
| Pepe downstream | 15 | 488 | 7.83 | 25.8 | 1.0 | 1.33 | 1.5 | 0.017 | 0.12 | 0.038 | 80,400 | 54.10 | 1773 |
| Pepe upstream | 55 | 686 | 7.65 | 9.1 | 1.2 | 0.31 | 1.0 | 0.164 | 0.13 | 0.034 | 67,700 | 71.80 | 67.80 |
| Pepe midstream | 19 | 525 | 7.75 | 32.2 | 1.0 | 1.13 | 1.8 | 0.032 | 0.18 | 0.085 | 79,900 | 60.30 | 1787 |

To assess the appropriateness of river water to become a clean water resource, we refer to the Republic of Indonesia Government Regulation Number 82 of 2001 concerning Water Quality Management and Water Pollution Control. Based on these regulations, the classification of water quality that is required to be used as drinking water is the Class 1 classification.

**Table 2.** Class 1 water quality criteria for drinking water [12]

| **Parameters** | **TSS (mg/L)** | **TDS (mg/L)** | **pH** | **BOD (mg/L)** | **DO (mg/L)** | **Phosphate (mg/L)** | **Nitrate (mg/L)** | **Nitrite (mg/L)** | **Chlorine (mg/L)** | **Sulphur (mg/L)** | **Total Coli/100ml** | **COD (mg/L)** | **Detergent (ug/L)** |
|---|---|---|---|---|---|---|---|---|---|---|---|---|---|
| Baseline | 50 | 1000 | 6-9 | 2 | 6 | 0.2 | 10 | 0.06 | 0.03 | 0.002 | 1000 | 10 | 200 |

Based on the criteria in Table 2, the three rivers in Surakarta are categorized as polluted. There is no river water that is included in class I as a source of drinking water. The condition of river water, which is still far from the class I quality standard, indicates that stakeholders need to intervene to improve the quality of river water, both for fulfilling urban drinking water sources, as well as for environmental sustainability. This paper then describes the land use conditions around the river as input for making groupings to facilitate problem identification, intervention, and decision making.

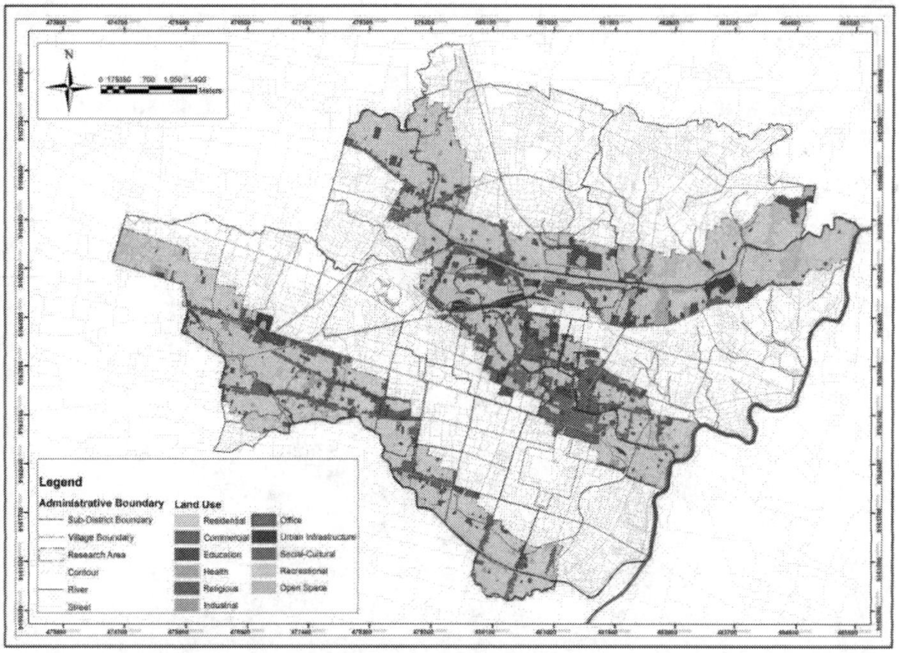


**Figure 1.** Land use of Surakarta's three major river basin

Pepe, Anyar, and Premulung rivers divide the city of Surakarta into areas with a high density of land use. Figure 1 illustrates how the density and variety of land uses in each watershed. Spatially, the majority of the river is surrounded by residential land uses that produce greywater which flowed into the river. Pepe river downstream, there is a mixed-use pattern, whereas, in Anyar river, there is still quite a lot of open green space. Quantitatively in Table 3, the percentage area of each land use in the watershed

IOP Conf. Series: Earth and Environmental Science 1016 (2022) 012048 doi:10.1088/1755-1315/1016/1/012048

of each river is presented, which shows that the majority of land uses in the river basin are settlements. This indicates that there is a potential for river water pollution from household waste.

**Table 3.** Land use area on river watershed, 2018

| Land Use | Pepe | | Anyar | | Premulung | |
|---|---|---|---|---|---|---|
| | ha | % | ha | % | ha | % |
| Residential | 408.79 | 46.68 | 293.71 | 46.39 | 402.42 | 63.84 |
| Commercial | 128.28 | 14.65 | 25.11 | 3.97 | 69.19 | 10.98 |
| Road | 122.79 | 14.02 | 112.71 | 17.80 | 45.48 | 7.22 |
| Open space | 69.63 | 7.95 | 93.16 | 14.71 | 38.76 | 6.15 |
| Educational | 43.50 | 4.97 | 43.01 | 6.79 | 26.02 | 4.13 |
| Office | 28.15 | 3.21 | 15.51 | 2.45 | 17.31 | 2.75 |
| River | 22.13 | 2.53 | 22.86 | 3.61 | 8.13 | 1.29 |
| Social-Cultural | 20.14 | 2.30 | 0.11 | 0.02 | 4.82 | 0.76 |
| Urban Infrastructure | 16.46 | 1.88 | 7.62 | 1.20 | 1.70 | 0.27 |
| Industrial | 7.14 | 0.82 | 6.64 | 1.05 | 10.45 | 1.66 |
| Religious | 6.79 | 0.78 | 1.32 | 0.21 | 2.07 | 0.33 |
| Health | 1.99 | 0.23 | 11.26 | 1.78 | 3.20 | 0.51 |
| Recreational | 0 | 0 | 0.18 | 0.03 | 0.75 | 0.12 |
| TOTAL | 875.79 | 100 | 633.20 | 100 | 630.31 | 100 |

River water quality and land use data were then used for grouping analysis which resulted in the four groups as shown in Figure 2.

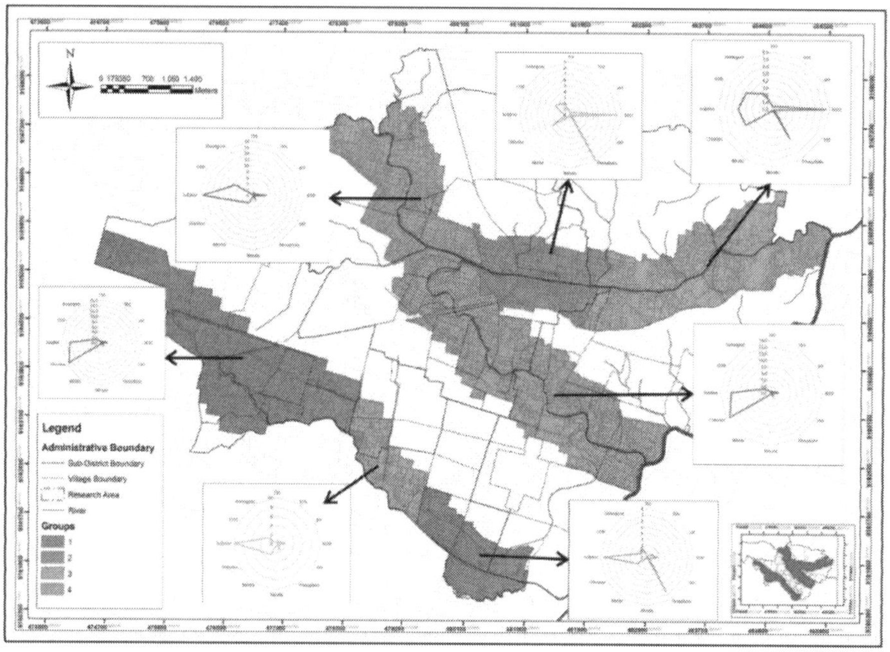


**Figure 2.** Group of three major river watershed characteristics

The four groups have different characteristics. Group 1, which includes the upstream, midstream 1, and downstream Premulung river segments, shows that the majority of land use characters in the first layer of the river are settlements. However, this group includes textile industries in their river basin. Some textile industries have Waste Water Treatment Plant (WWTP) but are inactive. The water quality in this group is very high, exceeding the criteria is the total coli, sulphur, chlorine, and phosphate content.

Group 2 contains only one river segment, Anyar upstream. Even though the river water quality does not meet the baseline class 1, this river segment has better water quality than other river segments except for the phosphate content and total coli. Some land uses in group 2 consist of urban infrastructure managed by the government.

Group 3 is the border area of the Pepe upstream and Anyar Tengah to Anyar downstream rivers. Although this area is dominated by settlements, it still has many green open spaces. The green open space is located not only in a 500m radius from the river but also in the river basin. Even though it has a lot of green open space the quality of river water in this segment is not better. The levels of BOD, COD and Total Coli of group 3 were high. Even in the midstream to downstream Anyar segment, the total coli content is up to 90 times the water quality standard required for drinking water.

For Group 4 members, the river segment Premulung midstream 2 with Pepe midstream to downstream, there are similarities in character. Group 4 has a variety of land-use types but has very little green open space. In terms of water quality, the midstream to downstream Pepe segments have water quality far exceeding the class 1 quality standard the total coli is 80 times the quality standard. However, in Premulung midstream 2, although the land use is the same as the other two segments in group 4, the water quality is still better. This is possible because the upstream part of the Premulung river has better land use and water quality characteristics.

Each group has different characters both in terms of land use and river water quality. The river water content in group 1, especially total coli, is closely related to industrial land use around the river because many industries do not have WWTP and drain their waste into the rivers. In the middle stream to downstream of Anyar river which has the highest total coli content, even though it has large green open spaces, it is dominated by settlements with high populations. For residential land and the activities of the population in it, of course, it is necessary to further investigate its relation to the water content that flows into the river. E. coli concentrations increase with strong population growth, limited environmental controls and large changes in the climate [13,14]. Therefore, intervention does not only need to be done on the river but also land-use activities related to the river.

## 4. Conclusions

The resulting grouping can assist relevant stakeholders to formulate surface water management policies. To process the river water into raw drinking water, it is not only necessary to use drinking water processing technology, but also regulation for land use and community activities in the river basin. The results show that the required interventions vary based on the characteristics of river water quality and land use. In groups that have residential and industrial land uses, the total coli content is higher than other groups. This indicates the need for management in land use and community activities. Industrial WWTP needs to be reactivated and greywater processing from residential activities needs to be encouraged. Future research is needed to find more about the root causes of river water quality.

## References

[1] UN-ESC 2018 *Sustainable cities, human mobility and international migration* (New York City, United States; Geneva, Switzerland)

[2] Hagemann S, Chen C, Clark D B, Folwell S, Gosling S N, Haddeland I, Hanasaki N, Heinke J, Ludwig F, Voss F and Wiltshire A J 2013 Climate change impact on available water resources obtained using multiple global climate and hydrology models *Earth System Dynamics* **4** 129–44

[3] Harding R J, Weedon G P, van Lanen H A J and Clark D B 2014 The future for global water assessment *Journal of Hydrology* **518** 186–93

[4] Rahayu P, Rini E F and Soedwiwahjono 2019 Domestic water adequacy of Surakarta, Indonesia: Is it prone to vulnerability? *Environment and Urbanization ASIA* **10** 81–98
[5] Lee S 2015 *Resource efficiency for green and resilient urban development in the Asia-Pacific Region (The case of water)* (Bangkok)
[6] Iles C E and Hegerl G C 2015 Systematic change in global patterns of streamflow following volcanic eruptions *Nature Geoscience* **8** 838–42
[7] Wu C, Ji C, Shi B, Wang Y, Gao J, Yang Y and Mu J 2019 The impact of climate change and human activities on streamflow and sediment load in the Pearl River basin *International Journal of Sediment Research* **34** 307–21
[8] Tong S T Y, Sun Y, Ranatunga T, He J and Yang Y J 2012 Predicting plausible impacts of sets of climate and land use change scenarios on water resources *Applied Geography* **32** 477–89
[9] Khalid R M, Mokhtar M bin, Jalil F, Rahman S A and Spray C 2018 Legal framing for achieving 'good ecological status' for Malaysian rivers: Are there lessons to be learned from the EU Water Framework Directive? *Ecosystem Services* **29** 251–9
[10] Yang W, Long D and Bai P 2019 Impacts of future land cover and climate changes on runoff in the mostly afforested river basin in North China *Journal of Hydrology* **570** 201–19
[11] DLH 2019 *Profil Dinas Lingkungan Hidup Tahun 2019* (Surakarta: Dinas Lingkungan Hidup Kota Surakarta)
[12] Republik Indonesia 2001 *Peraturan pemerintah tentang pengelolaan kualitas air dan pengendalian pencemaran air* (Indonesia: Republik Indonesia)
[13] Islam M M M, Iqbal M S, Leemans R and Hofstra N 2018 Modelling the impact of future socio-economic and climate change scenarios on river microbial water quality *International Journal of Hygiene and Environmental Health* **221** 283–92
[14] Iqbal M S, Islam M M M and Hofstra N 2019 The impact of socio-economic development and climate change on E. coli loads and concentrations in Kabul River, Pakistan *The Science of The Total Environment* **650** 1935–43

**Acknowledgement**
This research is funded by non-tax state revenue of Universitas Sebelas Maret, Surakarta, Indonesia.

# Reservoir to ecotourism and its climate change mitigation type development: Challenges and opportunities in North Sumatra's Botanic Garden Samosir Areas

**Netty Demak H Sitanggang[1,2]*, Zuhud M A Ervizal[3], Masy'us Burhanuddin[3] and Soekmadi Rinekso[3]**

[1]Indraprasta PGRI University,TB. Simatupang, Jl. Nangka Raya No.58 C, Jagakarsa, South Jakarta, 12530, Indonesia
[2]Postgraduate School of Forestry Campus IPB Darmaga, Bogor 16680 Indonesia IPB University of Bogor, Indonesia
[3]Faculty of Forestry Campus IPB Darmaga, Bogor 16680 Indonesia IPB University of Bogor, Indonesia
*Corresponding author: nettybiologi@gmail.com

**Abstract.** Optimizing reservoirs for ecotourism development is one way to increase local or national economics without impairing environmental conditions, alongside as an adapting or mitigating process to climate change, which is part of sustainable development goals, especially in the Aek Natonang'reservoir, where there is also a botanic garden and a conservation zone function. Hence, this research tried to explore, investigate, and analyze the role of reservoirs in ecotourism and their coping with climate change impacts, as well as their hopefully positive welfare impact on the welfare of the local people. The direct observation and in-depth individual interviews with some of the respondents collected data in the field, alongside assessment of environmental aspects or conditions. The result of this research has shown that the reservoir of Aek Natonang areas as a water conservation paradigm that would support the local activities in their daily lives as well as act as an adapting and mitigating process to climate change impacts. Hence, ecotourism development is one way to cope with it, and it is part of the challenges and opportunities in Aek Natonang areas management to ecotourism development in the framework of local or national economic growth based on the reservoir of potential.

**Keywords** : climate change, ecotourism, local economic, sustainable development, water conservation

## 1. Introduction

Climate change is one big issue, and it is part of the challenges to development processes that land use change brings. That could be caused by problems such as symptoms of degradation and their impacts, such as the loss of vegetation, the loss of biodiversity, and so forth [1,2]. Hence, developing tour areas based on natural potential, which leads to ecotourism development goals, is one way to maintain biodiversity, alongside as a means to mitigate or adapt to climate change impacts [3,4] and it leads to the welfare of the local people kindly and sustainably. Ecotourism development based on reservoirs is part of the trend in this era, which has been able to show some beautiful natural views and other functions (ecological function). The importance of the reservoir as water conservation and its functions [5] and

[6] are part of one whole to mitigate or adapt to climate change impacts, alongside its ability to local economic or welfare kindly and sustainability.

It is part of the water conservation development and it has a function [6]. Alongside water supply, it supports local people's daily activities For example, it supports aquaculture and aqua-plant development [7,8] and, of course, irrigation [9]. Alongside it, it will be created and developed into ecotourism areas based on reservoir water management [10]. This is one natural resource potential (water resource) that can support local or regional economic development goals for welfare or wellbeing kindly and sustainably. Hence, preserving and maintaining water quality (management) in building tourism based on natural or ecological principles is needed, alongside it is part of one criterion for sustainability or sustainable development goals [11]. On the other hand, land use change is unfriendly and destructive. Which one is cutting down plants or trees in the surroundings destructively? Hence, it needs to be careful in developing a tour of the site (ecotourism), so it is necessary to consider that, of course, without impairment to ecological systems [12], which have multi-functions or ecological functions, which one is as an ecosystem service to others (around).

Referring to the description above, reservoir water is part of the living heart of the surroundings (humans, plants, animals, and so on), and it has multiple functions (social, economic, and ecological), so maintaining it is needed, in the framework of sustainability or sustainable tourism and water security shaped kindly and sustainably [13,14]. Reservoir water and botanic garden Samosir is one site that has potential for ecotourism development, alongside reviving local or regional economies based on natural potential (ecological). Indirectly or directly, ecotourism development has brought land-use change, and, of course, it would be an ecological disturbance if it's not responded to effectively and kindly. Hence, evaluating ecotourism areas' development is needed to maintain and preserve their rich natural resources, chiefly the botanic garden samosir, alongside it hopefully to benefit the welfare of local people and sustainably develop conservation areas [15].

## 2. Methods

Administratively, that botanic garden in Samosir, North Sumatra, has multi-functions (Figure 1), and indirectly it would be created to create equilibrium (stability of ecosystems) kindly, sustainably, and lead to welfare or wellbeing. In general, the botanic garden at Samosir serves an ecological function in the form of a key role of equilibrium in ecological systems (ecosystem stability). On the other hand, because the Samosir areas are part of biodiversity or conservation hotspot areas, identifying and inventories of natural resource potential in Samosir (reservoir water and botanic garden) is critical to local economic development while not jeopardizing environmental aspects, one of which ecotourism development is part of the government's program to create and empower local people in natural resources management, chiefly in water management with botanic gardens elaborated for a tour of the site (tourism of object to ecotourism paradigm development).

There is some data collection in this research, which is based on both to direct observation and in depth individual interviews that have been used for data collection in this field of research, which is at least 25 people that have explored the advantages and disadvantages of ecotourism development. The importance of this research would be focused on utilizing reservoir water management in supporting ecotourism development, alongside it as an adapting and mitigating type of development to climate change impacts [3], alongside policing to appear necessary [16], and purposed to minimize impacts and vulnerability [17]. Hence, this research will describe the role of ecotourism in coping with climate change impacts.

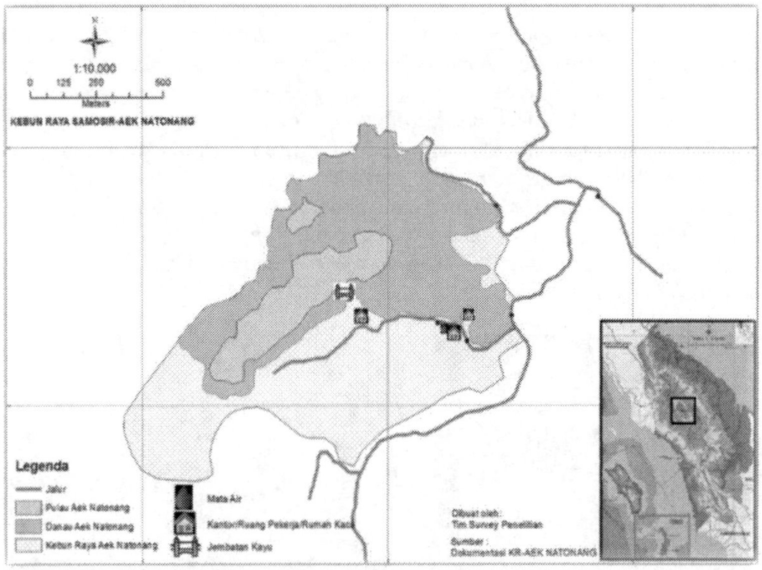


**Figure 1.** Site investigation

Referring to the description above, ecotourism development based on reservoir (dam) management is part of one targeted at welfare or (adding) income for the local government and people around it. It would be hoped to utilize nature's potential kindly, friendly, and sustainably (a rural sustainable or social resilience system), and it would hopefully be a way to mitigate or adapt to climate change impacts kindly.

## 3. Result and discussion

A tour area development based on natural resources (reservoir and botanic garden) has become a developing trend in this era, which is part of the economic creativity that would be local or national economic growth. This is one investment based on a rich natural environment that could create new jobs for local people, which is the river and botanic gardens' potential as a tour area, or ecotourism development, alongside as a means of mitigating and adapting to climate change impacts kindly. These two research sites are conservation hot spots, where each is able to respond to water availability (dam or reservoir) sustainability as well as ecological functions in their process. Hence, it is more interesting to explore and analyze whether the role of ecotourism in climate change impacts has decreased, but on the other hand, ecotourism development has also brought about biodiversity change (loss of biodiversity), which is a land use change factor for the tour area expansion relationship.

### *3.1. Reservoir and botanic gardens-based ecotourism development*

The botanic garden of Samosir is one of the areas that has potential and challenges with natural beauty and its impact on the environment. If managed kindly, it would support the welfare of local people. Hence, creating and developing a tour must be done kindly and sustainably, alongside it hopefully to carry out stability in society, economy, and the environment (sustainable development paradigm). Hence, the botanic garden has existed alongside rivers, so it would be designed, created, and developed into a tour area based on natural beauty and empirical studies of reservoir water in Aek Natonang, which is part of the biodiversity hotspot relationship in botanic garden areas, which would increase carrying capacity and ecosystem services, which leads to equilibrium (ecosystem stability), which is part of botanical garden function [18,19]. Hence, Aek Natonang has ecological functions, and it could be developed for local economic development, which one is ecotourism area development (Figure 2).

There were beautiful views (natural), biodiversity of plants and animals, freshwater abundance (reservoir water), and so on. Those are part of the sustainable development criteria and goals, which are used to take a tour of sites based on natural (ecologically) development, alongside increasing capability in local economics. Hence, Aek Natonang's rivers as water conservation and their multifunction, which is part of the water supply or stock that is used to irrigate or meet the daily needs of local or community people, also serve as a process of adaptation to drought, which is caused by climate change impacts [7,8]. On the other hand, it used to be secondary income for the local government and, of course, it was for local people, so there were some changes in botanic gardens and samosir areas, which in Figure 1 shows the cases that happened. Land use change that happened in the botanic garden is one step towards land modification in the destination's tour, alongside supported by the reservoir view, so ecotourism development based on the reservoir and botanic garden has considered the ability to achieve social, economic, and ecological equilibrium and sustainability. The Botanic Garden of Samosir has now changed to ecotourism areas with reservoir potential, so indirectly this development has brought many advantages, without impairing environmental aspects, and directly to where there were some facilities where you could look at Figures 2 and 3.

**Figure 2.** Dynamic land use change

In Figure 2 above, it is disclosed that there has been a change in where botanic garden samosir has changed, which one is a cutting of plant, which land has changed, and so on. Land use change has caused primary issues that have impacted the environment, such as the loss of biodiversity (plants and animals), trash (plastic and so on), and so on [20,21] while it has brought to the local economy (the economy), it has also increased. Hence, it is necessary to explore, understand, and analyze from all perspectives the botanic gardens' development and the evaluation needed to maintain conservation (the land, the water, and others) continually and consistently.

**Figure 3.** Ecotourism area developments and its natural view

*3.2. Ecotourism development as a mitigating tool for climate change impacts*

According to the description above, relating to land use change to a tour area development based on ecological prospective (Eco), with the nature of beautiful views and reservoir water management [dam]. Those are part of the ecotourism type development that, hopefully, leads to equilibrium (ecosystem stability), as well as the welfare of the local people, and as part of kindly and dynamically mitigating or coping with climate change impacts. Hence, ecotourism is one way to cope with climate change, and it is part of sustainable development [22]. Moreover, it would be one strategy to preserve biodiversity [23], or conservation goals [24], while maintaining the equilibrium of ecological systems kindly, in order to increase ecosystem service, which leads to the wellbeing of local people continually.

Inherently, that botanic garden of Samosir areas has natural potential, which has created a tour of areas of object development based on biodiversity, while also creating and advancing local economic development (Figure 3), and it has brought new jobs for others. Hence, it refers to the direct observation (field observation) and according to the description formerly (above), the Aek Natonang areas naturally have natural potential (ecological function) with a site of high altitude at least 1.300 *masl*, which are reservoir area and its functions, plant and animal diversity (biodiversity), and a beautiful natural view. These aspects is criterion to ecotourism areas development and automatically it brought to local economic development kindly and ecologically, which one is trader with various of snack, drinks, foods, and so on (Figure 4).

In Aek Natonang, for example, is one of the places that has some natural potential. For example, it is part of the water or conservation areas' function, which is used to fulfill the needs of around a hundred communities. Hence, it is necessary to pay attention to all academics and stake holders in order to actualize it, consistently and sustainable with involves involving local people.

**Figure 4.** Trader in Samosir botanic garden areas developed

There were some public facilities of service to tourists, which are part of the ecotourism development areas, alongside there were some rest areas, villas, home stays, and so on (Figure 5). This is part of the advantage of adding to income secondary for local or regional economic growth, as a representative of a development type based on natural resources without sacrificing environmental aspects. According to the description above, or formerly, ecotourism is one example of maintaining and preserving biodiversity as conservation spot areas development [25], which has multiple functions that must be guarded and, of course, involves involving local people in managing it. Those are also part of one way of both mitigating and adapting to climate change impacts kindly, and effectively [3].

**Figure 5.** (A) the river that developed to rest area with natural view in around, and (B) Villa development in river or botanic garden areas as used to stay overnight and others

## 4. Conclusion

Development of ecotourism in botanic gardens area is one way to increase local economies and as conservation areas function, which is one goal of sustainability, and it would be a welfare goal ecologically. Hence, the botanic garden of Samosir is one of many examples of how this is changing to ecotourism based on reservoirs in the area. On the other hand, that is part of the mitigate and adapt process to climate change impacts, which is capability in water storage and sustainability, although it would be in dry season occurrence, alongside it would be creating an ecological nuance developed that is more comfortable, cool, and the others.

## References

[1] Newbold T 2018 Future Effects of Climate and Land-use Change on Terrestrial Vertebrate Community Diversity under Different Scenarios *Proc. R. Soc. B Biol. Sci.* **285** 1–9

[2] Jung M, Rowhani P and Scharlemann J P W 2019 Impacts of Past Abrupt Land Change on Local Biodiversity Globally *Nat. Commun.* **10** 1–8

[3] Agyeman Y B 2020 Ecotourism as an Adaptation Strategy for Mitigating Climate Change Impacts on Local Communities Around Protected Areas In Ghana *Handb. Clim. Chang. Resil.* 1–20

[4] Hakim L and Nakagoshi N 2014 Ecotourism and Climates Changes: the Ecolodge Contribution in Global Warming Mitigation *J. Trop. Life Sci.* **4** 26–32

[5] Harada J and Yasuda N 2004 Conservation and Improvement of the Environment in Dam Reservoirs *Int. J. Water Resour. Dev.* **20** 77–96

[6] Kawashima S 2007 Conserving Reservoir Water Storage: An Economic Appraisal *Water Resour. Res.* **43** 1–9

[7] Kristiyanto 2020 Traditional Development of Water Conservation based on Local Knowledge: Coping with Climate Change Impacts in Rural Areas *IOP Conf. Ser. Earth Environ. Sci.* **423** 1–8

[8] Kristiyanto 2021 The "Jamban Systems" as Coping to both of Land Use and Climate Change Impacts in Rural Area *IOP Conf. Ser. Earth Environ. Sci.* **724** 1–8

[9] Ward F A and Pulido-Velazquez M 2008 Water Conservation in Irrigation can Increase Water Use *Proc. Natl. Acad. Sci. U. S. A.* **105** 18215–20

[10] Muryani C, Santoso S and Utomowati R 2018 Potential Analysis and Development of Reservoir Water for Ecotourism at Gajah Mungkur Wonogiri *IOP Conf. Ser. Earth Environ. Sci.* **145** 1–9

[11] Tang Q, Fu B, Collins A L, Wen A, He X and Bao Y 2018 Developing a Sustainable Strategy to Conserve Reservoir Marginal Landscapes *Natl. Sci. Rev.* **5** 8–10

[12] Tian M, Min Q, Lun F, Yuan Z, Fuller A M, Yang L, Zhang Y and Zhou J 2015 Evaluation of Tourism Water Capacity in Agricultural Heritage Sites *Sustainability* **7** 15548–69

[13] Enriquez J, Tipping D C, Lee J J, Vijay A, Kenny L, Chen S, Mainas N, Holst-Warhaft G and Steenhuis T S 2017 Sustainable Water Management in the Tourism Economy: Linking the Mediterranean's Traditional Rainwater Cisterns to Modern Needs *Water* **9** 1–118

[14] Gössling S, Peeters P, Hall C M, Ceron J P, Dubois G, Lehmann L V and Scott D 2012 Tourism and Water Use: Supply, Demand, and Security. An International Review *Tour. Manag.* **33** 1–15

[15] Indrawardani K F, Maryunani, Multifiah and Irawanto R 2019 Purwodadi botanic garden: conservation area with sustainable ecotourism potential *Int. J. Business, Econ. Law* **20** 206–17

[16] Jamaliah M M and Powell R B 2018 Ecotourism Resilience to Climate Change in Dana Biosphere Reserve, Jordan *J. Sustain. Tour.* **26** 519–36

[17] Grimm I J, Alcântara L C S and Sampaio C A C 2018 Tourism under Climate Change Scenarios: Impacts, Possibilities, and Challenges *Brazilian J. Tour. Res.* **12** 1–22

[18] Miller J S, Lowry P P, Aronson J, Blackmore S, Havens K and Maschinski J 2016 Conserving Biodiversity through Ecological Restoration: the Potential Contributions of Botanical Gardens and Arboreta *Candollea* **71** 97–8

[19] Sharrock S L and Group B G I S M D 2011 The Biodiversity Benefits of Botanic Gardens *Trends Ecol. Evol.* **26** 433

[20] Butarbutar R and Soemarno S 2013 Environmental Effects of Ecotourism in Indonesia *J. Indones. Tour. Dev. Stud.* **1** 97–107

[21] Sangpikul A 2017 Ecotourism Impacts on the Economy, Society and Environment of Thailand *J. Rev. Glob. Econ.* **6** 302–12

[22] Mondino E and Beery T 2019 Ecotourism as a Learning Tool for Sustainable Development. The Case of Monviso Transboundary Biosphere Reserve, Italy *J. Ecotourism* **18** 107–21

[23] Sridhar K R 2015 Ecotourism and Biodiversity : A New Perspective *Voyag. A Travel Tour. Exped.* 4–6

[24] Stronza A L, Hunt C A and Fitzgerald L A 2019 Ecotourism for Conservation? *Annu. Rev. ofEnvironment Resour.* **44** 229–53

[25] Hvenegaard G T and Dearden P 1998 Linking Ecotourism and Biodiversity Conservation: A Case Study of Doi Inthanon National Park, Thailand *Singap. J. Trop. Geogr.* **19** 193–211

IOP Conf. Series: Earth and Environmental Science 1016 (2022) 012050 doi:10.1088/1755-1315/1016/1/012050

# Effectiveness of solar panel positioning and grid tie inverter (GTI) in generating solar energy in Indonesian houses

**Dodik Siswantoro**

Department of Accounting, Universitas Indonesia, Kampus Baru UI Depok, West Java, 16424 Indonesia

Corresponding author: dodik.siswantoro@ui.ac.id

**Abstract**. In Indonesia, most electricity powers are based on non-renewable resources, which is not good for climate change. Therefore, the use of solar panels based on rooftops is an excellent alternative way to reduce electricity use. Thus, the objective of this paper is to analyze the effectiveness of solar panel positioning and Grid Tie Inverter (GTI) in generating solar energy for small- to medium-sized houses in Indonesia. Solar panel usage by residents living in the city may be limited, as the rooftop space in city houses is limited. GTIs can generate different types of power as they have different resources, this is good for climate change as no battery is needed In this research, two different solar panels with different positioning and angles, and connected to different types of GTI, were analyzed. The period of study was from 23rd July 2020 to 5th May 2021. The analysis was based on index comparison and movement of the sun during the said period. The result of the study indicates that the position and angle of solar panels affect the electricity generated by them. Solar panels placed in a horizontal position were observed to generate more power, and imported model inverters were observed to provide better electricity supply to homes.

**Keywords** : Solar panel, inverter, sun, rooftop, resident

## 1. Introduction

The development toward using solar energy for generating electricity in roof top-type residential houses in Indonesia has been quite rapid. Despite receiving only a small incentive for exported electricity (only 65% of the total electricity power), the residents of Indonesia still prefer to have rooftop solar panels as their alternative source of electricity, as it is more economical. In the long run, rooftop solar panels provide higher benefits to users, as they serve as a free source of electricity. They have an average payback period of around 7–12 years, depending on the type of solar panel used, such as grid tie (which runs on battery), off-grid (which does not run on battery), and hybrid (which runs both with and without battery). Of these types, the grid tie type of solar panel is the most economical. This type of solar panel is not battery-supported and so can be used only in the afternoons.

The use of rooftop solar panels for electricity generation is a new trend that has been developing slowly. The Indonesian government did not provide a good incentive for developing solar energy in home solar panels, and the export electricity generated by solar panels is at only 65% of the normal price. To generate electricity from a solar panel, the power meter has to be replaced, which can cost around IDR 1.655 million. Therefore, thorough feasibility and profitability studies must be conducted before starting the usage of rooftop solar panels to generate electricity. Indonesia is well-positioned to generate considerable amounts of power from rooftop solar panels [1]. Its location on the equator is ideal for the absorption of substantial amounts of sunlight.

A majority of the houses in Indonesia measure an area of 24–70m2. If rooftop solar panels are installed in such limited spaces, they may not be able to generate maximum electricity, as their capacity would be diminished by the lack of space. So far, not much research has been conducted on the effectiveness of solar panel positioning and Grid Tie Inverter (GTI) usage in residential units in Indonesia. This research paper will attempt to fill this gap and suggest the best location and position for installing solar panels in small- to medium-sized residential units in Indonesia. The research will focus on small-scale homes only.

## 2. Literature review

Rooftop solar panels can be highly potent when they are installed on a large scale in big cities [2]. Currently, many Indonesian households do not use solar panels to generate electricity. Therefore, there is significant potential for installing rooftop solar panels not only in big cities but also in the rural areas of Indonesia, which do not get stable electricity [3]. It reports that countries such as Switzerland have a high potential for rooftop solar panel power generation [4]. Although Switzerland is located far from the equator compared with Indonesia, solar panels are still expected to have high potential in sun-dominated areas of the country.

Research that studied the performance of rooftop solar panels in Thailand states that the efficiency of solar energy generation will differ from one region to another based on altitude and longitude [5]. However, since Thailand is located in the north of the equator and falls under sun movement, the sunlight received by different regions within Thailand would not vary much. Other studies argue that all rooftops may not be suitable for solar panel power generation [6,7]. In Thailand, only 30% of residential rooftops meet the standards required for rooftop solar panel installation [8].

Rooftop solar panels have different feasibilities for electricity generation as proved by a cost benefit analysis performed [9,10]. Rooftop solar panels also provide economic benefits over investment [11]. Today, many techniques, such as the skymap, can be used to generate maximum power from solar panels [12]. Many factors can aid the efficient installation of rooftop solar panels, such as Photovoltaic (PV) devices [13], mathematic algorithms [14], and Solar3D technology [15].

## 3. Research method

The research was conducted by analyzing observations using quantitative methods. Two grid tied rooftop solar panels installed on residential units were observed from 23rd July 2020 to 5th May 2021. The one-year observation was not possible due to certain electrical issues. Each solar panel comprised two packages of rooftop solar panels with 340wp and was connected to a GTI inverter of 1000VA. The two solar panels were of different brands. One inverter was imported from China, as it was not available in the Indonesian market. The electricity generation of each package was measured manually using a power meter at 06:00 AM every day. Previously, a Blynk platform was used to record the activity automatically through WiFi, but since it was encountering a problem, it had to be discontinued.

The first solar panel was positioned horizontally and was attached to a horizontal cement roof, while the second solar panel was positioned in a tilted position following the angle of the roof (see figure 1). The electricity generated by both inverters was compared and analyzed in relation to the movement of the sun, using Sun Surveyor. On 21st June 2020, the sun was located at the top north of the solar panels. On 5th October 2020, the sun was located above the solar panels. On 22nd December 2020, the position of the sun changed to the top south of the solar panels. On 6th March 2021, it moved back to above the solar panels. The sun was located at the south of the solar panels for only five months. For the remaining seven months of the study period, it was located the north of the solar panels (see figure 2).

On 23rd January 2021, we changed the solar panel connected to a different inverter to check the consistency of electricity generation. In this manner, we were able to assess the effectiveness of both the inverters.

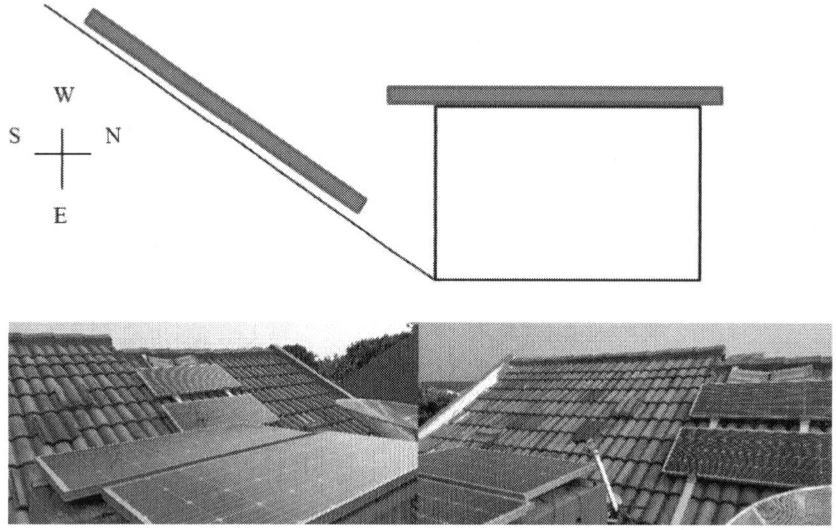


**Figure 1.** Position of the rooftop solar panels
Source: Research data

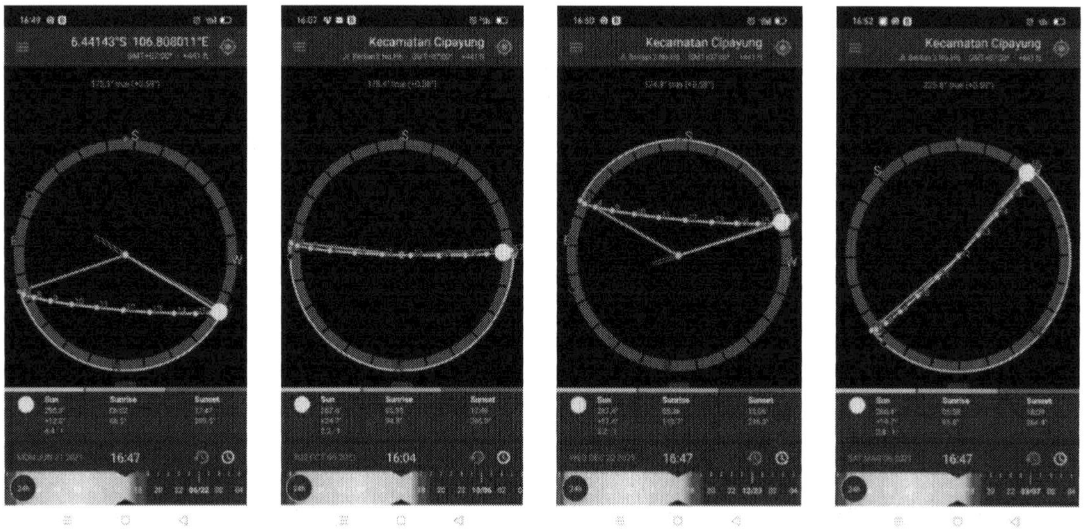

a b c d

**Figure 2.** Sun position at top north on 21st June (a), above the solar panels on 5th October (b) at top south on 22nd December (c), and above the solar panels on 6th March (d)
Source: Sun Surveyor

## 4. Analysis

From Figure 3 given below, we can see that the electricity generation ratio has increased slightly from Package 1 to Package 2. This is consistent with the movement of the sun from above the solar panels on 5th October 2020 to the south of the solar panels towards the end of the year. Thereafter, it kept increasing until 3rd December 2020. Then it started decreasing again until 22nd January 2021. The sun absorption of Package 2 decreased after 5th October 2020 because the solar panel was positioned toward the north, while the sun was positioned toward the south. Total watts generated by Package 1 was 128.64kWatt, while that by Package 2 was 139.43kWatt for the period ranging from 23rd July to 3rd

December 2020. Based on this data, we can infer that although the positioning of Package 1 is better, Package 2 will be more viable as it generates more electricity.

For the period between 4th December 2020 and 22nd January 2021, the total watt generated by Package 1 was 39.2kWatt and that by Package 2 was 39.4kWatt. During this period, the sun moved to the top of south on 22nd December 2020 and then moved back to the north. Between 4th December and 22nd December 2020, as the sun moved towards the south, both packages generated the same amount of electricity, that is, 15kWatt.

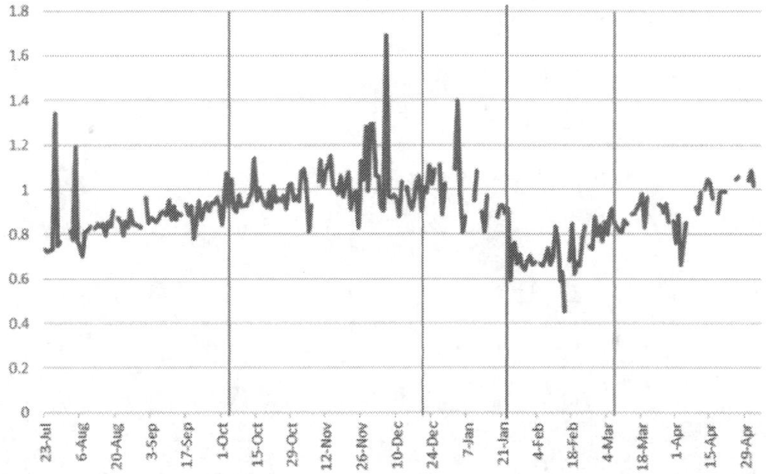


**Figure 3.** Ratio of electricity generation by Package 1 versus Package 2
Source: Research data

On 23rd January 2021, we changed the connection of the solar panels to different inverters. This made the ratios reduce from the previous recording. This meant that the inverter connected to Package 2 gave better electricity generation. The trend was sharper than what was recorded during similar movement of the sun after 5th October and before 6th March on the reverse position, when the sun moved to the south (after 5th October) and to above the solar panel (before 6th March). Similar events were also recorded before 5th October and after 6th March on the reverse position, when the sun moved to above the solar panel (before 5th October) and then to the north (after 6th March) (see Figures 3, 4 and 5). This data proves that inverter Package 2 provided better electricity generation.

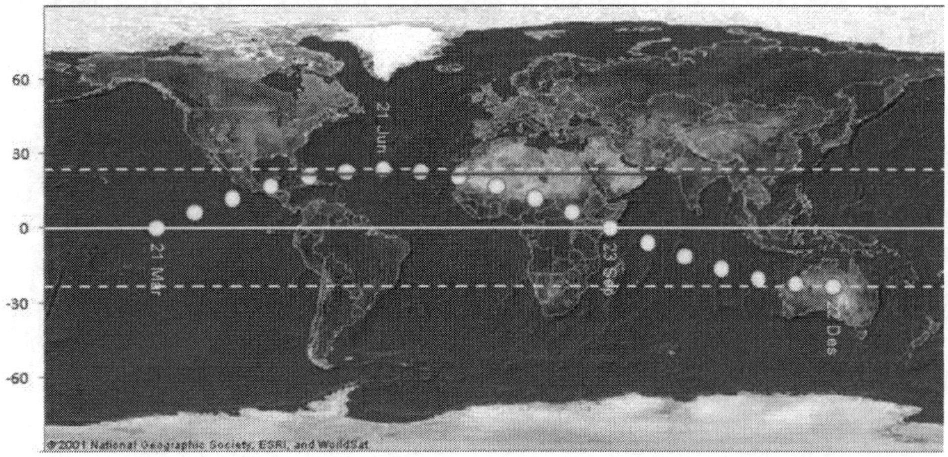


**Figure 4.** Movement of the sun in one year, Source: Google.com

From 23rd January until 6th March 2021, inverter Package 1 produced 27.5kWatt of electricity, while inverter Package 2 produced 37.7kWatt of electricity. The ratio between the two was 0.73. Between 7th March and 5th May 2021, inverter Package 1 produced 61.8kWatt of electricity, while inverter Package 2 produced 66kWatt of electricity. The ratio between the two was 0.93. The ratio improved between the two time periods because the solar panel of Package 1 was placed in a tilted position to the sun, which kept moving north until 21st June 2021. This research is important as previous research only focused on the location of the solar panel [5–7,9]. Other researchers only discussed the potency of solar panel rooftop for electricity [2,4], but if it does not treat well, it would be less efficient.

**Figure 5.** Sun movement above solar panel on 5th October and 6th March
Source: Google Maps, Sun Surveyor

## 5. Conclusion

The study proved that the position and location of solar panels can affect the amount of electricity generated by them. In this research, limited rooftop space may have been a limitation for solar panel positioning. It would be better to position solar panels such that they face the sun for a longer period of time during a year. In addition, the type of inverter used may also affect the amount of electricity generated. The more efficient the solar panel generation it would be better for climate change mitigation. Efficiency could be a significant problem for solar panel electricity power to mitigate climate change.

The limitation of this research was that it was restricted to only one location and only two types of inverters were used. Further research can be conducted in different locations using similar rooftop solar panels.

## References

[1] Hidayatno A, Setiawan A D, Wikananda Supartha I M, Moeis A O, Rahman I and Widiono E 2020 Investigating policies on improving household rooftop photovoltaics adoption in Indonesia *Renew. Energy* **156** 731–42

[2] Mansouri Kouhestani F, Byrne J, Johnson D, Spencer L, Hazendonk P and Brown B 2019 Evaluating solar energy technical and economic potential on rooftops in an urban setting: the city of Lethbridge, Canada *Int. J. Energy Environ. Eng.* **10** 13–32

[3] Porse E, Fournier E, Cheng D, Hirashiki C, Gustafson H, Federico F and Pincetl S 2020 Net solar generation potential from urban rooftops in Los Angeles *Energy Policy* **142** 111461

[4] Assouline D, Mohajeri N and Scartezzini J L 2018 Large-scale rooftop solar photovoltaic technical potential estimation using Random Forests *Appl. Energy* **217** 189–211

[5] Yoomak S, Patcharoen T and Ngaopitakkul A 2019 Performance and economic evaluation of solar rooftop systems in different regions of Thailand *Sustain.* **11**
[6] Hong T, Lee M, Koo C, Jeong K and Kim J 2017 Development of a method for estimating the rooftop solar photovoltaic (PV) potential by analyzing the available rooftop area using Hillshade analysis *Appl. Energy* **194** 320–32
[7] Chaves A and Bahill A T 2010 Locating Sites for Photovoltaic Solar. Pilot study uses DEM derived from LiDAR *ArcUser* 24–7
[8] Kerdoum P and Premrudeepreechacharn S 2020 Analysis of PV penetration level on low voltage system in Chiang Mai Thailand *Energy Reports* **6** 754–60
[9] Shukla A K, Sudhakar K and Baredar P 2016 Simulation and performance analysis of 110 kWp grid-connected photovoltaic system for residential building in India: A comparative analysis of various PV technology *Energy Reports* **2** 82–8
[10] Shukla A K, Sudhakar K and Baredar P 2016 Design, simulation and economic analysis of standalone roof top solar PV system in India *Sol. Energy* **136** 437–49
[11] Vimpari J and Junnila S 2019 Estimating the diffusion of rooftop PVs: A real estate economics perspective *Energy* **172** 1087–97
[12] Yousuf M U and Shere S M 2020 A novel computational methodology to estimate solar energy on building rooftops *Environ. Prog. Sustain. Energy* **39** e13385
[13] Xu L, Long E and Wei J 2020 Study on the limiting height of rooftop solar energy equipment in street canyons under the cityscape constraints *Sol. Energy* **206** 1–7
[14] Li Y, Si T and Liu C 2018 Geographical variation in energy yields of rooftop photovoltaic projects in Australia *Adv. Build. Energy Res.* **14** 1–18
[15] Liang J, Gong J, Xie X and Sun J 2020 Solar3D: An Open-Source Tool for Estimating Solar Radiation in Urban Environments *ISPRS Int. J. Geo-Information* **9** 524

# Annual characteristics of gross primary productivity (GPP) in mangrove forest during 2016-2020 as revealed by Sentinel-2 remote sensing imagery

**Linda Sri Rahayu Romadhoni[1], Abd. Rahman As-syakur*[23], Zainul Hidayah[1], Dwi Budi Wiyanto[2], Rahma Safitri[1], Raden Yusuf Satriyana Utama[1], I Made Sara Wijana[3], Alfandy Putra Anugrah[1] and I Made Oka Guna Antara[3]**

[1]Marine Science and Fisheries Department, Faculty of Agriculture, Trunojoyo University of Madura, Jalan Raya Telang No 02, Kamal-Bangkalan, East Java 69162, Indonesia.
[2]Marine Science Department, Faculty of Marine and Fisheries, Udayana University, Bukit Jimbaran Campus, Bali, 80361 Indonesia.
[3]Research Centre for Environment (PPLH), Udayana University, Sudirman Campus, Denpasar, Bali 80232, Indonesia.
*Corresponding author: ar.assyakur@pplh.unud.ac.id

**Abstract**. Carbon dioxide ($CO_2$) is the most responsible gas in the phenomenon of global warming on earth because of its greatest concentration and longevity in the long atmosphere. Meanwhile mangrove is one of the *blue carbon* parameters that can take $CO_2$ for photosynthesis and store it into biomass and sediment, so the existence of Mangrove plays a key role in the balance of the global carboncycle. Gross Primary Productivity (GPP) is one of the key variables in conducting a carboncycle study because the GPP values constitute the total value of carbon fixation by terrestrial ecosystem through vegetation photosynthesis. The aim of this study is to recognize the annual characteristics of GPP values in the mangrove ecosystem using remote sensing satellite Sentinel-2 during the period of 2016 to 2020 in TAHURA Ngurah Rai, Bali, Indonesia. The vegetation photosynthesis model (VPM) model is used to calculate annual GPP by using remote sensing indices of Enhanced Vegetation Index (EVI) and Land Surface Water Index (LSWI). Both remote sensing indices are supported by temperature and solar radiation data to determine photosynthesis fraction. The results of the current study indicated that the annual GPP values in the mangrove forest of TAHURA Ngurah Rai have decreased during the observation year. The total value of GPP in 2016 is reached to 28790 tC $m^{-2}$ $year^{-1}$, while the total amount of GPP value in 2020 decrease to 26223 tC $m^{-2}$ $year^{-1}$. The decline in GPP values may be due to changes of land cover and mangrove mortality that occurred around Benoa port in 2018.

**Keywords** : Sentinel-2, gross primary productivity, mangrove forest, remote sensing

## 1. Introduction

Global warming is one of the world's issues which is marked by an increase in the earth's surface temperature related to greenhouse gases (GHG). One of the causes of the increasing number of greenhouse gases is $CO^2$

emissions in the atmosphere through land burning, motor vehicle emissions and factory waste [1]. Anthropogenic global warming is known to have caused major changes in temperature, rainfall, wind patterns and sea level height on the earth in the last hundred years and has an impact on regional and global climate change [2]. The impact of climate change can be minimized by decreased carbon emissions in the atmosphere or increasing carbon sequestration by primary producers [1].

Indonesia targets to decreased carbon emissions in 2030 by 834 million tons of $CO^2$e or 29% of BAU (business as usual). This target has been achieved by 7.55% in 2018 through better sustainable management practices and increasing the role and area of tree cover [3]. According to the Ministry of Environment and Forestry of the Republic of Indonesia [4,5], Indonesia's forest area in 2020 is 95.6 million ha or 50.9% of the land area, of which 3.31 million ha is mangrove forest. The role of mangrove forests in decreased carbon emissions is considered more effective when compared to tropical rain forests such as peatlands. This is evidenced by the incidence of mangrove deforestation covering an area of 0.7% of the total forest area which has an emission impact of 0.02-0.12 Pg carbon per year, which is equivalent to 10% of global deforestation emissions [6].

On the other hand, to understand the role of mangrove forest areas in reducing carbon emissions in the atmosphere, further studies are needed. In this case, one important variable that needs to be known is the value of primary productivity (GPP). GPP is the total value of carbon fixation by terrestrial ecosystems through vegetation photosynthesis in a certain period [7]. GPP is considered to be very efficient to know and understand the variation of the coral productivity of vegetation in carbon sequestration in the atmosphere and its relationship to changes around mangrove forests, the increase in $Co^2$ in the atmosphere and climate change [8].

Temporal estimation of the GPP value of mangrove forests can be known using remote sensing technology. In this case remote sensing provides repeated and systematic satellite imagery data in a temporal scale that allows characterizing the vegetation structure including leaf area index (LAI) and photosynthetic active radiation absorption (APAR) [9,10]. Remote sensing technology is highly effective in monitoring GPP at regional and local scales by linking the absorbed photosynthetic active radiation fraction (*f*APAR) driven by the vegetation index [11,12]. In calculating the estimated GPP value in this mangrove forest, using Sentinel 2A satellite image data level 1C. This is because the Sentinel 2 Level 1C image is included in the medium resolution image (10 x 10 meters) and the study area has a heterogeneous landscape. Based on these considerations, the Sentinel 2 level 1C image is a helpful solution when compared to the use of Landsat 8 imagery which has a resolution of 30 x 30 meters [13].

The calculation of the estimated GPP value is carried out using an Efficiency model of using LUE light which has been commonly used by using a simple algorithm in estimating the GPP value [14]. The LUE model states that carbon exchange is a function of the amount of light energy absorbed by vegetation and the efficiency of light energy used to sequester carbon [15]. one of the efficiency models of using light to be used in estimating the GPP value is the vegetation production model (VPM) [16]. VPM is a light efficiency model that utilizes remotely sensed imagery to estimate GPP based on the impact of temperature, water pressure, and phenology [17]. Based on what has been described, this study was conducted to determine the annual characteristics of the gross primary productivity value in the TAHURA Ngurah Rai mangrove forest, Bali using Sentinel 2 imagery from 2016 to 2020.

## 2. Materials and method

### *2.1. Research location*

The study is located in the Ngurah Rai Bali forest park. By administration lies in Denpasar district, and Kuta district, Bali. The Ngurah Rai forest park administration is part of UPT. Tahura Ngurah Rai's district of forestry service in Bali Province. Astronomically at 115⁰10'-115⁰15' E and 8⁰41'-8⁰47' S (Figure 1).

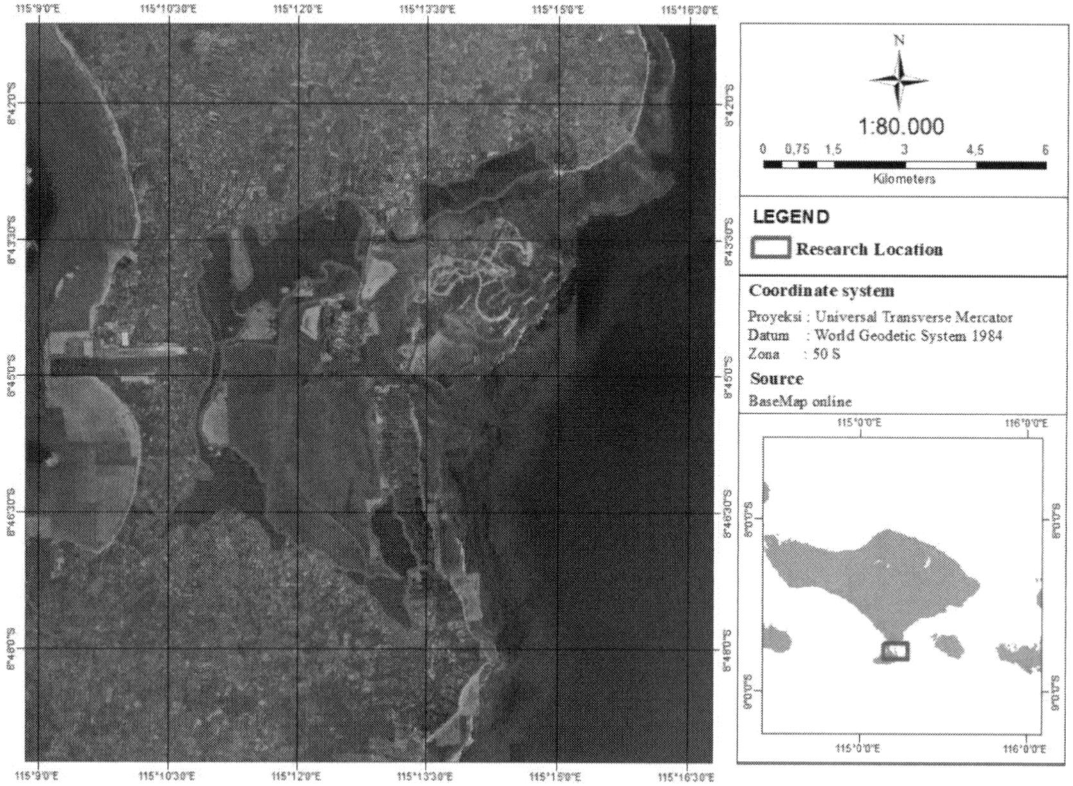


**Figure 1.** Research location

*2.2. Materials*

The remote sensing data used is the sentinel-2 level 1-c image file from 2016 to 2020 were obtained from Copernicus through google earth engine. In the VPM models, two remote sensing indexes are created by the enhancement index of vegetation (EVI) and the land surface water index (LSWI). Enhanced Vegetation Index (EVI) is linearly correlated with the green leaf index based on multispectral data. EVI index is sensitive to variations in the canopy of vegetation [15]. The EVI employs an additional blue band for atmospheric correction and variable soil and canopy background reflectance and is calculated from blue, red and NIR bands [18] EVI values range from -1 to +1. The EVI is calculated as Equation 1:

$$\text{EVI} = 2.5\, x \frac{\rho nir - \rho red}{\rho nir + (6\, x\, \rho red - 7.5\, x\, \rho blue) + 1} \quad (1)$$

LSWI is calculated Equation 2:

$$LSWI = \frac{\rho nir - \rho swir}{\rho nir + \rho swir} \quad (2)$$

Because of Short-Wave Spectral Infrared (SWIR) is sensitive to water content, vegetation, and moisture land, so a combination between NIR and SWIR should be used to know water sensitive vegetation indices. The higher water level in a leaf, or an increased moisture land, making the absorption of SWIR increases and the SWIR reflectance decreases, so that the LWSI can be increased. LSWI values range from -1 to + 1 [16].

*2.3 Overview of the VPM model*

The canopy vegetation consisting of photosynthesis active (PAV, chloroplasts) and non-photosynthesis active (NPV, mostly branches and stems). Based on the concept of PAV dan NPV, VPM model was developed to estimate the value of GPP [16]. A brief description of VPM is given below (Equation 3):

$$GPP = \varepsilon g \times fPAR_{chl} \times PAR \quad (3)$$

where $\varepsilon g$ is the light used efficiency (μmol m$^{-2}$ s$^{-1}$, photosynthetic photon flux density, PPFD). $FPAR_{chl}$ is fraction of photosynthetic active radiation (PAR) that absorbed by chlorophyll and PAR is Photosynthetically Active Radiation (μmol photosynthetic photon flux density, PPFD) calculated as 0.45 x R (R, incoming global solar radiation). In this research, daily PAR values are added up for one year to get the annual GPP value. The light used efficiency is affected by temperature, water and leaf phenology which is defined as follows (Equation 4):

$$\varepsilon g = \varepsilon_0 \times T_{scalar} \times W_{scalar} \times P_{scalar} \quad (4)$$

$\varepsilon_0$ is the light used efficiency maximum which is equal to 0,40 gC mol$^{-1}$ [19] used in this research, and $T_{scalar}$ is temperature values, $W_{scalar}$ is water values dan $P_{scalar}$ is phenology leaf values, all that matter is included to determine the effect of leaf age on photosynthesis at the canopy level. The calculation of the $P_{scalar}$ value depends on the longevity of the leaves (deciduous, green leaves all year round). For a leaf-dominated canopy with a life expectancy of one year, $P_{scalar}$ is calculated using two distinct phases as a linear function [13]. In this research the $P_{scalar}$ value used is 1 because mangroves are vegetation that remains green throughout the year. $T_{scalar}$ is a parameter used to represent photosynthetic temperature. Using Equations 5, developed for the Terrestrial Ecosystem Model [15]:

$$Tscalar = \frac{(T-Tmin)\ x\ (T-Tmaks)}{[(T-Tmin)\ x\ (T-Tmax)]-(T-Topt)^2} \quad (5)$$

where T is the average monthly temperature acquired from the meteorological, climatological, and geophysical services Ngurah Rai, Bali. The $T_{min}$, $T_{max}$, and $T_{opt}$ parameters for photosynthesis used in the study are 0,48,28.

$W_{scalar}$ is the parameters used to know the effect of water on photosynthesis vegetation. The equation used to calculate $W_{scalar}$ was as follows (Equation 6) [16]:

$$Wscalar \frac{1+LSWI}{1+LSWImax} \quad (6)$$

where the $LSWI_{max}$ value is the maximum LSWI value in growing seasons for individual pixels.

In the VPM model, $FPAR_{chl}$ in photosynthesis is assumed to be a linear function of EVI and A coefficient set at 1.0 (Equation 7) [16].

$$fPARchl = EVI\ x\ A \quad (7)$$

## 3. Result and discussion

VPM Model uses the EVI index and the LSWI index to determine the annual GPP value. The GPP value shows different results every year (Figure 2). The total value of GPP in 2016,2017,2018,2019,2020 is 28790 tC m$^{-2}$ year$^{-1}$, 25442 tC m$^{-2}$ year$^{-1}$, 25769 tC m$^{-2}$ year$^{-1}$, 26522 tC m$^{-2}$ year$^{-1}$, 26223 tC m$^{-2}$ year$^{-1}$ (Figure 3). In 2016 the GPP had the highest value which then continued to decline over the years. In 2017 the value of GPP has decreased by as much as 3348 tC m$^{-2}$ year$^{-1}$, and this is because the value of the fraction of photosynthesis active in 2017 was lower than 2016, in 2016 the fraction of the photosynthetic active for 15954.6835 mol photosynthetic photon flux density, PPFD and in 2017 it was 147738.1522 mol photosynthetic photon flux density, PPFD. The value of the active photosynthetic fraction is very influential on the total annual GPP value so that the total annual GPP value in 2017 is smaller compared to 2016. In addition, in 2017 there was a phenomenon *dieback* due to reclamation by PT. PELINDO so that the mangrove area decreased by 2.43 ha in Benoa Cape.

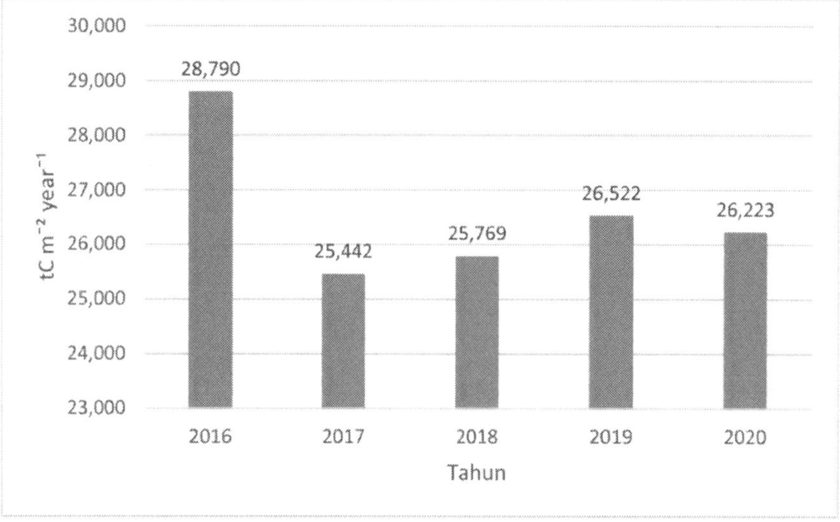


**Figure 2**. Total of annual gross primary productivity in TAHURA Ngurah Rai, Bali

In 2018 the value of GPP has increased from 2017, GPP total value in 2017 increased by 329 tC m$^{-2}$ year$^{-1}$. This is maybe due by the reforestation process and low rainfall in 2018. In October 2018 PT. PELINDO carried out rehabilitation by planting mangrove species of *Rhizopora apicullata* and *Rhizopora mucronata* as many as 100 thousand trees in the Benoa Bay area. In addition, based on data from the BMKG Ngurah Rai, Bali in 2018, the average rainfall in the TAHURA area was 99.06 mm/year, which is lower than the rainfall in 2017 which was 187.3 mm/year on average. When rainfall is high, the intensity of sunlight that is useful for plant photosynthesis will decrease, thereby reducing plant productivity [20]. In 2019 the value of GPP increased by 753 tC m$^{-2}$ year$^{-1}$. The increase in the value of GPP in 2019 are the result of rehabilitation by PT. PELINDO in Benoa Bay since October 2018. In 2020 the value of GPP decreased by 299 tC m$^{-2}$ year$^{-1}$.

GPP value varies each year, the value of the lowest GPP namely in 2017 amounted to 25442 tC m$^{-2}$ year$^{-1}$ and the highest value of GPP in the year 2016 amounted to 28 790 tC m$^{-2}$ year$^{-1}$. During the period of 5 years from 2016 to 2020 the value of GPP decreased by 2567 tC m$^{-2}$ year$^{-1}$ (Figure 4). The GPP value has decreased a lot in the northwest area of Benoa Cape marked in red on the map, then on the north side of Benoa Cape there was also a decreased in value of the GPP (Figure 5). Which the largest decrease occurred between 2017 to 2018 were caused by the presence of reclaimed land up to a total reduction in the mangrove area of 17 Ha, Changes in mangrove density also affect the total carbon uptake so that the GPP value decreases.

IOP Conf. Series: Earth and Environmental Science 1016 (2022) 012051 doi:10.1088/1755-1315/1016/1/012051

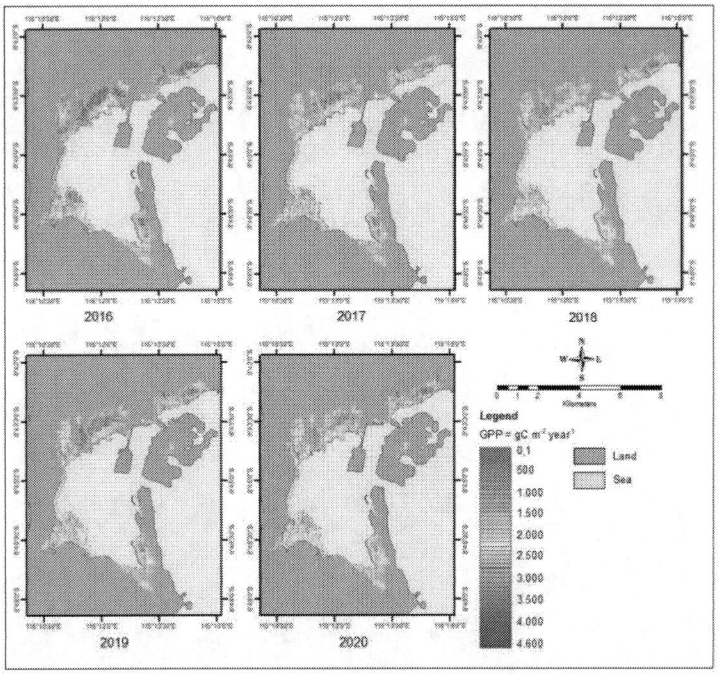


**Figure 3.** Annual gross primary productivity in TAHURA Ngurah Rai, Bali

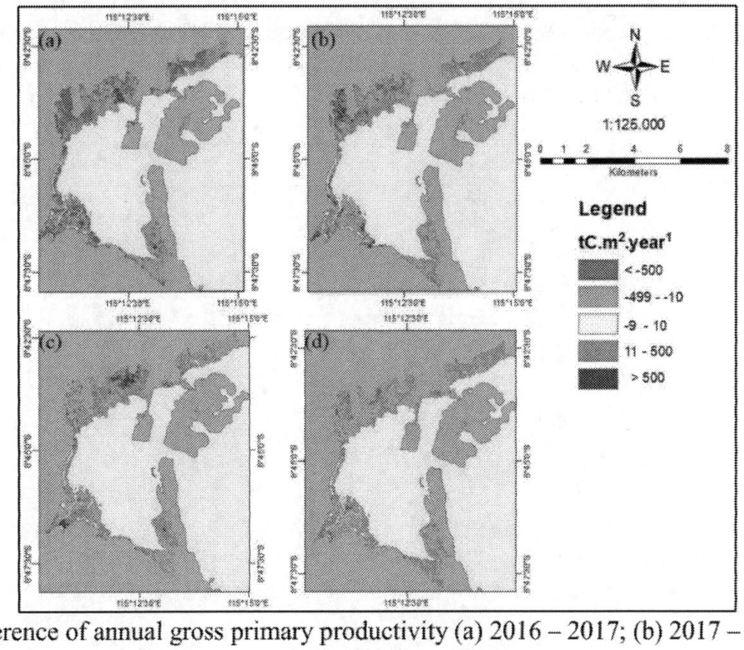


**Figure 4**. Difference of annual gross primary productivity (a) 2016 – 2017; (b) 2017 – 2018; (c) 2018 - 2019; (d) 2019-2020

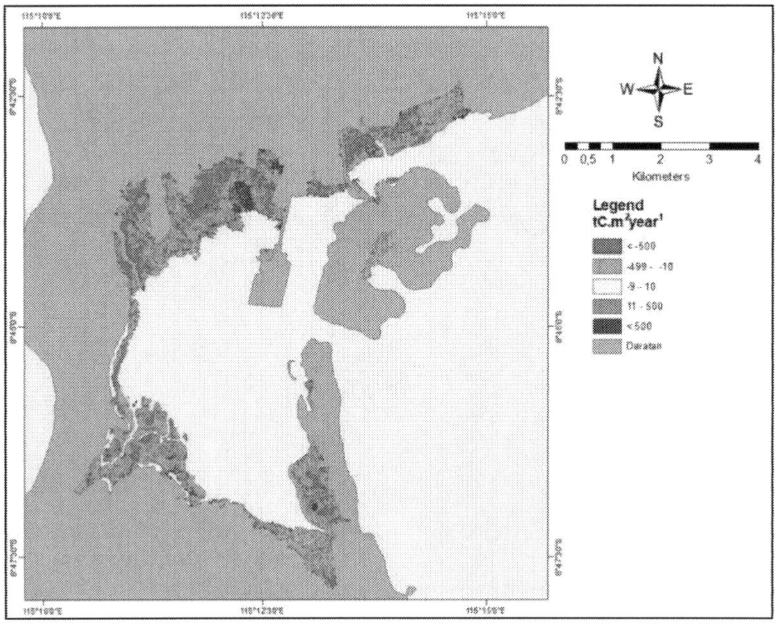


**Figure 5.** Difference of annual gross primary productivity 2016-2020

This study also shows the average annual GPP value as analyzed using Sentinel-2 image data. The lowest average GPP value was in 2017 which was 2211 gC m$^{-2}$ year$^{-1}$ and the highest average GPP value was in 2020, which was 2505 gC m$^{-2}$ year$^{-1}$. Meanwhile, the average annual GPP value increased in 2018 and 2019 by 2277 gC m$^{-2}$ year$^{-1}$ and 2323 gC m$^{-2}$ year$^{-1}$, respectively where the average annual GPP value again declined in 2020 to 2318 gC m$^{-2}$ year$^{-1}$ (Figure 6).

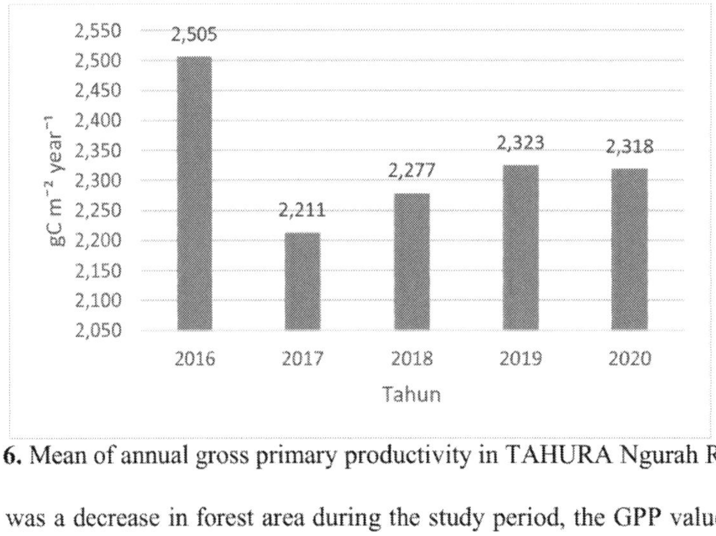


**Figure 6.** Mean of annual gross primary productivity in TAHURA Ngurah Rai, Bali

Although there was a decrease in forest area during the study period, the GPP value increase in 2018. This study shows the importance of rehabilitation and reforestation of mangrove forests in maintaining the carbon cycle. It is important to plant mangroves to increase carbon sequestration by vegetation. Mangrove ecosystems can sequester carbon much higher than tropical rain forests [21]. Mangroves can store carbon up to three times the average carbon storage per hectare by mainland tropical forests, where the optimal function of mangrove carbon absorption can reach 77.9% [22]. In addition, anthropogenic activities must also be controlled to maintain the balance of the mangrove ecosystem. Reducing the reclamation area is one of the efforts to maintain and preserve the mangrove forest in TAHURA Ngurah Rai, Bali.

## 4. Conclusion

Gross Primary Productivity calculation using the VPM model results in varying annual GPP values. GPP highest total value in 2016 and the total value of the lowest GPP which in 2017. For 5 years the total value of GPP in TAHURA Ngurah Rai, Bali experienced a decline of up to 2567 tC $m^{-2}$ $year^{-1}$. This is due to the reclamation activity that occurred from 2017 to 2018 covering an area of 17 ha, causing a lot of mangrove deaths in Benoa Bay. In addition, the GPP value decreased, influenced by the PAR value absorbed by plants, which decreased in 2017. This study shows the importance of rehabilitation and reforestation of mangrove forests in maintaining the carbon cycle. It is important to plant mangroves to increase carbon sequestration by vegetation, so that it can reduce the impact of global warming on the earth.

## References

[1] Prakoso T B, Afiati N and Suprapto D 2018 Biomassa Kandungan Karbon dan Serapan CO2 pada Tegakan Mangrove di Kawasan Konservasi Mangrove Bedono, Demak *Manag. Aquat. Resour. J.* **6** 156–63

[2] Cook J, Nuccitelli D, Green S A, Richardson M, Winkler B, Painting R, Way R, Jacobs P and Skuce A 2013 Quantifying the consensus on anthropogenic global warming in the scientific literature *Environ. Res. Lett.* **8** 024024

[3] Ruslan 2021 Status Pemanfaatan Energi Baru Terbarukan dan Opsi Nuklir dalam Bauran Energi Nasional *J. Pengemb. Energi Nukl.* **23** 39–49

[4] KLHK 2021 Deforestasi Indonesia Turun, Terendah Dalam Sejarah - Kementerian LHK *https://www.menlhk.go.id/site/single_post/3640/deforestasi-indonesia-turun-terendah-dalam-sejarah*

[5] KLHK Kick Off Penanaman Mangrove Tahun 2021 - Kementerian LHK *https://www.menlhk.go.id/site/single_post/3653/kick-off-penanaman-mangrove-tahun-2021*

[6] Donato D C, Kauffman J B, Murdiyarso D, Kurnianto S, Stidham M and Kanninen M 2011 Mangroves among the most carbon-rich forests in the tropics *Nat. Geosci.* **4** 293–7

[7] Takarina N D and Runtuwene D O 2019 Primary Productivity of Blanakan Fishpond at Subang, West Java *J. Phys. Conf. Ser.* **1245** 012034

[8] Nuarsa W, As-Syakur A R, Gunadi I G A and Sukewijaya I M 2018 Changes in Gross Primary Production (GPP) over the Past Two Decades Due to Land Use Conversion in a Tourism City *ISPRS Int. J. Geo-Information* **7** 57

[9] Lele N, Kripa M K, Panda M, Das S K, Nivas A H, Divakaran N, Naik-Gaonkar S, Sawant A, Pattnaik A K, Samal R N, Thangaradjou T, Saravanakumar A, Rodrigues B F and Murthy T V R 2021 Seasonal variation in photosynthetic rates and satellite-based GPP estimation over mangrove forest *Environ. Monit. Assess.* **193** 1–20

[10] Inoue Y, Peñuelas J, Miyata A and Mano M 2008 Normalized difference spectral indices for estimating photosynthetic efficiency and capacity at a canopy scale derived from hyperspectral and CO2 flux measurements in rice *Remote Sens. Environ.* **112** 156–72

[11] Turner D P, Gower S T, Cohen W B, Gregory M and Maiersperger T K 2002 Effects of spatial variability in light use efficiency on satellite-based NPP monitoring *Remote Sens. Environ.* **80** 397–405

[12] Iwata H, Ueyama M, Iwama C and Harazono Y 2013 Variations in fraction of absorbed photosynthetically active radiation and comparisons with MODIS data in burned black spruce forests of interior Alaska *Polar Sci.* **7** 113–24
[13] Yuan W, Cai W, Liu S, Dong W, Chen J, Arain M A, Blanken P D, Cescatti A, Wohlfahrt G, Georgiadis T, Genesio L, Gianelle D, Grelle A, Kiely G, Knohl A, Liu D, Marek M V., Merbold L, Montagnani L, Panferov O, Peltoniemi M, Rambal S, Raschi A, Varlagin A and Xia J 2014 Vegetation-specific model parameters are not required for estimating gross primary production *Ecol. Modell.* **292** 1–10
[14] Sims D A, Luo H, Hastings S, Oechel W C, Rahman A F and Gamon J A 2006 Parallel adjustments in vegetation greenness and ecosystem CO2 exchange in response to drought in a Southern California chaparral ecosystem *Remote Sens. Environ.* **103** 289–303
[15] Xiao X, Hollinger D, Aber J, Goltz M, Davidson E A, Zhang Q and Moore B 2004 Satellite-based modeling of gross primary production in an evergreen needleleaf forest *Remote Sens. Environ.* **89** 519–34
[16] Xiao X, Zhang Q, Saleska S, Hutyra L, De Camargo P, Wofsy S, Frolking S, Boles S, Keller M and Moore B 2005 Satellite-based modeling of gross primary production in a seasonally moist tropical evergreen forest *Remote Sens. Environ.* **94** 105–22
[17] Kanniah K D, Kang C S, Sharma S, Amir A A, Kanniah C :, Kang K D ;, Sharma C S ;, Amir S ; and Remote A A 2021 Remote Sensing to Study Mangrove Fragmentation and Its Impacts on Leaf Area Index and Gross Primary Productivity in the South of Peninsular Malaysia *Remote Sens.* **13** 1427
[18] Rahman As-syakur A, Osawa T and Adnyana I W S 2010 Medium Spatial Resolution Satellite Imagery to Estimate Gross Primary Production in an Urban Area *Remote Sens.* **2** 1496–507
[19] Hirano T, Segah H, Harada T, Limin S, June T, Hirata R and Osaki M 2007 Carbon dioxide balance of a tropical peat swamp forest in Kalimantan, Indonesia *Glob. Chang. Biol.* **13** 412–25
[20] Harun R A, Irsal and Jonis 2019 Pengaruh Curah Hujan dan Hari Hujan terhadap Produksi Tanaman Karet Umur 13, 16, dan 19 Tahun di PT. Socfin Indonesia Kebun Lima Puluh *J. Online Agroekoteknologi* **7** 20–8
[21] Windarni C, Setiawan A and Jurusan Kehutanan Fakultas Pertanian Universitas Lampung Jl Sumatri Brojonegoro No R 2018 Carbon Stock Estimation of Mangrove Forest in Village Margasari Sub-District Labuhan Maringgai District East Lampung *J. Sylva Lestari* **6** 66–74
[22] Bachmid F, Sondak C F A and Kusen J D 2018 Estimasi penyerapan karbon hutan mangrove Bahowo Kelurahan Tongkaina Kecamatan Bunaken *J. Pesisir dan Laut Trop.* **6** 8–13

**Acknowledgment**
The authors acknowledge to the The Directorate General of Higher Education, Ministry of Education, Culture, Research, and Technology for the grant funding for the competence program of the Independent Campus of Marine Science Study Program, Faculty of Agriculture, Trunojoyo University, Madura. The authors also acknowledge to the Research Centre for Environment (PPLH), Udayana University, which has facilitated and assisted in the implementation and completion of this research.

# Understorey species diversity and its effect on soil temperature and relative humidity in the Riparian Zone of Pusur River

**Alvian Febry Anggana[1] and Mujiyo[2]**

[1]Department of Environmental Science, Graduate Faculty, Universitas Sebelas Maret Jl. Ir Sutami 36 A, Kentingan, Surakarta, 57126 Indonesia
[2]Department of Soil Science, Faculty of Agriculture, Universitas Sebelas Maret Jl. Ir Sutami 36 A, Kentingan, Surakarta, 57126 Indonesia
*Corresponding author: angga.kshe43@gmail.com

**Abstract.** Climate change has had a negative impact on all of life aspects. Currently, we have seen frequent occurrences of droughts, rising temperatures, floods, landslides. Therefore, mitigation and adapatation efforts are needed to reduce climate change. One of these efforts is Identify the diversity of riparian understorey plants todetermine their role in the riparian microclimate.This study aims to determine the diversity of understorey plants and their effect on temperature and soil moisture in the Riparian Pusur. It was carried out along the Pusur River, from the upstream to the downstream (administratively included in Boyolali Regency and Klaten Regency) in August 2021. A vegetation analysis method was employed. Sample was obtained from nine observation stations in which 10 plots (2 m x 2 m in size) were established in each station, at elevation 100-800 m asl. This study identifies 9364 individuals that belong to 74 species and 35 families. The lower Upstream has the highest species diversity (30 species), while the mid Downstream has the lowest (14 species). Based on the measurement of the soil temperature, the lowest temperature is in the upper Upstream of 27°C, while the highest is in the lower Downstream of 33°C. The average relative humidity ranges from 39 to 60%. This study concludes that the maximum H' is identified at soil temperature of 27-28.8°C and relative humidity of 59-61% with the species diversity of riparian understorey ranges from 2.9 to 3.08.

**Keywords :** riparian, diversity, vegetation, soil, relative humidity

## 1. Introduction

Climate change occurs due to the increasing concentration of $CO_2$ in the atmosphere. Riparian understorey is one of the forest components that has the potential as a natural $CO_2$ absorber and is able to stabilize the microclimate. The riparian vegetation in transitional ecosystems, however, provides significant ecological benefits and functions for the sustainability of living organisms[1]. The riparian zone in rivers has been characterized by the presence of vegetation that has a positive effect on the river ecosystem. Mitsch and Gosselink [2] suggest that vegetation is able to maintain soil and water temperatures. Similarly, Anggana and Susanti [3] affirm that riparian vegetation significantly affects erosion adsorption and increased DO levels. Understorey vegetation is one of the components that make up a land ecosystem. It is defined as underlying vegetation of 5-10 cm in height [4]. Its presence has a positive impact on the riparian ecosystem.

Maisyaroh [5] suggests that understorey helps reduce water velocity, preventing erosion. In addition, it regulates the microclimate conditions, maintaining the stability of soil temperature and humidity[6]. These two factors are influential in plant growth. Lubis [7] confirms the effect of soil temperature on water uptake. The lower the temperature, the smaller the amount of water is absorbed by roots. Meanwhile, relative humidity refers to soil moisture content, which is the water content of the soil [8]. It affects soil erosion rate and water quality [9]. The Pusur River is a meandering alluvial river that flows from the Mount Merapi and ends at the Bengawan-Solo River. It is located between Klaten Regency and Boyolali Regency. The Upstream is in Boyolali Regency, while and the Mid- and Downstream are at Klaten Regency. Afandi et al [10] emphasize the potential of Pusur River that ranges from irrigation to eco-tourism. The present study aimed to investigate the diversity of understorey species and its effect on soil temperature and relative humidity in the riparian zone of Pusur River.

## 2. Methodology

### 2.1. Time and study site

The present study was carried out in the riparian zone of Pusur River; from the upstream (Boyolali) to the downstream (Klaten), Central Java Province, Indonesia. The data were collected in August, 2021.

### 2.2. Data collection technique

A survey method was employed, a total of 9 plots (90 sub-plots) were established in the upstream, midstream, and downstream of the Pusur River. Each plots consisted of 10 sub-plots of 2m x 2m for understorey class. Subsequently, the collected data were examined using the vegetation analysis to obtain the important value index (IVI). This index is used to determine the dominance of a species in a given ecosystem. It is calculated based on the formula in Kusmana [11]. IVI = RD + RF.

$$\text{Density (D)} = \frac{\text{Number of individuals of species}}{\text{Sample plot area}}$$

$$\text{Relative density (RD)} = \frac{\text{Density of a species (D)}}{\text{density of all species}} \times 100\%$$

$$\text{Frequency (F)} = \frac{\text{Number of plots of a species}}{\text{Sum of all plots all species are identied}}$$

$$\text{Relative frequency (FR)} = \frac{\text{Freqency of a species (F)}}{\text{Frequency of all species}} \times 100\%$$

### 2.3. Species diversity index (H')

Diversity serves as a community-level indicator of vegetation stability. The species diversity index is

$$H' = -\sum \frac{ni}{N} \log \frac{ni}{N}$$

where:

H' = Diversity index
Ni = Number of individuals of each species
N = Total number of individuals in the site

calculated using the Shannon-Wiener Index Ludwig and Reynolds [12]. The formula is:

Meanwhile, the criteria are:

1. $H < 1$ : indicates low species diversity,

2. $1 < H < 3$ : indicates moderate species diversity,
3. $H > 3$ : indicates high species diversity.

*2.4. Species richness index (R)*
Species richness index is obtained from the total number of individuals of all species in an ecosystem. It is calculated using the formula in Odum [13]:

$$R = \frac{S-1}{\ln (N)}$$

where:
R= Species richness index
S= Number of species
N= Total number of individuals

Evenness Index (E)
Evenness refers to how evenly the individuals in a community among the different species. Magurran [14] suggests that species evenness ranges from zero to one, with zero signifying no evenness and one signifying a complete evenness.

$$E = H'/\ln S$$

Where:
H' = Vegetation diversity index
S = Species richness

*2.5. Soil temperature and relative humidity*
Soil temperature and relative humidity are estimated using the formula in Sabaruddin [15] as follows:

$$T = \frac{2T\ morning + T\ midday + T\ afternoon}{4}$$

$$RH = \frac{2RH\ morning + RH\ midday + RH\ afternoon}{4}$$

where:
T = Temperature of soil
RH = Relative humidity

## 3. Result and discussion

*3.1. Pusur River habitat*
The analysis of riparian understorey vegetation in the Pusur River was divided into three parts of the river, namely: upstream, midstream, and downstream. Each part was subdivided into three 3 observation stations (upper, middle, lower) thus totally, nine stations were established (Table 1). The understorey vegetation analysis identified as many as 9364 individuals that belong to 74 species and 35 families in the research location. The understorey plants in the riparian zone of Pusur River varied. The highest number of individuals was found in the upper Upstream (1777 individuals), while the lowest was in the middle Downstream (688 individuals). Subsequently, the highest number of species was identified in the lower Upstream (30 species), while the lowest was in the middle Downstream (14 species). According to Zega [16], understorey development is mainly affected by light reaching the forest floor and elevation of the area.

**Table 1.** Number of individual and understorey plant Species in the Riparian Zone of Pusur River

| **Observation Station** | **Σ Individuals** | **Σ Species** |
|---|---|---|
| Lower Upstream | 710 | 30 |
| Upper Upstream | 1777 | 26 |
| Upper Midstream | 416 | 24 |
| Middle Upstream | 1487 | 27 |
| Middle Midstream | 1117 | 22 |
| Upper Downstream | 1022 | 21 |
| Lower Midstream | 945 | 18 |
| Lower Downstream | 1202 | 16 |
| Middle Downstream | 688 | 14 |

The species dominance at stand level is determined using the IVI. A type of vegetation that has a high IVI is able to adapt well to its environment. In addition, it is able to grow optimally. The results of vegetation analysis show that *Sphagnetiloca trilobata* (L.) Pruski, *Ficus septica*, *Brachiaria mutica*, *Ageratum conyzoides* (L.) L, and *Cyperus rotundus* are the species that dominate the riparian zone along the upstream to downstream of Pusur River. It indicates that these species are able to adapt well in the riparian ecosystem. It confirms Wicaksono [17] on the capacity of dominant species in optimizing the environmental factors as well as in thriving and regenerating itself. The list of the understorey plants that dominate the study site from the upstream to the downstream of Pusur River is presented in Table 2:

Based on the indices in Table 2, the observation stations in the Upstream are dominated by waderan (*Oplismenus compositus*), namely the upper Upstream (19.34%) and middle Upstream (23.88%), while the lower Upstream is dominated by wedelia (*Sphagnetiloca trilobata* (L) Pruski) (26.58%). The riparian zone of the middle Midstream is dominated by awar-awar (*Ficus septica*) of 19.37%, wedelia (*Sphagnetiloca trilobata* (L) of 41.06%, and rumput gajah (*Brachiara mutica*) of 49.58%. Subsequently, the upper Downstream is dominated by wedelia (*Sphagnetiloca trilobata (L)* of 37.89%, while the middle and lower Midstream are dominated by bandotan (*Ageratum conyzoides* (L) L of 41.23% and 42.34%, respectively.

### *3.2. Species diversity index (H')*

The diversity analysis of understorey vegetation in the riparian zone of Pusur River reveals the species diversity index ranges from 2.37 to 3.08 (Table 3). The lowest H' is obtained by the middle Downstream while the highest is obtained by the lower Upstream. Based on the index, the species diversity of the riparian zone of Pusur River varies, ranging from moderate to high. The highest is found in the Upstream, followed by the Midstream and Downstream. It is possibly related to the elevation of the area. It has been suggested by Dalmanik et al [18] on the influence of elevation on type of vegetation. Nevertheless, other factors may contribute to species diversity.

IOP Conf. Series: Earth and Environmental Science 1016 (2022) 012052 doi:10.1088/1755-1315/1016/1/012052

**Table 2.** Dominant understorey plants in the Riparian Zone of Pusur River

| No | Observation station | Local name - *Latin name* | Description | | | |
|---|---|---|---|---|---|---|
| | | | N | RD | RF | IVI |
| a | Upper Upstream | Waderan - *Oplismenus compositus* | 207 | 11.65 | 7.69 | 19.34 |
| | | Rumput Gajah - *Brachiaria mutica* | 201 | 11.31 | 7.69 | 19 |
| | | Putri Malu - *Mimosa pudica* | 132 | 7.43 | 673 | 14.16 |
| | | Pegagan - *Centella asiatica* (L) Urb. | 127 | 7.15 | 5.77 | 12.92 |
| | | Teh-tehan - *Acalypha siamensis* | 151 | 8.5 | 385 | 12.34 |
| b | Middle Upstream | Waderan - *Oplismenus compositus* | 265 | 17.82 | 6.06 | 23.88 |
| | | Wedelia - *Sphagnetiloca trilobata* (L) Pruski | 149 | 10.02 | 12.12 | 22.14 |
| | | Rumput Teki - *Cyperus rotundus* | 214 | 14.39 | 6.06 | 20.45 |
| | | Tumpang air - *Peperomia pellucida* (L.) Kunth | 128 | 8.61 | 9.09 | 17.7 |
| | | Bandotan - *Ageratum Conyzoides* (L) L | 187 | 12.57 | 4.54 | 17.12 |
| c | Lower Upstream | Wedelia - *Sphagnetiloca trilobata* (L) Pruski | 138 | 19.43 | 7.14 | 26.58 |
| | | Pacing Air Tawar - *Cheilocostus speciosus* (J.Koenig) C.D.Specht | 47 | 6.69 | 11.43 | 18.48 |
| | | Rumput gajah - *Brachiaria mutica* | 64 | 9.01 | 5.71 | 14.73 |
| | | Putri Malu - *Mimosa pudica* | 44 | 6.19 | 5.71 | 11.91 |
| | | Teh-tehan - *Acalypha siamensis* | 60 | 8.45 | 2.857 | 11.31 |
| d | Upper Midstream | Awar-awar - *Ficus septica* | 35 | 8.14 | 10.95 | 19.37 |
| | | Teh-tehan - *Acalypha siamensis* | 22 | 5.28 | 10.95 | 16.25 |
| | | Pegagan - *Centella asiatica* (L) Urb. | 47 | 11.298 | 4.11 | 15.41 |
| | | Bayam Duri - *Amaranthus spinosus* L. | 56 | 13.46 | 1.37 | 14.83 |
| | | Semanggi - *Marselia drummondii* L. | 44 | 10.58 | 1.369 | 11.95 |
| e | Middle Midstream | Wedelia - *Sphagnetiloca trilobata* (L) Pruski | 345 | 30.89 | 10.17 | 41.06 |
| | | Rumput gajah - *Brachiaria mutica* | 175 | 15.66 | 11.86 | 27.53 |
| | | Talas - *Homalomena rubescens* (ROXB.) Kunth | 58 | 5.192 | 11.86 | 17.06 |
| | | Jarong - *Achyranthes aspera* | 97 | 8.68 | 5.08 | 13.76 |
| | | Awar-awar - *Ficus septic* | 32 | 2.86 | 10.16 | 13.034 |
| f | Lower Midstream | Rumput Gajah - *Brachiaria mutica* | 311 | 32.91 | 16.67 | 49.58 |
| | | Cyperus - *Cyperus papyrus* L. | 163 | 17.25 | 11.11 | 28.35 |
| | | Bandotan - *Ageratum Conyzoides* (L) L | 153 | 16.19 | 11.11 | 27.3 |
| | | Rumput Teki - *Cyperus rotundus* | 99 | 10.48 | 8.33 | 18.81 |
| | | Sidoguri - *Sida rhombifolia* | 4 | 0.42 | 8.33 | 8.76 |
| g | Upper Downstream | Wedelia - *Sphagnetiloca trilobata* (L) Pruski | 248 | 24.26 | 13.33 | 37.59 |
| | | Rumput Teki - *Cyperus rotundus* | 148 | 14.48 | 13.33 | 27.81 |
| | | Bandotan - *Ageratum Conyzoides* (L) L | 129 | 12.62 | 11.11 | 23.73 |
| | | Ara Sungsang - *Asystasia gangetica* | 49 | 4.79 | 6.66 | 11.46 |
| | | Rumput Gajah - *Brachiaria mutica* | 69 | 6.75 | 4.44 | 11.196 |
| h | Middle Downstream | Bandotan - *Ageratum Conyzoides* (L) L | 175 | 25.436 | 15.789 | 41.23 |
| | | Rumput Teki - *Cyperus rotundus* | 90 | 13.081 | 18.42 | 31.5 |

| | | | | | | |
|---|---|---|---|---|---|---|
| | | Rumput gajah - *Brachiaria mutica* | 127 | 18.459 | 10.52 | 28.99 |
| | | Cyperus - *Cyperus papyrus* L. | 44 | 6.395 | 7.894 | 14.29 |
| | | Wedelia - *Sphagnetiloca trilobata* (L) Pruski | 58 | 8.43 | 5.26 | 13.69 |
| I | **Lower Downstream** | Bandotan - *Ageratum Conyzoides* (L) L | 280 | 23.29 | 19.04 | 42.34 |
| | | Waderan - *Oplismenus composites* | 295 | 24.54 | 9.5 | 34.07 |
| | | Kangkung Pagar - *Ipomea carnea* | 59 | 4.908 | 11.9 | 16.81 |
| | | Tulangan - *Eleusine indica* (L) | 100 | 8.139 | 7.142 | 15.46 |
| | | Daun Putri - *Centrosema pubescens* | 125 | 10.399 | 4.761 | 15.16 |

### *3.3. Evenness index (E)*

The evenness index (E) provides information about the dominance of each species in a community. The evenness index of understorey plants in the riparian zone of Pusur River ranges from 0.83 to 0.93 (Table 3). The lowest index is obtained by the upper Downstream (0.83), while the highest is obtained by the upper Midstream (0.93). It signifies the absence of dominant species in the riparian zone of Pusur River. According to Magurran [14], species evenness ranges from zero to one, with zero signifying no evenness and one signifying a complete evenness. The species evenness between observation stations varies, showing the dominant species or vegetation with high number of individuals. It is in accordance with the observation in the upper Downstream, which indicates the dominance of particular species that causes the evenness index of this area is lower than others. The species that dominate the upper Downstream are: wedelia (*Sphagnetiloca trilobata* (L) Pruski of 62,000/ha, rumput teki (*Cyperus rotundus*) of 37,000/ha, and bandotan (*Ageratum Conyzoides* (L) L of 32,250/ha.

**Table 3.** Species diversity index (H') and evenness index (E)

| No | Observation Station | H' | E |
|---|---|---|---|
| 1 | Upper Upstream | 3.01 | 0.92 |
| 2 | Middle Upstream | 2.9 | 0.88 |
| 3 | Lower Upstream | 3.08 | 0.91 |
| 4 | Upper Midstream | 2.96 | 0.93 |
| 5 | Middle Midstream | 2.6 | 0.87 |
| 6 | Lower Midstream | 2.46 | 0.85 |
| 7 | Upper Downstream | 2.6 | 0.83 |
| 8 | Middle Downstream | 2.37 | 0.93 |
| 9 | Lower Downstream | 2.46 | 0.89 |

### *3.4. Species richness index (R)*

Species richness index is used to determine the species diversity based on the total number of individuals of species in an ecosystem. The result of the calculation using the Margalef index shows that the species richness in the upper Upstream has the highest value compared to other observation stations (Figure 1). This value is also proportional to the number of species found in the area. It verifies Nahlunisa et al [19], that the species richness index depends on the number of species found in a plot.

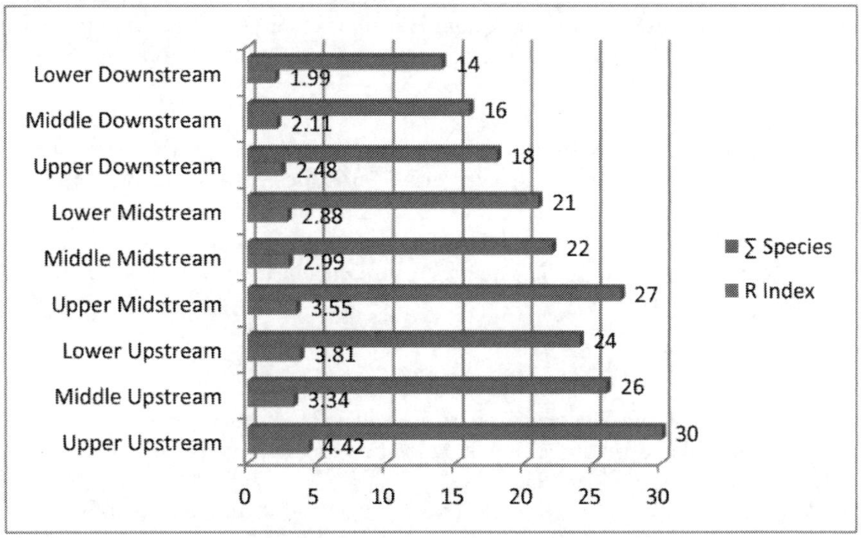


**Figure 1**. The species richness index

**Table 4.** Soil temperature and relative humidity in the Riparian Zone of Pusur River

| No | Observation Station | T | RH | H' |
|---|---|---|---|---|
| 1 | Upper Upstream | **27** | **60** | **3.01** |
| 2 | Middle Upstream | **27.25** | **59** | **2.9** |
| 3 | Lower Upstream | **27.75** | **61** | **3.08** |
| 4 | Upper Midstream | **28.8** | **61** | **2.96** |
| 5 | Middle Midstream | 29.5 | 58 | 2.6 |
| 6 | Lower Midstream | 30.1 | 56 | 2.46 |
| 7 | Upper Downstream | 30.9 | 56 | 2.6 |
| 8 | Middle Downstream | 32.75 | 40 | 2.37 |
| 9 | Lower Downstream | 33 | 39 | 2.46 |

In general, soil temperature and relative humidity are the abiotic factors that influence the understorey development and its diversity in the riparian zone of Pusur River (Table 4). Based on the measurement of the average soil temperature at a depth of 5 cm below the ground, the lowest temperature is in the upper Upstream (27°C) while the highest is in the lower Downstream (33°C). The effect of soil temperature on vegetation is obvious on tiller development and microorganism activities. It confirms Karyati et al. [20] in the study of the impact of soil temperature on plant revegetation in post-mining land. The average relative humidity at a depth of 5 cm below the ground in the riparian zone of Pusur River ranges from 39 to 60%. The upper Upstream has the highest relative humidity of 60%, while the lower Downstream has the lowest of 39%. The variations in soil temperature and relative humidity significantly affect the species diversity of riparian understorey in Pusur River. Optimal soil temperature and relative humidity affect the diversity level of understorey plants. The variations in elevation from the Upstream to Downstream also affect the level of soil temperature. High moisture content as indicated by relative humidity is also influenced by the presence of riparian litter. Based on the observations, the upper Midstream to the upper Upstream has very diverse litter. The observations also show that the maximum H' value is identified at soil temperature of 27-28,8°C and relative humidity of 59-61% with the species diversity of riparian understorey ranges from 2.9 to 3.08.

## 4. Conclusion

Based on the analysis of understorey vegetation in the riparian zone of Pusur River, 9,364 individuals can be identified in which they belong to 74 species and 35 families. The variations in elevation from the Upstream to Downstream also affect the level of soil temperature. High moisture content as indicated by relative humidity is also influenced by the presence of riparian litter. The contribution of riparian understorey in reducing climate change is capable of influencing the stability of temperature and micro humidity in riparian ecosystem.

## References

[1] Bental W P, Siahaan R and Maabuat P V 2017 Keanekaragaman vegetasi riparian sungai polimaan, minahasa selatan–sulawesi utara (riparian vegetation diversity of polimaan river, south minahasa-sulawesi utara) *J. Bioslogos* **7** 26–31
[2] Mitsch W and Gosselink J 1993 *Wetlands* (New York: Van Rostrand Reinhold)
[3] Anggana A and Susanti P 2020 Evaluation of water quality in The Swamp River border using water quality index *J. Degrad. Min. L. Manag.* **7** 2373–9
[4] LIPI 2016 *Ekologi Vegetasi* (Jakarta: LIPI Press)
[5] Maisyaroh W 2010 Struktur Komunitas Tumbuhan Penutup Tanah di Taman Hutan Raya R. Soerjo Cangar, Malang Structure of Ground Cover Plant Community R. Soerjo Grand Forest Malang *J. Pembang. dan Alam Lestari* **1** 1–9
[6] Kunarso A and Azwar F 2013 Keragaman jenis tumbuhan bawah pada berbagai tegakan hutan tanaman di Benakat, Sumatera Selatan *J. Penelit. Hutan Tanam.* **10** 85–98
[7] Lubis K S 2007 Aplikasi Suhu dan Aliran Panas Tanah *USU Repos. Medan*
[8] Suyono and Sudarmadi 1997 *Hidrologi Dasar* (Yogyakarta: Fakultas Geografi, Universitas Gadjah Mada)
[9] Djumali D and Mulyaningsih S 2014 Pengaruh Kelembaban Tanah Terhadap Karakter Agronomi, Hasil Rajangan Kering Dan Kadar Nikotin Tembakau (Nicotiana tabacum L; Solanaceae) Temanggung Pada Tiga Jenis Tanah *Ber. Biol.* **13** 1–11
[10] Afandi M, Zakaria R, Wardoyo A, Kusumastuti A and Saputro R 2018 Pendekatan Integratif Dalam Pelestarian Sub Das Pusur Kabupaten Klaten *Prosiding Seminar Nasional Geografi UMS IX* (Prosiding Seminar Nasional Geografi UMS IX 2018) pp 349–61
[11] Kusmana C 2018 *Metode survey dan interpretasi data vegetasi* (PT Penerbit IPB Press)
[12] Ludwig J A, Reynolds J F, Quartet L and Reynolds J F 1988 *Statistical ecology: a primer in methods and computing* (John Wiley & Sons)
[13] Odum E P 1998 *Dasar-Dasar Ekologi* (Yogyakarta: Gadjah Mada University Press)
[14] Magurran A E 1988 *Ecological diversity and its measurement* (New Jersey: Princeton university press)
[15] Sabaruddin L 2012 Agroklimatologi Aspek-aspek Klimatik untuk Sistem Budidaya Tanaman *Alf. Bandung*
[16] Zega C A 2017 *Keanekaragaman Jenis Vegetasi di Hutan Desa Rambang Kecamatan Rungan Barat Kabupaten Gunung Mas Provinsi Kalimantan Tengah* (Palangkaraya University)

[17] Wicaksono F B 2015 Komposisi jenis pohon dan struktur tegakan hutan mangrove di desa pasar banggi kabupaten rembang provinsi jawa tengah *J. Hutan Mangrove* **5** 55–62
[18] Damanik A J, Kartikawati S M and Prayogo H 2018 Studi Keanekaragaman Jenis Anggrek (Orchidaceae) Berdasarkan Ketinggian Tempat di Bukit Wangkang Kabupaten Kubu Raya *J. Hutan Lestari* **6** 447–55
[19] Nahlunnisa H, Zuhud E A M and Santosa Y 2019 Keanekaragaman spesies tumbuhan di areal Nilai Konservasi Tinggi (NKT) Perkebunan Kelapa Sawit Provinsi Riau *Media Konserv.* **21** 91–8
[20] Karyati K, Putri R O and Syafrudin M 2018 Suhu Dan Kelembaban Tanah Pada Lahan Revegetasi Pasca Tambang Di Pt Adimitra Baratama Nusantara, Provinsi Kalimantan Timur **27** 103–14

# Effect of various types and doses of biochar on hybrid maize growth

**M Rahayu*[1], A I Nurmalasari[1] and N N Aini[2]**

[1]Lecturer of Agrotechnology Study Program, Faculty of Agriculture, Sebelas Maret University, Surakarta, Indonesia
[2]Undergraduate Student of Agrotechnology Study Program, Agriculture Faculty, Sebelas Maret University, Surakarta, Indonesia
*Corresponding author: mujirahayu@staff.uns.ac.id

**Abstract**. Climate change also has an impact on agriculture, especially at Mediterranean red-yellow soil which is characterized with low fertility. The application of biochar is an alternative to increase soil fertility, as well as promoting the growth and yield of maize in red-yellow Mediterranean soil. The study aimed it determining the appropriate type and dose of biochar in red-yellow Mediterranean soil to support the growth of hybrid maize. The study used a factorial Randomized Completely Block Design (RCBD) with two factors. The first factor was the type of biochar which consisted of 3 levels, namely coconut shell biochar (B1), rice husk biochar (B2), and maize cob biochar (B3). The second factor was the dose of biochar which consists of 5 levels, namely 0 ton.ha$^{-1}$ (D0), 5 ton.ha$^{-1}$ (D1), 10 ton.ha$^{-1}$ (D2), 15 ton.ha$^{-1}$ ( D3), and 20 ton.ha$^{-1}$ (D4). The analysis of variance (ANOVA) continued with the Duncan's Multiple Range Test (DMRT) at level 5% were employed for data analysis. The results showed that the application of biochar had a significant effect on the growth of hybrid maize, especially on some observational variables. The application of biochar rice husks significantly affected the stem diameter, leaves number, and dry weight of 7 week after plant (WAP). The dose of 15 tons.ha$^{-1}$ significantly affected the dry weight of 4 WAP, while the interaction of rice husk biochar with a dose of 15 tons.ha$^{-1}$ significantly affected the leaves width.

**Keywords:** husk biochar, soil fertility, ANOVA, rice husk

## 1. Introduction

Climate change in Indonesia is causing a decrease in water availability that has an effect on the adequacy of crop needs [1]. The yellow red Mediterranean is a marginal soil that has low infiltration and drainage characteristic [2]. This condition has a bad impact on the growth of cultivated crops, especially food crops [3]. Hybrid maize is a food crop that is widely cultivated in Indonesia for food and animal feed [4]. The growing need for hybrid maize every year should be accompanied by increased productivity of hybrid maize [5]. Hybrid maize cultivation in marginal land has not provided good hybrid maize productivity, hybrid maize productivity can increase if accompanied by good growth [6]. Improper processing of marginal soil can degrade soil quality causing soil fertility to decrease [7]. One way to improve marginal soil fertility can be done with the use of biochar [8]. Biochar is an agricultural organic waste obtained through pyrolysis [9].

The use of biochar has a good impact on the availability of water, improving soil nutrients that will help increase plant growth. Based on research conducted by Verdiana [10], biochar implementation can increase the growth of hybrid maize, especially in the addition of crop height and nutrient absorption in the soil. Some studies show an interaction between biochar and fertilizer at the fresh weight of hybrid maize. This is because the provision of biochar is able to increase the quality and quantity of soil so that it affects plant growth [11]. Based on the description above, this study was conducted to obtain the appropriate type and dose of biochar to be applied in yellow-red Mediterranean soil so that it is expected to improve soil fertility and increase the growth of hybrid maize.

## 2. Materials and methods

The study was conducted from April to September 2021. Location for pots/polybag experiments at screen house Faculty of Agriculture Universitas Sebelas Maret Surakarta, while laboratory analysis was carried out in the Laboratory of Ecology and Plant Production Management; and the Laboratory of Plant Physiology and Biotechnology, Faculty of Agriculture Sebelas Maret University, Surakarta. The materials used included hybrid maize seeds, inorganic fertilizers (SP36, KCl, and urea fertilizer), soil, biochar, pesticides, and fungicides. The soil, which was yellow-red Mediterranean were taken from the Menggoran, Playen, Yogyakarta, and Gunung Kidul forest areas; while the biochar were derived from rice husk, coconut shell and maize cobs. The study used a factorial Randomized Completely Block Design (RCBD) with two experimental factors. The first factor, namely the type of biochar consisting of 3 levels, namely coconut shell biochar (B1), rice husk biochar (B2), and maize cob biochar (B3). The second factor, namely the biochar dose consisting of 5 levels, namely 0 (D0), 5 (D1), 10 (D2), (D3), and 20 tons.ha$^{-1}$ (D4).

Parameters observed included plant height, number of leaves, stem diameter, leaf length, leaf width, leaf area, plant fresh weight, and plant dry weight. The research data obtained were analyzed using analysis of variance (ANOVA), followed by the Duncan's Multiple Range Test (DMRT) with 5% level [12].

## 3. Results and discussion

### *3.1. Plant height*

The application of different types and doses of biochar showed a varied influence on the height of hybrid maize plant growth from the beginning of growth to the maximum vegetative at 9 WAP (Figure 1). The application of rice husk biochar at a dose of 15 ton.ha$^{-1}$ produced the highest hybrid maize plants compared to other treatments. In this study, the giving of various biochars resulted in higher plants compared to those without biochar. According to Widyantika (2019), the addition of biochar can improve soil properties and support the growth of hybrid maize crops [13].

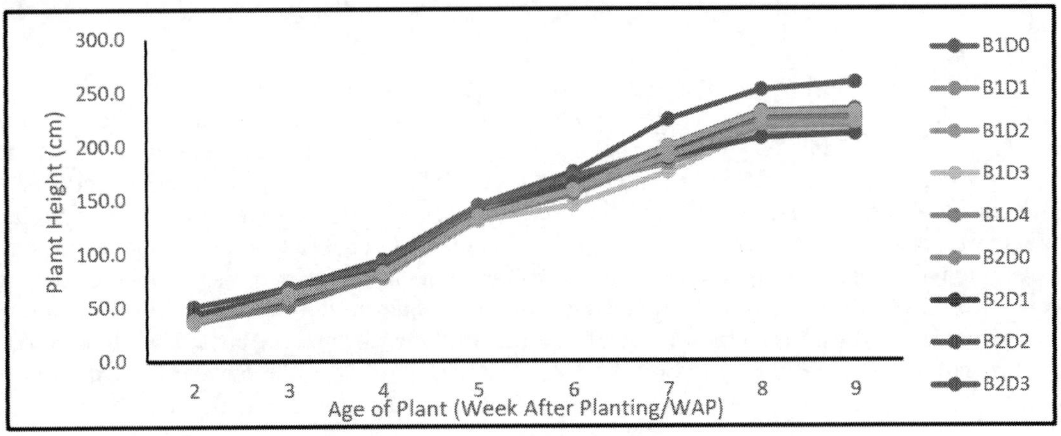


**Figure 1.** Influence of biochar type and doses on the height of hybrid maize plant
(Remarks: B1= coconut shell biochar; B2= rice husk biochar; B3= maize cob biochar; D0= 0 ton.ha$^{-1}$; D1=5 ton.ha$^{-1}$; D2= 10 ton.ha$^{-1}$; D3= 15 ton.ha$^{-1}$ D4= 20 ton.ha$^{-1}$)

**Table 1.** The effects of biochar application on plant height, leaves number and stem diameter of hybrid maize at 9 WAP

| Treatment | Plant height (cm) | | Leaves number | | Stem diameter (mm) | |
|---|---|---|---|---|---|---|
| **Type of Biochar** | | | | | | |
| Coconut Shell Biochar | 220.93 | a | 13.07 | a | 15.13 | a |
| Rice Husk Biochar | 223.87 | ab | 13.93 | b | 15.98 | b |
| Maize Cob Biochar | 232.87 | b | 13.13 | a | 14.92 | a |
| **Biochar Dose** | | | | | | |
| 0 (ton.ha$^{-1}$) | 215.11 | a | 13.11 | a | 15.02 | a |
| 5 (ton.ha$^{-1}$) | 225.00 | ab | 13.22 | a | 14.92 | a |
| 10 (ton.ha$^{-1}$) | 225.22 | ab | 13.33 | a | 15.59 | a |
| 15 (ton.ha$^{-1}$) | 232.56 | b | 13.78 | a | 15.86 | a |
| 20 (ton.ha$^{-1}$) | 231.56 | b | 13.44 | a | 15.33 | a |
| Average | 225.89 | | 13.38 | | 15.34 | |
| Interaction | - | | - | | - | |

Note: The number followed by the same letter shows no significant difference based on DMRT at the level of 5%; (-):no interaction.

Table 1 showed biochar type and biochar dose significantly affected the plant height at 9 WAP, but there was no interaction between the type and the dose of biochar. The maize cobs biochar produced the highest height at 9 WAP with an average value of 232.87 cm, but not different with that of rice husk biochar treatment. Meanwhile, the biochar with dose of 15 tons.ha$^{-1}$ produced the biggest plant height but not significantly different from the dose of 20 tons.ha$^{-1}$. The bigger the dose of biochar applied would increase the plant height. Rice husk biochar provided the highest K element of about 0.90% as well as high Cation Exchange Capacity (CEC) 29.27 me/100 g [14]. The utilization of hybrid maize cobs biochar can help increase C-organic in the soil. The high content of C-organic in the soil is due to the decomposition process by decomposing microbes in the soil in biochar so that the addition of hybrid maize cob biochar can increase soil organic matter useful for plant growth [15] [16]. The application of biochar in drought periods can act as a booster to plant growth and hold nutrients in the soil which ultimately affects plant growth [17]. The higher the dose of biochar, the higher the increase in C-organic content in the soil, hence increase plant growth [18].

### *3.2. Stem diameter*

The stem is one of the organs of plants which plays an important role to support the plant performance. In addition, the stem is the storage area for food reserves to support the development of other organs in plants [19]. Based on Table 1, type of biochar had a significant effect on stem diameter at 9 WAP, but not with the dose of biochar. The type of rice husk biochar produced the largest diameter of hybrid maize stem with an average value of 15.98 mm. The need for nutrients is quite supportive of the growth of hybrid maize, especially the increase in stem diameter that plays a role to support the standing plant and the path of nutrient spread [20]. The provision of biochar rice husks can help increase the capacity to hold water. This has a good influence on the increase in the diameter of the trunk [21]. The treatment of biochar dose at 9 WAP has no significant effect on the diameter of the stem, but the best biochar dose is 15 tons.ha$^{-1}$ with an average of 15.86 (Table 1). The content of N in the biochar needed for plant growth. N content plays a role in stimulating the growth of plants, especially stems, branches and leaves [22].

### 3.3. Leaves number

Table 1 confirmed there was no interaction between the type and the dose of biochar on the number of hybrid maize leaves at 9 WAP. The treatment of rice husk biochar produced the highest leaves number at 9 WAP with an average value of 13.93. Increasing the number of leaves will increase the fresh weight of plant because the leaves are organs for photosynthesis [23]. The treatment of biochar doses of 15 tons.ha$^{-1}$ resulted in the highest leaves number with an average of 13.78. Leaf formation is inseparable from the role of nutrients such as nitrogen, the nutrient content in biochar for plants plays a role in the formation of new cells and the constituent components of organic compounds in plants that affect the vegetative growth of plants, especially the increase in the number of leaves [24]. Biochar rice husks affect the formation of leaves which are affected by the absorption and availability of nutrients [25].

### 3.4. Leaf length

Optimal growth and development of leaves will have a larger cross-sectional area for the absorption of light that the plant needs. The longer the leaves. the larger the place where chlorophyll can absorb light [26].

**Tabel 2.** Leaf length, leaf width, and leaf area of hybrid maize on various types and doses of biochar

| Treatment | Leaf Length (cm) At 9 WAP | | Leaf Width (cm) At 9 WAP | | Leaf Area (cm$^2$) at 4 WAP | | Leaf Area (cm$^2$) at 7 WAP | |
|---|---|---|---|---|---|---|---|---|
| **Type of Biochar** | | | | | | | | |
| Coconut Shell Biochar | 74.07 | a | 6.33 | a | 888.34 | a | 3843.66 | a |
| Rice Husk Biochar | 81.10 | a | 6.64 | a | 995.82 | a | 4593.37 | a |
| Maize Cob Biochar | 79.33 | a | 6.40 | a | 798.36 | a | 3927.70 | a |
| **Biochar Dose** | | | | | | | | |
| 0 (ton.ha$^{-1}$) | 79.22 | a | 6.22 | a | 853.17 | ab | 4092.69 | a |
| 5 (ton.ha$^{-1}$) | 78.11 | a | 6.38 | a | 620.40 | a | 3934.13 | a |
| 10 (ton.ha$^{-1}$) | 79.22 | a | 6.38 | a | 866.70 | ab | 4171.52 | a |
| 15 (ton.ha$^{-1}$) | 82.06 | a | 6.85 | b | 1051.89 | b | 41116.52 | a |
| 20 (ton.ha$^{-1}$) | 72.56 | a | 6.44 | ab | 1078.71 | b | 4293.03 | a |
| Average | 77.66 | | 6.45 | | 894.18 | | 8746.57 | |
| Interaction | - | | + | | - | | - | |

Note: The number followed by the same letter shows no significant difference based on DMRT at the level of 5%; (-): there is no interaction. (+): there is an interaction.

Table 2 shows there was no interaction between biochar type and biochar dose to the length of hybrid maize leaves at 9 WAP. The treatment of biochar type and biochar dose also had no significant effect on leaf length. The longest leaves are produced on the biochar treatment of rice husks with an average value of 81.10 cm (Table 2). Photosynthetic resulting from photosynthesis will be reused during respiration and produce the energy required by the cell for division and enlargement causing the leaves to reach maximum length and width [27]. Biochar dose treatment also has no significant effect on leaf length. but the best biochar dose is 15 tons.ha$^{-1}$ with an average of 82.06 cm. The provision of biochar is able to increase leaf length. because biochar increases the availability of macro and micro elements that support plant growth. and can improve soil characteristics. such as increasing the capacity of the soil to hold water. The development of plant tissues will be influenced by the availability of micro and macro nutrients. plants that meet the needs of nutrients will stimulate the growth and development of their leaves [28].

### 3.5. Leaf area

Leaf area is a parameter that provides an explanation of the process and rate of photosynthesis in plants [29]. Table 2 presents there was no interaction between the type of biochar and the dose of biochar against the area of hybrid maize leaves at 4 WAP and 7 WAP. The treatment of biochar type has no significant effect on the area of hybrid maize leaves aged 4 WAP and 7 WAP. but the best type of biochar is biochar rice husk with an average value of 995.82 cm$^2$ and 4593.37 cm$^2$ (Table 2). The higher the leaf area, the process of sunlight capture and $CO_2$ fixation is higher so that the photosynthesis process runs well and the photosynthetic produced more and more [30]. The treatment of biochar doses has a significant effect on the leaf area of 4 WAP. The biochar dose that produces the largest leaf area at a dose of 20 tons.ha$^{-1}$ with an average of 1078.71 cm$^2$ but is not significant different from the dose of 15 tons.ha$^{-1}$. The higher the leaf area. the process of sunlight capture and $CO_2$ fixation is higher so that the process of photosynthesis runs well and the photosynthetic produced more and more [31].

### *3.6 Leaf width*

The wide leaf size illustrates that the nutrients available can meet the needs of the plant. so that the plant can grow and develop optimally [32]. Table 3 shows there was an interaction between the type of biochar and the dose of biochar to the width of hybrid maize leaves (Table 2).

**Table 3.** The application of various types and doses of biochar on maize leaf width

| Type of Biochar | Biochar Dose (ton.ha$^{-1}$) | Leaf Width (cm) | |
|---|---|---|---|
| Coconut Shell Biochar | 0 | 6.500 | a |
| Coconut Shell Biochar | 5 | 6.167 | a |
| Coconut Shell Biochar | 10 | 5.833 | a |
| Coconut Shell Biochar | 15 | 6.667 | a |
| Coconut Shell Biochar | 20 | 6.500 | a |
| Rice Husk Biochar | 0 | 6.167 | a |
| Rice Husk Biochar | 5 | 6.333 | a |
| Rice Husk Biochar | 10 | 6.667 | a |
| Rice Husk Biochar | 15 | 7.733 | b |
| Rice Husk Biochar | 20 | 6.333 | a |
| Maize Cob Biochar | 0 | 6.000 | a |
| Maize Cob Biochar | 5 | 6.667 | a |
| Maize Cob Biochar | 10 | 6.667 | a |
| Maize Cob Biochar | 15 | 6.167 | a |
| Maize Cob Biochar | 20 | 6.500 | a |
| Average | | 6.46 | |

Note: The number followed by the same letter shows no significant difference at $\alpha$= 5%

Table 3 shows the biochar of rice husks with a dose of 15 tons.ha$^{-1}$ produces the largest leaf width of 7.733 cm and is in stark contrast to other treatments. This is because the provision of biochar to plants can cause the addition of header width in plants. The wide increase of leaf width can have a good influence on crop production [33]. Leaves with a wide size means that having nutrients given fertilizer can meet the needs of plants. Wide leaves indicate the nutrient demands are fulfilled. then the plant can grow and develop optimally which will affect the growth and development of plants [34].

### *3.7. Fresh weight plant*

The fresh weight of plants is one of the parameters that describe the process of photosynthesis that takes place. High fresh weights indicate the process of photosynthesis goes well, as well as the accumulation of assimilate. The amount of asymptotic hoarding will affect the increase in fresh weight of plants [35]. Table 4 shows there was no interaction between biochar type and biochar dose on the fresh and dry weights at 4 and 7 WAP. Type of biochar significantly affected the dry weight at 7 WAP, while the dose of biochar significantly affected the fresh and dry weights at 4 and 7 WAP (Table 4).

**Table 4.** The application of various types and doses of biochar on fresh and dry weight at 4 and 7 WAP

| Treatment | Fresh Weight Plant 4 WAP (g) | Fresh Weight Plant 7 WAP (g) | Dry Weight Plant 4 WAP (g) | Dry Weight Plant 7 WAP (g) |
|---|---|---|---|---|
| **Type of Biochar** | | | | |
| Coconut Shell Biochar | 29.23 a | 180.53 a | 3.78 a | 17.07 a |
| Rice Husk Biochar | 37.89 a | 188.01 a | 4.29 a | 22.39 b |
| Maize Cob Biochar | 29.52 a | 184.39 a | 3.42 a | 18.48 a |
| **Biochar Dose** | | | | |
| 0 (ton.ha$^{-1}$) | 31.70 ab | 174.63 a | 3.65 ab | 18.97 a |
| 5 (ton.ha$^{-1}$) | 22.05 a | 174.04 a | 2.21 a | 18.40 a |
| 10 (ton.ha$^{-1}$) | 28.85 ab | 190.86 a | 3.32 ab | 18.80 a |
| 15 (ton.ha$^{-1}$) | 41.27 b | 197.50 a | 5.58 b | 20.01 a |
| 20 (ton.ha$^{-1}$) | 37.19 ab | 184.51 a | 4.40 ab | 20.40 a |
| Average | 32.21 | 184.31 | 3.93 | 19.31 |
| Interaction | - | - | - | - |

Note: The number followed by the same letter shows no significant difference $\alpha$= 0.05; (-):no interaction.

Table 4 shows the application of biochar rice husks produced the highest dry weight at 7 WAP, and differs markedly from coconut shell biochar and maize cob biochar with an average of 22.39 g. This suggests that biochar can improve the physical properties of the soil. Biochar applications can also increase the yield of hybrid maize crops in sour soils through increased pH and soil CEC [36]. Biochar application with the dose 15 tons.ha$^{-1}$ produced fresh weight and was the highest with an average of 197.50 g at 7 WAP, but was not significantly different with those of 0, 10, and 20 tons.ha$^{-1}$. This suggests that the addition of organic matter to the soil can increase the soil's absorption of water and aid the absorption of added chemical nutrients. Optimal absorption of nutrients will increase the photosynthetic of the plant's final process of photosynthesis then translocated to all plant organs to create wet biomass [37].

### *3.8. Dry weight plant*

The dry weight of plants indicates the efficiency of the photosynthetic process that takes place in the plant body. It will also affect the growth of hybrid maize. as it can show nutrients and moisture content used as metabolic materials in plants. Table 4 shows there was no interaction between biochar type and biochar dose to the dry weight of hybrid maize at 4 and 7 WAP. Type of biochar rice husk at 7 WAP significantly affected the dry weight of plants (22.39 g). Biochar dose of 15 tons.ha$^{-1}$ at 4 WAP also had a significant effect on the dry weight of the plant, with an average of 5.58 g.

The dry weight of hybrid maize plants on each treatment is presented in Figure 2. The dry weight of the plant affects the growth of hybrid maize crops because it indicates the nutrients and moisture content used as plant metabolism. The vegetative phase is the phase in which element N is brought to the growing point of the plant. so this vegetative phase occurs proper nutrient absorption [38]. At 7 WAP the highest dry weight of the plant is produced on the biochar treatment of rice husks at a dose of 10 tons.ha$^{-1}$ with an average dry weight of 25.11 g (Figure 2). The application of biochar with the appropriate dose in plants will help increase microbial activity in the soil and will have an effect on increasing the availability of organic matter in the soil. Optimal organic matter in the soil helps growth and development in plants [39].

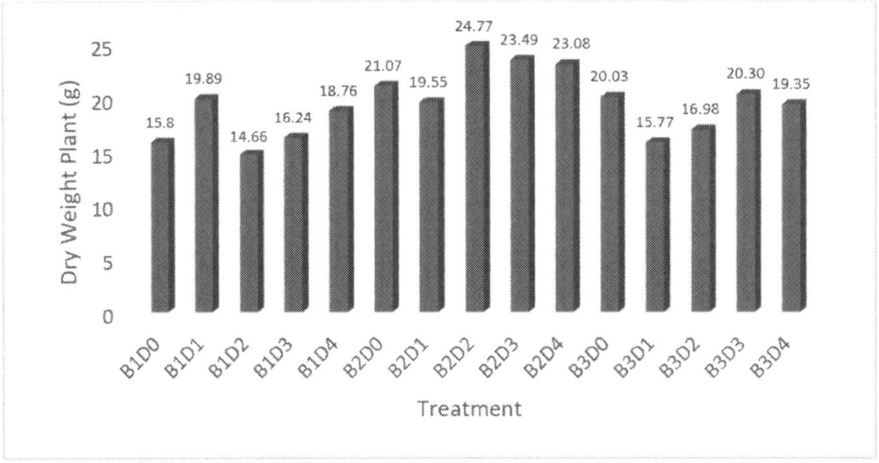


**Figure 2.** Effect of biochar type and dosage on dry weight of hybrid maize at 7 WAP
(Remarks: B1= coconut shell biochar; B2= rice husk biochar; B3= maize cob biochar; D0= 0 ton.ha$^{-1}$; D1=5 ton.ha$^{-1}$; D2= 10 ton.ha$^{-1}$; D3= 15 ton.ha$^{-1}$ D4= 20 ton.ha$^{-1}$)

## 4. Conclusions

Climate change affects the decline in land fertility, and biochar application is an effort to improve soil fertility. This study concluded application of rice husk biochar increased the growth of hybrid maize including plant height, number of leaves, stem diameter and dry weight at 7 weeks after planted. The provision of biochar with a dose of 15 tons.ha$^{-1}$ promoted the highest plant height, leaf width, leaf area of 4 WAP, fresh and dry weight of hybrid maize plant at 4 weeks after planted. The interaction between biochar type and biochar dose can be seen at leaf width, especially biochar rice husks with a dose of 15 tons.ha$^{-1}$.

## References

[1] Edgerton M D 2009 Increasing crop productivity to meet global needs for feed, food, and fuel *Plant Physiol.* **149** 7–13
[2] Ippolito J A, Laird D A and Busscher W J 2012 Environmental benefits of biochar *J. Environ. Qual.* **41** 967–72
[3] Sutch R 2011 The Impact of the 1936 Corn Belt Drought on American Farmers' Adoption of Hybrid Corn. u: Libecap Gary D., Steckel Richard H.[ur.] The Economics of Climate Change: Adaptations Past and Present
[4] Purwono M S and Hartono R 2007 Bertanam Jagung Unggul *Penebar Swadaya, Jakarta*
[5] Li G, Zhao B, Dong S, Zhang J, Liu P, Ren B, Lu D and Lu W 2019 Morphological and Physiological Characteristics of Maize Roots in Response to Controlled-Release Urea under Different Soil Moisture Conditions *Agron. J.* **111** 1849–64
[6] Oliveira F R, Patel A K, Jaisi D P, Adhikari S, Lu H and Khanal S K 2017 Environmental application of biochar: Current status and perspectives *Bioresour. Technol.* **246** 110–22
[7] Septiana L M, Djajakirana G and Darmawan D 2018 Characteristics of Biochars from Plant Biomass Wastes at Low-Temperature Pyrolysis *Sains Tanah-Journal Soil Sci. Agroclimatol.* **15** 15–28
[8] Rinaldi M E and Yunis M 2013 Pertumbuhan dan hasil tanaman jagung (Zea Mays L.) yang ditumpangsarikan dengan kedelai (Glycine Max L.) *Skripsi. Fak. Pertan. Jur. Agroteknologi Univ. Tamansiswa Padang. Padang*
[9] Koide R T, Nguyen B T, Skinner R H, Dell C J, Adler P R, Drohan P J, Licht M, Matthews M B, Nettles R and Ricks K 2018 Comparing biochar application methods for switchgrass yield

and C sequestration on contrasting marginal lands in Pennsylvania, USA *BioEnergy Res.* **11** 784–802

[10] Verdiana M A, Sebayang H T and Sumarni T 2017 Pengaruh Berbagai Dosis Biochar Sekam Padi dan Pupuk NPK terhadap Pertumbuhan dan Hasil Tanaman Jagung (Zea mays L.) *J. Produksi Tanam.* **4**

[11] Praing M W, Situmeang Y P and Mahardika I B K 2018 Penggunaan berbagai jenis biochar dan jenis pupuk dalam meningkatkan pertumbuhan dan hasil tanaman jagung manis *Gema Agro* **23** 176–81

[12] Gomez K and Gomez A 1984 *Statistical procedures for agricultural research.* (New York: John Wiley and Sons Inc.)

[13] Widyantika S D and Prijono S 2019 Pengaruh biochar sekam padi dosis tinggi terhadap sifat fisik tanah dan pertumbuhan tanaman jagung pada typic kanhapludult *J. Tanah dan Sumberd. Lahan* **6** 1157–63

[14] Lu S-G, Sun F-F and Zong Y-T 2014 Effect of rice husk biochar and coal fly ash on some physical properties of expansive clayey soil (Vertisol) *Catena* **114** 37–44

[15] Anon 2016 *Organic Fertilizers - From Basic Concepts to Applied Outcomes*

[16] Rodriguez A, Lemos D, Trujillo Y T, Amaya J G and Ramos L D 2019 Effectiveness of biochar obtained from corncob for immobilization of lead in contaminated soil *J. Heal. Pollut.* **9**

[17] Kurniawan A, Haryono B, Baskara M and Tyasmoro S Y 2016 Pengaruh penggunaan biochar pada media tanam terhadap pertumbuhan bibit tanaman tebu (Saccharum officinarum L.) *J. Produksi Tanam.* **4**

[18] Githinji L 2014 Effect of biochar application rate on soil physical and hydraulic properties of a sandy loam *Arch. Agron. Soil Sci.* **60** 457–70

[19] Feng W, Yang F, Cen R, Liu J, Qu Z, Miao Q and Chen H 2021 Effects of straw biochar application on soil temperature, available nitrogen and growth of corn *J. Environ. Manage.* **277** 111331

[20] Sombo Y, Astutik A and Widowati W 2016 Evaluasi residu jenis biochar dan waktu pemupukan nitrogen dan kalium pada tanaman jagung (Zea Mays L.) musim tanam kedua *Publ. Ilm. Mhs. Fak. Pertan. Univ. Tribhuwana Tunggadewi* **4**

[21] Islam S J M, Mannan M A, Khaliq Q A and Rahman M M 2018 Growth and yield response of maize to rice husk biochar *Aust. J. Crop Sci.* **12** 1813–9

[22] Bahri S, Juanda B R and Maulida H 2018 Pengaruh Jenis Biochar dan Pupuk Za Terhadap Pertumbuhan dan Produksi Tomat (Lycopersicum Esculentum Mill.) *J. Penelit. Agrosamudra* **5** 46–60

[23] Polii M G M 2009 Respon produksi tanaman kangkung darat (ipomoea reptans poir) terhadap variasi waktu pemberian pupuk kotoran ayam *J. Soil Environmnt* **1** 18–22

[24] Haryadi D, Yetti H and Yoseva S 2015 Pengaruh pemberian beberapa jenis pupuk terhadap pertumbuhan dan produksi tanaman kailan (Brassica alboglabra L.)

[25] Surachman S and Santoso E 2020 Aplikasi Biochar Sekam Padi Dan Pupuk NPK Terhadap Serapan N, P, K Dan Komponen Hasil Jagung Manis Di Lahan Gambut *J. Ilm. Hijau Cendekia* **5** 42–9

[26] Varela Milla O, Rivera E B, Huang W-J, Chien C and Wang Y-M 2013 Agronomic properties and characterization of rice husk and wood biochars and their effect on the growth of water spinach in a field test *J. soil Sci. plant Nutr.* **13** 251–66

[27] Cahyani S, Sudirman A and Azis A 2016 Respons pertumbuhan vegetatif tanaman tebu (Saccharum officinarum L.) ratoon 1 terhadap pemberian kombinasi pupuk organik dan pupuk anorganik *J. Agro Ind. Perkeb.* 69–78

[28] Salawati S, Basir-cyio M, Kadekoh I and Thaha A R 2016 Potensi biochar sekam padi terhadap perubahan pH, KTK, C organik dan P tersedia pada tanah sawah inceptisol *Agrol. J. Ilmu-ilmu Pertan.* **23** 101–9

[29] Hussien Ibrahim M E, Adam Ali A Y, Zhou G, Ibrahim Elsiddig A M, Zhu G, Ahmed Nimir N E and Ahmad I 2020 Biochar application affects forage sorghum under salinity stress *Chil. J. Agric. Res.* **80** 317–25

[30] Sari P, Meri R, Maghfoer M D and Koesriharti K 2016 Pengaruh frekuensi penyiraman dan dosis pupuk kandang ayam terhadap pertumbuhan dan hasil tanaman pakchoy (Brassica rapa L. Var. Chinensis) *J. Produksi Tanam.* **4** 342–51

[31] Krisnadhi J D, Yurlisa K and Sudiarso S 2020 Pengaruh Pemberian Plant Growth Promoting Rhizobacteria (PGPR) dan Dosis Pupuk Kandang Kambing terhadap Pertumbuhan dan Hasil Jagung Manis (Zea mays L. var. Saccharata) *J. Produksi Tanam.* **8** 234–40

[32] Jabborova D, Annapurna K, Al-Sadi A M, Alharbi S A, Datta R and Zuan A T K 2021 Biochar and Arbuscular mycorrhizal fungi mediated enhanced drought tolerance in Okra (Abelmoschus esculentus) plant growth, root morphological traits and physiological properties *Saudi J. Biol. Sci.* **28** 5490–9

[33] Akmal S and Simanjuntak B H 2019 Pengaruh pemberian biochar terhadap pertumbuhan dan hasil tanaman sawi pakchoy (Brassica rapa Subsp. chinensis) *Agril. J. Ilmu Pertan.* **7** 168–74

[34] Aditiameri A 2016 Respon pemberian macam pupuk organik dan dosis pupuk daun terhadap pertumbuhan dan hasil tanaman pakcoy (Brassica rapa L.) *Agrisia-Jurnal Ilmu-Ilmu Pertanian-Jurnal Ilmu-Ilmu Pertan.* **8**

[35] Zhang L, Sun X, Tian Y and Gong X 2014 Biochar and humic acid amendments improve the quality of composted green waste as a growth medium for the ornamental plant Calathea insignis *Sci. Hortic. (Amsterdam).* **176** 70–8

[36] Major J, Rondon M, Molina D, Riha S J and Lehmann J 2010 Maize yield and nutrition during 4 years after biochar application to a Colombian savanna oxisol *Plant Soil* **333** 117–28

[37] Kusumaningrum I, Hastuti R B and Haryanti S 2007 Pengaruh Perasan Sargassum crassifolium Dengan Konsentrasi Yang Berbeda Terhadap Pertumbuhan Tanaman Kedelai (Glycine max L. Merrill) *Bul. Anat. dan Fisiol.* **15**

[38] Maynita S, Wawan and Amri A I 2015 Uji Efektivitas Mimba (Azadirachta Indica a. Juss) dan Pemberian Nitrogen terhadap Pelindian Nitrogen dan Pertumbuhan Tanaman Jagung Manis (Zea Mays Saccharata Sturt) pada Tanah Gambut *J. Online Mhs. Fak. Pertan. Univ. Riau* **2** 1–12

[39] Warnock D D, Lehmann J, Kuyper T W and Rillig M C 2007 Mycorrhizal responses to biochar in soil–concepts and mechanisms *Plant Soil* **300** 9–20

# The bidirectional interaction between climate change and type 2 diabetes burden

**Ratih Dewi Yudhani*[1,3], Eti Nurwening Sholikhah[2] Dwi Aris Agung Nugrahaningsih[2], and Widana Primaningtyas[4]**

[1]Department of Pharmacology, Faculty of Medicine, Universitas Sebelas Maret, Jl. Ir. Sutami No.36 A, Kentingan, Jebres, Surakarta, Central Java, Indonesia, 57126
[2]Department of Pharmacology and Therapy, Faculty of Medicine, Public Health and Nursing, Universitas Gadjah Mada, Jl. Farmako Sekip Utara, Sleman, Yogyakarta, 55281
[3]Doctoral Programs in Health and Medicine, Faculty of Medicine, Public Health and Nursing, Universitas Gadjah Mada, Jl. Farmako Sekip Utara, Sleman, Yogyakarta, 55281
[4]Medicine Study Program, Faculty of Medicine, Universitas Sebelas Maret, Jl. Ir. Sutami No.36 A, Kentingan, Jebres, Surakarta, Central Java, Indonesia, 57126
*Corresponding author: ratihyudhani@staff.uns.ac.id

**Abstract**. Nowadays, type 2 diabetes (T2D) and climate change are the most urgent issues that being the focus worldwide. Interestingly, T2D directly or indirectly affects climate change and vice versa, climate change also contributes to the increased risk and the mortality of various non-communicable diseases (NCD), including T2D. This review aims to synthesis and give the comprehensive understanding related to interaction between climate change and T2D burden, which still being a limited subject in the Indonesian healthcare sector. It also explores those mechanisms for considering the right strategies and policies addressing those issues, especially the proper strategies for the prevention and management of diabetes. This review was conducted by screening from Google Scholar, Science Direct and PubMed with keywords "diabetes AND climate change", "diabetes AND extreme temperature", "diabetes AND cold weather". Various predisposing factors for the rapid growth of diabetes prevalence (urbanization, physical inactivity, unhealthy diets, and obesity) are also act as a risk vector that triggering climate change. While extreme temperature increases the susceptibility of the diabetic subjects to get life-threatening complications like dehydration and heatstroke. There is bidirectional interaction between climate change and T2D burden, which should be a priority for the government's attention to achieve the Sustainable Developmental Goal (SDG), especially for NCD including type 2 diabetes.

**Keywords** : bidirectional interaction, climate change, diabetes burden

## 1. Introduction

Nowadays, diabetes especially type 2 diabetes (T2D) and climate change are the most urgent issues that are being faced globally. The accelerating of these challenges not only affects the safety and welfare of the global population but also human health and human cost expenditure [1,2]. Diabetes mellitus has emerged as an epidemic and a global crisis since the sufferers at 2017 reached 425 million people and is expected to reach 693 million by 2045 [3]. Moreover, the report from the global burden of diseases

(GBD) indicated that in 2016, 72.3% of death causes were related to non-communicable diseases with 1.4 million death was caused by diabetes [4]. The rise of diabetes burden is mostly in low and middle-income countries due to there being an increase in diagnosed cases of diabetes in children and young adults [3,5]. Moreover, four of five diabetes sufferers are found in low and middle-income countries and South-East Asia is estimated to be one of the regions with the most diabetes prevalence in the next 20 years [2].

In the 21st century, along with diabetes burden, climate change is also act as a serious threat worldwide [6]. Climate is the average change in the weather over a period-specific time, from months to thousands of years. Since the last 1950s there has been an increase in the global mean temperature of $0.6^0$ C [1]. The rapid development of industrialization triggers the massive release of greenhouse emission into the atmosphere, thereby increasing the global temperature. This condition not only affects the environment but also adversely affects human life and health [6]. Interestingly, diabetes directly or indirectly affects climate change and vice versa, climate change also contributes to the increased risk and the mortality of various non-communicable diseases (NCD), including diabetes [2,7].

The elevating of global temperature possibly plays a role in increasing the diabetes epidemic due to higher environmental temperatures negatively affecting glucose metabolism through suppressing the brown adipose tissue activity. There was an increase in the diabetes incidence by 0.31 % for every degree Celsius rising in global temperature [8]. Epidemiological study shows that T2D sufferers have a higher risk (reach of 56%) for hospitalization as well as an increased risk of morbidity at extremely high temperatures environment or during exposure to heatwaves [9,10]. Other evidence indicated that the elderly diabetics with autonomic neuropathy complications are at a higher risk of developing hypothermia in cold environments [1,10]. On the other hand, the increased burden of T2D with obesity as the main risk factor is likely to trigger climate change indirectly. The greater obesity prevalence in the population will raise the greenhouse emissions, especially from the escalation of food production. A population with 40% obesity was estimated to require dietary energy 19% more than a normal body mass index distribution in population [2,11].

Diabetes mellitus and climate change as the main global issues are not only interconnected but also share the same vectors, therefore they possibly have the same solution [2,6]. A review focusing on the bidirectional interaction between type 2 diabetes and climate change are not established yet. This review aims to describe the comprehensive overview on the bidirectional interaction between type 2 diabetes and climate change. It also explores those mechanisms for considering the right strategies and policies addressing those issues, especially the proper strategies for the prevention and management of diabetes.

## 2. Method

This narrative review was conducted by screening from Google Scholar, Science Direct and PubMed with keywords "diabetes AND climate change", "diabetes AND extreme temperature", "diabetes AND cold weather".

## 3. Results and discussion

### *3.1. The effect of climate change on T2D burden*

*3.1.1. The effect extreme heat exposure on T2D burden.* Blauw et al. (2017) concluded that the global temperature rising is plays a role in the recent epidemic of T2D [8]. It supporting by the study in Canada that found in urban areas with the hottest outdoor temperatures there is a 3% increase in relative risk for gestational diabetes compared to the coldest areas. It showed the direct interaction between the risk of gestational diabetes and the ambient temperature among almost 400.000 women [12].

This role is possibly related to the decrease in brown adipose tissue (BAT) activity causes by climate change [13]. Brown adipose tissues (BAT) have the capacity for combusting stored fatty acid from triglyceride pools to produce heat [14]. BAT is equipped with a certain mitochondrial uncoupling protein (UCP1). UCP1 has the ability to produce high amount of heat immediately after its activation [13].

Physiologically, BAT is activated in cold condition and its activation not only decreasing body mass fat but also promoting glucose oxidation[14]. The increasing rate of fatty acids influx into activated BAT will induce the translocation and uptake of glucose to other metabolically active tissues as compensation. Thereby, improving insulin sensitivity [8]. The 10 days acclimatization of T2D patients to moderate cold temperature (14-15$^0$ C) increase the insulin sensitivity without altering body weight [15]. BAT acts as potential tissues that inhibit the development of diabetes since it has been shown to increase energy expenditure, affect glucose homeostasis and increase insulin sensitivity in humans [16]. This collecting evidence suggests that decreased BAT activity at higher ambient temperatures will increase the propensity for the development of diabetes mellitus in the population.

Furthermore, individuals with underlying diseases, including diabetes are at higher risk for the negative impacts of climate change. At high temperatures, diabetes patients are more susceptible to get dehydration and heatstroke, thereby rising their morbidity and mortality risk [2]. Epidemiological data showed people with diabetes are at higher risk of both developing temperature-related illness and death in heatwaves periods compared to the general population. In addition, during heatwave exposure, the hospitalization rate for diabetic patients significantly increased by 30% [9].

Diabetes results in several complications, including cardiovascular, metabolic, and neurological dysfunction which may tend to disrupt the thermoregulatory response during heat exposure [17]. In diabetic patients, microvascular complications are likely to decrease the vasodilation ability of the skin's capillaries, hence reducing blood transport to the skin's surface to dissipate heat [17,18]. In addition, peripheral neuropathy and autonomic dysfunction in diabetes might disrupt the sweating response, especially in the limbs while resting or doing physical activities at a high temperature. This condition placing patients at higher risk for developing heat stress and hyperthermia [17,19]. In normal individuals, heat exposure will be dissipated by the body through the skin surface and sweating mechanisms [18]. Both mechanisms were modulated by sympathetic nervous system through hypothalamus respond [20]. This reason underlies the disruption of heat loss from the skin blood flow and sweating in diabetic patients will increase the risk of heat-related illness.

Moreover, exercise is one of the non-pharmacological therapies recommended for diabetic patients, but prolonged exposure to high temperatures results in fluid loss and contributes to dehydration and a rapid increase in core body temperature. Dehydration is further exacerbated in patients with uncontrolled hyperglycaemia [19].

*3.1.2. The effect extreme cold exposure on T2D burden.* Beside extreme heat exposure, many studies also documented the contribution of extreme cold exposure to the higher risk for morbidity and mortality in diabetes patients [20]. An ecological study between 2006-2013 recruiting more than 1 million T2D patient in Taiwan revealed that low temperature associated with the increasing risk for hypoglycaemic complications and the peak incidence of hypoglycaemia in T2D patients is always found in the January-February period. According to the Taiwan Central Weather Bureau, this period is listed as the winter period in this country [21]. Other epidemiological study in Finnish population involving more than 5000 participants showed that T2D individuals experienced more cardiovascular symptoms (dyspnoea, chest pain, and respiratory symptoms) associated with cold weather compared to normal individuals. This vulnerability is also associated with the worsening of glycaemic control [22]. The longitudinal follow-up study involving 4399 T2D patients in Taiwan documented that every 1°C decrease in temperature significantly increases the risk of having HbA1c levels higher than 7% with an odds ratio 1.01. This study also showed that T2D patients with HbA1c levels above 7% had a higher risk of worsening diabetes control in winter and spring weather compared to summer [23].

The susceptibility of T2D individuals to be more at risk of experiencing the cardiovascular symptoms in cold weather may be mediated through various factors such as dysfunction of the autonomic nervous system (sympathetic over-reactivity) in both prediabetic and diabetic individuals. It improves vasoconstriction and blood pressure which lead to the myocardial oxygen demand and supply imbalance, hence increasing cardiac symptoms. The decreased of arterial compliance which also found in diabetics can impair extremity blood flow. Peripheral neuropathy associated with diabetes results in sensory and thermoregulatory dysfunction, especially in the extremities. In addition, the response to heat production by brown adipose tissues in cold weather conditions is also impaired by the reduction of insulin sensitivity [22].

### *3.2. The shared vectors of T2D burden and climate change*

The rising burden of T2D and its main risk factor, obesity is expected to play a major role in triggering climate change. This is due to the higher prevalence of obesity in the population will increase greenhouse gas emissions (GHG), especially from massive food production. The over-release of GHG is also caused by the increasing demands on the health care system related to the burden of non-communicable diseases including diabetes as well as failures to prevent and to manage diabetes properly [2]. Moreover, T2D indirectly influences climate change through several shared global vectors that exist between them, including massive urbanization, sedentary lifestyle, and the increasing of food consumption [7].

In the 21st century, there is a rapid increase in the pattern of urbanization. More than 50% of the world's population lives in urban areas and by 2030 it is expected to reach up to 5 billion [2]. Some previous studies reported the positive correlation between urbanization and the higher prevalence of T2D [24]. The rapid increase in urbanization is more pronounced in developing countries, thus optimizing its role as a vector for the higher prevalence of T2D and increasing GHG. This is possibly underlined by unplanned urbanization that triggers the expansion of slums, massive industrialization, increased pollution, while decreasing the resources. This massive urbanization also triggers the rapid lifestyle shifting, such as unhealthy diet and lack of physical activity due to increasing number of private cars use which are known to have strong correlation with the risk for the development of T2D [2]. Urbanization has also been shown to have a strong correlation with rising temperatures and currently being one of the important research focus [25]. Studies involving several metropolitan cities in China (Beijing-Tianjin-Hebei) show that urbanization triggers an increase in surface temperature, reduce wind speed and simultaneously increase the convective precipitation [26].

Other shared global vectors that interlink T2D and climate change in this recent era is the changing in global food system (GFS) which triggers by globalization, economic transition and massive urbanization [2]. Globalization and economic progress have had a crucial impact on the shifting of dietary pattern (nutrition transition) from traditional diets (fruits, vegetables, and grain) toward processed food and unhealthy diets, such as saturated-fatty acid, sugar, and low food-fibre which were known correlated to NCD, especially obesity and T2D [2,27,28]. This nutritional transition is based on higher demand for lower-priced processed foods and increased consumption of animal products [6]. The rapid development of food production increases the availability and affordability of processed foods. However, processed foods are more prone to trigger diabetes and obesity than traditional diets. The food processing steps also have a negative impact on the environment due to those several steps is known as a major contributor for the elevated level of GHG [2].

### *3.3. The proposed strategies addressing the burden of climate change and T2D focusing on those bidirectional interaction*

Through a comprehensive understanding of bidirectional interaction as well as the shared vectors of T2D and climate change as illustrated in Figure 1, it can be a good way and create an opportunity to direct policies that can simultaneously address both threats [29]. The health sector should play a major role in developing plans to address health threats related to climate change, including educating health care providers and the public to increase awareness and identify individuals who are more vulnerable in areas at higher risk of heat waves [30].

*3.3.1. Proposed strategies to reduced T2D mortality risk in countries with extreme temperature.* The increasing population of diabetics living in countries with relatively high temperatures such as the Middle east, India, Sub-Saharan Africa and several states in the United States, especially around the Southwest desert requires efforts to increase public awareness [19]. The globally action in supporting low-income countries with enormous vulnerable populations is needed for adapting and reducing health risk related to climate change [6]. This can be pursued through education to increase knowledge of preventing the increased risk of morbidity in diabetics during periods of high temperatures environmental conditions [19].

Diabetes patients, especially who has cardiovascular complication may require a certain individualized management plan since they should be considered as a vulnerable groups related to the climate change [7]. Comprehensive campaigns for increasing public awareness, education, and advising to prepare patients when facing extreme temperatures are urgent issues, especially for diabetes patients. Educational topics related to basic weather knowledge such as temperature, humidity, and heat index are needed, hence they can understand the weather conditions that they are facing [31]. During extreme heat periods, there needs an interventions to ensure that vulnerable T2D patients stay indoors and have access to cooling systems safely. For reducing the risk of patient morbidity during this heat periods, doctors should also identify T2D patients who have susceptibilities such as those with cardiovascular disease, or chronic kidney disease [20].

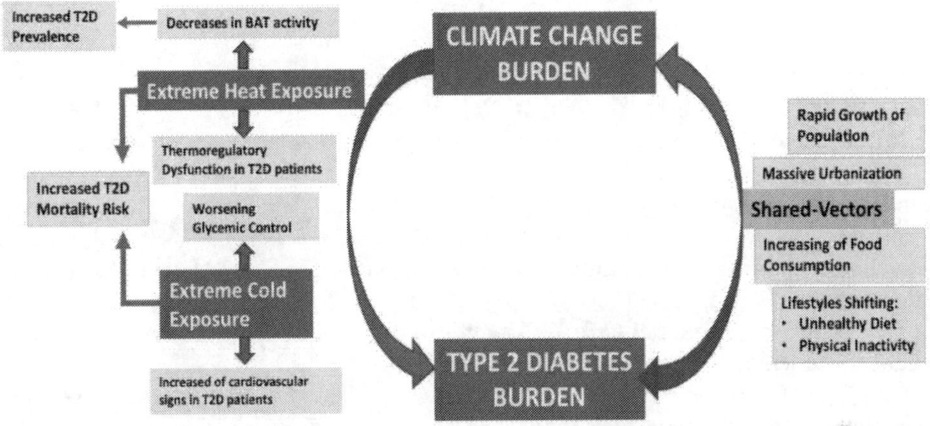


**Figure 1.** The scheme of bidirectional interaction between climate change and T2D burden

*3.3.2. Proposed strategies addressing the burden of climate change* and *T2D related to the urbanization policies.* The rapid increase in urbanization concentrated in developing countries combined with unplanned urbanization is a strong shared vector between T2D and increasing GHGs as climate change drivers. This is based on urbanization triggers lifestyle modification such as unhealthy diets, sedentary lifestyles, and increasing the number of private cars use that closely related to physical inactivity. Policies to reduce GHG and create conditions that encourage physical activity such as walking and cycling need to be implemented as an effort to overcome these problems [2].

The offered strategies that possibly also be applied to overcome these problems related urban planning is optimizing "active action" in population. Switching transportation modes to low-carbon transportation has great potential to reduce carbon emissions. In addition, optimizing "active travel" movements such as cycling, e-cycling or walking can be one of the promising strategies to reduce carbon dioxide emissions, especially from the transportation field [32]. It also can be supported by building infrastructures that facilitated active action in the population, such as increasing bicycle lanes and parking areas in urban as well as policies to make the bicycle more affordable for the community [2]. In addition, designing urban population completed with accessible welfare community facilities especially in preventive care will be an effort to optimize the prevention of chronic diseases including T2D [27].

*3.3.3. Proposed strategies addressing the burden of climate change and T2D related to the global food policies.* The massive growth of the world's population accompanied by urbanization and the shifting in diet pattern has emerged an increase in production, industrialization, and globalization especially in the food production sector. It can promote the increased burden of T2D and climate change. Strategies that might be able to suppress this shared vector is reforming a global food system aimed for optimizing the availability of food worldwide, in line with the concern in minimizing the environmental effects caused by the processing of the global food system [2]. The efforts in improving global food security can be pursued through the promotion of the International agricultural trade system from areas with abundant water resources into water-scarce areas. In addition, a breakthrough is needed in ensuring the availability of additional food through increasing food production apart from the agricultural sector, such as expanding the production function of the marine sector [33].

## 4. Conclusions

There is bidirectional interaction between climate change and T2D burden. Through a comprehensive and briefly understanding of the bidirectional interaction as well as the shared vector of T2D and climate change creates an opportunity to direct policies that can simultaneously address both threats. This issues should be a priority for the government's attention to achieve the Sustainable Developmental Goal (SDG), especially for NCD including type 2 diabetes.

## References

[1] Cuschieri S and Agius J C 2021 The interaction between diabetes and climate change–A review on the dual global phenomena *Early Hum. Dev.* **155** 105220

[2] IDF 2012 Diabetes and climate change report international diabetes federation *Int. Diabetes Fed.* 1–17

[3] International Diabetes Federation 2017 *IDF Diabetes Atlas, 8th edition*

[4] Naghavi M, Abajobir A A, Abbafati C, Abbas K M, Abd-Allah F, Abera S F, Aboyans V, Adetokunboh O, Ärnlöv J, Afshin A, Agrawal A, Kiadaliri A A, Ahmadi A, Ahmed M B, Aichour A N, Aichour I, Aichour M T E, Aiyar S, Al-Eyadhy A, et al 2017 Global, regional, and national age-sex specifc mortality for 264 causes of death, 1980-2016: A systematic analysis for the Global Burden of Disease Study 2016 *Lancet* **390** 1151–210

[5] Dunachie S and Chamnan P 2019 The double burden of diabetes and global infection in low and middle-income countries *Trans. R. Soc. Trop. Med. Hyg.* **113** 56–64

[6] Dain K and Hadley L 2012 Diabetes and climate change-Two interconnected global challenges *Diabetes Res. Clin. Pract.* **97** 337–9

[7] Zilbermint M 2020 Diabetes and climate change *J. Community Hosp. Intern. Med. Perspect.* **10** 409–12

[8] Blauw L L, Aziz N A, Tannemaat M R, Blauw C A, de Craen A J, Pijl H and Rensen P C N 2017 Diabetes incidence and glucose intolerance prevalence increase with higher outdoor temperature *BMJ Open Diabetes Res. Care* **5**

[9] Semenza J C, McCullough J E, Flanders W D, McGeehin M A and Lumpkin J R 1999 Excess hospital admissions during the July 1995 heat wave in Chicago *Am. J. Prev. Med.* **16** 269–77

[10] Kenny G P, Sigal R J and McGinn R 2016 Body temperature regulation in diabetes *Temperature* **3** 119–45

[11] Edwards P and Roberts I 2009 Population adiposity and climate change *Int. J. Epidemiol.* **38** 1137–40

[12] Booth G L, Luo J, Park A L, Feig D S, Moineddin R and Ray J G 2017 Influence of environmental temperature on risk of gestational diabetes *Cmaj* **189** E682–9

[13] Symonds M E, Farhat G, Aldiss P, Pope M and Budge H 2019 Brown adipose tissue and glucose homeostasis–the link between climate change and the global rise in obesity and diabetes *Adipocyte* **8** 46–50

[14] Hoeke G, Kooijman S, Boon M R, Rensen P C N and Berbeé J F P 2016 Role of Brown Fat in Lipoprotein Metabolism and Atherosclerosis *Circ. Res.* **118** 173–82

[15] Hanssen M J W, Hoeks J, Brans B, Van Der Lans A A J J, Schaart G, Van Den Driessche J J, Jörgensen J A, Boekschoten M V., Hesselink M K C, Havekes B, Kersten S, Mottaghy F M, Van Marken Lichtenbelt W D and Schrauwen P 2015 Short-term cold acclimation improves insulin

sensitivity in patients with type 2 diabetes mellitus *Nat. Med.* **21** 863–5
[16] Chondronikola M, Volpi E, Børsheim E, Porter C, Annamalai P, Enerbäck S, Lidell M E, Saraf M K, Labbe S M, Hurren N M, Yfanti C, Chao T, Andersen C R, Cesani F, Hawkins H and Sidossis L S 2014 Brown adipose tissue improves whole-body glucose homeostasis and insulin sensitivity in humans *Diabetes* **63** 4089–99
[17] Kenny G P, Yardley J, Brown C, Sigal R J and Jay O 2010 Heat stress in older individuals and patients with common chronic diseases *Cmaj* **182** 1053–60
[18] Yardley J E, Stapleton J M, Sigal R J and Kenny G P 2013 Do heat events pose a greater health risk for individuals with type 2 diabetes? *Diabetes Technol. Ther.* **15** 520–9
[19] Westphal S A, Childs R D, Seifert K M, Boyle M E, Fowke M, Iñiguez P and Cook C B 2010 Managing diabetes in the heat: Potential issues and concerns *Endocr. Pract.* **16** 506–11
[20] Vallianou N G, Geladari E V., Kounatidis D, Geladari C V., Stratigou T, Dourakis S P, Andreadis E A and Dalamaga M 2020 Diabetes mellitus in the era of climate change *Diabetes Metab.* 101205
[21] Lai S W, Chang W C, Lin C L, Chou I C, Tsai F J, Lai Y J and Nakajima K 2020 Low ambient temperatures correlate with increased risk of hypoglycemia in patients with type 2 diabetes: An ecological study in Taiwan *Med. (United States)* **99**
[22] Ikäheimo T M, Jokelainen J, Hassi J, Hiltunen L, Keinänen-Kiukaanniemi S, Laatikainen T, Jousilahti P, Peltonen M, Moilanen L, Saltevo J and Näyhä S 2017 Diabetes and impaired glucose metabolism is associated with more cold-related cardiorespiratory symptoms *Diabetes Res. Clin. Pract.* **129** 116–25
[23] Tien K J, Yang C Y, Weng S F, Liu S Y, Hsieh M C and Chou C W 2016 The impact of ambient temperature on HbA1c in Taiwanese type 2 diabetic patients: The most vulnerable subgroup *J. Formos. Med. Assoc.* **115** 343–9
[24] Gassasse Z, Smith D, Finer S and Gallo V 2017 Association between urbanisation and type 2 diabetes: An ecological study *BMJ Glob. Heal.* **2** 1–8
[25] Chapman S, Watson J E M, Salazar A, Thatcher M and McAlpine C A 2017 The impact of urbanization and climate change on urban temperatures: a systematic review *Landsc. Ecol.* **32** 1921–35
[26] Wang M, Zhang X and Yan X 2013 Modeling the climatic effects of urbanization in the Beijing-Tianjin-Hebei metropolitan area *Theor. Appl. Climatol.* **113** 377–85
[27] Hawkes C 2006 Uneven dietary development: Linking the policies and processes of globalization with the nutrition transition, obesity and diet-related chronic diseases *Global. Health* **2** 1–18
[28] WHO 2002 Diets and Noncommunicable Diseases 1–185
[29] Shubair 2013 Climate Change and Type 2 Diabetes *J. Endocrinol. Diabetes Mellit.* **1** 22–6
[30] Frumkin H, Hess J, Luber G, Malilay J and McGeehin M 2008 Climate change: The public health response *Am. J. Public Health* **98** 435–45
[31] Nassar A A, Childs R D, Boyle M E, Jameson K A, Fowke M, Waters K R, Hovan M J and Cook C B 2010 Diabetes in the desert: What do patients know about the heat? *J. Diabetes Sci. Technol.* **4** 1156–63
[32] Brand C, Götschi T, Dons E, Gerike R, Anaya-Boig E, Avila-Palencia I, de Nazelle A, Gascon M, Gaupp-Berghausen M, Iacorossi F, Kahlmeier S, Int Panis L, Racioppi F, Rojas-Rueda D, Standaert A, Stigell E, Sulikova S, Wegener S and Nieuwenhuijsen M J 2021 The climate change mitigation impacts of active travel: Evidence from a longitudinal panel study in seven European cities *Glob. Environ. Chang.* **67**
[33] Hanjra M A and Qureshi M E 2010 Global water crisis and future food security in an era of climate change *Food Policy* **35** 365–77

**Acknowledgement**
The authors would like thanks to the Universitas Sebelas Maret which has provided funding for this research through the PDD grant No: 260/UN27.22/HK.07.00/2021

# Review on the effect of climate change on ecosystem services

**Berhanu Zawude Bakure[1]*, Kitessa Hundera[2] and Magarsa Abara[1]**

[1]Department of Environmental Science, Wollega University, P.O. Box 395, Nekemte, Ethiopia
[2]Department of Biology, Jimma University, P.O. Box: 378, Jimma, Ethiopia
*Corresponding author: berzew5@gmail.com

**Abstract.** Currently, ecosystem services (the benefits society drive from the ecosystem) are under pressure from climate change. With increasing climate change over time, the influence that it can cause ecosystem service attracted the attention of the world more than ever. In our daily life, directly or indirectly we rely on benefits gained from the ecosystem. This review paper was aimed to address the effects of climate change on ecosystem services with its possible mitigation and adaptation measures by analyzing articles, books, and reports collected from trusted journals and websites. The range and extents of ecosystem service can be affected in quality and quantity both directly and indirectly due to climate change over time. Supporting services of the ecosystem like biomass production, nutrient cycling, soil formation, atmospheric oxygen, production, and the water cycle were affected. Similarly, provisioning services such as food, drinking water, timber, wood fuel, fodder, wood, fibers, plants, and animal species are affected. Regulating services like cross-pollination, seed dispersal, decomposition, water regulation, flood control, carbon sequestration, and climate regulation are affected. Furthermore, cultural services like tourism and recreation, aesthetic values, cultural heritage, spiritual and religious values, educational values, social relations, and ecotourism are going to decline. Generally, understanding the effects of climate changes on ecosystem services became fundamentally important adaptation and mitigation of effect. Hence, conservation, protection, restoration, and appropriate management ecosystem are required for adaptation and mitigation of climate change effect.

**Keywords:** ecosystem services, effect of climate change, climate change adaptation

## 1. Introduction

Currently, ecosystem services are under the pressure of climate change [1, 2]. The concept of ecosystem service was much older, although it became popularized after the last three decades [2]. Ecosystem service refers to benefits that people drive from the ecosystem. The most popular definition for ecosystem services is that used by Millennium Ecosystem Assessment (MEA) [3]. The MEA defines ecosystem services as those benefits that society attains from ecosystems. The advantages that society drives from the ecosystem were categorized as provisioning services, cultural services, regulating services, and supporting services [3, 4].

There are several benefits that human beings obtain from the ecosystem that can be categorized under the scheme as utilitarian and non-utilitarian values. Utilitarian values are provisioning, regulating, and supporting aspects, whereas non-utilitarian values include ethical, religious, cultural, and philosophical aspects [5, 6]. Utilitarian values can be further classified as direct, indirect, and option use-values. Direct use values are benefits that people consume directly or non-consumptive values like aesthetic and recreational values. Consumptive values include the exploitation of fish and other aquatic flora. Indirect use values can be water regulation, ecosystem production, and climate regulation. Other ecosystem values are option values that are deemed to be used in the future, these include preserving threatened species for next-generation [4, 5, 6, 7].

It was believed by many authors that ecosystem services were accepted as the integrating systems of community and ecosystems [8, 9]. As a result, ecosystem services was defined jointly with social and ecosystems. The range and extents of ecosystem service varies with altering ecosystems and societies through periods [10, 11]. Ecosystems free of disturbance can afford services that are the basis for human well-being [12]. Therefore, several authors have reported that general definition for the term Ecosystem services refers to benefits society gain from ecosystems including provisioning, regulating, cultural and supporting services like cycling of minerals and nutrient [9, 10, 13].

It was commonly noted that certain groups of ecosystem services have shown synergistic variation positive co-variation) and antagonistic variation (negative co-variation); which leads to concept of the ecosystem service packages [14, 15]. These marvels of ecosystem service packages may be due to co-delivered amenities initiate from a mutual primary situation. The great increase in provisioning services of sustenance and integrity in current world have commonly come at overhead of services such as water quality and landscape amenity [2, 16].

Climate change have significant effect on ecosystems services that may be direct and indirect [5, 6]. The effect can be due to escalating $CO_2$ concentration, land use modification, environmental pollution (like ambient air and water bodies), increasing natural resource use, and loss of biological diversity [17]. Climate change also affects ecosystem services stock share a compassion as it alters underlying drivers, like increase in inhabitants and depletion. Therefore, understanding the effects of climate changes on ecosystem services became fundamentally important in order to adapt and mitigate the effect [1, 2]. It was reported that the potential of ecosystem to provide necessary services to society is under stress [17]. Among the stressors population growth, overexploitation, urban expansion, pollution, land use change and climate change have played fundamental roles [2, 5]. According to [6] from 1997 to 2011 the economic benefits earned from ecosystem service became globally declined by $4.3–20.2 trillion per year due to climate change.

The effects of climate change on ecosystem service is being increased today more than ever [5]. As a result of climate change, most freshwater ecosystems have been totally restructured, oceans have been severely altered, and coral reefs are severely endangered, and little consideration was given for ecosystem services that are being lost [1, 2]. Climate changes also affected ecosystem services through challenges that species face by behavioral change and disruption of mutualism. In some habitats, species extinctions observed where migrations are necessary for survival but pathways not available for successful movement in the fragmented. These implies the services derived from the species is being lost as result of climate change [16, 18].

Climate change can affect quality and quantity of freshwater ecosystems negatively [17, 18]. Warming temperature have direct effects on precipitation patterns that lead to amplified extent and irregularity of flooding and scarcities in numerous regions [6]. Reductions in snow cause reduced stream flow, warming cause seasonal hydrology of receiving streams and freshwater availability decrease because of vaporization and human consumption. Smaller water bodies are expected to be affected more by climate change due to lower buffering potential for thermal and hydrological cycles [5, 18].

High temperatures aggravate the concentration of water contaminants due to high water loss that reduces the amount of water and effectively increases pollutants concertation and decrease flushing of these materials [19]. Additionally, climate change will also alter seasonal patterns in lakes and extend the ice-free season and period of summer stratification in higher latitudes. High-temperature causes change in circulation patterns and stretched stratification which leads to an increased chance of anoxia in cooler, deepwater habitats that cold-water species need to survive.

Biodiversity loss is also one of the consequences of climate change since it affects ecosystem function and stability. The growth rate, size, fertility, reproduction rate, and species distribution of many aquatic organisms are affected by change in temperature. Warming can desynchronize life cycles and seasonal phenology of consumers and resources, similarly, the patterns of flowers and pollinators can be affected [18, 19].

This review paper was aimed to address the effects of climate change on ecosystem services, mitigation and adaptation measures of the effects and finally future directions. To review this paper, articles, books and reports were collected from trusted journals and website. Search engines like google, google scholar, science direct, book and other websites were used to assess materials. After materials relevant to the purpose of the paper were collected and reviewed, the analyzed result was written and presented in this paper.

## 2. Effect of climate change on ecosystem service

The effect of climate change on ecosystem services was expected as the main challenge of this century [3]. According to Intergovernmental Panel on Climate Change report, increased temperature up to 3°C causes considerable alterations on ecosystem process, structure and functions both spatially and temporally. The symptom was currently manifested in several regions of the world, and frustrated the future of ecosystem services. The future predictions of climate change by scientists have stated that it will become 10 times faster than the average rate of the past glacial period [1, 2, 19, 20].

Several authors have reported that climate change causes significant damage to natural systems from which peoples drive benefit [5, 6]. These implies that, fast changing climate is responsible for the extinction of most valuable species that play key role in ecosystem services. The rates of species extinction today is 10–100 times greater than that of fossil record, and even expected to increase in the future. According to [3] predictions, extinction rates of biodiversity in the twenty-first century will become 100 times greater than today. Thus, climate change aggravates the ecosystem services damage directly and indirectly in terms of both quantity and quality.

The most obvious effect of climate changes like rainfall distribution, temperature, flooding and sea level rise have caused serious challenge to ecosystem services [15, 19] particularly species with small fragmented populations, or populations restricted to small areas are susceptible to these climatic shifts [16]. The four main classes of ecosystem services; supporting services, provisioning services, regulating services and cultural services were reviewed in the next sections.

### *2.1. Effect on supporting services*

Supporting services sustain environments for life on earth, but it have indirect benefits for peoples as opposed to other ecosystem services [18;21]. Ecosystem by itself cannot be constant and continued without the consistency of underlying natural manners, for example biomass production (photosynthesis, primary production), nutrient cycling, soil formation, atmospheric oxygen production and the water cycle [22]. These processes allow the Earth to sustain basic life forms. Therefore, other ecosystem services will not exist without supporting services.

*2.1.1. Biomass production (photosynthesis, primary production).* Biomass production is the basic underlying ecosystem services that can be affected by climate changes [10, 12]. All higher trophic levels i.e. herbivores and carnivores are dependent on primary production or photosynthesis. Conversion of light energy to chemical energy and biomass through photosynthesis is the central processes in terrestrial and aquatic ecosystems. Change in climate affects the rate of photosynthesis and hence biomass production. These leads to loss of ecosystem services like availability of food, water and clean air for society [8, 9]. Primary productivity refers to changes in biomass of primary producer per unit area and time [9, 12], which is the outcome of photosynthesis, respiration, and other metabolic processes. The relationship between ecosystem process and biodiversity was studied by many authors, largely as total biomass of primary producers or accumulation rate [23, 24] It was commonly believed that the better producer species richness can improve efficiency of conversion into standing biomass by plants and algae. The stored biomass by primary producers is used as a source of energy (biofuels), timber, food or decomposed by decomposers. Since biological diversity are sensitive to climate change, there is direct negative effect on biomass production of ecosystem services [10, 11].

*2.1.2. Nutrient cycling and soil formation.* Nutrient cycling and soil formation is very important process in ecosystems [25, 26], which is involves mineralization of nutrients, absorption, storing in their biomass that pass through food chain and finally goes to soil after death. The cycling of basic elements like phosphorus, nitrogen, sulfur and carbon plays significant role in ecosystem services as the constituent of biomass and chemical form of energy [27]. The climate change affects the rate of nutrients cycling and soil formation process. If soil formation and cycling of minerals are not in a needed amount and rate it causes plausibly effect on ecosystem services [28]. Because organisms cannot get energy and nutrients for survival from the food they eat. Therefore, the loss of primary producers can significantly degrade ecosystem efficiency to assimilate inorganic resources [29]. Nutrient cycling and soil formation greatly support ecosystem health and biological diversity of aquatic, terrestrial and soil organisms [29, 30]. The structural and functional diversity of plant species plays a key role to control evapotranspiration and runoff [26]. Additionally, the diversity of soil organisms (such as earthworms, micro-arthropods, and microbes) contributes to soil physical properties like presence of macropores which help for penetration. Several authors have reported the direct effects of climate change on these process and, hence it affects ecosystem services [27, 31].

### *2.2. Effect on provisioning services*

MEA defines provisioning services as the physical assets or things that society gain from ecosystems [3] which can be limited, renewable, directly consumed, appropriated and traded. It includes food, water for domestic consumption, timber products, and energy from biomass, fossil fuels, and oils, plants body parts for clothes, materials, and medicinal remunerations. The provision of food was considered as the fundamental services of ecosystem for society. This provision can be gained through agricultural practices, farming of fishes, wildlife hunting and gathering. There are also another services like forage, wood and a diversity of living and non-living resources gained from both terrestrial, wetland and aquatic ecosystems [2, 4].

*2.2.1. Provision of food.* The effect of climate change on food provision is commonly negative with increasing temperature [19, 24]. It was reported that the provision of food for world is little affected at moderate level and negatively affected nearly at 1°C temperature. If temperature becomes greater than 2°C, the net effect becomes increasingly negative. Therefore, there is expectation of growing need and struggle for consumption of food and water resources on earth [20, 23].

*2.2.1.1 Agricultural Crops.* According to [32], climate change have direct effect on crop cultivation. Other studies have also reported decrease in growth and production of cereal crop with increasing temperature [33]. As a result, decline of crop yield was reported at dry places; especially in the subtropical semi-arid areas of the world [30, 33]. Conversely, high harvest and production was reported for regions with optimum moisture and moderate temperature [34, 35]. These implies that increased climate change can affect crop production and leads to starvation and food insecurity. Most studies have reported the significant decline of food production for global mean warming above 2°C [19, 20]. The integrated factors of the need and supply [33] were directly correlated with climate changes like inter annual climate variability. These may be the reason for crop failure in the region known for the bulk production of staple crops [32, 34]. Different regions of the world is facing problems of climate changes that declines productivity and harvest of crops.

*2.2.1.2. Livestock and dairy product.* The excessive amount of agricultural landscape was taken over by requirements of domestic animals. Livestock's like fowls, goats, cattle need huge amount of land for grazing or crop cultivation for their consumption [24]. Accordingly, the higher plant production appropriated for human use is consumed by domesticated species [4]. In most cases grazing lands allotted for domestic animals are not fertile areas, but too dry and soils not appropriate for agriculture [33]. Therefore, animal production decreased and hence community livelihoods. Even though the influences of climate change on animal nourishment development are related to those delineated for crops, animal-based production has extra temperature sensitivity; most animals have naturally very strict range of body temperature [33]. Their productivity declines as their body temperature gets closer with environmental temperature [34]. The comprehensive rangelands of the planet, where incomes and beliefs are frequently related to pastoralism are now quite near to this limit throughout the world [4].

*2.2.1.3. Wild Fisheries.* The food provision of aquatic ecosystem, particularly fish can be rigorously affected by climate change [19]. Changes in climate changes water temperature, dissolved oxygen, and concentration of pollutants in the water [33]. Thus, these conditions affect directly and indirectly fish population and diversity. The productivity of ocean is expected to escalate discreetly below a 2°C, but retort of fish production is complicated by changes in biome, body size and ocean acidification [24]. Some authors have forecasted shift in location of many marine fisheries as a result of climate change or of ocean temperature deviations [32]. Additionally, climate change is directly related to species richness and distributions; hence it affect their interaction, and breeding periods, mating time and seasons of maturity [33].

Acidification of ocean is an obvious concern of rising $CO_2$ levels with numerous negative effects on living organisms [5]. Organisms those build their shells from calcium carbonate can be negatively affected at acidic pH values [6]. Although the effect is manifested directly on wild-harvested like mollusks' and crustacean, other planktonic organisms on which oceanic food web is based are also seriously affected. Other organisms with an internal skeleton like fish were also susceptible to the complications [32]. Similarly, coral reefs were seriously subjected to the synergetic effects of high temperature and eutrophication [33] Theses affects organisms those directly dependent on coral reef habitat as well as fisheries and tourism services. Furthermore, there was report from recent study showing combined effect of other stressors (like overharvesting, water pollution, and perturbations) with climate changes [5;35]. These synergetic effect causes additional problem that impose variation to coastal species, including key habitat-forming species (like algae) and the associated ecological functioning of temperate coasts.

Therefore, appropriate monitoring at both national and international level pressures can escalate adaptation of aquatic communities to pressures of climate change [32]

*2.2.2. Provision of forest products.* Climate change have significant effect on forest, since it increase tree mortality and decrease heterogeneity and diversity of forest [18]. It was reported by [22] that observation over past half century have shown reasonable degradation. There are numerous reasons for the death of trees that cooperate with climate change. These includes spread and outbreaks disease, nuisance, fire, flooding and storms [36]. Most of the time, this kind of configuration happened if complex biome where subjected to climate conversion [21]. The vibrant observation of biosphere reactions to a gradually changing climate is that the biome advances progressively and increasingly at its principal verge, and found being pressurized in an orderly fashion [36, 37]. The conviction is that, movement of a biome when it includes numerous interrelated, long-lasting species, is manifest through an increasing frequency of disturbances leading to unexpected, disordered and discontinuous community reorganization [21, 38].

*2.2.3. Timber provision.* Another effect that climate change cause to benefits derived from ecosystem service as provision is the challenges imposed on timber productions. The developing countries produce timber for several purposes like for construction material used by community at local level and raw materials that can be used in industry [21, 37]. Presently, total timber products are derived from plantation forests worldwide. The portion of firewood delivered from plantations is expected to rise to three-quarters in mid-century. Appropriately managed and conserved forests resources are a well-accepted measure of numerous traditional landscapes. Even though plantation forest production rates per unit area increased recently, there is large variability from region to region [39].

*2.2.4. Energy provision.* Ecosystem is considered as the fundamental storage and source of energy. As the primary source of energy, biomass (like wood and charcoal) was the major supply for both developed and developing countries [40]. In the serous scenarios for mitigation of climate change, the great dependency on biomass energy was reported. The biomass of plants have high potential of carbon sequestration and storage [38]. Forest and woody plants are the major source of these biomass, and hence grown in short rotations on agricultural land [41]. Beside biomass, other abiotic components of ecosystem such as water and wind also provide massive amount energy like hydropower and wind energy. Several authors have reported that the overall productions of ecosystem works under a lot of problems without unsuitable conditions [40, 41]. Thus, diverting a much higher fraction of biomass to bioenergy certainly affects agricultural practices or semi-natural areas providing other ecosystem services like habitat for living organisms.

*2.2.5. Water provision.* Water provides many benefits for human being; like domestic consumptions, industrial purpose, power generation, irrigated agriculture and urban sanitation [42]. The quantity and quality of freshwater became problematic to our today's world as climate change increased [41]. Several authors agree that freshwater provisions are under stress in many dry parts of the world, due to rising demand and climate changes [43]. Consequently, the relationship between alterations in rainfall and vaporization, runoff is expected to decline in most tropical deserts and escalation in the wet tropics [40]. Most authors reported that worldwide climate change aggravates current inclinations toward poor water quality due to a number of reasons [42, 43]. High water temperature and nutrient concentrations boost up algal growth and pathogens. The augmentation of nutrient concentration and sediment can be resulted from disturbance of terrestrial ecosystems like biome movement and land use changes [40, 41]. Excessive storms and intensive water use are another combined effects that affects water provision with climate changes.

*2.2.6. Medicinal provision.* It is obvious that plants and animals or their body parts have been collected and used for a long period of time by society, for their medicinal properties [44]. The interest of people towards medicinal and aromatic plants become increased worldwide throughout developing and developed countries [45, 46]. Many authors were agreed that medicines derived from plants, animals and microbes were foundation ground for today's medical treatment. Most of these species with medicinal importance were basically harbored in grasslands and forest ecosystems [36, 46]. Similarly, plants and animals of medicinal importance also found in aquatic ecosystems. For example, traditional medicine were extracted from sea horse, star fish, sea urchin and sea cucumber [45]. Chemicals like beta carotene and other carotenoids extracted from algae are widely used in pharmaceuticals. Other medicines like Quinine, which efficiently fights malaria are extracted from plants [47]. Currently, climate change is affecting plant and animal species abundance and distribution by aggravating pollution, degradation and alteration of habitat [36]. Hence, to attain appropriate conservation and management of these medicine pool alive, adaptation and mitigation measures to climate change is needed.

### *2.3. Effect on regulating services*

It is commonly believed that regulating services of ecosystems are groups that mostly depend on relationship between biodiversity and ecosystem. Consequently, it is vulnerable to biodiversity loss and ecosystem degradation due to climate change [3]. Regulating services of ecosystem is result of multiple ecosystem processes and ecosystem functioning, particularly, regulation human habitat, soil conditions, water quality and atmosphere [47, 49]. The concept of ecosystem regulation comprises of broad ranges like seed dispersal and pollination, decomposition and mineralization, flood control, storage of carbon and climate regulation. The capacity of ecosystem to maintain natural balance can be affected by extreme climate changes and other disturbances from both natural process and human activities.

*2.3.1. Regulation of climate.* Climate regulation refers to ecosystem service that normalize processes related to atmospheric compositions, greenhouse effect, ozone, precipitation, air quality, and moderation of temperature and weather patterns at global and local scales [49, 50]. Ecosystem contribute in moderating climate changes through sequestration of carbon, trapping particulate matters and other pollutants [36]. Climate change affects significantly potentials of ecosystem to regulate atmosphere; air quality, precipitation and maintain natural balance of chemical compounds found in air. Currently, both terrestrial and aquatic ecosystem moderate rate of climate change through sequestration of $CO_2$ released from human activities [47, 49] . The capacity of terrestrial ecosystem to absorb $CO_2$ is restricted and time dependent, due to respiration of deposited carbon that finally balances or exceeds the uptake.

The rate of respiration rises with temperature, whereas sequestration rate saturates with rising $CO_2$ and decease as temperature drops [36, 49]. Moreover, the occurrence of wildfire was increasing with the regions climate change. Air quality can be regulated as plants can remove contaminants and pollutions by capturing and absorption [51]. The ability of various species to capture particulate matter may be different for different particulate matter [36], most studies have reported the significance of plant species diversity for health benefits and regulation of air quality [44]. For example, it was believed that coniferous tree species (*Pinus sylvestris*) have high capacity in removing particulate matters than broadleaved species. Similarly, the complexity of leaf structure and combinations of species can enhance removal of particulate matters [52, 53].

*2.3.2. Regulation of water.* Climate change affects ability to regulate water quantity and quality. Integral riparian and coastal vegetation reduce flooding and erosion [53]. Climate change affected these service worldwide, but there is synergetic effect with riparian development [54]. For instance, areas under appropriate management have better cost-effective mechanism of defense against rising sea level due to climate change [55]. Climate change escalate heavy rainfall, disaster of flooding and drought [52]. The hazard of flooding can be controlled by riparian forest management and percolation of rainfall into soil [53]. Hence, areas covered by vegetation can contribute significant role in adaptation and mitigation of flood disaster both for long period and short period of time [44, 55]. Ecosystem can regulate the quality of water through removing of pollutants [55]. These could happen when riparian forest diverts rainfall and decrease transfer to ground, it can also minimize the flooding risk [52]. Additionally, wetland ecosystems can function of basic role of regulation improving air and water quality [54]. These potential can also help to store waters and restrict the risk of ecosystems eutrophication [51].

*2.3.3. Regulation of soil.* There is natural ability of ecosystem to self-regulate through removal of contaminant [49;57]. For example, bioremediation is used to remove pollutants from soil and water by changing environmental conditions that facilitate growth of microbes to decompose (convert contaminants into $CO_2$, biomass and water), transform to nonhazardous products and trap contaminants [24] . These process is biological mechanisms to remove contaminant, which is safe and environmentally friend process [32;33]. Climate change can affect negatively these process as a result of sensitivity of organisms. Organisms suitable for the process are those with physiological abilities to transform by degrading and detoxifying pollutants to nontoxic compounds [28;57]. The efficiency is extremely reliant on environmental parameters that encourage the growth of microorganisms [29]; in another way the process needs to have appropriate microorganisms at appropriate place under convenient environmental conditions for degradation [57]. Moreover, soil nutrient content, moisture, pore, air and ability to harbor soil organisms depends on climatic conditions [56]. Therefore, climate change affects regulation of soil and its formation. Since climate change influences regulation of soil, crop diversity and heterogeneity can improve resistance to change and facilitate retention of soil, nutrients cycles, decrease erosion and runoff [56]. Crop residue can be used to increase soil organic nutrients for improvement of water retention and nutrient provision for crop [28].

*2.3.4. Regulation of health and wellbeing.* Current evidence and past records show the spread in level, rate, and harshness of epidemics of unusual diseases of animals and plants [10]. The spread and invasion of 'alien invasive species' is also documented at different parts of the world [2]. Although many reasons and justifications like improved movement of society, high chance for transmission of disease, larger and more densely packed human and livestock populations are reflected as cause of the problems, the underlying and basic cause of problem is climate change [4]. Furthermore, there are reasonable influences that ecosystems are highly susceptible to the occurrence of single species populations, when they decreased areas and species richness with climate prompted stress [11]. There is inferential suggestion that situations of extraordinary climate change will be complemented by abating of the ecosystem service to pest and disease outbreak which can suppress provision services.

*2.3.5. Pollination.* Pollination is the mutualistic interaction of living organisms that can be affected by climate changes since most of insects responsible for carrying pollen grains are sensitive to climate changes [21]. These can affect food security of nation, decrease distribution and richness of biodiversity, and integrity of natural ecosystems [10]. Recently, the declination of the number of pollinator abundance and richness was reported universal. There are reports from other authors showing that the problem was also resulted from frequently used insecticides and outbreak of disease and habitat [16]. Hence, pollinator and its target flower respond to climate change and other disturbance differently they have chance to become decoupled [18]. Detected variations in flowering phenology is the first biological markers clearly endorsed to climate change [58]. Other factors determining the efficiency of pollination is structure and composition of landscape. These land features are sensitive to climate variability and their efficiency in pollination can be negatively affected by fast change in climate [59]. Consequently, predictions of the effect of climate change on regulation of pollination service ranges from more pessimistic to less pessimistic [59]. Roaming organisms pollinate numerous vital crops species that plays crucial role both socially and commercially [21]. Pollination can be made rarely by introduced species, artificially by human, wild animals; but more commonly by insects like bees, butterflies, beetles and birds other invertebrates [59, 60].

### *2.4. Effect on cultural services*
According to MEA [3], cultural services refers to nonmaterial profits that society gain from ecosystems, and specifically values like tourism and recreation, spiritual and religious, aesthetic, cultural heritage, knowledge systems, educational, inspiration, social relations, sense of place, cultural heritage values, recreation and ecotourism. According to [48], cultural ecosystem services was considered as the most difficult services to characterize. Similarly, other authors reported that their connections to human emotion, and motivation can make them complex but important for human well-being [44].

*2.4.1. Tourism and recreation.* Ecosystem gives services of recreation and tourism that can fundamentally provide support for greater incomes [10;12]. Almost recreational services are extremely susceptible to universal warming; particularly outdoor winter sports and tropical holidays [44]. Many peoples use outdoor entertainment as an opportunity for interaction with other community, religion, culture and beliefs, and similarly, they use it as a chance to interact with their environment [48]. These offers them recreational activities, like mobility, and nature study [10] . Climate change is affecting the ecosystem services of tourism and recreation internationally [11], not because of their unique characters and specific to certain region but are not viable to society who created them.

*2.4.2. Landscape aesthetics.* Aesthetics are regularly used as an example of cultural ecosystem service, but more specific operational definitions to guide assessments are rarely provided. According to MEA, it refers to the "beauty or aesthetic value in various aspects of ecosystems, as reflected in the support for parks, 'scenic drives,' and the selection of housing locations"[3]. More recently, [9] represent aesthetic services based on "appreciation of natural scenery," and [12] link aesthetic values in rural areas with "the amount or configuration of open space in agricultural or forested (land use/land cover) types." These conceptualizations emphasize visual landscape aesthetics, especially scenic beauty [10; 11]. The broad range of landscape aesthetics was studied under context of different disciplines [14]. Qualitative and quantitative surveys have shown that aesthetic values are affected by climate changes at local and regional level [9], and hence service that society gains from ecosystem is being threatened. Variations in aesthetic inclinations through society, ethnics and nations are commonly assumed, and variances in aesthetic principles have been verified, especially in the context of culturally modified landscapes [12]. Conversely, perceptual evaluations of primarily natural landscapes have frequently revealed agreement than disagreement [59], features normally explain the largest share of variance in measures of perceived aesthetic quality within a context of ecosystem services.

*2.4.3. Cultural heritage.* According to millennium ecosystem assessment [3], cultural heritage is explained as the legacy of biophysical features, physical artifacts, and intangible attributes of a group or society that are inherited from past generations, maintained in present, and bestowed for benefit of future generations. Consequently, several articles reviewed in this area reveal a broad range of biocultural relationships, ranging beyond visual focus in the former section [60]. Cultural landscapes are significant constituents of cultural heritage characterized by long-term interaction between site conditions and human influences [61]. As it was indicated by several studies, cultural heritage is inseparably linked with community [61, 62]. It contributes to the identity of the community and may exemplify a whole region and act as a significant characteristic for touristic offers and product promotion [61, 62]. Culture is not static and hence can drive ecosystem change. Thus, conserving cultural heritage can extensively integrate with preserving other ecosystem services. Ecosystem services that can play a crucial role in cultural heritage are frequently mutual goods that are shared rather than owned [63]. They typically lack appropriate and fair prices as signals of value, which may be more clearly (but still imperfectly) expressed via politics. Nonmarket economic evaluation methods have been positively applied to cultural heritage objects; nevertheless, estimates of some aspects, such as regional identity or sense of place, largely remain vague [61]. One confirmed method is expertly facilitated deliberation, which can elicit and refine relevant cultural heritage values and the ecosystem features with which they are associated.

*2.4.4. Spiritual and religious services.* Concern in spiritual and religious importance and beliefs credited to certain aspects of nature has been emerged, as replicated in their inclusion as a subcategory of cultural ecosystem service [11, 61]. Environment protection specialists have discussed about the ways in which spiritual and religious values can be involved in promoting biodiversity conservation with some risk for underestimating the difficulties of lived involvements of spirituality and religiosity [61]. Diverse religious groups and conservationists have tried to strengthen the link between religion and environmental conservation, promoting the concept of "environmental stewardship". Relationships between environments and religion include moral and symbolic concepts but can also center on very measurable concerns, such as staking claim to land contested by immigrants, invading states, or development agencies [63]. Language is among the most powerful ways cultures map meanings through which the world is made more intelligible [11]. For instance, the variety of names for a single site points to shared histories in an increasingly multicultural world. Spiritual and religious amenities do not generalize well across communities and they are difficult to assess in economic or monetary terms [16, 19]. However, there are more comprehensive methods for studying spiritual and religious ecosystem service, the way they are constructed and perceived, and their relation to land use and resource management. Many historical and anthropological studies demonstrate the complexities of spiritual services, and hence may contribute to policies that avoid the trap of overgeneralization [15, 61]. Ecologists and ecological economists are increasingly adopting methods derived from history and social sciences to include spiritual and religious services in their analyses.

## 3. Mitigation and adaptation measures of the effects

Ecosystem services play an important role in strategies for tackling climate change: mitigation and adaptation [16, 19]. Mitigation aims at reducing emissions sources or enhancing sinks of greenhouse gases, and adaptation aims at adjusting natural or human systems to moderate harm or exploit beneficial opportunities from climate variations [20]. Because of their different rationales, these strategies have different priority sectors and locations: mitigation prioritizes larger emission sources or stronger potential sinks, whereas adaptation prioritizes vulnerable people, ecosystems and activities. While some sectors are mostly concerned by one of the two strategies, ecosystems and their services are clearly relevant to both [14, 15].

Ecosystems contribute to mitigation because of their capacity to remove carbon from the atmosphere and to store it [19]. Ecosystems contribute also to adaptation because they provide services that can help people adapt to both current climate hazards and future climate change. While ecosystem services are part of the solution to climate change, they are also affected by changing climatic conditions [20, 61]. Ecosystem-based approaches to climate change should recognize the multiple links between ecosystem services and climate change: management can enhance the contribution of ecosystem services to adaptation and mitigation and, as climate change will affect ecosystems and their services, adaptation measures are needed to reduce negative impacts and maintain ecosystem functions [14, 16].

Responding effectively to climate change requires urgent action to halt net greenhouse gas emissions and to adapt to changes that cannot be prevented [64]. The Paris Agreement of the United Nations Framework Convention on Climate Change has committed governments to the following: keeping global temperature rise below 2°C, pursuing efforts to limit it to 1.5°C, and adapting to reduce the vulnerability of people and ecosystems to the damaging consequences of a changing climate [19, 20]. The appropriate protection, restoration, or management of natural and semi-natural ecosystems can make critical contributions to climate change mitigation and help people adapt to climate change [14, 15, 16]. Ecosystem themselves are vulnerable to climate change, but by restoring natural ecosystem processes, resilience can be built, and a wide range of adaptation strategies can ameliorate the impacts.

## 4. Future prospects

Predictions about future effects of climate change on ecosystem services are being made by several authors. It was clearly discovered that climate change have significant effect on ecosystem services [34, 35]. Climate change affects both abiotic and biotic components of ecosystem directly or indirectly; hence ecosystem process and functions [64]. Thus, services derived from ecosystem can be declined in quantity and quality. Further investigations are needed for appropriate adaptation and mitigation measures in future. From this review it is possible to realize that ecosystem services was being under stress not only due to climate changes but also there are other underlying human activities like over exploitation and land use changes [61]. Therefore, the synergetic effects of both climate change and human activity can aggravate loss of ecosystem services with increasing human population growth, industrial and technological development for exploitations.

Due to the complex nature of ecosystem, the short-term effects of simple and minor changes may not be revealed. For example, soil formation takes place over millennia through the interaction of climate and the biota, thus the rapid migration of organisms that will be stimulated by climate change will not only result in new combinations of species but also perhaps mismatches of the biota with the substrates upon which they evolved [33]. But it can seriously affect sensitive species that are important to society. Hence, the long term and short-term effects of climate change and other human activities needs further investigations by different experts.

## 5. Conclusion

The climate change affect ecosystem services directly or indirectly. These is through land use change, rising $CO_2$, air and water pollution, intensifying natural resource use, and loss of biodiversity. Therefore, understanding the effects of climate changes on ecosystem services became fundamentally important in order to adapt and mitigate the effect. The change affects supporting services of ecosystem like biomass production (photosynthesis, primary production), nutrient cycling, soil formation, atmospheric oxygen production and the water cycle. Without supporting services, provisional, regulating, and cultural services wouldn't exist. Appropriate adaptation and mitigation strategies can reduce effects of climate changes. These can be through ecosystem protection, restoration, or appropriate management. It helps people adapt to climate change and minimize problems they face from loss of ecosystem services.

**References**

[1] Locatelli B 2018 Ecosystem Services and Climate Change *Routledge Handb. Ecosyst. Serv.* 481–90
[2] Scholes R J 2016 Climate Change and Ecosystem Services *Wiley Interdiscip. Rev. Clim. Chang.* **7** 537–50
[3] Mooney H A, Cropper A, Capistrano D and Carpenter S 2005 *Millennium Ecosystem Assessment*
[4] Alcano J, Ash N, Butler C, Callicot B and Capistrano D 2003 *Ecosystems and Human Well-Being* (Washington, Covelo, and London: Island Press)
[5] Haines-Young R and Potschin M 2012 The Links Between Biodiversity, Ecosystem Services and Human Well-Being *Ecosyst. Ecol.* **1** 110–39
[6] Lele S, Springate-Baginski O, Lakerveld R, Deb D and Dash P 2013 Ecosystem Services: Origins, Contributions, Pitfalls, and Alternatives *Conserv. Soc.* **11** 343–58
[7] Zegeye H 2018 Climate Change Mitigation and Adaptation *Int. J. Res. Environ. Stud.* **2** 105–21
[8] Jabbour R 2016 Coffee Agroecology: A New Approach to Understanding Agricultural Biodiversity, Ecosystem Services and Sustainable Development *Am. Entomol.* **62** E1–E1
[9] Colvin S A R, Sullivan S M P, Shirey P D, Colvin R W, Winemiller K O, Hughes R M, Fausch K D, Infante D M, Olden J D, Bestgen K R, Danehy R J and Eby L 2019 Headwater Streams and Wetlands are Critical for Sustaining Fish, Fisheries, and Ecosystem Services *Fisheries* **44** 73–91
[10] Reyers B, Biggs R, Cumming G S, Elmqvist T, Hejnowicz A P and Polasky S 2013 Getting the Measure of Ecosystem Services: A Social-Ecological Approach *Front. Ecol. Environ.* **11** 268–73
[11] Tuvendal M and Elmqvist T 2011 Ecosystem Services Linking Social and Ecological systems: River Brownification and the Response of Downstream Stakeholders *Ecol. Soc.* **16**
[12] Malhi Y, Franklin J, Seddon N, Solan M, Turner M G, Field C B and Knowlton N 2020 Climate Change and Ecosystems: Threats, Opportunities and Solutions *Philos. Trans. R. Soc. B Biol. Sci.* **375** 1–8
[13] Chiabai A, Quiroga S, Martinez-Juarez P, Higgins S and Taylor T 2018 The Nexus between Climate Change, Ecosystem Services and Human Health: Towards a Conceptual Framework *Sci. Total Environ.* **635** 1191–204
[14] Biggs R, Schlüter M and Schoon M L 2015 *Principles for Building Resilience: Sustaining Ecosystem Services in Social-Ecological Systems*
[15] Lescourret F, Magda D, Richard G, Adam-Blondon A F, Bardy M, Baudry J, Doussan I, Dumont B, Lefèvre F, Litrico I, Martin-Clouaire R, Montuelle B, Pellerin S, Plantegenest M, Tancoigne E, Thomas A, Guyomard H and Soussana J F 2015 A Social-Ecological Approach to Managing Multiple Agro-Ecosystem Services *Curr. Opin. Environ. Sustain.* **14** 68–75
[16] Folke C 2007 Social-Ecological Systems and Adaptive Governance of the Commons *Ecol. Res.* **22** 14–5
[17] De Groot R S, Wilson M A and Boumans R M J 2002 A Typology for the Classification, Description and Valuation of Ecosystem Functions, Goods and Services *Ecol. Econ.* **41** 393–408
[18] Carpenter S R, Mooney H A, Agard J, Capistrano D, Defries R S, Diaz S, Dietz T, Duraiappah A K, Oteng-Yeboah A, Pereira H M, Perrings C, Reid W V., Sarukhan J, Scholes R J and Whyte A 2009 Science for Managing Ecosystem Services: Beyond the Millennium Ecosystem Assessment *Proc. Natl. Acad. Sci. U. S. A.* **106** 1305–12
[19] Dawson T P, Jackson S T, House J I, Prentice I C and Mace G M 2011 Beyond Predictions: Biodiversity Conservation in a Changing Climate *Science (80-. ).* **332** 53–8

[20] Urry J 2015 *Climate Change and Society* (London: Palgrave Macmillan)
[21] Kremen C 2005 Managing Ecosystem Services: What do We Need to Know about Their Ecology? *Ecol. Lett.* **8** 468–79
[22] Dale V H and Polasky S 2007 Measures of the Effects of Agricultural Practices on Ecosystem Services *Ecol. Econ.* **64** 286–96
[23] Cardinale B J, Matulich K L, Hooper D U, Byrnes J E, Duffy E, Gamfeldt L, Balvanera P, O'Connor M I and Gonzalez A 2011 The Functional Role of Producer Diversity in Ecosystems *Am. J. Bot.* **98** 572–92
[24] Vitousek P M and Hooper D U 1994 Biological Diversity and Terrestrial Ecosystem Biogeochemistry *Biodiversity and Ecosystem Function* pp 3–14
[25] Richter D de B, Oh N H, Fimmen R and Jackson J 2007 *The Rhizosphere and Soil Formation*
[26] Schulz S, Brankatschk R, Dümig A, Kögel-Knabner I, Schloter M and Zeyer J 2013 The Role of Microorganisms at Different Stages of Ecosystem Development for Soil Formation *Biogeosciences* **10** 3983–96
[27] Bormann F . and Likens G . 1967 Nutrient Cycling *Science (80-. ).* **155** 424–9
[28] Burke D J, Weintraub M N, Hewins C R and Kalisz S 2011 Relationship Between Soil Enzyme Activities, Nutrient Cycling and Soil Fungal Communities in a Northern Hardwood Forest *Soil Biol. Biochem.* **43** 795–803
[29] Hickman J E, Ashton I W, Howe K M and Lerdau M T 2013 The Native-Invasive Balance: Implications for Nutrient Cycling in Ecosystems *Oecologia* **173** 319–28
[30] Jiao S, Chen W, Wang J, Du N, Li Q and Wei G 2018 Soil Microbiomes with Distinct Assemblies Through Vertical Soil Profiles Drive the Cycling of Multiple Nutrients in Reforested Ecosystems *Microbiome* **6** 1–13
[31] Azcón-Aguilar C and Barea J M 2015 Nutrient Cycling in the Mycorrhizosphere *J. Soil Sci. Plant Nutr.* **15** 372–96
[32] Challinor A J, Ewert F, Arnold S, Simelton E and Fraser E 2009 Crops and Climate Change: Progress, Trends, and Challenges in Simulating Impacts and Informing Adaptation *J. Exp. Bot.* **60** 2775–89
[33] Reilly J, Tubiello F, McCarl B, Abler D, Darwin R, Fuglie K, Hollinger S, Izaurralde C, Jagtap S, Jones J, Mearns L, Ojima D, Paul E, Paustian K, Riha S and Rosenberg, N. Rosenzweig C 2003 US Agriculture and Climate Change: New Results *Clim. Change* **57** 43–67
[34] Lobell D B and Gourdji S M 2012 The Influence of Climate Change on Global Crop Productivity *Plant Physiol.* **160** 1686–97
[35] Roudier P, Sultan B, Quirion P and Berg A 2011 The Impact of Future Climate Change on West African Crop Yields: What Does the Recent Literature Say? *Glob. Environ. Chang.* **21** 1073–83
[36] Patterson T M and Coelho D L 2009 Ecosystem Services: Foundations, Opportunities, and Challenges for the Forest Products Sector *For. Ecol. Manage.* **257** 1637–46
[37] Buongiorno J, Zhu S, Zhang D, Turner J and Tomberlin D 2003 The Global Forest Products Model: Structure, Estimation, and Applications *Glob. For. Prod. Model Struct. Estim. Appl.* 1–301
[38] Hodder K H, Newton A C, Cantarello E and Perrella L 2014 Does Landscape-Scale Conservation Management Enhance the Provision of Ecosystem Services? *Int. J. Biodivers. Sci. Ecosyst. Serv. Manag.* **10** 71–83
[39] Shackleton C and Shackleton S 2004 The Importance of Non-Timber Forest Products in Rural Livelihood Security and as Safety Nets: A review of Evidence from South Africa *S. Afr. J. Sci.* **100** 658–64
[40] Spangenberg J H, Görg C, Truong D T, Tekken V, Bustamante J V and Settele J 2014 Provision of Ecosystem Services is Determined by Human Agency, not Ecosystem Functions. Four Case studies

*Int. J. Biodivers. Sci. Ecosyst. Serv. Manag.* **10** 40–53

[41] Brockerhoff E G, Barbaro L, Castagneyrol B, Forrester D I, Gardiner B, González-Olabarria J R, Lyver P O B, Meurisse N, Oxbrough A, Taki H, Thompson I D, van der Plas F and Jactel H 2017 Forest Biodiversity, Ecosystem Functioning and the Provision of Ecosystem Services *Biodivers. Conserv.* **26** 3005–35

[42] Jewitt G 2002 Can Integrated Water Resources Management Sustain the Provision of Ecosystem Goods and Services? *Phys. Chem. Earth* **27** 887–95

[43] Olguín E J, Sánchez-Galván G, Melo F J, Hernández V J and González-Portela R E 2017 Long-Term Assessment at Field Scale of Floating Treatment Wetlands for Improvement of Water Quality and Provision of Ecosystem Services in a Eutrophic Urban Pond *Sci. Total Environ.* **584** 561–71

[44] Khan S M, Page S, Ahmad H, Shaheen H, Ullah Z, Ahmad M and Harper D M 2013 Medicinal Flora and Ethnoecological Knowledge in the Naran Valley, Western Himalaya, Pakistan *J. Ethnobiol. Ethnomed.* **9**

[45] Speak A F, Mizgajski A and Borysiak J 2015 Allotment Gardens and Parks: Provision of Ecosystem Services with an Emphasis on Biodiversity *Urban For. Urban Green.* **14** 772–81

[46] Maes J, Paracchini M-L and Zulian G 2011 *A European assessment of the provision of ecosystem services - Towards an atlas of ecosystem services* vol JRC63505

[47] Zhang W, Ricketts T H, Kremen C, Carney K and Swinton S M 2007 Ecosystem Services and Dis-Services to Agriculture *Ecol. Econ.* **64** 253–60

[48] Bennett E M, Peterson G D and Gordon L J 2009 Understanding Relationships among Multiple Ecosystem Services *Ecol. Lett.* **12** 1394–404

[49] West P C, Narisma G T, Barford C C, Kucharik C J and Foley J A 2011 An Alternative Approach for Quantifying Climate Regulation by Ecosystems *Front. Ecol. Environ.* **9** 126–33

[50] Ma L, Bicking S and Müller F 2019 Mapping and Comparing Ecosystem Service Indicators of Global Climate Regulation in Schleswig-Holstein, Northern Germany *Sci. Total Environ.* **648** 1582–97

[51] Beckett K P, Freer-Smith P H and Taylor G 2000 The Capture of Particulate Pollution by Trees at Five Contrasting Urban Sites *Arboric. J.* **24** 209–30

[52] Chen L, Liu C, Zhang L, Zou R and Zhang Z 2017 Variation in Tree Species Ability to Capture and Retain Airborne Fine Particulate Matter (PM2.5) *Sci. Rep.* **7** 1–11

[53] Broadmeadow S and Nisbet T R 2004 The Effects of Riparian Forest Management on the Freshwater Environment: A Literature Review of Best Management Practice *Hydrol. Earth Syst. Sci.* **8** 286–305

[54] Hupp C R 2000 Hydrology, Geomorphology and Vegetation of Coastal Plain Rivers in the South-Easthern USA *Hydrol. Process.* **14** 2991–3010

[55] Shepard C C, Crain C M and Beck M W 2011 The Protective Role of Coastal Marshes: A Systematic Review and Meta-Analysis *PLoS One* **6** e27374

[56] Margesin R, Zimmerbauer A and Schinner F 2000 Monitoring of Bioremediation by Soil Biological Activities *Chemosphere* **40** 339–46

[57] Diels L, De Smet M, Hooyberghs L and Corbisier P 1999 Heavy Metals Bioremediation of Soil *Appl. Biochem. Biotechnol. - Part B Mol. Biotechnol.* **12** 149–58

[58] Schulp C J E, Lautenbach S and Verburg P H 2014 Quantifying and Mapping Ecosystem Services: Demand and Supply of Pollination in the European Union *Ecol. Indic.* **36** 131–41

[59] Häussler J, Sahlin U, Baey C, Smith H G and Clough Y 2017 Pollinator Population Size and Pollination Ecosystem Service Responses to Enhancing Floral and Nesting Resources *Ecol. Evol.* **7** 1898–908

[60] Daniel T C, Muhar A, Arnberger A, Aznar O, Boyd J W, Chan K M A, Costanza R, Elmqvist T, Flint C G, Gobster P H, Grêt-Regamey A, Lave R, Muhar S, Penker M, Ribe R G, Schauppenlehner

T, Sikor T, Soloviy I, Spierenburg M, Taczanowska K, Tam J and Von Der Dunk A 2012 Contributions of Cultural Services to the Ecosystem Services Agenda *Proc. Natl. Acad. Sci. U. S. A.* **109** 8812–9
[61] Tengberg A, Fredholm S, Eliasson I, Knez I, Saltzman K and Wetterberg O 2012 Cultural Ecosystem Services Provided by Landscapes: Assessment of Heritage Values and Identity *Ecosyst. Serv.* **2** 14–26
[62] Plieninger T, Dijks S, Oteros-Rozas E and Bieling C 2013 Assessing, Mapping, and Quantifying Cultural Ecosystem Services at Community Level *Land use policy* **33** 118–29
[63] Hølleland H, Skrede J and Holmgaard S B 2017 Cultural Heritage and Ecosystem Services: A Literature Review *Conserv. Manag. Archaeol. Sites* **19** 210–37
[64] Morecroft M D, Duffield S, Harley M, Pearce-Higgins J W, Stevens N, Watts O and Whitaker J 2019 Measuring the Success of Climate Change Adaptation and Mitigation in Terrestrial Ecosystems *Science (80-. ).* **366**

# Assessment of current solid waste management practices, community perceptions, and contributions in the City of Kigali, Rwanda

**Fidele Iraguha[1], Ari Handono Ramelan[2] and Prabang Setyono[3]**
[1]Environmental Science Department, Postgraduate program, Sebelas Maret University, Surakarta 57126, Indonesia
[2]Environmental Science Department, Faculty of Mathematics and Natural Sciences, Sebelas Maret University, Surakarta 57126, Indonesia
[3]Environmental Science Department, Postgraduate program, Sebelas Maret University, Surakarta 57126, Indonesia
*Corresponding author: iraguhafidele@gmail.com

**Abstract.** Socioeconomic development, urbanization, an uncontrollable massive population rate, and higher living standards in cities increase the amount and variety of solid waste generated. If it accumulates, it degrades the urban environment, affects natural resources, and causes health problems. Dumping sites are becoming an increasing source of concern for Rwanda municipalities, particularly in Kigali City, and they are a significant source of pollution in the air, water, and land. This article aims to assess current solid waste treatment practices in the City of Kigali and evaluate the local community attitudes, participation, and perception in waste management. This research uses a mixed method to understand better how the waste management system performs in Kigali City. The data was collected by distributing a questionnaire to respondents, conducting interviews with communities, and contacting all companies that collect and recycle waste, government institutions, and other sources such as national household surveys and reports. The quantity and type of municipal solid waste (MSW) varies in each identified district. The findings indicate that personal attitude and perception are essential while engaging in waste management practices. There are poorly implemented WM laws and regulations. Due to incomplete combustion, open landfill fire produces smoke, hazardous gases, and particles that pollute the air. In addition, methane concentrations can reach up to 50 percent of landfill gas composition at the maximum anaerobic decomposition. These gases participate in the increase of greenhouse gas effect, global warming, and climate change.

**Keywords:** solid waste management, air pollution, public attitudes, organic law

## 1. Introduction

Socioeconomic development, urbanization, an uncontrollable massive population rate, and improved urban living standards increase the amount and complexity of solid waste generated. If it accumulates, it degrades the urban environment, affects natural resources, and causes health problems [1]. Solid waste management is becoming an environmental and public health concern in many countries, including

Rwanda, where appropriate systems for managing waste, from its source to its final disposal (landfill) or processing, are still poor and unsustainable [2]. It does not meet the best practices of waste management [3]. Kigali City's solid waste management system is still an open dumping landfill. Dumping sites are becoming an increasing source of concern for Rwanda municipalities, particularly in Kigali City, and they are a significant source of pollution in the air, water, and land. Due to incomplete combustion, open landfill fire produces smoke, hazardous gases, and particles that pollute the air.

Such a rapid increase in population and development produces an increasing volume of solid waste. As a result, it increases its management's need for infrastructure, institutional setup, and community involvement. But, the city municipality that runs the solid waste management activities could not fulfill the required approach to ensure effective solid waste management. Waste collection in Kigali City consists primarily of organic wastes associated with the market and household waste, as well as broken glasses, scrap metal, plastics, polyethylene, rubber, paper, wood, and other inorganic wastes associated with the industrial and commercial sectors [4,5]. The citizens are expected to store domestic wastes using sack or other temporary storage inside the home and hand them over to private waste collectors. However, there are wastes dumped on open areas, roadsides, gullies, burn-in their backyards, and thrown in sewerages by the households.

According to the RIO Declaration (1992) [6], it has been emphasized that all countries worldwide must seek and invest in environmental beautification and safe waste management practices. Cities cannot have clean and healthy living conditions unless waste is collected regularly and disposed of properly. Solid waste management reduces or eliminates adverse environmental and human health impacts while promoting economic development and a higher quality of life. Some processes are involved in waste management for a municipality, and personal attitude, perception, and participation are essential while engaging in waste management practices [7].

Even today, waste collection and management remain a problem in most developing countries [8]. Developing countries' waste sectors have failed to provide citizens with sustainable and adequate solid waste management services [9]. In Kigali City, population growth and economic development have led to increased MSW output; due to this expansion, more hazardous waste generated increases pollution and public health risks. Methane gas is produced during the decomposition of organic compounds, which causes fires and contributes to the increased greenhouse gas effect, global warming, and climate change. Kigali city does not yet have a solid strategy for waste management [10]. In Kigali, some studies on solid waste management were carried out, so only two peer-reviewed documents were available, focused only on landfill and composting, recycling systems [11]. This article aims to assess current solid waste treatment practices in Kigali city, evaluate the community's attitudes, participation, and perception in solid waste management, explore environmental law, and analyze the effects of solid waste management on climate change.

## 2. Materials and method

This research uses a mixed method to understand better how Kigali City's waste management system performs. The data was collected by distributing a questionnaire to respondents, conducting interviews with communities, and contacting all waste collection and recycling companies, as well as government entities such as the City of Kigali, Rwanda Environmental Management Authority (REMA), Sector officials, and other sources such as national household surveys and reports.

The households were selected by using systematic random sampling. The research did not seek to generalize the entire population. It looked at the case of Kigali city, attempted to capture the attitude, perception, and participation in solid waste management practices among the small size sampling population, and how it accounts for their practices. This study used statistical data analysis with an assistant of Jamovi software.

## 3. Discussion

*3.1Kigali city SWM system evolution*

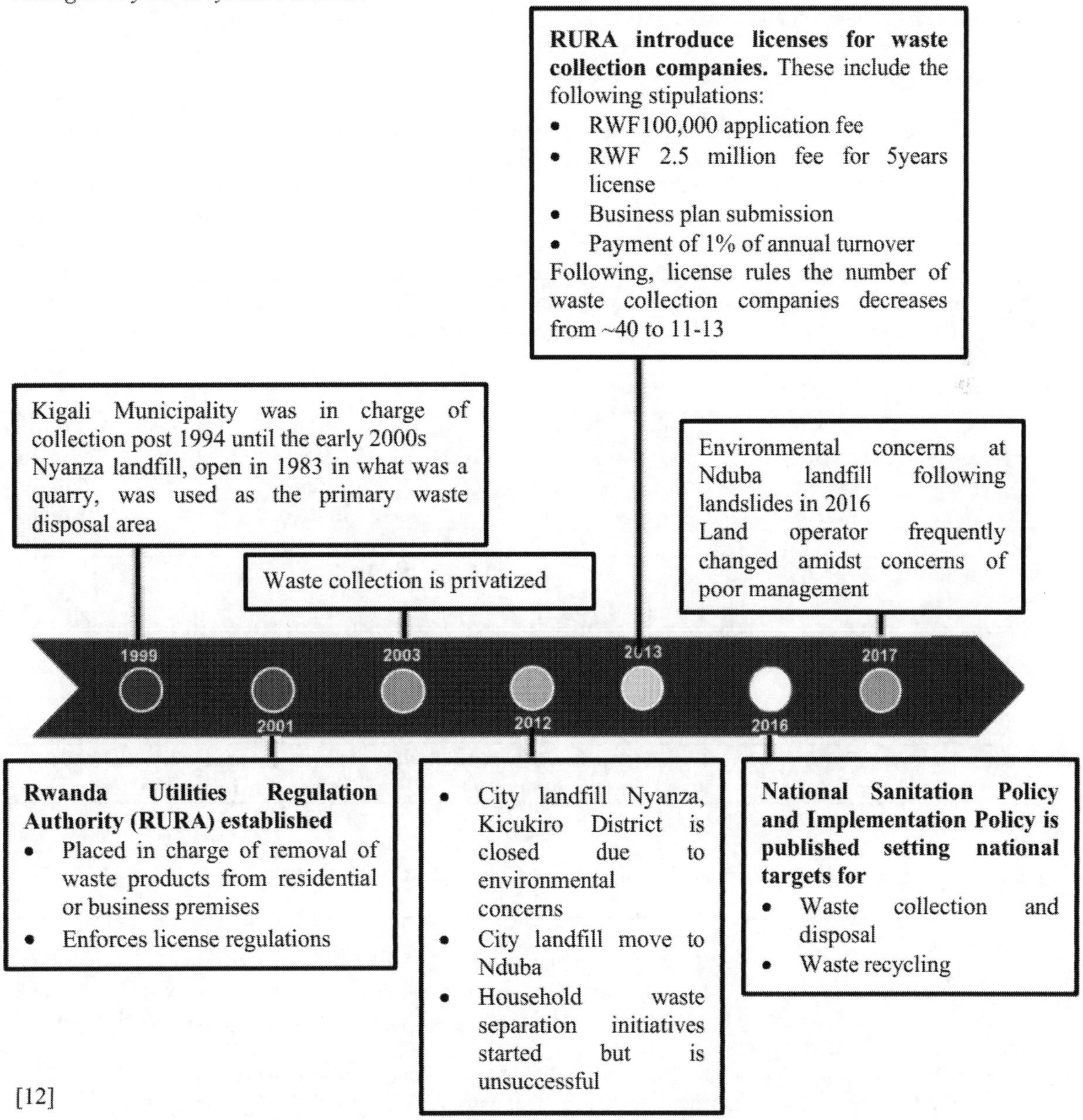


[12]

**Figure 1.** The evolution of waste management in Kigali city since 1994

Figure 1 depicts the evolution of waste management in Kigali city since 1994; Kigali's garbage management system has improved significantly during the previous ten years. Before 2010, there was no government policy or policy or a standardized plan of action for regulating SWM, resulting in SWM activities carried out by households, communities, nongovernmental organizations (NGOs), and local government officials. All of whom were working with limited technical and financial resources. SWM has been formed by gradual modifications rather than a single, comprehensive endeavor.

*3.2 Solid waste management status in Kigali City*

*3.2.1 Waste generation.* Table 1 shows the estimated amount of solid waste generated per capita and citywide from different sources. Annual waste generation in Kigali city was determined by weighing cars transporting SW to the main dumpsite provided information(Nduba landfill). It was estimated to be 800 tons per day and 292,000 tons per year. In comparison, all waste generation estimates a waste generation rate of 0.57 kg on average per capita per day or 205 kg per year. Gasabo generates the most garbage at the district level, followed by Kicukiro and Nyarugenge. Gasabo has a larger population as expected by EICV 5 and generates roughly 0.51 kilograms per capita per day. Gasabo will produce 459 tons per day, compared to 207 tons in Kicukiro and 135 tons per day in Nyarugenge.

**Table 1.** Solid waste generation (per capita and citywide) estimated from several sources.

| Waste Generation per day | UNEP GWMO/ EICV5 (2018) | Rapid Planning (2018) | Wilson, Kabera (2017) | Auditor General (2015) | Kigali Master Plan (2013) | Mucyo (2013) | SWM strategic Plan for Kigali (2012) |
|---|---|---|---|---|---|---|---|
| Per capita (kg) | 0.49 | 0.5 | 0.57 | 0.6 | 0.6 | 0.6 | 0.47 |
| Province wide (tons) | 800 | 808 | 638 | - | 640 | 450 | 408 |

**Table 2.** The total quantity of solid waste generated by households in Kigali city.

| | Waste generated by 204 households in Kigali City for 24 days | Waste generated by all 309,947 households in Kigali City |
|---|---|---|
| **Solids waste collected (24 days)(kg)** | 14,042.379 | 21,335,261.12 |
| **Solids waste collected per day (kg)** | 585,099 | 888,969.208 |

This research shows that 204 households generate 14,042.379 kg of solid waste during 24days and 585,099 kg per day; Kigali city has 309,947 total households, which these households generate 21,335,261 kg in 24 days. The total amount of solid waste generated in Kigali city by households is 888,9 tons per day, as presented in table 2.

Two hundred four households generate 10,272.8 kg of biodegradable waste during 24 days and 428,033kg per day, and one household generates 50,356kg in 24days and 2,098kg per day as presented in table 3.

**Table 3.** The total amount of biodegradable organic garbage generated in Kigali.

| | **Biodegradable waste generated by 204 households for 24 days** | **Biodegradable waste generated by one household** |
|---|---|---|
| **Biodegradable waste collected (24 days)(kg)** | 10,272.8 | 50,356 |
| **Biodegradable waste collected per day (kg)** | 428,033 | 2,098 |

**Table 4.** Waste composition in Kigali City.

| **Type** | | **Frequency (%)** |
|---|---|---|
| **Organic** | Food, garden<br>Wood | 75.3 |
| | Paper and Cardboard | 6.6 |
| **Non-Organic** | Plastic | 3.7 |
| | Metal | 1.6 |
| | Glass | 1.1 |
| | Other (including textiles etc.) | 11.7 |
| | **Total** | **100** |

Table 4 shows that biodegradable waste (organic waste) is the most generated in Kigali, 78% consisting of food, garden, and wood, while 6.6% of paper and cardboards. Non-Organic waste generated is plastic at a rate of 3.7%, metal (1.6%), glass (1.1%), and metal(1.6%)

*3.2.2 Waste collection, transportation, disposal, and treatment in Kigali City.* A sectoral approach is used in coordination with the districts of Kigali city to collect solid waste and communicate with the public. Cooperatives or companies that collect solid waste try to impose a pricing framework, efficiency control, and even sanctions on households reluctant to comply with the agreement while sectors coordinate solid waste collection. Kigali city has only one landfill (Nbuba landfill), waste collected transported by trucks and disposed at Nduba dumpsite as presented in table 3.

In areas that lack collection, low-income communities generally dump their rubbish at the closest empty lots, water channels (ruhurura), public spaces, creeks, or rivers and pollute the environment. The results showed that some people in the Gasabo district burn waste in their backyards, producing severe hazardous gases and particulates polluting the air. People living around the Nyabugogo watershed dump their rubbish at this watershed. Some people do not pay collection fees and dump their waste in the water channel; during the rainy season, that waste is transported by water toward the watershed, pollutes water, harms aquatic species, and causes eutrophication. These results agree with Manaf et al. [13], who mention irregular waste collection services, insufficient waste collection equipment, and the weakness of legal provisions as significant obstacles to waste recycling in Kigali today.

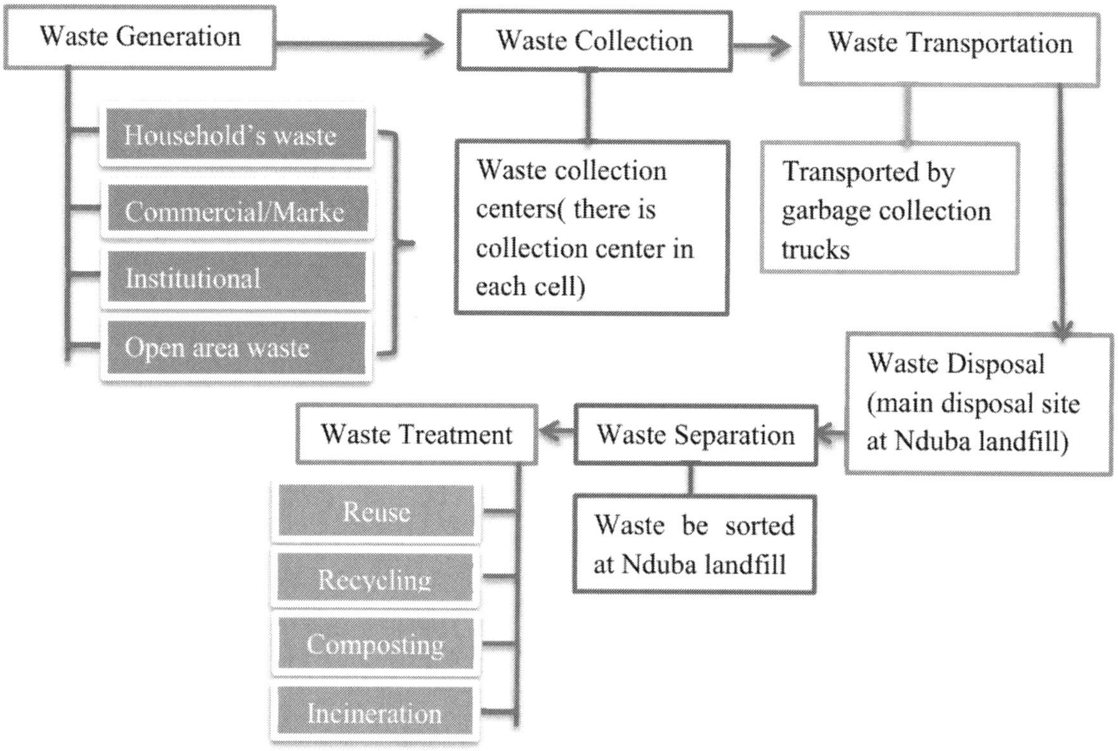


**Figure 2.** The flow of collection of solid wastes in Kigali City municipality

*3.3 Waste treatment in Kigali city*

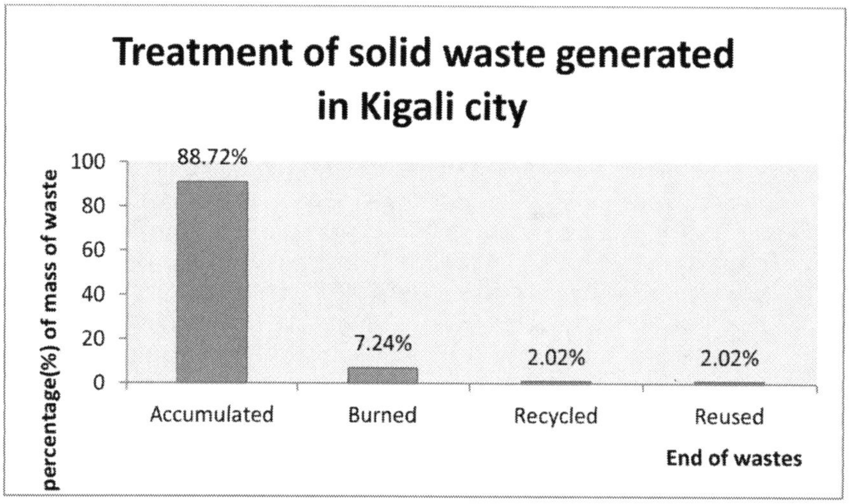


**Figure 3.** Treatment of SW in Kigali City.

The graph in Figure 3 shows the methods used for MSWM in Kigali city 88.72% of waste accumulated at dumpsite (35.5% are composted and used as fertilizers, while 52.33% are accumulated at the dumpsite). A small quantity (7.24%) burned without energy recovery. Recycling and reuse only used for a small quantity of waste; only 2.02% of wastes consisting of glass and plastic bottles and cans reused, 2.02% % are recycled (mainly consisting of metals, plastic bottles, and sawdust to make briquettes to heat bricks and toils). Waste treatment has not yet been developed in Kigali. Composting only a tiny amount of organic waste and inorganic waste like plastic bottles, glass, papers are recycled and reused at a low level. The findings showed a low level of participation among citizens, weakness in policy-making, poor private sector performance, and lack of integrated management. One essential solid waste treatment method used in Kigali city is composting and other techniques, including incineration without recovering energy, mainly used at different hospitals in Kigali city. All waste generated wastes are dumped at Nduba landfill and produce significant environmental and health problems. Methane gas is produced during the decomposition of organic compounds, which causes fires and contributes to the increased greenhouse gas effect, global warming, and climate change.

### 3.4 Solid waste and climate change

*3.4.1 The effect of solid waste on climate change and potential Remedies.* Solid waste can contribute to climate change by emitting greenhouse gases, and climate change can also impact solid waste management. Globally, most municipal solid waste (MSW) is disposed of in unmanaged landfills, where LFG (landfill gas) is produced as a by-product. Landfill gas is formed when organic material decomposes anaerobically; it comprises 45 percent to 60 percent $CH_4$, 40 percent to 60 percent $CO_2$, and 2 percent to 9 percent other gases released mainly into the atmosphere [14]. Organic waste decomposition emits $CO_2$ and $CH_4$, the leading greenhouse gases [15]. Although both $CO_2$ and $CH_4$ are greenhouse gases that contribute to climate change and global warming, the relative impact of solid waste on climate change is minimal.

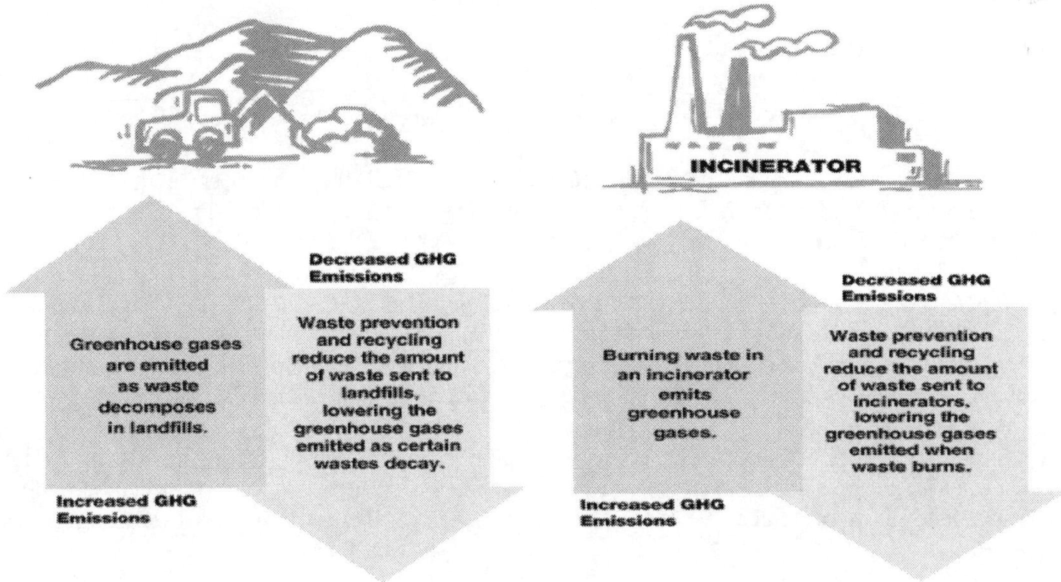


**Figure 4.** How to decrease the greenhouse gases emitted by incinerators and waste decomposition

Waste prevention and recycling help Kigali city better manage solid waste and help reduce $CO_2$ emissions and greenhouse gases. Recycling saves power; made from recycled materials often uses less energy than made from raw materials, as presented in Figure 4 [16]. And waste minimization is even more effective.

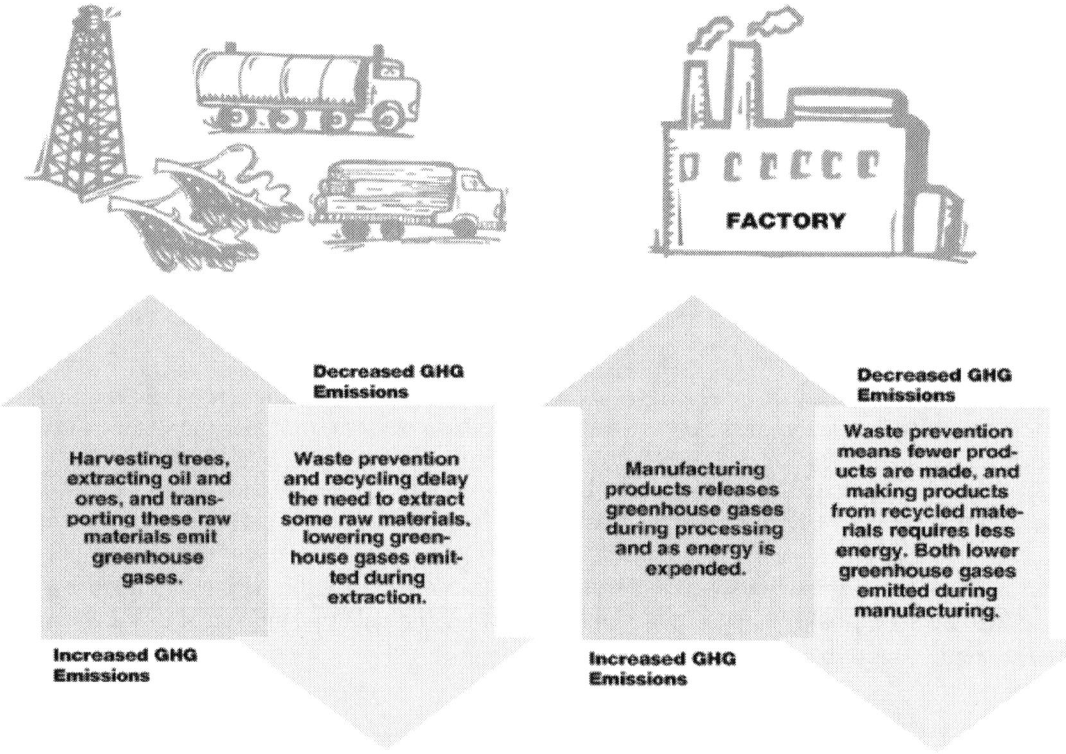


**Figure 5.** How to decrease the greenhouse gases emitted by factories

When individuals reuse things or make products with less material, they need less energy to extract, transport, process, and produce them. Less energy demand means less fossil fuel burning and less $CO_2$ emissions, as shown in Figure 5.

*3.4.2 Climate change and their effects on solid waste management in Kigali city.* Table 5 shows how high temperatures exacerbate fire risk from combustible garbage. Residents of areas with waste treatment or dumping sites complain of foul odors [17]. Temperature influences the biological, chemical, physical, and mechanical properties and behavior of wastes and landfill liner materials, as well as waste decomposition and degradation [18].

When it rains a lot, open garbage storage containers might become flooded, affecting processing waste. Heavy rain can alter the rate and frequency of waste decomposition and increase leachate strength, both of which impact landfills as presented in table 5 [19]. Intense rainfall increases the risk of erosion at the dump [19].

**Table 5.** Climate factors and their effects on solid waste management.

| Variables of climate | |
|---|---|
| **Temperature** | - Affects biological processes (composting, anaerobic digestion)<br>- It affects waste degradation and decomposition |
| **Precipitation** | - Intense rainfall deteriorates a landfill without proper water collection systems, creating cracks in the containment structure that allow debris and leachate to escape and damage surrounding resources.<br>- Alter waste decomposition rate |

Source: [15]

*3.5 Community attitude, knowledge, and perception on SWM in Kigali City*

Inadequate waste management policies and services provided by the state are not the only causes of improper waste management. Individual attitudes also play a role in waste management [20]. cited public perception and behavior as a barrier in waste management.

Rwanda's Law No. 17/2019 says that the manufacturing, importing, or selling plastic carry bags and single-use plastic items is prohibited. Article 3. For an exceptional reason, people who use banned products may be authorized in writing. Article 4. [21] Plastic bags would be prohibited because they are harmful to the environment. Following the ban, there was a public controversy. Some people supported the ban on plastic bags, while others opposed it.

The research showed that 98% of the respondents do not do the solid waste separation process. Citizens have a limited understanding of proper waste management and its consequences and lack information and awareness. Even though some respondents expressed support for the plastic bag ban, the interviews revealed that situational and factors are at work, resulting in a gap between people's thoughts and actions regarding environmental issues. As a result, the importance of environmental education and knowledge is the critical foundation for raising public awareness of problems of the environment.

The study looked into factors impacting waste separation at the households level. Individuals who care about the environment and want to help solve problems are influenced by awareness efforts to sort waste. Some households separate waste to be paid for by recycling companies. Due to the rising cost of these secondary resources, communities have sprouted cooperatives for recycling materials. These factors seem to have an improvement at household-level separation. And, sharing responsibility for the city's waste system improves separation.

*3.6 Socio-cultural and SWM in the city of Kigali*

*3.6.1 Community participation in solid waste management payment services.* In Kigali, people pay for waste collection services according to the municipality's Ubudehe categorizations (4 different social classes). There are four different types. The first category is the most impoverished, the fourth category is the most prosperous. People of the first category do not have to pay; however, the others do that. The amount to be paid varies depending on the sector, but it is often in the range of 1000 Rwf (USD 0.99) per month for the second category to 5000 Rwf (4.96 USD) per month for the third category. Once a week, rubbish collection is carried out in houses, whereas waste collection is done daily in restaurants, markets, and hotels. Even though private waste collection companies complain about the waste collection fee and say it is insufficient, they continue to generate income.

In many areas, There were no routine assessments of garbage collection companies to ensure the quality of their services, and companies were allowed to operate autonomously in many cases. While household objections were either communicated directly to corporations or stated during Umuganda, there were few precise mechanisms in place to ensure that waste providers collected rubbish from all residences within their territory on schedule.

*3.6.2 Contribution of Rwandan culture to MSW management.* A statewide communal workday known as Umuganda can be translated as getting together for a common goal, a practice that has long existed in Rwandan culture but has just lately become mandated. On the last Saturday of each month, from 8:00 a.m. to 11:00 a.m., businesses close, and traffic comes to a halt as residents around the country head to their communities, shovels and hoes in hand, to engage in various community projects.

All citizens are expected to participate, and those who do not participate must pay a fine or risk arrest. Generally, during Umuganda day, people clean streets, plant trees, cut grass, and trim bushes alongside the roads. When asked about the impact of Umuganda, more than half of the people said it improved neighborhood cleanliness, and 16.3 percent said it improved air quality and overall environmental quality. A substantial portion of Umuganda's activities is devoted to environmental conservation and general cleanup [22].

*3.7 Environmental organic law*

Waste management in Kigali and Rwanda is influenced by environmental, sustainability, and sanitation principles. As of April 8, 2005, the Rwandan government implemented statute No 4/2005, Organic Law Determining the Modalities of Protecting, Conserving, and Promoting the Environment in Rwanda [23]. Conservation and sustainable use of natural resources is based on these five principles [23]. Lays forth the main concepts that regulate the entire environmental sector, such as protection and polluter pays.

*3.7.1 Cooperation between stakeholders and the city of Kigali on SWM.* The Rwandan government is in charge of monitoring stakeholder input on SWM systems. Environmental committees exist at all levels of government, all levels of Kigali city(from the cell level, sector level, and district level). Citizens' committees attempt to deal with environmental issues such as waste and report to authorities the obstacles they encounter during Umuganda Day (day of community service) [22].

*3.7.2 Legal and institutional framework.* Private and public sector partners now provide services for waste management. But for this cooperation to work and achieve full-service coverage and environmental cleanliness, the government must adopt, enforce and maintain formal norms, policies, laws, and regulations [24].

Solid waste management in Kigali still has some challenges. In general, stakeholders seek equipment to handle their concerns. This research shows that a functional system relies on technology solutions and environmental, socio-cultural, and economic linkages that should be present to enable the overall approach to function. Figure 6 shows how Kigali's waste management organization includes national policy-making entities and local implementation and regulatory authorities. MININFRA, MoE, MoH, and MINICOM draft policies for local administrative organizations, including Kigali City and sectoral, local government authorities. RURA and REMA set the guidelines for all implementation activities [23]. WASAC also conducts SWM through landfill management. Finally, with Kigali's assistance, districts and sector officials are participating in contracts of waste management providers and ensuring the protection of the environment [25].

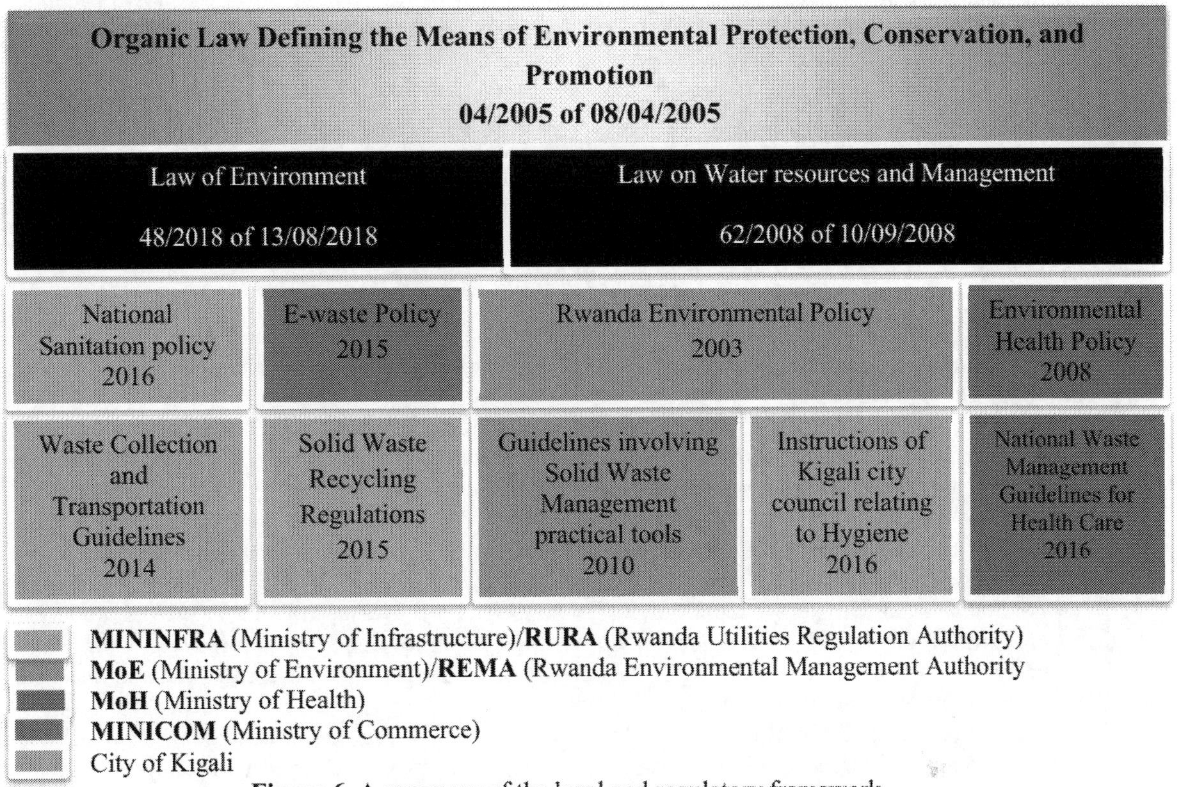


**Figure 6.** A summary of the legal and regulatory framework

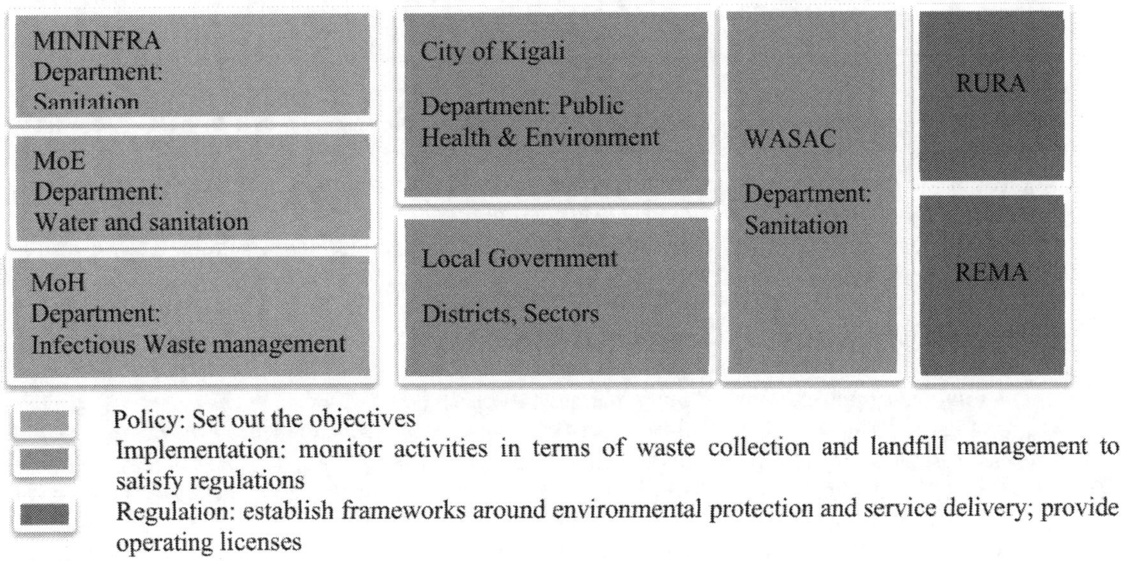


**Figure 7.** Roles of institutions in the SWM system

Waste management involves various stakeholders with varying areas of expertise, as shown in Figure 7. They all contribute to developing a city's system by establishing policies and solid waste management systems. Nonetheless, it is sometimes regarded as solely the duty of local governments. The primary recognized or formal stakeholders comprised the local authority (Kigali city, districts, sectors), government institutions (WASAC, RURA, REMA), several government ministries (MoE, MINENFRA, MoH), and private service providers. In the best-case scenario, citizens are regarded as co-responsible with the municipality. A thorough understanding of the stakeholders and their roles is essential for establishing a productive and efficient system.

## 4. Conclusion

This research shows that the essential solid waste treatment method used in Kigali city is composting. Other techniques, including incineration without recovering energy, recycling, and reuse, are still low. The manufacturing, importing, or selling of plastic carry bags and single-use plastic items is prohibited except for an exceptional reason.

Citizens have a limited understanding of proper waste management and its consequences and lack information and awareness. Some households separate waste to be paid for by recycling companies. People participate in waste management by paying for waste collection services. All citizens are expected to join during Umuganda day, clean streets, and collect solid waste. Numerous waste management legislation and regulations exist but are rarely enforced. The Rwandan government should enforce waste management laws more strictly.

The results showed that some people in the Gasabo district burn waste in their backyards, producing severe hazardous gases and particulates polluting the air. Methane gas is produced during the decomposition of organic compounds, which causes fires and contributes to the increased greenhouse gas effect, global warming, and climate change.

The results of this study reinforce SWM status, the effect of waste on climate change; the findings indicated a close relationship between climate change and solid waste. However, this result is necessary further investigated to get the whole picture. Further research can calculate the amount of $CO_2$ emitted during the decomposition process at Nduba landfill and the factors affecting the solid waste management system in Kigali.

## References

[1] Singh A 2019 Managing the uncertainty problems of municipal solid waste disposal *J. Environ. Manage.* **240** 259–65

[2] Office of the Auditor General of State Finances 2016 *Performance Audit Report on Management of Solid and Liquid (Sewage) Waste in City of Kigali* (Kigali, Rwanda)

[3] United Nations Environment Programme 2005 *Solid Waste Management* (Nairobi, Kenya: United Nations Environment Programme)

[4] Erasu D, Feye T, Kiros A and Balew A 2018 Municipal solid waste generation and disposal in Robe Town, Ethiopia *J. Air Waste Manag. Assoc.* **68** 1391–7

[5] Kabera T 2019 Solid Waste Management in Rwanda *Sustainable Waste Management Challenges in Developing Countries* (Pennsylvania: IGI Global) pp 287–305

[6] Koy T 2019 The Rio Declaration on Environment and Development *The Earth Summit Agreements: A Guide and Assessment* (London: Routledge) pp 85–96

[7] Bowan P A and Mumuni A I 2013 Urban solid waste management in Ghana: an assessment of zoomlion's approach to waste management in the wa municipality *J. Environ. Earth Sci.* **3** 73–80

[8] Tanskanen J H 2000 Strategic planning of municipal solid waste management *Tanskanen, Juha-Heikki* **30** 111–33

[9] Laurent A, Bakas I, Clavreul J, Bernstad A, Niero M, Gentil E, Hauschild M Z and Christensen T H 2014 Review of LCA studies of solid waste management systems - part i: lessons learned and perspectives *Waste Manag.* **34** 573–88
[10] Isugi J and Niu D 2016 Research on landfill and composting guidelines in Kigali City, Rwanda based on China's experience *Int. Proc. Chem. Biol. Environ. Eng.* **32** 12–6
[11] Kabera T, Wilson D C and Nishimwe H 2019 Benchmarking performance of solid waste management and recycling systems in East Africa: Comparing Kigali Rwanda with other major cities *Waste Manag. Res. J. a Sustain. Circ. Econ.* **37** 58–72
[12] Rajashekar A, Bowens A and Gatoni A S 2019 *Assessing Waste Management Services in Kigali* (London)
[13] Manaf L A, Samah M A A and Zukki N I M 2009 Municipal solid waste management in malaysia: practices and challenges *Waste Manag.* **29** 2902–6
[14] Aljaradin M and Persson K M 2012 Comparison of different waste management technologies and climate change effect—jordan *Am. J. Clim. Chang.* **1** 57–63
[15] Olayide O E and Ikpi A E 2010 Political system, budget performance and economic growth in Nigeria, 1970-2004 *J. Sustain. Dev. Africa* **12** 90–9
[16] Eneh A E O and Oluigbo S N 2012 Mitigating the impact of climate change through waste recycling *Res. J. Environ. Earth Sci.* **4** 776–81
[17] Winne S, Horrocks L, Kent N, Miller K, Christopher H, Benzie M and Power R 2012 *Increasing the Climate Resilience of Waste Infrastructure* (Harwell, Oxfordshire)
[18] Zhao Y R, Liu T J, Chen X S, Xie Q and Huang L P 2016 The effect of temperature on the biodegradation properties of municipal solid waste *Waste Manag. Res.* **34** 265–74
[19] Watson G V R and Powrie W 2015 The impact of climate-related environmental change on the uk solid waste sector *Infrastruct. Asset Manag.* **2** 84–94
[20] Byamba B and Ishikawa M 2017 Municipal solid waste management in ulaanbaatar, mongolia: systems analysis *Sustainability* **9** 896
[21] Official Gazette 2019 *Official Gazette n° 37 bis of 23/09/2019* (Gazette)
[22] Rwanda Governance Board 2017 *Impact Assessment of Umuganda 2007-2016* (Kigali, Rwanda: Rwanda Governance Board)
[23] Government of Rwanda 2005 Official Gazette of the Republic of Rwanda No . 04 / 2005 of 08 / 04 / 2005 Organic Law Determining the Modalities of Protection , Conservation and Promotion of the Environment in Rwanda
[24] Fobil J N, Armah N A, Hogarh J N and Carboo D 2008 The influence of institutions and organizations on urban waste collection systems: an analysis of waste collection system in Accra, Ghana (1985-2000) *J. Environ. Manage.* **86** 262–71
[25] Twagiramungu F 2006 *Environmental Profile of Rwanda* (Kigali, Rwanda)

# Patterns of *Tetragonula Laeviceps* (Smith) colony activity in KHDTK Gunung Bromo UNS

**Ana Agustina[1*], Retno Tanding Suryandari[2] and Tetuko Rawidyo Putro[2]**

[1]Department of Forest Management, Universitas Sebelas Maret, Jl. Ir. Sutami 36 A, Kentingan, Surakarta, 57126, Indonesia
[2]Department of Management, Universitas Sebelas Maret, Jl. Ir. Sutami 36 A, Kentingan, Surakarta, 57126, Indonesia
*Corresponding author: ana.agustina2018@staff.uns.ac.id

**Abstract.** Due to climate change, many impacts occur on the sustainability of organisms, including the trigona bee. The occurrence of climate change not only affects decreasing honey production but also reduces the number of colonies. Presume that it can change the activity pattern of the trigona bee. This study aims to determine the pattern of activity demonstrated by *Tetragonula laeviceps* (Smith) colony in KHDTK Gunung Bromo UNS. This research was conducted in July 2021 at KHDTK Gunung Bromo UNS. The observation involved 128 colony boxes placed into four cages, namely cages A, B, C, and D. Observations on the activity of trigona bees starting from 06.00 WIB to 18.00 WIB, assuming trigona bee activity begins in the morning and end the activity outside the hive after sunset. The results showed that of the cages A, B, C, and D, the worker bees tend to be active in the morning (6-10 am) to collect nectars and resin. Before noon (10 to 12 noon), bee activities were decreasing. The activity of the worker bees increased between 12 to 2 pm and decreased again after that time slot. The food source most favored by trigona bees: *Pinus merkusii* (rp), *Artocarpus heterophyllus* (r), *Spathiphyllum wallisii* (np), and *Xanthostemon chrysanthus* (np). Resin extractions are very needed for bees as batumen materials, and this is due to natural enemies (lizards, ants, vespa wasps, and spiders) that appear around the cage.

**Keywords**: activity pattern, food source, honey, natural enemy, trigona bee

## 1. Introduction

Stingless bees are one of the types of insects that belong to the tribe Meloponini with the Family Apidae (Hymenoptera). Characteristics of stingless bees Forewing with reduced distal venation, marginal cell frequently open at apex; pretarsal claws simple; penicillum present in worker; auricle absent [1]. There are more than 40 types of stingless bees in Indonesia, one of which is T. laeviceps [2]. Stingless bees are known by different names in each region, including teuweul (West Java), klanceng (Central Java, Yogyakarta, East Java), kelulut (Kalimantan), and galo-galo (West Sumatra) [3, 4].

The pollination of flowers in plants can occur naturally or assist by intermediaries such as wind, insects, bats, birds, and water. One type of insect that can be an intermediary in the pollination process is the stingless bee which belongs to the Apidae family. It is estimated that about 66% of plant pollinators are bees, including stingless bees. The types of plants that stingless bees can pollinate are very diverse, not only types of plants that have economic value but also have an essential role in the process of pollinating forest plants so that the process of forest regeneration continues [5].

Stingless bees can produce bee pollen, honey, and propolis. Bee pollen serves as antioxidant, anti-inflammatory, antimicrobial, and antimutagenic activities [6]. Honey is a source of nutrition and provides health to the body [7]. Propolis is a source of antioxidants [8], antibacterial [9], anti-inflammatory, immunomodulatory activities, and antitumor activity [10]. Economically, honey and propolis produced by stingless bees can be an alternative economic source for rural to urban areas. There have been many community economic activities by cultivating stingless bees, both individually and in groups.

The issue that has developed over the past few years is about climate change, where one of the impacts can cause changes in the life cycle and habits of stingless bees so that they need to adapt to their environment. Because of climate change stingless bees would experience a significant reduction in their climatically suitable areas and this phenomenon will affect to agriculture and rural livelihoods [11]. Adaptations made by stingless bees include all aspects such as the presence of natural enemies, pests and diseases, and food sources [12]. In this study, observations were made on the activity pattern of the stingless bee of T. laeviceps and correlated it with air humidity, temperature, and food sources. In addition, it is necessary to observe food sources and natural enemies for the survival of stingless bees.

## 2. Material and methods

The study was conducted by observing the T. laeviceps colony at KHDTK Mount Bromo UNS held in July 2021. The observed stingless bee colonies were divided into four main cages, namely cages A, B, C, and D, with 128 colonies. From 128 colonies, only three boxes were selected in each cage by purposive sampling. Observation activities were carried out on worker bees that carried out activities in and out of the hive (carrying pollen, carrying resin, disposing of garbage, and looking for food). Observations were made from 06.00-18.00 WIB with time intervals every 2 hours and measured within 5 minutes (a modification of the study [4]). In addition to counting the number of bees, light intensity measurements were also carried out using a Krisbow brand Light meter 10176593, measuring humidity and temperature using a Krisbow brand Thermo-hygrometer 10207853.

The activity data analysis of *T. laeviceps* was tested using One way ANOVA and Duncan's New Multiple Range Test (DMRT) to provides significance levels for the different between activity time and cage. Data processing and statistical testing using Microsoft Excel and SPSS version 22 for windows.

## 3. Results and discussion

### 3.1. *Hive exit activity*

Activity out of the hive has a pattern (Figure 1) where activities out of the hive are most often carried out in the morning. This activity is related to the availability of nectar in the morning more so that the activity of bees when leaving the hive tends to be high. Activity outside the hive increased at 10.00-14.00 WIB, not only taking nectar but also taking resin, pollen, and throwing garbage. Activities, when bees carry garbage will be seen from the mandible, which brings something dark [4]. As the evening progresses, the activity of getting out of the hive decreases until sunset.

The results of statistical tests showed that there were significant differences at each time counting the number of individuals leaving the hive. After it was found that there was a significant difference between time intervals, DMRT was carried out, the results are as presented in Table 1. Based on DMRT result, it can be seen that there are significant differences in each time interval except during the day, which is in the same subset. DMRT was also carried out on cages where it was found that cage A had a significant difference from cages B, C, and D (Table 2). This was most likely influenced by shade, the presence of food sources, natural enemies, and the position of cages B, C, and D, which were close together compared to the location of cages A.

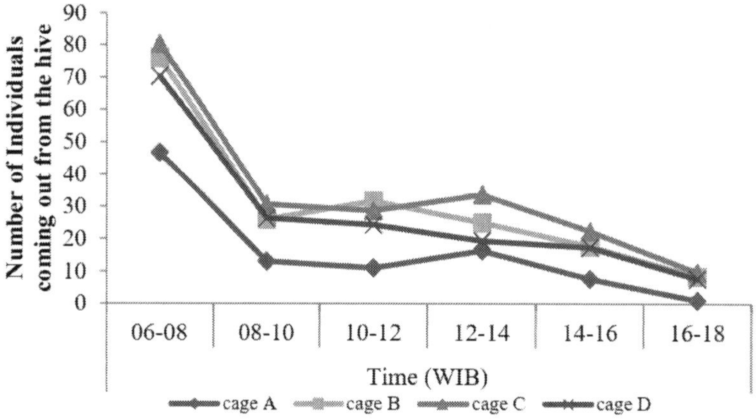


**Figure 1.** The activity of *T. laeviceps* leave the hive

**Table 1.** Duncan's new multiple range test on observation time interval

| Time activity | Subset | | | | |
|---|---|---|---|---|---|
| | 1 | 2 | 3 | 4 | 5 |
| 16-18 WIB | 12.000 | | | | |
| 14-16 WIB | | 24.833 | | | |
| 12-14 WIB | | | 38.250 | | |
| 10-12 WIB | | | 41.833 | | |
| 6-8 WIB | | | | 59.500 | |
| 8-10 WIB | | | | | 84.583 |
| Sig. | 1.000 | 1.000 | .573 | 1.000 | 1.000 |

Means for groups in homogeneous subsets are displayed.
Based on observed means.
The error term is Mean Square(Error) = 238.889
a. Uses Harmonic Mean Sample Size = 12.000
b. Alpha = .05
c. WIB = West region of Indonesian Time

**Table 2.** Duncan's new multiple range test by cage

| Cage | N | Subset | |
|---|---|---|---|
| | | 1 | 2 |
| A | 18 | 31.944 | |
| C | 18 | | 47.167 |
| D | 18 | | 47.222 |
| B | 18 | | 47.667 |
| Sig. | | 1.000 | .928 |

Means for groups in homogeneous subsets are displayed.
Based on observed means.
The error term is Mean Square(Error) = 238.889
a. Uses Harmonic Mean Sample Size = 18.000
b. Alpha = .05

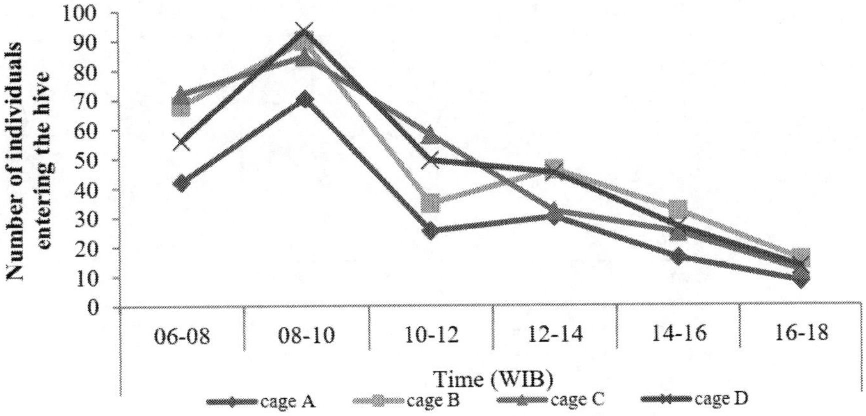


**Figure 2.** The activity of *T. laeviceps* enter the hive

*3.2. Activity back to the hive*

The activity of returning to the hive, the climax occurs at 08.00-10.00 WIB; this is related to the activity of bees when they leave the hive, which mainly occurs in the morning at 06.00-08.00 WIB. This activity is carried out after the worker bees get feed ingredients in pollen, honey, or resin obtained from plants around the cage. Bees that come home with pollen will show a yellowish tinge to the corbicula, or if they bring resin, the corbicula will appear white/glossy brown [4]. The activity of entering the hive again increased at 12.00-14.00 WIB, and after that, it continued to decrease until the afternoon (Figure 2).

DMRT was carried out to see the extent of the differences between variables as presented in Table 3 and Table 4. Based on the results of statistical tests, there was a significant difference between the activity of entering the hive on the observation time interval and the type of cage. Based on the time interval of bee activity when returning to the hive at 8 am to 2 pm were in the same subset, until late afternoon the number of active bees decreases, while activity peaks in the morning. DMRT for each cage showed a subset difference between cages A, while cages B, C, and D showed a relationship with each other.

**Table 3.** Duncan's new multiple range test based on observation time interval

| Time activity | N | Subset | | | |
|---|---|---|---|---|---|
| | | 1 | 2 | 3 | 4 |
| 16-18 WIB | 12 | 6.667 | | | |
| 14-16 WIB | 12 | | 16.250 | | |
| 12-14 WIB | 12 | | | 23.583 | |
| 10-12 WIB | 12 | | | 23.917 | |
| 8-10 WIB | 12 | | | 24.000 | |
| 6-8 WIB | 12 | | | | 68.333 |
| Sig. | | 1.000 | 1.000 | .913 | 1.000 |

Means for groups in homogeneous subsets are displayed.
Based on observed means.
The error term is Mean Square(Error) = 74.792
a. Uses Harmonic Mean Sample Size = 12.000
b. Alpha = .05
c. WIB = West region of Indonesian Time

**Table 4.** Duncan's new multiple range test by cage

| Cage | N | Subset | | |
|---|---|---|---|---|
| | | 1 | 2 | 3 |
| A | 18 | 15.944 | | |
| D | 18 | | 27.556 | |
| B | 18 | | 30.778 | 30.778 |
| C | 18 | | | 34.222 |
| Sig. | | 1.000 | .269 | .238 |

Means for groups in homogeneous subsets are displayed.
Based on observed means.
The error term is Mean Square(Error) = 74.792
a. Uses Harmonic Mean Sample Size = 18.000
b. Alpha = .05

### 3.3. *Activity of bees leaving and entering the hive*

Stingless bees carry out activities in and out of the hive alternately and simultaneously. Activities in and out of the hive occur because of several activities such as collecting honey, pollen, and resin. It can be in the form of activities to remove garbage from the hive. The data presented is a combination of data leaving the hive and returning to the hive. Figure 3 shows the activity of the bees that took place from the morning (6 am) and the highest activity in cages A, B, C, and D. Towards the afternoon, the bee activity tends to decrease, then increases again at 12.00-14.00 WIB. Again there was a decrease in the number of activities until the afternoon.

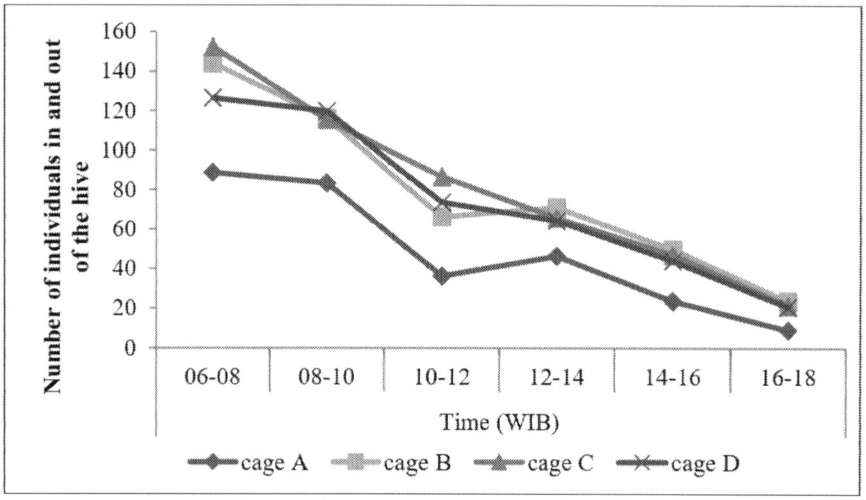


**Figure 3.** The activity of *T. laeviceps* in and out of the hive

### 3.4. *Relationship between daily activities and environmental conditions*

The results of daily weather elements measurements around the hive of *T. laeviceps* are presented in Figure 4. The average air temperature at 06.00 WIB was 25.6°C in cage A, 26.3°C in cages B and C, and 26.5°C in cage D. Then it increased during the day with a peak at 12.00-14.00 WIB (33.3°C) and then decreased again at 16.00-18.00 WIB (26.07-26.10°C). In light intensity, changes occurred in each cage with a maximum light intensity of 21.15 Klux in cage A, 27.19 Klux in cage B, 25.32 Klux in cage C, and 26.43 Klux in cage D, all of which occurred at 12.00-14.00 WIB. Air humidity decreases with increasing temperature during the day and increases again in the afternoon.

### 3.5. *Food source around the cage of T. laeviceps*

Stingless bees have a narrower home range compared to *Apis* sp., such as *T. carbonaria*, which has a home range of 333 to 712 m [13]. In other hand the type of *T. laeviceps* only has the ability to fly up to 100 m from the hive. The food resources most favored by the colony were *Pinus merkusii* (rp), *Artocarpus heterophyllus* (r), *Spathiphyllum wallisii* (np), *Xanthostemon chrysanthus* (np), *Chrysogonum* sp. (p), *Hydrangea macrophylla* (n), *Antigonon leptopus* (np), and *Chrysanthemum* sp. (np) (added information: n = nectar; r = resin; p = pollen). The availability of food sources in the form of nectar, pollen, and resin needs to be considered in cultivating stingless bees for the colony's survival.

### 3.6. *Natural enemy to stingless bees*

Every insect certainly has natural enemies that can threaten the survival of the colony. The natural enemies of stingless bees include fruit flies (*Drosophila* sp.), beetles (Family Nitidulidae), ants, lizards, spiders, termites, cockroaches, and sun bears [14]. At the location of Mount Bromo KHDTK, natural enemies that can be found include lizards, ants, vespa wasps, and spiders (Figure 5). All-natural enemies around the hive can cause reduced honey production due to the number of individuals dying. If it continues, the colony will disappear, which is detrimental to the stingless bees colony.

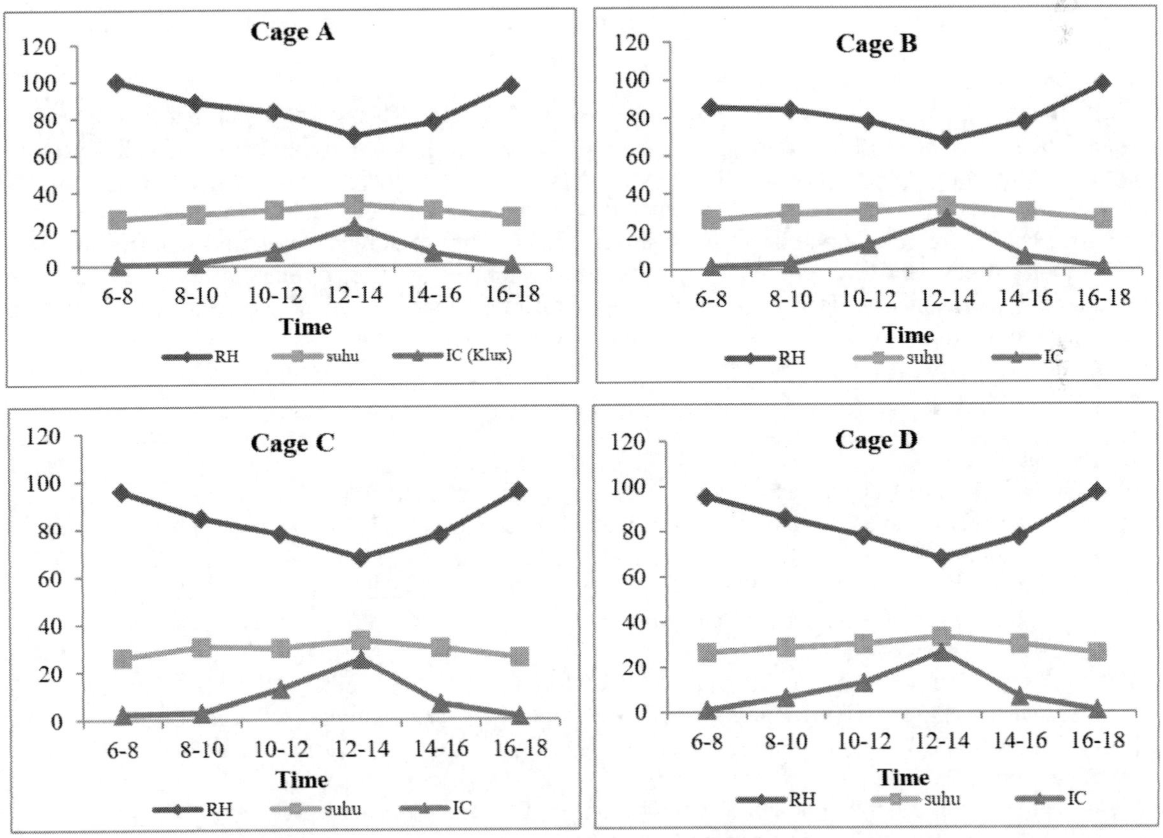


**Figure 4.** Daily weather element data around the hives of *T. laeviceps*

**Figure 5.** Natural enemies that found around the colonies (a. Ants; b. Lizards; c. Vespa wasp)

## 4. Conclusion

The activity pattern of *T. laeviceps* started from sunrise to sunset. This study found that the highest amount of activity was in the morning and increased-decreased phenomenon activity periodically occur until the afternoon. Stingless bees build a complex relationship with their thermal environment, which means solar radiation affects the body temperature of bees and their foraging activities in tropical environments. Climate change impact to time changing activity of stingless bee, and will impact to pollination process. The ideal conditions for *T. laeviceps* colonies to develop optimally are in the vicinity of stands or plants that produce food resources in the form of nectar, pollen, and resin. In KHDTK Gunung Bromo, food resources for stingless bees available from forest area and flower plants such as *Pinus merkusii* (rp), *Artocarpus heterophyllus* (r), *Spathiphyllum wallisii* (np), and *Xanthostemon chrysanthus* (np).

## References

[1] Engel M S 2012 The honey bees of Indonesia (Hymenoptera: Apidae) *Treubia* **39** 41–9
[2] Gratzer K, Susilo F, Purnomo D, Fiedler S and Brodschneider R 2019 Challenges for beekeeping in Indonesia with autochthonous and introduced bees *Bee World* **96** 40–4
[3] Syafrizal, Kusuma I W, Saud O R, Wiandany R, Yahya M F and Harmonis 2020 Conservation of kelulut (stingless bee) in East and North Kalimantan, Indonesia *IOP Conference Series: Earth and Environmental Science* **591**
[4] Yustia I P J, Rauf A and Maryana N 2017 Ritme aktivitas penerbangan harian Tetragonula laeviceps (Smith) (Hymenoptera: Apidae) di Bogor *Jurnal Entomologi Indonesia* **14** 117–25
[5] Erniwati 2013 Kajian biologi lebah tak bersengat (Apidae: Trigona) di Indonesia *Fauna Indonesia* **12** 29–34
[6] Guine R P F 2015 Bee pollen: Chemical composition and potential beneficial effects on health *Current Nutrition & Food Science* **11** 301–8
[7] Bogdanov S, Jurendic T, Sieber R and Gallmann P 2008 Honey for nutrition and health: A review *Journal of the American College of Nutrition* **27** 677–89
[8] Khairunnisa K, Mardawati E and Putri S H 2020 Karakteristik fitokimia dan aktivitas antioksidan ekstrak propolis lebah Trigona Sp *Jurnal Industri Pertanian* **2** 124–9

[9] Sabir A 2005 Aktivitas antibakteri flavonoid propolis Trigona sp terhadap bakteri Streptococcus mutans (in vitro) (In vitro antibacterial activity of flavonoids Trigona sp propolis against Streptococcus mutans) *Dental Journal (Majalah Kedokteran Gigi)* **38** 135–41
[10] Lotfy M 2006 Biological activity of bee propolis in health and disease *Asian Pacific Journal of Cancer Prevention* **7** 22–31
[11] Gonzalez V H, Cobos M E, Jaramillo J and Ospina R 2021 Climate change will reduce the potential distribution ranges of Colombia's most valuable pollinators *Perspectives in Ecology and Conservation* **19** 195–206
[12] Conte Y Le and Navajas M 2008 Climate change : impact on honey bee populations and diseases *Revue scientifique et technique (International Office of Epizootics)* **27** 499–510
[13] Smith J P, Heard T A, Beekman M and Gloag R 2016 Flight range of the Australian stingless bee Tetragonula carbonaria (Hymenoptera: Apidae) *Austral Entomology* 1–4
[14] Pangestika N W, Atmowidi T and Kahono S 2018 Additional nest structures and natural enemies of stingless bees (Hymenoptera: Apidae: Meliponinae) *Jurnal Sumberdaya Hayati* **4** 42–7

# Vulnerability analysis of coastal communities in Merauke as impact of climate change

**Untari[1,2], Rahim Darma[3]*, Philipus Betaubun[4], Letty Fudjaja[3] and Andi Adri Arief[6]**

[1]PhD. Student on Agricultural System, Graduate School, Hasanuddin University, South Sulawesi Province, Indonesia.
[2]Department of Agribusiness, Faculty of Agriculture, Musamus University, Papua Province, Indonesia.
[3]Agribusiness Study Program, Faculty of Agriculture, Hasanuddin University, South Sulawesi Province, Indonesia.
[4]Department of Civil Engineering, Faculty of Engineering, Musamus University, Papua Province, Indonesia.
[6]Socio Economic of Fisheries Study program, Faculty of Marine Science and Fisheries, Hasanuddin University, South Sulawesi Province, Indonesia.
*Corresponding author: rdarma@unhas.ac.id

**Abstract.** The coastal area of Merauke is already experiencing the impact of climate change which causes community vulnerability which can lead to structural poverty. The purpose of the study conducted in the coastal area of Merauke Regency with a sample of 2 observation areas namely Merauke and Naukenjerai Districts using a qualitative research approach with survey methods. Research data were collected using questionnaires and in-depth interviews with expert respondents who have competencies related to the topic being studied. Data were analyzed using vulnerability analysis which measures exposure, sensitivity, and resilience in community groups who work as fishermen, farmers, and become fishermen/farmers, traders, and transportation workers. The results of the study indicate that the groups of people who have vulnerability as a result of climate change are fishermen and farmers. The level of vulnerability is caused by pressure from natural resources, namely tidal water disasters and floods due to high rainfall intensity which has an impact on the loss of income and work of the community. While the highest level of pressure caused by human activities, namely from sand excavation and land conversion into settlements and public facilities, population growth, and activities to dispose of garbage and household waste on the coast have the biggest impact on decreasing the carrying capacity of the region, coastal ecosystems, reducing livelihoods. livelihoods and sources of income.

**Keywords:** resilience, vulnerability, climate change, coastal

## 1. Introduction

The management of coastal areas in Indonesia has been mandated in the 1945 Constitution that the Republic of Indonesia provides protection and empowerment of fishermen, independence, usefulness, togetherness, integration, openness, efficiency-justice, sustainability, welfare, local wisdom, and the preservation of environmental functions [1]. Management of coastal areas can promote the values embedded in indigenous peoples and local communities who have made natural resources a part of their lives [2]. However, the results of the study Voyer et al., (2017) shows that there are groups of people who use coastal areas that often ignore local cultural values [3] that prioritize the environment to be part of the social order of society. Natural resources are used to increase people's prosperity while still prioritizing environmental balance so that natural resources have a role as a capital for people's economic growth and as a support for the living system [4] and a sustainable environment [5] to reduce social and economic vulnerability [6].

Community-based natural resource management is one form that is widely used in several regions to improve the adaptive capacity of the community to environmental changes by looking at various endogenous and exogenous factors such as technical, financial, legal factors, which can be seen from livelihood, political, and environmental factors. Social, economic, scale, knowledge, community, and cultural aspects involve values and ethics involving multiple stakeholders [7–11]. The social and economic development of the community cannot be separated from the contribution of maintaining coastal ecosystems. One of the models that can be used is Integrated Coastal Area Management by emphasizing strengthening the capacity of development actors and local communities to remain relevant in formulating eco-centric policies in coastal resource management [12]. The development of coastal communities to face socio-economic changes in the future by increasing the capacity of local communities through community empowerment by improving the partnership system can overcome socio-economic pressures so that the community can be independent [13,14].

Coastal zones can be considered inherently vulnerable because of their sensitive ecosystems, geomorphological, and complexities, and the clustering of many social, economic, and sometimes conflicting activities. The tourism sector in urban areas gets more attention than in rural areas which can increase the demographic pressure of coastal areas which are very sensitive to change. By using a matrix that relates the level of exploitation and adaptive capacity of vulnerability. This approach contributes to the effective use of metrics and Integrated Coastal Vulnerability Assessment Diagrams and can guide coastal management actions [15]. Meanwhile, according to Pricope, Joanne and Lauren, (2019) to overcome the impact of a change in coastal areas in the future, an integrated social and economic vulnerability model at the geographic area level can be developed using census data to measure the populations-based on social variability [16].

Based on the above phenomenon, Merauke is one of the areas facing the risk of global climate change, especially coastal communities as a result of carbon emissions by reducing mangrove forests [17,18]. Data on the area of Merauke mangrove forest in 2005 covering an area of 216,196 hectares based on satellite imagery [19] Meanwhile, according to 2009 data from Bakorsurtanal, Merauke Regency has 293,061.159 hectares of mangrove land. However, different data revealed in the 2011 RTRWK data contained in the Land-use Planning book supporting low emission development of Merauke Regency in 2017 that the Merauke mangrove forest area is 216,001.95 Ha or 6.98% of the Merauke forest area. One of the reasons for the fluctuating area of the Merauke mangrove forest is the area management system that has not been integrated between related parties [20]. Damage to mangrove areas disrupts natural coastal ecosystems which can result in a decrease in fishery resources and cause low incomes for traditional fishermen [21,22]. Efforts to develop a low emission development strategy in Merauke using the principle of sustainability have not been able to run optimally [23]. For this reason, the purpose of the research carried out in this paper is analysing the level of vulnerability in community groups on the Merauke coast which was taken in the Merauke and Naukenjerai districts. This paper describes the level of vulnerability and impact on the ecological, social, and economic dimensions of coastal community groups and the causes that occur so that they can be used as recommendations for implementing sustainable development in coastal areas to improve community welfare.

## 2. Method

The research framework uses qualitative research with a survey method approach. The research was carried out in the coastal area of Merauke by taking 2 sample areas in Merauke District and Naukenjerai District as research observation areas. Research data were collected using questionnaires and in-depth interviews using a structured list of questions from expert respondents.

Vulnerability analysis using the vulnerability index published by the IPCC (2001), which is used in determining the vulnerability index to disturbances or shocks related to climate change [24] by reviewing the conduct of the community in the culture of the coastal community. The vulnerability analysis is formulated as an accumulation of the level of exposure, the level of sensitivity, and adaptive capacity, which is formulated as follows:

$$Vulnerability = f\ (Exposure, sensitivity, Adaptive\ Capacity) \quad (1)$$

Based on the phenomena that occurred in Merauke Regency, the determinants of vulnerability, the main components, and indicators used in the study are presented in Table 1.

**Table 1.** Determinants, main components and indicators of vulnerability index assessment

| No | Determinants of Vulnerability | Main component | Indicator |
|---|---|---|---|
| 1 | *Exposure* | Inundation rise<br>Resource Degradation<br>Population income decline | Inundation rate rise<br>Land degradation rate<br>The rate of decline in population productivity |
| 2 | *Sensitivity* | Livelihood dependence on the coast<br>Land use rights<br>The dependence of social life on the coast | Percentage of population whose livelihoods depend on coastal resources<br>Percentage of the population claimed as individuals such as rice fields, fields, and yards<br>The ratio of the population relying on water transportation: land |
| 3 | *Resilience* | Interaction factor | • Learn to live with change and uncertainty<br>• Nurturing diversity for resilience<br>• Combine different types of knowledge to learn<br>• Create opportunities for socio-ecological self-organization towards sustainability. |

The exposure value is measured by the level of exposure between 0-3, namely 0 (not exposed); 1 (slightly exposed); 2 (sufficiently exposed); 3 (highly exposed). The sensitivity value is measured by the level of synchronicity between 0-3, namely 0 (not sensitive); 1 (slightly sensitive); 2 (quite sensitive); 3 (very sensitive). Meanwhile, adaptation capacity refers to the four interaction factors developed by Folke related to socio-ecological adaptation [25]. The value of adaptive capacity is measured by the level of adaptation between values 0 – 3, namely 0 (not adaptive); 1 (slightly adaptive); 2 (quite adaptive); 3 (highly adaptive).

## 3. Result and discussion

### *3.1. Level of the vulnerability of coastal communities*

Developing countries are under threat from unsustainable global and national changes and can create vulnerabilities that change social, economic, and ecological conditions. Social and ecological conditions of the community [26]. Vulnerability analysis is carried out to identify the characteristics of the community both individually/groups in anticipating/dealing/fighting against environmental changes such as natural disasters and natural damage due to human activities. In addition, it can also identify the community's inability to cope with losses, damage, and disturbances that arise as a result of a threat, both short-term and long-term. Vulnerability is the degree of loss or damage that may occur due to extreme events that cause disasters or ecological conditions that cannot perform their normal functions. Vulnerability analysis measures exposure, sensitivity, and resilience factors in livelihood groups that are vulnerable to direct changes in natural resources. The results of the vulnerability analysis in coastal areas in the respondent group in the coastal areas of Merauke and Naukenjerai Districts are presented in Table 2.

**Table 2.** Vulnerability analysis on the coast of Merauke and Naukenjerai Districts, Merauke District.

| No | Factor | District | | | | | |
|---|---|---|---|---|---|---|---|
| | | Merauke | | | Naukenjerai | | |
| | | 1 | 2 | 3 | 1 | 2 | 3 |
| 1 | *Exposure* | | | | | | |
| | Rob | 0.55 | 0.55 | 0 | 0.55 | 0.45 | 0.45 |
| | Land resource degradation | 0.45 | 0.55 | 0 | 0.55 | 0.45 | 0.45 |
| | Population Income Decrease | 0.45 | 0.55 | 0.45 | 0.45 | 0.45 | 0.55 |
| | Avarage Amount | 0.48 | 0.55 | 0.15 | 0.52 | 0.45 | 0.48 |
| 2 | *Sensitivities* | | | | | | |
| | Coastal Livelihood Dependence | 0.55 | 0 | 0.45 | 0 | 0.54 | 0 |
| | Residents who claim the location of the waters | 0.55 | 0 | 0.55 | 0.45 | 0.45 | 0.55 |
| | The ratio of the population relying on transportation | 0 | 0 | 0.55 | 0.55 | 0.55 | 0 |
| | Average Amount | 0.37 | 0 | 0.52 | 0.33 | 0.51 | 0.18 |
| 3 | *Resilience* | | | | | | |
| | Ability to learn to live with change and uncertainty | 0.45 | 0.55 | 0.45 | 0.55 | 0.55 | 0.55 |
| | Ability to develop various means for reorganization and renewal | 0.45 | 0.55 | 0.45 | 0.45 | 0.45 | 0.55 |
| | Ability to coordinate various kinds of knowledge | 0 | 0.45 | 0.54 | 0.45 | 0.55 | 0.45 |
| | Ability to create opportunities for self-organization | 0.45 | 0.55 | 0 | 0.55 | 0.55 | 0.45 |
| | Average Amount | 0.34 | 0.53 | 0.36 | 0.50 | 0.53 | 0.50 |
| Total Vulnerability (E + S – Ac) | | 0.51 | 0.03 | 0.31 | 0.35 | 0.44 | 0.17 |

Note: 1. Fishermen; 2. Farmers; 3. Services (farmers/fishermen, traders, transportation).

Based on the results of data analysis on the level of vulnerability to community groups in the coastal areas of Merauke and Naukenejrai Districts which is presented in Table 2, it shows the level of vulnerability in community groups based on the dominant occupation in the two different coastal areas. Community groups in coastal areas with the highest level of vulnerability are fishermen, then community groups who work in the service sector such as farm labourers and fishermen, and traders. The fishing communities in Merauke District depend on coastal natural resources for their livelihoods, which are influenced by the condition of the coastal ecosystem itself. This is because the condition of the coastal ecosystem provides valuable ecosystem services for the community [27–29]. It is like what happened in Bulgaria [30]. Seasonal fish production causes fishermen's income to be uncertain. Fishermen in Merauke District do not farm because Merauke District is an urban area, so land for agriculture in the coastal area is limited and land has been used for development as a result of the expansion of the urban area.

Meanwhile, the people in Naukenjerai District can anticipate environmental changes. People always live side by side with nature and manage natural resources by prioritizing local wisdom from culture handed down from their ancestors. In addition, the Naukenjerai area is included in a protected area, which is designated as a national park area. The management of the national park area prioritizes local wisdom to maintain the natural ecosystem within the Wasur National Park area itself. Based on the results of the analysis, that in Naukenjerai, coastal community groups that have a level of vulnerability to change and the ability to face environmental uncertainty are farmers and fishermen (rice cultivation and gardening). This is in line with what happened in Kupang City [31].

The condition and dynamics of coastal waters in both districts are strongly influenced by human activities that accelerate land degradation in the coastal areas, thus affecting the aquatic ecosystems on the coast. For this reason, increasing knowledge of coastal communities and ease of access to facilities and infrastructure need to be improved [32]. However, the high dependence on ecosystems by communities can change and degrade ecosystems at an alarming rate [29].

### *3.2. Identification of pressure and disturbances causing vulnerability*

The process of interaction of ecological or natural and social systems in the research area in addition to supporting the development of the social system of the people who live and depend on their natural resources, but also produces various forms of pressure that can limit and even destroy the social system during the lives of coastal communities.

Natural pressures will affect the lives of people living in coastal areas. The impacts felt by the community such as loss of assets, jobs, income, increasing fishing operational costs, and business uncertainty. Some of the natural pressures identified on the coast of the research area are as follows: 1) tidal water disaster, 2) high rainfall intensity floods, 3) land-use change, 4) fishing area conflicts. The risk impact of vulnerable groups against natural pressures in coastal and marine areas on the coastal communities of the research area, namely Merauke and Naukenjerai districts is presented in Table 3.

**Table 3.** Risk impact of vulnerable groups on coastal and marine natural stresses on coastal communities in Merauke and Naukenjerai Districts.

| No. | Vulnerable group | Natural Pressure | | | |
|---|---|---|---|---|---|
| | | tidal water disaster | Flood | Land-use change | Territorial conflict |
| 1 | Fisherman/farmer | 1,2,3,5,6,7 | 1,2,3,4,5,6,7 | 1,2,3,7 | 2,7 |
| 2 | Farm laborer/fisherman labor | 2,3,5,7 | 2,5,7 | 2 | 0 |
| 3 | Service | 2,5,6,7 | 2,5,6,7 | 3,6,7 | 0 |
| 4 | Trader | 2,5,6,7 | 2,3,6,7 | 5,6 | 2 |
| 5 | General public | 2,5,7 | 1,2,3,4,5,6,7 | 3,6,7 | 0 |

Note : Risk impact : 0. No impact; 1. Losing a job; 2. Loss of income; 3. Loss of assets; 4. Losing life; 5. Increase the risk of work; 6. Increased operational costs; 7. Increase uncertainty.

The results of field observations show that natural pressure poses a threat to vulnerable groups of coastal communities. Table 3. shows that the groups that are most vulnerable to pressure from changes in coastal natural conditions are fishermen and farmers, especially during tidal waves and floods caused by high-intensity rains in coastal areas. Natural pressures such as tidal water disasters, floods, land-use changes, and regional conflicts have the highest risk impact, namely job loss, loss of income, and increased uncertainty. This is because the people in the research area depend on their families for their livelihood from coastal water resources, both fishery, and non-fishery. Meanwhile, the vulnerable groups who will feel the impact of risk from the pressure of natural change are the labour group. Coastal communities with the main job as fishermen are greatly affected by the fishing season. Community labour groups in coastal areas based on survey results in the field that natural pressures in coastal areas will have an impact on loss of community income, especially for groups of farmworkers and fishermen, namely loss of income. Farmworkers and fishermen will lose their income if there is a high tidal wave on the mainland and floods because they cannot find fish and agricultural land is flooded during high rainfall and high tides. In addition, changes in land use in the coastal area of Merauke from the results of field surveys have a direct impact on groups of fishermen, farmers, and labourers. Changes in the use of coastal land, especially mangrove forests as one of the protected areas that have an important function to become an ecology in and behind the mangrove forest area. Changes in mangrove forests on the coast of Merauke for changes in land use by the community for the expansion of population growth settlements and infrastructure development for the community. The same thing is in line with the results of research in the coastal village of Subang Regency, that the causes of changes in coastal land use are the increase in population, settlements, and infrastructure [33].

Natural pressures that often occur and become an interesting issue in Papua in general and Merauke, in particular, are regional conflicts. The territorial conflicts in question are conflicts overfishing areas and land ownership. This is the trigger for developing into an ecological-political issue that becomes sensitive in the implementation of development. Ecology-politics can develop as a result of the complexity of the problem of the relationship between humans and nature [34]. In addition to natural pressures that occur on the coast, there are human pressures both individually and in groups on resources in the coastal areas in the research area, namely the Merauke and Naukenjerai districts which cause coastal environmental degradation which will reduce the quality of life of the people living around them. Some of the human activities that can have a risk impact on nature are presented in Table 4.

**Table 4.** Community pressure on natural resources in the coastal areas of Merauke and Nukenjerai Districts.

| **Impact on natural resources** | | **Public Pressure** | | | |
|---|---|---|---|---|---|
| **Function** | **Interest** | **1** | **2** | **3** | **4** |
| Ecology | The decline in the carrying capacity of the region | √ | √ | √ | √ |
| | Decreased carrying capacity for coastal ecosystem life | √ | √ | √ | √ |
| | Decreased land resources | - | √ | | √ |
| | Declining water quality | - | - | √ | √ |
| | The Decline in coastal aesthetics | √ | - | √ | √ |
| | Decreased biodiversity | √ | √ | - | √ |
| Economy | Decreased sources of livelihood | √ | √ | √ | √ |
| | Decreased sources of income | √ | √ | √ | √ |
| Social | Public welfare | - | √ | - | √ |
| | Life system support | - | √ | - | √ |
| | Law enforcement | √ | √ | √ | √ |

Note*: 1. Population growth; 2. Land conversion; 3. Dispose of garbage and waste to the coast; 4. Sand excavation

Human pressure both individually and in groups in the coastal area that has an impact on all ecological, economic and social functions, namely the excavation of sand on the coast in both Merauke and Naukenjerai districts. Past sand excavation activities still have an impact that is felt by the community today. Sand excavation in coastal areas is not only carried out on the coast but is carried out in the yard of houses containing sand. These community activities are dominated by local communities who have customary land rights on the coast. This happened in both study areas, but in Naukenjerai it caused the resettlement of settlements to the mainland. This was stated by the head of Kampung Onggaya:

> *"....In the past, this village was on the beach which you are now looking at. We have moved the pole or honey on the beach because the location had been eroded by the waves. I think that every year the abrasion can be 3 meters and maybe the village we live in now will be exhausted if we don't take precautions from now on..."*

The increase in population, both from the rate of increase in family members and the addition of new residents, will have an impact on the increasing competition for livelihoods which has an impact on the economic sector and the ecology of the region. Likewise, the activity of throwing garbage on the coast will cause the fastest environment to pile up garbage. The research coastal area has the potential to develop the coastal tourism sector which can bring in a source of community income by selling on the coast and even selling catches to visitors who come for tours. There is a need for integrated waste management education so that garbage and household waste are not dumped into the sand which will be carried to the sea during high tides.

## 4. Conclusion

The research concludes that climate change has a major impact on fishing communities and farmers. Dependence on land and water resources on the coast causes the high vulnerability. Natural pressure factors that cause high are tidal water disasters, high-intensity floods, and land conversion due to tidal water degradation which causes loss of income and jobs, as well as increasing fishery business risks. Meanwhile, human pressure that causes vulnerability is the excavation of sand and the conversion of land into settlements and infrastructure development for the community. To reduce the risk of the vulnerability of coastal communities as a result of climate change, it is necessary to carry out several activity scenarios, namely the rehabilitation of coastal areas by planting mangroves by involving the community and diversifying livelihood sources based on natural resources.

## References

[1] Kemenkumham RI 2016 *Undang-Undang Republik Indonesia Nomor 7 Tahun 2016 tentang perlindungan dan pemberdayaan nelayan, pembudi daya ikan, dan petambak garam* (Indonesia)

[2] Sangha K K, Stoeckl N, Crossman N and Costanza R 2019 A state-wide economic assessment of coastal and marine ecosystem services to inform sustainable development policies in the Northern Territory, Australia *Marine Policy* **107** 103595

[3] Voyer M, Barclay K, McIlgorm A and Mazur N 2017 Using a well-being approach to develop a framework for an integrated socio-economic evaluation of professional fishing *Fish and Fisheries* **18**

[4] Sutrisno E 2014 Implementasi pengelolaan sumber daya pesisir berbasis pengelolaan wilayah pesisir secara terpadu untuk kesejahteraan nelayan (studi di Perdesaan Nelayan Cangkol Kelurahan Lemahwungkuk Kecamatan Lemahwungkuk Kota Cirebon) *Jurnal Dinamika Hukum* **14** 1–12

[5] Arie Nurzaman, Rajib Shaw M S R P 2020 Measuring community resilience against coastal hazards: A Case study in Baron Beach, Gunungkidul Regency Arie *Progress in Disaster Science*

[6] Hadipour V, Vafaie F and Kerle N 2019 An indicator-based approach to assess social vulnerability of coastal areas to sea-level rise and flooding: A case study of Bandar Abbas city, Iran *Ocean and Coastal Management* 105077

[7] Armitage D 2005 Adaptive capacity and community-based natural resource management *Journal of Environmental Management* **35** 703–15

[8] Wiber M, Berkes F, Charles A and Kearney J 2004 Participatory research supporting community-based fishery management *Journal of Marine Policy* **28** 459–68

[9] Masozera M K, Alavalapati J R R, Jacobson S K and Shrestha R K 2006 Assessing the suitability of community-based management for the Nyungwe Forest Reserve, Rwanda *Journal of Forest Policy and Economics* **8** 206–16

[10] Pomeroy R S 1995 Community-based and co-management institutions for sustainable coastal fisheries management in Southeast Asia *Journal of Ocean and Coastal Management* **27** 143–62

[11] Thompson P M, Sultana P and Islam N 2003 Lessons from community based management of floodplain fisheries in Bangladesh *Journal of Environmental Management* **69** 307–21

[12] Nagabhatla N, T.Hung N, Tuyen L T, Cam V T N, Dhanraj J, T.T N and Swierczek F W 2019 Ecosystem-based approach for planning research and capacity development for integrated coastal zone management in Southeast Asia *APN Science Bulletin* **9** 3–9

[13] Barker A 2005 Capacity building for sustainability: Towards community development in coastal Scotland *Journal of Environmental Management* **75** 11–9

[14] Hariyanto S 2014 Analisis pemberdayaan masyarakat nelayan di pantai Prigi Kecamatan Watulimo Kabupaten Trenggalek *Jurnal Universitas Tulungagung BONOROWO* **2**

[15] Lins-de-Barros F M 2017 Integrated coastal vulnerability assessment: A methodology for coastal cities management integrating socioeconomic, physical and environmental dimensions - Case study of Região dos Lagos, Rio de Janeiro, Brazil *Ocean and Coastal Management* **149** 1–11

[16] Pricope N G, Halls J N and Rosul L M 2019 Modeling residential coastal flood vulnerability using finished-floor elevations and socio-economic characteristics *Journal of Environmental Management* **237** 387–98
[17] Untari, Witdarko Y and Sembiring J 2018 Analysis of carbon emission level on Merauke Regency land cover *E3S Web of Conferences* **73** 0–4
[18] Untari, Melmambessy E H P and Simatupang D O 2018 Carbon emissions and mitigation actions in Merauke *E3S Web of Conferences* **73** 1–6
[19] WWF Kabupaten Merauke 2005 *Luasan mangrove Kabupaten Merauke. dengan menggunakan data citra.*
[20] Untari, Darma R, Betaubun P and Arief A A 2020 Review of the use of mangrove forests in supporting the socio-economic life of fishing communities *IOP Conference Series: Earth and Environmental Science* **575** 012042
[21] Imbanop Y P, Widiastuti M M D and Fachriza R 2019 Analisis kelayakan usaha penangkapan ikan pada musim peralihan oleh nelayan di kampung onggaya distrik naukenjerai kabupaten merauke *Musamus Journal of Agribussines* **1** 60–6
[22] Untari, Chintia A, Malino P, Ginting N M, Fachrizal R, Betaubun P and Arief A A 2021 Sicio-economic conditions of small-scale traditional fishermen: a case study in Payum Village, Merauke District, Papua, Indonesia *Jurnal Ilmiah Pertanian* **18** 20–8
[23] Rahail E, Untari, Herdjiono I, Saadah, Wahyuni C and Dariati T 2019 Low-emission strategy through effective planning and public participation in Merauke *IOP Conference Series: Earth and Environmental Science* **235**
[24] IPPC 2001 Reference document on best available techniques in the ferrous metals processing industry *Integrated Pollution Prevention and Control (IPPC)*
[25] Folke C, Carpenter S, Elmqvist T, Gunderson L, Holling C S and Walker B 2002 Resilience and sustainable development: building adaptive capacity in a world of transformations *AMBIO: A J. of the Human Environment* **31** 437–40
[26] Nel R, Mearns K F, Jordaan M and Goethals P 2021 Ecological informatics towards understanding the role of islandness in shaping socio-ecological systems on SIDS : The socio-ecological islandscape concept *Journal of Ecological Informatics* **62** 101264
[27] Heckwolf M J, Peterson A, Jänes H, Horne P, Künne J, Liversage K, Sajeva M, Reusch T B H and Kotta J 2021 From ecosystems to socio-economic bene fi ts : A systematic review of coastal ecosystem services in the Baltic Sea *Science of the Total Environment* **755** 142565
[28] Fukamachi K 2020 Building resilient socio-ecological systems in Japan : Satoyama examples from Shiga Prefecture *Journal of Ecosystem Services* **46** 101187
[29] Berrio-giraldo L, Villegas-palacio C and Arango-aramburo S 2021 Understating complex interactions in socio-ecological systems using system dynamics : A case in the tropical Andes *Journal of Environmental Management* **291** 112675
[30] Gocheva K, Lü Y, Li F, Bratanova-doncheva S and Chipev N 2019 Ecosystem restoration in Europe : Can analogies to Traditional Chinese Medicine facilitate the cross-policy harmonization on managing socio- ecological systems ? *Science of the Total Environment* **657** 1553–67
[31] Ledoh L Y, Satria A and Hidayat R 2018 Kerentanan penghidupan masyarakat pesisir perkotaan terhadap variabilitas iklim (studi kasus di Kota Kupang) *Journal of Natural Resources and Environmental Management* **9** 758–70
[32] Hoque S, Andrioti D, Pike K, Hasan M, Arifur M, Akter S, Mitchell D and Holliday E 2021 Knowledge, practice, and economic impacts of COVID-19 on small-scale coastal fishing communities in Bangladesh : Policy recommendations for improved livelihoods *Marine Policy* **131** 104647
[33] Muliani, Adrianto L, Soewardi K and Hariyadi S 2018 Sistem sosial ekologi kawasan desa pesisir Kabupaten Subang *Jurnal Ilmu dan teknologi Kelautan Tropis* **10** 575–88
[34] Dharmawan A H 2007 Dinamika sosio-ekologi pedesaan : perspektif dan *Jurnal Transdisiplin Sosiologi, Komunikasi, dan Ekologi Manusia* **01** 1–40

**Acknowledgement**
Acknowledgement to the Ministry of Education, Culture, Research, and Technology; the Directorate General of Higher Education as the funder for the 2021 Doctoral Dissertation Grant; Hasanuddin University Research and Community Service Institute which has facilitated this activity, as well as parties who contributed to the research that we did not mention one by one.

# The effect of indigenous vegetations on the biological control of oil palm basal stem rot (BSR) disease caused by *Ganoderma* in peatlands

**Supriyanto[1,3]*, Purwanto[2], S H Poromarto[2] and Supyani[2]**

[1]Department of Agricultural Science, Graduate School, Sebelas Maret University, Surakarta, Indonesia
[2]Faculty of Agriculture, Universitas Sebelas Maret, Jl. Ir. Sutami 36A Kentingan, Surakarta, Central Java, 57126 Indonesia.
[3]Department of Agriculture Cultivation, Faculty of Agriculture, Universitas Tanjungpura, Jl. Prof. Dr. H. Hadari Nawawi, Pontianak, Indonesia
*Corresponding author: supriyantountan2013@gmail.com

**Abstract.** Oil palm is an important plantation commodity in Indonesia. Oil palm cultivation in Indonesia is always associated with global warming because some of its plantation's areas are on peatlands and its management not yet environmentally friendly. A serious problem of oil palm cultivation on peatlands is the basal stem rot disease caused by the *Ganoderma* fungi. There is no effective method to control this disease. Biological control is an alternative method that is environmentally friendly and is the focus of current development, however, in peatland, its development was limited by the lack of information about the influence of peat environmental factors. This study aims to determine the effect of indigenous vegetation on the biological control of *Ganoderma* on oil palm on peat soil. The research has been conducted by examining the effect of 4 species of peat indigenous plants on the effectiveness of *Ganoderma* biological control on peat soil in the experimental garden. The results showed that the presence of peat indigenous plants affected the effectiveness of *Ganoderma* control in oil palm on West Kalimantan peat soils. The single indigenous plant *Nephrolepis biserrata* was able to reduce the intensity of the disease. However, the use of mixed indigenous plants *N. biserrata*, *Asystasia intrusa*, *Melastoma affine*, and *Stenochlaena palustris* actually increased the intensity of the disease. These results indicate that the management of existing peat indigenous vegetations around oil palm plantations can help reduce the *Ganoderma* attacks on oil palm on peat soils.

**Keywords:** biological control, *Ganoderma*, indigenous vegetation oil palm, peatland

## 1. Introduction

Indonesia is a major producer of palm oil in the world [1]. Currently, Indonesia is also utilizing peatlands as part of its oil palm cultivation area. The proportion has 14.58% of the total oil palm plantations [2]. Until 2017, only 16.7% of plantations were certified Indonesian Sustainable Palm Oil (ISPO) [3]. This condition makes Indonesian palm oil is always associated with global warming. The main problem in oil palm cultivation on peatlands that have not been resolved is the basal stem rot (BSR) disease caused by the *Ganoderma* sp. In the West Kalimantan peatlands, the intensity of the *Ganoderma* attack reaches 76% [4]. There has not been found an effective way to control BSR disease in peatlands [5]. Based on the initial studies, it is known that there are indigenous antagonist fungi and bacteria from peatlands that have been able to control *Ganoderma* in vitro. These antagonists also can act as acid-tolerant fungi and bacteria, so they have the potential to be developed as biological control agents for *Ganoderma* in peatlands. However, currently, the biological control of *Ganoderma* in the field has not been effective. Allegedly, there are environmental factors that affect the effectiveness of these antagonists in the peat soil environment [6;7].

In previous studies, it was found that differences in dominant plant species influenced the percentage of antagonistic fungi in peat soil [4]. Allegedly, there is a close relationship between plant species around the palm and the effectiveness of biological control against *Ganoderma* on peat soil. The effect of vegetation around oil palm plantations on the effectiveness of *Ganoderma* biological control on peatlands is possible considering that the plant rhizosphere affects the type of microorganism community around it. Differences in plant species affect the type of root exudate produced and affect the type of microorganism community in the rhizosphere [8]. This research was conducted to determine the effect of indigenous plants on the effectiveness of *Ganoderma* control on oil palm on peat soil. This information is important considering that indigenous vegetation is relatively easy to manage, is abundant in resources, and has not been used properly in the process of oil palm cultivation on peatlands.

## 2. Materials and methods

### 2.1. Experimental design

The test was carried out in a natural peat soil situation, consisting of hemic peat soil and sapric peat soil, as well as mineral soil as a comparison. A total of 3 isolates of *Ganoderma*-antagonistic fungi, namely 13EJ15, B3J19, and E4J8 isolates [6], and 2 isolates of *Ganoderma*-antagonistic bacteria, namely E4B6 and E2B12 isolates from peatlands of West Kalimantan [7] were tested in this study. Antagonists were applicated as a consortium of antagonistic fungi (13EJ15+B3J19+E4J8 isolates), a consortium of antagonistic bacteria (E4B6+E2B12 isolate), and a consortium of fungi and bacterial antagonists (13EJ1+B3J19+E4J8+E4B6+E2B12 isolates). All antagonistic treatments were tested in situations of cultivation without vegetations, cultivation with single plants, and cultivation with mixed plants, in order to obtain 9 antagonistic fungi consortium treatments, 9 antagonistic bacterial consortium treatments, and 9 fungal and bacterial antagonist consortium treatments. All treatments were applied to control *Ganoderma* which were inoculated on oil palm seedlings in a polyethylene plastic bag (polybag). As the pathogen, *Ganoderma* from peatland was used, namely, *Ganoderma* G301 isolate. As a comparison, 9 control treatments of oil palm seedlings grown in the same situation but not inoculated with antagonists (positive control), and 9 control treatments of oil palm seedlings that were not inoculated with *Ganoderma* and not applicated with antagonists (negative control). All treatments were arranged in a factorial completely randomized design and 3 times replication.

### 2.2. Peat soil preparation

The soil was made friable, stirred, and sieved with a 2 cm-hole sieve to obtain a homogenous aggregate size. Peat soil was weighed in air-drying condition, and as much as 5 kg of peat soil were placed in a 30 x 40 cm polybag. Mineral soils were measured by volume to equal the volume of peat soil in polybag in air-drying conditions due to the difference in their bulk density. The soil was kept covered for 2 weeks before use.

### 2.3. Preparation of oil palm seedlings

This research used commercial oil palm seeds of Marihat DxP 239 variety from PPKS Medan, Indonesia. Oil palm sprouts were grown in 10 x 20 cm polybags containing a mixture of peat soil and mineral soil in a 1:1 ratio. The soil mixture was homogenized by making it friable and sieving with a 2 cm-hole sieve. Before it was used, the medium was sterilized by steaming for 6 hours. Seedlings were cultivated for two months under 50% shade.

*2.4. Inoculum preparation and inoculation of oil palm seedlings with Ganoderma*

*Ganoderma* inoculum preparation followed the method of [9] with modifications. The *Ganoderma* inoculum was prepared by growing the *Ganoderma* isolate G301 onto 6 x 6 x 12 cm rubber woodblocks. Before it was used, the rubber woodblocks were soaked in tap water for a night, thoroughly washed and wrapped in an autoclaved-resistant plastic bag, sterilized in an autoclave for 1 hour, cooled overnight, and then re-sterilized in the same way. After that, 10 ml of liquid PDA were added to the plastic and flattened over the entire surface of the rubber wood block. After the PDA achieved room temperature, aseptically, a quarter part of 1 Petri culture of 7 days-age *Ganoderma* isolate was put on the surface of rubberwood, and then closed again and incubated for 2 months. The inoculum was ready to use after the mycelium has covered all the rubber woodblock surfaces.

*Ganoderma* inoculation of oil palm seedlings follows the method of [10] with modifications. Rubber woodblocks that have been overgrown with *Ganoderma* mycelium were buried in the soil in polybags 1 week before the oil palm seedlings were planted at a depth of 10 cm. The 2-month old oil palm seedling was planted on the medium, placed on a rubber woodblock and the main root was made sure to sit on the rubber woodblock, and then backfilled. Oil palm seedling was then cultivated for 24 weeks. The growing medium humidity was maintained in a condition of field capacity.

*2.5. Preparation and inoculation of antagonists onto peat soil medium*

Three isolates of antagonistic fungi and 2 isolates of antagonistic bacteria were rejuvenated. The antagonistic fungal inoculum was prepared in a corn-water medium (1:1 ratio) for 2 weeks. The bacterial inoculum was prepared in a bacterial suspension with a minimum density of 108 cfu/ml. The antagonist inoculum was given into the medium according to the predetermined treatment. The fungal consortium treatment, 10 g of corn-water medium inoculum was used for each isolate so that for each treatment 30 g of the antagonist was obtained. Isolates were buried and mixed into the planting medium 3 days after *Ganoderma* inoculation. The polybags were incubated under plastic sheeting to sunlight avoided. Antagonistic bacteria were given in a bacterial suspension by soil drenching. Each isolate was taken as much as 15 ml using a sterile syringe so that 30 ml of bacterial suspension was obtained for consortium bacterial treatment. The bacterial suspension was splashed evenly onto the surface of the planting medium and then incubated a day before planting oil palm seedlings.

*2.6. Preparation of indigenous plants*

The single plant treatment used the dominant indigenous plant on peatlands, namely *Nephrolepis biserrata*. The mixed plants used 4 indigenous plant species, namely *N. biserrata, Asystasia intrusa, Melastoma affine* and *Stenochlaena palustris*. The root of about 5 cm high seedling indigenous plants were washed with sterile water before use. The seedlings of the indigenous plants were planted one day after planting the oil palm seedlings, about 5 cm from the oil palm. The treatment for a single indigenous plant, each polybag planted with 4 individual plants of *N. biserrata*. The treatment for mixed indigenous plants, in each polybag, was planted with one individual *N. biserrata, A. intrusa, M. affine* and *S. palustris*, planted evenly within 5 cm of the oil palm plantation.

*2.7. BSR disease intensity observation*

Observation of the disease was carried out every 4 weeks for 24 weeks, by observing the symptoms on the shoot of oil palm seedlings. Observation of disease intensity was carried out using scoring criteria. The disease symptom scoring criteria used was shown in Table 1.

**Table 1.** Criteria for scoring symptoms of the disease on oil palm seedling [9]

| Score | Disease symptoms |
|---|---|
| 0 | Healthy plants, green leaves, not any fungal mass on the seedlings. |
| 1 | 1-3 chlorotic leaves without showing any fungal mass on the seedlings |
| 2 | Seedlings reveal a mass of fungi in any part |
| 3 | More than 3 chlorotic leaves, necrotic leaves with or without fungal mass on seedlings |
| 4 | At least 50% of leaves have chlorosis and necrosis with or without a fungal mass on the seedlings |
| 5 | Dead seedlings with or without the appearance of a fungal mass on the seedlings |

### *2.8. Statistical analysis*

Data on disease severity, in the form of a scoring number, were calculated using the formula of [11] to determine the success rate of disease control (1).

$$\mathrm{IP} = \frac{\sum_{i=1}(k\ x\ nk)}{\mathrm{Z\ x\ N}}\ x\ 100\% \tag{1}$$

Where: IP= Disease intensity, nk= number of plants with a score of k (k=0,1,2,3,4), k= Score used, Z= Highest score, and N= Number of plants observed.

## 3. Results and discussion

In hemic peat, the lowest symptom intensity was obtained from single plant treatment. The highest symptom intensity was obtained from the treatment with mixed plants and slightly higher than the uncontrolled plants (positive control). The intensity of symptoms in plants treated with the consortium of fungi was 40%, higher than in control plants (33.3%). The intensity of symptoms in the application of antagonists without the addition of plants was lower than those given mixed plants (Figure 1). These results indicate that plant management on hemic peat can be beneficial in the biological control of *Ganoderma* using indigenous antagonists. *N. biserrata* is a plant with a dense canopy so that it can act as a ground cover plant. Its presence reduces sunlight that reaches the soil surface while increasing soil moisture so that the environment becomes conducive to soil microorganisms in the rhizosphere. In addition, the very shallow roots of *N. biserrata* also reduce the level of competition with oil palm roots. Another possibility is that there is a specific interaction between *N. biserrata*, with certain microorganisms present in hemic peat, which greatly affects soil suppression of *Ganoderma* as stated by [12].

Observation of disease intensity on mineral soils showed that the use of single plants in biological control was able to reduce disease symptoms to zero percent compared to 13.3% in the positive control treatment. In plants that were controlled with antagonists and applicated mixed plants, the intensity of the disease was more severe than the control treatment. Meanwhile, those that were not applicated plants showed the same symptoms as the control (in the consortium fungal treatment) and slightly lower in the bacterial consortium and fungal and bacterial consortium treatments (Figure 1). These results are almost the same as those for hemic peat soils.

Single indigenous plants had a positive effect on the effectiveness of biological control of *Ganoderma* by indigenous antagonists on hemic peat soils, however, negatively on sapric peat soils (Table 2). It is assumed that, apart from its influence on the microclimate around oil palm, it is also related to the possible influence of several different biological mechanisms on the two types of peat soil. First, the influence of microorganism communities in hemic peat soils is different from sapric peat. Research by [13] in Kubu Raya shows that each level of peat maturity contains different amounts and types of fungi. The more mature the peat, the higher the level of diversity and density. The same case was found in observations of mycorrhizae [14] and cellulolytic bacteria [15]. The existence of a single plant is thought to have provided a conducive environment for the innate microorganism community so that it inhibited the given indigenous antagonists. The cover plants have a significant impact on the energy transformation process and microclimate formation for innate microorganisms [16].

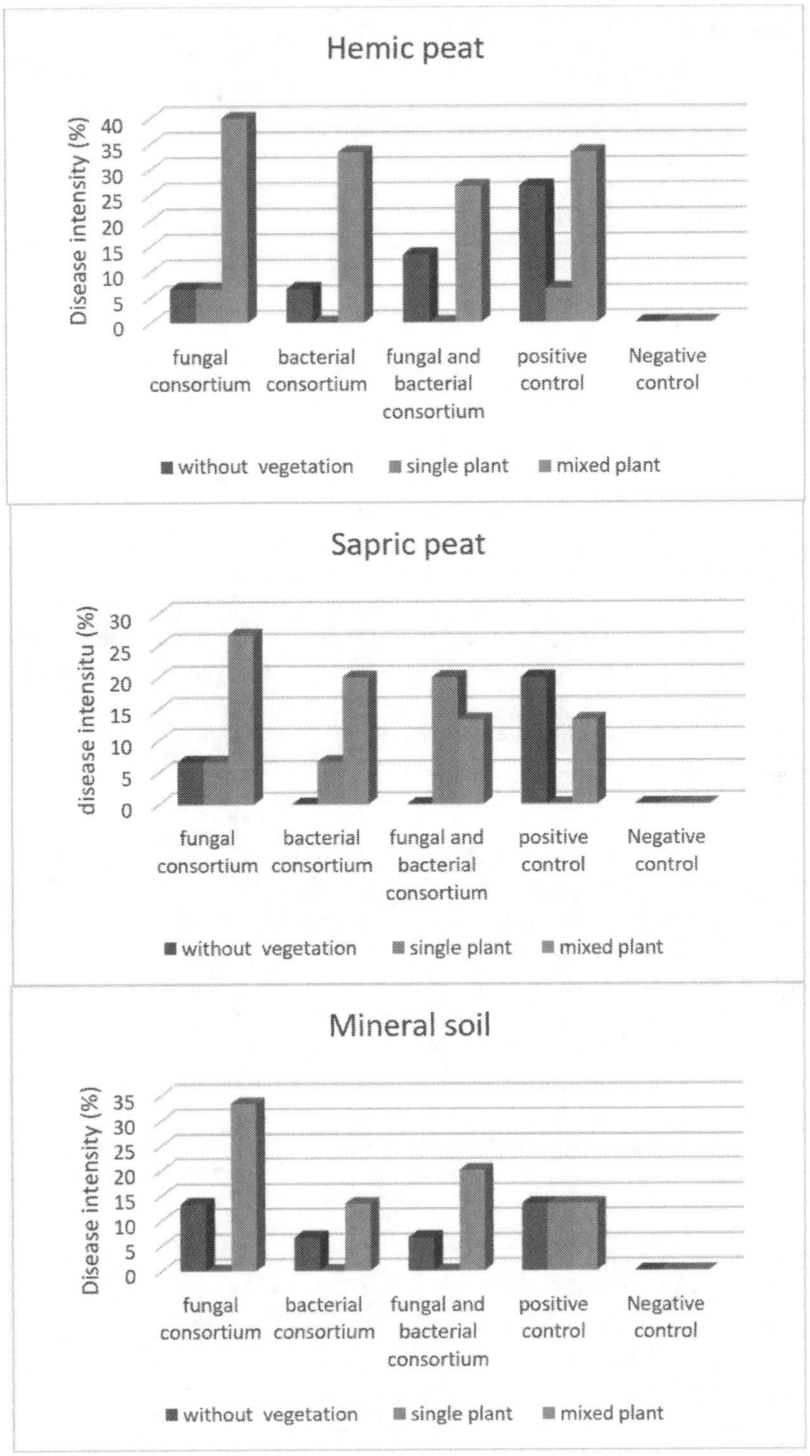


**Figure 1.** Effect of indigenous vegetations and antagonists on the intensity of disease symptoms in oil palm seedlings.

**Table 2.** Effect of plants on the effectiveness of biological control of indigenous antagonists of peatland

| Treatment | Control effectivenes (%) * |
|---|---|
| Without vegetation | |
| Hemic peat | 50 |
| Sapric peat | 100 |
| Mineral soil | 49.96 |
| Single plant | |
| Hemic peat | 100 |
| Sapric peat | 0 |
| Mineral soil | 100 |
| Mixed Plant | |
| Hemic peat | 19.98 |
| Sapric peat | 0 |
| Mineral soil | 0 |

Note: *Control effectiveness is the percentage of the difference in the ratio between disease intensity in treated oil palms and control oil palms

The second possibility is that there is a specific relationship between the indigenous plant *N. biserrata* and these antagonists because it comes from the same environmental site. It is assumed that to occur because based on field observations, there is a correlation between the SDR of plants and the number of types of antagonistic fungi in hemic peat soils [4]. In addition, there is a phenomenon that in the process of succession of microorganisms in peatlands, at the initial weathering level, it is dominated by microorganisms of the symbiotic type than those of the saprotrophic type [17]. Also, the plant community under certain conditions is more influential than the type of soil [18]. In the peat soil situation, it is presumed that there are specific hosts for certain rhizosphere fungi to support their growth [19]. Further studies on this phenomenon are urgently needed to explain why the use of single plants can increase the effectiveness of control by indigenous antagonists in hemic peat soils.

## 4. Conclusions

The presence of indigenous vegetations affects the effectiveness of *Ganoderma* control in oil palm on West Kalimantan peat soils. The management of indigenous plants *N. biserrata* alone was able to reduce the intensity of the disease better than the control. However, the use of mixed indigenous plants *N. biserrata, A. intrusa, M. affine* and *S. palustris* actually increased the intensity of the disease. The results of this study indicate that the proper management of indigenous vegetation in oil palm cultivation can improve the health of oil palm as well as increase its role in reducing greenhouse gas emissions which causes global warming through the use of indigenous vegetation as a cover crop. To understand further the role of plants in supporting the biological control of *Ganoderma* on peatlands, further research is necessary. An important aspect that needs to be studied is the possibility of a paired relationship that can create soil suppression against *Ganoderma* on peatlands. Likewise with regard to the mechanism that occurs in it.

## References

[1] USDA 2021 Oilseeds: World Markets and Trade *Foreign Agric. Serv.* https://www.fas.usda.gov/data/grain-world-markets-
[2] Miettinen J, Shi C and Liew S C 2016 Land Cover Distribution in the Peatlands of Peninsular Malaysia, Sumatra and Borneo in 2015 with Changes Since 1990 *Glob. Ecol. Conserv.* **6** 67–78
[3] BPDPKS 2017 Riset Kelapa Sawit BPDPKS Periode 2017 *BPDPKS*
[4] Supriyanto, Purwanto, Poromarto S H and Supyani 2020 The Relationship of Some Characteristics of Peat with Oil Palm Basal Stem Rot (BSR) Caused by Ganoderma in Peatlands *IOP Conference Series: Earth and Environmental Science* vol 423 p 012064
[5] Muniroh M S, Nusaibah S A, Vadamalai G and Siddique Y 2019 Proficiency of Iocontrol Agents

as Plant Growth Promoters and Hydrolytic Enzyme Producers in Ganoderma Boninense Infected Oil Palm Seedlings *Curr. Plant Biol.* **20** 1–23

[6] Supriyanto, Purwanto, Poromarto S H and Supyani 2020 Evaluation of In Vitro Antagonistic Activity of Fungi from Peatlands Against Ganoderma Species Under Acidic Condition *Biodiversitas* **21** 2935–45

[7] Supriyanto, Purwanto, Poromarto S H and Supyani 2021 Evaluation of In Vitro Activity of Ganoderma-Antagonistic Bacteria from Peatland Under Acidic Condition *IOP Conference Series: Earth and Environmental Science* vol 724 p 012013

[8] Bakker M G, Manter D K, Sheflin A M, Weir T L and Vivanco J M 2012 Harnessing the Rhizosphere Microbiome Through Plant Breeding and Agricultural Management *Plant Soil* **360** 1–13

[9] Rakib M R M, Bong C F J, Khairulmazmi A and Idris A S 2015 Aggressiveness of Ganoderma Boninense and G. Zonatum Isolated from Upper - and Basal Stem Rot of Oil Palm (Elaeis guineensis) in Malaysia *J. Oil Palm Res.* **27** 229–40

[10] Idris A S, Kushairi D, Ariffin D and Basri M 2006 Technique for Inoculation of Oil Palm Germinated Seeds with Ganoderma *MPOB Infomation Ser.* **314** 1–4

[11] Arwiyanto T, Goto M, Tsuyumu S and Takikawa Y 2014 Biological Control of Bacterial Wilt in South East Asia *J. Perlindungan Tanam. Indones.* **18** 55–64

[12] Latz E, Eisenhauer N, Rall B C, Scheu S and Jousset A 2016 Unravelling Linkages between Plant Community Composition and the Pathogen-Suppressive Potential of Soils *Sci. Rep.* **6** 23584

[13] Efa Rosita, Riza Linda S K 2014 Kapang pada Tingkat Kematangan Gambut yang Berbeda di Kawasan Hutan Lindung Gunung Ambawang Kabupaten Kubu Raya *Protobiont* **3** 10–6

[14] S.Friska, R.Linda S K 2005 Kepadatan Spora Jamur Mikoriza Vesikular Arbuskular Pada Tiga Tingkat Kematangan Gambut Di Kawasan Hutan Lindung Gunung Ambawang Kabupaten Kubu Raya Friska *Protobiont* **4** 96–102

[15] Wahyuni D, Khotimah S and Linda R 2015 Eksplorasi Bakteri Selulolitik pada Tingkat Kematangan Gambut yang Berbeda di Kawasan Hutan Lindung Gunung Ambawang *Probiont* **4** 69–76

[16] Procházka J, Brom J, Šťastný J and Pecharová E 2011 The Impact of Vegetation Cover on Temperature and Humidity Properties in the Reclaimed Area of a Brown Coal Dump *Int. J. Mining, Reclam. Environ.* **25** 350–66

[17] Kang H, Kwon M J, Kim S, Lee S, Jones T G, Johncock A C, Haraguchi A and Freeman C 2018 Biologically Driven DOC Release from Peatlands During Recovery from Acidification *Nat. Commun.* **9** 1–7

[18] Reese A T, Lulow K, David L A and Wright J P 2018 Plant Community and soil Conditions Individually Affect Soil Microbial Community Assembly in Experimental Mesocosms *Ecol. Evol.* **8** 1196–205

[19] Supriyanto, Priyatmojo A and T.Arwiyanto 2009 Penapisan PGPF untuk Pengendalian Penyakit Busuk Lunak Lidah Buaya (Aloe vera) di Tanah Gambut *J. Perlindungan Tanam. Indones.* **15** 71–82

## Acknowledgement

The author would like to thank the Directorate General of Research and Development, Ministry of Research and Higher Education, Indonesia for providing funding for this research according to contract No. 107/ADD/SP2H/LT/DRPM/VIII/2017.

## AUTHOR INDEX